Measuring Process Capability

Measuring Process Capability

A Reference Handbook for Quality and
Manufacturing Engineers

Davis R. Bothe

Landmark Publishing, Inc.
Cedarburg, WI

Library of Congress Cataloging-in-Publication Data

Bothe, Davis R., 1950–
    Measuring process capability : a reference handbook for
quality and manufacturing engineers / Davis R. Bothe.
        p.   cm.
    Includes index.
    ISBN 0-07-006652-3 (alk. paper)
    1. Manufacturing processes.   2. Quality control–Statistical
methods.   I. Title.
TS176.B67 2001
658.5–dc21

96-30087
CIP

1 2 3 4 5 6 7 8 9 0   DOC/DOC   9 0 2 1 0 9 8 7

ISBN 0-07-006652-3

*The sponsoring editor for this book was Steven T. Bothe, the editing supervisor was
Berdine G. Bothe, and the production supervisor was Betsy K. Zera. It was set in Times
Roman by Davis R. Bothe.*

Printed and bound by Van Lanen Printing Company, Inc.

Landmark Publishing books are available at special quantity discounts to use as
premiums and sales promotions, or for use in corporate training programs. For more
information, please write to the Director of Special Sales, International Quality
Institute, W5030 Landmark Drive, Cedarburg, WI 53012-2910 or visit the Web site
www.I-Q-I.com.

This book is dedicated to the two people who have helped me so greatly throughout my entire life, Mom and Dad.

Contents

Preface

In recent years there has been a growing interest in quantifying the ability of a process to satisfy customer requirements. The size of this interest is demonstrated by the increasing number of articles appearing in the various quality magazines and technical journals where authors praise (or in some cases, condemn) a particular measure of process capability. Unfortunately, these authors usually present different versions of capability formulas, definitions, notation, terminology and, worst of all, offer varying opinions on what information about the process these measures actually convey.

Over time, these inconsistencies naturally led to much confusion on the shop floor as well as in the front office. Quality practitioners lacked a single reliable source that described, compared, and discussed the advantages and disadvantages of all these various measures in an unbiased and consistent fashion. In view of the need for such a source, an idea was born for creating an easy to read reference book explaining the common measures of process capability in a clear and concise manner, as well as introducing several new approaches for measuring capability in special situations.

The end result of that idea is this text, which attempts to integrate all capability measures into one cohesive, unified system. It is written primarily for those who plan and run capability studies, analyze measures of capability as part of their daily work, assess the outcome of process modifications, and make machine scheduling decisions. A secondary audience consists of managers who prepare (or read) capability reports and need to thoroughly understand the information contained therein so they can clearly and correctly communicate these results to top management and/or customers. Thus, a full listing of personnel who would find this book beneficial includes quality engineers, SPC facilitators, manufacturing engineers, supplier quality engineers, manufacturing department supervisors, quality control managers, quality assurance managers, plant managers, directors of quality assurance, as well as all other mid- and top-level managers who request capability reports.

Extensive knowledge of statistics is not required to successfully use this book because the bulk of it is written for "shop floor" personnel. Readers with a good understanding of basic control charting methodology will have no problem comprehending any new ideas presented. For those readers interested in a more advanced study of a specific topic, an extensive list of references is given at the conclusion of each chapter. In addition to numerous case studies, where all required calculations are fully illustrated, there are many practice exercises to further help readers master the concepts. You are encouraged to take advantage of these learning opportunities as they offer valuable insights about how to select an appropriate capability measure, properly apply it, and then correctly analyze and interpret the results. Please note that although all exercises are based on real-life scenarios, all associated data and conclusions regarding capability have been altered so no proprietary information is disclosed.

The organization of this book follows a (hopefully) logical progression in presenting capability measures for almost all types of manufacturing situations. After a brief introduction to the subject of assessing process capability in Chapter 1, Chapter 2 stresses the importance of stabilizing the process output before any attempt is undertaken to measure its capability. A brief review of several popular variable-data control charts (along with their control limit formulas) is offered. Readers already familiar with this topic and its related terminology may wish to only skim this chapter.

Chapter 3 explains how the process parameters needed to measure capability (*e.g.*, μ, σ, and $\bar{p}$) are estimated from the various control charts. The crucial difference between short- and long-term process variation is identified and examined at this point. Process parameters are then used in Chapter 4 to help define the concept of process capability for both bilateral and unilateral specifications. Next, the difference between potential capability (what a process could possibly do under ideal conditions) and performance capability (what a process is actually doing under existing conditions) is presented. The chapter ends by explaining how to select critical process characteristics, establish relevant capability goals, and properly conduct (and when to update) a process capability study.

The next two chapters cover more than 50 different measures of capability for variable data. Chapter 5 deals with measures of potential capability, whereas Chapter 6 concentrates on measures of performance capability. An abundance of case studies highlighting the application of these various measures in numerous manufacturing environments is provided, along with a list of advantages and disadvantages for each measure. A flowchart is given for helping select the most appropriate potential capability measure for a process.

All measures of capability presented up to this point assume that the process output is normally distributed. Chapter 7 describes three separate methods (histograms, probability paper, goodness-of-fit tests) for checking this vital assumption. For a process having a non-normally distributed output, Chapter 8 explains two different procedures for estimating an "equivalent" capability measure.

Occasionally, the output of a process involves some form of attribute data, having either a Poisson or binomial distribution. Chapter 9 first reviews four common attribute-data control charts (c, u, np, p), then demonstrates how some of the more popular measures of capability are calculated for this special category of industrial data. The distinction between defects per unit (*dpu*) and defects per opportunity (*dpo*) is discussed along with guidelines for estimating capability at both unit and opportunity levels.

As there appears to be substantial interest in the subject of conducting machine capability studies (many companies now specify some minimum capability level in all contracts for buying new equipment), Chapter 10 is devoted exclusively to that topic. Two approaches (control chart and sequential s test) are offered for estimating how much of the total process variation is created by just the machine. Readers also learn how to establish reasonable machine capability goals based on process capability goals and how to eliminate the effect of gage variation from an estimate of machine capability.

Because estimates of process parameters are incorporated in their calculation, all capability measures are estimates of the true process capability. To fully appreciate the variation involved with a particular capability estimate, one should really consider the confidence interval associated with that measure. In Chapter 11, an appropriate confidence bound is derived for almost every measure of capability contained in this book.

Sometimes product received by the customer comes from the merged output of several similar process streams, *e.g.*, a sixteen-cavity die. Chapter 12 describes how a single capability measure (Average C_{PK}) can estimate the combined capability of such a process. At other times, a single, overall measure (Product C_{PK}) is needed for assessing the capability of a process (or an assembly line) producing multiple characteristics on the same product.

To date, very little developmental work has focused on how to apply capability measures to processes involving tool wear, autocorrelated data (usually generated by continuous processing operations), short production runs, geometric tolerances, within-piece variation (*e.g.*, taper, out of round, surface finish), dimensions without tolerances, and several other special circumstances. Chapter 13 presents techniques for control charting these processes, as well as some unique alternatives for assessing both their potential and performance capability.

The final chapter discusses the increasingly popular concept of "six-sigma" quality and reveals the related capability metrics and goals (for both variable and attribute data) needed to quantify this intriguing, but demanding, quality philosophy.

Several statistical tables are contained in the Appendix. Of note are Table II, which lists c_4 factors from 2 through 500 (by 1); Table III, giving Z values from 0 through 7.79 (by .01); Table VI, which has χ^2 values from 1 through 500 (by 1) degrees of freedom; and Table VII, containing Student's t values from 1 to 500 (by 1) degrees of freedom.

Some definitions and explanations are repeated in several chapters, thus allowing this text to function as a reference book. A reader may zero in on just the one capability measure of interest without having to wade through every preceding page in order to understand the terminology associated with that measure. Exercises are included after most chapters so this book may also serve as an undergraduate textbook.

Considerable care was taken to be as consistent as possible with ANSI, AIAG, and ISO 9000 current, or proposed, standards. Regrettably, complete agreement is not always possible because there are some differences between these standards. In addition, much research was done to provide references with differing viewpoints about particular capability measures so readers can be fully informed of any potential concerns.

Despite these precautions, certain definitions and methods for measuring process capability contained in this book may be considered controversial by some, and I take full responsibility for all opinions expressed. Readers who take exception are encouraged to contact me so their reservations can be incorporated into future editions, thus making this book an even more valuable reference for those in the quality profession.

Writing this book has been an extremely enjoyable (and lengthy) learning experience for me. I hope it is as enjoyable and worthwhile for you (my customer), the reader. As always, best of luck in all your quality-improvement efforts.

Davis R. Bothe
Cedarburg, WI
drbothe@sprynet.com

Measuring Process Capability

Chapter 1

Introduction

*"To understand, you
must first measure."*

Most books on statistical process control (SPC) emphasize defining a manufacturing process, applying the correct control chart, interpreting patterns on the chart and working to achieve stability of the process. This is definitely an important first step in quality improvement because stability means there is at least consistency in the process output. However, having a stable output does not necessarily imply the process is producing parts that satisfy customers.

At the top of Figure 1.1 is an $\overline{X}$, R chart displaying a good state of control for a process machining the depth of an O-ring groove. Below this chart is a histogram of the individual depth measurements collected to create the chart. Print specifications defining the minimum and maximum allowable measurements for groove depth are added to this histogram so that the process output can be compared to the customer requirement. As many grooves have their depth outside these specification limits[1], this in-control (stable) process is *consistently* producing a large percentage of nonconforming product, as well as a substantial percentage of unhappy customers.

This situation is fairly common because control charts are typically placed on "problem" processes, the ones where nonconforming parts are being produced. This makes economical sense, as there is very little chance of saving money by reducing scrap and rework on a process that is manufacturing all conforming parts. Thus, process performance like that presented in Figure 1.1 should be expected quite often when conducting capability studies. One useful outcome of these studies is a prediction of the percentage of nonconforming parts, which enhances a company's ability to make better business decisions about process management.

---

[1] Throughout this text, the lower specification limit is abbreviated LSL (ISO 3534-2.2 uses LTL, for lower tolerance limit), whereas USL represents the upper specification limit (ISO 3534-2.2 uses UTL).

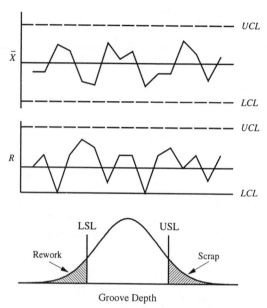

Fig. 1.1 An in-control process can consistently produce a large number of nonconforming parts.

As Ford and Motorola clearly demonstrated throughout the 1990s, there is a strong correlation between profitability and customer satisfaction. Measuring process capability is one of the best methods for quantifying how well customer requirements are being fulfilled. A process capability study provides valuable insight on how well an existing process is performing with regard to those requirements, what needs to be done to improve performance, and after these changes are made, how much of an improvement actually occurred (Goh). Without this knowledge, how can management have any hope of fully understanding their operations? In fact, Dr. Deming comments if he were a banker, he would not lend money to any company for new equipment until they demonstrated they were utilizing their current equipment to its fullest capability. Companies that know their capacity to delight customers have a powerful advantage over those that don't.

Process capability studies also offer product development engineers valuable information on whether or not current manufacturing equipment has the ability to accurately produce their proposed designs. This knowledge of how well a process can hold a given tolerance helps them choose between alternative methods of production, which aids them in designing for manufacturability and which ultimately leads to manufacturing superior products at lower cost.

Messina discloses the importance of capability studies for companies operating under just-in-time manufacturing systems. This kind of process information lowers work-in-process inventories by reducing work stoppages, and thereby improving the flow of material through the plant. In addition, capability studies can help to:

- schedule critical jobs to machines having appropriate capability;

- establish a target for the process average that minimizes total operating costs (see Feigenbaum);

- identify qualified operators;

- plan for efficient equipment usage;

- establish labor and material content based on process yield rates;

- make manufacturing problems visible and quantifiable;

- prioritize areas in need of quality improvement;

- verify that process-improvement efforts are successful;

- track capability improvement over time;

- communicate information concerning product quality to customers;

- accept new, or reconditioned, equipment;

- benchmark processes in other industries;

- develop realistic tolerances based on actual process performance;

- plan production methods for new products;

- set pricing of products: higher markups are needed to cover the costs associated with incapable processes;

- select qualified suppliers;

- determine sampling frequency for control charts: with higher capability, the interval between subgroups may be increased;

- decide if 100 percent inspection is necessary: part containment and 100 percent inspection must be implemented for every process not meeting its mandated capability goal;

With all these benefits, it's no wonder the popularity of capability measures continues to grow. However, like all tools, they must be used wisely. This means the concept of process capability should be familiar to almost everyone in your organization, including both technical and non-technical personnel. Any terminology associated with this subject must therefore be relatively easy to understand and provide a common language for discussing quality on the shop floor, in the corporate boardroom, with suppliers, as well as with customers. Unfortunately, many diverse measures of process capability have been proposed over the years, usually with different mathematical symbols and inconsistent terminology. A small handful of these measures are listed in Figure 1.2. This proliferation of techniques has caused quite a bit of confusion on how to properly conduct a capability study, as well as how to correctly interpret (and report) the results.

By identifying the basic principles and strategies involved with measuring capability, this book offers quality practitioners a means of translating this "Tower of Capability Babel" so they may fully understand (and appreciate) the more commonly-used capability indexes. The entire discipline of properly measuring process capability is integrated into one cohesive and unified system, with step-by-step instructions listed for assessing the capability of almost every type of manufacturing in a wide variety of industries. By helping manage and reduce process variability, capability measures significantly diminish the level of uncertainty in making business decisions.

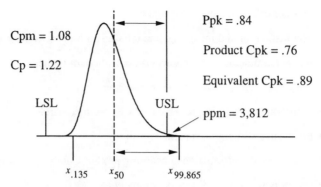

Fig. 1.2 Many different measures of process capability exist.

In addition to covering all the conventional measures, several new, and hopefully worthwhile, measures are presented for unique circumstances where standard procedures do not readily apply. However, the objective is not to have quality practitioners employ all measures in every study, but, by gaining an awareness of available alternatives, to be better able to select the most pertinent ones, apply them correctly, and fully comprehend their meaning. In support of that objective, a list of both advantages and disadvantages is supplied for each measure.

Although most capability indexes are easily calculated, they may not always measure the right aspect of a particular process. Because the yardstick selected to assess capability exerts a strong influence on how manufacturing operations are conducted, choosing the wrong capability measure may lead to such undesirable outcomes as suboptimal process performance, misallocated resources, increased waste, unhappy customers, and lost profits.

But before any discussion about specific capability measures can begin, the vital issue of process stability must be addressed. Control of a process must be achieved first, *before* any attempt is made to measure capability or estimate the percentage of nonconforming product. Always keep in mind that capability implies stability. Without stability, no reliable predictions can be formulated about process capability, nor any other aspect of the process. That's why the entire next chapter is devoted exclusively to the extremely important topic of process stability.

1.1 References

Deming, W. Edwards, *Out of the Crisis*, 1986, Massachusetts Institute of Technology, Cambridge, MA, p. 13

Feigenbaum, Armand V., *Total Quality Control*, 3rd edition, 1983, McGraw-Hill Book Co., New York, NY, p. 787

Goh, T.N., "Essential Principles for Decision-making in the Application of Statistical Quality Control," *Quality Engineering*, Vol. 1, No. 3, 1989, p. 247-263

Messina, William S., "JIT and SQC: Partners in Productivity," *42nd ASQC Annual Quality Congress Transactions*, May 1988, Dallas, TX, pp. 783-787

Chapter 2

Importance of Process Stability

*"To be perfectly honest, we don't care if the process is
stable or not, we just want to know what the capability is."*

Quote from a Quality Control Manager
in the aerospace industry.

Capability studies involve forecasting the future state of the process output. This is an
impossible task if past process performance doesn't provide a sound basis for prediction.
Thus, before any type of meaningful capability study can be undertaken, the process being
studied must be stable (Gunter). An unstable process is not well-defined and behaves in an
unpredictable fashion, as is seen in Figure 2.1. It is extremely difficult to summarize how
this process performed in the past, and there is absolutely no way of knowing what the process
output will look like in the future. Because of this instability, there is no way of assessing
its current, or future, ability to satisfy customer requirements.

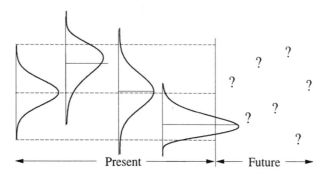

Fig. 2.1 An unstable process has no predictable behavior.

When a process is stable, it is repeatable, well-defined, and predictable, as illustrated in
Figure 2.2. Reliable estimates of process parameters can be made to help calculate meaningful
measurements of current process capability. Because the process is predictable, these results

5

can be expected to reflect process performance in the future as well, just as long as the process remains stable (Wheeler and Chambers, 1986). Promises can be made, and kept, to customers about expected levels of quality.

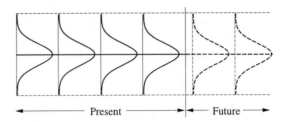

Present ◄────►◄──── Future ────►

Fig. 2.2 Future performance can be predicted for a stable process.

Process stability furnishes a high degree of assurance that the future will closely resemble the past. But how does one know if a process is stable? One of the best ways is provided by statistical process control (SPC) charts.

2.1 Definition of Stability

A process output is considered stable when it consists of only *common-cause* variation. Common-cause variation originates from the basic elements of a manufacturing process, which are typically the machine, manpower (operator), material, work method, and measurement system (the 5 Ms). The first step in verifying stability is to plot measurements of the process output on an appropriate control chart, then interpret the results.

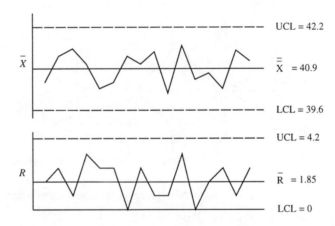

Fig. 2.3 Example of an $\overline{X}, R$ control chart displaying good control.

For an $\overline{X}, R$ chart like the one displayed in Figure 2.3, stability means all subgroup averages and ranges are between their respective control limits and display no evidence of *assignable-source* variation. Assignable-source variation is unplanned and originates outside of the expected operating conditions for the process. Examples are a plugged coolant line, a chip wedged beneath the fixture, a broken tool, or a spike in air pressure. These unpredictable events usually act to change process variation by an unknown amount.

The appearance of nonrandom patterns on the control chart provides strong evidence that assignable-source variation is present in the process. Several types of nonrandom patterns (called "out-of-control" conditions) are listed here:

- Points outside a control limit

- Runs: 7 or more consecutive points on one side of the chart's centerline

- Trends: 7 or more points moving consecutively upward (or downward)

- Cycles: repeated patterns

- Hugging the center line: 12 consecutive points lying in the middle third of the control chart

- Hugging the control limits: a large number of points near the control limits and very few near the centerline of the chart

The presence of assignable-source variation makes the correct assessment of process capability, which evaluates only common-cause variation, difficult, if not impossible. Below are quotes by several prominent statisticians on the importance of achieving process stability *before* attempting to measure any aspect of process capability.

> One point cannot be overemphasized: *the process capability cannot be estimated until a state of statistical control has been achieved.* (Pyzdek)

> The presence of special causes makes capability evaluation not meaningful. (Kane, p. 279)

> Process capability . . . refers to the product uniformity resulting from a process which is in a state of statistical control, *i.e.*, in the absence of time-to-time "drift" or other assignable causes of variation. (Juran *et al*, p. 9-16)

> The process must be stable in order to produce a reliable estimate of process capability. (Montgomery, p. 287)

> For a process-capability study to be meaningful, the process being analyzed should be . . . *in a state of statistical control*. (Feigenbaum, p. 779)

> A process has a capability only if it is stable. (Deming, p. 314)

> There is no process, no capability, and no meaningful specifications, except in statistical control. (Deming, p. 404)

Because stability is so important when measuring capability, it is worthwhile to spend some time reviewing basic control charting concepts to understand how assignable-source variation is separated from common-cause variation. Note that the following is not meant to be a full course on control charting theory or methodology. For a more detailed explanation, see Montgomery, Grant and Leavenworth, or AIAG.

2.2 Review of Variable-Data Control Charts

The output of every stable process follows some type of distribution. The "bell-shaped" curve, or normal, distribution displayed in Figure 2.4 is a common model for many manufacturing situations involving variable data.

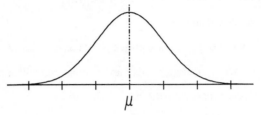

Fig. 2.4 Distribution of individual measurements, *X*.

The two process parameters describing a particular normal distribution are called μ (pronounced "mew") and σ (pronounced "sigma"). μ describes the location (the process average), while σ indicates spread (the process standard deviation). They are calculated with the following formulas, where *N* is the total number of items in the process and X_i represents the measurement of an individual item.

$$\mu = \frac{\sum\limits_{i=1}^{N} X_i}{N} \qquad \sigma = \sqrt{\frac{\sum\limits_{i=1}^{N} (X_i - \mu)^2}{N}}$$

Areas underneath the normal curve have been compiled for these two parameters with some of the more popular ones listed in Table 2.1 (Appendix Table III has a complete listing).

Table 2.1 Areas under the normal curve.

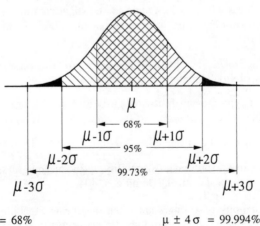

$\mu \pm 1\sigma = 68\%$	$\mu \pm 4\sigma = 99.994\%$
$\mu \pm 2\sigma = 95\%$	$\mu \pm 5\sigma = 99.99994\%$
$\mu \pm 3\sigma = 99.73\%$	$\mu \pm 6\sigma = 99.9999998\%$

Subgroup Statistics Versus Process Parameters

μ and σ are calculated with measurements from *every* part produced by the process. In most cases it is not possible to collect measurements from all items of a process's output, usually because of cost. As all N measurements are not available, the process parameters cannot be directly calculated. Luckily, μ and σ can be accurately *estimated* from a smaller subset of sample measurements.

With control charts, subgroups of several consecutive pieces are collected from a process at different times throughout the day. From the measurements in each subgroup, several statistics are computed, usually the subgroup average and a measure of the within-subgroup variation. Shortly, it will be explained how these subgroup statistics help estimate the unknown process parameters.

Some typical subgroup statistics associated with variable-data control charts are given in Figure 2.5, where n denotes the subgroup size, $\overline{X}$ (read X bar) is called the subgroup average, R is the subgroup range, and S is the subgroup standard deviation.

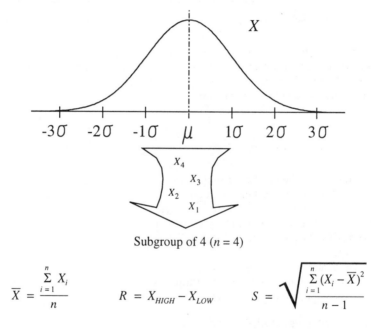

Subgroup of 4 $(n = 4)$

$$\overline{X} = \frac{\sum\limits_{i=1}^{n} X_i}{n} \qquad R = X_{HIGH} - X_{LOW} \qquad S = \sqrt{\frac{\sum\limits_{i=1}^{n}(X_i - \overline{X})^2}{n-1}}$$

Fig. 2.5 Typical statistics calculated from subgroup data.

To help monitor a process, these subgroup statistics are usually plotted on the appropriate type of control chart, and then analyzed for control (stability).

Example 2.1 The $\overline{X}, R$ Chart

One of the most common variable-data charts is the $\overline{X}, R$ chart. The subgroup ranges are plotted on the R chart, while the subgroup averages are plotted on the $\overline{X}$ portion, as is illustrated in Figure 2.6 for a subgroup size of four.

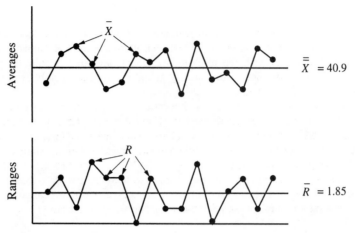

Fig. 2.6 Subgroup averages and ranges plotted on a control chart for $n = 4$.

When subgroup statistics from at least twenty subgroups are plotted, the subgroup statistics are first averaged and then used to estimate the process parameters μ and σ, as is shown below. The "^" symbol appearing above a process parameter indicates this quantity is an *estimate* for the true value of that parameter.

$$\hat{\mu}_{\bar{X}} = \overline{\overline{X}} = \frac{\sum\limits_{i=1}^{k} \overline{X}_i}{k} \qquad \hat{\sigma} = \frac{\overline{R}}{d_2} \quad \text{where} \quad \overline{R} = \frac{\sum\limits_{i=1}^{k} R_i}{k}$$

k represents the number of subgroups plotted on the chart (some textbooks use m instead of k), while d_2 is a constant based on the subgroup size, and is found in Appendix Table I.

The centerline of the $\overline{X}$ chart, $\overline{\overline{X}}$ (called X double bar), estimates the average of the $\overline{X}$s, which will later be shown to help estimate the process average. Comparing $\overline{\overline{X}}$ to the desired average for the process provides an indication of the accuracy of this process to produce parts on target.

The centerline of the range chart, $\overline{R}$, when divided by d_2 provides an estimate of the process standard deviation. This estimate furnishes information about the precision (spread) of the process output around its average.

Distribution of X vs. Distribution of $\overline{X}$

When a process is in control, the distribution of all subgroup averages collected from it has three important properties. First, the average of all subgroup averages, $\mu_{\bar{X}}$, is equal to the average of the individuals, μ (Grant and Leavenworth, p. 59). In this text, a process parameter with no subscript attached signifies that it pertains to the distribution of individuals.

$$\mu_{\bar{X}} = \mu = \frac{\sum\limits_{i=1}^{N} X_i}{N}$$

Thus, $\overline{\overline{X}}$, which is an estimate of $\mu_{\bar{X}}$, is therefore also an estimate of μ.

$$\hat{\mu} = \hat{\mu}_{\bar{X}} = \bar{\bar{X}} = \frac{\sum\limits_{i=1}^{k} \bar{X}_i}{k}$$

The second property is perhaps the most crucial of the three as it is the reason why control charts are so very powerful in detecting process changes. The standard deviation of the $\bar{X}$s (labeled $\sigma_{\bar{X}}$) is directly related to the standard deviation of the individuals, σ, as indicated by the following formula (Grant and Leavenworth, p. 59).

$$\sigma_{\bar{X}} = \frac{\sigma}{\sqrt{n}}$$

The next section describes in detail why this relationship between the variation of subgroup averages and the variation of individuals is so vitally important to control charts.

A third property involves the central limit theorem, which specifies that subgroup averages tend to form a normal distribution (Wheeler and Chambers, 1992, p. 78). This is true no matter what shape the distribution of individuals has. Because of this theorem, control limits of an $\bar{X}$ chart are symmetrical about $\bar{\bar{X}}$.

To better understand the impact of these three properties on control charts, consider the distribution of individuals (X) displayed in Figure 2.7. Assuming a normal distribution, these individuals are centered at the average, μ. Between $\mu - 3\sigma$ and $\mu + 3\sigma$ lie the middle 99.73 percent of the process output. Because of the first property mentioned above, the distribution of subgroup averages ($\bar{X}$) has its average of $\mu_{\bar{X}}$ also centered at μ.

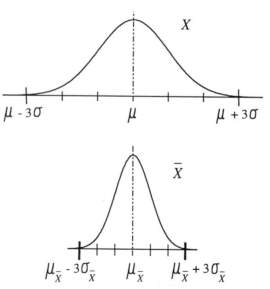

Fig. 2.7 Relationship between distribution of individuals and subgroup averages.

The upper control limit (*UCL*) for subgroup averages is set at $3\sigma_{\bar{X}}$ above $\mu_{\bar{X}}$, whereas the lower control limit (*LCL*) is set at $3\sigma_{\bar{X}}$ below $\mu_{\bar{X}}$. Thus, 99.73 percent of all subgroup averages should fall between these two limits when the process is in control.

$$LCL_{\overline{X}} = \overline{\overline{X}} - 3\sigma_{\overline{X}} \qquad\qquad UCL_{\overline{X}} = \overline{\overline{X}} + 3\sigma_{\overline{X}}$$

In practice, $A_2\overline{R}$ replaces $3\sigma_{\overline{X}}$ in these formulas. The A_2 factors are a function of the subgroup size, and are listed in Appendix Table I for n of 1 through 10.

$$3\hat{\sigma}_{\overline{X}} = \frac{3\hat{\sigma}}{\sqrt{n}} = \frac{3\overline{R}}{d_2\sqrt{n}} = A_2\overline{R} \qquad \text{where } A_2 = \frac{3}{d_2\sqrt{n}}$$

This makes the $\overline{X}$ control limit calculations as follows:

$$LCL_{\overline{X}} = \overline{\overline{X}} - A_2\overline{R} \qquad\qquad UCL_{\overline{X}} = \overline{\overline{X}} + A_2\overline{R}$$

Control limits for the range chart are developed in a similar manner. D_3 and D_4 for n equal to 2 through 10 are also given in Appendix Table I.

$$LCL_R = D_3\overline{R} \qquad\qquad UCL_R = D_4\overline{R}$$

As an example of these calculations, recall that the chart in Figure 2.6 has $\overline{R}$ equal to 1.85, $\overline{\overline{X}}$ equal to 40.9, and n equal to 4. The $\overline{X}$ chart control limits are calculated with these formulas:

$$UCL_{\overline{X}} = \overline{\overline{X}} + A_2\overline{R} = 40.9 + .729(1.85) = 42.2$$

$$LCL_{\overline{X}} = \overline{\overline{X}} - A_2\overline{R} = 40.9 - .729(1.85) = 39.6$$

The range chart control limits for this process are determined as follows:

$$UCL_R = D_4\overline{R} = 2.282(1.85) = 4.2$$

$$LCL_R = D_3\overline{R} = 0(1.85) = 0$$

These limits and the completed chart are displayed in Figure 2.8. One of the chart's purposes is to assist in maintaining a product characteristic (or a process variable) at a prescribed level, while simultaneously preserving process consistency (variation).

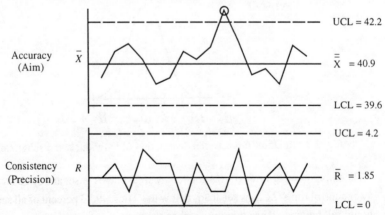

Fig. 2.8 Example of an $\overline{X}, R$ chart with a point out of control.

A subgroup average should rarely fall outside a control limit when the process is in control (only a .27 percent chance). If one does, it is assumed the process is no longer in control. Some type of assignable-source variation has apparently changed the process in some manner to disrupt the naturally occurring pattern of points on the $\overline{X}$ chart. Work should begin immediately to identify the source of this abnormal variation.

2.3 Why Control Charts Are Better Than Inspection

As demonstrated by this next example, the relationship between σ and $\sigma_{\overline{x}}$ is the main reason why control charts are better than inspection for detecting process changes. Imagine a process being monitored by inspection where one part is checked each hour. Initially, the process average is 0 and the standard deviation is 2, as is shown in Figure 2.9. As long as the inspector finds measurements within the specification limits of ± 6, there is no call for alarm and production is allowed to continue. If a measurement appears outside these limits, production is halted and prompt corrective action taken.

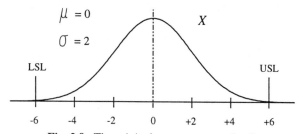

Fig. 2.9 The original process centered at 0.

Suppose an assignable source of variation causes the process average to shift upward a distance of 6 units (a 3σ change). Due to being centered at the upper specification limit, half the process output is now outside the USL. As Figure 2.10 clearly illustrates, this is certainly a very grim situation.

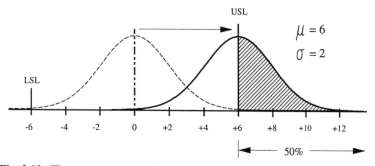

Fig. 2.10 The process output after an upward shift of 6 in the process average.

What is the probability that an inspector would catch this enormous change with the next piece checked? Because half of the pieces are still within specification, there is a 50 percent chance of selecting such a part and mistakenly deciding to continue running this modified process. Not very good odds, yet many manufacturing companies have rules similar to this:

"check one part every hour," or "check one part out of every 20."

What happens if a control chart is chosen to monitor this process instead of inspection? Since the average of the $\overline{X}$s equals the average of the individuals ($\mu_{\overline{x}} = \mu$), the centerline of the $\overline{X}$ chart equals zero, the average of the process output.

$$\mu_{\overline{x}} = \mu = 0$$

Assuming a subgroup size of 4 is taken once an hour, $\sigma_{\overline{x}}$ can be calculated from the second property given in the previous section.

$$\sigma_{\overline{x}} = \frac{\sigma}{\sqrt{n}} = \frac{2}{\sqrt{4}} = \frac{2}{2} = 1$$

For an in-control process, $3\sigma_{\overline{x}}$ is similar to $A_2\overline{R}$ (Podolski). Thus, the *UCL* of the $\overline{X}$ chart is located at the centerline plus $3\sigma_{\overline{x}}$, while the *LCL* is $3\sigma_{\overline{x}}$ below the centerline.

$$LCL_{\overline{x}} = \overline{\overline{X}} - 3\sigma_{\overline{x}} = 0 - 3(1) = -3 \qquad UCL_{\overline{x}} = \overline{\overline{X}} + 3\sigma_{\overline{x}} = 0 + 3(1) = 3$$

As long as the process remains stable, the vast majority of future subgroup averages (99.73 percent) will fall between these control limits of ± 3 and a decision will be made to continue running the process (see Figure 2.11).

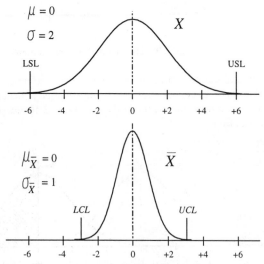

Fig. 2.11 The $\overline{X}$ distribution is also centered at 0, but with less variation than the X distribution.

Suppose the distribution of individuals shifts up a distance of six units (3σ) as before, and is now centered at the USL of 6. When the process average shifts, so does the average of the subgroup averages (recall that $\mu_{\overline{x}}$ is always equal to μ). Thus, the new $\mu_{\overline{x}}$ is now also 6.

$$\mu_{\overline{x}} = \mu = 6$$

As there is no change in the process standard deviation or subgroup size, $\sigma_{\bar{X}}$ is still 1. Thus, the lower $3\sigma_{\bar{X}}$ tail for the shifted distribution is located at +3 and the upper $3\sigma_{\bar{X}}$ tail at +9.

$$\text{Lower tail} = \mu_{\bar{X}} - 3\sigma_{\bar{X}} = 6 - 3(1) = +3 \qquad \text{Upper tail} = \mu_{\bar{X}} + 3\sigma_{\bar{X}} = 6 + 3(1) = +9$$

This new condition is displayed in Figure 2.12.

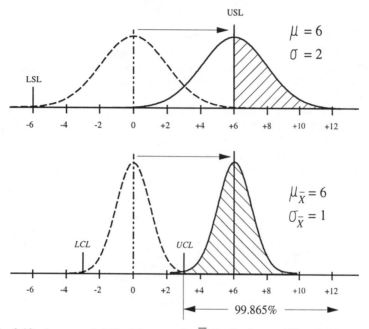

Fig. 2.12 An upward shift of 6 causes the $\bar{X}$ distribution to shift up by 6 as well.

Suppose the first subgroup taken after this change in process average consists of these four measurements.

$$X_1 = 3 \qquad X_2 = 4 \qquad X_3 = 4 \qquad X_4 = 3$$

The average of this subgroup is 3.5.

$$\bar{X} = \frac{3 + 4 + 4 + 3}{4} = 3.5$$

Even though all four individual measurements from the new distribution are within print specifications, the subgroup average of 3.5 falls *above* the *UCL* of 3. This provides the operator with a signal to stop the process and investigate why a change has caused the process average to shift significantly higher.

What percentage of the subgroup averages from this new process will fall above the *UCL* of 3 and alert the operator of a change in the process average? Because the lower $3\sigma_{\bar{X}}$ tail of the shifted $\bar{X}$ distribution is equal to the *UCL*, only .135 percent of the subgroup averages would fall within the control limits (review Figure 2.12). Conversely, this means 99.865

percent will be above the *UCL*. When the first subgroup is collected after this change and its subgroup average plotted on the $\overline{X}$ chart, there is almost a 100 percent chance this shift in the process average will be detected. Compare this result to the 50 percent chance of detecting the change with inspection. Because the variation of subgroup averages is much less than that for individuals, control charts are much more powerful at detecting process changes than any general inspection plan.

Power of the $\overline{X}$ Chart to Detect Changes

The $\overline{X}$ chart helps detect changes in the process average. Large shifts in the process average are caught quite readily as just demonstrated, however, very *small* changes in the process average have only a slight chance of being noticed (Kane, p. 230). There is a direct relationship between the size of the change and the probability of an out-of-control point on the $\overline{X}$ chart (Heyes).

This relationship is displayed in Figure 2.13 for a subgroup size of 4 (the code for a BASIC software program to calculate this curve is given by Hurayt). To allow this graph to apply to any process, the size of the shift is expressed in standard deviations on the *x* axis. For example, in the last section a change of 6 units occurred in the process average. Since the process standard deviation is 2, this change is expressed as a 3σ shift in the process average $(3 = 6/2)$. On the *y*-axis, the probability of the next subgroup average being outside a control limit is read as 99.865 percent for this magnitude of shift.

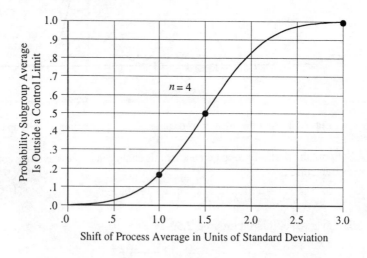

Fig. 2.13 Probability of detecting a shift in process average.

However, the $\overline{X}$ chart has only a 50 percent chance of detecting a change of 1.5σ in the process average and just a 16 percent chance of catching a 1σ shift. Since changes smaller than 1.5σ have less than a 50 percent chance of being caught, it is quite probable, that over time, many minor changes in the process average will come and go without being noticed on the control chart!

Because these changes in the process average between subgroups may go unrecognized on the $\overline{X}$ chart, process variation occurring over long periods of time will usually be slightly greater than variation happening over a very short period of time, *even for a chart appearing to be in control* (Burns). This important difference in variation becomes extremely critical when evaluating capability measures, as will be explained in Chapter 5.

The sensitivity of a control chart to detect changes in the process average can be improved by increasing the subgroup size. Notice how the curves displayed in Figure 2.14 become steeper as *n* increases. The steeper the curve, the greater the probability a shift of a given magnitude is discovered.

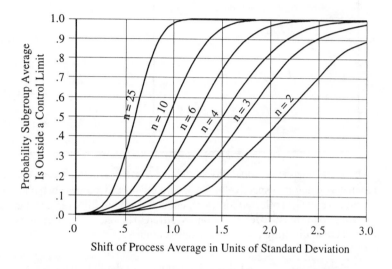

Fig. 2.14 Probability of detecting a shift in process average for different subgroup sizes.

Unfortunately, a larger subgroup size requires more time and money to collect, measure, and have its plot points calculated. Note that the sensitivity of a chart to detect a given size change increases by the square root of *n*.

$$\sigma_{\bar{X}} = \frac{\sigma}{\sqrt{n}}$$

More importantly, the cost of sampling increases directly with *n*. Thus, there is a diminishing marginal return to increasing the subgroup size. Sometimes this is a worthwhile trade-off when the characteristic under study is very critical, and a small shift in the process average could cause nonconforming parts to be produced. As an alternative to increasing *n*, the time between subgroups could be decreased to bolster chart sensitivity. The application of more out-of-control rules also helps enhance the chart's ability to detect process changes. Keats *et al* describe an effective method for determining an appropriate subgroup size and sampling frequency for $\bar{X}$ charts.

2.4 Control Limit Formulas for Variable-Data Charts

This section presents control limit formulas for three common variable-data charts, beginning with the very popular $\bar{X}, R$ chart, whose formulas are given in Table 2.2. The control limit constants for all variable-data charts are listed in Appendix Table I for *n* from 1 to 10. As before, *n* is the subgroup size and *k* is the number of subgroups. All formulas assume the measurements of individual pieces are independent of each other.

Table 2.2   Formulas for the $\overline{X}, R$ chart.

R Chart	$\overline{X}$ Chart
$R = X_{HIGH} - X_{LOW}$ $\overline{R} = \dfrac{\sum\limits_{i=1}^{k} R_i}{k} \qquad \dfrac{\overline{R}}{d_2} = \hat{\sigma}$ $UCL_R = D_4 \overline{R}$ $LCL_R = D_3 \overline{R}$	$\overline{X} = \dfrac{\sum\limits_{i=1}^{n} X_i}{n}$ $\overline{\overline{X}} = \dfrac{\sum\limits_{i=1}^{k} \overline{X}_i}{k} = \hat{\mu}_{\overline{X}} = \hat{\mu}$ $UCL_{\overline{X}} = \overline{\overline{X}} + A_2 \overline{R}$ $LCL_{\overline{X}} = \overline{\overline{X}} - A_2 \overline{R}$

Of special interest for capability studies are $\overline{\overline{X}}$ and $\overline{R}$, which are used to estimate the process parameters μ and σ, respectively. To accurately portray process behavior on a control chart, subgroups must be collected for a long enough period to fully represent the process under all operating conditions: different operators; different batches of material; high and low operating temperatures; differing levels of ambient humidity.

Table 2.3   Formulas for the $\overline{X}, S$ chart.

S Chart	$\overline{X}$ Chart
$S = \sqrt{\dfrac{\sum\limits_{i=1}^{n} (X_i - \overline{X})^2}{n-1}}$ $\overline{S} = \dfrac{\sum\limits_{i=1}^{k} S_i}{k} \qquad \dfrac{\overline{S}}{c_4} = \hat{\sigma}$ $UCL_S = B_4 \overline{S}$ $LCL_S = B_3 \overline{S}$	$\overline{X} = \dfrac{\sum\limits_{i=1}^{n} X_i}{n}$ $\overline{\overline{X}} = \dfrac{\sum\limits_{i=1}^{k} \overline{X}_i}{k} = \hat{\mu}_{\overline{X}} = \hat{\mu}$ $UCL_{\overline{X}} = \overline{\overline{X}} + A_3 \overline{S}$ $LCL_{\overline{X}} = \overline{\overline{X}} - A_3 \overline{S}$

Due to the difficulty in manually calculating the subgroup standard deviation, the $\overline{X}, S$ chart is employed almost exclusively where charting has been automated. However, it should be used whenever n is greater than 10, as the subgroup range loses some efficiency in assessing process variation for large subgroup sizes (Montgomery, p. 230). Formulas for the $\overline{X}, S$ chart are presented in Table 2.3. Of particular interest for capability studies are the centerlines $\overline{\overline{X}}$ and $\overline{S}$.

Even though the subgroup standard deviation provides a slightly better estimate of process variation, $\overline{X}, R$ charts are more popular because when charts are manually prepared, subgroup ranges are much easier to calculate.

The *IX & MR* chart is implemented in cases where the logical subgroup size is 1. The plot point on the individual X chart is just the measurement of this one part. However, two consecutive measurements are grouped together to calculate the moving range. The moving range (*MR*) is defined as being the absolute value of the difference between two consecutive *IX* plot points.

$$MR \text{ Plot Point } = |IX_i - IX_{i-1}|$$

Formulas for the centerlines and control limits of this chart are displayed in Table 2.4. Since a moving range cannot be calculated for the first individual measurement, k minus 1 is used in the denominator when calculating the average moving range. Of interest for capability studies are the centerlines, $\overline{X}$ and $\overline{MR}$, which are utilized in estimating μ and σ, respectively.

Table 2.4 Formulas for the *IX & MR* chart.

MR Chart	IX Chart
$MR = |IX_i - IX_{i-1}|$ $\overline{MR} = \dfrac{\sum\limits_{i=1}^{k-1} MR_i}{k-1}$ $\qquad \dfrac{\overline{MR}}{1.128} = \hat{\sigma}$ $UCL_{MR} = 3.27\,\overline{MR}$ $LCL_{MR} = 0$	$IX = IX_i$ $\overline{X} = \dfrac{\sum\limits_{i=1}^{k} IX_i}{k} = \hat{\mu}$ $UCL_{IX} = \overline{X} + 2.66\,\overline{MR}$ $LCL_{IX} = \overline{X} - 2.66\,\overline{MR}$

Because the *MR* plot-point calculation involves two *IX* values, the d_2, D_3, and D_4 factors for the moving range chart are based on a subgroup size of 2.

Within vs. Between Subgroup Variation

Control limits for an $\overline{X}$ chart are calculated from either $\overline{R}$ or $\overline{S}$. These two averages are derived from the variation occurring only *within* a subgroup. Under the premise of "rational" subgrouping (Montgomery, p. 113), pieces in a subgroup must be purposely selected to be as similar as possible. They should be produced under identical ("homogeneous") manufacturing conditions, meaning consecutive pieces are collected with no operator, tool, or material changes allowed while these pieces are made (Wheeler and Chambers, 1992, p. 99).

Therefore, any estimates of process variation derived from only within-subgroup variation (such as $\overline{R}$ or $\overline{S}$) reflect the process operating under these special, ideal conditions. This preferential sampling procedure generates an estimate of inherent (common-cause) process variation. Process variability may be reduced to this level, but no further without significant changes to the process. In this book, predictions of process variability based on only within-subgroup variation are called estimates of "short-term" process variation and provide an optimistic view ("best case") of product variability (AT&T; Montgomery, p. 215; Wheeler and Chambers, 1992, p. 100).

Suppose subgroups are collected every hour. Changes in operators, tools, material, temperature, and humidity will definitely occur *between* subgroups. Some of these changes may result in shifting the process average. As seen earlier in this chapter (Figure 2.13), a few of the smaller shifts may not be detected on the $\overline{X}$ chart, and therefore, no corrective action is taken. Missing these movements in the process average causes estimates of process variation based on both within- and between-subgroup variation to be equal to, or greater than, the short-term estimate, even when the process is in control. This more representative, or realistic estimate is referred to as "long-term" process variation (Fellers, p. 45).

$$\sigma_{LT} \geq \sigma_{ST}$$

The subscript "ST" is for short term, while "LT" is for long term. Chapter 5 reveals how this difference in estimating the process standard deviation can greatly influence estimates of process capability.

Notice that short-term process variation is used to calculate control limits on the $\overline{X}$ chart ($\overline{\overline{X}} \pm A_2 \overline{R}$ or $\overline{\overline{X}} \pm A_3 \overline{S}$). If plot points on the $\overline{X}$ chart, which display between-subgroup variation, fall between the control limits, which are based on within-subgroup variation, then the process is said to have achieved a state of statistical control, since it can manufacture parts hour after hour (long term) as consistently as it can produce consecutive parts (short term). σ_{LT} and σ_{ST} are approximately equal when the process is in statistical control. When the process is out of control, σ_{LT} will be substantially larger than σ_{ST} (Podolski, see also Figure 3.5).

It is incorrect to base control limits on the standard deviation derived by pooling all individual measurements collected for a control chart (Wheeler). The formula for this type of standard deviation is presented below, where kn is the total number of measurements collected, and S_{TOT} is the standard deviation of all these measurements.

$$S_{TOT} = \sqrt{\frac{\sum\limits_{i=1}^{kn} (X_i - \hat{\mu})^2}{kn - 1}}$$

This measure of process variation combines variability occurring both within and between subgroups. To preserve their statistical validity, control limits must be based solely on within-subgroup variation.

Another important consideration in the calculation of control limits is the handling of out-of-control subgroups. Measurements contained in these subgroups reflect more than just common-cause variation. In order for a control chart to properly detect assignable-cause variation, these measurements must be excluded from all calculations of control limits and centerlines. Because process parameters (μ and σ) are estimated from these centerlines, they reflect process performance under stable operating conditions.

$$\hat{\mu} = \overline{\overline{X}} = \frac{\sum\limits_{i=1}^{k} \overline{X}_i}{k} \qquad \hat{\sigma}_{ST} = \frac{\overline{R}}{d_2} = \frac{\sum\limits_{i=1}^{k} R_i}{k d_2}$$

Even though omitted from the calculation of control limits and centerlines, out-of-control points (and their explanations) are still left on the chart to guide problem-solving efforts.

2.5 Operational Definition of "In Control"

As this chapter has repeatedly emphasized, before a process parameter can be estimated from measurements collected on a control chart, the process must display evidence of stability, *i.e.*, demonstrate that it is in a good state of statistical control. Does this requirement mean every subgroup plotted on a chart must be in control before estimating μ or σ? As charts are typically applied to processes with a history of stability problems, there should be no surprise when an occasional subgroup does go out of control. In fact, that's why charts are kept on a process, to alert personnel when corrective action is needed. Thus, the concept of "statistically in control" is an ideal state that is rarely, if ever, obtained in real life.

So, if it is unlikely for any control chart to ever be in a "perfect" state of control, does this mean estimates of the process parameters can never be made? For control charts to be relevant, an operational definition of "in control" must be formulated. The inventor of control charts, Dr. Walter Shewhart, concluded a person would not be justified in stating a process is in control until obtaining a least 25 subgroups between out-of-control conditions. Thus, in practice, if there is an average of 25 in-control subgroups for each out-of-control condition, the process can be considered stable. This means no more than about 4 percent of the points can be out of control.

A similar percentage is recommended by Fellers (p. 40) and by McNeese and Klein (5 percent). Obviously, this definition may vary for different industries and, perhaps, even for different characteristics, depending their criticality. Data plotted on the control chart should also be collected over a long enough period of time (at least several days) to represent the full span of operating conditions, such as different operators, batches of materials, tools, setups, and temperatures.

2.6 Specification Limits vs. Control Limits

Quite often, confusion exists about the difference between control limits and specification limits. Perhaps this is because both terms include the word "limits." However, these two concepts have absolutely no mathematical relationship whatsoever to each other (Goh).

Specification limits help inspectors decide if measurements of a particular dimension are within print requirements by comparing each measurement to the specification limits. These limits help separate nonconforming parts from conforming, as is illustrated in Figure 2.15. When a measurement is between the LSL and USL, the part is considered to be conforming. If the measurement is either below the LSL, or above the USL, the part is judged to be nonconforming for this dimension.

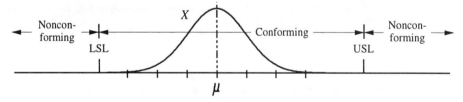

Fig. 2.15 Specification limits help determine if a part is conforming or nonconforming.

Control limits help operators determine if a process is stable (in control) or not by helping to separate assignable-cause variation from common-cause variation. When a subgroup statistic falls between the control limits, and there are no non-random patterns present, the process is assumed to be in control. Figure 2.16 shows this situation for the $\overline{X}$ chart. If a subgroup average is outside these control limits, the process is considered to be unstable.

Whether measurements collected to calculate a plotted statistic are in or out of specification has absolutely no effect on the interpretation of the control chart for stability. Recall from Tables 2.2 through 2.4 that specification limits are not used in the calculation of plot points, centerlines, or control limits.

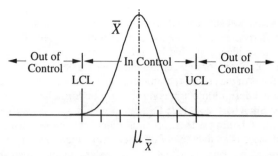

Fig. 2.16 Control limits help determine if a process is in or out of control.

Furthermore, an in-control point does not mean conforming product has been produced. The control chart only reveals if the process output is stable or not. A process can be in control and producing all parts outside the print specifications (Deming, p. 334). Conversely, it is possible for a process to be out of control, but still be making all parts within specification. A stable process does not imply there is an absence of variation, or that the variation is small, or even that the process output is within specification. It only means the process output is consistent.

In addition, as subgroup averages are not directly comparable to the specification limits (which are for individual measurements), it makes absolutely no sense to draw specification limits on the $\overline{X}$ portion of a control chart. This mixing of control limits and specification limits on one chart is perhaps the most common mistake made by companies just beginning their implementation of SPC.

Control charts help strike a balance between "tampering" with the process (making adjustments when none is required) and reacting to assignable causes when they are present. For an excellent example of this trade-off, read about the funnel experiment conducted by Dr. Deming, as described by Boardmen and Boardmen.

Once stability is achieved, the consistent process output must be compared to specifications for determining how well customer requirements are being met. Making and evaluating this comparison are the principal objectives for undertaking a capability study. But before any measure of capability can be computed, process parameters must first be correctly estimated, which is the main topic of the next chapter.

2.7 Summary

The output from every manufacturing process varies to some extent. When the observed variation is a result of only common (inherent) causes, the process is said to be stable. Owing to the separation of assignable and common-cause variation, control charts are one of the best tools available for demonstrating stability. Done correctly, control charting is more effective at detecting process changes than inspection.

To be statistically valid, measurements must be selected to minimize variation within the subgroup, while allowing variation to occur between subgroups. In addition, the measurements making up each subgroup must be independent of each other. One method for verifying this assumption is the Durbin-Watson test (Dielman).

Although nothing is certain in life except death, taxes, and process variation, at least through control charts we are able to gain some control over the latter. In addition, control charts also help track and improve quality in the following three ways.

- *Stabilize the process*. Control charts identify when out-of-control conditions occur in the process. This provides clues for identifying causes of these special conditions.

- *Improve the process*. Once a good state of control is attained, changes to the process can be tried and their effect on the process seen on the chart. If the proposed change is effective, it can be incorporated into the process for long-term improvement.

- *Measure process capability*. When control is established, meaningful estimates of process parameters can be made to quantify how well a process meets customer expectations. The natural variation of the process may now be compared to specification limits to reliably estimate the percentage of conforming parts.

For all subsequent discussion about measuring process capability in this book, all processes analyzed are assumed to have attained a good state of statistical control or the few out-of-control subgroups have been removed from the analysis. In presenting a case study on the zinc plating of wire, Clifford emphasizes the important role control charts play when conducting capability studies.

All product specifications listed on the engineering drawings are also assumed to be correct and to adequately reflect true customer requirements. If not, any comparison of the process output to these specifications is meaningless, and there is little sense in attempting to estimate, or evaluate, any measure of process capability.

Stability is always a prime concern, however, simply having a stable process does not assure customers' needs are being met. To make this evaluation, the capability of a process must be accurately assessed, correctly compared to its goal, and verified to equal, or exceed, this goal. But before any type of capability can be measured, estimates of process parameters must be properly obtained. Fortunately, this is the subject of the next chapter.

2.8 Exercises

2.1 Center Gap Size

An $\overline{X}, R$ chart is started on a process making corrugated boxes. The critical characteristic being monitored is the center gap between two folded flaps on the bottom of the box (Figure 2.17). After 24 subgroups of size 4 ($n = 4$) are collected over a two-day period, the following sums are obtained.

$$\sum_{i=1}^{24} R_i = .165 \, cm \qquad \sum_{i=1}^{24} \overline{X}_i = 7.6368 \, cm$$

Fig. 2.17

Calculate the appropriate centerlines and control limits for this chart.

2.2 Keyway Width

An $\overline{X}, S$ chart is monitoring the width of a keyway cut on a milling machine. After 20 subgroups ($k = 20$) of size 5 ($n = 5$) are collected, the sum of the 20 subgroup standard deviations is .28952 mm, while the sum of the 20 subgroup averages is 99.827 mm. Calculate control limits for this chart.

$$\sum_{i=1}^{20} S_i = .28952 \text{ mm} \qquad \sum_{i=1}^{20} \overline{X}_i = 99.827 \text{ mm}$$

2.3 Particle Contamination

Readings are recorded on an *IX & MR* chart ($n = 1$) for particle contamination (in *ppm*) of a cleaning bath for printed circuit boards after they pass through a wave soldering operation. When 25 readings are plotted, the following totals are derived. Use these to determine control limits for this chart.

$$\sum_{i=1}^{24} MR_i = 327 \text{ } ppm \qquad \sum_{i=1}^{25} X_i = 3183 \text{ } ppm$$

Remember that there is one less moving range than *IX* values.

2.4 Solder Pull Test

Following assembly, a sample of PC boards (Figure 2.18) is taken to an inspection station for several tests. During the solder pull test, one of the wire leads that was soldered to a board is pulled until it breaks free. The force required to remove the wire is charted on an $\overline{X}, R$ chart ($n = 4$).

After three days, 23 subgroups have been collected, with the sum of their $\overline{X}$s equal to 1086.98 grams, while the sum of their ranges is 211.6 grams. Compute $\overline{\overline{X}}$ and $\overline{R}$, then use these to calculate control limits for both charts.

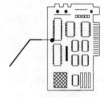

Fig. 2.18

2.5 Shaft Length

The tapered steel shaft displayed in Figure 2.19 is made from bar stock cut to the desired length by a sawing operation. As shaft length is important for the proper functioning of this part, management has decided to monitor this characteristic with an $\overline{X}, R$ chart. A subgroup of five consecutive shafts is taken every half hour, with the individual length measurements recorded in Table 2.5 (data are deviations from the nominal length requirement).

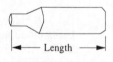

Fig. 2.19

Calculate the subgroup averages and ranges, and then plot them on a control chart. When all points are plotted, calculate control limits and interpret the results. Save this chart as it will be used in several future exercises.

Table 2.5 Shaft length measurements.

Subgroup Number	Length Measurements	$\overline{X}$	R
1	0, 1, -2, 0, -1		
2	-2, 0, 1, 3, 2		
3	-3, -1, 0, -2, -2		
4	-3, 1, 3, 0, -2		
5	1, -1, -2, 0, 1		
6	2, 3, 1, 0, 1		
7	1, -2, 1, -1, -2		
8	0, -3, -1, 2, 1		
9	2, 0, 1, -2, -1		
10	3, -1, -1, 0, 2		
11	1, 0, 0, -1, -1		
12	0, -3, -1, -1, 0		
13	1, -1, 0, -2, -1		
14	0, 2, -1, 2, 1		
15	-2, -3, 0, -1, -1		
16	-1, 0, 0, -2, 2		
17	1, -2, -1, 0, 1		
18	0, 0, -3, 1, -2		
19	-1, 0, 3, 0, 2		
20	-4, 1, -1, 0, -1		
21	-1, 1, 1, 0, 0		
22	2, 0, -2, -2, 0		
23	-3, -1, 0, 2, 1		
24	0, 3, 0, -1, 0		
25	0, -2, 1, -1, -1		

2.6 Rib Width

A French supplier in the aerospace industry produces metal "ribs" for use in the construction of wings for military aircraft. A key characteristic is the width of the rib at the front end. As the machining cycle is quite long, the department supervisor decides to monitor the process output for rib width with an *IX & MR* chart.

Help the supervisor plot the 30 width measurements recorded in Table 2.6 (on the next page), calculate the control limits, and interpret the chart. Construct a histogram of these measurements. Save the control chart and histogram as they will be used in later exercises.

Table 2.6 Rib width measurements (cm).

Rib Number	IX	Rib Number	IX
1	1.572	16	1.576
2	1.576	17	1.575
3	1.574	18	1.571
4	1.580	19	1.577
5	1.577	20	1.577
6	1.570	21	1.578
7	1.579	22	1.572
8	1.577	23	1.570
9	1.573	24	1.575
10	1.574	25	1.581
11	1.578	26	1.582
12	1.568	27	1.578
13	1.591	28	1.578
14	1.575	29	1.575
15	1.576	30	1.573

2.9 References

AIAG (Automotive Industry Action Group), *Fundamental Statistical Process Control Reference Manual*, 2nd printing, 1995, AIAG, Southfield, MI

AT&T, *Statistical Quality Control Handbook*, 2nd edition, 1984, Delmar Printing Co., Charlotte, NC, p. 61

Boardmen, Thomas J., and Boardmen, Eileen C., "Don't Touch that Funnel!," *Quality Progress*, Vol. 23, No. 12, December 1990, pp. 65-69

Burns, Clarence R., "SPC Training and Decision Making with OC Curves," *46th Annual Quality Congress Transactions*, Nashville, TN, May, 1992, p. 515

Clifford, Paul C., "A Process Capability Study Using Control Charts," *Journal of Quality Technology*, Vol. 3, No. 3, July 1971, pp. 107-111

Deming, W. Edwards, *Out of the Crisis*, 1986, Massachusetts Institute of Technology, Cambridge, MA

Dielman, T.E., *Applied Linear Regression*, 1991, PWS-Kent Publishing Co., Boston, MA

Feigenbaum, Armand, *Total Quality Control*, 3rd edition, 1983, McGraw-Hill, New York, NY, p. 779

Fellers, Gary, *SPC for Practitioners: Special Cases and Continuous Processes*, 1991, ASQC Quality Press, Milwaukee, WI

Goh, T.N., "Essential Principles for Decisionmaking in the Application of Statistical Quality Control," *Quality Engineering*, Vol. 1, No. 3, 1989, p. 256

Grant, Eugene L., and Leavenworth, Richard S., *Statistical Quality Control*, 6th edition, 1988, McGraw-Hill Book Co., New York, NY

Gunter, Bert, "Part 1: What is a Process Capability Study?," *Quality Progress*, Vol. 24, No. 2, February 1991, pp. 97-100

Heyes, Gerald B., "Control Chart Sensitivity," *Quality*, 26, 8, August 1987, p. 81

Hurayt, Gerald, "Operating Characteristic Curves for the X-bar Chart," *Quality*, Vol. 31, No. 9, September 1992, p. 59

Juran, Joseph M., Gryna, Frank M., Jr., and Bingham, R. S., Jr., *Quality Control Handbook*, 3rd edition, 1979, McGraw-Hill, New York, NY, p. 9-16

Kane, Victor E., *Defect Prevention*, 1989, Marcel Dekker, New York, NY

Keats, J. Bert, Miskulin, John D., and Runger, George C., "Statistical Process Control Scheme," *Journal of Quality Technology*, Vol. 27, No. 3, July 1995, pp. 214-225

McNeese, William H., and Klein, Robert A., *Statistical Methods for the Process Industries*, 1991, ASQC Quality Press, Milwaukee, WI, p. 427

Montgomery, Douglas C., *Introduction to Statistical Quality Control*, 1985, John Wiley & Sons, New York, NY

Podolski, Garry, "Standard Deviation: Root Mean Square versus Range Conversion," *Quality Engineering*, Vol. 2, No. 2, 1989-90, pp. 156-157

Pyzdek, Thomas, *Pyzdek's Guide to SPC, Volume Two*, 1992, Quality Press, Milwaukee, WI, p. 57

Shewhart, Walter A., *Statistical Method from the Viewpoint of Quality Control*, 1986, Dover Publications, Inc., New York, NY, p. 37

Wheeler, Donald J., and Chambers, David S., *Understanding Statistical Process Control*, 1986, Keith Press, Knoxville, TN, p. 130

Wheeler, Donald J., and Chambers, David S., *Understanding Statistical Process Control*, 2nd edition, 1992, SPC Press, Inc., Knoxville, TN

Wheeler, Donald J., "Charts Done Right," *Quality Progress*, 27, 5, May 1994, pp. 65-68

Chapter 3

Estimating Process Parameters from Variable-Data Charts

Measuring is knowing.

Once a process demonstrates stability, reliable estimates can be made of the various process parameters required for capability studies. The parameters of interest for variable-data charts (such as the $\overline{X}, R; \overline{X}, S;$ or $IX\&MR$) are the process average, μ, and process standard deviation, σ. For normal distributions, estimates of these two provide the necessary information to predict the expected process output, as is demonstrated in Figure 3.1.

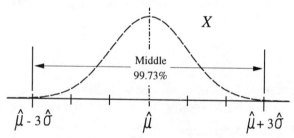

Fig. 3.1 Estimated distribution of individual measurements.

All measures of process capability require an estimate of σ, while most require estimates of both σ and μ. Although there is only one method for estimating μ, there are several ways of estimating σ. The method chosen depends on the type of capability being studied, as will be fully explained in Chapter 5.

3.1 Estimating the Process Average, μ

μ is the average, or mean, of the process output distribution. It is calculated by averaging all individual measurements for a characteristic produced by a given process.

$$\mu = \frac{\sum\limits_{i=1}^{N} X_i}{N}$$

Since rarely, if ever, is it feasible to collect *all* possible measurements for a characteristic, μ is usually estimated from the sample measurements contained in each in-control subgroup on the control chart. This estimate of μ is labeled $\hat{\mu}$, which is read "mew hat." For an $\overline{X}$ chart, μ is most often estimated with the following formula, where k is the number of in-control subgroups.

$$\hat{\mu} = \overline{\overline{X}} = \frac{\sum\limits_{i=1}^{k} \overline{X}_i}{k}$$

μ can also be estimated by summing all individual measurements of each in-control subgroup. As there are n measurements in each subgroup and k subgroups, there is a total of kn measurements available. These are summed and divided by kn to estimate μ. This method is used by many SPC software packages as it minimizes any rounding-off problems associated with calculating the subgroup averages.

$$\hat{\mu} = \frac{\sum\limits_{i=1}^{kn} X_i}{kn}$$

For an *IX* chart, μ is estimated from the individual plot points with this formula:

$$\hat{\mu} = \overline{X} = \frac{\sum\limits_{i=1}^{k} X_i}{k}$$

Remember, measurements contained in any out-of-control subgroup must *not* be included in the estimate of any process parameter, as the following example illustrates.

Example 3.1 Center Gap Size

An $\overline{X}, R$ chart with 25 subgroups of size 4 ($n = 4$) for the center gap between the two bottom flaps folded in a process making corrugated boxes is displayed in Figure 3.2. Notice there is one $\overline{X}$ plot point below the lower control limit of the $\overline{X}$ chart. This subgroup average should *not* be included in the summation of subgroup averages used to estimate μ because it does not represent the normal operating conditions of this process (Pignatiello and Ramberg). For this same reason, out-of-control points are also excluded when calculating centerlines and control limits for a chart.

So instead of summing all 25 subgroup averages recorded on this chart, only the 24 in-control points are added to get a total of 7.6368 cm.

$$\sum\limits_{i=1}^{24} \overline{X}_i = 7.6368 \, \text{cm}$$

From this sum, an estimate of .3182 cm is made for the average center gap size.

$$\hat{\mu} = \overline{\overline{X}} = \frac{\sum\limits_{i=1}^{24} \overline{X}_i}{k} = \frac{7.6368 \, \text{cm}}{24} = .3182 \, \text{cm}$$

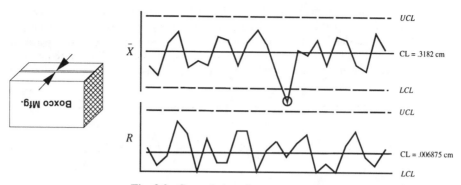

Fig. 3.2 Control chart for center gap size.

Alternatively, μ could be estimated from the 96 individual measurements (with $n = 4$, and $k = 24$, $kn = 96$). This alternative method always produces an identical answer to that for the first, within round off.

$$\hat{\mu} = \frac{\sum\limits_{i=1}^{96} X_i}{kn} = \frac{30.55 \text{ cm}}{24(4)} = .3182 \text{ cm}$$

Is this average gap size too large? Too small? Just right? Without specifying a target for the average, there is no way of knowing. When the specification limits are available, they may be compared to this estimated average gap size of .3182 cm to determine if the process is properly centered or not.

Typically the calculation for this estimate of the process average does not have to be made as it is just the centerline of the $\overline{X}$ chart (see Exercise 2.1). This centerline is needed to determine control limits and should already be computed and displayed on the chart. When deriving control limits, the one out-of-control subgroup would have been dropped from the calculation because it represents an assignable source of variation that is not a normal part of this flap-folding process.

The units of measurements, cm in this case, are just as much a part of the answer as the .3182. It will be seen later that for most measures of process capability, these units cancel out, leaving a dimensionless result. This is a useful feature since comparisons of capability may then be made between processes producing different characteristics.

Example 3.2 Particle Contamination

After being populated with components, circuit boards (Figure 3.3) pass over a wave soldering operation, then through a cleaning solution to remove solder debris and flux.

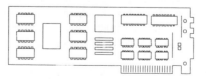

Fig. 3.3 A printed circuit board.

Cleanliness of this solution is designated a critical process variable and is checked by measuring the particle contamination in *ppm* (parts per million) of the cleaning solution. To better understand the behavior of this variable, the supervisor of this department begins an *IX & MR* chart (*n* = 1) to monitor particle contamination. Following one month of data collection, with one reading taken per day, the chart presented in Figure 3.4 is created.

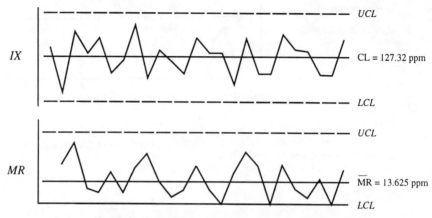

Fig. 3.4 *IX & MR* chart for particle contamination.

To calculate $\overline{X}$, the supervisor added up all 25 *IX* plot points for a total of 3183 *ppm*, then divided this sum by 25. As all points are in control, in addition to being the centerline of the *IX* chart, this value is also an estimate of the average particle contamination for this period of time.

$$\hat{\mu} = \overline{X} = \frac{\sum\limits_{i=1}^{25} X_i}{k} = \frac{3183 \, ppm}{25} = 127.32 \, ppm$$

Again, units are associated with the answer, 127.32 *ppm*. This average may be compared to the process requirements for particle contamination and a decision made on whether or not improvements to the cleaning bath are needed.

3.2 Estimating the Process Standard Deviation

Unfortunately, estimating the process standard deviation is somewhat more involved than estimating the process average. First of all, there are two distinct types of process variation, short-term and long-term (Koons; Montgomery, p. 385). Then, there are several methods for estimating the short-term standard deviation, depending on the type of control chart chosen to monitor the process. The remainder of this section discusses the differences between short- and long-term variation, while the following two sections demonstrate how each type is estimated.

For control charts to function properly, a sampling plan must be established that minimizes process variation *within* a single subgroup. This is why consecutive pieces produced under as similar operating conditions as possible (homogeneous) are taken to form one subgroup. When collecting a subgroup size of 5, it is not a good idea to collect the first two pieces, switch operators, then collect the last three pieces of that subgroup. Nor is it a good practice to take one piece, change to a different lot of material, then collect the remaining four pieces of that subgroup. Likewise, stopping the process in the middle of collecting a subgroup to take a lunch break should be avoided. Every effort ought to be made to eliminate the chance of introducing any type of assignable-cause variation during the production of pieces selected for a subgroup.

By taking these precautions, the variation present between pieces in a single subgroup reflects only *within*-subgroup variation. This has also been referred to as "instantaneous" (Montgomery, p. 365), "inherent," or "short-term" (Fellers) variation. This type of variation is not influenced by shifts in the process average due to different operators, unlike lots of material, tool wear, or changes in ambient temperature and humidity, which take place primarily *between* subgroups. A standard deviation estimated from only within-subgroup variation represents the most consistent a process can be over a short period of time and is therefore labeled σ_{ST}, the subscript "ST" denoting short term (AT&T).

$$\sigma_{ST} = \text{Short-term process standard deviation} = \sigma_{WITHIN}$$

Subgroups are then collected at predetermined intervals that allow the possibility of assignable-cause variation to occur *between* them, *e.g.* temperature fluctuations, dissimilar batches of material, voltage spikes, different operators, air pressure surges, tool wear, or unequal setups. A standard deviation estimated from the combination of within-subgroup *and* between-subgroup variations more accurately reflects the performance of the entire process over longer periods of time (Shunta; Hare).

$$\sigma_{LT}^2 = \sigma_{WITHIN}^2 + \sigma_{BETWEEN}^2 \quad \Rightarrow \quad \sigma_{LT} = \sqrt{\sigma_{WITHIN}^2 + \sigma_{BETWEEN}^2}$$

The notation for this long-term standard deviation is σ_{LT}, with the subscript "LT" representing long term.

$$\sigma_{LT} = \text{Long-term process standard deviation}$$

In a perfect state of control, variation in the process average between subgroups is zero and, thus, the long-term and short-term standard deviations are equal (Podolski).

$$\sigma_{LT} = \sqrt{\sigma_{WITHIN}^2 + \sigma_{BETWEEN}^2} = \sqrt{\sigma_{WITHIN}^2 + 0^2} = \sigma_{WITHIN} = \sigma_{ST}$$

Unfortunately, the process average is seldom in a "perfect" state of control. Any change in the process average causes the variation between subgroups to be greater than zero, resulting in the long-term standard deviation being somewhat greater than the short-term standard deviation (Pignatiello and Ramberg).

$$\sigma_{BETWEEN}^2 \geq 0$$

$$\sigma_{BETWEEN}^2 + \sigma_{WITHIN}^2 \geq \sigma_{WITHIN}^2$$

$$\sqrt{\sigma_{BETWEEN}^2 + \sigma_{WITHIN}^2} \geq \sigma_{WITHIN}$$

$$\sigma_{LT} \geq \sigma_{ST}$$

To better understand why this difference occurs, look at the individual measurements contained in the subgroups of size 4 displayed in Figure 3.5. Initially the process average and spread are quite stable. At about the halfway mark, the process average is subjected to a sustained upward drift, causing the subgroup averages to become larger. However, the subgroup ranges before and after this change remain very similar. Because the within-subgroup variation is almost equivalent, the estimate of short-term variation for the period before the change would be approximately the same as the estimate made with subgroups collected during the drift. This is because each subgroup range represents variation occurring only during the time required to produce four consecutive parts.

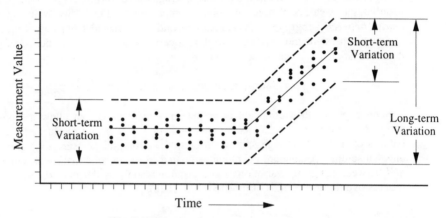

Fig. 3.5 Short-term versus long-term variation.

Measurements from all subgroups are combined together to estimate variation occurring both within and between subgroups. This measure of long-term variation will certainly be much greater than the one for just within-subgroup variation due to the dramatic drift in process average. If this process were being charted, it's very likely this upward trend would be detected and the appropriate corrective action implemented. As a result of being identified as part of an out-of-control condition, measurements from these subgroups would be removed from any estimates of process parameters. This assistance in generating reliable estimates is certainly one of the major benefits of control charts.

But what if the change in the process average is not as pronounced as that presented in Figure 3.5? What if the largest movements occurring over time are only plus or minus $.5\sigma$ shifts, as is depicted in Figure 3.6?

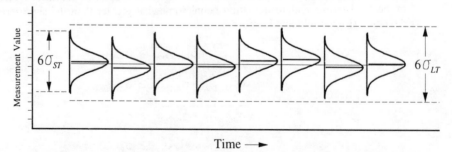

Fig. 3.6 Small shifts in the process average may go undetected on a control chart.

The graph in Figure 2.13 reveals there is only a 3 percent chance of detecting this magnitude of shift with a subgroup size of 4. It is unlikely a subgroup taken while the process average was shifted higher by .5σ would ever indicate the process is out of control. Thus, there would be no reason to exclude measurements from these subgroups when estimating the long-term standard deviation. Because the between subgroup variation is no longer zero, an estimate of σ_{LT} will be slightly larger than an estimate of σ_{ST}, which is based on only within-subgroup variation. Thus, it becomes vitally important to specify which standard deviation is used when estimating measures of process capability (Wheeler and Chambers, page 127).

The concept of "rational subgrouping" is based on this difference between short- and long-term variation. A measure of short-term process variation, like $\overline{R}$, is used to calculate control limits for the $\overline{X}$ chart ($\overline{\overline{X}} \pm A_2 \overline{R}$). The long-term variation seen from subgroup average to subgroup average is then compared to the control limits on the $\overline{X}$ chart, which are based on short-term variation. If an $\overline{X}$ plot point is outside a control limit, like the one shown in Figure 3.7, variation between subgroups, σ_{LT}, will be significantly greater than variation within subgroups, σ_{ST}. As a result, the process is judged out of control because it has changed significantly between subgroups.

The $\overline{X}$ chart being in control implies σ_{LT} is very similar to σ_{ST}, and the process output is considered stable. As long as a process remains in control, it performs just as consistently over a long period of time as it does in a short period of time. Notice that σ_{LT} may not be *exactly* equal to σ_{ST}, just fairly close.

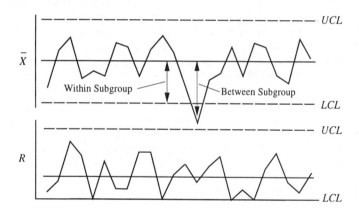

Fig. 3.7 An out-of-control point on the $\overline{X}$ chart implies variation between subgroups is greater than variation within subgroups.

Once the process is in a good state of control, a capability study can be undertaken to predict the ability of the process to meet certain requirements. Choosing the proper capability measure depends in part whether interest is in short-term or long-term process performance. Using σ_{ST} will provide an indication of process capability under ideal conditions, whereas using σ_{LT} will estimate capability under more normal operating conditions. This is why many quality practitioners prefer incorporating the long-term standard deviation into most capability studies (Koons; Grant and Leavenworth). Whatever your preference, it is extremely important to recognize the difference between these two types of process variation and clearly specify which is to be, or has been, incorporated in a given capability study.

Estimating σ_{ST} from $\overline{R}$, $\overline{S}$, or $\overline{MR}$

If either an R, S, or MR chart is monitoring the process, σ_{ST} is estimated with the formulas below, once the chart shows the process has achieved good stability. The symbol $\hat{\sigma}_{\overline{R}}$ is defined as an estimate of the process standard deviation derived from $\overline{R}$.

$$\hat{\sigma}_{ST} = \hat{\sigma}_{\overline{R}} = \frac{\overline{R}}{d_2} \quad \text{where } \overline{R} = \frac{\sum\limits_{i=1}^{k} R_i}{k}$$

Likewise, $\hat{\sigma}_{\overline{S}}$ is derived from $\overline{S}$, and $\hat{\sigma}_{\overline{MR}}$ is based on $\overline{MR}$. As before, k is the number of subgroups, while d_2 and c_4 are constants determined by the subgroup size, n.

$$\hat{\sigma}_{ST} = \hat{\sigma}_{\overline{S}} = \frac{\overline{S}}{c_4} \quad \text{where } \overline{S} = \frac{\sum\limits_{i=1}^{k} S_i}{k}$$

$$\hat{\sigma}_{ST} = \hat{\sigma}_{\overline{MR}} = \frac{\overline{MR}}{d_2} \quad \text{where } \overline{MR} = \frac{\sum\limits_{i=1}^{k-1} MR_i}{k-1}$$

Due to each R (or S) being calculated exclusively from measurements within a single subgroup, they represent only *short-term* process variation. As a result of rational subgrouping, variation in the process average occurring between subgroups is not comprehended by R nor S and, therefore, is not reflected in $\overline{R}$ or $\overline{S}$. Because these measures of within-subgroup variation capture just short-term process variation, estimates of σ derived from them are considered estimates of short-term process variation, and are accordingly labeled $\hat{\sigma}_{ST}$. This is the "best," *i.e.*, the least variation, one can expect from the current process. Using $\hat{\sigma}_{ST}$ to measure process performance results in an optimistic estimate of the process's ability to meet specifications as it represents the ideal obtainable when the process average is in a perfect state of statistical control (Grant and Leavenworth).

The examples given on the next two pages explain how to estimate the short-term standard deviation from information contained on the various variable-data control charts.

Example 3.3 Center Gap Size

In Exercise 2.1, an $\overline{X}, R$ chart was created for a process making corrugated boxes. The critical characteristic being monitored is the center gap between two folded flaps on the bottom of the box. $\overline{R}$ is calculated as .006875 cm from 24 subgroups of size 4 ($n = 4$) and is the centerline of the range chart displayed in Figure 3.2. An estimate of σ_{ST} is derived from $\overline{R}$, as is done here, where d_2 for n equal to 4 is found as 2.059 in Appendix Table I:

$$\hat{\sigma}_{ST} = \hat{\sigma}_{\overline{R}} = \frac{\overline{R}}{d_2} = \frac{.006875 \text{ cm}}{2.059} = .003339 \text{ cm}$$

Because this estimate of the process standard deviation is computed from only within-subgroup variation, it depicts the best the current process is capable of doing.

Example 3.4 Solder Pull Test

In Exercise 2.4, an $\overline{X}, R$ chart ($n = 4$) for the pull test of a soldering operation (Figure 3.8) was finally brought into a good state of control, with $\overline{R}$ equal to 9.20 grams and $\overline{X}$ equal to 47.260 grams. Estimates of μ and σ_{ST} are found as follows:

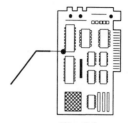

$$\hat{\mu} = \overline{X} = 47.260 \text{ gm}$$

$$\hat{\sigma}_{ST} = \hat{\sigma}_{\overline{R}} = \frac{\overline{R}}{d_2} = \frac{9.20 \text{ gm}}{2.059} = 4.468 \text{ gm}$$

Fig. 3.8

This estimate of the short-term process standard deviation provides a picture of the "instantaneous" variability of this process. Just as for estimates of the process average, units are attached to estimates of the process standard deviation (in this example, grams).

Assuming the process output follows a normal distribution, 68 percent of the pull test results are predicted to fall between 42.792 and 51.728 grams. These numbers are generated by adding and subtracting one short-term standard deviation to the process average.

$$\hat{\mu} - 1\hat{\sigma}_{ST} = 47.260 \text{ gm} - 1(4.468 \text{ gm}) = 42.792 \text{ gm}$$

$$\hat{\mu} + 1\hat{\sigma}_{ST} = 47.260 \text{ gm} + 1(4.468 \text{ gm}) = 51.728 \text{ gm}$$

Likewise, about 95 percent of the pull-test results will be between the average minus two standard deviations and the average plus two standard deviations.

$$\hat{\mu} - 2\hat{\sigma}_{ST} = 47.260 \text{ gm} - 2(4.468 \text{ gm}) = 38.324 \text{ gm}$$

$$\hat{\mu} + 2\hat{\sigma}_{ST} = 47.260 \text{ gm} + 2(4.468 \text{ gm}) = 56.196 \text{ gm}$$

Example 3.5 Keyway Width

For an $\overline{X}, S$ chart monitoring the width of a keyway cut on a milling machine, k is 20 while n is 5. The sum of the subgroup standard deviations from all 20 subgroups is .28952 mm. An estimate of σ_{ST} is derived from $\overline{S}$, as is shown below:

$$\overline{S} = \frac{\sum_{i=1}^{20} S_i}{k} = \frac{.28952 \text{ mm}}{20} = .014476 \text{ mm}$$

$$\hat{\sigma}_{ST} = \hat{\sigma}_{\overline{S}} = \frac{\overline{S}}{c_4} = \frac{.014476 \text{ mm}}{.9400} = .01540 \text{ mm}$$

c_4 is from Appendix Table I for n equal to 5, because that is how many measurements were used to calculate each subgroup standard deviation. As with calculating $\overline{R}$, the centerline of the S chart is $\overline{S}$, based on measurements from only in-control subgroups. Again, units of measurement (mm) are associated with this estimate of short-term process variation.

If the average keyway width is estimated as 4.99135 mm, then in the short-term operation of this process, approximately 99.73 percent of all widths should fall between 4.94515 mm and 5.03755 mm, assuming the process output for width is normally distributed.

$$\hat{\mu} - 3\hat{\sigma}_{ST} = 4.99135 \text{ mm} - 3(.01540 \text{ mm}) = 4.94515 \text{ mm}$$

$$\hat{\mu} + 3\hat{\sigma}_{ST} = 4.99135 \text{ mm} + 3(.01540 \text{ mm}) = 5.03755 \text{ mm}$$

This is the expected output as long as the process remains stable. With a LSL of 4.92500 mm and an USL of 5.05500 mm, both $3\sigma_{ST}$ tails are well within print specifications. This indicates the process has the ability to machine the proper width on at least 99.73 percent of the keyways.

Example 3.6 Particle Contamination

From an *IX & MR* chart (*k* is 25 readings) monitoring a cleaning solution for printed circuit boards, the sum of the 24 moving ranges is 327 *ppm*. This chart indicates the process is in a reasonably good state of control, as was shown in Figure 3.4.

An estimate of σ_{ST} for particle contamination of this process is determined from the average moving range. When calculating the average moving range, remember to use *k* minus 1 in the denominator.

$$\overline{MR} = \frac{\sum\limits_{i=1}^{k-1} MR_i}{k-1} = \frac{327 \text{ ppm}}{24} = 13.625 \text{ ppm}$$

d_2 is found in Appendix Table I to be 1.128 for an *n* of 2, which is the subgroup size for these moving ranges since they are based on the absolute value of the difference between *two* consecutive *IX* plot points. If a contamination reading is taken every hour, then σ_{ST} estimates the variation observed in contamination levels from hour to hour.

$$\hat{\sigma}_{ST} = \hat{\sigma}_{\overline{MR}} = \frac{\overline{MR}}{d_2} = \frac{13.625 \text{ ppm}}{1.128} = 12.079 \text{ ppm}$$

In Example 3.2, the process average for particle contamination was estimated as 127.32 *ppm*. Assuming these readings are normally distributed, approximately 99.73 percent are predicted to lie between 91.08 *ppm* and 163.56 *ppm*.

$$\hat{\mu} - 3\hat{\sigma}_{ST} = 127.32 \text{ ppm} - 3(12.079 \text{ ppm}) = 91.08 \text{ ppm}$$

$$\hat{\mu} + 3\hat{\sigma}_{ST} = 127.32 \text{ ppm} + 3(12.079 \text{ ppm}) = 163.56 \text{ ppm}$$

Usually in statistics, larger sample sizes are better. However, moving ranges are one of the few cases where *smaller* samples sizes are preferred. To minimize the influence of time-to-time changes, the *MR* values should be calculated from only two consecutive *IX* plot points. If more than two are used, more long-term variation is included with each additional *IX* value, and the resulting moving ranges are less likely to represent solely short-term process variation. This concept is extremely important for calculating proper *IX* control limits.

$$LCL_{IX} = \overline{X} - 2.66\,\overline{MR} \qquad\qquad UCL_{IX} = \overline{X} + 2.66\,\overline{MR}$$

In order to preserve their statistical validity, control limits on the *IX* chart must be based on just short-term process variation (review Chapter 2). As the control limits are calculated from $\overline{MR}$, the individual moving ranges used in determining $\overline{MR}$ must therefore represent only short-term variation. Using two *IX* measurements for each moving range best accomplishes this objective.

Estimating σ_{LT} from S_{TOT}

If the process average is in a "perfect" state of control, long-term and short-term variation are identical (Gunter, 1989*b*). However, some shifts in the process average are so small they may not be detected by the chart, as was demonstrated in Figure 2.13. Detecting these small movements is difficult unless very large subgroup sizes are used. Any undetected changes in μ occurring between subgroups will make the estimate of σ_{LT} slightly greater than the one for σ_{ST}, even when the chart appears to be in control (Burr; Koons).

The long-term process standard deviation, σ_{LT}, is estimated by first calculating a sample standard deviation using *all* individual measurements from in-control subgroups on the control chart. In effect, all measurements are treated as if they belong to one, large subgroup. This calculation combines variation among pieces within a subgroup with the variation between pieces in different subgroups. Because *k* is the number of subgroups and *n* is the subgroup size, *kn* is the total number of measurements included in calculating this standard deviation (total is abbreviated by "*TOT*").

$$S_{TOT} = \sqrt{\frac{\sum_{i=1}^{kn} (X_i - \hat{\mu})^2}{kn - 1}}$$

Once S_{TOT} is calculated, divide by c_4 to get $\hat{\sigma}_S$, which is an estimate of the long-term process standard deviation, $\hat{\sigma}_{LT}$.

$$\hat{\sigma}_{LT} = \hat{\sigma}_S = \frac{S_{TOT}}{c_4}$$

Division by c_4 is necessary to provide an unbiased estimate of σ_{LT} (Duncan). c_4 is determined by *kn*, the number of individual measurements used in calculating S_{TOT}. c_4 values for *kn* up to 500 are given in Appendix Table II. If such a table is not readily available, c_4 values may be approximated from the following formula (ASQC).

$$c_4 \cong \frac{4kn - 4}{4kn - 3}$$

This estimate of the long-term process standard deviation is more difficult to calculate manually than those for the short-term standard deviation, but offers a much better prediction of the actual process variation observed by a customer receiving the parts produced on this process over a period of time. Because capability studies indirectly measure customer satisfaction (and the availability of computer software to calculate σ_{LT}), many quality practitioners now base their capability measures on σ_{LT} rather than σ_{ST} (Charbonneau and Webster; Hart and

Hart). This preference exists because σ_{LT} more precisely reflects the true process variation, thus presenting a more realistic picture of actual process performance (Cryer and Ryan), and therefore, of actual customer satisfaction.

Example 3.7 Valve O.D. Size

The sums given below are calculated from in-control subgroups on an $\overline{X}, R$ chart monitoring a grinding operation for the outer diameter (O.D.) of a collar located on an automotive exhaust valve (see arrows in Figure 3.9).

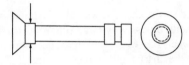

Fig. 3.9 An automotive exhaust valve.

An estimate of the process average is found from the sum of the subgroup averages.

$$\sum_{i=1}^{20} \overline{X}_i = 200.082 \text{ cm} \quad \Rightarrow \quad \hat{\mu} = \overline{\overline{X}} = \frac{\sum\limits_{i=1}^{k} \overline{X}_i}{k} = \frac{200.082 \text{ cm}}{20} = 10.0041 \text{ cm}$$

The sum of the subgroup ranges is used to calculate the average range.

$$\sum_{i=1}^{20} R_i = .722 \text{ cm} \quad \Rightarrow \quad \overline{R} = \frac{\sum\limits_{i=1}^{k} R_i}{k} = \frac{.722 \text{ cm}}{20} = .0361 \text{ cm}$$

The short-term process standard deviation is estimated from the average range to be about .01552 cm.

$$\hat{\sigma}_{ST} = \hat{\sigma}_R = \frac{\overline{R}}{d_2} = \frac{.0361 \text{ cm}}{2.326} = .01552 \text{ cm}$$

The sum of the squared deviations from the process average is needed to calculate S_{TOT} ($k = 20$, $n = 5$, $kn = 100$).

$$\sum_{i=1}^{100} \left(X_i - \overline{\overline{X}} \right)^2 = .025534 \text{ cm}^2$$

This sum is substituted into the formula for S_{TOT}.

$$S_{TOT} = \sqrt{\frac{\sum\limits_{i=1}^{kn} (X_i - \overline{\overline{X}})^2}{kn - 1}} = \sqrt{\frac{.025534 \text{ cm}^2}{100 - 1}} = \sqrt{.0002579 \text{ cm}^2} = .01606 \text{ cm}$$

Once S_{TOT} is determined, σ_{LT} can be estimated. From Appendix Table II, c_4 for 100 is .9975.

$$\hat{\sigma}_{LT} = \hat{\sigma}_S = \frac{S_{TOT}}{c_4} = \frac{.01606 \text{ cm}}{.9975} = .01610 \text{ cm}$$

Owing to c_4 being so close to 1.00 for large values of kn (.9975 in this example), many authors choose to omit it when estimating σ_{LT} (Feigenbaum), thus simplifying the long-term standard deviation formula.

$$\hat{\sigma}_{LT} \cong S_{TOT}$$

In this example, dividing by c_4 changes the estimate of σ_{LT} from .01606 cm (biased) to .01610 cm (unbiased), a difference of only .25 percent. Although these differences are generally quite small, and perhaps do not warrant the extra time to look up the appropriate c_4 factor, this book always incorporates c_4 in this calculation to provide an unbiased estimate of σ_{LT}. It is left up to each reader to decide whether or not to include it. Fortunately, most capability calculations are done by software programs, which should have no difficulty in providing the best possible estimate by incorporating the c_4 factor, no matter how many measurements are used.

Notice how the sum of the squared deviations has units of cm^2. When the square root of this sum is taken during the calculation of S_{TOT}, the units become cm again. This is now consistent with the units of the process average, as well as the specification limits. For example, to predict where approximately 95 percent of the outer diameters will be produced over the long term, $\pm 2\sigma_{LT}$ is added directly to the process average.

$$\hat{\mu} - 2\hat{\sigma}_{LT} = 10.0041 \text{ cm} - 2(.01610 \text{ cm}) = 9.9719 \text{ cm}$$

$$\hat{\mu} + 2\hat{\sigma}_{LT} = 10.0041 \text{ cm} + 2(.01610 \text{ cm}) = 10.0363 \text{ cm}$$

This is why σ (the standard deviation), rather than σ^2 (the variance), is seen most often in quality control work. The reason most statisticians prefer working with variances will be explained in later chapters (see Section 10.4).

Observe the difference between estimates of the short-term (.01552 cm) and long-term (.01610 cm) standard deviations. This difference of about 4 percent is fairly typical and ought to be expected due to the different method of calculating each type of standard deviation. Obviously, the choice of standard deviation affects estimates of process capability. Thus, to prevent confusion, a company should standardized on one or the other, and avoid reporting or comparing both short- and long-term capability measures.

Remember this example, as it appears extensively in Chapters 5 and 6 to illustrate how the various capability measures are calculated and, more importantly, interpreted.

Example 3.8 Particle Contamination

σ_{LT} is estimated from the individual X measurements of the IX & MR chart monitoring particle contamination (Figure 3.4), where k is 25 and n is 1, making kn equal to 25. The deviations of the individual measurements from the estimated process average, $\overline{X}$, have been calculated, squared, and summed. The resulting total is 3932.34 ppm^2.

$$\sum_{i=1}^{25} (X_i - \overline{X})^2 = 3932.34 \ ppm^2$$

With this sum, S_{TOT} is calculated as is done here.

$$S_{TOT} = \sqrt{\frac{\sum\limits_{i=1}^{25}(X_i - \overline{X})^2}{kn - 1}} = \sqrt{\frac{3932.34\ ppm^2}{25 - 1}} = 12.8003\ ppm$$

An estimate of σ_{LT} is derived when S_{TOT} is divided by c_4. Notice how the c_4 factor of .9896 makes only a small difference in the result for $\hat{\sigma}_{LT}$ (a little over 1 percent).

$$\hat{\sigma}_{LT} = \frac{S_{TOT}}{c_4} = \frac{12.8003\ ppm}{.9896} = 12.9348\ ppm$$

Compare this $\hat{\sigma}_{LT}$ of 12.9348 *ppm* to the $\hat{\sigma}_{ST}$ of 12.079 *ppm* computed previously in Example 3.6, which is about a 7 percent difference. As expected, the estimate of long-term process variation is somewhat larger than the one for short-term variation.

The process average for particle contamination was estimated earlier as 127.32 *ppm*. Assuming these readings are normally distributed, approximately 99.73 percent of all contamination readings are expected to range from 91.08 and 163.56 *ppm*, based on short-term process variation. According to the long-term standard deviation estimate, this same percentage of contamination readings is predicted to fall between 88.52 and 166.12 *ppm*. The spread of this last interval is about 7 percent wider than the one based on $\hat{\sigma}_{ST}$, as displayed in Figure 3.10.

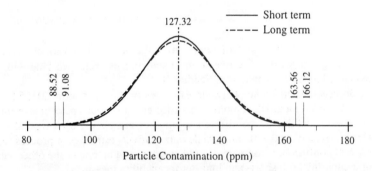

Fig. 3.10 The process spread based on σ_{ST} is slightly less than the one estimated from σ_{LT}.

Although the spreads of these two distributions appear very similar, the difference between σ_{ST} and σ_{LT} often has a substantial effect on estimates of process performance.

3.3 Why Process Parameters Cannot Be Correctly Estimated for an Unstable Process

Some authors recommend attempting to estimate process parameters of an unstable (out-of-control) process. These types of estimates are sometimes referred to as "performance" parameters of the process (AIAG). However, very little, if any, useful information about a process can be gained from analyzing an unstable process (Wheeler and Chambers, page 130). First of all, there is no well-defined output distribution. By definition, out of control means either the process average, or standard deviation, or both, are changing. How can one characterize a moving target that is changing shape in a capricious manner and attempt to make any type of prediction about its past, much less future, performance?

A few authors recognize this predicament and qualify their use of these parameter estimates by stating the values only summarize what has just happened, *i.e.*, the past "performance" of the process. In this section, it is clearly demonstrated that even this is not necessarily true. These pseudo estimates have the potential of dangerously misleading decision makers. No serious attempt at describing any aspect of a process output (either its parameters, or later on, its capability) can be attempted until the process is in control.

The output distribution displayed in Figure 3.11 came from a process whose average was not stable. Most of the time it was running at μ_1, but occasionally the average would jump up to μ_2. When all these measurements are averaged, the "performance" process average for this period of time is calculated as μ_3, which is slightly higher than the middle of the tolerance.

Assuming the middle of the tolerance is best for the process average, should this operator raise or lower the process aim? The bulk of the process output is centered below the target, but the average of all these measurements is slightly higher than the target due to the contribution of measurements contained in the out-of-control subgroups. If a decision is based on only the overall "performance" average, the operator will *wrongly* decide to lower the process aim, pushing the bulk of the output distribution even further from the target average, and perhaps even cause some parts to be produced below the LSL.

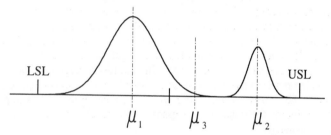

Fig. 3.11 The output of a process with an unstable average.

In Figure 3.12, the majority of the process output is correctly centered at the middle of the tolerance (assuming this is best). However, thanks to the contribution of measurements from several out-of-control subgroups, the overall average "performance" of this process is μ_1, located somewhat higher than the target. An operator making a decision about process aim based on this reported "performance" average would again wrongly conclude to lower the process average.

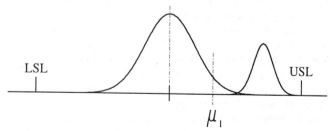

Fig. 3.12 The output of a process with an unstable average.

The process average should represent the most commonly occurring measurements, but in this situation, very few parts are at μ_1. Without drawing a histogram of the measurements, this pseudo average provides absolutely no benefit for decision making and may actually cause incorrect decisions to be made. Deming (p. 312) stresses that process parameters, such as μ, serve no useful purpose for process improvement unless the data used to estimate them are in a state of control. If a control chart indicates a process is unstable, few valid conclusions can be stated for the parts contained in subgroups collected for the chart, much less for the parts produced between subgroups. Thus, the first step in any analysis of process data is checking for control (Wheeler).

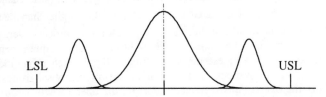

Fig. 3.13 The output of a process with an unstable average.

Figure 3.13 shows a process centered at the target with "flyers" occurring on both sides of the process average. Some measurements were collected when the process average was centered much lower than usual, while others were taken when the average was centered much higher. If all measurements are included in a calculation of the process "performance" standard deviation, a greatly inflated estimate of σ results due to the added variation between the different averages (Harding *et al*). In fact, it is so large that some of the output from this process is predicted to be outside of the specification limits (Figure 3.14), when in fact, no parts are. Exercises 3.7 and 3.8 provide additional numerical examples of similar interpretation problems for unstable processes, while several more are presented in an article by Gunter (1989*a*).

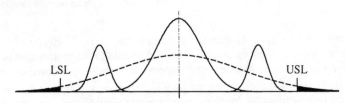

Fig. 3.14 The suspected process output.

The *Manual on Presentation of Data* (ASTM) states, that although estimates of the average and standard deviation can be calculated for any set of measurements, the results may be of little value from the standpoint of interpretation unless the measurements arise from a stable system. Merging measurements produced under significantly different conditions leads to a condensed summary that may be of little practical worth. Deming (p. 340) emphasized that estimates of the process average, standard deviation, and output distribution serve no useful purpose for process improvement unless the measurements originate from a stable process. If not, the estimates tell nothing about the process under study.

When a problem-solving team is assigned to help reduce variation in this process, the members will most likely concentrate on common causes of process spread when they should be finding the assignable sources responsible for major shifts in the process average. This misdirection of effort results in a misallocation of resources, adding cost to the product and very little value. If a control chart is used, the team would at least have an idea of which type of variation to address first.

If these last three examples, with relatively few out-of-control subgroups, demonstrate how misleading "performance" estimates are, what about situations where the process is extremely unstable? Unfortunately, there is no method available to make a good analysis of "bad" data. Some believe that eventually, after collecting enough data (bad or good), somehow it will all average out to the right answer. More data is worse than a little bad data because more time and money have been wasted. There is no "central limit" theorem for bad data. In fact, having no data is better than having bad data. With bad data you will often be deceived and make incorrect decisions, all the time believing you understand what's going on. Deming (p. 313) called this walking "into a bear trap." At least with no data, you realize you don't know what's going on. This is why all assignable sources of variation must be identified and eliminated prior to estimating process parameters.

Control charts are the best method for identifying assignable causes of variation that foul estimates of process parameters. Because estimates of process parameters are an integral part of process capability studies, there is no hope of measuring the capability (or "performance") of an unstable process in any meaningful manner. Any attempt to do so is just a waste of time. No capability study can be performed prior to control charting, not even machine capability or short-term studies. It makes better economic sense to invest resources in stabilizing a process, than to waste them on calculating flawed numbers and generating misleading reports. Until stability is achieved, management decisions about a process will be just pure guesswork.

Once control is established, process performance can be guaranteed only as long as stability is maintained. This is why charting must always be continued after completing a capability study, for if stability is lost, so is capability.

Several authors understand how unreliable these "performance" estimates are but propose tracking them over time as an indicator of process improvement or deterioration (AIAG). Unfortunately, tracking meaningless numbers is also a waste of resources, as will be demonstrated using the last example (Figure 3.14). The standard deviation calculated from all these measurements is inflated due to an unstable process average. This inflated standard deviation makes it appear that parts are being produced outside the print specifications, when in fact, none are.

Engineers seeing these "performance" estimates would immediately implement changes to reduce process variation. $4,000 is allocated for new tooling which, it is hoped, will accomplish this objective. However, unknown to the engineers, the new tooling actually *increases* the process variation. After installation, this "improvement" generates the process output of Figure 3.15. When measurements are gathered from this new process output and analyzed, the process average remains the same as before, and it appears the jumps in the process average have even gone away. But best of all, just as the engineers hoped for, the newly calculated process standard deviation is much smaller than before, thus justifying their expenditures. There may still be a few nonconforming parts, but the situation is certainly much improved compared to the old tooling, or so they think.

Note that with the old tooling (Figure 3.14), *no* nonconforming parts were actually produced. The engineers only thought so because their decision was based on a questionable statistic, a standard deviation miscalculated from the mixture of in- and out-of-control data. With the newer (and more expensive) tooling, there are now many nonconforming parts that require extra processing time to rework or that must be scrapped entirely. The engineers have unwittingly worsened the process, and what's even worse, they're celebrating because they think they made it better! Obviously, ignorance is bliss. It certainly is costly.

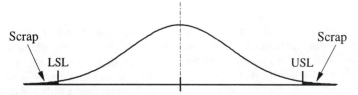

Fig. 3.15 The "improved" process output.

Even if a meaningful measure of capability could be derived for an unstable process, of what use is it? If no deliberate changes are made to the process and the study repeated tomorrow, a different outcome (maybe better, maybe worse) will mostly likely occur. If deliberate changes are made in hopes of improving the process, and the study repeated, a different result (maybe better, maybe worse) will be seen because process performance is unpredictable. Was the intentional change the cause of this outcome? Did it make process capability better? Worse? There is no way to tell if this planned change had any effect, much less the anticipated effect.

The result you get depends on which data you use. And since the process is out of control, the data collected from day to day is guaranteed to be different. Reported capability results will vary all over the scale, some months you'll feel good about the process, while you'll be greatly disappointed in others. Considerable time will be squandered trying to explain why the numbers are worse. Customers will never know what to expect. The quality level of each shipment will be a surprise, which means they are not likely to remain customers for long. Unless capability studies are based on reliable estimates of process parameters, results are suspicious at best. So what can one learn from an unstable process? As Dr. Wheeler (p. 135) states so succinctly, "Not much."

About all that can be done when a process is badly out of control is to present a histogram of the collected measurements, showing the spread of the process output and how many parts were out of specification. A statement must be included that identifies this process as being out of control, what is being done to fix the problems, and when these corrective actions will be completed. Do not make any promises about how this process will perform tomorrow, other than it could be worse, could be better, or could be the same.

The remainder of this book assumes the process is in a reasonably good state of control before attempting any type of process capability study. But before capability can be estimated, it must first be defined, which is the topic of the next chapter.

3.4 Summary

This chapter explained how parameters describing a process output distribution are estimated from measurements collected for constructing a variable-data control chart. For a process output having a normal distribution, the two important parameters are the average, μ, and standard deviation, σ.

The process average is estimated from the centerline of the $\overline{X}$ or IX chart and provides an indication of where the process output is located. This estimated average can be compared to the target average to determine if any adjustments are necessary to better center the process.

If a measure of short-term variation is desired, the process standard deviation is derived from the centerline of the R, S, or MR chart. This parameter is labeled σ_{ST}, with the subscript "ST" representing short term. When an estimate of long-term variation is preferred, the process standard deviation is based on the sample standard deviation of individual measurements from all in-control subgroups on the chart. This parameter is designated σ_{LT}, where the subscript "LT" is for long term.

Distinguishing between the short-term standard deviation (based on only within-subgroup variation) and the long-term standard deviation (based on within- and between-subgroup variation) is very important when undertaking capability studies. Even for processes exhibiting good statistical control, σ_{LT} is typically greater than σ_{ST}. Because of small unde-tected changes occurring in the process average between subgroups, σ_{ST} will, by design, underestimate the actual amount of total process variation. Taylor estimates that σ_{LT} is typically about 14 percent larger than σ_{ST}. Failure to recognize this difference leads to problems with downstream operations; estimates of the percentage of nonconforming product; estimates of scrap and rework costs; and finally, how well customer expectations are truly being met.

As the purpose of any statistical study is to enlighten, and not confuse or mislead, the importance of process stability was again emphasized. The exclusion of data from out-of-control subgroups is critical to obtaining correct estimates of process parameters. Including data collected when a process was out-of-control in the estimation of process parameters often leads to false conclusions concerning the true performance of that process. Estimates made from a mixture of measurements from both in-control and out-of-control subgroups can lead to overestimating or underestimating process capability, with either outcome causing severe repercussions.

Underestimating capability causes money to be spent needlessly improving an acceptable process. Personnel hesitate quoting on new business due to not knowing how good the process really is. Capacity is lost as jobs will be routed to other, seemingly more capable machines (Podolski). On the other hand, overestimating process capability generates a false sense of security. Jobs requiring a much more capable process are assigned to this one, resulting in unexpected scrap and rework. False promises are made to customers concerning anticipated quality levels and costs.

Recognizing that a stable process is required to provide meaningful estimates of process parameters and that these estimates are a prerequisite for correctly assessing capability, we are finally now ready to formally discuss the concept of process capability.

3.5 Exercises

3.1 Connecting Force

The force required to join the male and female parts of an electrical connector should be between 1.30 and 1.50 kilograms. A subgroup of three consecutive connectors is collected every hour, with the average and range of each subgroup plotted on an $\overline{X}$, R chart. After eliminating several assignable sources of variation, the most recent 26 subgroups indicate the process is in a good state of control. The following sums are derived from these 26 in-control subgroups. Estimate the process average and short-term standard deviation for connector joining force.

$$\sum_{i=1}^{26} \overline{X}_i = 36.820 \text{ kg} \qquad \sum_{i=1}^{26} R_i = 1.32 \text{ kg}$$

Assuming joining force has a normal distribution, estimate where the middle 68 percent of the process output is expected to fall.

3.2 Bolt Diameter

A major manufacturer of fasteners for the aerospace industry determines that bolt diameter is extremely important to customers and should be considered a key characteristic. Because of automatic gaging, data recording, and control charting, an $\overline{X}$, S chart ($n = 4$) is selected to monitor this process.

Once in control, $\overline{S}$ is calculated as .01198 cm, and $\overline{\overline{X}}$ as 2.221 cm. After estimating σ_{ST}, predict where the middle 99.73 percent of bolt diameters are expected to fall as long as this process remains in control (assume bolt diameter has a normal distribution). If the print specification for bolt diameter is 2.200 $\pm$ .050 cm, how is this process performing as far as meeting this requirement?

3.3 Relay Voltage

The voltage required to lift an armature in a relay is being monitored on an *IX & MR* chart. When control is established, the sum of 32 individual voltage readings is 947.2 volts, while the sum of the 31 associated moving ranges for this period is 74.7 volts. The manufacturer has decided on a target average of 32 volts, with a USL of 40 volts.

Assuming a normal distribution, is the estimated upper $3\sigma_{ST}$ tail of the process output for voltage below the USL? Is the estimated upper $4\sigma_{ST}$ tail of the process output for voltage below the USL? How many $\hat{\sigma}_{ST}$s could fit between the estimated process average and the USL? How many could fit between the target average and the USL?

3.4 Piston Ring Diameter

As a top producer of piston rings, a European automotive supplier is notified that the inside diameter of these rings is now considered a critical characteristic by its main customer. The first decision made by a quality-improvement team formed to work on this part is to monitor the involved machining operation with an $\overline{X}$, R chart ($n = 6$). Wanting to verify various calculations performed by their SPC software program, the team collects individual measurements from the first 22 in-control subgroups. With n equal to 6, this results in 132 (22 $\times$ 6) inside diameter readings, the sum of which is given below. Help them derive an estimate of the process average.

$$\sum_{i=1}^{132} X_i = 1190.64 \text{ cm}$$

There turns out to be good agreement between this estimate of the process average and that generated by the software. The team now wishes to check the calculation for σ_{LT}. They sum up the squared deviations between the individual measurements and the estimated process average.

$$\sum_{i=1}^{132} (X_i - \hat{\mu}) = .06144 \text{ cm}^2$$

Use this result to estimate σ_{LT}, and compare it to the software's estimate of .0216564 cm.

3.5 Shaft Length

In Exercise 2.5, a control chart was constructed for the length of a tapered metal shaft. Using data from only in-control subgroups, estimate the process average, the short-term standard deviation, and the long-term standard deviation for shaft length.

3.6 Rib Width

In Exercise 2.6, an *IX & MR* control chart was constructed for the width of ribs used in constructing aircraft wings. First, estimate the process average, short-term standard deviation and long-term standard deviation for rib width using all 30 measurements. Then estimate these process parameters again, but this time exclude all out-of-control measurements. Why the difference in results? Which set of parameter estimates best predicts the expected process output under stable operating conditions?

The specifications for rib width are 1.580 ± .016 cm. Are any width measurements out of specification?

3.7 Core Weight

After assembly, transformer cores are checked for weight. This is an important characteristic, as any fluctuation in core weight from its target will adversely affect the transformer's electrical performance. To study the output distribution for weight, 36 cores are selected at random from a week's worth of production and weighed to the nearest gram. These readings are presented in Table 3.1. Use these measurements to estimate the process average and long-term standard deviation for core weight. Predict what the output distribution looks like for this characteristic by estimating the $\pm 3\sigma_{LT}$ tails. Make a sketch of this distribution, and draw on the specifications limits of 136 ± 7 grams. Are most of the core weights on the target of 136 grams? How well does the process spread fit within the tolerance?

Table 3.1 Transformer core weights (grams).

135	133	137	138
132	136	140	132
139	140	138	137
137	133	131	134
133	139	133	138
141	134	136	135
139	134	139	138
138	139	132	133
132	138	133	134

Now construct a histogram from these measurements. How many core weights actually exceed one of the specification limits? Why is this such a different answer than your previous one based on the $\pm 3\sigma_{LT}$ tails?

3.8 Film Sag

Film used in an industrial X-ray machine can sag because of accumulated part tolerances in the machine. As film sag leads to distorted images, a team is promptly formed to investigate the extent of this problem. The first action taken is to collect 25 consecutive sag measurements from the assembly process. Because only two or three units are produced per day, nearly two full weeks are needed to amass this amount of process performance data.

As implementing a quick solution to this problem has now become the team's major concern, a decision is made not to waste time charting this data, but to immediately discover what the process output distribution looks like. Assist this team in their task by first estimating the process average and long-term standard deviation for the 25 sag measurements listed in Table 3.2. Then sketch the expected process output by determining the $\pm 3\sigma_{LT}$ tails. Add the specification limits of 15 ± 5; then help the team formulate specific recommendations on how to improve this process.

Make a histogram of these individual sag measurements. How does this compare to your sketch based on the $\pm 3\sigma_{LT}$ tails? Use an *IX & MR* chart to check the stability of this process.

Table 3.2 Amount of film sag.

14	19	12	14	13
13	14	15	18	12
15	16	13	15	22
12	13	11	15	14
12	13	15	13	14

3.6 References

AIAG, *Fundamental Statistical Process Control: Reference Manual*, 1991, Automotive Industry Action Group, Southfield, MI, p. 80

ASTM, *ASTM Manual on Presentation of Data and Control Chart Analysis*, STP 15D, 1976, American Society for Testing and Materials, Philadelphia, PA, pp. 3-4

ASQC Statistics Division, *Glossary & Tables for Statistical Quality Control*, 1983, ASQC Quality Press, Milwaukee, WI, p. 133

AT&T, *Statistical Quality Control Handbook*, 2nd edition, 1984, Delmar Printing Co., Charlotte, NC, p. 61

Burr, Irving W., *Statistical Quality Control Methods*, 1976, Marcel Dekker, New York, NY, p. 84

Charbonneau, Harvey C., and Webster, Gordon L., *Industrial Quality Control*, 1978, Prentice-Hall, Englewood Cliffs, NJ, p. 120

Cryer, Johnathan D., and Ryan, Thomas P., "The Estimation of Sigma for an X Chart, $\overline{MR}/d_2$ or S/c_4," *Journal of Quality Engineering*, Vol. 22, No. 3, July 1990, pp. 187-192

Deming, W. Edwards, *Out of the Crisis*, 1986, Massachusetts Institute of Technology, Cambridge, MA

Duncan, Acheson J., *Quality Control and Industrial Statistics*, 5th edition, 1986, Irwin, Homewood, IL, p. 61

Feigenbaum, Armand, *Total Quality Control*, 3rd edition, 1983, McGraw-Hill, New York, NY, p. 360

Fellers, Gary, *SPC for Practitioners: Special Cases and Continuous Processes*, 1991, ASQC Quality Press, Milwaukee, WI, pp. 44-45

Grant, Eugene L., and Leavenworth, Richard S., *Statistical Quality Control*, 6th edition, 1988, McGraw-Hill Book Co., New York, NY, p. 155

Gunter, Bert, "The Use and Abuse of Cpk, Part 2," *Quality Progress*, Vol. 22, No. 3, March 1989a, pp. 108-109

Gunter, Bert, "The Use and Abuse of Cpk, Part 3," *Quality Progress*, Vol. 22, No. 5, May 1989b, p. 79

Harding, Arved J., Jr., Lee, Kwan R., and Mullins, Jennifer L., "The Effect of Instabilities on Estimates of Sigma," *46th ASQC Annual Quality Congress Transactions*, Nashville, TN, May 1992, pp. 1037-1043

Hare, Lynne, "In the Soup: A Case Study to Identify Contributors to Filling Variability," *Journal of Quality Technology*, Vol. 20, January 1988, pp. 36-43

Hart, Marilyn K., and Hart, Robert F., *Quantitative Methods for Quality and Productivity Improvement*, 1989, ASQC Quality Press, Milwaukee, WI, p. 215

Koons, George F., *Indices of Capability: Classical and Six Sigma Tools*, 1992, Addison-Wesley Publishing Co., Readings, MA, pp. 18-20

Montgomery, Douglas C., *Introduction to Statistical Quality Control*, 1985, John Wiley & Sons, New York, NY

Pignatiello, Joseph J., and Ramberg, John S., "Process Capability Indices: Just Say No!," *47th ASQC Annual Quality Congress Transactions*, Boston, MA, May 1993, pp. 101-102

Podolski, Garry, "Standard Deviation: Root Mean Square versus Range Conversion," *Quality Engineering*, Vol. 2, No. 2, 1989-90, pp. 156-157

Shunta, Joseph P., *Achieving World Class Manufacturing through Process Control*, 1995, Prentice-Hall, Englewood Cliffs, NJ, pp. 28-29

Taylor, Wayne A., *Optimization & Variation Reduction in Quality*, 1991, McGraw-Hill, Inc., New York, NY, p. 173

Wheeler, Donald J., and Chambers, David S., *Understanding Statistical Process Control*, 2nd edition, 1992, SPC Press, Inc., Knoxville, TN

Wheeler, Donald J., *Beyond Capability Confusion*, 1999, SPC Press, Inc., Knoxville, TN, p. 41

Chapter 4

Defining Process Capability

*"The true quality of a manufactured product is
ultimately determined by customer satisfaction."*

Process capability is broadly defined as the ability of a process to satisfy customer expectations. Some processes do a good job of meeting customer requirements and therefore are considered "capable," while others do not and are designated "not capable." This definition provides a helpful overall concept about process capability, but in practice, capability must be defined more specifically (and in more quantifiable terms) so it can be consistently observed, measured, analyzed, and reported.

The most popular quantitative definition of process capability is the width of the middle 99.73 percent of the process output distribution under statistically stable operating conditions (Kane, 1989, p. 267; Montgomery, 1991, p. 366; Wheeler and Chambers, p. 119). For an output with a normal distribution, this width is the distance from $\mu - 3\sigma$ to $\mu + 3\sigma$ or, as it is commonly referred to, the "6σ" spread (see Figure 4.1). These 3σ "tails" are often called the natural process limits because they define the "voice" of the process.

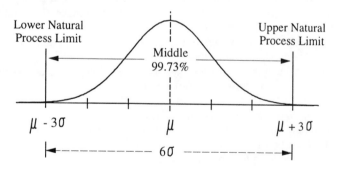

Fig. 4.1 The 6σ spread of a normal distribution.

A *measure* of process capability summarizes some aspect of a process's ability to satisfy customer requirements. Most measures compare the stable output distribution for a particular product characteristic to the specification limits for that characteristic. Because specification limits are assumed to reflect customer desires, capability measures are said to relate the "voice" of the process to the "voice" of the customer. In some cases this customer could be internal (the next operation), but more often is external (the end user).

The ability of a process to meet a particular customer requirement depends in part if a bilateral (double-sided) or unilateral (single-sided) specification is involved. Therefore, an operational definition of process capability is necessary for each of these two cases.

4.1 Process Capability Defined for Bilateral Specifications

To fulfill the *minimum* criterion for process capability, at least 99.73 percent of a process's output must fall within bilateral specifications. Figure 4.2 shows a process just meeting this minimum requirement, as the middle 99.73 percent of its output falls exactly within the specification limits (Deleryd, 1998).

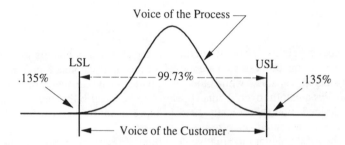

Fig. 4.2 The minimum capability requirement for characteristics with bilateral specifications.

Achieving this minimum stipulation for capability does not mean a process is making 100 percent conforming parts. Up to a maximum of .27 percent nonconforming parts could be produced (100 percent minus 99.73 percent) by a process satisfying this definition of capability. In practice, capability goals are quite often set much higher than this minimum, as is explained further in Section 4.5.

An analogy may be drawn to the current educational system. In order to pass from one grade to the next, a student must achieve a minimum of a "C" average. Although a "C" average is all that is necessary for advancement, few parents are completely satisfied when their child just meets this minimum prerequisite. Most children are encouraged to be a "B," or even an "A," student. Obviously, to be a "B" or "A" student, one must at least fulfill the minimum requirement for passing. Likewise, many companies want their products to exceed the minimum quality requirements so they can be more successful in the marketplace.

The percentage in this capability definition was originally selected because it corresponds to a specific area under a normal curve, *i.e.*, 99.73 percent of the process output is expected to be within $\mu \pm 3\sigma$. In the case of non-normally distributed variable data, it will be helpful to apply the more general definition of capability as being the width of the middle 99.73 percent of the process output. Attribute-data capability studies are also based on a requirement of having no more than a certain percentage of nonconforming parts when in a state of statistical control. More details about these special situations are presented in later chapters (Chapters 8 and 9).

If a process with a normally distributed output is centered at the middle of the tolerance, and just meets the minimum capability requirement, 99.73 percent of its output is within specification. Half of the .27 percent of nonconforming parts (.135 percent) falls below the

LSL, while the other .135 percent lies above the USL. Thus, this minimum condition for process capability could be restated as having no more than .135 percent nonconforming parts (or "fallout") beyond either specification limit.

> *A process producing a characteristic with a bilateral specification meets the minimum requirement of capability when it is stable, and has no more than .135 percent of its output for this characteristic outside either specification limit.*

This alternative definition is very useful when considering capability for characteristics with unilateral specifications.

4.2 Process Capability Defined for Unilateral Specifications

The definition of capability for unilateral specifications allows a process to produce a maximum of .135 percent nonconforming parts. Because there is only one specification limit, this means the process must be able to produce at least 99.865 percent (100 percent minus .135 percent) of its output within specification, as is illustrated in Figure 4.3.

> *A process producing a characteristic with a unilateral specification meets the minimum requirement of capability when it is stable, and has no more than .135 percent of its output for this characteristic outside the single specification limit.*

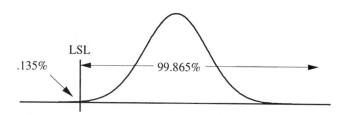

Fig. 4.3 The minimum capability requirement for characteristics having a unilateral specification.

These particular percentages were originally chosen because they correspond to areas of a normal distribution below and above $\mu - 3\sigma$ when working with a characteristic having just a LSL (as in Figure 4.3), or above and below $\mu + 3\sigma$ if only an USL is given.

4.3 Need for Different Capability Measures

The above definition of process capability is fairly straightforward, but many diverse methods have been devised to measure the various aspects of how well a process complies to it. As most processes are quite complex, any single number is unlikely to adequately summarize all pertinent information about process performance (Nelson). Those people obsessed with having all information reduced to just a single number are said to be suffering from either "mononumerosis" or "monomania." Either affliction can prove fatal.

An engineering drawing for a complex part displays many different views (front, top, side) to help explain what the finished part is to look like. In a similar manner, several different capability measures are often required to adequately describe the complicated process which manufactures that part.

Given that a process is in control, there are three ways it can fail to meet a customer requirement. Suppose the output voltage for a newly designed audio amplifier has a specification of 60 ± 4 volts. The process assembling amplifiers may not be capable for the voltage requirement as a result of either:

a. the process variation (spread) is too large (Figure 4.4a),

b. the process average is not properly centered (Figure 4.4b), or

c. the process average is not properly centered *and* the process variation is too large (Figure 4.4c).

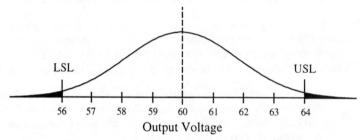

Fig. 4.4a Process variation (spread) is too large.

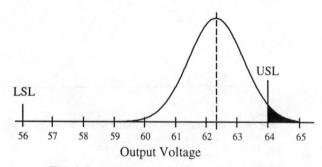

Fig. 4.4b Process average is centered too high.

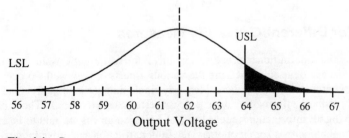

Fig. 4.4c Process average is too high *and* process variation too large.

Over the years, different capability measures were developed to cover each of these three conditions. Some (such as C_P and P_P) consider only process variation, others deal with just process centering (the k factor), while still others (like C_{PK} and P_{PK}) look at both variation and centering. A few (C_{PMK} or C_{PG}) assign a heavy penalty when the process average is not on target, while some do not even consider the specification limits at all (6σ spread). Several, like C'_{RL}, P'_{STU}, were created to work exclusively with unilateral specifications. As the definition of capability involves the percentage of nonconforming parts, many quality practitioners prefer a measure that is somehow related to this value, which is sometimes referred to as the process yield. Thus, a few measures are primarily concerned with the amount of nonconforming product, *e.g.*, *ppm* and equivalent $P^{\%}_{PK}$. Yet a small number place little significance on this aspect of capability, for instance, C_{PM}.

In addition to the above differences, some capability measures provide an insight to only short-term process capability (such as C_P), while others track long-term process capability (like P_P). Several are concerned with *potential* capability (C_{ST}), but most reflect *performance* capability (Z_{MIN}). The distinction between short- and long-term variation was discussed in the last chapter, whereas the difference between potential and performance capability is explained in the next section.

Potential vs. Performance Capability

Measures considering only process spread are called measures of potential capability, while those comprehending both spread and centering are designated measures of performance capability. Potential capability reveals what could happen *if* the process is properly centered (Kane, 1989, pp. 267-271; Montgomery, 1991, p. 374). Quite often, a process is considered to possess potential capability if its 6σ spread is equal to (or less than) the width of the tolerance. Nothing can be said about the actual percentage of conforming parts being produced because this depends on where the process average is centered, as is illustrated in Figure 4.5 for the output distributions of processes A and B.

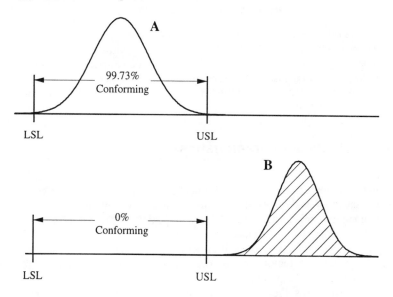

Fig. 4.5 Processes A and B have the same potential, but differ greatly in performance.

Both these processes have the same *potential* capability (based on process spread) of manufacturing parts within specification, but their actual *performance* is significantly different. Performance capability measures how well the process output actually conforms to the specification. Process A is producing practically 100 percent conforming parts, while process B is making all nonconforming parts. Customers receiving these parts would certainly notice the great difference in quality, and thus, rate the performance capability of process A much higher than that for B. Any capability measure based solely on the process standard deviation is blind to the disparity between A and B, whereas a measure incorporating both the process standard deviation *and* average would easily recognize the difference in quality level performance.

Changing the process average does not affect potential capability, but greatly affects performance capability (Kane, 1986, p. 45). For example, if process B's average is lowered so it is centered at the middle of the tolerance, its potential capability remains the same, but its performance (in terms of satisfying the customer) changes dramatically. For bilateral specifications, measures of potential capability answer the question, "*Could* the process produce at least 99.73 percent conforming parts?" Performance capability measures answer the question, "*Is* the process currently producing at least 99.73 percent conforming parts?"

> *A process is said to possess potential capability if it could meet the minimum requirement for capability by shifting the process average.*

As will be seen in the next two chapters, measures of potential capability do not incorporate the process average in their calculation, while performance measures do. Thus, potential measures are best suited for operations where process centering is not very important, or for indicating how well customer requirements might be met *if* the process output can be correctly centered. By including both the process average and standard deviation, measures of performance capability reveal how well customer expectations are actually being met.

> *A process is said to possess performance capability if it is actually meeting the minimum requirement for capability.*

No matter which definition of capability is chosen, no organization has enough manpower to concurrently chart and conduct capability studies on *all* product characteristics. A company must develop procedures for identifying critical characteristics so improvement efforts can initially focus on just those exerting a major influence on customer satisfaction, reduced operating costs, and/or improved delivery times. The next section describes several such selection methods.

4.4 Selecting Critical Characteristics

Due to limited resources, organizations cannot simultaneously undertake quality-improvement efforts on every product characteristic and process variable. In view of this limitation, it makes economic sense to begin with those features deemed most critical (sometimes called "significant," "key," "special," or "major") for subsequent machining or assembly operations, product function, and/or customer satisfaction. On occasion, features may be chosen because of their effect on reliability, durability, safety, or even compliance with government regulations. Somerton and Mlinar offer an approach for identifying the key characteristics of a process.

In many cases, the customer selects these characteristics by listing them in the contract with the supplier. For example, the automotive industry identifies important characteristics with the use of certain symbols on their engineering drawings (Chrysler *et al*, 1995*a*). Ford Motor Company places an inverted delta (∇) on its drawings to highlight critical dimensions. Chrysler Corporation relies on three different symbols: a shield for safety-related items; a diamond for features responsible for quality, reliability, or durability; a pentagon for critical tooling characteristics. General Motors uses a diamond for characteristics where variation could adversely affect customer satisfaction or the ability to properly build the product.

In the aerospace industry, Boeing designates key characteristics with a symbol similar to *a* in Figure 4.6. According to the ASME standard Y14.5M-1994 on dimensioning and tolerancing (page 38), a symbol like *b* in Figure 4.6 denotes critical features which are statistically toleranced and must be produced with statistical process controls. In addition to denoting critical product features, customers may also identify crucial process *variables* which need monitoring. When in doubt about how critical a characteristic (or process variable) may be, a supplier should not be afraid to ask the customer to define what is important. A good place for a supplier to start is learning what the customer checks at incoming inspection. Undoubtedly, these product characteristics are quite important to the customer. Johnson Controls identifies critical characteristics on its engineering drawings with a triangle containing the letter "c." Up to three numbers (1 through 3) are placed beneath this symbol. If a 1 appears, this feature is crucial to product performance and the end user's satisfaction. If a 2 is present, the feature is very important with regards to how it is manufactured. The presence of a 3 signals suppliers of a critical characteristic, one that will most likely be checked at incoming inspection.

a $\quad\quad\quad\quad\quad\quad$ b

Fig. 4.6 Symbols used to identify key characteristics.

If the product is designed internally, your own product/design engineers (and even marketing personnel) can be asked for their opinion on which product dimensions, based on the product's intended use and operating environment, are likely candidates for being critical characteristics. Frequently, these characteristics are identified by reviewing the design and process failure mode and effects analyses (FMEA) created for a product. Chrysler *et al* (1995*b*) offers an example of how conducting a process FMEA on the manual application of wax to the inside of a door revealed several important process variables. An example of applying this procedure for selecting critical characteristics in the food-processing industry is provided by Puri, while Gevirtz offers a case study of using an FMEA during the development of a cylinder head to identify critical product features and their associated process variables. All such identified features should be documented and clearly labeled on the engineering drawings in some manner, with the method of identification communicated to all personnel. This will also help create a meaningful manufacturing control plan.

There may be customer complaints (low tensile strength, excessive burrs, oversize counterbores) which result in product being returned to the supplier. Unquestionably, these are characteristics in need of quality improvement and present obvious opportunities for reducing manufacturing costs, as well as increasing customer satisfaction.

When there are few, or no, customer problems or returned material, examine reports on internal scrap, rework, and test rejects. Most shops keep very detailed records on these types of problems, and some even prepare cost of quality reports to flag problem areas (Campanella). Although these quality problems are caught internally and the customer is not aware of them, they do add cost to the product. The added costs for scrap, rework, and inspection are passed on to the customer in the form of higher prices, which results in the supplier being less competitive and losing business. If a fixed-priced contract is in force, the supplier must absorb the added cost and suffer lower profit margins. Neither option is very attractive. And sooner or later, a nonconforming part will slip through inspection and result in either a customer complaint, lost business, or even a lawsuit concerning product liability.

In addition to these formal quality reports, ask operators and other shop floor personnel about which characteristics they must watch most often, or cause the most problems. Machine uptime may also be crucial in some operations. Improving these elements of the process helps "sell" the concept of quality improvement to the shop floor personnel as they will witness first hand how their jobs become easier. It also helps if the chosen characteristic can be quickly measured and corrective action is easy to administer when problems arise.

As critical characteristics become surrogates for customer satisfaction, considerable care must be taken to ensure that they are in fact closely tied to what customers are really concerned about, *i.e.*, that they are indeed true indicators of customer satisfaction.

Relating Customer Problems to Specific Product Features

Sometimes customer complaints are very specific: "low weld strength on the left side of cross brace number 3," "hole number 9 is .2 mm off location in the *x* axis." These characteristics can ordinarily be easily traced to the responsible department for resolution. Decisions can then be made concerning the type of control chart required, subgroup size, sampling interval, measurement system, who will collect data, who will analyze data, corrective action plan when process goes out of control, who will interpret chart, who will take corrective action, when 100 percent inspection should be initiated, and finally, what type of capability study is best suited for the process.

But there are often times when customer complaints are very general, like "excessive vibration," or "low horsepower." In these cases, the generalized complaint must be traced back to specific product characteristics (or process variables) before a capability study can be effectively conducted. To best accomplish this, a cross-functional team of subject experts must be assembled. One of the team's first efforts should be drawing a flowchart to map out the entire process by describing all process inputs and their effect on the output. This chart helps everyone on the team learn how the process operates by identifying critical control points and areas where variation may be introduced. As an illustration, customers are complaining about finding wire breaks on an automotive voltage regulator coil. The flowchart for winding coils is depicted in Figure 4.7, using different symbols to denote the various activities performed in this process.

Once the process is fully understood, the team can brainstorm which characteristics may be responsible for this complaint. After the brainstorming session, the list of critical characteristics should be organized on a cause-and-effect diagram to aid the analysis.

Example 4.1 Lens Cracks

Optical lenses for telescopes are carefully ground to the proper dimensions depending upon customer requests. Recently, several customers have complained about discovering cracks in lenses they have received. A team is quickly assembled to resolve this customer concern. After creating a process flowchart, the team members decide to identify all characteristics that may lead to this type of crack. During a one-hour brainstorming session, quite a few ideas are generated for potential causes of variation in the grinding process. To organize all these suspected causes, the team carefully arranges them on the cause-and-effect diagram exhibited in Figure 4.8. These are the characteristics that should definitely be considered

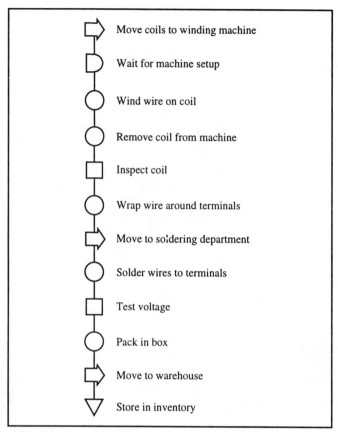

Fig. 4.7 A process flowchart for wire breaks on regulator coils.

for the application of control charts. Once control is established, capability studies can be conducted to determine if the process has the ability to consistently meet the requirements, or if the requirements need to be changed.

If variation in these product characteristics may result in several dissimilar customer complaints, a matrix analysis can be applied to help identify the most crucial characteristics. Each customer complaint is written on a separate row of the matrix, as is illustrated in Figure 4.9. The product characteristics (or process variables) suspected of causing these complaints are listed across the top of the matrix, one per column.

Example 4.2 Injector Complaints

Several customer complaints have been received by a company manufacturing electronic fuel injectors. The majority of complaints involve "leaks," followed by "low output," "smoking," and "seized." Product characteristics identified in a brainstorming session as being responsible for these complaints are written as column headings across the top of the matrix in Figure 4.9. Each characteristic in turn is evaluated for its effect on the various

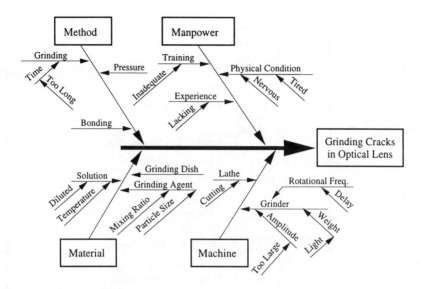

Fig. 4.8 The cause-and-effect diagram approach of identifying likely candidates for a capability study.

customer complaints. A single dot indicates a likely cause, whereas a dot within a circle represents a possible strong cause-and-effect relationship. For example, variation in "plunger to bushing clearance" is suspected of causing injector leaks, excessive exhaust smoke, or even injector seizure. However, as the dot within a circle in the "seized" row indicates, the team believes variation in plunger to bushing clearance is very closely related to seized injectors.

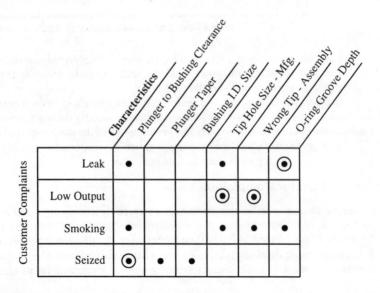

Fig. 4.9 A matrix approach for identifying potential candidates for a capability study.

On the other hand, some characteristics are thought to be responsible for generating just one particular customer complaint. For example, variation in "plunger taper" is suspected of causing only seized injectors.

Note that some of these characteristics involve variable data (bushing I.D. size), while others (wrong tip) are attribute data. The type of data determines the sort of control chart and capability study required. Capability studies involving both variable (Chapters 5 and 6) and attribute data (Chapter 9) are covered in this book.

If desired, weights may be assigned to each customer concern based on its severity. Most often a scale of 1 to 10 is used, with 10 being reserved for the most serious complaint. In this example, "seized" is thought to be the most severe problem, and is thus given a weight of 10. Using this first assignment as a baseline, "low output" complaints are considered to be an 8, "smoking" a 5, while "leak" is assigned a weight of only 4.

After weights are allocated, each characteristic is assigned a rating proportional to the strength of its effect on a specific customer complaint. The rating scale typically ranges from 0 to 10, with 10 indicating the strongest relationship. For instance, "plunger to bushing clearance" is rated a strength of "6" for causing a "leak," "0" (no relationship) for "low output," "3" for "smoking," and "10" for "seized" (see Figure 4.10).

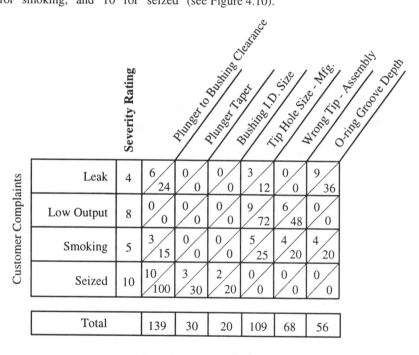

Fig. 4.10 Complaints can be weighted for severity and each characteristic rated for its contribution to the complaint.

When all characteristics are rated for each customer complaint, the weight of the customer complaint is multiplied by the strength rating. These results are summed for each characteristic and recorded at the bottom of the matrix in the "total" row. When the results are summed for all characteristics, a Pareto diagram is constructed to rank characteristics by their totals.

Applying this type of analysis to the injector example, "plunger to bushing clearance" is identified as the most critical product characteristic and becomes the prime candidate for a detailed capability study (see the Pareto diagram in Figure 4.11). In addition, this characteristic would most likely be assigned a capability goal much higher than the minimum capability requirement.

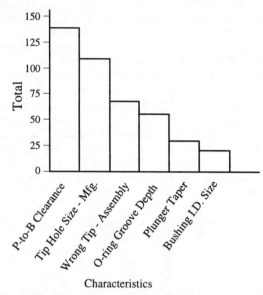

Fig. 4.11 A Pareto diagram of the weighted totals identifies critical characteristics.

Example 4.3 Paper Processing

As another example of this procedure for identifying critical characteristics based on customer requirements, a matrix similar to the one displayed in Figure 4.12 was developed by Shenoy for the paper-processing industry.

Customer Requirements / Paper Characteristics	Brightness	Tensile	Burst Index	Tear Index	Coating Thickness
Grade	●	▲	▲	▲	●
Thickness				▲	●
Whiteness	⋇				▲
Folds			●	●	
Durability		⋇	⋇	⋇	
Smoothness	▲				

Relationship
⋇ Strong
● Medium
▲ Weak

Fig. 4.12 A matrix analysis of customer complaints in the paper-processing industry.

This matrix analysis may be applied to internal scrap and rework, test rejects, or even engineering concerns. For control purposes, it is best to identify the process variables responsible for the observed variation in the product characteristics, as this represents the truest sense of statistical *process* control. Thus, a second matrix analysis can be done but with product characteristics now in the rows and process variables in the columns. See Cohen's book on quality function deployment (QFD) for additional information on this topic. Anderson reveals how a division of Allied Signal employed QFD to identify relationships between product characteristics and process variables to optimize the production of spark plug insulators. McLaughlin offers a list of other statistical techniques that help discover the interrelationship of key process variables.

4.5 Establishing Capability Goals

To judge how "good" the process capability is for an identified critical characteristic, the capability measure generated by a capability study must be compared to some desired goal. These goals must be established, so that when achieved, the product demonstrates superior performance in the marketplace. Obviously, various industries will have different goals for capability, and within a given industry, different companies will have dissimilar goals, and inside of a single company, different product characteristics will have diverse capability requirements. Thus it is extremely difficult to establish one goal that is suitable for all manufacturing situations (Davis *et al*). Nevertheless, general guidelines for goals have been recommended by several authors. These are listed in the appropriate sections of the next two chapters where specific measures of capability are discussed.

Quite often, capability goals are expressed as either a maximum percentage of nonconforming product, or conversely, a minimum percentage of conforming product. In today's global economy, the goal for the percentage of conforming product must be remarkably high. Although 99 percent conformance may sound acceptable to some organizations, Harry points out how achievement of this goal would result in the following events occurring *daily*:

- No electricity for 14 minutes.
- Unsafe drinking water for 15 minutes.
- 2 plane crashes at *each* major airport.
- 548 wrong drug prescriptions.
- 714 incorrect surgical procedures.
- 480,000 lost articles of U.S. mail.
- 5,280,000 checks deducted from the wrong bank accounts.
- 18,921,600 phone calls misplaced by telecommunications services.

Many customers have established functional specifications and capability goals for purchased products (Fabian). According to Pang, Cummins/Onan Corporation has made attainment of capability goals one of their major criteria for selecting suppliers. However, Pennucci recommends jointly setting capability goals with suppliers, based on cost considerations, process ability, and the functionality of the part. Always keep in mind that a goal easily achieved by one supplier may not be so easy for another.

Other companies aren't so concerned about the current level of capability, but instead look for continuous improvement in capability (Sullivan). Some companies eliminate, or significantly reduce, inspection of a process's output when it exceeds a predetermined capability goal (Skellie and Ngo). Still others, like Alcoa, have incorporated this capability goal philosophy as part of their company's decision-making process for capital expenditures

(Rosenfeld).

In addition to setting detailed capability goals, Runger suggests also specifying proper data collection procedures, requirements for gage capability studies and calibration, minimum sample sizes, desired confidence bounds, degree of process control, and methods for determining the shape of the process output distribution. These requirements should be listed on the capability study form.

Juran (1993) takes the position that top management is responsible for establishing quality goals and determining appropriate means for comparing actual quality results against these goals. Capability goals should be based on past product performance, current and future customer needs (accounting for rising customer expectations), current and future competition (predicted through benchmarking), market sensitivity to quality, and the overall company's strategic business objectives. To be realistic and achievable, all goals must consider life cycle costs, material purchase costs, cost of assembly or fabrication, serviceability, ease of repair, and durability. Capability studies must be performed and reviewed on a regular, on-going basis to ensure that goals are met. Where current performance falls short of its goal, management must develop plans, and provide resources, for improving the involved process.

Many companies set an identical capability goal for *all* characteristics of *every* product, including those supplied from completely different industries. This is frequently done because a company doesn't wish to invest the necessary time scrutinizing each process to determine its critical characteristics and their true capability needs. However, all characteristics are not equally important, and therefore do not have the same capability requirements. A hypothetical ranking of the capability level needed for several critical product characteristics is presented in Figure 4.13.

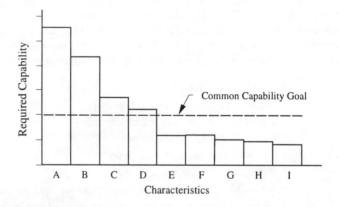

Fig. 4.13 Capability needed by critical characteristics to satisfy customer's requirements.

Although a simple method of setting goals, this "common-goal" policy is very wasteful because it leads to an improper allocation of resources. Time and money are spent improving all product characteristics whose current capability falls short of the goal. Those characteristics that don't require capability as high as the common goal are worked on with the same intensity as those requiring capability higher than this goal. Efforts expended on improving characteristics whose true capability goals are less than the common one would be better invested on those whose true capability needs exceed the general goal.

Making matters worse, those characteristics truly needing higher capability are improved only until they reach the common goal. At this point, improvement stops as attention and resources are diverted to other characteristics whose capability is currently less than the universal goal. The final result of this inefficient goal-setting policy is a shop where all characteristics have about the same level of capability, as is depicted in Figure 4.14. Obviously, capability goals must not be arbitrarily set.

Schneider *et al* have authored an insightful article on how the misapplication of capability measures results in misleading conclusions regarding supplier quality. They also offer several cautions as well as some alternative capability measures for special situations.

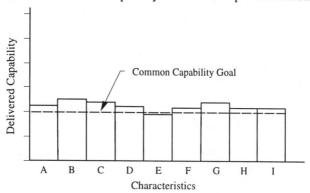

Fig. 4.14 All characteristics end up with about the same level of capability.

Capability goals must also take into account process volatility. If a particular process is often subjected to rapidly occurring changes, it becomes very difficult to control. Subgroup sizes should be increased, and time between subgroups decreased, for charts monitoring these kinds of processes so that changes are detected sooner. To ensure customer requirements are consistently met in this environment, substantially higher capability goals must be specified to provide a sufficient buffer for these changes. For example, consider the outputs for processes A and B displayed in Figure 4.15.

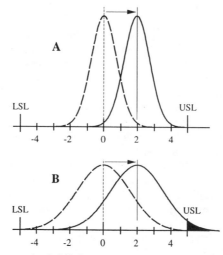

Fig. 4.15 The same-sized shift in average affects process B much more than A.

Process A can withstand a two-unit shift in the process average and still produce nearly all conforming parts because of its high capability. On the other hand, a similar two-unit shift in the average of process B causes a significant number of nonconforming parts to be produced due to its marginal capability. To provide a sufficient "safety margin," a process frequently experiencing significant shifts in its process average (or increases in process spread) must be assigned a higher capability goal than would a process with reasonably good stability. Chapter 14 explains how this concept of a shifting process average plays a major role in determining capability goals under the six-sigma quality philosophy.

Managers establishing capability goals must also evaluate the costs associated with scrap, rework, assembly problems, test rejects, and field failures. Product characteristics with high costs in any of these areas should be designated as critical and given an appropriately higher capability goal. Of course, any characteristic considered to be safety related should also have an elevated capability goal.

4.6 Process Capability Studies

A process capability study is a formal procedure for undertaking a systematic analytical investigation to reliably assess a process's ability to consistently meet a specific requirement. This may be a customer's requirement for a critical product characteristic, or an internal requirement for a important process variable. After selecting a characteristic for analysis, and then verifying the accuracy and precision of the measurement system, there are six major activities that must be completed during the course of a process capability study:

1. verify process stability,

2. estimate process parameters,

3. measure process capability,

4. compare actual capability to desired capability,

5. make a decision concerning process changes, and

6. report results of the study along with recommendations.

As mentioned in the last chapter, the remainder of this text assumes the first step is done and the process is in a relatively good state of statistical control. This allows for a meaningful estimation of process parameters (step 2) and measures of process capability (step 3). From these valid estimates, actual process capability may be compared to a specified capability goal so a decision can be made about what actions need to be taken (step 5).

These studies are often done during the development phase of new processes, as this type of information often helps manufacturing and product engineering agree on realistic tolerances. Sometimes capability studies are focused on a specific component of a process, *e.g.* a gage capability study or a machine capability study (see Chapter 10). Occasionally, some studies are conducted to quantify the ability of a process to satisfy a prospective customer's needs. At other times, they are employed to help make a decision as to which of several competing part suppliers should be chosen. However, for most companies, the majority of capability studies are performed on current manufacturing operations to evaluate how well they meet existing customer requirements. To be successful, a capability study must be well planned and carefully organized (Deleryd, 1997).

Obviously, there are many different definitions of what a process capability study entails. Some, like AT&T's (pp. 34-36 and 45-46), are much broader, including the selection of critical characteristics, implementation of control charts, identifying and reacting to out-of-control conditions, recalculating control limits, measuring process capability, and reducing common-source variation by means of designed experiments. Gunter provides an example of this broader definition applied to a blending operation. DeRosa *et al* include flowcharting the process and call this detailed approach a "process characterization study." They also present an administrative case study involving a government contract for office automation equipment. Lorenzen *et al* provide an example of how quality function deployment, design of experiments, and SPC are combined to improve quality in manufacturing electronic multichip modules. For more electronics examples, see Fasser and Brettner. Stevick explains how Eaton Corporation developed an organized approach for successfully applying statistical methods to improve a process producing engine components. An interesting case study of attempting to make a process capable before control charts are started is presented by Ackroyd and Turner.

This book purposely limits the scope of the study to focus on the capability measurement aspect, not because SPC and design of experiments are not important, they are. However, both topics are already adequately covered in the literature (Box *et al*; Montgomery, 1984, 1991; Wheeler and Chambers).

Capability Study Form

An example of a capability study form is displayed in Figure 4.16. Clearly, no single form can fit all applications in every type of industry, but many of the items on this one should be included in any detailed capability report. Note that all setup and processing parameters must be recorded so the estimated process capability can be linked to a specific set of operating conditions. An example of a form designed for foundries is offered by Smith, while one for machine shops is provided by Feigenbaum. A more general list that is less industry specific is given by Hradesky.

It is important that the time frame for conducting a capability study capture as many of the known common sources of variation as possible. These include effects of setups, changes in daily temperature or humidity, different operators, material from different lots and/or suppliers, tooling (cutting speeds and feed rates), and shut downs for lunch, breaks, and scheduled maintenance activities. If any of these items cause out-of-control conditions to appear on the control chart, they must be addressed before attempting to estimate process parameters or assess capability.

For certain types of studies, *e.g.*, machine or gage capability, a time frame may be selected to purposely *exclude* some of the above factors. However, this must be prominently stated on the capability study report. Machine capability studies are covered in Chapter 10. Explanations of, and forms for, gage capability studies are given by AIAG.

Almost all authors agree there must be an absolute minimum of 30 measurements in any type of study. Many prefer a minimum of 100 measurements. Obviously, a larger number of measurements will improve the confidence one has that the sample data properly described the true state of process capability (more on this topic in Chapter 11). However, larger sample sizes require more time to collect and will cost more as well.

Updating a Capability Study

After a capability study has been successfully completed for a particular characteristic of a given process, when and how often should the capability measure be updated? Several occasions for redoing a capability study have been proposed and are discussed in the next five bulleted items.

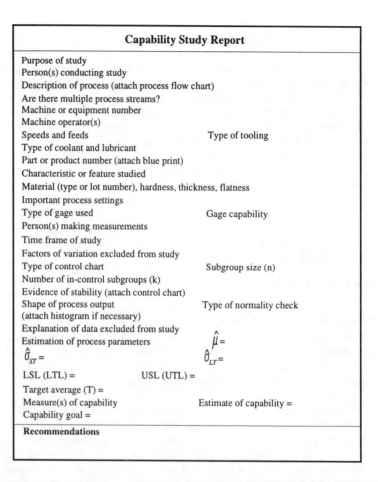

Capability Study Report

Purpose of study
Person(s) conducting study
Description of process (attach process flow chart)
Are there multiple process streams?
Machine or equipment number
Machine operator(s)
Speeds and feeds   Type of tooling
Type of coolant and lubricant
Part or product number (attach blue print)
Characteristic or feature studied
Material (type or lot number), hardness, thickness, flatness
Important process settings
Type of gage used   Gage capability
Person(s) making measurements
Time frame of study
Factors of variation excluded from study
Type of control chart   Subgroup size (n)
Number of in-control subgroups (k)
Evidence of stability (attach control chart)
Shape of process output   Type of normality check
(attach histogram if necessary)
Explanation of data excluded from study
Estimation of process parameters $\hat{\mu} =$
$\hat{\sigma}_{ST} =$ $\hat{\sigma}_{LT} =$

LSL (LTL) =   USL (UTL) =
Target average (T) =
Measure(s) of capability   Estimate of capability =
Capability goal =

Recommendations

Fig. 4.16 A process capability study from for a machining operation.

- *End of each control chart*. Many companies update capability measures when a sheet of control chart paper has been completely filled out. The results of this chart are used for calculating new control limits for the next sheet of chart paper, as well as making an estimate of the current process capability. However, if a process is still in control, it hasn't changed significantly and there is no need to update control limits nor the estimate of capability (Gitlow *et al*). Recalculating control limits and process capability when not necessary is a waste of valuable time.

- *After each run (or batch) is completed*. This a common approach in job shops where short runs of many different part numbers are routine. However, this approach really measures *part number* (or batch) capability rather than *process* capability. A summary of the last run is provided, but there is no way to predict process capability of future runs.

- *End of month (or quarter)*. Some quality practitioners wait until the end of some stated time interval before updating their capability measures. Typically, this is done every month (in some companies, every quarter) for all critical characteristics. Usually this list of characteristics, and their updated capability measures, is sent to the customer at the end of each reporting period. But what if the process, and therefore its capability, changes on the second day of the month? This event is not discovered until the end of the month. If capability has decreased, this is definitely too long a period to wait before initiating corrective action. What if no change to process capability occurs during the month? Then updating capability measures at the end of the month is an investment of resources with no return. Companies should avoid all such non-value-added activities.

- *Every time a new plot point is added to the control chart*. Many SPC software programs advertise the ability to automatically recalculate control limits and capability measures as each plot point is added to the chart. This is an extremely dangerous practice for control limits since it may mask out-of-control conditions. Suppose an upward trend develops on the $\overline{X}$ chart. As subgroup averages increase, the control limits also creep upward as a result of being recalculated with the most recent plot points. These moving limits make detection of the upward trend more difficult. If process variation gradually increases, the recalculated control limits will widen, reducing the chance new plot points will fall outside a control limit. In addition, points previously outside the control limits may now be within the changed limits, and vice versa. Is this point in or out of control? Recalculating limits with each new subgroup seriously decreases a chart's effectiveness in separating assignable causes of variation from common, which, of course, is the principal purpose of any control chart.

- *The night before a customer visit*. Although this is perhaps the most common reason triggering the updating of control charts and capability measures, there is little merit in doing so at these times.

The above circumstances are all inappropriate reasons for updating a capability measure. Capability studies should be repeated only when the process changes. How does one know whether the process has changed or not? By the control chart. If the chart is still in control, there has been no significant change to the process, or to the process parameters. Repeating a capability study will produce the same, or a very similar, estimate of process capability (Gitlow *et al*). As one major objective of a company should be to maximize quality at *minimum* cost, time should not be squandered doing unnecessary tasks.

If a process changes, there should be some evidence of an out-of-control signal on the control chart. A run below the centerline of an $\overline{X}$ chart monitoring tensile strength implies the process average has shifted lower (Figure 4.17). Because lower tensile strength is a deterioration of the process, effort should be expended on isolating the assignable source of variation and eliminating it, not wasted on recalculating capability measures. When the process returns to being in control, the prior capability is also restored, so again there is no need to recalculate capability measures.

In some cases a downward shift in the process average may be desirable, *e.g.*, a decrease in the average amount of taper for a shaft. In these situations, the cause should be found and made a *permanent* part of this process. When this is done, control limits and centerlines ought be recalculated after 20 subgroups are collected. Once control at this new level is demonstrated, all measures of capability must be updated because the process average (and perhaps the standard deviation) will have changed significantly (Tickel *et al*).

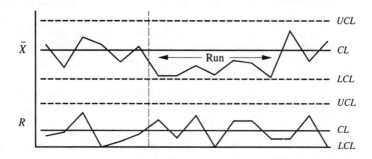

Fig. 4.17 $\bar{X}$ chart shows a significant downward shift in the process average.

A similar reasoning applies to the range chart. If there is a run above the centerline, the process variation has increased (Figure 4.18). The assignable cause of variation responsible should be identified and corrective action taken to prevent its reoccurrence in the future.

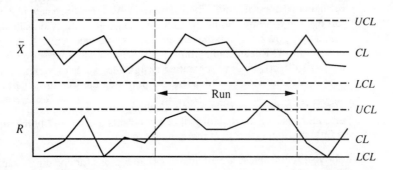

Fig. 4.18 Range chart indicates an increase in the process standard deviation.

When the assignable source of variation cannot be removed (the more-experienced Bill retires and the newly hired Hillary takes over running the operation), then the deterioration of the process is permanent. After a minimum of 20 subgroups are collected under these new operating conditions and plotted on the chart, recalculate the control limits and centerlines. When control is reestablished at this new level, update the capability study because the process output parameters have changed significantly. Work now begins on reducing process variation to at least what it was before the change.

On the other hand, if the run is *below* the centerline of the range chart, discover what changed and incorporate this cause of improvement as a *permanent* part of the process. After 20 subgroups, recalculate control limits, reestablish control, and finally, redo the capability study since process variation is considerably less than before, which means process capability should be significantly improved.

Always let the process be your guide for updating a capability study. Do not waste valuable resources performing worthless tasks.

Importance of Histograms in Capability Studies

Histograms are very helpful for understanding the results of a capability study by graphically displaying the relationship of the process output to the specification limits. As Yogi Berra is often credited with stating, "It's amazing how much you can see just by looking." However, histograms cannot replace control charts for demonstrating process stability (Deming, p. 312). They are unable to detect runs, trends, cycles, or other time-related out-of-control conditions. As explained in Chapter 3, any estimate of process parameters made when one (or more) of these conditions are present is invalid and will most likely lead to faulty decision making.

Even a histogram appearing to be somewhat normally distributed is *not* sufficient evidence the process is in control. Figure 4.19 displays a running record of the individual measurements for each subgroup of size 5 collected hourly from an operation turning the outer diameter of an automotive piston. As the tool wears and becomes shorter, the finished diameter of the piston increases gradually over time, as evidenced by the upward trend on this chart.

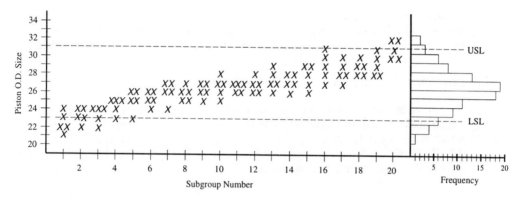

Fig. 4.19 A "bell-shaped" histogram does not mean the process is in control.

The obvious upward trend indicates the process average is unstable, yet the histogram of individual diameter measurements displayed at the right-hand side of Figure 4.19 appears as the typical "bell-shaped" curve expected from a process with a normal distribution. If an evaluation of process capability is made from only this histogram, a false conclusion would be reached that there is too much process spread, when in fact the process could easily meet the diameter requirement if the tool is changed more often or a switch is made to another style of tool that wears less.

Unless the measurements making up a histogram originate from a stable process, a histogram will *not* provide a reliable summary of that process's performance (AT&T, p. 61). However, if stability is first demonstrated with a control chart, a histogram becomes a very helpful device for explaining process capability, as it is easier to understand for those with little statistical training (Zaciewski). In addition, specification limits may be drawn on, and the number of sample measurements falling out of specification quickly determined. When properly constructed, a histogram presents an excellent visual assessment of the relationship between the process output and the customer requirement, as is clearly seen in Figure 4.20.

In this example, metal mounting brackets are run through an anodizing bath. The feature being studied is the thickness of the anodizing coating. The histogram discloses how the process is currently producing some brackets with too thick a coating. However, it may be able to consistently meet the print specifications for thickness if the process average is shifted slightly lower.

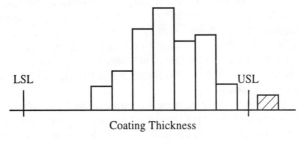

Fig. 4.20 A histogram of the process output compared to print specifications.

Furthermore, histograms are sometimes employed to help determine the shape of the process output distribution, as explained in Chapter 7. This is quite important, as one major assumption of most capability measures is that the process output is normally distributed.

4.7 Flowchart for Conducting a Process Capability Study

The important steps and decisions involved with conducting a process capability study are diagrammed on the flowchart displayed in Figure 4.21 on the next page. Spencer and Tobias present a case study featuring a process capability flowchart created by SEMATECH specifically for the semiconductor industry (see Deleryd *et al* for a list of critical factors for successfully implementing a capability study).

4.8 Summary

In today's competitive marketplace, customers assume manufacturers will provide on-time delivery of high-quality, low-priced products. By assisting companies in improving internal operations, capability studies enable manufacturers to improve productivity, reduce costs, and enhance their strategic advantage over competitors. Because of these benefits, many managers now request capability reports from their process and quality engineers. Some companies use the results of these capability studies as a report card to track the progress of quality improvement.

Process capability is determined by variation originating from *only* common causes, which is the minimum achievable variation after all assignable causes are eliminated. Thus, capability always implies stability.

Process capability studies analyze the relationship between actual process output and acceptable process output in an effort to measure customer satisfaction. The actual process output is represented by the 6σ spread (for normal distributions in a state of control) and the acceptable process output by the print specifications. For features with bilateral specifications, the process is considered to meet the minimum requirement of capability if at least 99.73 percent of the output is conforming. Characteristics with unilateral specifications must have at least a 99.865 percent conformance rate to be classified as capable.

Many measures of capability exist. The ones presented in the next chapter are based solely on process spread and measure potential capability. Those in Chapter 6 consider both process spread *and* process centering, thus measuring performance capability. Some measures consider only short-term process variation, while others focus on long-term variation. Still others are interested in only the percentage of nonconforming parts. However, one must suppress the urge to track every measure of capability. Select just those which are pertinent to good decision making, but don't feel obligated to keep the same set of measures forever. When circumstances change, adjust the set of measures accordingly so the needed process insight is always provided.

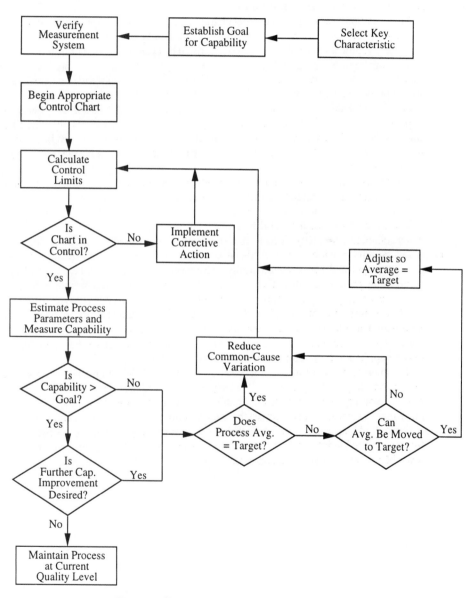

Fig. 4.21 Flowchart for process capability studies.

Be aware that no one particular measure is "perfect", as all have certain advantages and disadvantages. The "best" capability measure most often depends on the purpose of the study. For example, is the main interest on short- or long-term capability? In many cases, a more complete understanding of process capability is obtained by evaluating *several* of these measures along with a histogram of the process output. To be meaningful, the measurements on which any capability measure is based must represent the entire process, be independent of each other, and originate from a stable process. Many measures also assume that the process output is normally distributed.

Most customers now specify critical characteristics for which suppliers must achieve (and maintain) a minimum level of capability. Additional key product characteristics (and more importantly, process variables) are often identified by customer complaints, test rejects, scrap and rework reports, operator difficulties, as well as manufacturing and product engineering concerns. These characteristics must be prioritized in some meaningful manner so that efforts are properly assigned in resolving the most important quality problems first.

Goals established for process capability must consider the customer's needs, cost of improvements, available resources, volatility of the process, as well as the criticality of the characteristic. This usually involves a tremendous amount work, but it must be done to assure scarce resources are effectively and efficiently allocated to maximize customer satisfaction at minimum cost.

Capability studies must be updated whenever there is a permanent change to the process. These changes are detected by a shift in the centerline of a control chart. As long as the chart is in control, there is no need to recalculate control limits or update the capability study.

The development of a capability study form assures important items aren't overlooked when conducting the study. This form also serves as a permanent record of the study's results.

Histograms provide a helpful visual summary for perceiving the relationship between process output and specification limits. However, they cannot replace control charts for determining process stability. Even though other methods have been proposed for measuring process capability, only the control chart approach is rigorously correct. Any other approach is, at best, just an approximation (Juran *et al*, 1979). Control chart analysis leads to the elimination of assignable-cause variation, thus allowing capability studies to accurately assess the amount of common-cause variation. Once stability is achieved, other statistical methods may be applied to help reduce common-cause variation.

The main objective of measuring process capability is to better understand how adequately the current output of a process will satisfy customer requirements. Properly designed and run, a capability study provides a wealth of vital information concerning a process. This knowledge establishes a benchmark by which future process improvement, or deterioration, is tracked. Measuring capability before and after the implementation of process changes furnishes valuable information about continuous process improvement, which is the driving force behind all quality-enhancement efforts.

4.9 Exercises

4.1 Connecting Force

In Exercise 3.1, the process average for connecting force was estimated as 1.41615 kg, while the short-term standard deviation was predicted to be .02999 kg. With a LSL of 1.300 kg and an USL of 1.500 kg, do the $3\hat{\sigma}_{ST}$ tails fit within the tolerance (assume a normal distribution)? If not, what can be done to help make them fit?

4.2 Bolt Diameter

The average bolt diameter and short-term standard deviation were estimated in Exercise 3.2. Given a specification of 2.200 ± .050 cm, and assuming a normal distribution, does this process meet the minimum requirement of capability for a bilateral specification? If not, on what should process improvement efforts concentrate?

4.3 Relay Voltage

In Exercise 3.3, the average voltage to lift a relay armature was estimated as 29.6 volts, while the short-term standard deviation (σ_{ST}) was estimated to be 2.136 volts. With an USL of 40 volts, does this process meet the minimum requirement of capability for a unilateral specification (assume the process output has a normal distribution)?

4.4 Piston Ring Diameter

In Exercise 3.4, the process average and long-term standard deviation were estimated for the inside diameter of a piston ring.

$$\hat{\mu} = 9.02 \text{ cm} \qquad \hat{\sigma}_{LT} = .02170 \text{ cm}$$

With a LSL of 8.9250 cm and an USL of 9.0750 cm, could one estimated $6\sigma_{LT}$ spread fit between these specification limits? Given where the process is centered, do both the estimated lower and upper $3\sigma_{LT}$ tails fit within these specifications?

4.5 Shaft Length

In Exercise 3.5, estimates of the process average and short-term standard deviation were made for the length of a tapered steel shaft shown in Figure 4.22. Assuming a normal distribution, how does this process output compare to the customer specifications of ±3 (answer in part by sketching a histogram)? Does this process meet the minimum requirement of short-term process capability for a bilateral specification?

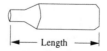

Fig. 4.22 A tapered steel shaft.

The long-term standard deviation was also estimated in Exercise 3.5. Does this process meet the minimum requirement for long-term process capability?

4.6 Rib Width

In Exercise 3.6, the process average and short-term standard deviation were estimated from an *IX & MR* chart. Given a specification of $1.580 \pm .016$ cm, how many σ_{ST}s fit between the process average and the LSL? Between the process average and the USL? Which of these two results is more critical?
Repeat this analysis using the long-term standard deviation.

4.7 Connector Length

Electrical connectors are cut to the specified length on a shearing operation. Because connector length is critical to the proper functioning of the finished connector, an $\overline{X}, R$ control chart ($n = 4$) is kept on this characteristic to help stabilize the process. When control is finally established, measurements from the last 25 in-control subgroups of the chart are arranged into the frequency tally displayed in Table 4.1. As the subgroup size is 4, there is a total of 100 (25 × 4) length measurements.

Table 4.1   Frequency tally of 100 connector length measurements.

Lower Class Boundary	Class Midpoint	Upper Class Boundary	Freq.
66.775	66.800	66.825	0
66.825	66.850	66.875	1
66.875	66.900	66.925	2
66.925	66.950	66.975	4
66.975	67.000	67.025	16
67.025	67.050	67.075	21
67.075	67.100	67.125	30
67.125	67.150	67.175	12
67.175	67.200	67.225	7
67.225	67.250	67.275	5
67.275	67.300	67.325	2
67.325	67.350	67.375	0

Construct a histogram of these measurements. Draw on the specification limits, and determine what percentage of this sample is nonconforming. The LSL for connector length is 66.700, while the USL is 67.300. Does this process satisfy the minimum requirement for process capability?

4.8 Roller Waviness

An $\overline{X}, R$ chart ($n = 4$) is monitoring a critical characteristic of an automotive machining operation which determines the waviness of the surface finish for a cam roller (Figure 4.23). After three weeks of observing the process and making several improvements, the chart finally shows good control. Before estimating process capability, the department supervisor wishes to construct a histogram of the process output for waviness.

To do this, the last 31 in-control subgroups (representing 124 waviness measurements) are taken from the chart and the frequency tally shown in Table 4.2 is made. Construct a histogram for this data and draw on the USL of 75 microinches. What effect would the shape of this output distribution have on estimating a measure of process capability?

Fig. 4.23

Table 4.2   Frequency tally for 124 measurements of cam roller waviness.

Class Midpoint	Freq.
5	0
10	3
15	16
20	27
25	22
30	20
35	14
40	8
45	6
50	3
55	2
60	1
65	0
70	1
75	0
80	0
85	1
90	0

The profilometer used to measure waviness can only read in intervals of 5. Thus, in the class with a midpoint of 10, the 3 waviness measurements were in fact equal to 10. With this additional information about the process, estimate the process average and long-term standard deviation.

4.9 Fan Test

Electric fans for cooling personal computers are assembled and tested. Unfortunately, quite a few fans fail this functional test. A quality-improvement team obtains a list of reasons from the inspection department regarding why fans fail this test. Next, a brainstorming session is held to generate ideas on which product characteristics might be responsible for these various modes of failures. After assigning a number between 1 and 10 to each failure mode based on its severity, the team rates every product characteristic (from 0 to 10) based on its perceived role in causing each type of test failure. All this information is presented in the matrix displayed in Table 4.3.

Table 4.3   Matrix analysis for electrical fan rejects.

Test Failure Mode	Severity	Oversize Armature	Loose Shroud	Dry Bearings	Bent Blades	Wrong Blade Assembly
Low RPM	6	7	0	6	2	4
High RPM	3	3	0	0	0	3
Noisy	7	0	4	7	6	8
High Temp.	10	0	0	9	0	0
Total						

Help the team create a Pareto diagram that ranks these characteristics by their overall effect on test failures.

4.10 References

Ackroyd, Peter, and Turner, Tom, "Forget About Tolerances! It's Your Cpk That Counts," *Modern Machine Shop*, Vol. 65, No. 11, April 1993, pp. 56-63

AIAG, *Measurement Systems Analysis: Reference Manual*, 1990, Automotive Industry Action Group, Southfield, MI

Anderson, Larry H., "Controlling Process Variation is Key to Manufacturing Success," *Quality Progress*, Vol. 23, No. 8, August 1990, pp. 91-93

AT&T, *Statistical Quality Control Handbook*, 2nd edition, 1984, Delmar Printing Co., Charlotte, NC

Boeing, *Key Characteristics: The First Step to Advanced Quality*, D6-55596 TN, Boeing Commercial Airplane Group, Quality & Surveillance Department., Materiel Division., Seattle, WA, p. 4

Box, George E.P., Hunter, William G., and Hunter, J. Stuart, *Statistics for Experimenters*, 1978, John Wiley & Sons, New York, NY

Campanella, Jack, editor, *Principles of Quality Costs: Principles, Implementation, and Use*, 2nd edition, 1990, Quality Press, Milwaukee, WI

Chrysler, Ford, and General Motors, *Potential Failure Mode and Effect Analysis: FMEA*, 1995*b*, Chrysler Corp., Ford Motor Company, General Motors Corporation (available through AIAG), Southfield, MI, pp. 29-43

Chrysler, Ford, and General Motors, *Quality Systems Requirements: QS-9000*, 1995*a*, Chrysler Corporation, Ford Motor Company, General Motors Corporation (available through AIAG), Detroit, MI, p. 82

Cohen, Lou, *Quality Function Deployment: How to Make QFD Work for You*, 1995, Addison-Wesley Publishing Co., Reading, MA

Davis, E.J., Scott, G.F., and Obray, C.D., "Process Capability for Manufacturing Processes," *Quality Forum*, Vol. 19, No. 3, September 1993

Deleryd, Mats, "A Strategy for Mastering Variation," *51st ASQ Annual Quality Congress Transactions*, May 1997, pp. 760-768

Deleryd, Mats, "On the Gap Between Theory and Practice of Process Capability Studies," *International Journal of Quality and Reliability Management*, Vol. 15, No. 2, 1998, pp. 178-191

Deleryd, Mats, Deltin, Johan, and Klefsjo, Bengt, "Critical Factors for Successful Implementation of Process Capability Studies," *Quality Management Journal*, Vol. 6, No. 1, 1999, pp. 40-59

Deming, W. Edwards, *Out of the Crisis*, 1986, Massachusetts Inst. of Tech., Cambridge, MA

DeRosa, Dominick A., Ashley, Karen M., and Berstein, Abe J., "Process Characterization: The Key to Quality Planning," *44th ASQC Annual Quality Congress Transactions*, May 1990, pp. 159-168

Fabian, Robert, "SPC-Guided Purchasing," *Quality*, Vol. 26, No. 11, Nov. 1987, pp. 32-34

Fasser, Yefim and Brettner, Donald, *Process Improvement in the Electronics Industry*, 1992, Wiley-Interscience, New York, NY

Feigenbaum, Armand V., *Total Quality Control*, 3rd edition, 1983, McGraw-Hill, NY, NY, pp. 780-781

Gevirtz, Charles, "The Fundamentals of Advanced Quality Planning," *Quality Progress*, Vol. 24, No. 4, April 1991, pp. 49-51

Gitlow, Howard, Gitlow, Shelly, Oppenheim, Alan, and Oppenheim, Rosa, *Tools and Methods for the Improvement of Quality*, 1989, Irwin, Homewood, IL, p. 432

Gunter, Bert, "Process Capability Studies, Part 4: Applications of Experimental Design," *Quality Progress*, Vol. 24, No. 8, August 1991, pp. 123-132

Harry, Mikel J., *The Nature of Six Sigma Quality*, 1987, Technical Report, Government Electronics Group, Motorola, Inc., Scottsdale, AZ, p. 1

Hradesky, John L., *Productivity and Quality Improvement*, 1988, McGraw-Hill, NY, NY, pp. 104-108

Juran, Joseph M., "Made in the USA - A Renaissance in Quality," *Harvard Business Review*, July-August 1993

Juran, Joseph M., Gryna, Frank M., Jr. and Bingham, R. S., Jr., *Quality Control Handbook*, 3rd edition, 1979, McGraw-Hill, New York, NY, p. 9-18

Kane, Victor E., *Defect Prevention: Use of Simple Statistical Tools*, 1989, Marcel Dekker, Inc., New York, NY

Kane, Victor E., "Process Capability Indices," *Journal of Quality Technology*, Vol. 18, No. 1, January 1986, p. 45

Lorenzen, Jerry A., Iqbal, Asif, Erz, Karen, and Rosenberger, Lisa M., "QFD, DOE and SPC in a Process for Total Quality," *47th ASQC Annual Quality Congress Transactions*, May 1993, pp. 421-427

McLaughlin, Gregory C., "Designing an Effective Process Capability Study," *42nd ASQC Annual Quality Congress Transactions*, Dallas, TX, May 1988, pp. 524-529

Montgomery, Douglas C., *Design & Analysis of Experiments*, 2nd edition, 1984, John Wiley & Sons, New York, NY,

Montgomery, Douglas C., *Introduction to Statistical Quality Control*, 2nd edition, 1991, John Wiley & Sons, New York, NY,

Nelson, Peter R., Editorial, *Journal of Quality Technology*, 24, 4, October 1992, p. 175

Pang, Vera K., "Scientifically Selecting Suppliers," *Quality Progress*, Vol. 25, No. 2, February 1992, pp. 43-45

Pennucci, Nicholas J., "Supplier SPC and Process Capability," *Quality*, Vol. 26, No. 11, November 1987, pp. 43-44

Puri, Subhash C., "Food Safety and Quality Control: SPC with HACCP," *44th ASQC Annual Quality Congress Transactions*, San Francisco, CA, May 1990, pp. 729-735

Rosenfeld, Manny, "A Quality-Based Capital Decision Process," *Quality Progress*, Vol. 24, No. 10, October 1991, pp. 75-79

Runger, George C., "Designing Process Capability Studies," *Quality Progress*, Vol. 26, No. 7, July 1993, pp. 31-32

Schneider, Helmut, Pruett, James, and Lagrange, Cliff, "Uses of Process Capability Indices in the Supplier Certification Process," *Quality Engineering*, Vol. 8, No. 2, 1995-1996, pp. 225-235

Shenoy, Mulalidhar K., "Machine Monitoring for Quality Assurance," *48th ASQC Annual Quality Congress Proceedings*, Las Vegas, NV, May 1994, p. 440

Skellie, Joel, and Ngo, Phung, "Use Statistical Process Monitoring Instead of Inspection," *Quality Progress*, Vol. 23, No. 6, June 1990, pp. 99-100

Smith, Kenneth M., *The Key Minimum-Maximum Standards Process Control System*, 1983, American Foundrymen's Society, Des Plaines, IL, pp. 125-127

Somerton, Diana G., and Mlinar, Sharyn E., "What's Key? Tool Approaches for Determining Key Characteristics," *50th ASQC Annual Quality Congress Transactions*, May 1996, pp. 364-369

Spencer, William J., and Tobias, Paul A., "Statistics in the Semiconductor Industry: A Competitive Necessity," *The American Statistician*, Vol. 49, No. 3, August 1995, pp. 245-249

Stevick, G.E., "Preventing Process Problems," *Quality Progress*, Vol. 23, No. 9, Sept. 1990, pp. 67-73

Sullivan, Larry P., Reply in "Letters" Column, *Quality Progress*, Vol. 18, No. 4, April 1985, pp. 7-8

Tickel, Craig M., Constable, Gordon K., and Cleary, Michael J., "Communicating Quality Improvement to Your Customers," *42nd ASQC Annual Quality Congress Transactions*, May 1988, p. 713

Wheeler, Donald J., and Chambers, David S., *Understanding Statistical Process Control*, 2nd edition, 1992, SPC Press, Inc., Knoxville, TN

Zaciewski, Robert, D., "Versatile Visual Aids," *Quality Progress*, Vol. 26, No. 1, January 1993, p. 112

Chapter 5

Measuring Potential Capability

Capability measures are like golf clubs:
more than one is needed to be successful.

There are many different methods of measuring process capability for variable-data control charts. Selecting the appropriate one depends on whether interest is in:

1. short-term process variation,

2. long-term process variation,

3. only process spread,

4. process spread *and* centering.

Most measures of process capability can be grouped by these four main categories, as is presented in Table 5.1. There are some measures that could be placed into more than one category (for example, *ppm*) depending on how they are calculated. This chapter covers both short- and long-term measures of potential capability, all of which are listed in the bold boxes of Table 5.1.

Even though many different types of capability measures exist, most companies rely on only one or two in practice. However, there is no one "best" measure of capability that will work well in all circumstances. Each presents a unique view of process capability and has its own advantages and disadvantages. To fully comprehend the manufacturing ability of any process, several measures should be analyzed during the course of a capability study, along with a histogram of the individual measurements.

Throughout this book, capability measures beginning with the letter "C" are calculated using short-term process variation (σ_{ST}), whereas those beginning with the letter "P" are based on long-term process variation (σ_{LT}). For those measures in which either short- or long-term variation could apply, a subscript of "ST" or "LT" identifies which is meant. For example, $Z_{MIN.ST}$ is the Z_{MIN} measure calculated with the short-term standard deviation, while $Z_{MIN.LT}$ represents the Z_{MIN} measure calculated with the long-term standard deviation.

This chapter introduces capability measures for situations where only process spread is of interest. These measures indicate the capability which is possible *if* the process average is centered at the desired target value. Due to this assumption, they are called measures of *potential* (possible) capability, and are typically applied in those situations where the process

Table 5.1 Categorizing the various measures of process capability.

Type of Process Variation	When Concern Is With:	
	Only Process Spread (Potential Capability)	Process Spread and Centering (Performance Capability)
Short Term	$6\sigma_{ST}$ Spread $C_R, C^*_R, C'_{RL}, C'_{RU}$ $C_P, C^*_P, C'_{PL}, C'_{PU}$ $C_{ST}, C^*_{ST}, C'_{STL}, C'_{STU}$	$Z_{LSL,ST}, Z_{USL,ST}$ $Z_{MIN,ST}, Z_{MAX,ST}$ $Z^*_{LSL,ST}, Z^*_{USL,ST}, Z^*_{MIN,ST}$ $C_{PL}, C_{PU}, C^*_{PL}, C^*_{PU}$ $C_{PK}, C^*_{PK}, C_{PM}, C^*_{PM}$ $C_{PG}, C^*_{PG}, C_{PMK}, C^*_{PMK}$ $ppm_{LSL,ST}, ppm_{USL,ST}$ $ppm_{MAX,ST}, ppm_{TOTAL,ST}$
Long Term	$6\sigma_{LT}$ Spread $P_R, P^*_R, P'_{RL}, P'_{RU}$ $P_P, P^*_P, P'_{PL}, P'_{PU}$ $P_{ST}, P^*_{ST}, P'_{STL}, P'_{STU}$	$Z_{LSL,LT}, Z_{USL,LT}$ $Z_{MIN,LT}, Z_{MAX,LT}$ $Z^*_{LSL,LT}, Z^*_{USL,LT}, Z^*_{MIN,LT}$ $P_{PL}, P_{PU}, P^*_{PL}, P^*_{PU}$ $P_{PK}, P^*_{PK}, P_{PM}, P^*_{PM}$ $P_{PG}, P^*_{PG}, P_{PMK}, P^*_{PMK}$ $ppm_{LSL,LT}, ppm_{USL,LT}$ $ppm_{MAX,LT}, ppm_{TOTAL,LT}$

average is easily adjusted and concern is only about variation around the average (Clements). Measures considering both spread and actual process centering are called performance measures and are covered in the next chapter.

One application for potential capability measures is an operation cutting the length of metal electrical connectors in a job shop environment. Various types of connectors are run on this process, each having a different length requirement. It's relatively easy to set the machine to the desired cut-off length, but once located there, variation around the average determines how well customer requirements are met. As interest is focused solely on process spread and not centering, potential capability measures are appropriate for this situation as they will reveal the best this process can do without fundamental changes to reduce common-cause variation.

Another application area is for operations having automatic gaging with feedback control to adjust the process average. With this system, the average should remain on target, and concern is focused on variation around this average. By measuring potential capability, management becomes aware of the lower limit for this variation under current manufacturing conditions.

The discussion about potential capability measures begins with one called simply just the "process spread."

5.1 Short-term Process Spread, $6\sigma_{ST}$ Spread

One of the easiest capability measures to calculate (and to comprehend) is the short-term process spread, labeled the $6\sigma_{ST}$ spread. This measure of potential capability is derived by multiplying the short-term process standard deviation by 6 (Jamieson).

$$\text{Short-term process spread} = 6\sigma_{ST}$$

Since the definition of process capability for a normal distribution involves the six standard deviation spread of the process output (Figure 5.1), this first measure of process capability is fairly easy to understand.

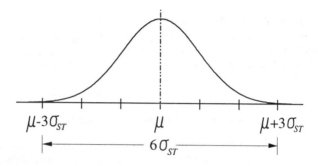

Fig. 5.1 The $6\sigma_{ST}$ spread of the process output.

$6\sigma_{ST}$ simply represents the distance from the upper $3\sigma_{ST}$ tail to the lower $3\sigma_{ST}$ tail. This is where the middle 99.73 percent of the process output (for a normal distribution) will fall over an extended period of time.

$$
\begin{aligned}
6\sigma_{ST} \text{ Spread} &= \text{Upper } 3\sigma_{ST} \text{ tail minus lower } 3\sigma_{ST} \text{ tail} \\
&= \mu + 3\sigma_{ST} - (\mu - 3\sigma_{ST}) \\
&= \mu + 3\sigma_{ST} - \mu + 3\sigma_{ST} \\
&= 3\sigma_{ST} + 3\sigma_{ST} \\
&= 6\sigma_{ST}
\end{aligned}
$$

This $6\sigma_{ST}$ spread represents the smallest interval in which the process can produce 99.73 percent of its output in the short term. Such a limit is helpful for design engineers to have in mind when establishing tolerances for new products, and for sales managers when quoting on new business.

For most capability studies, the actual short-term standard deviation is not known, so it must be estimated from subgroup data collected on a control chart. Thus, the following formula is often applied in practice to estimate the $6\sigma_{ST}$ spread. Recall that the symbol "^" above a process parameter indicates it is an estimate rather than the true value of that parameter.

$$\text{Estimate of } 6\sigma_{ST} \text{ Spread} = 6\hat{\sigma}_{ST} \text{ Spread}$$

Depending on the type of control chart monitoring the process, σ_{ST} is estimated in one of the following three ways, as was explained in Section 3.2 (page 36).

$$\hat{\sigma}_{ST} = \hat{\sigma}_{\bar{R}} = \frac{\bar{R}}{d_2} \quad \text{if an } \bar{X}, R \text{ chart is used, or}$$

$$= \hat{\sigma}_{\bar{S}} = \frac{\bar{S}}{c_4} \quad \text{if an } \bar{X}, S \text{ chart is used, or}$$

$$= \hat{\sigma}_{\overline{MR}} = \frac{\overline{MR}}{1.128} \quad \text{if an } IX \ \& \ MR \text{ chart is used.}$$

In order to meet the minimum requirement of short-term potential capability (the ability to produce 99.73 percent conforming parts), the $6\sigma_{ST}$ spread must be no greater than the tolerance. Sometimes this tolerance is referred to as the specification width, or the engineering tolerance, and is defined as the distance between the USL and LSL.

$$\text{Tolerance} = \text{USL - LSL}$$

Thus, a "rating scale" for this measure is developed as follows:

$6\sigma_{ST}$ Spread < Tolerance means the minimum requirement for short-term potential capability is exceeded and the process has the ability to produce at least 99.73 percent conforming parts with proper centering,

$6\sigma_{ST}$ Spread = Tolerance implies the minimum requirement for short-term potential capability is just met and the process has the ability to produce at most 99.73 percent conforming parts with correct centering,

$6\sigma_{ST}$ Spread > Tolerance indicates the minimum requirement for short-term potential capability is *not* met and the process lacks the ability to produce 99.73 percent conforming parts, even with proper centering.

Sometimes in manufacturing, "in-process" specifications are developed for certain characteristics. These tend to be tighter than the original print specifications, and are typically imposed by the manufacturer to help downstream operations in the production (or assembly) of this part. If the purpose of the capability study is to evaluate how well this process satisfies the requirements of the downstream operations (the internal customer), the in-process specifications should be used to rate the capability of the process. If the study is conducted to determine how well the final customer's requirements are being met (the external customer), then the $6\sigma_{ST}$ spread must be compared to the print specifications.

Although the $6\sigma_{ST}$ spread may be computed for a characteristic with a unilateral specification, it cannot be judged acceptable or not as the tolerance is undefined.

Example 5.1 Fan Housing Weight

An $\bar{X}, R$ chart (subgroup size of 5) monitoring the weight of fan housings produced on a plastic injection molding process is finally brought into a good state of control. A histogram of the individual weight measurements for the last 20 in-control subgroups is displayed in Figure 5.2.

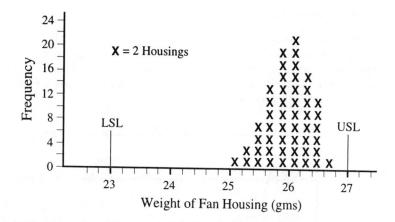

Fig. 5.2 Weight measurements for plastic fan housings.

The $\bar{R}$ value from the range chart is .848 grams while the tolerance for part weight is 4.00 grams (25 ± 2 grams). Substituting these values into the formula below yields an estimate of the short-term process spread.

$$6\hat{\sigma}_{ST} \text{ Spread } = 6\hat{\sigma}_{ST} = 6\,(\bar{R}\,/\,d_2) = 6\,(.848\,\text{gms}\,/\,2.326) = 2.187\,\text{gms}$$

This calculation estimates the width of the $6\sigma_{ST}$ spread for part weight to be 2.187 grams. This interval is required by the process to produce the middle 99.73 percent of its output distribution for part weight (assuming a normal distribution). Because this $6\sigma_{ST}$ spread of 2.187 grams is quite a bit less than the tolerance of 4.00 grams, this process exceeds the minimum requirement for potential capability on part weight. Whether this span of weights falls completely within, only partially within, or totally outside the tolerance is not addressed and cannot be readily ascertained from this measure of capability. However, a quick look at the histogram of weight measurements discloses the process is centered closer to the USL than the LSL. The majority of the process output is located in the upper half of the tolerance. This lack of proper centering may cause some fan housings to be molded overweight. Note that this process spread measure does not reveal the percentage of nonconforming product.

Because control limits for the $\bar{X}$ chart are based on short-term variation, the $6\sigma_{ST}$ spread may also be estimated directly from the difference between these limits. Recall that these control limits are set at $\pm 3\sigma_{\bar{X}}$ from the centerline. Thus, the difference between the limits is equal to $6\sigma_{\bar{X}}$.

$$6\hat{\sigma}_{\bar{X}} = UCL - LCL$$

$$\frac{6\hat{\sigma}_{ST}}{\sqrt{n}} = UCL - LCL$$

$$6\hat{\sigma}_{ST} = \sqrt{n}\,(UCL - LCL)$$

$$6\hat{\sigma}_{ST} \text{ Spread } = \sqrt{n}\,(UCL - LCL)$$

The next example illustrates how this capability measure is applied in practice.

Example 5.2 Rough Turning

The inner diameter (I.D.) of a transmission gear for heavy-duty construction equipment undergoes a rough turning operation on a lathe. As I.D. size is important for subsequent machining operations, this characteristic is charted on an $\overline{X}$, R chart, where n is five. Once the process is in a good state of control (as evidenced by the chart in Figure 5.3), it is desired to assess potential capability.

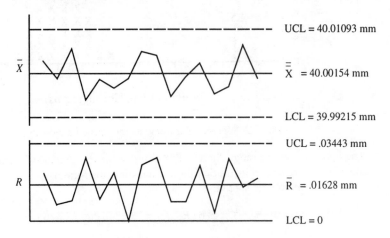

Fig. 5.3 The control chart for gear I.D. size.

The short-term process spread can be estimated from σ_{ST} or from the difference between control limits on the $\overline{X}$ chart. First, based on an estimate of σ_{ST}:

$$6\hat{\sigma}_{ST}\ \text{Spread}\ =\ 6\hat{\sigma}_{ST}\ =\ 6\frac{\overline{R}}{d_2}\ =\ \frac{6\,(.01628\,\text{mm})}{2.326}\ =\ .04199\,\text{mm}$$

A second estimate is provided by the control chart method.

$$6\hat{\sigma}_{ST}\ \text{Spread}\ =\ \sqrt{n}\,(UCL - LCL)\ =\ \sqrt{5}\,(40.01093\,\text{mm} - 39.99215\,\text{mm})\ =\ .04199\,\text{mm}$$

The specifications for I.D. size are 40.00 mm $\pm$ .02 mm, making the tolerance .04 mm. As the estimated $6\sigma_{ST}$ spread of .04199 mm is greater than the tolerance, this roughing operation lacks short-term potential capability. Work must begin to identify, and reduce, sources of common-cause variation. In addition, whenever process capability doesn't meet its goal, the output must be contained and 100 percent inspection implemented. These safeguards continue until process capability is improved to the required level.

The $6\hat{\sigma}_{ST}$ spread presents an idea of how a process can perform in the short term (instantaneous capability), as it is calculated from variation occurring only within subgroups. Due to the manner of collecting rational subgroups, this variation represents the absolute best this process can do without drastic changes. Quite often, interest exists for a more realistic idea of what the process spread will actually look like when variation happening between subgroups, as well as within subgroups, is also considered. This is accomplished by replacing σ_{ST} with σ_{LT} when estimating the process spread.

5.2 Long-term Process Spread, $6\sigma_{LT}$ Spread

This measure of capability provides an idea of the range of measurements expected over a period of time for the process to produce the middle 99.73 percent of its output. It is calculated by multiplying the long-term process standard deviation by 6.

$$6\sigma_{LT} \text{ Spread} = 6\sigma_{LT}$$

As the definition of process capability (for a normal distribution) involves the six standard deviation spread of the process output, this measure of process capability is fairly easy to understand. It is simply the smallest width in which the process can produce the middle 99.73 percent of its output over the long term, as is shown in Figure 5.4.

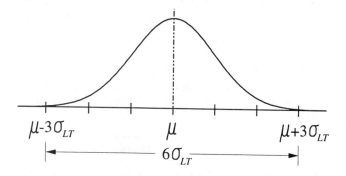

Fig. 5.4 The $6\sigma_{LT}$ spread of the process output.

For most capability studies, the true long-term process variation is seldom known so it must be estimated from data collected on a control chart. In practice, the following formula is used to estimate the long-term process spread, where $\hat{\sigma}_{LT}$ is the estimated long-term process standard deviation. Confidence bounds for the true value of this capability measure are covered in Section 11.4.

$$\text{Estimate of } 6\sigma_{LT} \text{ Spread} = 6\hat{\sigma}_{LT}$$

Recall from Section 3.2 (page 39) that no matter which type of variable-data chart is monitoring the process, σ_{LT} is always estimated from the individual measurements of all in-control subgroups on the chart. Thus, the c_4 factor is based on a total sample of kn measurements, where k is the number of in-control subgroups and n is the number of individual measurements in each subgroup.

$$\hat{\sigma}_{LT} = \frac{S_{TOT}}{c_{4_{kn}}} \qquad \text{where } S_{TOT} = \sqrt{\frac{\sum_{i=1}^{kn}(X_i - \hat{\mu})^2}{kn - 1}}$$

The c_4 factors for kn are found in Appendix Table II. The rating scale for the long-term process spread is the same as for the short-term process spread.

Example 5.3 Fan Housing Weight

In the injection-molding example (Example 5.1), S_{TOT} is calculated as .3937 grams from the individual measurements contained in 20 (k) in-control subgroups of the $\overline{X}, R$ chart monitoring housing weight. As n is 5, the number of individual measurements included in this calculation is equal to 100 $(kn = 20 \times 5)$. The long-term process spread is estimated as follows:

$$6\hat{\sigma}_{LT} \text{ Spread } = 6\hat{\sigma}_{LT} = \frac{6S_{TOT}}{c_{4_{kn}}} = \frac{6(.3937 \text{ gm})}{.9975} = 2.368 \text{ gm}$$

This capability measure estimates the $6\sigma_{LT}$ spread for this process output to be 2.368 grams, which is the range required by the process to produce 99.73 percent of the part weights over the long term (assuming a normal distribution). Because this is less than the tolerance of 4.00 grams, the process is said to possess long-term potential capability.

$$6\hat{\sigma}_{LT} = 2.368 \text{ gm} < \text{Tolerance} = 4.00 \text{ gm}$$

Notice how this range is about 8.3 percent larger than the 2.187 grams required for the estimated $6\sigma_{ST}$ spread (review Example 5.1). This difference is caused by the small changes in process centering occurring between subgroups that go undetected by the control chart, but are included in σ_{LT}. Recall from Section 3.2 that:

$$\sigma_{LT} = \sqrt{\sigma_{WITHIN}^2 + \sigma_{BETWEEN}^2}$$

The $6\sigma_{ST}$ spread assumes the ideal case, that $\sigma_{BETWEEN}$ is zero.

$$\sigma_{ST} = \sqrt{\sigma_{WITHIN}^2 + 0^2} = \sigma_{WITHIN}$$

Thus, the $6\sigma_{ST}$ spread becomes:

$$6\sigma_{ST} \text{ Spread} = 6\sigma_{WITHIN}$$

On the other hand, the $6\sigma_{LT}$ spread assumes that $\sigma_{BETWEEN}$ is small because the process is in control, but that it is not equal to zero. This makes the $6\sigma_{LT}$ spread greater than the $6\sigma_{ST}$ spread.

$$\sqrt{\sigma_{WITHIN}^2 + \sigma_{BETWEEN}^2} > \sqrt{\sigma_{WITHIN}^2 + 0^2}$$

$$\sigma_{LT} > \sigma_{WITHIN}$$

$$6\sigma_{LT} > 6\sigma_{WITHIN}$$

$$6\sigma_{LT} \text{ Spread} > 6\sigma_{ST} \text{ Spread}$$

Capability Goals for $6\sigma_{ST}$ and $6\sigma_{LT}$

A very common goal for these particular capability measures is to be equal to, or less than, 75 percent of the tolerance (Juran and Gryna). This provides a safety margin which allows the process spread to increase slightly but still retain a high potential for producing conforming product. With a tolerance of 4.00 grams for the fan housing weight example, the goal for the

6σ spread becomes 3.00 grams (4.00 grams × .75), or less. As the estimates for $6\sigma_{ST}$ (2.187 grams) and $6\sigma_{LT}$ (2.368 grams) are both significantly less than 3.00 grams, the process is considered to have very good potential capability for meeting this goal.

Suppose a different customer expresses a capability goal requiring these measures to be less than 50 percent of the tolerance. 50 percent of 4.00 grams is 2.00 grams. As both short-term and long-term process spread measures are greater than this second goal, the process is classified as lacking potential capability for this customer. Work must begin as quickly as possible to reduce process variation for this characteristic in order to realize this more stringent requirement.

If there are at least two customers with different capability goals, the process may be capable for one, but not the other. Always work to the most restrictive goal since meeting this one will satisfy all other customers.

Example 5.4 Exhaust Valves

In Example 3.7 (page 40), σ_{ST} and σ_{LT} were estimated for the O.D. size of an exhaust valve run on a grinding machine (Figure 5.5).

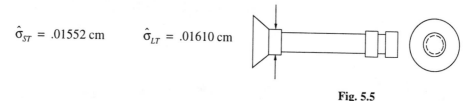

$$\hat{\sigma}_{ST} = .01552 \text{ cm} \qquad \hat{\sigma}_{LT} = .01610 \text{ cm}$$

Fig. 5.5

Assuming the process output has a normal distribution, the potential capability of the grinding process to produce this characteristic can be estimated with the $6\sigma_{ST}$ (for short term) and $6\sigma_{LT}$ (for long term) spreads.

$$6\hat{\sigma}_{ST} \text{ Spread } = 6(.01552 \text{ cm}) = .09312 \text{ cm}$$

$$6\hat{\sigma}_{LT} \text{ Spread } = 6(.01610 \text{ cm}) = .09660 \text{ cm}$$

If print specifications for O.D. size are a LSL of 9.9400 cm and an USL of 10.0600 cm, the tolerance is .1200 cm (10.0600 cm - 9.9400 cm). As the estimated short-term process spread of .09312 cm is less than this tolerance of .1200 cm, this grinding process surpasses the minimum requirement for short-term potential capability. When properly centered, it has the potential to produce at least 99.73 percent of the valve O.D. sizes well within tolerance.

What if a customer specifies the goal for short-term potential capability for valve O.D. size to be at most 75 percent of the tolerance? To meet this goal, the estimate of $6\sigma_{ST}$ must be less than .0900 cm (.75 × .1200 cm). With the $6\sigma_{ST}$ spread estimated as .09312, the process does not meet this goal no matter where the process average is centered. One hundred percent inspection and product containment should be initiated, and continued, until process variation is reduced enough to meet this goal.

The long-term process spread of .09660 cm is also less than the tolerance of .1200 cm, revealing that the process possesses long-term potential capability for producing O.D. size as well. However, it would not meet a goal of being less than 75 percent of the tolerance.

Advantages and Disadvantages

The short- and long-term process spreads are perhaps the easiest capability measures to calculate, requiring knowledge about only σ_{ST} or σ_{LT}. As process capability is expressed in the same units of measurement as the characteristic under study, these measures are readily understood by personnel on the shop floor as well as in upper management.

The same formula applies to unilateral and bilateral specifications. As will be seen later, some measures of capability require different formulas for these two cases, causing some confusion. Although the process spread is easily estimated for characteristics with a unilateral specification, it is difficult to rate the process's capability because the tolerance is undefined. If a target value for the process average (T) is given, a "tolerance" may be defined as follows when only a LSL is provided.

$$Tolerance = 2(T - LSL)$$

As can be seen in Figure 5.6a, this pseudo tolerance is just twice the length of the interval from the LSL to the target average.

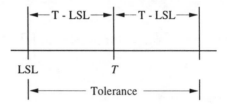

Fig. 5.6a Creating a surrogate tolerance for a unilateral specification.

If only an USL is available, switch to this formula:

$$Tolerance = 2(USL - T)$$

In addition, these capability measures cannot be directly compared to those for other processes due to these very same units of measurement. For example, which process has a better capability: process A with a $6\sigma_{ST}$ spread of 2.187 grams or process B with a $6\sigma_{ST}$ spread of 1152.9 newton-meters? There is no easy way to answer this question with this first measure of process capability.

Comparisons are difficult even when studying the differences between two similar machines producing the same characteristic of the same part number. Assume the process spread of machine A is identical to that for machine B ($6\sigma_{ST.A} = 6\sigma_{ST.B}$), as is displayed in Figure 5.6b. Machine A is currently doing a much better job producing this characteristic within tolerance than B, yet both are classified equal in potential capability by the process spread measure. As emphasized previously, examining more than one measure of capability, along with a histogram, should be mandatory before making any conclusions concerning how well customer requirements are being fulfilled.

In order for the 6σ spread to accurately represent 99.73 percent of the process output, the output must have a normal distribution, while individual measurements collected from the process should be independent. These will be implicit assumptions throughout this chapter (as well as the next) and is quite important for obtaining valid estimates of capability (Gunter). Chapter 7 explains how this assumption is verified, while Chapter 8 presents several capability measures for process outputs having non-normal distributions.

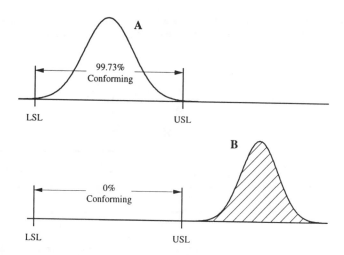

Fig. 5.6b The process spreads for machines A and B.

As indicators of potential capability, these measures apply to only those situations where the location of the process average is of little, or no, concern (Barnett). For the majority of processes, *both* spread and centering are important. In fact, for some processes centering is even more important than spread.

There is no way to estimate the percentage of conforming (or nonconforming) parts from these measures. Neither is it possible to determine if the process is expected to produce any parts below the LSL or above the USL. This makes it extremely difficult to tell if customer requirements, as represented by the specification limits, are really being met or not. Thus, these first two capability measures have little correlation to actual customer satisfaction. Constructing a histogram and drawing on the specification limits will shed some light on these questions.

Since meeting customer requirements is a vital part of any assessment of process capability, it makes sense to somehow incorporate the specification limits into the capability measure itself. Such a measure is presented next.

5.3 Short-term Potential Capability Ratio, C_R

One of the earliest methods for measuring short-term potential process capability is to relate the process spread for a given characteristic (typically represented by $6\sigma_{ST}$) directly to the tolerance for that characteristic (Charbonneau and Webster). This relationship, called the short-term potential capability ratio, produces a dimensionless number that may be compared to a customer-specified goal, as well as to the capability of other processes.

$$C_R = \frac{6\sigma_{ST}}{\text{Tolerance}} = \frac{6\sigma_{ST}}{\text{USL - LSL}}$$

The C_R measure calculates a ratio of the short-term process spread to the tolerance for the characteristic under study. This ratio represents what percentage of the tolerance is required to accommodate one $6\sigma_{ST}$ spread. Assuming the process output is stable and normally distributed, $6\sigma_{ST}$ represents the width of the middle 99.73 percent of the output distribution that the process can maintain over a short period of time. Hopefully this width is substantially less than the tolerance, which represents the customer's requirement for this characteristic (see Figure. 5.7).

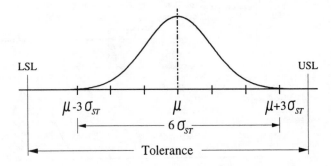

Fig. 5.7 Comparing the $6\sigma_{ST}$ spread to the tolerance.

If $6\sigma_{ST}$ is less than the tolerance, this ratio is less than 1.0.

$$6\sigma_{ST} < \text{Tolerance}$$

$$\frac{6\sigma_{ST}}{\text{Tolerance}} < \frac{\text{Tolerance}}{\text{Tolerance}}$$

$$C_R < 1$$

This implies that the process has the *potential* of producing at least 99.73 percent conforming parts in the short term, and therefore, meets the minimum requirement of potential capability. As with the $6\sigma_{ST}$ spread measure, the actual percentage of conforming parts depends on where the process average is centered.

On the other hand, if $6\sigma_{ST}$ is greater than the tolerance, this ratio is greater than 1.0, and the process fails to meet the minimum criteria for short-term potential capability for this characteristic. No matter where the process is centered, it lacks the ability to produce at least 99.73 percent conforming parts.

$$6\sigma_{ST} > \text{Tolerance}$$

$$\frac{6\sigma_{ST}}{\text{Tolerance}} > \frac{\text{Tolerance}}{\text{Tolerance}}$$

$$C_R > 1$$

In most cases the actual (true) short-term process standard deviation is not known, so it must be estimated from subgroup data collected on the control chart monitoring this characteristic. In practice the following formula (which replaces σ_{ST} with $\hat{\sigma}_{ST}$) is used to estimate C_R.

$$\hat{C}_R = \frac{6\hat{\sigma}_{ST}}{\text{Tolerance}} = \frac{6\hat{\sigma}_{ST}}{\text{USL - LSL}}$$

Depending on the type of control chart being used, σ_{ST} is estimated from one of the formulas listed here. d_2 and c_4 for n from 1 to 10 are found in Appendix Table I.

$$\hat{\sigma}_{ST} = \frac{\overline{R}}{d_2} \qquad \hat{\sigma}_{ST} = \frac{\overline{S}}{c_4} \qquad \hat{\sigma}_{ST} = \frac{\overline{MR}}{1.128}$$

A rating scale for this measure is established as given below.

$C_R < 1$ means the minimum requirement is exceeded and the process has the ability to produce at least 99.73 percent conforming parts with proper centering,

$C_R = 1$ indicates the minimum requirement is just met and the process has the ability to produce at most 99.73 percent conforming parts with proper centering,

$C_R > 1$ implies the minimum requirement for short-term potential capability is *not* satisfied, and the process lacks the ability to produce 99.73 percent conforming parts, even with the correct centering.

Example 5.5 Fan Housing Weight

An $\overline{X},R$ chart (subgroup size of 5) monitoring the weight of fan housings produced on a plastic injection-molding process is finally brought into a good state of control. The tolerance for part weight is 4.00 grams, and the $\overline{R}$ value from the chart is .848 grams. Substituting these values into the formula below yields an estimate of C_R.

$$\hat{C}_R = \frac{6\hat{\sigma}_{ST}}{\text{Tolerance}} = \frac{6(\overline{R}/d_2)}{\text{Tolerance}} = \frac{6(.848 \text{ gms}/2.326)}{4.00 \text{ gms}} = \frac{2.187 \text{ gms}}{4.00 \text{ gms}} = .547$$

Notice how the gram units cancel out, resulting in a dimensionless measure of capability. This C_R estimate of .547 indicates the process has the potential to fit one full $6\sigma_{ST}$ spread into only 54.7 percent of the tolerance. If properly centered with a normal distribution, this process should have no difficulty producing at least 99.73 percent conforming parts, thus exceeding the minimum requirement for short-term potential capability. This agrees with the above rating scale, as this ratio of .547 is much less than 1.0.

Example 5.6 Roller Width

After molding and curing, hard rubber rollers for guiding and driving paper sheets through copying machines are ground to a final diameter. Rollers are then sliced with a rotating knife to the proper width, with a finished roller shown in Figure 5.8. As excessive variation in roller width results in paper jams, this dimension must be held to within plus or minus .0127 cm of the nominal length of 31.2250 cm.

Fig. 5.8 A hard rubber roller for a copying machine.

A subgroup size of 10 is chosen, so an $\overline{X}$, S chart is selected to monitor the slicing process. When a good state of control is demonstrated, the last 35 in-control subgroups are selected to calculate $\overline{S}$, which turns out to be .005169 cm. Based on this result, the short-term potential capability ratio is estimated as 1.25.

$$\hat{C}_R = \frac{6\hat{\sigma}_{ST}}{\text{Tolerance}} = \frac{6(\overline{S}/c_4)}{\text{Tolerance}} = \frac{6(.005169\,\text{cm}/.9727)}{.0254\,\text{cm}} = 1.25$$

An estimated C_R ratio of 1.25 means the $6\sigma_{ST}$ spread is 25 percent greater than the tolerance. Even if properly centered, the slicing operation cannot come close to cutting all rollers to the desired width. Efforts to reduce common-cause variation must begin immediately. After improvements are implemented, process control for width must be reestablished and the capability study repeated. This cycle continues until the desired capability goal is met.

While these variation-reduction activities are being carried out, the rollers should be subjected to 100 percent inspection for width. Note that a C_R of 1.25 does not mean 25 percent of these rollers have their width out of specification. Unfortunately, there is no way to estimate the actual percentage of nonconformance from the C_R ratio.

With an $\overline{X}$ of 31.2306, the UCL on the $\overline{X}$ chart is computed as 31.23564 cm, while the LCL is 31.22556 cm.

$$UCL = \overline{\overline{X}} + A_3\overline{S} = 31.2306\,\text{cm} + .975(.005168\,\text{cm}) = 31.23564\,\text{cm}$$

$$LCL = \overline{\overline{X}} - A_3\overline{S} = 31.2306\,\text{cm} - .975(.005168\,\text{cm}) = 31.22556\,\text{cm}$$

C_R may also be estimated from the difference between these control limits.

$$\hat{C}_R = \frac{\sqrt{n}\,(UCL - LCL)}{\text{Tolerance}} = \frac{\sqrt{10}\,(31.23564\,\text{cm} - 31.22556\,\text{cm})}{.0254\,\text{cm}} = 1.25$$

This yields a result identical to the first estimate of C_R.

5.4 Long-term Potential Capability Ratio, P_R

This next measure calculates the ratio of the long-term process spread, as measured by $6\sigma_{LT}$, to the customer requirement, as measured by the tolerance. P_R reveals the percentage of tolerance required to accommodate one $6\sigma_{LT}$ spread.

$$P_R = \frac{6\sigma_{LT}}{\text{Tolerance}} = \frac{6\sigma_{LT}}{\text{USL - LSL}}$$

$6\sigma_{LT}$ represents the distance from the upper $3\sigma_{LT}$ tail to the lower $3\sigma_{LT}$ tail. This is where the middle 99.73 percent of the process output (for a normal distribution) will fall over an extended period of time.

$$6\sigma_{LT} \text{ Spread} = \text{Upper } 3\sigma_{LT} \text{ tail minus lower } 3\sigma_{LT} \text{ tail}$$

$$= \mu + 3\sigma_{LT} - (\mu - 3\sigma_{LT})$$

$$= \mu + 3\sigma_{LT} - \mu + 3\sigma_{LT}$$

$$= 3\sigma_{LT} + 3\sigma_{LT}$$

$$= 6\sigma_{LT}$$

In view of this being the best that can be expected of this process over the long term, hopefully the estimated $6\sigma_{LT}$ spread is considerably less than the tolerance, as is illustrated in Figure 5.9. If so, the process has the potential of producing at least 99.73 percent conforming parts, and therefore, meets the minimum requirement for long-term potential capability.

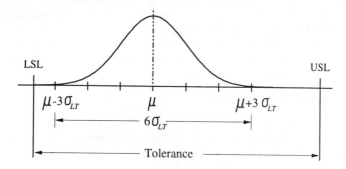

Fig. 5.9 P_R compares the $6\sigma_{LT}$ spread to the tolerance.

Due to being unknown in most cases, the true long-term process standard deviation must be estimated from subgroup data collected for the control chart. In practice, this formula is employed most often to estimate P_R.

$$\hat{P}_R = \frac{6\hat{\sigma}_{LT}}{\text{Tolerance}} = \frac{6\hat{\sigma}_S}{\text{USL - LSL}}$$

No matter which type of variable-data chart is chosen to control the process, σ_{LT} is always estimated by first calculating the standard deviation of individual measurements (S_{TOT}) from all in-control subgroups on the chart. This result is then divided by c_4, which is based on a total of kn measurements, to derive an unbiased estimate of σ_{LT}.

$$\hat{\sigma}_{LT} = \frac{S_{TOT}}{c_{4_{kn}}} \qquad \text{where} \quad S_{TOT} = \sqrt{\frac{\sum_{i=1}^{kn} (X_i - \hat{\mu})^2}{kn - 1}}$$

The rating scale for P_R is the same as the one for C_R.

Example 5.7 Fan Housing Weight

In the previous injection-molding example, S_{TOT} is calculated as .3937 grams from the individual measurements contained in the 20 (k) in-control subgroups of the $\overline{X}, R$ chart monitoring part weight. As n is 5, the number of individual measurements included in this calculation is equal to 100 ($kn = 20 \times 5$). Given a tolerance for part weight of 4.00 grams, an estimate of P_R is calculated as .592.

$$\hat{P}_R = \frac{6\hat{\sigma}_{LT}}{\text{Tolerance}} = \frac{6(S_{TOT}/c_4)}{\text{Tolerance}} = \frac{6(.3937 \text{ gms}/.9975)}{4.00 \text{ gms}} = .592$$

The $6\sigma_{LT}$ spread of the process output is estimated as having the potential to take up only 59.2 percent of the tolerance for weight. This estimate of potential process capability is slightly different (about 8 percent higher) from the one estimated by C_R because the estimate of long-term process variation is usually greater than the one for short-term variation.

As the estimate of P_R is quite a bit less than 1.0, this process easily meets the minimum requirement for long-term potential capability. However, there is no way of knowing what percentage of housings actually have their weight within tolerance. All that can be surmised, is that with proper centering, the process *could* mold at least 99.73 percent of the fan housings within the specified weight tolerance.

Capability Goals for C_R and P_R

A very common goal for these potential capability measures is to be equal to or less than .75. Meeting this goal requires the process spread to take up no more than 75 percent of the tolerance. This is identical to the common goal for the process spread measure.

$$\frac{6\sigma_{ST}}{\text{Tolerance}} = .75 \quad \Rightarrow \quad 6\sigma_{ST} = .75\,(\text{Tolerance})$$

In the last example where the estimates for C_R of .547 and P_R of .592 are less than .75, the injection-molding process for fan housing meets this goal for both short- and long-term potential capability.

Suppose a different customer expresses a capability goal requiring these measures to be less than .50, which now requires the 6σ spread to occupy no more than 50 percent of the tolerance. As the estimates of both C_R and P_R are greater than .50, the process fails to meet the capability goal for this customer. Work must begin as quickly as possible to reduce process variation for housing weight. While these improvement efforts are proceeding, all housings must be contained and undergo 100 percent inspection to assure customers receive the desired level of quality.

Example 5.8 Exhaust Valve

σ_{ST} and σ_{LT} were both estimated in Example 3.7 for the O.D. size of an exhaust valve (Figure 5.10) run on a grinding machine.

$$\hat{\sigma}_{ST} = .01552 \text{ cm} \qquad \hat{\sigma}_{LT} = .01610 \text{ cm}$$

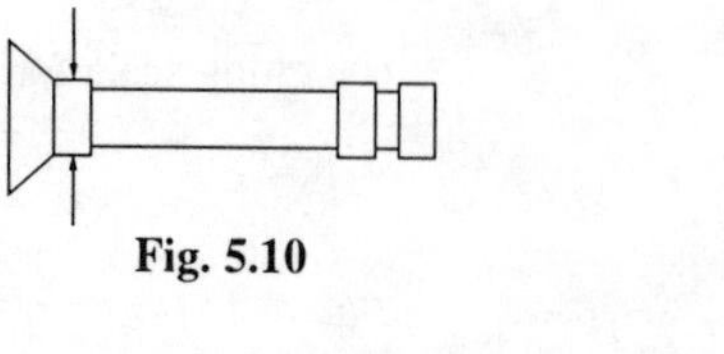

Fig. 5.10

If print specifications for O.D. size are a LSL of 9.9400 cm and an USL of 10.0600 cm, the short-term potential capability of the grinding operation for this characteristic can be estimated with the C_R capability ratio.

$$\hat{C}_R = \frac{6\hat{\sigma}_{ST}}{\text{USL - LSL}} = \frac{6(.01552 \text{ cm})}{10.0600 \text{ cm} - 9.9400 \text{ cm}} = .776$$

An estimated C_R ratio of .776 means one $6\sigma_{ST}$ spread of the process output should easily fit within print specifications, as it would require only 77.6 percent of the total tolerance, assuming normality and proper centering. Therefore, the grinding process for the valve O.D. size readily exceeds the minimum requirement of short-term potential capability for a bilateral specification.

In a similar manner, the long-term potential capability of the grinding operation for this characteristic is estimated with the P_R capability ratio.

$$\hat{P}_R = \frac{6\hat{\sigma}_{LT}}{\text{USL - LSL}} = \frac{6(.01610 \text{ cm})}{10.0600 \text{ cm} - 9.9400 \text{ cm}} = .805$$

The estimated P_R ratio of .805 is less than 1.0, implying one $6\sigma_{LT}$ spread of the process output will fit within print specifications. Therefore, this process also meets the minimum requirement for long-term potential capability because it has the ability to grind at least 99.73 percent conforming parts (with proper centering).

Although both these estimated measures exceed the minimum capability requirements, neither meets the common capability goal of .75.

Additional Interpretation of C_R and P_R Ratios

To more fully understand the information contained in these ratios, consider the three cases discussed on the next few pages. All three assume process stability has been achieved, the process output is normally distributed, and the process average is centered at the middle of the tolerance, which is 50 in this example.

Case #1 Minimum Potential Capability

The process output for a machine cutting the length of an electrical connector has its upper 3σ tail on the USL of 54 and its lower 3σ tail on the LSL of 46, as is displayed in Figure 5.11. As long as process stability is maintained, 99.73 percent of the connectors are expected to have their length within this tolerance. Conversely, no more than .27 percent will be out of tolerance.

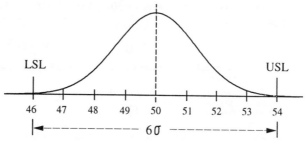

Fig. 5.11 Example of minimum potential capability.

Because 6σ is equal to the tolerance in this example, the C_R (or P_R) ratio equals 1, identifying this process as just meeting the *minimum* requirement of potential capability for a bilateral specification.

$$\hat{C}_R = \frac{6\hat{\sigma}_{ST}}{\text{Tolerance}} = \frac{6\hat{\sigma}_{ST}}{6\hat{\sigma}_{ST}} = 1 \qquad \hat{P}_R = \frac{6\hat{\sigma}_{LT}}{\text{Tolerance}} = \frac{6\hat{\sigma}_{LT}}{6\hat{\sigma}_{LT}} = 1$$

In particular, a C_R of 1 indicates a process has the potential to produce 99.73 percent conforming parts in the short term, given that the process average is properly centered. Nevertheless, the process may not have potential capability in the *long* term due to the many small, undetected time-related changes acting on the process (recall that $\sigma_{ST} \leq \sigma_{LT}$). On the other hand, a P_R of 1 would imply the process has potential capability to produce 99.73 percent of the connectors within the length tolerance over the long term.

Although meeting the minimum requirement for capability is encouraging, it is certainly a very precarious position. Even a small increase in variation, or change in process average, will result in less than 99.73 percent of the connector lengths being cut to an acceptable length, and the process no longer can be considered capable. Recall from the graph in Figure 2.13 (page 16) that for *n* equal to 4, there is only a 50 percent chance of detecting a 1.5σ shift in the process average. A movement in the average of just one standard deviation would greatly increase the number of nonconforming parts. Enlarging the subgroup size to increase the sensitivity of the control chart to notice smaller changes may prove worthwhile in situations of marginal capability. The time interval between subgroups may also be decreased to help catch process changes sooner.

Case #2 Lack of Potential Capability

If the 6σ spread is greater than the tolerance, the process cannot produce a minimum of 99.73 percent conforming parts (regardless of process centering), as is clearly shown in Figure 5.12. Much more than .27 percent of the connector lengths will be outside of tolerance and, therefore, this cut-off process definitely lacks potential capability for connector length.

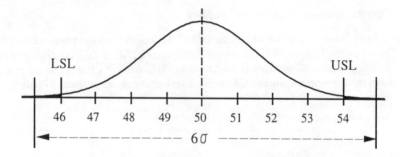

Fig. 5.12 A process lacking potential capability.

In this case the C_R (or P_R) ratio is greater than 1, indicating the process fails to meet the minimum potential capability requirement.

$$\hat{C}_R = \frac{6\hat{\sigma}_{ST}}{\text{Tolerance}} > 1 \qquad\qquad \hat{P}_R = \frac{6\hat{\sigma}_{LT}}{\text{Tolerance}} > 1$$

This is a very expensive situation as costs are incurred for the scrap and rework being produced, the 100 percent inspection needed to catch the nonconforming product, the warranty claims and customers returns from the nonconforming items that get by the 100 percent inspection, the lost business, and finally, any product liability lawsuits.

A C_R ratio greater than 1 quite often leads to operators over adjusting the process. Suppose a part is produced below the LSL, which is quite likely to happen when a process is not capable. After detecting such a nonconforming part, most operators become concerned and shift the process average *higher* to prevent making another undersize part. If the process is already centered at the middle of the tolerance, this reaction pushes an even greater percentage of parts above the USL. When the operator catches one of these, he or she becomes alarmed and responds by adjusting the process average *lower*, thinking this will prevent anymore oversize parts from being produced. This may be true, but unfortunately, this reaction causes parts to be produced below the LSL. When one of these undersize parts is detected, the operator again reacts by shifting the process average, but this time moving it higher. Of course, this means oversize parts will again be found and the cycle repeated. The net result of all this tampering is to increase process variation and actually cause more nonconforming parts to be produced than if the process were left alone, as is illustrated in Figure 5.13. And to make matters even worse, valuable production time is lost making all these unnecessary (and detrimental) adjustments. This is why a control chart must be maintained on a process after stability is demonstrated. Operators should react only to out-of-control conditions, *not* out-of-specification measurements.

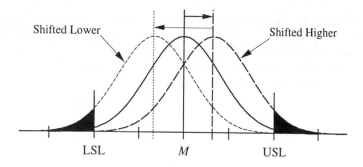

Fig. 5.13 The process spread is increased by making unnecessary adjustments.

Instead of wasting energy shifting the process average higher and lower, efforts must focus on reducing the inherent process variation contributed by the five "M"s (machine, manpower, method, measurement, material). After process changes are made to reduce process variation, verify their effectiveness on the control chart, recalculate control limits, reestablish control, then recalculate the C_R ratio to determine the amount of improvement in potential capability.

When P_R is greater than 1, the process lacks potential for producing at least 99.73 percent of the connectors within the allowable length tolerance over the long term. Actions targeted at reducing process variation must be undertaken as soon as possible. Note that this process may have short-term potential capability, as σ_{LT} is usually greater than, or equal to, σ_{ST}. The only way of knowing is to calculate the C_R ratio and compare its value to 1.

Juran *et al* (p. 9-57) list the six choices available to management when a process is not capable of producing the required percentage of conforming product.

1. *Reduce process variation.* Rebuild existing, or purchase new, equipment. Provide better operator training. Seek out better raw materials, work methods, tooling, gaging, and other process components to help reduce common-cause variation.

2. *Revise tolerances.* Convince the customer to accept a lower quality standard. Note that changing the specification is an administrative decision that does not directly improve product quality.

3. *Change the design.* A different design may make the product less sensitive to variation in this characteristic.

4. *Suffer and sort.* Endure the increased cost of scrap, rework, and customer complaints. Perform 100 percent inspection in an attempt to sort out the nonconforming product.

5. *Subcontract the work.* Find another company which has the required capability to produce this product.

6. *Get out of the business.* Stop selling this product altogether.

In today's competitive marketplace, option 1 offers the best hope of successfully remaining in business. By properly analyzing measures of capability, managers gain valuable insight that helps them make better decisions on how to efficiently run their manufacturing operations.

Case #3 Good Potential Capability

When the 6σ spread is less than the tolerance the C_R (or P_R) ratio for the process is less than 1.0 (Figure 5.14). Such a process has the potential of producing more than 99.73 percent of connector lengths within specification. Ratios less than 1.0 indicate the minimum requirement for potential capability is exceeded, but process capability may not meet capability goals that are set significantly below 1.0.

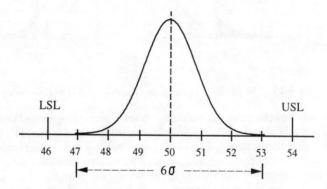

Fig. 5.14 A process with good potential capability.

Assuming that the specifications accurately reflect customer requirements and that the process is properly centered, this process will produce only a negligible amount of scrap and rework. Under these conditions, product containment and 100 percent inspection are not necessary, and there should be practically no warranty claims nor customer returns. With satisfied customers, there is no lost business and no lawsuits.

$$\hat{C}_R = \frac{6\hat{\sigma}_{ST}}{\text{Tolerance}} < 1 \qquad \hat{P}_R = \frac{6\hat{\sigma}_{LT}}{\text{Tolerance}} < 1$$

When the C_R ratio is less than 1, the process has good potential for short-term process capability. However, it may not be capable in the long term because of the many small, undetected time-related changes acting on the process. If P_R is also less than 1, the process has long-term potential capability to produce at least 99.73 percent of the connectors within the allowable length tolerance. This is an ideal situation because the minimum requirement for process capability is met, and there is still a safety margin to protect the process from any future changes in spread or centering. A large safety margin is desirable as it insulates the process from external shocks.

To demonstrate this concept of robustness, Figure 5.15 displays an $\overline{X}$, R chart with the distribution of individuals directly above it. A safety margin allows this process to change (as it most likely will), be detected on the chart, corrective action taken, and the process returned to its capable state *before* any nonconforming units are produced. This is the true purpose of SPC: to *prevent* problems from occurring rather than waiting until after they happen to take corrective action.

In section A of Figure 5.15, the process displays a good state of control and has a C_R ratio significantly less than 1.0. If the process average shifts upward for some reason, individual measurements become larger than before. The subgroup averages collected during this period will also be higher than before, resulting in a run above the centerline of the $\overline{X}$ chart, or even in points above the upper control limit, as is shown in section B. When this out-of-control condition is detected and corrective action taken, the process is brought back into control (section C) before any nonconforming parts are produced. Notice how in section B the upper tail for the distribution of individuals remains *below* the USL due to the safety margin provided when a process has good capability.

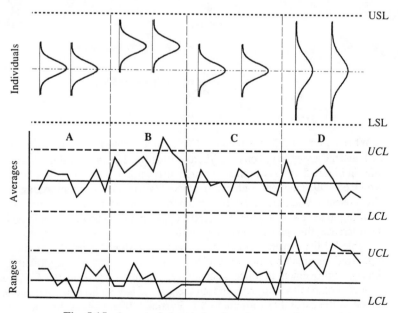

Fig. 5.15 A control chart helps detect process changes.

If the process spread increases, the highest measurement observed in each subgroup will tend to be larger than before. Likewise, the lowest measurement will tend to be smaller than before. Because they are determined by the difference between the highest and lowest measurements in each subgroup, ranges now tend to be larger than before, resulting in a run above the centerline of the R chart (or even in points above the UCL), as is seen in section D of the chart. As a result of the safety margin, this change in process spread can happen, be detected on the chart, corrective action implemented and the process brought back into control *before* any nonconforming parts are produced.

It should now be evident why C_R equaling 1.0 is just the bare minimum requirement for capability. To make product quality robust to unforeseen changes in the process average and standard deviation, larger safety margins are needed. This is why process improvement efforts must continue, even after the minimum requirement is met. Improving the process increases the probability all parts are produced within specification, even when the process goes out of control.

Just how much of a safety margin is needed? The exact answer depends upon many factors: criticality of the characteristic, volatility of the process, ease of making adjustments, cost of scrap and rework, customer requirements, industry quality levels. Historically, many companies have chosen a goal of .75.

$$C_{R.GOAL} = \frac{6\sigma_{ST}}{\text{Tolerance}} \le .75 \qquad\qquad P_{R.GOAL} = \frac{6\sigma_{LT}}{\text{Tolerance}} \le .75$$

If C_R equals .75, then $6\sigma_{ST}$ equals 75 percent of the tolerance. This is the same as $8\sigma_{ST}$ equaling 100 percent of the tolerance, as is demonstrated here:

$$C_R = .75$$

$$\frac{6\sigma_{ST}}{\text{Tolerance}} = .75$$

$$6\sigma_{ST} = .75\,(\text{Tolerance})$$

$$\frac{6}{.75}\sigma_{ST} = \text{Tolerance}$$

$$8\sigma_{ST} = \text{Tolerance}$$

When C_R equals .75, it is possible (with proper centering) to have the upper 4σ tail on the USL and the lower 4σ tail on the LSL, as is illustrated by the distribution with the long-dashed lines in Figure 5.16. Assuming a normal distribution, the process has the inherent ability to produce 99.9937 percent conforming parts (.0063 percent nonconforming, or 63 *ppm*). Other often-selected goals are: $C_R \le .60$, which means the $\pm 5\sigma$ tails could fit within the tolerance, resulting in as few as .6 *ppm*; $C_R \le .50$, suggesting the $\pm 6\sigma$ tails should be able to fit within the tolerance, thus generating as few as .002 *ppm*.

Table 5.2 lists several ranges of C_R and their assigned capability status. However, many organizations aren't so much interested in a specific numerical goal, but prefer to see continuous improvement in process capability, *i.e.*, smaller and smaller C_R ratios (Sullivan).

Some companies who have chosen the goal of having 8σ equal the tolerance will change the 6σ in the capability formula to 8σ. Pyzdek (1992a) labels this measure C_M. When the ratio of this revised formula equals 1.0, the goal of 8σ has been achieved.

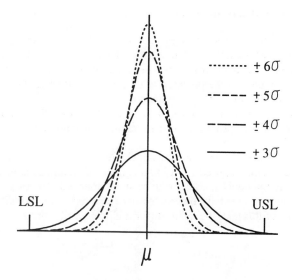

Fig. 5.16 Different potential capability goals.

Table 5.2 Capability ratings for various C_R values.

C_R Range	Capability Rating
$.00 < C_R \leq .50$	Terrific
$.50 < C_R \leq .60$	Excellent
$.60 < C_R \leq .75$	Good
$.75 < C_R \leq 1.00$	Fair
$1.00 < C_R \leq 1.25$	Poor
$1.25 < C_R$	Terrible

$$C_M = C_{R,8\sigma} = \frac{8\sigma_{ST}}{\text{Tolerance}} \qquad P_M = P_{R,8\sigma} = \frac{8\sigma_{LT}}{\text{Tolerance}}$$

This practice is not recommended due to the confusion created when comparing the C_R ratio for one process to another with a different capability goal. Which process is better: A with a C_R of .63 (using the 6σ formula), or B with a C_R of .84 (using the 8σ formula)? It is certainly not obvious that A and B have identical short-term potential capability (the tolerance equals $9.523\,\sigma_{ST}$ for each process). Think of the total bewilderment created for the customer. Some suppliers report capability based on one formula, while other suppliers rely on a different formula. The less-confusing approach is for all to agree on the same formula for measuring capability and change the goal instead.

Advantages and Disadvantages

The C_R and P_R ratios are easy to calculate and express variability in terms understandable to shop floor personnel as well as top management and customers: the percentage of tolerance equal to one 6σ spread (either short or long term) of the process output. If operators are told a process output has the potential to take up only 60 percent of the tolerance, they have an intuitive feeling this is a good situation.

These ratios have a dimensionless rating scale that allows a direct comparison to ratios calculated for other processes. But this rating scale is not consistent with those for the majority of capability measures.

There are also some other drawbacks to these ratios. First, a different formula (and interpretation) is required for a unilateral specification, as is explained in Section 5.6. In addition, if potential capability is to be an accurate indicator of *actual* capability, the process output must be normally distributed and centered at the *middle* of the tolerance. In the remainder of this text, the letter *M* (for midpoint) will represent the middle of the tolerance.

$$M = \frac{\text{LSL} + \text{USL}}{2}$$

There are times when it is desired to center the process output at some target (T) which is *not* equal to *M*. A target average is the most suitable value for the product characteristic because it optimizes product performance and/or minimizes manufacturing costs. Situations in which T is not equal to M are identified by specifications such as 65, plus 3, minus 5. For example, circuit board manufacturers drill holes in the substrate used for mounting electronic components. They prefer to have each hole slightly oversize (at 65 instead of 64) as this lessens the chances of bent leads when components are machine inserted into the board.

If attempts are made to center the process output at the target of 65, rather than at *M* (which is 64), C_R and P_R overestimate true potential capability. To understand why, assume C_R is 1.0, which means the $6\sigma_{ST}$ spread will just fit into the tolerance.

$$C_R = 1.0$$

$$\frac{6\sigma_{ST}}{\text{Tolerance}} = 1$$

$$6\sigma_{ST} = \text{Tolerance}$$

In order to fit one $6\sigma_{ST}$ spread into the tolerance, the upper $3\sigma_{ST}$ tail must be right on the USL, while the lower $3\sigma_{ST}$ tail is on the LSL. Due to the symmetry of a normal distribution, the process average under these conditions must be located halfway between these two specification limits (at 64, as Figure 5.17 illustrates) if the full potential capability is to be realized. However, efforts are undertaken by those running this process to center the process average at the target of 65. When the process output for hole size is finally located at 65, the upper $3\sigma_{ST}$ tail is pushed beyond the USL, and more than .135 percent of the holes will be drilled oversize (Figure 5.18). This drilling operation is unable to achieve the full potential capability promised by the C_R ratio because the process output will not be centered at *M* under normal operating conditions. Due to its low value, a ranking of C_R ratios for all critical characteristics in this shop would not identify this operation as a problem. Yet, over 2 percent of hole sizes is anticipated to be nonconforming when its output is centered at *T*.

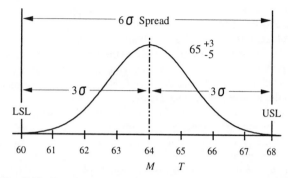

Fig. 5.17 Process centered at 64, the middle of the tolerance.

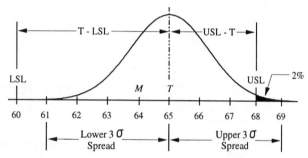

Fig. 5.18 Process centered at 65, the target average.

Fortunately, there are few characteristics where the target average is not the middle of the tolerance. But for those infrequent cases where T is not equal to M, basing conclusions on C_R will result in faulty decision making as this ratio overstates expected potential capability. Other measures that remedy this problem (C_R^* and P_R^*) are introduced in Section 5.5.

Even in those circumstances where T is the middle of the tolerance, there may be a problem assuming the process can be centered there. As Figure 5.19 illustrates, a process can be in control, have a very low C_R ratio, which indicates excellent potential capability, but still produce nearly all nonconforming parts as a result of not being centered exactly at M. Thus, a low C_R ratio is not always a sure guarantee of a low nonconformance rate.

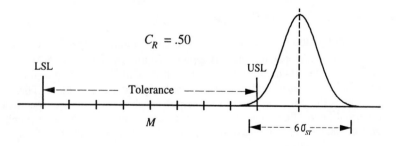

Fig. 5.19 A process with good potential capability, but producing many nonconforming parts.

A quick check of the control charts on any shop floor will reveal most processes are *not* centered exactly at M, even though it is the target average. The centerline of the $\overline{X}$ chart is usually close, but not equal to this value ($\hat{\mu} = \overline{\overline{X}} \neq M$). As it appears difficult for many processes to be properly positioned within their tolerances, this centering assumption becomes quite important and must always be checked. The performance capability measures presented in the next chapter don't have this problem because, in addition to process spread, they also factor in the location of the process average.

Another important condition for correctly interpreting these ratios involves the assumption of normality for the process output distribution. 6σ does not represent the middle 99.73 percent of the process output for all distributions. As an example, consider a distribution skewed to the right, like the one presented in Figure 5.20. The span from $\mu + 3\sigma$ to $\mu - 3\sigma$ does not capture the middle 99.73 percent of this process output.

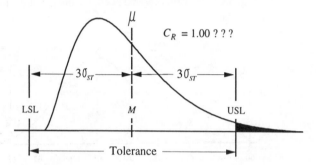

Fig. 5.20 The C_R ratio may be misleading for non-normal distributions.

Even though the process average is centered at M, a C_R ratio of 1.0 does not imply the process has the potential to produce 99.73 percent conforming parts. Comparing the 6σ spread of a non-normal distribution to the tolerance could yield an erroneous assessment of potential capability, leading to flawed decision making. Chapter 7 explains several methods for checking the normality assumption, while Chapter 8 offers procedures for estimating capability when the process output is non-normally distributed.

In addition, even if the normality assumption is satisfied, these ratios cannot predict the actual percentage of nonconforming parts. If a C_R ratio is 1.63, what percentage of the process output is potentially out of tolerance? It is not 63 percent. Unfortunately, there is no direct relationship between these ratios and the percentage of nonconforming parts. Other measures of process capability presented later in this book will readily divulge this important aspect of process performance.

Relationship of C_R to the Process Average

As the C_R ratio involves only σ_{ST} in its calculation, changes in the process average have absolutely no effect on its value, as depicted on the graph in Figure 5.21. C_R remains constant at .75 no matter where the process average is located (even when it is moved outside the print specifications) because this ratio is completely independent of μ. This independent relationship is true for all measures of potential capability for bilateral specifications, as will be observed later when other such measures are plotted on this same graph. However, the performance capability measures introduced in the next chapter will respond to changes in the process average.

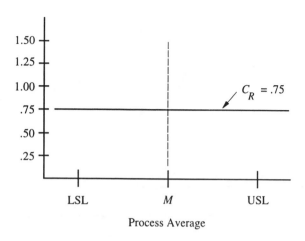

Fig. 5.21 Relationship of C_R to the process average.

Example 5.9 Floppy Disks

A rash of complaints from unhappy customers has resulted in decreasing sales for a major manufacturer of floppy disks (Figure 5.22). The quality control manager for this company decides to form a problem-solving team to resolve this serious problem.

Fig. 5.22 A floppy disk with the diameter of the inner hole indicated.

The cause of customer dissatisfaction is traced to excessive variation in the diameter of the inner hole of the floppy disk. If the hole is cut too large, the floppy disk fails to rotate properly when inserted into a disk drive, and therefore, cannot store or retrieve information. The customer must return the floppy for a replacement. On the other hand, undersize holes will cause a jam and occasionally damage the delicate disk drive, resulting a major warranty claim (a replacement floppy *plus* repairing the disk drive).

As a first step, the problem-solving team decides to implement an $\overline{X}, R$ chart on the hole-cutting process. After several days of charting, the process is finally brought into control, as evidenced by the chart presented in Figure 5.23. Important chart statistics and other information needed for estimating potential process capability are listed here:

$$\overline{\overline{X}} = 40.9 \qquad \overline{R} = 1.85 \qquad S_{TOT} = .913 \qquad k = 21$$

$$n = 4 \qquad LSL = 39.5 \qquad USL = 44.5$$

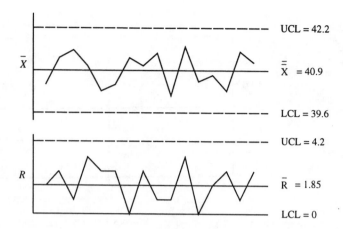

Fig. 5.23 The control chart for hole I.D. size.

As mentioned in Chapter 3, measurements from all out-of-control subgroups must be excluded from the calculation of $\overline{X}$ and $\overline{R}$. Only data representing the process operating in a stable mode are included in the calculation of control limits and in any measurement of process capability. If there are many out-of-control points, problem-solving efforts must first be directed at eliminating assignable-cause variation. Once these are resolved and process stability is demonstrated, a formal capability study can be initiated.

As their first attempt at evaluating capability, some members of the team want to draw the specification limits of 39.5 and 44.5 for hole size on the $\overline{X}$ chart, then compare them to the control limits for hole size of 39.6 and 42.2. Because these control limits lie within the specification limits, they are ready to conclude the process is capable. Unfortunately, this is not necessarily true, since specification limits are for *individuals* and control limits are for subgroup *averages* (review Section 2.6, page 21). Process capability is correctly assessed by calculating some measure of capability, in this case, the C_R and P_R ratios.

Before estimating C_R, an estimate of σ_{ST} must be obtained. When an $\overline{X}, R$ chart is monitoring the process, σ_{ST} is estimated from $\overline{R}$ over d_2 (for a subgroup size of 4).

$$\hat{\sigma}_{ST} = \hat{\sigma}_{\overline{R}} = \frac{\overline{R}}{d_2} = \frac{1.85}{2.059} = .898$$

With this estimate of the short-term standard deviation, C_R is estimated as 1.078.

$$\hat{C}_R = \frac{6\hat{\sigma}_{ST}}{\text{Tolerance}} = \frac{6(.898)}{44.5 - 39.5} = 1.078$$

Assuming a normal distribution for the process output, the resulting C_R ratio of 1.078 indicates the estimated $6\sigma_{ST}$ spread for the hole-cutting operation is 7.8 percent *greater* than the tolerance. Even if this process output is centered at the middle of the tolerance, there is no way it can produce at least 99.73 percent conforming holes and, thus, definitely lacks short-term potential capability. Serious consideration must be given to initiating 100 percent inspection for hole size. In the meantime, the team must attempt to reduce common-cause process variation for the hole-cutting operation. When completed, another study should be run to measure the amount of improvement in capability. There may be times when the ability to reduce common-cause variability comes to a standstill. At this point, the team's focus must shift to redesigning the product and/or the process.

To gain additional insight about this process, the team decides to estimate the P_R ratio, which requires obtaining an estimate of σ_{LT}. No matter which variable-data chart is chosen, σ_{LT} is always estimated from S_{TOT} over c_4. Because S_{TOT} is calculated from the individual measurements of all in-control subgroups, the c_4 factor from Appendix Table II is based on kn. For this capability study, k is 21 and n is 4, making kn equal to 84.

$$\hat{\sigma}_{LT} = \frac{S_{TOT}}{c_4} = \frac{.913}{.9970} = .916 \qquad \hat{P}_R = \frac{6\hat{\sigma}_{LT}}{\text{Tolerance}} = \frac{6(.916)}{44.5 - 39.5} = 1.099$$

The estimated P_R ratio of 1.099 exposes this hole-cutting process as *not* meeting the minimum requirement for long-term potential capability. The tolerance would have to be 9.9 percent larger in order for one $6\sigma_{LT}$ spread to be able to fit within it. This conclusion should really be no surprise since the C_R ratio has already revealed that the process is not capable, even in the short-term.

Even though classified as not capable, the control chart monitoring this process must be maintained. Not capable means a certain amount of nonconforming hole sizes are guaranteed to be produced. Without a chart, operators discovering a floppy with an out-of-specification hole size will attempt to adjust the process. As previously explained, this type of tampering results in the production of a larger number of nonconforming floppies. The best an operator can do is leave an in-control process alone while the team works on improving it. The control chart is also needed to help verify the effectiveness of the team's attempts at improvement.

Note that these ratios leave a lot of unanswered questions: What's the total percentage of nonconforming hole diameters? What percentage of the hole diameters are above the USL? Below the LSL? Answers to these key questions are extremely important for sound decision making about how to operate a process, but are not provided by these two measures of potential capability. As will be discovered in the next chapter, other capability measures readily furnish these crucial pieces of information.

5.5　C^*_R and P^*_R for Cases Where $T \neq M$

The C_R and P_R ratios assume the target (sometimes called the aim, or set point) for the process average, T, is the middle of the tolerance, M. This is usually, but not always, true. There are times when the average is targeted away from M to either: facilitate downstream processing; optimize product performance; minimize costs; or help with the assembly of mating parts (Kane, 1989). Potential capability measures are designed to indicate how well the output distribution of a characteristic could fit within the tolerance *when properly centered*. In those cases where T is not equal to M, the C_R and P_R ratios overestimate the true potential process capability, as demonstrated in this next example.

Suppose the print specifications are given as 65 plus 3, minus 5. In Figure 5.24a, the process is centered at 64 (the middle of the tolerance), with the 3σ tails just fitting within these specifications of 60 and 68. Due to the 6σ spread being equal to the tolerance, the C_R ratio is 1.0, and the process just meets the minimum requirement for potential capability. However, assume this process is centered at the desired average of 65 instead of 64, as is shown in Figure 5.24b. When centered at 65, the upper 3σ tail is above the USL and approximately 2 percent of the parts exceed the USL of 68. If the process is to be kept at 65, it does not have the potential of producing at least 99.73 percent conforming parts and therefore, is not able to meet the minimum requirement for potential capability.

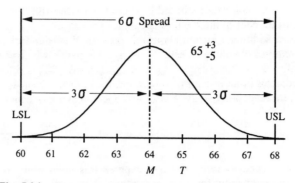

Fig. 5.24a Process centered at 64, the middle of the tolerance.

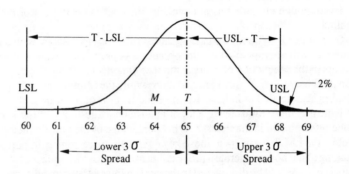

Fig. 5.24b Process centered at the target average of 65.

When T is not the middle of the tolerance, each 3σ tail of the process output distribution must be considered separately. The distance from the process average to the lower 3σ tail is just 3σ.

$$\mu - (\mu - 3\sigma) = \mu - \mu + 3\sigma = 3\sigma$$

This length of this 3σ spread is compared to the distance from T to the LSL. If σ_{ST} is being studied, this ratio is called C_{RL}^T, with the superscript "T" for "target," and the subscript "L" for "LSL."

$$C_{RL}^T = \frac{3\sigma_{ST}}{T - LSL}$$

When centered at T, 3σ must be less than T minus LSL if this process is to meet the minimum requirement of potential capability that no more than .135 percent nonconforming parts are below the LSL.

$$3\sigma_{ST} \leq T - LSL$$

If this happens, the C_{RL}^T ratio will be less than 1.0.

$$3\sigma_{ST} \le T - \text{LSL}$$

$$\frac{3\sigma_{ST}}{T - \text{LSL}} \le \frac{T - \text{LSL}}{T - \text{LSL}}$$

$$C_{RL}^{T} \le 1$$

Likewise, the distance from the process average to the upper $3\sigma_{ST}$ tail must be less than the distance between the USL and T.

$$\mu - (\mu - 3\sigma_{ST}) \le \text{USL} - T$$

$$3\sigma_{ST} \le \text{USL} - T$$

Thus, the ratio of $3\sigma_{ST}$ to USL minus T (called C_{RU}^{T}) must be less than 1.0 if the process is to satisfy the minimum requirement for potential capability.

$$3\sigma_{ST} \le \text{USL} - T$$

$$\frac{3\sigma_{ST}}{\text{USL} - T} \le \frac{\text{USL} - T}{\text{USL} - T}$$

$$C_{RU}^{T} \le 1$$

The worse case, the *larger* of these two, determines the potential capability for this process. This is expressed by the following equations, where C_{R}^{*} is pronounced "CR star":

$$C_{R}^{*} = \text{Maximum} \, (C_{RL}^{T}, C_{RU}^{T}) = \text{Maximum} \left(\frac{3\sigma_{ST}}{T - \text{LSL}}, \frac{3\sigma_{ST}}{\text{USL} - T} \right)$$

$$P_{R}^{*} = \text{Maximum} \, (P_{RL}^{T}, P_{RU}^{T}) = \text{Maximum} \left(\frac{3\sigma_{LT}}{T - \text{LSL}}, \frac{3\sigma_{LT}}{\text{USL} - T} \right)$$

Throughout this book, the notation * added to a measure indicates it is for the case where T is not the middle of the tolerance. Sometimes the letter "T" is added at the end of the measure's subscript. Under this convention, C_{R}^{*} is labeled C_{RT}, while P_{R}^{*} becomes P_{RT}.

When the process standard deviation is estimated from sample data, estimate C_{R}^{*} and P_{R}^{*} with this next set of formulas:

$$\hat{C}_{R}^{*} = \text{Maximum} \left(\frac{3\hat{\sigma}_{ST}}{T - \text{LSL}}, \frac{3\hat{\sigma}_{ST}}{\text{USL} - T} \right) \qquad \hat{P}_{R}^{*} = \text{Maximum} \left(\frac{3\hat{\sigma}_{LT}}{T - \text{LSL}}, \frac{3\hat{\sigma}_{LT}}{\text{USL} - T} \right)$$

Defined in this manner, the rating scale for C_{R}^{*} and P_{R}^{*} is identical to that for C_{R} and P_{R}.

- > 1 discloses the process fails to satisfy the minimum requirement for potential capability,
- $= 1$ means the process just meets the minimum requirement and has the ability to produce 99.73 percent conforming parts,
- < 1 indicates the process exceeds the minimum requirement.

Example 5.10 Contact Closing Speed

A South Korean manufacturer checks circuit breakers for contact motion to verify they close at the proper speed. Closing too quickly causes premature wear out of the contacts, while closing too slowly could burn or weld the contacts so the breaker could not be reopened. The engineering department has developed the following window for closing speed to optimize the life of this product.

$$LSL = .0140 \text{ seconds} \qquad USL = .0250 \text{ seconds} \qquad T = .0170 \text{ seconds}$$

The middle of the tolerance, M, is .0195 seconds.

$$M = \frac{LSL + USL}{2} = \frac{.0140 \text{ seconds} + .0250 \text{ seconds}}{2} = .0195 \text{ seconds}$$

These requirements are portrayed graphically in Figure 5.25.

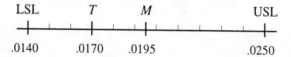

Fig. 5.25 Closing time requirements (in seconds) for a circuit breaker.

From the control chart monitoring closing speed, σ_{ST} is estimated as .00129 seconds. C_R is then estimated as .70, which indicates excellent short-term potential capability.

$$\hat{C}_R = \frac{6\hat{\sigma}_{ST}}{USL - LSL} = \frac{6(.00129)}{.0250 - .0140} = .70$$

Unfortunately, this measure of capability has little meaning for this process as it assumes the average will be centered at .0195 seconds. Efforts will be made to center the process average at the target closing time of .0170 seconds, which is considerably less than the middle of the tolerance. This desired operating condition makes a substantial difference in the evaluation of potential capability, as is disclosed when C_R^* is estimated.

$$\hat{C}_R^* = \text{Maximum}\left(\frac{3\hat{\sigma}_{ST}}{T - LSL}, \frac{3\hat{\sigma}_{ST}}{USL - T}\right)$$

$$= \text{Maximum}\left(\frac{3(.00129)}{.0170 - .0140}, \frac{3(.00129)}{.0250 - .0170}\right)$$

$$= \text{Maximum}(1.29, .48) = 1.29$$

Although the $6\sigma_{ST}$ spread is only 70 percent of the tolerance, this process is not capable for closing time under the added restriction of having to center the process average at .0170 seconds. Work must be done to reduce process variation or a design change made to move the target average closer to the middle of the tolerance or to lower the LSL.

Relationship of C_R^* to C_R

When T equals the middle of the tolerance, the formula for C_R^* yields a result identical to C_R. In this special situation, T divides the tolerance into two equal halves (Figure 5.26).

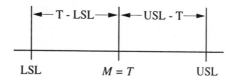

Fig. 5.26 T is located at the middle of the tolerance.

$$T - LSL = \frac{1}{2} \text{Tolerance} = USL - T$$

The formula for C_R^* can be simplified due to T minus LSL and USL minus T both being equal to one half of the tolerance.

$$C_R^* = \text{Maximum}\left(\frac{3\sigma_{ST}}{T - LSL}, \frac{3\sigma_{ST}}{USL - T} \right)$$

$$= \text{Maximum}\left(\frac{3\sigma_{ST}}{1/2\,\text{Tolerance}}, \frac{3\sigma_{ST}}{1/2\,\text{Tolerance}} \right)$$

$$= \text{Maximum}\left(\frac{6\sigma_{ST}}{\text{Tolerance}}, \frac{6\sigma_{ST}}{\text{Tolerance}} \right) = \frac{6\sigma_{ST}}{\text{Tolerance}} = C_R$$

The C_R^* ratio applies when T is equal to M as well as in cases where T is not. Thus, C_R^* is just a more generalized form of C_R; one that isn't burdened with the assumption that T is located at the middle of the tolerance.

Example 5.11 Gold Plating

The LSL for the thickness of gold being plated on a silicon wafer is 50 angstroms and the USL is 100 angstroms. For cost reasons, a decision is made to target the average thickness at 65 ($T = 65$), as is displayed in Figure 5.27. If σ_{LT} is 2.5, estimates can be made for both P_R (assumes centering at M) and P_R^* (assumes centering at T).

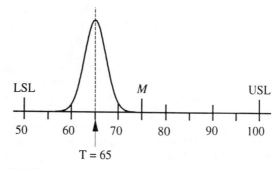

Fig. 5.27 The distribution of gold plating thickness when centered on target.

$$\hat{P}_R = \frac{6\hat{\sigma}_{LT}}{USL - LSL} = \frac{6(2.5)}{100 - 50} = .30$$

If the process is centered at 75, which is the middle of the tolerance, the process spread for thickness would occupy merely the middle 30 percent of the tolerance. But, this is unlikely to happen as efforts will be made to keep this process centered at 65 to minimize manufacturing costs. Reporting the potential capability of this plating process as .30 is misleading when the condition required to achieve that potential, *i.e.*, centering at M, is not likely to occur. A more realistic idea of true potential capability is acquired by replacing P_R with the P_R^* ratio whenever T does not equal M.

$$\hat{P}_R^* = \text{Maximum} \left(\frac{3\hat{\sigma}_{LT}}{T - LSL}, \frac{3\hat{\sigma}_{LT}}{USL - T} \right)$$

$$= \text{Maximum} \left(\frac{3(2.5)}{65 - 50}, \frac{3(2.5)}{100 - 65} \right)$$

$$= \text{Maximum} \left(\frac{7.5}{15}, \frac{7.5}{35} \right)$$

$$= \text{Maximum} \, (.50 , .21) = .50$$

When this gold-plating process is centered at the target of 65 angstroms, the lower $3\sigma_{ST}$ tail of the process output will extend from 65 to 57.5, a spread of 7.5 angstroms. This spread would take up 50 percent of the 15 angstroms distance between the process average and the LSL (65 minus 50). The resulting P_R^* estimate of .50 realistically reflects the actual potential capability of this process. Instead of comparing $3\sigma_{ST}$ to 50 (one half of the tolerance) as P_R does, P_R^* compares it to only 15, the distance from the target average to the LSL. This "loss" of available tolerance causes P_R^* to be less than P_R whenever T is different than M. P_R^* equals P_R only when T equals M.

5.6 C'_{RL}, C'_{RU}, P'_{RL}, and P'_{RU} for Unilateral Specifications

The C_R and P_R formulas must be modified to measure capability for characteristics with unilateral specifications. First of all, with only one specification, the tolerance (difference between the USL and LSL) is undefined. Secondly, the minimum requirement for potential capability is the ability to produce 99.865, not 99.73, percent conforming product.

These special formulas for measuring potential capability for characteristics with unilateral specifications require a target value for the process average (T) be specified. For a unilateral specification, a target value means the process average should be at this value, or *better*, than it. How can it be better? Suppose the critical characteristic being studied is the taper of a shaft. Excessive taper is undesirable, so an USL of 10 is set with a target average of 6, as is pictured in Figure 5.28. Specifying a target of 6 for this type of dimension really means a target *zone* of 6 or lower, as less taper is better. A process average located anywhere within this zone is considered "on target."

In a case like taper where only an USL is given, the following capability ratio measures how well a process meets this unilateral requirement. This special version of the C_R ratio is called C'_{RU} (pronounced "CRU prime") and is sometimes referred to as a "one-sided" potential capability ratio.

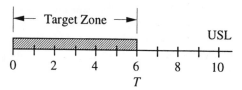

Fig. 5.28 A process average for taper between 0 and 6 is "on target."

$$C'_{RU} = \text{Minimum} \, (\, C^T_{RU}, C_{RU} \,) \qquad \text{where } C^T_{RU} = \frac{3\sigma_{ST}}{\text{USL} - T} \quad \text{and} \quad C_{RU} = \frac{3\sigma_{ST}}{\text{USL} - \mu}$$

In a similar fashion, C'_{RL} is defined for characteristics having just a LSL.

$$C'_{RL} = \text{Minimum} \, (\, C^T_{RL}, C_{RL} \,) \qquad \text{where } C^T_{RL} = \frac{3\sigma_{ST}}{T - \text{LSL}} \quad \text{and} \quad C_{RL} = \frac{3\sigma_{ST}}{\mu - \text{LSL}}$$

Notice that attention is focused exclusively on just one tail of the process output distribution, either the upper $3\sigma_{ST}$ tail when dealing with only an USL, or the lower when dealing with a LSL. In addition, the process average is needed when computing C_{RL} and C_{RU}.

If no target is specified, C^T_{RL} and C^T_{RU} are undefined, reducing the formulas for C'_{RL} and C'_{RU} to the following simplified versions:

$$C'_{RL} = C_{RL} \qquad\qquad C'_{RU} = C_{RU}$$

As explained in the next chapter, this actually makes C'_{RL} and C'_{RU} measures of performance, rather than potential, capability because they incorporate μ in their calculation.

When interest is in long-term potential capability, apply this formula to characteristics having only a LSL.

$$P'_{RL} = \text{Minimum} \, (\, P^T_{RL}, P_{RL} \,) \quad \text{where } P^T_{RL} = \frac{3\sigma_{LT}}{T - \text{LSL}} \quad \text{and} \quad P_{RL} = \frac{3\sigma_{LT}}{\mu - \text{LSL}}$$

In a similar manner, P'_{RU} is defined as follows for characteristics having just an USL.

$$P'_{RU} = \text{Minimum} \, (\, P^T_{RU}, P_{RU} \,) \quad \text{where } P^T_{RU} = \frac{3\sigma_{LT}}{\text{USL} - T} \quad \text{and} \quad P_{RU} = \frac{3\sigma_{LT}}{\text{USL} - \mu}$$

These unilateral capability measures have a rating scale identical to that for C_R and P_R.

> 1 means the process fails to meet the minimum requirement for potential capability,

= 1 indicates the process just satisfies the minimum requirement,

< 1 implies the process exceeds the minimum requirement.

Unfortunately, in order to maintain the same rating scale, the formulas for these unilateral situations are somewhat more complex than those for C_R and P_R. Figures 5.29 through 5.34 exhibit how this modified formula for C'_{RU} generates a measure of capability for a characteristic with only an USL, which is equivalent in interpretation to the C_R ratio calculated for bilateral specifications.

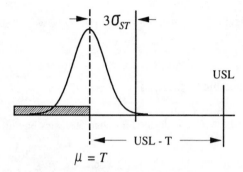

Fig. 5.29   Process has good potential capability and good performance.

In Figure 5.29, the process is centered at T ($\mu = T$). The upper $3\sigma_{ST}$ tail (below which lies 99.865 percent of the process output) is considerably less than the USL, visually indicating the process exceeds the minimum potential capability requirement for a unilateral specification. Thus, when μ equals T:

$$C'_{RU} = \text{Minimum}\left(\frac{3\sigma_{ST}}{\text{USL}-T}, \frac{3\sigma_{ST}}{\text{USL}-\mu}\right) = \text{Minimum}\left(\frac{3\sigma_{ST}}{\text{USL}-T}, \frac{3\sigma_{ST}}{\text{USL}-T}\right) = \frac{3\sigma_{ST}}{\text{USL}-T}$$

Because the $3\sigma_{ST}$ spread is less than USL minus T in this first example, the above C'_{RU} ratio is less than 1.0.

$$3\sigma_{ST} < \text{USL}-T$$

$$\frac{3\sigma_{ST}}{\text{USL}-T} < \frac{\text{USL}-T}{\text{USL}-T}$$

$$C'_{RU} < 1.0$$

Just as in the bilateral case, such a result signifies acceptable potential capability, since the process has the ability to produce at least 99.865 percent conforming parts.

If the process is centered higher than T, *and* the $3\sigma_{ST}$ spread is less than the distance between the USL and T (as is the case in Figure 5.30), the process still possesses good *potential* capability. Just as before, when the $3\sigma_{ST}$ spread is less than the distance between the USL and T, the C'_{RU} ratio is less than 1.0.

There may be less than 99.865 percent conforming parts actually produced, but just as in the bilateral case, the actual performance (percentage of conforming parts) still depends on where the process average is centered. If it can be located in the target zone without changing the standard deviation, this process would have the ability to produce at least 99.865 percent conforming parts.

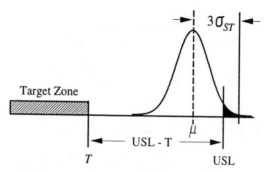

Fig. 5.30 Process has good potential capability, but poor performance.

Figure 5.31 reveals how this special measure for unilateral specifications equals 1.0 when the process average is centered at T, and the $3\sigma_{ST}$ spread equals the distance between the USL and the target average.

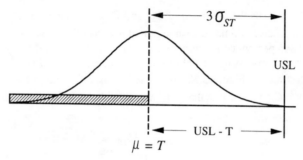

Fig. 5.31 Process has marginal potential capability.

Because μ equals T in this example, the formula for C'_{RU} can be simplified somewhat.

$$C'_{RU} = \text{Minimum}\,(\,C^T_{RU}\,,\,C_{RU}\,)$$

$$= \text{Minimum}\left(\frac{3\sigma_{ST}}{\text{USL}-T}\,,\frac{3\sigma_{ST}}{\text{USL}-\mu}\right)$$

$$= \text{Minimum}\left(\frac{3\sigma_{ST}}{\text{USL}-T}\,,\frac{3\sigma_{ST}}{\text{USL}-T}\right) = \frac{3\sigma_{ST}}{\text{USL}-T}$$

If $3\sigma_{ST}$ equals USL minus T, then:

$$C'_{RU} = \frac{\text{USL}-T}{\text{USL}-T} = 1.0$$

This particular process has the potential of producing at least 99.865 percent conforming parts (assuming normality), thus meeting the minimum requirement of short-term potential capability for a unilateral specification.

Had the process average been higher than T (with $3\sigma_{ST}$ still equal to USL minus T), C_{RU}^T would remain at 1.0 because it is not affected by changes in μ.

$$C_{RU}^T = \frac{3\sigma_{ST}}{\text{USL} - T} = \frac{\text{USL} - T}{\text{USL} - T} = 1.0$$

However, under these conditions C_{RU} would become greater than 1.0 because the distance from the USL to μ would now be less than $3\sigma_{ST}$.

$$3\sigma_{ST} > \text{USL} - \mu$$

$$\frac{3\sigma_{ST}}{\text{USL} - \mu} > \frac{\text{USL} - \mu}{\text{USL} - \mu}$$

$$C_{RU} > 1.0$$

As C'_{RU} equals the *minimum* of C_{RU}^T and C_{RU}, it remains equal to 1.0. This is appropriate since the process still has the same potential capability, although it would be producing less than 99.865 percent conforming parts. On the other hand, if the process average is lowered below T, the process has the ability to produce *more* than 99.865 percent conforming parts, as is illustrated in Figure 5.32. Because $3\sigma_{ST}$ is still equal to USL minus T, C_{RU}^T stays equal to 1.0, but C_{RU} is now less than 1.0 due to the distance from the USL to μ being greater than $3\sigma_{ST}$.

$$3\sigma_{ST} < \text{USL} - \mu$$

$$\frac{3\sigma_{ST}}{\text{USL} - \mu} < \frac{\text{USL} - \mu}{\text{USL} - \mu}$$

$$C_{RU} < 1.0$$

Being equal to the smaller of these two, C'_{RU} becomes less than 1.0, indicating this process exceeds the minimum capability requirement. Since the process has the ability to center μ below T, it has the capacity for producing more than 99.865 percent conforming parts.

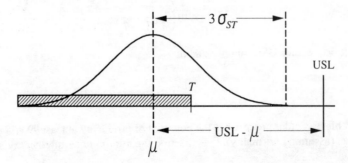

Fig. 5.32 Process exceeds minimum potential capability requirement.

Figure 5.33 shows a process centered at T where the $3\sigma_{ST}$ spread is greater than the distance between the USL and T.

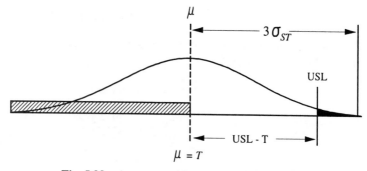

Fig. 5.33 A process with poor potential capability.

As before, when μ equals T:

$$C'_{RU} = C^T_{RU} = \frac{3\sigma_{ST}}{USL - T}$$

Because $3\sigma_{ST}$ is greater than USL minus T, C'_{RU} is greater than 1.0.

$$3\sigma_{ST} > USL - T$$

$$\frac{3\sigma_{ST}}{USL - T} > \frac{USL - T}{USL - T}$$

$$C'_{RU} > 1.0$$

Similar to the C_R ratio for bilateral specifications, a C'_{RU} ratio greater than 1.0 means the process does not meet the minimum requirement for potential capability, i.e., the ability to produce 99.865 percent conforming parts. The same conclusion would also be reached if the process average were centered above T. When μ is greater than the target average:

$$\mu > T$$

$$-\mu < -T$$

$$USL - \mu < USL - T$$

$$\frac{1}{USL - \mu} > \frac{1}{USL - T}$$

$$\frac{3\sigma_{ST}}{USL - \mu} > \frac{3\sigma_{ST}}{USL - T}$$

$$C_{RU} > C^T_{RU}$$

Because C'_{RU} is defined as the minimum of these two, it equals C^T_{RU} whenever μ is greater than T.

$$C'_{RU} = \text{Minimum}\,(\,C^T_{RU}\,,\,C_{RU}\,) = C^T_{RU}$$

When μ is greater than T, and $3\sigma_{ST}$ is greater than the distance between the USL and T (see Figure 5.33), C^T_{RU}, and therefore C'_{RU}, is greater than 1.0.

$$3\sigma_{ST} > \text{USL} - T$$

$$\frac{3\sigma_{ST}}{\text{USL} - T} > \frac{\text{USL} - T}{\text{USL} - T}$$

$$C^T_{RU} > 1.0$$

$$C'_{RU} > 1.0$$

On the other hand, when the process average is sufficiently *less than* T, the process may be able to make 99.865 percent conforming parts, as is revealed in Figure 5.34.

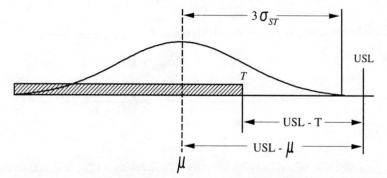

Fig. 5.34 Process has good potential capability because μ is less than T.

Anytime μ is located in the target zone, C'_{RU} equals C_{RU} because the distance USL minus μ is now greater than USL minus T. This makes C_{RU} less than C^T_{RU}.

$$\text{USL} - \mu > \text{USL} - T$$

$$\frac{1}{\text{USL} - \mu} < \frac{1}{\text{USL} - T}$$

$$\frac{3\sigma_{ST}}{\text{USL} - \mu} < \frac{3\sigma_{ST}}{\text{USL} - T}$$

$$C_{RU} < C^T_{RU}$$

Being the minimum of these two, C'_{RU} equals C_{RU} whenever μ is less than T. When $3\sigma_{ST}$ is less than the distance from the USL to the process average, C_{RU} is less than 1.0.

$$3\sigma_{ST} < USL - \mu$$

$$\frac{3\sigma_{ST}}{USL - \mu} < \frac{USL - \mu}{USL - \mu}$$

$$C_{RU} < 1.0$$

Because C'_{RU} equals C_{RU}, it is also less than 1.0, implying this process has the ability to produce at least 99.865 percent conforming parts, which the output distribution displayed in Figure 5.34 does.

In cases where T is not given, C_{RU}^T does not exist, so C'_{RU} equals C_{RU}. But remember, any estimate of capability made using $\hat{\mu}$ no longer represents *potential* capability, but rather *performance* capability (more on this in the next chapter). No measure can be calculated when the process average is equal to the specification, as division by zero is not allowed.

A similar interpretation applies to C'_{RL} when the characteristic targeted for a capability study has only a LSL.

Example 5.12 Particle Contamination

The following estimates of process parameters are made from an *IX & MR* chart monitoring particle contamination in *ppm* of a cleaning solution for printed circuit boards after they go through a wave soldering operation.

$$\hat{\mu} = 127.32\,ppm \qquad \hat{\sigma}_{ST} = 12.079\,ppm$$

If the USL for contamination is 165 *ppm* and T is 120 *ppm*, the C'_{RU} ratio is estimated as follows, assuming a normal distribution for particle contamination:

$$\hat{C}'_{RU} = \text{Minimum}\,(\,\hat{C}_{RU}^T, \hat{C}_{RU}\,)$$

$$= \text{Minimum}\left(\frac{3\hat{\sigma}_{ST}}{USL - T}, \frac{3\hat{\sigma}_{ST}}{USL - \hat{\mu}}\right)$$

$$= \text{Minimum}\left(\frac{3(12.079)}{165 - 120}, \frac{3(12.079)}{165 - 127.32}\right)$$

$$= \text{Minimum}\,(.81, .96) = .81$$

Although this process meets the minimum requirement for potential capability, it does not meet the imposed capability goal of .75. This sad situation is displayed in Figure 5.35a.

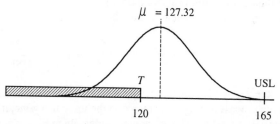

Fig. 5.35a Average particle contamination is 127 *ppm*, whereas the target is 120.

In an attempt to improve this process, a new type of filter is added to the cleaning bath. This causes a reduction in $\hat{\mu}$ to 108 *ppm*, with no change in $\hat{\sigma}_{ST}$, as is seen in Figure 5.35b.

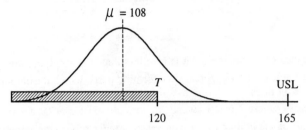

Fig. 5.35b Average particle contamination is 108 *ppm*.

The potential capability of this revised process is calculated as .64.

$$\hat{C}'_{RU} = \text{Minimum}\left(\frac{3\hat{\sigma}_{ST}}{USL - T}, \frac{3\hat{\sigma}_{ST}}{USL - \hat{\mu}} \right)$$

$$= \text{Minimum}\left(\frac{3(12.079)}{165 - 120}, \frac{3(12.079)}{165 - 108} \right)$$

$$= \text{Minimum}\,(\,.81\,,\,.64\,) = .64$$

As a result of being able to center the process average quite a bit below T, the potential capability becomes much better, and now exceeds the goal of .75. Note this happens even though the process variation was not reduced. But to maintain this level of potential capability, the process average must remain at 108 *ppm*.

Relationship of C'_{RL} and P'_{RU} to the Process Average

Consider a process where only an USL is given. For values of μ greater than T, C'_{RU} equals C_{RU}^{T}, which is independent of μ.

$$\text{When } \mu \geq T, \quad C'_{RU} = C_{RU}^{T} = \frac{3\sigma_{ST}}{USL - T}$$

Changes in the process average have no effect on C'_{RU} as long as μ remains above T, as depicted on the graph in Figure 5.36. C'_{RU} remains constant at .75 when the process average is greater than T, even if it is above the USL.

If μ ever drops below T, the C'_{RU} ratio becomes equal to C_{RU}, which is a function of μ.

$$\text{When } \mu < T, \quad C'_{RU} = C_{RU} = \frac{3\sigma_{ST}}{USL - \mu}$$

As μ decreases, USL minus μ becomes larger, making C_{RU} (and therefore, C'_{RU}) smaller. Anytime the process average is centered in the tolerance zone, it is rewarded with a "bonus" to its potential capability. This is shown graphically in Figure 5.36 by the downward sloping line for values of μ less than T.

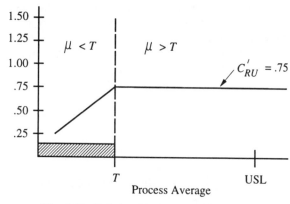

Fig. 5.36 Relationship between C'_{RU} and μ.

A similar relationship holds for the C'_{RL} ratio, as is displayed in Figure 5.37 and discussed in Example 5.13.

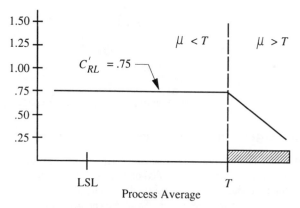

Fig. 5.37 Relationship between C'_{RL} and μ.

Example 5.13 Solder Pull Test

Wire leads are soldered onto printed circuit boards (Figure 5.38). Each hour, a "pull test" is performed on a lead by measuring the force required to remove it from the board. Acceptable leads must be able to withstand a minimum pull force of 25 grams.

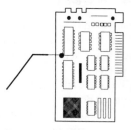

Fig. 5.38 Wire lead soldered to circuit board.

Test results are plotted on an *IX & MR* chart to check the stability of this soldering process. When in control, the average pull test is estimated as 35 grams, with the short-term standard deviation predicted to be 5 grams. With a target average pull test of 40 grams, the short-term potential capability of this process is estimated as 1.0. Unfortunately, the short-term potential capability goal for this pull test is .75.

$$\hat{C}'_{RL} = \text{Minimum}\left(\frac{3\hat{\sigma}_{ST}}{T - \text{LSL}}, \frac{3\hat{\sigma}_{ST}}{\hat{\mu} - \text{LSL}} \right)$$

$$= \text{Minimum}\left(\frac{3(5)}{40 - 25}, \frac{3(5)}{35 - 25} \right)$$

$$= \text{Minimum}\,(\,1.00\,,\,1.50\,) = 1.00$$

To improve the pull test results, a new type of solder is introduced. After control of this modified process is regained, its average is estimated as 41 grams, with a short-term standard deviation of 4 grams. C'_{RL} is now estimated as .75, which just meets the capability goal for this process.

$$\hat{C}'_{RL} = \text{Minimum}\left(\frac{3\hat{\sigma}_{ST}}{T - \text{LSL}}, \frac{3\hat{\sigma}_{ST}}{\hat{\mu} - \text{LSL}} \right)$$

$$= \text{Minimum}\left(\frac{3(4)}{40 - 25}, \frac{3(4)}{41 - 25} \right)$$

$$= \text{Minimum}\,(\,.80\,,\,.75\,) = .75$$

Establishing Target Values For Unilateral Specifications

Ford Motor Company has published a method for determining target values for unilateral specifications if they are not provided on the engineering drawings. When an estimate of the process standard deviation is computed, the target average is recommended to be set at a distance of four standard deviations from the specification limit.

$$T = \text{LSL} + 4\hat{\sigma} \quad \text{or} \quad T = \text{USL} - 4\hat{\sigma}$$

In specific applications, the factor of four may be changed (increased to 4.5, 5, or even 6) to provide a sufficient safety margin. Target averages established in this manner are subject to approval by quality and manufacturing engineering.

Example 5.14 Brake Shoe Hardness

An automotive company in the United Kingdom requires a minimum hardness for brake shoes (linings) installed into the vehicles it assembles. Brake shoe hardness is measured by the depth of an impression left by a metal contact pressed with a certain force on the surface of the shoe. As lower depth readings are associated with harder shoes and are thus desirable, an USL of 35.0 is specified on the engineering drawing. To determine the target average for hardness (none is given on the drawing), the supplier of these brake shoes finds the average range from the $\overline{X}, R$ chart ($n = 3$) tracking brake shoe hardness. With knowledge of the average range and subgroup size, σ_{ST} is estimated as 1.74.

$$\hat{\sigma}_{ST} = \frac{\overline{R}}{d_2} = \frac{2.946}{1.693} = 1.74$$

The quality-improvement team for this product decides a safety margin of 5 standard deviations is needed to provide sufficient protection against any upward shifts in the process average. Combining this desire with the estimate of σ_{ST}, a target average is calculated as 26.3. This proposed target average must be reviewed with both the supplier's and customer's engineering groups.

$$T = \text{USL} - 5\hat{\sigma}_{ST} = 35 - 5(1.74) = 26.3$$

From the $\overline{X}$ chart, the process average is estimated as 25.2. With this additional information, C'_{RU} may be estimated.

$$\hat{C}'_{RU} = \text{Minimum}\left(\frac{3\hat{\sigma}_{ST}}{\text{USL} - T}, \frac{3\hat{\sigma}_{ST}}{\text{USL} - \hat{\mu}}\right)$$

$$= \text{Minimum}\left(\frac{3(1.74)}{35.0 - 26.3}, \frac{3(1.74)}{35.0 - 25.2}\right)$$

$$= \text{Minimum}(.60, .53) = .53$$

As this estimate exceeds the goal of .60, this process is considered to possess excellent potential capability.

Advantages and Disadvantages

These ratios provide a measure of capability for unilateral specifications with the same rating scale as C_R and P_R. However, they are more complicated to calculate and difficult to understand. In addition, unilateral characteristics quite often have non-normal output distributions. Section 8.4 presents a procedure for estimating capability measures under these circumstances. Other approaches have been proposed for this situation (see Littig and Lam, or Flaig) but are much more complex and require special software programs.

5.7 Short-term Potential Capability Index, C_P

The 6σ process spread for a given characteristic can be compared to its tolerance in a manner other than that used with the C_R and P_R ratios. The capability measure presented in this section provides an index of how many "6σ spreads" will fit into the tolerance (Osuga). To meet the minimum potential capability requirement, *i.e.*, the ability to produce 99.73 percent conforming parts, at least one 6σ spread must be able to fit into the tolerance. In order to provide an adequate safety margin, more than one is required. For measuring short-term potential process capability, the C_P index is defined by Koons (pp. 18-19) as given below. Notice this formula is just the inverse of the one for the C_R ratio.

$$C_P = \frac{\text{Tolerance}}{6\sigma_{ST}} = \frac{\text{USL - LSL}}{6\sigma_{ST}} = \frac{1}{C_R}$$

Typically, σ_{ST} is not known and must be estimated from the control chart statistics (either $\overline{R}$ or $\overline{S}$). This estimate is labeled $\hat{\sigma}_{ST}$ and replaces σ_{ST} in the formula. Because $\hat{\sigma}_{ST}$ is now involved, the resulting measure becomes an estimate of the true capability measure.

$$\hat{C}_P = \frac{\text{Tolerance}}{6\hat{\sigma}_{ST}} = \frac{\text{USL - LSL}}{6\hat{\sigma}_{ST}} \qquad \text{where } \hat{\sigma}_{ST} = \frac{\overline{R}}{d_2} \text{ or } \hat{\sigma}_{ST} = \frac{\overline{S}}{c_4} \text{ or } \hat{\sigma}_{ST} = \frac{\overline{MR}}{1.128}$$

When the focus of a capability study is on the long-term aspects of the process, σ_{LT} is a more appropriate indicator of process behavior than is σ_{ST}.

5.8 Long-term Potential Capability Index, P_P

If interest is in acquiring a measurement of long-term potential process capability, then the P_P index applies since it is calculated from σ_{LT} (Koons, p. 20). This measure indicates how many $6\sigma_{LT}$ spreads could fit into the tolerance.

$$P_P = \frac{\text{Tolerance}}{6\sigma_{LT}} = \frac{\text{USL - LSL}}{6\sigma_{LT}} = \frac{1}{P_R}$$

Usually σ_{LT} is not known and must be estimated from S_{TOT}, which is calculated from the individual measurements of in-control subgroups collected for the control chart. This estimate is labeled $\hat{\sigma}_{LT}$ and replaces σ_{LT} in the above equation. As this capability measure is now an estimate, the symbol $\wedge$ is written above P_P.

$$\hat{P}_P = \frac{\text{Tolerance}}{6\hat{\sigma}_{LT}} = \frac{\text{USL - LSL}}{6\hat{\sigma}_{LT}} \qquad \text{where } \hat{\sigma}_{LT} = \frac{S_{TOT}}{c_4}$$

In order to meet the minimum requirements for long-term potential capability, at least one $6\sigma_{LT}$ spread must be able to fit within the tolerance. Munro recommends using this index as a preliminary assessment of potential capability.

Interpretation of C_P and P_P

The interpretation of these indexes is similar to that for the C_R and P_R ratios, as previously covered in Section 5.4. The only difference is that now *higher* numbers are better due to these new indexes being just the inverse of the C_R and P_R ratios.

$$C_P = \frac{1}{C_R} \qquad\qquad P_P = \frac{1}{P_R}$$

For example, if the $6\hat{\sigma}_{ST}$ spread is equal to half of the tolerance ($C_R = .50$), then two $6\hat{\sigma}_{ST}$ spreads should be able to fit within the tolerance ($C_P = 1/.50 = 2.0$). This is more intuitive in the sense that higher numbers mean higher capability, which implies a better match between the process output and customer requirements. Another way of interpreting these indexes, is as the width of the tolerance equal to one 6σ spread. If C_P equals .50, then the tolerance is one half the size of a $6\hat{\sigma}_{ST}$ spread. When C_P equals 1.00, the tolerance is equal to one $6\hat{\sigma}_{ST}$ spread. And when C_P equals 1.50, the tolerance is 1.5 times the size of one $6\hat{\sigma}_{ST}$ spread.

The inverse relationship to C_R and P_R also means the rating scale for C_P and P_P is the reverse of the one for C_R and P_R, with larger numbers now indicating higher potential capability.

< 1 implies the process fails to meet the minimum requirement for potential capability,

= 1 indicates the process just fulfills the minimum requirement,

> 1 means the process surpasses the minimum requirement.

Although similar in interpretation to the C_R and P_R ratios, the reversal of the rating scale can cause considerable confusion. This is especially true for companies supplying several customers, some of which want C_R values reported while others prefer to see C_P values. When customers visit, shop floor personnel don't know if they should show high capability numbers or low ones.

To help visualize the meaning of these new indexes, Figure 5.39 displays a process having C_P equal to 1.0. This means exactly one $6\hat{\sigma}_{ST}$ spread could fit within the tolerance when the process is centered at M.

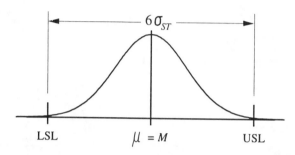

Fig. 5.39 A C_P of 1.0 means one $6\sigma_{ST}$ spread could fit within the tolerance.

Figure 5.40 shows a process with a C_P of 1.5, implying one and a half $6\hat{\sigma}_{ST}$ spreads could fit within the tolerance. Equivalently, one could say the tolerance equals $9\hat{\sigma}_{ST}$ ($1.5 \times 6 = 9$).

$$C_P = 1.5$$

$$\frac{\text{Tolerance}}{6\sigma_{ST}} = 1.5$$

$$\text{Tolerance} = 9\sigma_{ST}$$

The process output in Figure 5.41 has a C_P of 2.0. In this case, the tolerance has room for a full two $6\hat{\sigma}_{ST}$ spreads, meaning the tolerance is equal to a distance of $12\hat{\sigma}_{ST}$ ($2 \times 6 = 12$). Thus, higher C_P values also imply greater safety margins.

Goals

Just as the most common goal for C_R and P_R is .75, or lower, the most common goal for these two new measures of potential capability is just the inverse, 1.33, or higher.

$$C_{P,GOAL} = \frac{1}{C_{R,GOAL}} = \frac{1}{.75} = 1.33 \qquad\qquad P_{P,GOAL} = \frac{1}{P_{R,GOAL}} = \frac{1}{.75} = 1.33$$

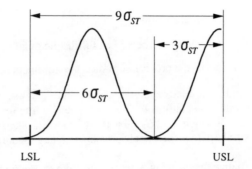

Fig. 5.40 A C_P of 1.5 means one and a half $6\sigma_{ST}$ spreads fit within the tolerance.

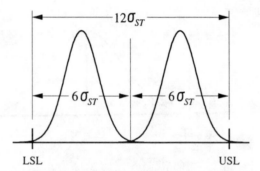

Fig. 5.41 A C_P of 2.0 means two $6\sigma_{ST}$ spreads will fit within the tolerance.

A process attaining this goal of 1.33 would be able to squeeze 1 and a third 6σ spreads (or a total of 8σ) within the tolerance. Alternately, this goal means the process has the ability to fit both 4σ tails within the tolerance when the process is centered at M (Figure 5.42).

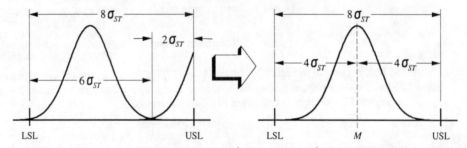

Fig. 5.42 When $C_P = 1.33$, one and a third $6\hat{\sigma}_{ST}$ spreads (or $8\hat{\sigma}_{ST}$) will fit in the tolerance.

A C_R goal of .60 ($\pm 5\hat{\sigma}_{ST}$) converts to a C_P goal of 1.67 (1/.60), while a C_R goal of .50 ($\pm 6\hat{\sigma}_{ST}$) translates into a C_P goal of 2.00. Juran *et al* (p. 9-22) proposed the rating scale for C_P presented in Table 5.3 on the facing page.

Table 5.3 Capability ratings for various C_P values.

C_P Range	Capability Rating
$2.00 \leq C_P$	Terrific
$1.67 \leq C_P < 2.00$	Excellent
$1.33 \leq C_P < 1.67$	Good
$1.00 \leq C_P < 1.33$	Fair
$.67 \leq C_P < 1.00$	Poor
$.00 \leq C_P < .67$	Terrible

To allow adequate room for shifts in the process average, Koons (p. 10) endorses a C_P goal of 4.0 for "important" characteristics, and as high as 10.0 for "critical" characteristics. Achievement of these higher goals provides a sufficient safety margin to decrease the chance of parts being produced outside of the specifications, should the process ever become unstable.

Example 5.15 Fan Housing Weight

An $\overline{X}, R$ chart (subgroup size of 5) monitoring the weight of fan housings produced on a plastic injection-molding process is finally brought into a good state of control. The tolerance for part weight is 4.00 grams and the $\overline{R}$ value from the chart is .848 grams. Substituting these values into the formula below yields an estimate of C_P.

$$\hat{C}_P = \frac{\text{Tolerance}}{6\hat{\sigma}_{ST}} = \frac{\text{Tolerance}}{6(\overline{R}/d_2)} = \frac{4.00 \text{ gms}}{6(.848 \text{ gms}/2.326)} = 1.828$$

The C_P estimate of 1.828 indicates the process has the potential to fit 1.828 $6\hat{\sigma}_{ST}$ spreads within the tolerance (an "excellent" rating by Juran's scale). The tolerance is almost equal to $11\hat{\sigma}_{ST}$ (6×1.828). As happened for the C_R and P_R ratios, the units of measurement cancel, leaving a dimensionless measure of process capability. This facilitates comparison of potential capability between different characteristics.

If properly centered with a normal distribution, this process should have no difficulty producing at least 99.73 percent conforming parts, thus exceeding the minimum requirement for short-term potential capability. This analysis agrees with the one made previously when measuring capability with the C_R ratio. In fact, this estimate of C_P is just the inverse of the .547 estimate for C_R obtained in Example 5.5.

$$\hat{C}_P = \frac{1}{\hat{C}_R} = \frac{1}{.547} = 1.828$$

Given a C_P goal of 1.33, this process is definitely capable. However, even though the capability goal is exceeded, the original control chart monitoring housing weight must be maintained to ensure that the current level of capability does not deteriorate (Koons, p. 25). To reduce operating costs, the subgroup size may be decreased, or the time between subgroups increased, but some type of chart must remain to prevent backsliding.

Example 5.16 Roller Width

After molding and curing, rubber rollers for guiding and driving paper sheets through copying machines are ground to a final diameter. Rollers are then sliced with a rotating knife to the proper width (Figure 5.43). As excessive variation in roller width results in paper jams, this dimension must be held to within plus or minus .0127 cm of the nominal length of 31.2250 cm, making the tolerance equal to .0254 cm.

Fig. 5.43 A roller for a copying machine.

A subgroup size of 10 is chosen, so an $\overline{X}$, S chart is selected to monitor the slicing process. When a good state of control is demonstrated, the last 35 in-control subgroups are used to calculate $\overline{S}$, which turns out to be .005169 cm. Based on this result, the short-term potential capability index is estimated as follows:

$$\hat{C}_P = \frac{\text{Tolerance}}{6\hat{\sigma}_{ST}} = \frac{\text{Tolerance}}{6(\overline{S}/c_4)} = \frac{.0254\,\text{cm}}{6(.005169\,\text{cm}/.9727)} = .80$$

An estimated C_P ratio of .80 means that even with proper centering, only 80 percent of one $6\hat{\sigma}_{ST}$ spread will fit within the allowed tolerance, resulting in a "poor" capability rating being assigned to this process. Even if properly centered, this slicing operation cannot come close to cutting all rollers to the desired width.

Notice that C_P can be estimated from the difference between the control limits of the $\overline{X}$ chart. With an $\overline{X}$ of 31.2306, the *UCL* of 31.23564 cm and the *LCL* of 31.22556 cm were previously calculated in Example 5.6. The result is identical to the above estimate of C_P.

$$\hat{C}_P = \frac{\text{Tolerance}}{\sqrt{n}\,(UCL - LCL)} = \frac{.0254}{\sqrt{10}\,(31.23564 - 31.22556)} = .80$$

Example 5.17 Exhaust Valves

The outer diameter at one location of the automotive exhaust valve displayed in Figure 5.44 is a critical characteristic. In Example 5.4, both the short- and long-term process standard deviations were estimated along with the C_R and P_R ratios.

$$\hat{\sigma}_{ST} = .01552\,\text{cm} \qquad \hat{\sigma}_{LT} = .01610\,\text{cm}$$

$$\hat{C}_R = .776 \qquad \hat{P}_R = .805$$

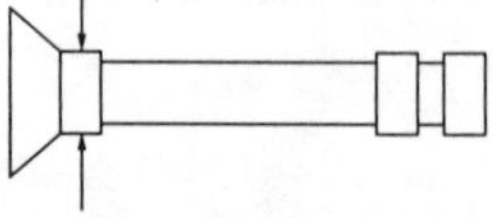

Fig. 5.44

Given print specifications for O.D. size of 10.000 cm $\pm$ .060 cm, both the short- and long-term potential capability of the grinding operation for this characteristic can be measured using the C_P and P_P indexes. The customer has specified a potential capability goal of 1.33 for this feature.

$$\hat{C}_P = \frac{\text{USL - LSL}}{6\,\hat{\sigma}_{ST}} = \frac{10.060\,\text{cm} - 9.940\,\text{cm}}{6(.01552\,\text{cm})} = 1.289$$

$$\hat{P}_P = \frac{\text{USL - LSL}}{6\hat{\sigma}_{LT}} = \frac{10.060\,\text{cm} - 9.940\,\text{cm}}{6(.01610\,\text{cm})} = 1.242$$

Both measures indicate that the process meets the minimum requirement for potential capability (greater than 1.0), but unfortunately, not the goal of 1.33. In fact, the process receives only a "fair" rating for potential capability. Similar to the decision made based on the C_R and P_R ratios, work must begin on reducing common-cause variation.

Because this process fails to meet its capability goal, the control chart for O.D. size must be continued for two important reasons. The first is for maintaining the process at its current level so potential capability does not become even worse. The chart helps in this regard by alerting the operator to the presence of any assignable sources of variation that might adversely affect the process, *e.g.*, a point above the UCL of the range chart.

The second reason for continuing the chart is to provide verification that the work done to reduce common-cause variation is successful. If these efforts are effective, a downward trend, or a run below the centerline, should be spotted on the range chart. Once the process is stabilized at this improved level, C_P and P_P should be estimated with the new process standard deviation and compared to the goal of 1.33.

Example 5.18 Acrylic Coating

A supplier to the aerospace industry produces integrated avionics systems for commercial aircraft. Many of the circuit boards inside these systems have surface mount components requiring an organic-acrylic coating to protect them from corrosion during their useful life (Figure 5.45). Boards are coated by passing through an automated spraying process, with coating thickness being one of the key output characteristics.

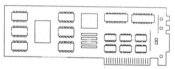

Fig. 5.45 A circuit board for an avionics system.

A new board is undergoing development and the engineering group is interested in establishing realistic specification limits for coating thickness so the spraying process should have little difficulty achieving a P_P goal of 2.00. From control charts monitoring coating thickness on current boards, the long-term standard deviation is estimated as .021 mm. With a desired target for average coating thickness of .190 mm on the new board, the required LSL and USL are derived by first determining how large a tolerance is needed to achieve a $P_{P.GOAL}$ of 2.00 (for a different approach, see Anand).

$$\frac{\text{Tolerance}}{6\hat{\sigma}_{LT}} = \hat{P}_{P.GOAL} \quad \Rightarrow \quad \text{Tolerance} = 6\hat{\sigma}_{LT} \times \hat{P}_{P.GOAL}$$

$$= 6(.021\,\text{mm})(2.00)$$

$$= .252\,\text{mm}$$

Assuming the target average of 1.90 mm is to be the middle of the tolerance, M, the LSL is found by subtracting one half of the required tolerance from this target.

$$\text{LSL} = M - .5\,(\text{Tolerance}) = 1.90\,\text{mm} - .5\,(.252\,\text{mm}) = 1.774\,\text{mm}$$

Likewise, the USL for the coating thickness of this board is found by adding one half the tolerance to M.

$$\text{USL} = M + .5\,(\text{Tolerance}) = 1.90\,\text{mm} + .5\,(.252\,\text{mm}) = 2.026\,\text{mm}$$

If the variation in coating thickness of the new board remains similar to that for current boards, the spraying operation should be able to realize a P_P index of 2.00 with these process-based specifications.

Relationship of C_P to the Process Average

By not including μ in its calculation, changes in the process average have no effect on C_P. This independence is illustrated on the graph in Figure 5.46. C_P remains constant at 1.33 no matter where the process average is, even if μ is located outside a print specification.

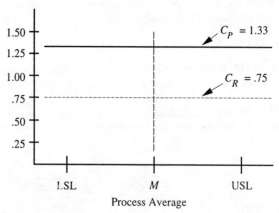

Fig. 5.46 Relationship between C_P and the process average.

Example 5.19 Floppy Disks

A serious field problem is traced to variation in the diameter of the inner hole of the floppy disk (Figure 5.47). If the hole is too big, the floppy disk fails to rotate properly and cannot store or retrieve information. On the other hand, if this hole is too small, it jams the delicate disk drive, causing a major warranty claim, involving both a replacement floppy plus repairing the disk drive. After four days of charting with an $\overline{X}, R$ chart, the process is finally brought into a good state of control and the following information is gathered:

Fig. 5.47

$$\text{Tolerance} = 5.0 \qquad \hat{\sigma}_{ST} = .898 \qquad \hat{\sigma}_{LT} = .916 \qquad \hat{C}_R = 1.078 \qquad \hat{P}_R = 1.099$$

To determine the potential capability of this hole-cutting operation, a team working on this problem decides to estimate both the C_P and P_P indexes.

$$\hat{C}_P = \frac{\text{Tolerance}}{6\hat{\sigma}_{ST}} = \frac{5.0}{6(.898)} = .928 \qquad \hat{P}_P = \frac{\text{Tolerance}}{6\hat{\sigma}_{LT}} = \frac{5.0}{6(.916)} = .910$$

C_P and P_P could also have been calculated by taking the inverse of C_R and P_R, respectively.

$$\hat{C}_P = \frac{1}{\hat{C}_R} = \frac{1}{1.078} = .928 \qquad \hat{P}_P = \frac{1}{\hat{P}_R} = \frac{1}{1.099} = .910$$

These two potential capability indexes tell the same story as the capability ratios, there is a serious capability problem producing the hole sizes. But this is certainly not new information. Everyone already knows there is a problem, but no one knows if it is because hole diameters are too big, too small, or both. Just as with the potential capability ratios, these indexes do not divulge what percentage of hole sizes are out of specification. Measures of potential process capability are quite limited in their information content. Knowledge about the actual process performance is desperately needed to understand what is happening in this process, and what must be done to improve its ability to satisfy customers. Capability measures presented in the next chapter will readily disclose this vital information.

Because C_P and P_P measure two different types of capability (short-term and long-term), most companies will avoid confusion by reporting one or the other, but not both.

Advantages and Disadvantages

These indexes are fairly easy to calculate and are dimensionless, which allows comparisons with other process characteristics. They have a familiar rating scale in that 1.0 indicates attainment of the minimum potential capability requirement. Unfortunately, the rating scale is reversed from the one for C_R, which leads to some confusion. Whereas *lower* C_R values mean better potential capability, *higher* values of C_P imply superior potential capability. However, this "higher is better" rating is used by the majority of capability measures.

C_P and P_P express the relationship between the tolerance and the process variation in terms somewhat less intuitive for shop floor personnel and many managers (how many 6σ spreads fit within the tolerance). This is why C_P is referred to as an "index" rather than a ratio. Many practitioners find this relationship harder to visualize than the one for C_R, which expresses capability as a percentage of the tolerance. For example, if new tooling for a process increases its C_P index from 1.24 to 1.30, how much of an improvement is made? Very few will realize this represents a potential 50 percent reduction in nonconforming parts.

Disadvantages are similar to those for the C_R and P_R ratios: different formulas are required for unilateral tolerances; assumption the process output can be centered at the middle of the tolerance; middle of tolerance is assumed to be best; process output must be normally distributed (Chan *et al*); measurements must be independent.

Even if all assumptions are met, these indexes cannot directly predict the percentage of nonconforming parts. If the C_P ratio is .93, what percentage of the process output is out of tolerance? There is no way of knowing. In fact, reductions in C_P can occur with no reduction in the percentage of nonconforming parts. The output for a process (curve A in Figure 5.48) has a C_P of only 1.00. Due to being centered at the USL, it is producing 50 percent nonconforming parts.

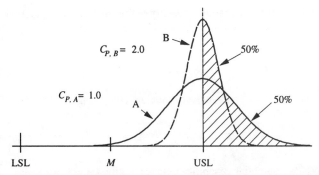

Fig. 5.48 An increase in C_P does not imply the nonconformance rate has decreased.

After several improvements are successfully implemented to diminish process spread, C_P increases to 2.00 (curve B). Unfortunately, this is not much of a cause for celebration as this "improved" process is still producing 50 percent nonconforming parts. Had the process average moved up as a result of the changes to reduce process spread, C_P would have become higher, while the nonconformance rate would have actually *increased*, as is demonstrated in Figure 5.49.

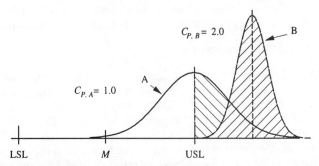

Fig. 5.49 Even though B has a higher C_P, its nonconformance rate is greater than A's.

Another disadvantage is that $\hat{P}_P$ is a "biased" estimator of P_P. Being a biased estimator means, that if a P_P index is estimated from each of many capability studies repeated on the same (stable) process, the average of all these estimates does not approach the true process P_P index as the number of studies increase. The deviation between the average of the estimates and the true process parameter is called the bias. Although this bias is quite small when kn is large, this undesirable trait bothers some practitioners. Fortunately for them, Pearn *et al* have developed a bias correction factor, called b_{kn}.

$$b_{kn} \cong 1 - \frac{.75}{kn - 1}$$

Multiplying $\hat{P}_P$ by this factor generates an unbiased estimate of P_P, labeled $\hat{P}_P^U$, with the "U" representing "unbiased."

$$\hat{P}_P^U = b_{kn} \hat{P}_P$$

Recall Example 5.17, where P_P was estimated as 1.242 for the O.D. size of an exhaust valve, based on a sample size (kn) equal to 100. The bias correction factor is calculated as:

$$b_{kn} \cong 1 - \frac{.75}{100 - 1} = 1.007576$$

Applying this to $\hat{P}_P$ generates an unbiased estimate of P_P.

$$\hat{P}_P^U = b_{kn} \hat{P}_P = 1.007576(1.242) = 1.251$$

Note that this unbiased estimate of 1.251 is less than 1 percent larger than the biased estimate of 1.242. Differences this small are not usually significant.

For an example of how capability is measured with the C_P index in the wood manufacturing industry, see Lyth and Rabiej. Tseng and Wu describe a statistical method for selecting the process with the largest C_P index from several competing manufacturing processes.

5.9 C_P^* and P_P^* for Cases Where $T \neq M$

The C_P and P_P indexes assume the target for the process average (T) is the middle of the tolerance (M). This is usually, but not always, true. In those few situations where T is not equal to M, these indexes overstate true potential process capability. Determination of an appropriate target is discussed by Arcelus and Rahim, Bisgaard *et al*, Hunter and Kartha, as well as Nelson.

As was done with the C_R and P_R ratios, the formulas for C_P and P_P are modified in similar manner to accurately assess potential capability in these cases. Again, the notation * is added to each measure's label to indicate this special condition where T does not equal M.

$$C_P^* = \text{Minimum}\,(C_{PL}^T, C_{PU}^T) = \text{Minimum}\left(\frac{T - \text{LSL}}{3\sigma_{ST}}, \frac{\text{USL} - T}{3\sigma_{ST}}\right)$$

$$P_P^* = \text{Minimum}\,(P_{PL}^T, P_{PU}^T) = \text{Minimum}\left(\frac{T - \text{LSL}}{3\sigma_{LT}}, \frac{\text{USL} - T}{3\sigma_{LT}}\right)$$

These special indexes report how many 3σ spreads could fit into the smaller "half" of the tolerance. The following relationships hold:

$$C_P^* = \frac{1}{C_R^*} \qquad P_P^* = \frac{1}{P_R^*}$$

When the process standard deviation is estimated from sample data, C_P^* and P_P^* are estimated with these two formulas.

$$\hat{C}_P^* = \text{Minimum}\left(\frac{T - \text{LSL}}{3\hat{\sigma}_{ST}}, \frac{\text{USL} - T}{3\hat{\sigma}_{ST}}\right) \qquad \hat{P}_P^* = \text{Minimum}\left(\frac{T - \text{LSL}}{3\hat{\sigma}_{LT}}, \frac{\text{USL} - T}{3\hat{\sigma}_{LT}}\right)$$

With the above formulas, the rating scale for C_P^* and P_P^* becomes the same as for the C_P and P_P indexes.

< 1 implies the process fails to satisfy the minimum potential capability requirement,

= 1 means the process just fulfills the minimum requirement,

> 1 reveals the process surpasses the minimum prerequisite.

In those situations where T equals M, C_P^* equals C_P.

$$C_P^* = C_P = \frac{1}{C_R} = \frac{1}{C_R^*}$$

If T is not the same as M, then C_P^* is less than C_P. The same relationship applies for P_P^* and P_P, as is demonstrated in the next example.

Example 5.20 Gold Plating

The LSL for the thickness of gold plated on a silicon wafer is 50 angstroms, with an USL of 100 angstroms. For cost reasons, a decision is made to target the average thickness at 65 ($T = 65$), as is shown in Figure 5.50. If $\hat{\sigma}_{LT}$ is 2.5, an estimate can be made for the P_P index.

$$\hat{P}_P = \frac{\text{USL} - \text{LSL}}{6\hat{\sigma}_{LT}} = \frac{100 - 50}{6(2.5)} = 3.33$$

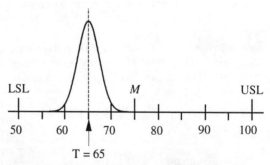

Fig. 5.50 The distribution of gold-plating thickness when centered on target.

Assuming the process is centered at 75 (the middle of the tolerance), it would be possible to fit as many as 3.33 $6\sigma_{LT}$ spreads into the tolerance. However, because care will be taken to keep the process centered at 65, this potential can never be achieved. The P_P measure is only legitimate under the assumption the process output will be centered at M. If this assumption is not valid, the P_P measure of 3.33 overstates the predicted potential capability of this process. Whenever T does not equal M, decisions concerning potential capability should be based on P_P^* instead of P_P.

$$\hat{P}_P^* = \text{Minimum}\left(\frac{T - \text{LSL}}{3\hat{\sigma}_{LT}}, \frac{\text{USL} - T}{3\hat{\sigma}_{LT}} \right)$$

$$= \text{Minimum}\left(\frac{65 - 50}{3(2.5)}, \frac{100 - 65}{3(2.5)} \right)$$

$$= \text{Minimum}\,(\,2.00\,,4.67\,) = 2.00$$

Notice that this is equal to the inverse of $\hat{P}_R^*$, which was calculated as .50 in Example 5.11.

$$\hat{P}_P^* = \frac{1}{\hat{P}_R^*} = \frac{1}{.50} = 2.00$$

When the process is centered at the target value of 65, the process average is closer to the LSL of 50 than the USL of 100. In light of this preferential operating condition, the P_P^* index discloses that only two $6\sigma_{LT}$ spreads will be able to fit into the tolerance with the process centered at 65. Even though quite a bit of "unused" tolerance exists at the upper end of the specification zone, it is unavailable to the process due to the added stipulation of centering μ at 65. Sections 8.5 and 8.9 explain how some non-normal distributions are able to take advantage of this "lost" tolerance.

If T is less than M, the available tolerance is reduced by bringing the USL down to a distance above T equal to the distance the LSL is below T, as is illustrated in Figure 5.51.

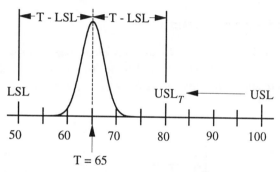

Fig. 5.51 The available tolerance is effectively reduced when T is not equal to M.

This artificially created USL is called USL_T, the subscript T indicating its association with the target average. Note that T is located half way between the LSL and USL_T. In this plating example, the LSL is 15 angstroms (65 minus 50) below the target of 65, so USL_T is placed 15 angstroms *above* the target. This makes USL_T equal to 80 angstroms (65 + 15).

$$\text{USL}_T = 65 + (65 - 50) = 80$$

In general, USL_T is positioned above T by a distance of T minus LSL.

$$\text{USL}_T = T + (T - \text{LSL}) = 2T - \text{LSL}$$

By requiring plating thickness to be centered at 65, the original tolerance is effectively compressed from 50 (100 minus 50) to only 30 (80 minus 50), a reduction of 40 percent. The "effective tolerance" for this example is defined as:

$$\text{Effective Tolerance} = \text{USL}_T - \text{LSL}$$

This loss in usable tolerance of 40 percent results in P_P^* being only 60 percent of P_P.

$$\hat{P}_P^* = \hat{P}_P \left(\frac{\text{Effective Tolerance}}{\text{Actual Tolerance}} \right) = \hat{P}_P \left(\frac{\text{USL}_T - \text{LSL}}{\text{USL - LSL}} \right) = 3.33 \left(\frac{80 - 50}{100 - 50} \right) = 2.00$$

P_P^* can also be estimated by dividing the effective tolerance by $6\sigma_{LT}$.

$$\hat{P}_P^* = \frac{\text{Effective Tolerance}}{6\hat{\sigma}_{LT}} = \frac{\text{USL}_T - \text{LSL}}{6\hat{\sigma}_{LT}} = \frac{80 - 50}{6(2.5)} = 2.00$$

If T is greater than M, the USL is left unchanged, while a surrogate LSL (called LSL_T) is established at a distance USL minus T *below* the target average.

$$\text{LSL}_T = T - (\text{USL} - T) = 2T - \text{USL}$$

Note that when T equals M, the effective tolerance is identical to the original tolerance, and P_P^* equals P_P.

$$P_P^* = P_P \left(\frac{\text{Effective Tolerance}}{\text{Actual Tolerance}} \right) = P_P(1) = P_P$$

Relationship Between C_P^* and μ

Whenever T is not equal to M, the effective tolerance is somewhat less than the full tolerance. As a result, C_P^* is less than C_P. As both these measures of potential capability are independent of μ, this relationship is true no matter where the actual process is centered, as is displayed in Figure 5.52.

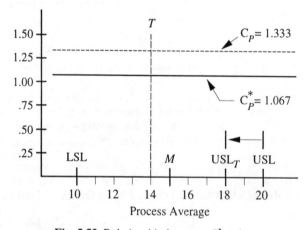

Fig. 5.52 Relationship between C_P^* and μ.

In this graph, the target average of 14 is one unit less than M, making USL_T equal to 18.

$$\text{USL}_T = 2T - \text{LSL} = 2(14) - 10 = 18$$

This lowering of the effective USL to 18 shrinks the effective tolerance to 80 percent of the full tolerance.

$$\frac{\text{Effective Tolerance}}{\text{Tolerance}} = \frac{\text{USL}_T - \text{LSL}}{\text{USL} - \text{LSL}} = \frac{18 - 10}{20 - 10} = .80$$

Thus, C_p^* is equal to 1.067, which is only 80 percent of C_p (.80 × 1.333 = 1.067). In general, the relationship between C_p^* and C_p can be expressed as follows, with the full derivation of k_T given at the end of this chapter.

$$C_p^* = C_p(1 - k_T) \qquad \text{where } k_T = \frac{|T - M|}{\text{Tolerance} / 2}$$

k_T represents how far T deviates from M as a percentage of half the tolerance. In this example, T is 14, and M is 15, making the absolute value of their difference equal to 1. As the tolerance is 10 (20 minus 10), half of this is 5. This makes k_T equal to .20, meaning T is one fifth of half the tolerance away from M.

$$k_T = \frac{|T - M|}{\text{Tolerance} / 2} = \frac{|14 - 15|}{(20 - 10)/2} = .20$$

This 20 percent reduction in usable tolerance causes C_p^* to be only 80 percent of C_p.

$$C_p^* = C_p(1 - k_T) = 1.333(1 - .20) = 1.067$$

C_p reveals the maximum available potential capability, which occurs when the process average is centered at M. k_T represents the loss in this maximum potential capability caused by specifying a target average that is not the middle of the tolerance.

In a similar manner, the relationship between P_p^* and P_p is expressed as:

$$P_p^* = P_p(1 - k_T) \qquad \text{where } k_T = \frac{|T - M|}{\text{Tolerance} / 2}$$

Because the standard deviation is not part of the calculation for k_T, this factor is identical for both short- and long-term indexes.

When *T* Equals One of the Specification Limits

There are some specifications such as 20 plus 6, minus 0. The design engineers would like to have all parts produced with this characteristic at 20. However, recognizing there is variation in manufacturing, they allow this characteristic to be somewhat larger than 20, but not lower. Thus, the engineers expect an output distribution for this characteristic to appear like curve A in Figure 5.53.

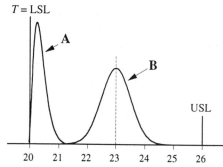

Fig. 5.53   Engineering wish (A) versus manufacturing reality (B).

Despite engineering's desire, manufacturing quite often centers the process output at the middle of the tolerance to provide an equal safety margin on both sides. This strategy results in the output average being located at 23, which is portrayed by curve B in Figure 5.53. The engineers who wished for *all* parts at 20, end up with *no* parts at 20. As all parts are well within "spec," manufacturing has a hard time understanding the engineers' complaints of not producing what they wanted. If manufacturing decides to respond to these complaints by making as many parts as possible at the desired target, it would have to center the process output at 20. Unfortunately, not every part can be made exactly at 20 due to inherent process variation. By centering the process average at *T*, half of the parts produced are below the LSL, adding considerable cost to this operation.

The C_P^* measure of potential capability is zero for situations where *T* equals LSL (or *T* equals USL). Note that the effective tolerance is also zero under these conditions.

$$C_P^* = \text{Minimum} \left(\frac{T - LSL}{3\hat{\sigma}_{LT}}, \frac{USL - T}{3\hat{\sigma}_{LT}} \right)$$

$$= \text{Minimum} \left(\frac{LSL - LSL}{3\hat{\sigma}_{LT}}, \frac{USL - LSL}{3\hat{\sigma}_{LT}} \right)$$

$$= \text{Minimum} \left(\frac{0}{3\hat{\sigma}_{LT}}, \frac{\text{Tolerance}}{3\hat{\sigma}_{LT}} \right)$$

$$= \text{Minimum} \left(0, 2C_P \right) = 0$$

The engineering group must consider the ease of manufacturability when designing new products. One of the best ways to avoid problems like this is having engineering and manufacturing work together during the design phase. This is called concurrent, or simultaneous, engineering. Rado gives an interesting example involving the application of capability indexes for enhancing new product development for memory storage systems.

5.10 C'_{PL}, C'_{PU}, P'_{PL}, and P'_{PU} for Unilateral Specifications

Formulas for the C_P and P_P indexes require modification in order to handle unilateral specifications, where the minimum requirement is to have the potential of 99.865 percent conformance (Kane, 1986). These modifications are similar to the changes made for the C_R and P_R ratios (Kume) covered in Section 5.6. Notice that attention is focused exclusively on just one tail of the process output distribution, either the upper 3σ tail when dealing with only an USL, or the lower when dealing with solely a LSL. For dimensions where just an USL is specified, use the following formula to calculate C'_{PU}, which is sometimes called a "one-sided" potential capability index (Vannman).

$$C'_{PU} = \text{Maximum} \left(C_{PU}^T, C_{PU} \right) \quad \text{where} \quad C_{PU}^T = \frac{USL - T}{3\sigma_{ST}} \quad \text{and} \quad C_{PU} = \frac{USL - \mu}{3\sigma_{ST}}$$

In a similar manner, C'_{PL} is defined for characteristics with only a LSL.

$$C'_{PL} = \text{Maximum} \left(C_{PL}^T, C_{PL} \right) \quad \text{where} \quad C_{PL}^T = \frac{T - LSL}{3\sigma_{ST}} \quad \text{and} \quad C_{PL} = \frac{\mu - LSL}{3\sigma_{ST}}$$

When no target is specified, C_{PL}^T and C_{PU}^T are undefined and dropped from the formulas for C'_{PL} and C'_{PU}. For certain circumstances, Section 5.6 explains how a target average may be established (see also Wheeler, 2000).

$$C'_{PL} = C_{PL} \qquad C'_{PU} = C_{PU}$$

As explained in the next chapter, this actually makes C'_{PL} and C'_{PU} measures of performance, rather than potential, capability. If interest is in long-term potential capability, the following formulas apply. For the case of only a LSL:

$$P'_{PL} = \text{Maximum} (P_{PL}^T, P_{PL}) \quad \text{where } P_{PL}^T = \frac{T - LSL}{3\sigma_{LT}} \text{ and } P_{PL} = \frac{\mu - LSL}{3\sigma_{LT}}$$

In a similar manner, P'_{PU} is defined as follows for characteristics with just an USL:

$$P'_{PU} = \text{Maximum} (P_{PU}^T, P_{PU}) \quad \text{where } P_{PU}^T = \frac{USL - T}{3\sigma_{LT}} \text{ and } P_{PU} = \frac{USL - \mu}{3\sigma_{LT}}$$

The rating scale for these indexes is the same as that for C_P and P_P.

    < 1 implies the process fails to meet the minimum potential capability requirement,

    = 1 indicates the process just meets the minimum requirement,

    > 1 reveals the process exceeds the minimum requirement.

Example 5.21 Particle Contamination

The following estimates of process parameters are taken from an *IX & MR* chart monitoring particle contamination of a cleaning solution for printed circuit boards after they pass through wave soldering.

$$\hat{\mu} = 127.32 \, ppm \qquad \hat{\sigma}_{ST} = 12.079 \, ppm$$

Given that the USL for contamination is 165 *ppm*, and *T* is 120 *ppm*, the C'_{PU} index is estimated as follows (a sketch of this situation is presented in Figure 5.54.):

$$\hat{C}'_{PU} = \text{Maximum} \left(\frac{USL - T}{3\hat{\sigma}_{ST}}, \frac{USL - \hat{\mu}}{3\hat{\sigma}_{ST}} \right)$$

$$= \text{Maximum} \left(\frac{165 - 120}{3(12.079)}, \frac{165 - 127.32}{3(12.079)} \right)$$

$$= \text{Maximum} (1.24, 1.04) = 1.24$$

Notice this result is equal to the inverse of the C'_{RU} ratio estimated in Example 5.12.

$$\hat{C}'_{PU} = \frac{1}{\hat{C}'_{RU}} = \frac{1}{.81} = 1.24$$

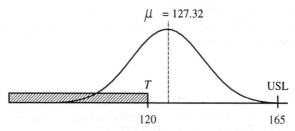

Fig. 5.54 Average particle contamination is 127 *ppm*.

Although this process has fulfilled the minimum requirement for potential capability, which assumes it can eventually be centered at 120 *ppm*, it does not meet a mandated capability goal of 1.33. In an attempt to improve this process, a new type of filter is added to the cleaning bath. This modification causes a reduction in $\hat{\mu}$ to 108 *ppm*, with no change in $\hat{\sigma}_{ST}$. The potential capability of this improved process is calculated as 1.57.

$$\hat{C}'_{PU} = \text{Maximum}\left(\frac{\text{USL} - T}{3\hat{\sigma}_{ST}}, \frac{\text{USL} - \hat{\mu}}{3\hat{\sigma}_{ST}} \right)$$

$$= \text{Maximum}\left(\frac{165 - 120}{3(12.079)}, \frac{165 - 108}{3(12.079)} \right)$$

$$= \text{Maximum}(1.24, 1.57) = 1.57$$

Because centering the process average a fair distance below *T* is now possible, potential capability is much higher, and now exceeds the goal of 1.33. But to maintain this increased level of potential capability, the process average must remain at, or below, 108 *ppm*.

Example 5.22 Acrylic Coating

A supplier to the aerospace industry produces integrated avionics systems for commercial aircraft. Many of the circuit boards used in these systems have surface mount components requiring an organic-acrylic coating to protect them from corrosion during their useful life. Boards are coated by passing through an automated spraying process, with coating thickness being one of the key output characteristics. Control of the spraying process has recently been established for a new board, with the average coating thickness estimated as 1.884 mm and the long-term standard deviation as .028 mm. As there is only a minimum thickness requirement of 1.774 mm, a decision is made to utilize the P'_{PL} index to measure potential capability. The target for coating thickness is 1.900 mm.

$$\hat{P}'_{PL} = \text{Maximum}(\hat{P}^T_{PL}, \hat{P}_{PL})$$

$$= \text{Maximum}\left(\frac{T - \text{LSL}}{3\hat{\sigma}_{LT}}, \frac{\hat{\mu} - \text{LSL}}{3\hat{\sigma}_{LT}} \right)$$

$$= \text{Maximum}\left(\frac{1.900 - 1.774}{3(.028)}, \frac{1.884 - 1.774}{3(.028)} \right)$$

$$= \text{Maximum}(1.50, 1.31) = 1.50$$

As the goal for P'_{PL} is 2.00, the capability team assigned to this process contemplates several alternatives for process improvements. After installing new spray nozzles, the process is once again brought into control. This time, the estimates for μ and σ_{LT} are 1.909 and .026, respectively. This single improvement increases the estimate of P'_{PL} to 1.73.

$$\hat{P}'_{PL} = \text{Maximum}\left(\frac{T-\text{LSL}}{3\hat{\sigma}_{LT}}, \frac{\hat{\mu}-\text{LSL}}{3\hat{\sigma}_{LT}}\right)$$

$$= \text{Maximum}\left(\frac{1.900-1.774}{3(.026)}, \frac{1.909-1.774}{3(.026)}\right)$$

$$= \text{Maximum}\,(1.62, 1.73) = 1.73$$

Because this estimate is still less than the goal of 2.00, additional improvements must be implemented to raise the average coating thickness, decrease thickness variation, or both.

Relationship of C'_{PU} and C'_{PL} to the Process Average

Consider a characteristic having only an USL. For values of μ greater than T, the C'_{PU} index equals C_{PU}^{T}, which is independent of μ. Changes in the process average have no effect on C'_{PU} as long as μ remains above T, as is depicted on the graph in Figure 5.55. C'_{PU} remains constant at 1.33 when the process average is greater than T, even if it is above the USL. If the process average ever drops below T, the C'_{PU} ratio equals C_{PU}, which is a function of μ. As μ decreases, C_{PU} increases, and therefore, C'_{PU} as well. Anytime the process average is centered in the target zone, it is rewarded with a "bonus" to its potential capability.

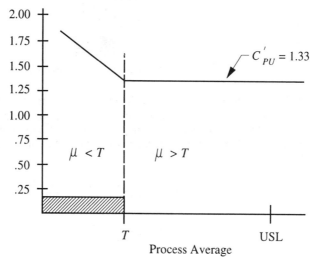

Fig. 5.55 Relationship between C'_{PU} and the process average.

As opposed to C'_{PU}, the graph for C'_{PL} has its upturn for process averages greater than T, as is seen in Figure 5.56.

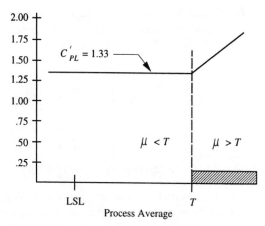

Fig. 5.56 Relationship between C'_{PL} and the process average.

Example 5.23 Solder Pull Test

Wire leads are soldered onto printed circuit boards (Figure 5.57). Each hour, a lead is checked by measuring the force required to pull it from the board (called a "pull test"). Acceptable leads must be able to withstand a minimum pull force of 25 grams.

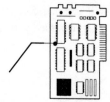

Fig. 5.57 Wire lead soldered to circuit board.

Test results are plotted on an *IX & MR* chart to check the stability of this soldering process. When in control, the average pull test is estimated as 35 grams, with the short-term standard deviation estimated as 5 grams. With a target average pull test of 40 grams, the short-term potential capability of this process is estimated as 1.00. Unfortunately, the short-term potential capability goal for this pull test is 1.33.

$$\hat{C}'_{PL} = \text{Maximum}\left(\frac{T - \text{LSL}}{3\hat{\sigma}_{ST}}, \frac{\hat{\mu} - \text{LSL}}{3\hat{\sigma}_{ST}} \right)$$

$$= \text{Maximum}\left(\frac{40 - 25}{3(5)}, \frac{35 - 25}{3(5)} \right)$$

$$= \text{Maximum}\,(1.00, .67) = 1.00$$

To improve the pull test results, a new type of solder is introduced. After control of this modified process is regained, its average is estimated as 41 grams, with a short-term standard deviation of 4 grams. C'_{PL} is now estimated as 1.33, which just meets the capability goal for this soldering process.

$$\hat{C}'_{PL} = \text{Maximum}\left(\frac{40 - 25}{3(4)}, \frac{41 - 25}{3(4)} \right) = 1.33$$

Note this result is equal to the inverse of the estimate for C'_{RL} obtained in Example 5.13.

$$\hat{C}'_{PL} = \frac{1}{\hat{C}'_{RL}} = \frac{1}{.75} = 1.33$$

Advantages and Disadvantages

These modified indexes provide a measure of capability for unilateral specifications with the same rating scale as C_P and P_P. However, they are more complicated to calculate and difficult to understand. In addition, unilateral characteristics quite often have non-normal output distributions. Section 8.4 presents a procedure for estimating capability measures under these circumstances. Other approaches have been proposed for this situation (see Littig and Lam, or Flaig) but are much more involved and require specialized software programs.

5.11 Short-term Specified Tolerance Index, C_{ST}

Because the C_P index is somewhat challenging to understand (how many 6σ spreads fit into the tolerance), some authors propose measuring, and reporting, just how many σs fit into the tolerance (Wheeler and Chambers). Bissell calls this the "standardized precision" index.

$$C_{ST} = \frac{\text{Tolerance}}{\sigma_{ST}}$$

If one $6\sigma_{ST}$ spread would just fit into the tolerance ($C_P = 1.0$), then C_{ST} would equal 6. If one $12\sigma_{ST}$ spread would fit into the tolerance ($C_P = 2.0$), then C_{ST} would equal 12. The following three measures of potential capability are related by a factor of six.

$$C_{ST} = 6C_P = \frac{6}{C_R}$$

When σ_{ST} is estimated, the following formula for estimating C_{ST} applies.

$$\hat{C}_{ST} = \frac{\text{Tolerance}}{\hat{\sigma}_{ST}}$$

Just as for the C_R and C_P measures, units of measurement cancel during the calculation of the specified tolerance index, leaving a dimensionless, or standardized, measure of process capability. This facilitates comparison of potential capability between different processes, as well as the establishment of a common capability rating system.

$C_{ST} < 6$ reveals the process fails to satisfy the minimum requirement for potential capability,

$C_{ST} = 6$ implies the process just meets the minimum criterion,

$C_{ST} > 6$ means the process exceeds the minimum requirement.

Instead of interpreting this index as the number of σ_{ST}s that could fit within the tolerance, it also expresses the size of one σ_{ST} as a function of the tolerance. For example, if C_{ST} equals 8, one σ_{ST} equals one eighth of the tolerance.

$$C_{ST} = 8$$

$$\frac{\text{Tolerance}}{\sigma_{ST}} = 8$$

$$\frac{1}{8}\text{Tolerance} = \sigma_{ST}$$

When C_{ST} equals 10, one σ_{ST} equals one tenth of the tolerance, and so on. In general:

$$\sigma_{ST} = \frac{1}{C_{ST}}\text{Tolerance}$$

Because they are practically the same, all discussion concerning interpretation, advantages and disadvantages for the C_P index also applies to the C_{ST} index. One advantage favoring C_{ST} over C_P is the ease of understanding the results of a capability study: how many standard deviations will fit into the tolerance. However, this is somewhat offset by the disadvantage of an unfamiliar rating scale.

Goals

The goal for C_{ST} is 6 times the goal for C_P.

$$C_{ST.GOAL} = 6C_{P.GOAL}$$

If $C_{P.GOAL}$ is 1.33, then $C_{ST.GOAL}$ is equal to 8.00, which is 6 times the $C_{P.GOAL}$ ($6 \times 1.33 = 8.0$). To meet this goal, the process must have the potential of fitting eight σ_{ST} within the tolerance ($8\sigma_{ST} = $ tolerance).

Example 5.24 Fan Housing Weight

An $\overline{X}, R$ chart (subgroup size of 5) monitoring the weight of fan housings produced on a plastic injection-molding process is finally brought into a good state of control. The tolerance for part weight is 4.00 grams and the $\overline{R}$ value from the chart is .848 grams. Substituting these values into the formula below yields an estimate of C_{ST}. Just as for C_R and C_P, the units of measurement cancel, leaving a dimensionless measure of process capability.

$$\hat{C}_{ST} = \frac{\text{Tolerance}}{\hat{\sigma}_{ST}} = \frac{\text{Tolerance}}{\overline{R}/d_2} = \frac{4.00\,\text{gms}}{.848\,\text{gms}/2.326} = 10.97$$

The C_{ST} estimate of 10.97 indicates the process has the potential to fit 10.97 σ_{ST}s within the tolerance. As this is much greater than the minimum requirement of 6, this process should have no difficulty producing at least 99.73 percent conforming parts (if properly centered with a normal distribution). This analysis agrees with the one made previously in Example 5.15, when the C_P index was estimated as 1.828.

$$\hat{C}_{ST} = 6\hat{C}_P = 6(1.828) = 10.97$$

5.12 Long-term Specified Tolerance Index, P_{ST}

If interest is in the long-term potential capability of a process, the P_{ST} index should be calculated.

$$P_{ST} = \frac{\text{Tolerance}}{\sigma_{LT}}$$

When σ_{LT} is estimated, this formula is modified as follows:

$$\hat{P}_{ST} = \frac{\text{Tolerance}}{\hat{\sigma}_{LT}}$$

This measure is similar to C_{ST}, only for long-term process variation. The rating scale and goals are also similar.

Example 5.25 Exhaust Valves

Print specifications for the O.D. size of an exhaust valve run on a grinding machine are a LSL of 9.9400 cm and an USL of 10.0600 cm. Several process parameters and measures of potential capability have already been estimated for this critical characteristic.

$$\hat{\sigma}_{ST} = .01552 \text{ cm} \qquad\qquad 6\hat{\sigma}_{ST} \text{ Spread} = .09312 \text{ cm}$$

$$\hat{\sigma}_{LT} = .01610 \text{ cm} \qquad\qquad 6\hat{\sigma}_{LT} \text{ Spread} = .09660 \text{ cm}$$

$$\hat{C}_R = .776 \qquad \hat{P}_R = .805 \qquad \hat{C}_P = 1.289 \qquad \hat{P}_P = 1.242$$

Two more measures for the potential capability of this grinding process are made by estimating both the short-term (C_{ST}) and long-term (P_{ST}) specified tolerance capability measures. The capability goal is for 8σ to equal the tolerance.

$$\hat{C}_{ST} = \frac{\text{USL - LSL}}{\hat{\sigma}_{ST}} = \frac{10.0600 \text{ cm} - 9.9400 \text{ cm}}{.01552 \text{ cm}} = 7.732$$

This estimated C_{ST} value means 7.732 σ_{ST}s of the grinding operation could fit between the USL and the LSL (or equivalently, 7.732 σ_{ST} equals the tolerance). This exceeds the minimum requirements for potential capability ($6\sigma_{ST}$ = tolerance), but does not meet the goal of $8\sigma_{ST}$ equaling the tolerance.

$$\hat{P}_{ST} = \frac{\text{USL - LSL}}{\hat{\sigma}_{LT}} = \frac{10.0600 \text{ cm} - 9.9400 \text{ cm}}{.01610 \text{ cm}} = 7.453$$

Note how this estimate of P_{ST} is 6 times the estimate of P_P.

$$\hat{P}_{ST} = 6\hat{P}_P = 6(1.242) = 7.452$$

The P_{ST} measure indicates 7.453 σ_{LT}s could potentially fit within the tolerance. Although this satisfies the minimum requirement for long-term potential capability, it does not meet the capability goal of 8.0.

Example 5.26 Floppy Disks

A field problem is traced to variation in the diameter of the inner hole of the floppy disk, which has a tolerance of 5.0. If the hole is too big, the floppy disk fails to rotate properly and cannot store or retrieve information. On the other hand, if this hole is too small, it jams the delicate disk drive, causing a major warranty claim. After the process is brought into a good state of control, the following information is gathered during the course of a capability study.

$$\hat{\sigma}_{ST} = .898 \qquad \hat{\sigma}_{LT} = .916 \qquad \hat{C}_R = 1.078 \qquad \hat{P}_R = 1.099 \qquad \hat{C}_P = .928 \qquad \hat{P}_P = .910$$

The C_{ST} and P_{ST} indexes are estimated as follows:

$$\hat{C}_{ST} = \frac{\text{Tolerance}}{\hat{\sigma}_{ST}} = \frac{5.0}{.898} = 5.568 \qquad\qquad \hat{P}_{ST} = \frac{\text{Tolerance}}{\hat{\sigma}_{LT}} = \frac{5.0}{.916} = 5.459$$

Notice the specified tolerance measures are simply 6 times their respective potential capability indexes.

$$\hat{C}_{ST} = 6\hat{C}_P = 6(.928) = 5.568 \qquad\qquad \hat{P}_{ST} = 6\hat{P}_P = 6(.910) = 5.460$$

The estimated C_{ST} value implies only 5.568 short-term standard deviations of the hole-cutting operation could fit between the USL and the LSL, or equivalently, $5.568\sigma_{ST}$ equals the tolerance. This result means the minimum requirement for potential capability is not met, *i.e.*, $6\sigma_{ST}$ will not fit within the tolerance, so this process is judged to be lacking in short-term potential capability.

Looking at long-term potential capability, the estimated P_{ST} measure indicates only 5.459 long-term standard deviations could possibly fit within the tolerance. This measure also signals a lack of potential capability. As expressed by all the other potential capability measures, a serious quality problem is also indicated by P_{ST}. However, there is no direct way to tell if it is because hole diameters are too big, too small, or both. In addition, there is no information about what percentage of holes are below the LSL or what percentage are above the USL. The specified tolerance indexes reveal only the existence of a capability problem, but offer no recommendation about what types of improvements are necessary to increase process capability.

5.13 C^*_{ST} and P^*_{ST} for Cases Where $T \neq M$

The C_{ST} and P_{ST} indexes assume the target for the process average (T) is the middle of the tolerance (M). In those few cases where T is not equal to M, these indexes overestimate the true potential process capability. When a target for the process average is not M, each "half" of the tolerance must be considered separately. The distance from T to the LSL is divided by σ, while the distance from the USL to T is also divided by σ. The *smaller* of these two determines the potential capability for this process, as is expressed by the following equations:

$$C^*_{ST} = 2 \times \text{Minimum}\,(C^T_{STL}, C^T_{STU}) = 2 \times \text{Minimum}\left(\frac{T-\text{LSL}}{\sigma_{ST}}, \frac{\text{USL}-T}{\sigma_{ST}}\right)$$

$$P^*_{ST} = 2 \times \text{Minimum}\,(P^T_{STL}, P^T_{STU}) = 2 \times \text{Minimum}\left(\frac{T-\text{LSL}}{\sigma_{LT}}, \frac{\text{USL}-T}{\sigma_{LT}}\right)$$

The multiplication by 2 is necessary so C_{ST}^* and P_{ST}^* have the same rating scale as do the C_{ST} and P_{ST} indexes.

< 6 means the process fails to meet the minimum prerequisite for potential capability,

$= 6$ discloses the process just barely fulfills the minimum requirement,

> 6 implies the process exceeds the minimum requirement.

C_{ST}^* is known as the modified short-term specified tolerance index, whereas P_{ST}^* is called the modified long-term specified tolerance index. They are related to other measures of potential capability as follows:

$$C_{ST}^* = 6C_P^* = \frac{6}{C_R^*} \qquad P_{ST}^* = 6P_P^* = \frac{6}{P_R^*}$$

When the process standard deviation is estimated from sample data, estimate C_{ST}^* and P_{ST}^* with these formulas:

$$\hat{C}_{ST}^* = 2 \times \text{Minimum}\left(\frac{T-\text{LSL}}{\hat{\sigma}_{ST}}, \frac{\text{USL}-T}{\hat{\sigma}_{ST}}\right)$$

$$\hat{P}_{ST}^* = 2 \times \text{Minimum}\left(\frac{T-\text{LSL}}{\hat{\sigma}_{LT}}, \frac{\text{USL}-T}{\hat{\sigma}_{LT}}\right)$$

In the special case where T equals M, the ensuing relationships hold true:

$$C_{ST}^* = C_{ST} = 6C_P^* = 6C_P = \frac{6}{C_R^*} = \frac{6}{C_R}$$

$$P_{ST}^* = P_{ST} = 6P_P^* = 6P_P = \frac{6}{P_R^*} = \frac{6}{P_R}$$

The next example demonstrates how these measures are applied and interpreted in an industrial setting.

Example 5.27 Gold Plating

The LSL for the thickness of gold being plated on a silicon wafer is 50 angstroms and the USL is 100 angstroms. For cost reasons, a decision is made to target the average thickness at 65 ($T = 65$), as is shown in Figure 58. If $\hat{\sigma}_{LT}$ is 2.5, P_{ST} may be estimated as follows:

$$\hat{P}_{ST} = \frac{\text{USL}-\text{LSL}}{\hat{\sigma}_{LT}} = \frac{100-50}{2.5} = 20$$

If the process is centered at 75 (the middle of the tolerance), it would be possible to fit twenty σ_{LT}s into the tolerance, which certainly meets the minimum capability requirement of 6. With 20 σ_{LT} equal to the tolerance, 10 σ_{LT} would fit from the middle of the tolerance (75 angstroms) to either specification limit. In particular, the number of σ_{LT}s from 75 to the LSL is calculated as is shown here:

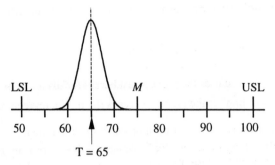

Fig. 5.58 The distribution of gold plating thickness when centered on target.

$$\text{Number of } \hat{\sigma}_{LT} \text{ from } M \text{ to LSL} = \frac{M - \text{LSL}}{\hat{\sigma}_{LT}} = \frac{75 - 50}{2.5} = 10$$

But for this process, care will be taken to keep the process centered at 65 angstroms, which is closer to the LSL of 50 angstroms than M is. Since a $10\sigma_{LT}$ spread will not fit into this reduced distance (65 minus 50 is less than 75 minus 50), the estimate of P_{ST} overstates the true potential capability of this plating process. The formula for P^*_{ST} compensates for T not being at the middle of the tolerance.

$$\hat{P}^*_{ST} = 2 \times \text{Minimum} \left(\frac{T - \text{LSL}}{\hat{\sigma}_{LT}}, \frac{\text{USL} - T}{\hat{\sigma}_{LT}} \right)$$

$$= 2 \times \text{Minimum} \left(\frac{65 - 50}{2.5}, \frac{100 - 65}{2.5} \right)$$

$$= 2 \times \text{Minimum} \left(6.0, 14.0 \right) = 2 \times 6.0 = 12.0$$

When the process is eventually centered at the target value of 65, only $6\sigma_{LT}$ will fit from this preferred average of 65 to the LSL of 50.

$$\text{Number of } \hat{\sigma}_{LT} \text{ from } T \text{ to LSL} = \frac{T - \text{LSL}}{\hat{\sigma}_{LT}} = \frac{65 - 50}{2.5} = 6.0$$

Thus, the P^*_{ST} estimate of 12.0 is a much more realistic characterization of the true potential capability available for this process given the centering objective.

5.14 C'_{STL}, C'_{STU}, P'_{STL}, and P'_{STU} for Unilateral Specifications

As done for C_P and P_P, the C_{ST} and P_{ST} measures of potential capability can be adapted to handle unilateral specifications. Attention is focused on how many standard deviations of the process output fit into the distance from the single specification limit to the target average, or the distance from the specification limit to the actual process average. When only an USL is given, the following formula is appropriate for calculating C'_{STU}. Note how the "2" is needed to make the scale for these measures identical to that for C_{ST}, where a "6" indicates the process meets the minimum requirement for potential capability.

$$C'_{STU} = 2 \times \text{Maximum}(C^T_{STU}, C_{STU}) \quad \text{where } C^T_{STU} = \frac{\text{USL} - T}{\sigma_{ST}} \quad \text{and} \quad C_{STU} = \frac{\text{USL} - \mu}{\sigma_{ST}}$$

In a similar manner, C'_{STL} is defined for characteristics having just a LSL.

$$C'_{STL} = 2 \times \text{Maximum}(C^T_{STL}, C_{STL}) \quad \text{where } C^T_{STL} = \frac{T - \text{LSL}}{\sigma_{ST}} \quad \text{and} \quad C_{STL} = \frac{\mu - \text{LSL}}{\sigma_{ST}}$$

If no target is specified, C^T_{STL} and C^T_{STU} are undefined, making the formulas for C'_{STL} and C'_{STU} as follows:

$$C'_{STL} = 2C_{STL} \qquad C'_{STU} = 2C_{STU}$$

Section 5.6 explains how a target average may be established when none is provided on the engineering drawing. If interest is in long-term potential capability, the following formulas apply. For the case of just a LSL:

$$P'_{STL} = 2 \times \text{Maximum}(P^T_{STL}, P_{STL}) \quad \text{where } P^T_{STL} = \frac{T - \text{LSL}}{\sigma_{LT}} \quad \text{and} \quad P_{STL} = \frac{\mu - \text{LSL}}{\sigma_{LT}}$$

In a similar manner, P'_{STU} is defined as shown here for characteristics with only an USL:

$$P'_{STU} = 2 \times \text{Maximum}(P^T_{STU}, P_{STU}) \quad \text{where } P^T_{STU} = \frac{\text{USL} - T}{\sigma_{LT}} \quad \text{and} \quad P_{STU} = \frac{\text{USL} - \mu}{\sigma_{LT}}$$

These measures are related to the others for unilateral specifications as follows:

$$P'_{STL} = 6P'_{PL} = \frac{6}{P'_{RL}} \qquad P'_{STU} = 6P'_{PU} = \frac{6}{P'_{RU}}$$

Example 5.28 Particle Contamination

From an *IX & MR* chart monitoring the particle contamination of a cleaning bath for printed circuit boards, the following estimates of process parameters are obtained:

$$\hat{\mu} = 127.32 \, ppm \qquad \hat{\sigma}_{ST} = 12.079 \, ppm$$

If the USL for contamination is 165 *ppm*, while *T* is 120 *ppm*, the C'_{STU} index is estimated as 7.45.

$$\hat{C}'_{STU} = 2 \times \text{Maximum}\left(\frac{\text{USL} - T}{\hat{\sigma}_{ST}}, \frac{\text{USL} - \hat{\mu}}{\hat{\sigma}_{ST}}\right)$$

$$= 2 \times \text{Maximum}\left(\frac{165 - 120}{12.079}, \frac{165 - 127.32}{12.079}\right)$$

$$= 2 \times \text{Maximum}(3.725, 3.119) = 7.45$$

This result is equal to 6 times the C'_{PU} index of 1.24 estimated in Example 5.21.

$$\hat{C}'_{STU} = 6\,\hat{C}'_{PU} = 6\,(1.24) = 7.44$$

Although this process has met the minimum requirements for potential capability, which assumes it can eventually be centered at 120 *ppm*, it does not meet the specified capability goal of 8.00. In an attempt to improve this process, a new type of filter is added to the cleaning bath. This causes a reduction in $\hat{\mu}$ to 108 *ppm*, with no change in $\hat{\sigma}_{ST}$. The potential capability of this modified process is calculated as demonstrated here:

$$\hat{C}'_{STU} = 2 \times \text{Maximum}\left(\frac{\text{USL} - T}{\hat{\sigma}_{ST}}, \frac{\text{USL} - \hat{\mu}}{\hat{\sigma}_{ST}} \right)$$

$$= 2 \times \text{Maximum}\left(\frac{165 - 120}{12.079}, \frac{165 - 108}{12.079} \right)$$

$$= 2 \times \text{Maximum}\,(3.725\,,\,4.719) = 9.44$$

Because it is possible to center the process average quite a bit below T, the potential capability is much better and now exceeds the goal of 8.00. But to achieve this level of potential capability, the process average must remain at, or below, 108 *ppm*.

Example 5.29 Solder Pull Test

Wire leads are soldered onto printed circuit boards like the one displayed in Figure 5.59. Each hour, a lead is checked by measuring the force required to pull it from the board, which is referred to as a "pull test." Acceptable leads must be able to withstand a minimum pull force of 25 grams (LSL = 25 gms).

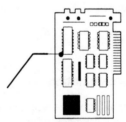

Fig. 5.59 Wire lead soldered to circuit board.

Test results are plotted on an *IX & MR* chart to check the stability of this soldering process. When in control, the average pull test is estimated as 35 grams, with the short-term standard deviation estimated as 5 grams. With a target average pull test of 40 grams, the short-term potential capability of this process is estimated as 6.00. Unfortunately, the short-term potential capability goal for this pull test is 8.00.

$$\hat{C}'_{STL} = 2 \times \text{Maximum}\left(\frac{T - \text{LSL}}{\hat{\sigma}_{ST}}, \frac{\hat{\mu} - \text{LSL}}{\hat{\sigma}_{ST}} \right)$$

$$= 2 \times \text{Maximum}\left(\frac{40 - 25}{5}, \frac{35 - 25}{5} \right)$$

$$= 2 \times \text{Maximum}\,(3.00\,,\,2.00) = 6.00$$

To improve the pull test results, a new type of solder is introduced. After control of this revised process is regained, its average is estimated as 41 grams, with a short-term standard deviation of 4 grams. C'_{STL} is now estimated as 8.00, which just meets the capability goal for this process.

$$\hat{C}'_{STL} = 2 \times \text{Maximum}\left(\frac{T - LSL}{\hat{\sigma}_{ST}}, \frac{\hat{\mu} - LSL}{\hat{\sigma}_{ST}}\right)$$

$$= 2 \times \text{Maximum}\left(\frac{40 - 25}{4}, \frac{41 - 25}{4}\right)$$

$$= 2 \times \text{Maximum}\left(3.75, 4.00\right) = 8.00$$

Advantages and Disadvantages

These ratios provide a measure of capability for unilateral specifications having the same rating scale as C_{ST} and P_{ST}. However, they are more complicated to calculate and difficult to understand. In addition, unilateral characteristics quite often have non-normal output distributions. Chapter 7 explains how to check the normality assumption, while Chapter 8 presents procedures for estimating capability measures when the process output is non-normally distributed.

5.15 Flowchart for Selecting Potential Capability Measures

With all the potential capability measures introduced in this chapter, one can easy become fairly confused when attempting to decide which one should be selected for a specific application, as is indicated in Figure 5.60.

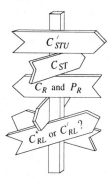

Fig. 5.60 *The multitude of capability measures offers many choices.*

The flowchart displayed in Figure 5.61 acts as a "road map" to assist in picking the correct potential capability measure for any situation. After following the arrows for either short- or long-term variation, unilateral or bilateral specification, and finally, whether or not the target average is the middle of the tolerance, users of this chart are led to a box listing the appropriate measures for that particular combination of choices. Notice that the 6σ spread is the most universal measure, as it appears in every box.

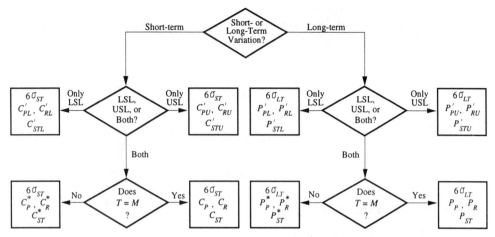

Fig. 5.61 Flowchart for selecting measures of potential capability.

5.16 Monitoring Potential Capability on a Control Chart

Some authors suggest tracking capability measures on an *IX & MR* chart so significant changes in process capability can be detected (DeRoeck, McCoy, Wheeler, 1997). However, most recommend watching the original control chart monitoring the characteristic under study (Hubele and Zimmer). Changes affecting potential capability will appear on this chart long before being evident on the control chart for the capability measure.

As long as the range chart is in control, the estimate of σ_{ST} will not change significantly, and thus, neither will any estimate of short-term potential capability. Any major change in process spread will manifest itself on the range chart as an out-of-control condition. Once detected, appropriate corrective action can be implemented to address this assignable cause of variation. The control chart monitoring a potential capability measure will not reveal this change since capability cannot be estimated for an out-of-control process.

When both the range and $\overline{X}$ charts are in control, the estimate of σ_{LT} will not change substantially from study to study, so neither will the estimate of long-term potential capability. Therefore, a control chart tracking this capability measure will also remain in control.

If either the range or $\overline{X}$ chart indicates the process has gone out of control, the person(s) watching the chart should react with a suitable, and timely, response to restore process stability. Again, the control chart for the capability measure would not change, as capability cannot be calculated when the process is out of control.

The performance capability measures introduced in the next chapter are a function of both the estimated process average and standard deviation. An increase in process spread could be offset by a shift in the process average, such that the capability measure is unaffected. If only the capability measure is monitored, this change could go unnoticed.

In addition, when capability measures are charted, there is a temptation to do away with the original chart monitoring the process. This would be a dangerous course of action because there is no way of knowing if the process is still stable or not. A capability measure by itself has no mechanism to reveal if the process is in or out of control. And once stability is lost, all measures of capability become meaningless and, like all misleading information, possibly even dangerous.

5.17 Summary

Measures of potential process capability express how well a process could perform under certain conditions. In addition to stability, the measures presented in this chapter assume the process output has a normal distribution. Concerned with only process spread and its relationship to the customer requirement, they do not incorporate process centering in their calculation and, thus, must also assume positioning the process average on a specified target average is fairly easy to accomplish.

Most measures require a special formula for unilateral specifications. The one exception is the 6σ process spread measure, which is also the only one to include the units of measurements for the characteristic under study in its result.

Potential capability depends on whether the goal is to center a process at the middle of the tolerance or at some other target, T. When T is not equal to M, modified measures of potential capability must be applied. These special measures are indicated with an * superscript, *e.g.*, C^*_p. An alternative method is calculating the USL_T (or LSL_T) to create a symmetrical tolerance centered around T. This reduces the original tolerance, thereby decreasing the process's potential capability.

Either short-term or long-term potential capability may be computed, based on how the process standard deviation is estimated. However, it is best to regularly report just one category, as comparing the C_p index of one process to the P_p index of another is quite difficult (if not impossible), and is certainly very confusing to those who have only a minimum understanding of capability measures.

Don't attempt to compute all measures of potential capability for every process. Select just those providing the needed information in the simplest manner, with the minimum amount of required effort. Different processes in one plant will no doubt require different measures. Don't be afraid to add and/or delete certain measures from consideration as circumstances surrounding a process change over time.

None of the potential measures reveals the actual percentage of nonconforming parts, which many believe is an important aspect of process capability (Pyzdek, 1992b). Thus, it is possible to obtain a "good" potential capability number, but still have a significantly high percentage of nonconforming parts being produced. Other capability measures are sorely needed to communicate more useful information about the *actual* state of a given process, as well as to recommend what can be done to improve the ability of this process to meet customer requirements or to make it more robust to unexpected changes. These types of "performance" capability measures form the main topic of the next chapter.

However, at least one potential capability measure should always be analyzed before looking at performance capability. If a process fails to meet its potential capability goal, there is no hope of achieving its performance capability goal. In this situation, efforts must concentrate on reducing process variation. When potential capability satisfies its goal, then performance capability should be checked. If this is not acceptable, improvement efforts will now focus on improving process centering.

5.18 Derivation of k_T Factor

The k_T factor expresses the loss in potential capability when a target not equal to M is specified for the process average. Assuming T is within the tolerance:

$$C_P^* = C_P(1 - k_T) \qquad \text{where } k_T = \frac{|T - M|}{\text{Tolerance} / 2}$$

Writing out the full formulas for C_P^* and C_P yields:

$$\frac{1}{3\sigma_{ST}} \text{Minimum} \, (T - \text{LSL}, \text{USL} - T) = \frac{\text{Tolerance}}{6\sigma_{ST}} \left(1 - \frac{|T - M|}{\text{Tolerance} / 2} \right)$$

First consider the case where T is greater than M, which means T minus LSL is greater than USL minus T.

$$\frac{1}{3\sigma_{ST}} \text{Minimum} \, (T - \text{LSL}, \text{USL} - T) = \frac{\text{Tolerance}}{6\sigma_{ST}} \left(1 - \frac{|T - M|}{\text{Tolerance} / 2} \right)$$

$$\frac{1}{3\sigma_{ST}} \text{USL} - T = \frac{\text{Tolerance}}{6\sigma_{ST}} \left(1 - \frac{T - M}{\text{Tolerance} / 2} \right)$$

$$\text{USL} - T = \frac{\text{Tolerance}}{2} \left(1 - \frac{T - M}{\text{Tolerance} / 2} \right)$$

$$\text{USL} - T = \frac{\text{Tolerance}}{2} - T + M$$

Half the tolerance is equal to USL minus M.

$$\text{USL} - T = \text{USL} - M - T + M = \text{USL} - T$$

Thus, the k_T factor works correctly for the case where T is greater than M. When T is less than M, T minus LSL is less than USL minus T. In addition, the quantity T minus M is now negative, which means M minus T is positive.

$$\frac{1}{3\sigma_{ST}} \text{Minimum} \, (T - \text{LSL}, \text{USL} - T) = \frac{\text{Tolerance}}{6\sigma_{ST}} \left(1 - \frac{|T - M|}{\text{Tolerance} / 2} \right)$$

$$\frac{1}{3\sigma_{ST}} (T - \text{LSL}) = \frac{\text{Tolerance}}{6\sigma_{ST}} \left(1 - \frac{M - T}{\text{Tolerance} / 2} \right)$$

$$T - \text{LSL} = \frac{\text{Tolerance}}{2} \left(1 - \frac{M - T}{\text{Tolerance} / 2} \right)$$

$$T - \text{LSL} = \frac{\text{Tolerance}}{2} - M + T$$

Half the tolerance is also equal to M minus LSL.

$$T - \text{LSL} = M - \text{LSL} - M + T = T - \text{LSL}$$

Thus, the formula for k_T is verified for all T values within the tolerance.

5.19 Exercises

5.1 Connecting Force

The force required to join the male and female parts of an electrical connector should be between 1.30 and 1.50 kilograms. In Exercise 3.1 (page 47), the short-term process standard deviation was estimated from the centerline of the range chart.

$$\hat{\sigma}_{ST} = \frac{\overline{R}}{d_2} = \frac{.05077 \text{ kg}}{1.693} = .02999 \text{ kg}$$

Assuming joining force has a normal distribution, estimate the short-term potential capability with the $6\sigma_{ST}$ spread, C_R, C_P, and C_{ST}. How well does the output of this process meet the minimum requirement for short-term potential capability for a bilateral specification? What about meeting the customer's requirement of having a C_P index of at least 1.33?

5.2 Bolt Diameter

A major manufacturer of fasteners for the aerospace industry determines that bolt diameter is extremely important to customers and should be considered a key characteristic. In Exercise 3.2 (page 48), σ_{ST} was estimated as .01300 cm. If the print specifications for bolt diameter are 2.200 cm $\pm$ .050 cm, how well is this process meeting the minimum requirement for short-term potential capability? Estimate the $6\sigma_{ST}$ spread, C_R, C_P, and C_{ST} measures to answer this question (assume bolt diameter has a normal distribution). The following capability goals have been established: .075 cm for the short-term process spread; .75 for C_R; 1.33 for C_P; 8.00 for C_{ST}. How much of a reduction in σ_{ST} is necessary for this process to meet its potential capability goals?

5.3 Relay Voltage

In Exercise 3.3, the average voltage to lift a relay armature was estimated as 29.6 volts, while the short-term standard deviation (σ_{ST}) was estimated to be 2.136 volts. With an USL of 40 volts and a target average of 32 volts, does this process meet the minimum requirement for short-term potential capability for a unilateral specification (assume normality)? Estimate the $6\sigma_{ST}$ spread, C'_{RU}, C'_{PU}, and C'_{STU} measures to answer this question.

Design changes to this relay increase the average voltage needed to lift the armature to 33.4 volts, but decrease the estimate of σ_{ST} to 1.975 volts. What effect do these changes have on the estimates of short-term potential process capability?

5.4 Piston Ring Diameter

In Exercise 3.4, the process average and long-term standard deviation were estimated for the inside diameter of a piston ring as 9.02 cm and .02170 cm, respectively. With a LSL of 8.9250 cm and an USL of 9.0750 cm, estimate the $6\sigma_{LT}$ spread, P_R, P_P, and P_{ST} measures of long-term potential capability. Does this process meet the minimum requirement for potential capability (assume normality)? Does it meet a goal of 1.33 for P_P?

Suppose a target for the process average is established at 9.0250 cm. How does this affect the assessment of potential capability? Use the long-term process spread, P_R^*, P_P^*, and P_{ST}^* measures to answer this question. What is the k_T factor?

5.5 Shaft Length

In Exercise 3.5 (page 49), estimates of the process average and short-term standard deviation were computed for the length of a tapered steel shaft. Assuming a normal distribution, how does this process output compare to the customer specifications of ± 3 mm? Compute the $6\sigma_{ST}$ spread, C_R, C_P, and C_{ST} measures to answer this question. Does this process meet the minimum requirement for short-term potential capability? Does the process meet a capability goal for C_P of 1.67? What percentage of shaft lengths are out of specification?

Assume a target for the process average of minus .1 mm is specified by the engineering department. How does this off-center target change the determination of potential capability (use C_R^*, C_P^*, and C_{ST}^*)? Compute the k_T factor. What percentage of potential capability is lost by requiring this off-center target for the process average?

Calculate LSL_T and USL_T. Combine these two values to estimate C_P^*. How large is the effective tolerance as a percentage of the original tolerance?

Assume the introduction of new tooling decreases the estimate of σ_{ST} by 68 percent. What effect does this substantial reduction in process spread have on the above estimates of short-term potential capability?

5.6 Rib Width

In Exercise 3.6, μ, σ_{ST}, and σ_{LT} were estimated from an *IX & MR* chart for the width of "ribs" used in constructing airplane wings. Given a specification of $1.580 \pm .016$ cm, estimate the $6\sigma_{ST}$ spread, C_R, C_P, and C_{ST}. Then estimate the $6\sigma_{LT}$ spread, P_R, P_P, and P_{ST}. Why the differences?

5.7 Connector Length

Electrical connectors are cut to the specified length on a shearing operation. From the length measurements given in Exercise 4.7 (page 78), the process average is estimated as 67.085, while the long-term standard deviation is estimated as .0883. With a LSL for connector length of 66.700 and an USL of 67.300, estimate the $6\sigma_{LT}$ spread, P_R, P_P, and P_{ST}. How does this conclusion about capability compare to that made in Exercise 4.7?

Assume a target of 67.100 for connector length is desired. How does this additional customer requirement alter this capability analysis (answer in part by estimating P_R^*, P_P^*, and P_{ST}^*)?

5.8 Cam Roller Waviness

Use the estimates of μ and σ_{LT} calculated in Exercise 4.8 (page 79) for the waviness of cam rollers to estimate the $6\sigma_{LT}$ spread, P'_{RU}, P'_{PU}, and P'_{STU}. The USL for waviness is 75 microinches, while T is 30 microinches.

The histogram of these measurements imply the process output for waviness may not be normally distributed. What effect would this have on the above estimates of potential capability?

5.9 Fan Test

Electric fans for cooling personal computers are assembled and tested. In Exercise 4.9 (page 80), oversize armature shafts (Figure 5.62) were identified as a major reason why fans fail on final test. A quality-improvement team is formed to conduct a capability study for the overall length of this part.

Fig. 5.62

After demonstrating stability with an $\overline{X}$, R chart ($n = 5$), the following information is taken from this chart.

$$\overline{\overline{X}} \;=\; 2.554 \text{ cm} \qquad \overline{R} \;=\; .0065 \text{ cm}$$

With a tolerance of 2.550 ± .012 cm for shaft length, estimate C_P and compare it to the goal of 1.33. If potential capability is so good, why should oversized shafts be such a large problem at final test (answer in part by sketching the process output distribution)?

5.10 Compressive Strength

Layers of woven glass-fiber fabric are laminated together with a unique type of heat-resistant resin. After curing, tests are performed to measure the compressive strength of this laminate, which must meet a minimum requirement of 50 ksi, with a target average of 62 ksi. The average and short-term standard deviation for compressive strength are estimated as 59.3 and 2.40, respectively. Armed with this information, estimate the short-term potential capability of this process with C'_{PL}. Does it meet a goal of 1.50?

The introduction of a new reformulated resin increases the average compressive strength to 63.8 ksi, but also increases the estimate of σ_{ST} to 2.76. What effect does this second resin have on the ability of this process to satisfy the customer's specification? Which resin would you recommend for regular production (assume equal costs)?

5.11 Pellet Weight

A process produces a certain size of ferrite pellets. Each pellet must weigh between a minimum of 20.5 grams and a maximum of 24.5 grams. After stability is established, the process average is estimated as 21.9 grams and the long-term standard deviation as .43 grams. With a target average of 23.1 grams, estimate the long-term process spread, P_R^*, P_P^*, and P_{ST}^* measures (assume normality). Determine the k_T factor. What percentage of potential capability is lost by requiring this off-center target for average pellet weight? Estimate P_P and compare it to P_P^* to verify this answer.

If it is possible to move the actual process average to 22.6 grams, with no change in σ_{LT}, how do these potential capability measures respond? If the process average can be shifted to 23.3 grams, how do these measures react?

5.20 References

Anand, K.N., "The Role of Statistics in Determining Product and Part Specifications: A Few Indian Experiences," *Quality Engineering*, Vol. 9, No. 2, 1996-97, pp. 187-193

Arcelus, F. J., and Rahim, M. A., "Simultaneous Economic Selection of a Variables and an Attribute Target Mean," *Journal of Quality Technology*, 26, 2, April 1994, pp. 125-133

Barnett, N.S., "Potentially Capable but Not Actual," *Proceedings from the Third International Conference on Teaching Statistics*, Vol. 2, August 1990, pp. 273-280

Bisgaard, Soren, Hunter, W. G., and Pallesen, "Economic Selection of Quality of Manufactured Product," *Technometrics*, Vol. 26, No. 1, February 1984, pp. 9-18

Bissell, Derek, *Statistical Methods for SPC and TQM*, 1994, Chapman & Hall, New York, NY, p. 240

Chan, L.K., Cheng, S.W., and Spiring, F.A., "The Robustness of the Process Capability Index, Cp, to Departures from Normality," *Proceedings of the 2nd Pacific Area Statistical Conference*, 1986, Tokyo, Japan

Charbonneau, Harvey C., and Webster, Gordon L., *Industrial Quality Control*, 1978, Prentice-Hall, Englewood Cliffs, NJ, p. 112

Clements, Richard B., *Handbook of Statistical Methods in Manufacturing*, 1991, Prentice-Hall, Inc., Englewood Cliffs, NJ, p. 59

DeRoeck, Richard, "Using a Graphical Technique for Determining and Displaying Process Capability," *ASQ Statistics Division Newsletter*, Vol. 16, No. 6, pp. 6-7

Flaig, John J., "A New Approach to Process Capability Analysis," *Quality Engineering*, Vol. 9, No. 2, 1996-97, pp. 205-211

Ford Motor Company, *Ford Body & Assembly Quality Manual*, Directive # SPC-501-B, issued March 1993, p. 1

Gunter, Bert, "The Use and Abuse of Cpk, Part 2," *Quality Progress*, Vol. 22, No. 3, March 1989, pp. 108-109

Hubele, Norma Faris, and Zimmer, Lora, "Average Run Lengths for a Cpm Control Chart," *48th ASQC Annual Quality Congress Proceedings*, Las Vegas, NV, May, 1994, pp. 577-582

Hunter, W.G., and Kartha, C. P., "Determining the Most Profitable Target Value for a Production Process," *Journal of Quality Technology*, Vol. 9, October 1977, pp. 176-181

Jamieson, Archibald, *Introduction to Quality Control*, 1982, Reston Publishing Co., Inc., Reston, VA, p. 72

Juran, Joseph M., and Gryna, Frank M., Jr., *Quality Planning and Analysis*, 2nd edition, 1980, McGraw-Hill, New York, NY, p. 285

Juran, Joseph M., Gryna, Frank M., Jr., and Bingham, R. S., Jr., *Quality Control Handbook*, 3rd edition, 1979, McGraw-Hill, New York, NY

Kane, Victor E., *Defect Prevention: Use of Simple Statistical Tools*, 1989, Marcel Dekker, Inc., New York, NY, p. 267

Kane, Victor E., "Process Capability Indices," *Journal of Quality Technology*, Vol. 18, No. 1, January 1986, pp. 42, 45

Koons, George F., *Indices of Capability: Classical and Six Sigma Tools*, 1992, Addison-Wesley Publishing Co., Reading, MA

Kume, Hitoshi, *Statistical Methods for Quality Improvement*, 1985, AOTS, Chosakai, Ltd., Tokyo, Japan, p. 63

Littig, Steven J., and Lam, C. Teresa, "Case Studies in Process Capability Measurement," *47th ASQC Annual Quality Congress Transactions*, Boston, MA, May 1993, pp. 569-575

Lyth, David M., and Rabiej, Roman J., "Critical Variables in Wood Manufacturing's Process Capability: Species, Structure, and Moisture Content," *Quality Engineering*, Vol. 8, No. 2, 1995-96, pp. 275-281

McCoy, Paul F., "Using Performance Indexes to Monitor Production Processes," *Quality Progress*, Vol. 24, No. 2, February 1991, pp. 49-55

Munro, Roderick A., "Using Capability Indexes in Your Shop," *Quality Digest*, Vol. 12, No. 5, May 1992, pp. 58-59

Nelson, Lloyd S., "Best Target Value for a Production Process," *Journal of Quality Technology*, Vol. 10, No. 2, April 1978, pp. 88-89

Osuga, Yutaka, "Process Capability Studies in Cutting Processes," *Reports of Statistical Applications Research*, Japanese Union of Scientists and Engineers, 11, 1, 1964, pp. 23-25

Pearn, W.L., Kotz, Samuel and Johnson, Norman L., "Distributional and Inferential Properties of Process Capability Indices," *Journal of Quality Technology*, 24, 4, Oct. 1992, p. 219

Pyzdek, Thomas, "Process Capability Analysis Using Personal Computers," *Quality Engineering*, Vol. 4, No. 3, 1992b, p. 438

Pyzdek, Thomas, *Pyzdek's Guide to SPC, Volume 2*, 1992a, ASQC Quality Press, Milwaukee, WI, p. 59

Rado, Leonard G., "Enhance Product Development by Using Capability Indexes," *Quality Progress*, Vol. 22, No. 4, April 1989, pp. 38-41

Sullivan, Larry P., "Japanese Quality Thinking at Ford," *Quality*, Vol. 25, No. 4, April 1986, pp. 32-34

Tseng, Sheng-Tsaing, and Wu, Tong-You, "Selecting the Best Manufacturing Process," *Journal of Quality Technology*, Vol. 23, No. 1, January 1991, pp. 53-62

Vannman, K., "Families of Capability Indices for One-Sided Specification Limits," *Statistics*, Vol. 31, 1998, pp. 43-66

Wheeler, Donald J., *The Process Evaluation Handbook*, 2000, SPC Press, Inc., Knoxville, TN, pp. 35-68

Wheeler, Donald J., "Capability Ratios Vary," *Quality Digest*, June 1997, pp. 49-51

Wheeler, Donald J., and Chambers, David S., *Understanding Statistical Process Control*, 2nd edition, 1992, SPC Press, Inc., Knoxville, TN

Chapter 6

Measuring Performance Capability

If better is possible, good is not good enough.

The last chapter dealt with measures of potential process capability, where attention is focused on the process spread and its relationship to print specifications. These measures provide a good indication of a process's ability to produce conforming product *if* centered at the target average. However, this assumes the process average is relatively easy to adjust, which is not always true. In those situations where the process average cannot be located on target, attention must shift to the relationship of the process's spread *and* centering to specification limits. Measures of capability incorporating both these process parameters are labeled "performance" measures (Meagher, Paulk *et al*; Pitt), as this is how the process is currently performing, without any assumptions about possible centering.

Just as there are measures for both short- and long-term potential capability, there are measures for short- and long-term performance capability as well. These are listed in the right-hand column of Table 6.1, with the first approach introduced in this chapter being the Z_{MIN} method.

6.1 The Z_{MIN} Method

Measures of potential capability ignore where the actual process average is located and simply assume it can be centered at the target average, which is usually the middle of the tolerance. The Z_{MIN} method eliminates this assumption by including both process spread and centering in its measure of performance capability. This is accomplished by calculating three "Z" values: Z_{LSL}, Z_{USL}, and Z_{MIN}, where Z represents the standardized normal deviate. Z_{LSL} and Z_{USL} are defined as:

$$Z_{LSL} = \frac{\mu - LSL}{\sigma} \qquad\qquad Z_{USL} = \frac{USL - \mu}{\sigma}$$

Z_{LSL} is the distance from the process average to the LSL, measured in terms of standard deviations. In Figure 6.1 on page 169, the distance between μ and the LSL is about 6σ. In the formula for Z_{LSL}, this distance is divided by σ, leaving just the 6, a dimensionless index which may easily be compared to Z_{LSL} values of other processes.

Table 6.1 Categorizing the various measures of process capability.

Type of Process Variation	When Concern Is With:	
	Only Process Spread (Potential Capability)	Process Spread and Centering (Performance Capability)
Short Term	$6\sigma_{ST}$ Spread C_R, C^*_R, C'_{RL}, C'_{RU} C_P, C^*_P, C'_{PL}, C'_{PU} C_{ST}, C^*_{ST}, C'_{STL}, C'_{STU}	$Z_{LSL,ST}$, $Z_{USL,ST}$ $Z_{MIN,ST}$, $Z_{MAX,ST}$ $Z^*_{LSL,ST}$, $Z^*_{USL,ST}$, $Z^*_{MIN,ST}$ C_{PL}, C_{PU}, C^*_{PL}, C^*_{PU} C_{PK}, C^*_{PK}, C_{PM}, C^*_{PM} C_{PG}, C^*_{PG}, C_{PMK}, C^*_{PMK} $ppm_{LSL,ST}$, $ppm_{USL,ST}$ $ppm_{MAX,ST}$, $ppm_{TOTAL,ST}$
Long Term	$6\sigma_{LT}$ Spread P_R, P^*_R, P'_{RL}, P'_{RU} P_P, P^*_P, P'_{PL}, P'_{PU} P_{ST}, P^*_{ST}, P'_{STL}, P'_{STU}	$Z_{LSL,LT}$, $Z_{USL,LT}$ $Z_{MIN,LT}$, $Z_{MAX,LT}$ $Z^*_{LSL,LT}$, $Z^*_{USL,LT}$, $Z^*_{MIN,LT}$ P_{PL}, P_{PU}, P^*_{PL}, P^*_{PU} P_{PK}, P^*_{PK}, P_{PM}, P^*_{PM} P_{PG}, P^*_{PG}, P_{PMK}, P^*_{PMK} $ppm_{LSL,LT}$, $ppm_{USL,LT}$ $ppm_{MAX,LT}$, $ppm_{TOTAL,LT}$

$$Z_{LSL} = \frac{\mu - LSL}{\sigma} = \frac{6\sigma}{\sigma} = 6$$

Z_{LSL} represents how many standard deviations fit into the distance from the process average to the LSL. In order to meet the minimum requirement of no more than .135 percent nonconforming parts below the LSL, Z_{LSL} must be at least 3 (assuming normality). For the process output displayed in Figure 6.1, this prerequisite is easily met as Z_{LSL} is 6. Note how this measure is a function of the process average as well as the standard deviation. When μ is shifted closer to the LSL, Z_{LSL} decreases. As μ moves further above the LSL, Z_{LSL} increases. If the average is stationary and σ increases, then Z_{LSL} decreases. Conversely, when σ decreases, Z_{LSL} increases.

In a similar fashion, Z_{USL} measures the distance from the process average to the USL in terms of standard deviations. In Figure 6.1, that distance is 3σ. When this distance is entered into the formula for Z_{USL}, the σs cancel out, leaving only the 3.

$$Z_{USL} = \frac{USL - \mu}{\sigma} = \frac{3\sigma}{\sigma} = 3$$

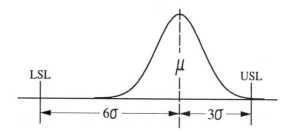

Fig. 6.1   Relationship of average to specification limits.

This result means 3 standard deviations will fit between the process average and the USL, or conversely, that the process average is located a distance of 3σ below the USL. Either way, this process just meets the minimum capability requirement for the USL, as there is .135 percent of its output above this specification.

Of these two Z values, the one causing more concern is the smaller, in this case the Z_{USL} of only 3. Any upward movement of the process average causes a greater increase in nonconforming parts above the USL than a similar-sized downward movement will cause below the LSL, as is demonstrated in Figure 6.2.

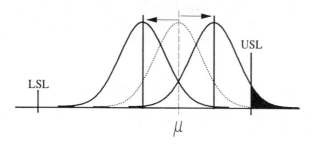

Fig. 6.2   An equal shift in centering affects the USL more than the LSL.

Likewise, any increase in the process standard deviation causes a greater increase in nonconforming parts above the USL than below the LSL, as is seen in Figure 6.3.

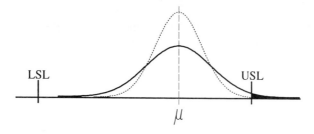

Fig. 6.3   An increase in spread impacts the USL more than the LSL.

To focus attention on the specification of greater concern, Z_{MIN} is defined as the smaller, or minimum, of Z_{LSL} and Z_{USL}.

$$Z_{MIN} = \text{Minimum}\,(Z_{LSL}, Z_{USL}) = \text{Minimum}\left(\frac{\mu - \text{LSL}}{\sigma}, \frac{\text{USL} - \mu}{\sigma}\right)$$

Wheeler and Chambers (p. 124) refer to Z_{MIN} as the distance from the process average to the nearest specification, expressed in "sigma" units. In order to meet the minimum requirement of performance capability, *i.e.*, no more than .135 percent of parts outside either specification limit, Z_{MIN} must be equal to, or greater than, 3. Satisfying this requirement forces both Z_{LSL} and Z_{USL} to be at least 3.

Estimating *Z* Values

Since the actual μ and σ are seldom known, estimates are often derived from data collected on a control chart monitoring the process. Formulas for estimating the three Z values based on estimates of the process parameters are given here:

$$\hat{Z}_{MIN} = \text{Minimum}\,(\hat{Z}_{LSL}, \hat{Z}_{USL}) = \text{Minimum}\left(\frac{\hat{\mu} - \text{LSL}}{\hat{\sigma}}, \frac{\text{USL} - \hat{\mu}}{\hat{\sigma}}\right)$$

The centerline of the $\overline{X}$ chart, $\overline{\overline{X}}$ (or $\overline{X}$ from an *IX* chart), always becomes an estimate of the process average. However, as seen in Chapter 3, σ may be estimated in several ways depending if interest is in short- or long-term performance capability.

Short- or Long-term *Z* Values

The standard deviation in the Z value formulas may represent either σ_{ST} or σ_{LT}, depending on the intent of the capability study. If σ_{ST} is chosen, then Z_{MIN} measures short-term performance capability, and the Z values are labeled as follows, with the subscript "*ST*" representing short term:

$$\hat{Z}_{MIN,ST} = \text{Minimum}\,(\hat{Z}_{LSL,ST}, \hat{Z}_{USL,ST}) = \text{Minimum}\left(\frac{\hat{\mu} - \text{LSL}}{\hat{\sigma}_{ST}}, \frac{\text{USL} - \hat{\mu}}{\hat{\sigma}_{ST}}\right)$$

$$\text{where}\ \ \hat{\sigma}_{ST} = \frac{\overline{R}}{d_2},\ \ \frac{\overline{S}}{c_4},\ \ \text{or}\ \ \frac{\overline{MR}}{d_2}$$

When an estimate of the long-term performance capability is desired, replace σ_{ST} with σ_{LT} in the above Z value formulas to estimate these following Z values, where the subscript "*LT*" denotes long term.

$$\hat{Z}_{MIN,LT} = \text{Minimum}\,(\hat{Z}_{LSL,LT}, \hat{Z}_{USL,LT}) = \text{Minimum}\left(\frac{\hat{\mu} - \text{LSL}}{\hat{\sigma}_{LT}}, \frac{\text{USL} - \hat{\mu}}{\hat{\sigma}_{LT}}\right)$$

$$\text{where}\ \ \hat{\sigma}_{LT} = \frac{1}{c_4}\sqrt{\frac{\sum\limits_{i=1}^{kn}(X_i - \hat{\mu})^2}{kn - 1}}$$

An estimate of the process average is calculated in the same manner whether short- or long-term performance capability is being studied.

$$\hat{\mu} = \overline{\overline{X}} = \frac{\sum\limits_{i=1}^{k} \overline{X}_i}{k} = \frac{\sum\limits_{i=1}^{kn} X_i}{kn}$$

There is no such thing as a short- or long-term process average. Identical answers, within round off, are obtained whether individuals or subgroup averages are chosen to estimate μ.

Example 6.1 Rough Turning

The inner diameter (I.D.) of a transmission gear for heavy-duty construction equipment made in Australia undergoes a rough turning operation on a lathe. As I.D. size is important for subsequent machining operations, this characteristic is charted on an $\overline{X}$, R chart, where n is 5. Once the process is in a good state of control, as evidenced by the chart in Figure 6.4, it is desired to evaluate both potential and performance capability.

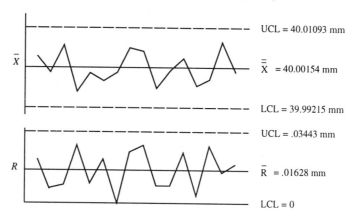

Fig. 6.4 The control chart for gear I.D. size.

With an $\overline{X}$, R chart, the estimate of σ_{ST} is based on $\overline{R}$.

$$\hat{\sigma}_{ST} = \frac{\overline{R}}{d_2} = \frac{.01628 \text{ mm}}{2.326} = .006999 \text{ mm}$$

The specification for I.D. size is 40.00 mm ± .03 mm, making the tolerance .06 mm. Assuming a normal distribution, the C_P and C_{ST} measures of potential capability are estimated from the relationship between $\hat{\sigma}_{ST}$ and the tolerance.

$$\hat{C}_P = \frac{\text{Tolerance}}{6\hat{\sigma}_{ST}} = \frac{.06 \text{ mm}}{6(.006999 \text{ mm})} = 1.429$$

$$\hat{C}_{ST} = \frac{\text{Tolerance}}{\hat{\sigma}_{ST}} = \frac{.06 \text{ mm}}{.006999 \text{ mm}} = 8.572$$

These indexes indicate this lathe has very good potential capability. In fact, the C_{ST} index implies a spread of over 8.5 σ_{ST}s could fit within the tolerance. However, this capability is attained only when the process average is centered at M, the middle of the tolerance. But in most industrial situations, the process is rarely located exactly at M. To evaluate how this turning operation is actually satisfying the customer's requirement for I.D. size, a measure of performance capability must be calculated and analyzed. As a first step, an estimate of the process average is obtained from the centerline of the $\overline{X}$ chart displayed in Figure 6.4.

$$\hat{\mu} = \overline{\overline{X}} = 40.00154 \, \text{mm}$$

The ability of this lathe to manufacture gears between the process average and the LSL is evaluated with the $Z_{LSL,ST}$ measure.

$$\hat{Z}_{LSL,ST} = \frac{\hat{\mu} - LSL}{\hat{\sigma}_{ST}} = \frac{40.00154 \, \text{mm} - 39.97 \, \text{mm}}{.006999 \, \text{mm}} = 4.506$$

This measure estimates the process average is about 4.5 short-term standard deviations above the LSL. As a minimum distance of 3 standard deviations is needed to assure no more than .135 percent of the holes are undersize, this turning process has excellent performance capability with respect to the LSL, as is exhibited in Figure 6.5.

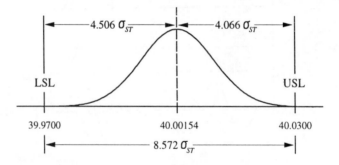

Fig. 6.5 The process output for gear I.D. size.

Despite this encouraging news about the LSL, there could be a problem producing hole sizes which conform to the USL. The extent of this difficulty is expressed by $\hat{Z}_{USL,ST}$.

$$\hat{Z}_{USL,ST} = \frac{USL - \hat{\mu}}{\hat{\sigma}_{ST}} = \frac{40.03 \, \text{mm} - 40.00154 \, \text{mm}}{.006999 \, \text{mm}} = 4.066$$

This measure estimates the process average is 4.066 short-term standard deviations below the USL, which also indicates respectable performance capability regarding this second customer requirement for gear I.D. size. As both specifications must be met, the smaller of these two measures limits the performance capability for this turning process.

$$\hat{Z}_{MIN,ST} = \text{Minimum} \, (\hat{Z}_{LSL,ST} , \hat{Z}_{USL,ST}) = \text{Minimum} \, (4.506 , 4.066) = 4.066$$

Notice how the sum of $Z_{LSL,ST}$ (the number of standard deviations from the LSL to the average) and $Z_{USL,ST}$ (the number of standard deviations from the average to the USL) equals C_{ST}, the number of standard deviations that could fit from the LSL to the USL.

$$\hat{C}_{ST} = \hat{Z}_{LSL,ST} + \hat{Z}_{USL,ST} = 4.506 + 4.066 = 8.572$$

Rating Scale for Z_{MIN}

Because the minimum level of performance capability requires at least three standard deviations fit between the process average and the nearest specification, the rating scale for Z_{MIN} is defined as follows:

$Z_{MIN} < 3.0$ indicates the process fails to meet the minimum requirement
for performance capability,

$Z_{MIN} = 3.0$ implies that the process just meets the minimum criterion,

$Z_{MIN} > 3.0$ means that the process exceeds the minimum prerequisite.

As with potential capability measures, just fulfilling the minimum capability requirement usually isn't good enough for today's competitive marketplace, thus most customers typically specify capability goals much higher than this.

Goals

A common goal is to have at least 4 standard deviations from the process average to the nearest print specification. This means Z_{MIN} must be equal to, or greater than, 4.0. If this goal is achieved, there would be no more than 32 *ppm* (parts per million) outside either specification, and therefore, no more than a total of 63 *ppm* outside both limits. Other often-used goals for Z_{MIN} are 5.0 (less than .6 *ppm* total) and 6.0 (less than .002 *ppm* total).

To put these various goals into perspective, assume a company is manufacturing 1,000 parts per day, 5 days per week, 50 weeks per year. At this rate of production, about 250,000 parts are made per year. In four years, one million parts are produced. For this situation, achieving a Z_{MIN} goal of 3 means one nonconforming part will be produced every 3 hours. Reaching a goal of 4 stretches this interval to one every 3 weeks, whereas attaining a goal of 5 pushes the interval to 6.7 *years*. Table 6.2 shows the relationship between these and several other goals for Z_{MIN}, along with the corresponding *ppm*, and the average production time between nonconforming parts. These calculations assume the process is centered at the middle of the tolerance and has a normal distribution.

Table 6.2   Relationship between Z_{MIN} goals, *ppm*, and time between nonconforming units.

Goal for Z_{MIN}	Total *ppm*	Time Between Nonconf. Units
3	2,700	3 hours
4	63	3 weeks
4.5	7	7 months
4.9	1	4 years
5	.6	6.7 years
6	.002	2,000 years

There are occasions when a Z_{MIN} goal less than 3 is chosen. In the chemical processing industry, yields for some processes are often in the range of 90 to 95 percent. A 95 percent yield for a characteristic with a bilateral specification equates to a Z_{MIN} goal of about 2.0. Recognizing that goals must be achievable, a goal of only 2.5 might be very reasonable for this process.

Example 6.2 Valve O.D. Size

In Example 3.7, estimates of the process average (10.0041 cm) and long-term process standard deviation (.01610 cm) were made for the O.D. size of an exhaust valve run on a grinding machine (Figure 6.6).

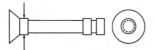

Fig. 6.6 An automotive exhaust valve.

Given the print specifications for O.D. size are a LSL of 9.9400 cm and an USL of 10.0600 cm, the long-term performance capability of the grinding operation for this characteristic is estimated by calculating $Z_{LSL,LT}$ and $Z_{USL,LT}$. Note how the units (in this case, cm) cancel out, leaving these Z values as dimensionless numbers.

$$\hat{Z}_{LSL,LT} = \frac{\hat{\mu} - LSL}{\hat{\sigma}_{LT}} = \frac{10.0041\,\text{cm} - 9.9400\,\text{cm}}{.01610\,\text{cm}} = 3.98$$

A $Z_{LSL,LT}$ of 3.98 means that there are almost 4 standard deviations between the process average and the LSL. If the process output has a normal distribution, there is substantially less than .135 percent of the valves with an O.D. size below 9.9400 cm. A later section explains how to convert Z_{LSL} into an estimate of the percentage of parts below the LSL.

$$\hat{Z}_{USL,LT} = \frac{USL - \hat{\mu}}{\hat{\sigma}_{LT}} = \frac{10.0600\,\text{cm} - 10.0041\,\text{cm}}{.01610\,\text{cm}} = 3.47$$

A $Z_{USL,LT}$ value of 3.47 implies that the process average is about 3.5 standard deviations below the USL (see Figure 6.7). This means fewer than .135 percent of the valves have their O.D. size above 10.0600 cm, again assuming the process output has a normal distribution. Chapter 8 explains how these percentages beyond the specification limits are determined when the process output is not normally distributed.

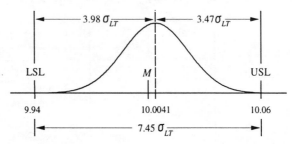

Fig. 6.7 The number of σ_{LT}s between the average and the specifications.

Adding $Z_{LSL,LT}$ and $Z_{USL,LT}$ together estimates how many long-term standard deviations fit into the tolerance.

$$\hat{Z}_{LSL,LT} + \hat{Z}_{USL,LT} = 3.98 + 3.47 = 7.45$$

This sum will always equal the P_{ST} index of potential capability.

$$\hat{P}_{ST} = \frac{\text{Tolerance}}{\hat{\sigma}_{LT}} = \frac{10.06 - 9.94}{.01610} = 7.45$$

P_{ST} is the total number of standard deviations that will fit from LSL to USL, while the Z values divide this total into the number of standard deviations from the LSL to the process average, and then the number from the process average to the USL.

Once $Z_{LSL,LT}$ and $Z_{USL,LT}$ are estimated, $Z_{MIN,LT}$ can be determined and compared to the goal so a decision about performance capability can be made.

$$\hat{Z}_{MIN,LT} = \text{Minimum}\,(\hat{Z}_{LSL,LT},\hat{Z}_{USL,LT}) = \text{Minimum}\,(3.98,3.47) = 3.47$$

A $Z_{MIN,LT}$ of 3.47 indicates that less than .135 percent nonconforming parts are predicted to be outside either specification limit, thus implying this process meets the minimum requirement of performance capability. However, the customer-imposed goal for this characteristic is a $Z_{MIN,LT}$ of 4.0 or more, meaning that at least $8\sigma_{LT}$s must fit within the tolerance. As this process is currently able to have at most $7.45\sigma_{LT}$s fit into the tolerance, work must be done to reduce process spread. Until then, part containment and 100 percent inspection should continue so this customer is assured of receiving the agreed upon quality level.

Because $Z_{USL,LT}$ is less than $Z_{LSL,LT}$, moving μ slightly lower will help increase $Z_{MIN,LT}$ somewhat, but not enough to reach the goal of 4.0. The best that can be done is to center μ at M, the middle of the tolerance, which is where $Z_{MIN,LT}$ is maximized. Just how large will $Z_{MIN,LT}$ become when μ is shifted to M, which in this example is 10.00 cm? There are several ways to find this answer, all of which assume the process output is normally distributed and σ_{LT} does not change when μ does. In the first method, the $Z_{LSL,LT}$ and $Z_{USL,LT}$ values are added together, then divided by 2 to obtain the maximum possible $Z_{MIN,LT}$ for this situation.

$$\text{Maximum } \hat{Z}_{MIN.\,LT} = \frac{\hat{Z}_{LSL,LT} + \hat{Z}_{USL,LT}}{2} = \frac{3.98 + 3.47}{2} = 3.725$$

The second approach is to divide the estimate of P_{ST} by 2.

$$\text{Maximum } \hat{Z}_{MIN.\,LT} = \frac{\hat{P}_{ST}}{2} = \frac{7.45}{2} = 3.725$$

A third alternative is multiplying P_P, the long-term potential capability index, by 3. For this case study, P_P is estimated as 1.242.

$$\hat{P}_P = \frac{\text{Tolerance}}{6\hat{\sigma}_{LT}} = \frac{10.06 - 9.94}{6(.01610)} = 1.242$$

Multiplying this result by 3 yields:

$$\text{Maximum } \hat{Z}_{MIN.\,LT} = 3\hat{P}_P = 3(1.242) = 3.726$$

A fourth approach is recalculating $Z_{MIN,LT}$ with the process average equal to 10.00 cm. When the process is centered at the middle of the tolerance, the distance from μ to the LSL is identical to that from μ to the USL. Thus, only one of the two Z values needs to be calculated, since $Z_{LSL,LT}$ equals $Z_{USL,LT}$.

$$\text{Maximum } \hat{Z}_{MIN,LT} = \text{Minimum} \left(\frac{\hat{\mu} - \text{LSL}}{\hat{\sigma}_{LT}}, \frac{\text{USL} - \hat{\mu}}{\hat{\sigma}_{LT}} \right)$$

$$= \text{Minimum} \left(\frac{10.00\,\text{cm} - 9.94\,\text{cm}}{.01610\,\text{cm}}, \frac{10.06\,\text{cm} - 10.00\,\text{cm}}{.01610\,\text{cm}} \right)$$

$$= \text{Minimum } (\,3.726\,,\,3.726\,) = 3.726$$

All four methods yield about the same result, 3.725. Thus, lowering the process average from 10.0041 so it is centered at the middle of the tolerance increases long-term capability from a $Z_{MIN,LT}$ of 3.470 to 3.725 (Figure 6.8). Unfortunately, as pointed out previously, this will not be enough of an increase to meet the customer's goal of 4.0.

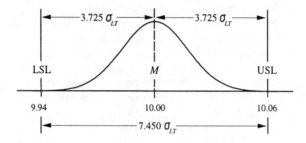

Fig. 6.8   Centering the process at 10.00 increases $Z_{MIN,LT}$ to 3.725.

So in addition to proper centering, the process standard deviation must also be reduced in order to satisfy the customer. But by how much? When the capability goal of $Z_{MIN,LT}$ being at least 4.0 is achieved, $Z_{LSL,LT}$ and $Z_{USL,LT}$ must both be at least 4.0. Thus, there must be at least $8\sigma_{LT}$ from the LSL to the USL, as is illustrated in Figure 6.9.

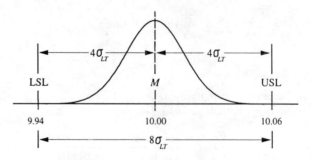

Fig. 6.9   Process centered at 10.00 and σ_{LT} reduced so $Z_{MIN,LT}$ is 4.0.

This condition allows a determination of how large σ_{LT} can be (call it $\sigma_{LT,GOAL}$) and still meet the capability goal of $Z_{MIN,LT}$ being greater than, or equal to, 4.0.

$$8\sigma_{LT,GOAL} = \text{USL} - \text{LSL}$$

$$8\sigma_{LT,GOAL} = 10.0600 \text{ cm} - 9.9400 \text{ cm}$$

$$\sigma_{LT,GOAL} = \frac{.1200 \text{ cm}}{8}$$

$$\sigma_{LT,GOAL} = .01500 \text{ cm}$$

This answer is easily checked by calculating $Z_{MIN,LT}$ for a process average of 10.00 cm and a long-term standard deviation of .01500 cm.

$$\hat{Z}_{MIN,LT} = \text{Minimum} \left(\frac{\hat{\mu} - \text{LSL}}{\hat{\sigma}_{LT}}, \frac{\text{USL} - \hat{\mu}}{\hat{\sigma}_{LT}} \right)$$

$$= \text{Minimum} \left(\frac{10.00 \text{ cm} - 9.94 \text{ cm}}{.01500 \text{ cm}}, \frac{10.06 \text{ cm} - 10.00 \text{ cm}}{.01500 \text{ cm}} \right)$$

$$= \text{Minimum} \ (4.0, 4.0) = 4.0$$

Because the estimated σ_{LT} for the current process is .01610 cm, a 6.8 percent reduction in process spread is necessary to reach the required goal of .01500 cm ($.01610 \times .932 = .01500$).

Example 6.3 Weight of Fan Housings

An $\overline{X}, R$ chart (subgroup size of 5) monitoring the weight of fan housings produced on a plastic injection molding process is finally brought into a good state of control. A histogram of the individual weight measurements is displayed in Figure 6.10.

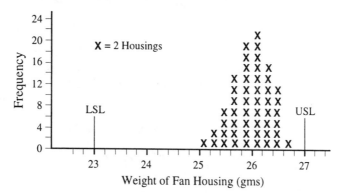

Fig. 6.10 Weight measurements for plastic fan housings.

$\overline{R}$ from the range chart is .848 grams and $\overline{\overline{X}}$ from the $\overline{X}$ chart is 26.11 grams. From these, estimates of the process parameters can be made.

$$\hat{\mu} = \overline{X} = 26.11 \text{ gm} \qquad\qquad \hat{\sigma}_{ST} = \frac{\overline{R}}{d_2} = \frac{.848 \text{ gm}}{2.326} = .365 \text{ gm}$$

As the tolerance for part weight is 25 grams $\pm$ 2 grams, the LSL is 23 grams, while the USL is 27 grams. Substituting these values into the formulas below yields estimates of the various short-term Z measures of performance capability.

$$\hat{Z}_{LSL,ST} = \frac{\hat{\mu} - \text{LSL}}{\hat{\sigma}_{ST}} = \frac{26.11 \text{ gm} - 23 \text{ gm}}{.365 \text{ gm}} = 8.52$$

Having 8.52 standard deviations between the process average and the LSL indicates the process has no problem meeting this specification. But this is a bilateral specification, so the distance from the process average to the USL must also be considered.

$$\hat{Z}_{USL,ST} = \frac{\text{USL} - \hat{\mu}}{\hat{\sigma}_{ST}} = \frac{27 \text{ gm} - 26.11 \text{ gm}}{.365 \text{ gm}} = 2.44$$

With only 2.44 standard deviations from the average to the USL, more than .135 percent of the housings will be overweight. Thus, this molding process does not meet the minimum prerequisite for performance capability, as is reflected by its $Z_{MIN,ST}$ value being less than 3.

$$\hat{Z}_{MIN,ST} = \text{Minimum} \, (\hat{Z}_{LSL,ST}, \hat{Z}_{USL,ST}) = \text{Minimum} \, (8.52, 2.44) = 2.44$$

This lack of performance capability occurs even though the potential capability of this process is quite high, with C_{ST} being almost 11.

$$\hat{C}_{ST} = \frac{\text{Tolerance}}{\hat{\sigma}_{ST}} = \frac{4.00 \text{ gm}}{.365 \text{ gm}} = 10.96$$

A C_{ST} of 10.96 means $Z_{MIN,ST}$ could be as high as 5.48 if μ can be centered at M.

$$\text{Maximum } \hat{Z}_{MIN.ST} = \frac{\hat{C}_{ST}}{2} = \frac{10.96}{2} = 5.48$$

Despite exceeding the minimum requirement for *potential* capability, this process fails to meet the minimum requirement for *performance* capability. Whenever potential capability is high, and performance capability is low, the first improvement efforts should focus on shifting the process average to the middle of the tolerance, in this case, 25 grams. Once located here, performance capability will be at its maximum.

$$\begin{aligned}
\hat{Z}_{MIN,ST} &= \text{Minimum} \, (\hat{Z}_{LSL,ST}, \hat{Z}_{USL,ST}) \\[2mm]
&= \text{Minimum} \left(\frac{\hat{\mu} - \text{LSL}}{\hat{\sigma}_{ST}}, \frac{\text{USL} - \hat{\mu}}{\hat{\sigma}_{ST}} \right) \\[2mm]
&= \text{Minimum} \left(\frac{25 \text{ gm} - 23 \text{ gm}}{.365 \text{ gm}}, \frac{27 \text{ gm} - 25 \text{ gm}}{.365 \text{ gm}} \right) \\[2mm]
&= \text{Minimum} \, (5.48, 5.48) = 5.48
\end{aligned}$$

If the average cannot be adjusted, or is too costly to adjust, attention must then focus on reducing process spread.

Example 6.4 Connector Length

Suppose an $\overline{X}, R$ chart monitoring the length of an electrical connector coming off a shearing operation is finally brought into a good state of control. From the chart data, μ is estimated as 3 and σ_{ST} as 1.

$$\hat{\mu} = \overline{\overline{X}} = +3 \qquad \hat{\sigma}_{ST} = \hat{\sigma}_{\overline{R}} = \frac{\overline{R}}{d_2} = 1$$

Length readings are given as deviations from a nominal length of 170. This method of coding makes the LSL equal to minus 6, and the USL equal to plus 6. The process output is displayed with these coded specifications in Figure 6.11

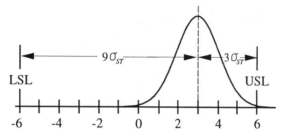

Fig. 6.11 A process centered close to the USL.

The $Z_{MIN.ST}$ measure of short-term performance capability is derived from these process parameter estimates as follows:

$$\hat{Z}_{MIN.ST} = \text{Minimum} \, (\hat{Z}_{LSL.ST}, \hat{Z}_{USL.ST})$$

$$= \text{Minimum} \left(\frac{\hat{\mu} - LSL}{\hat{\sigma}_{ST}}, \frac{USL - \hat{\mu}}{\hat{\sigma}_{ST}} \right)$$

$$= \text{Minimum} \left(\frac{3 - -6}{1}, \frac{6 - 3}{1} \right)$$

$$= \text{Minimum} \, (9, 3) = 3$$

An estimated $Z_{MIN.ST}$ of 3 means the process just barely meets the minimum requirement for performance capability. Because there is no safety margin, a constant vigil must be maintained over this process to detect any process changes as quickly as possible. Any upward drift in the process average, or increase in process spread, will result in more than .135 percent of the parts being above the USL, and the process would be considered not capable for this characteristic. In addition, because $Z_{MIN.ST}$ indicates marginal capability, the long-term performance of this process, as measured by $Z_{MIN.LT}$, will most likely not satisfy the minimum capability criterion since σ_{LT} is typically greater than σ_{ST}.

For this characteristic, the C_{ST} index is 12.00 and the C_P index is 2.00, both of which reveal tremendous short-term *potential* capability for this process.

$$\hat{C}_{ST} = \frac{\text{Tolerance}}{\hat{\sigma}_{ST}} = \frac{6--6}{1} = 12.00 \qquad\qquad \hat{C}_P = \frac{\text{Tolerance}}{6\hat{\sigma}_{ST}} = \frac{6--6}{6(1)} = 2.00$$

An estimated C_{ST} of 12.00 implies that one entire $12\sigma_{ST}$ spread could fit within the tolerance. This is the same result as adding the estimates of Z_{LSL} (the number of σ_{ST}s from μ to the LSL) and Z_{USL} (the number of σ_{ST}s from μ to the USL) together.

$$\hat{Z}_{LSL,ST} + \hat{Z}_{USL,ST} = 9 + 3 = 12$$

In spite of this tremendous potential capability, performance capability is scarcely marginal because the process average is not centered at the middle of the tolerance. Someone looking at just these two potential measures of capability could falsely assume this shearing operation has no difficulty in meeting the print specifications for connector length.

A comparison between potential and performance capability measures reveals information about how well the process average is centered near the middle of the tolerance. If potential capability is high, but performance capability low, then moving the process average closer to the middle of the tolerance will improve performance capability and, thereby, decrease the percentage of nonconforming parts (see Table 6.3).

Table 6.3 Relationship between performance and potential capability.

<table>
<tr><td></td><td></td><td colspan="2" align="center">Potential Capability</td></tr>
<tr><td></td><td></td><td align="center">Low</td><td align="center">High</td></tr>
<tr><td rowspan="2">Performance Capability</td><td>High</td><td align="center">Not Possible</td><td align="center">Desired State</td></tr>
<tr><td>Low</td><td align="center">Reduce Variation</td><td align="center">Move Average</td></tr>
</table>

When both indexes are low, work must concentrate on reducing process variation. If both are high, the process is properly centered and has small variation compared to the tolerance, which is the ideal condition. It is impossible to simultaneously have high performance capability and low potential capability, as performance can never exceed potential.

For this particular example, C_{ST} is 12.00, indicating it is possible to fit exactly one $12\sigma_{ST}$ spread within the tolerance. If the process is centered at the middle of the tolerance, there would be $6\sigma_{ST}$s from the process average to either specification limit, meaning $Z_{MIN.ST}$ could be as high as 6.

$$\text{Maximum } \hat{Z}_{MIN.ST} = \frac{\hat{C}_{ST}}{2} = \frac{12.00}{2} = 6.00$$

This maximum may also be derived directly from the individual Z values.

$$\text{Maximum } \hat{Z}_{MIN.ST} = \frac{\hat{Z}_{LSL.ST} + \hat{Z}_{USL.ST}}{2} = \frac{9 + 3}{2} = 6$$

Because the current estimate of $Z_{MIN,ST}$ is only 3, a substantial improvement in performance capability is possible by moving the process average closer to 0 from its present location of 3. In general, the size of this move is discovered through this formula, where M is the middle of the tolerance:

$$M = \frac{LSL + USL}{2} = \frac{-6 + 6}{2} = 0$$

$$\text{Shift in Process Average} = M - \hat{\mu} = 0 - 3 = -3$$

This shift may also be determined directly from the estimated Z values.

$$\text{Shift in Process Average} = \frac{\hat{\sigma}_{ST}}{2}(\hat{Z}_{USL,ST} - \hat{Z}_{LSL,ST}) = \frac{1}{2}(3 - 9) = -3$$

The minus 3 answer means $Z_{MIN,ST}$ can be made as high as 6.0 by *lowering* the process average 3 units.

Moving the Process Average

To improve the short-term performance capability of this process, the process average should be moved closer to 0, which is the middle of the tolerance. A process modification to lower the average must be implemented. After making this change, its effectiveness is verified by observing a run below the centerline of the $\overline{X}$ chart monitoring this process, as is depicted in Figure 6.12. This nonrandom pattern provides sufficient evidence the process average has been shifted significantly lower.

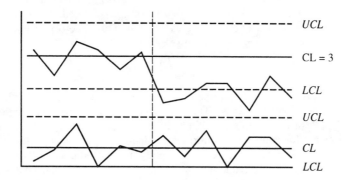

Fig. 6.12 A permanent change in μ causes a run to appear on the $\overline{X}$ chart.

Notice how the range chart remains in control, signifying there is no difference in the process spread from before the change to after. Once control limits are recalculated for this modified process, and stability established at this improved level, the revised estimates of μ and σ_{ST} listed below are computed from the newly calculated centerlines.

$$\hat{\mu} = \overline{\overline{X}} = 1 \qquad\qquad \hat{\sigma}_{ST} = \frac{\overline{R}}{d_2} = 1$$

With these process parameter estimates, the process output is predicted to look like the one displayed in Figure 6.13. Visually, this appears to be a definite improvement in performance capability, as the upper $3\sigma_{ST}$ tail is much further below the USL. In fact, $Z_{USL,ST}$ has increased by 2, from 3 up to 5.

$$\hat{Z}_{USL,ST} = \frac{USL - \hat{\mu}}{\hat{\sigma}_{ST}} = \frac{6 - 1}{1} = 5$$

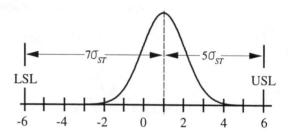

Fig. 6.13 The modified process output is centered closer to the middle of the tolerance.

However, this increase in $Z_{USL,ST}$ comes at the expense of $Z_{LSL,ST}$, which is reduced by 2.

$$\hat{Z}_{LSL,ST} = \frac{\hat{\mu} - LSL}{\hat{\sigma}_{ST}} = \frac{1 - (-6)}{1} = 7$$

Due to the downward shift in the process average, $Z_{USL,ST}$ increases from 3 to 5, while $Z_{LSL,ST}$ decreases from 9 to 7. When the process average is moved, one Z value always increases, while the other decreases by an identical amount. This trade-off is summarized in the updated estimate of $Z_{MIN,ST}$.

$$\hat{Z}_{MIN,ST} = \text{Minimum} \, (7, 5) = 5$$

By lowering the process average to 1, $Z_{MIN,ST}$ increases to 5, quite a bit higher than the $Z_{MIN,ST}$ of 3 for the original process. This improved process is producing significantly more than 99.73 percent of its output within print specifications and greatly exceeds the minimum requirement for short-term performance capability. With this safety margin, the process average (or variability) may change slightly and the process will still make at least 99.73 percent good parts. To reduce operating costs, the quality team in charge of this process may consider reducing the subgroup size of the control chart monitoring this process, or increasing the time between subgroups (Burns).

Even though the average has moved from 3 to 1, C_{ST} is still 12.00 and C_P stays at 2.00. These two potential capability measures remain unchanged and do *not* identify this significant performance capability improvement because they are not influenced by changes in the process average. They measure only the potential of a process to produce nonconforming parts and that potential hasn't been altered, as $Z_{MIN,ST}$ can still be as high as 6. But the actual performance has improved substantially, as the $Z_{MIN,ST}$ index of 5 is now much closer to this maximum of 6. Note that $Z_{LSL,ST}$ and $Z_{USL,ST}$ have changed, but still sum to 12 (7 + 5 = 12).

To further improve performance capability, the process average could be shifted even lower. Again, a change is made to the process in hopes of lowering the average, its effectiveness verified on the control chart, revised control limits calculated, and estimates of the new process average and standard deviation calculated with the following results:

$$\hat{\mu} = \overline{\overline{X}} = -5 \qquad \hat{\sigma}_{ST} = \frac{\overline{R}}{d_2} = 1$$

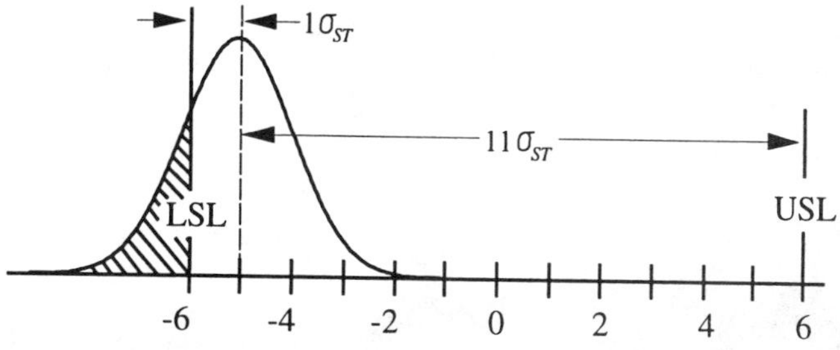

Fig. 6.14 Process average is shifted too close to the LSL.

The process average was definitely lowered, this time all the way down to minus 5, as is seen in Figure 6.14. This change makes $Z_{USL,ST}$ look terrific, as it's all the way up to 11.

$$\hat{Z}_{USL,ST} = \frac{USL - \hat{\mu}}{\hat{\sigma}_{ST}} = \frac{6 - (-5)}{1} = 11$$

Unfortunately, this drastic downward shift has placed the process average dangerously close to the LSL, as the low $Z_{LSL,ST}$ value warns.

$$\hat{Z}_{LSL,ST} = \frac{\hat{\mu} - LSL}{\hat{\sigma}_{ST}} = \frac{(-5) - (-6)}{1} = 1$$

With less than $3\sigma_{ST}$ between the process average and the LSL, much more than .135 percent of the connectors will have their length cut below the LSL. This lack of performance capability is expressed by a $Z_{MIN,ST}$ value of only 1, meaning the process is just 1 standard deviation away from a specification limit.

$$\hat{Z}_{MIN,ST} = \text{Minimum} \, (\, 1, 11 \,) = 1$$

Anytime $Z_{MIN,ST}$ is less than 3, the process is judged *not* capable of fulfilling the minimum requirement for performance capability. To improve this situation, either the process average must be moved, the process standard deviation decreased, or both.

As before, $\hat{C}_{ST}$ still equals 12.00 while $\hat{C}_p$ stays equal to 2.00. These two potential capability measures remain unchanged because $\hat{\sigma}_{ST}$ and the tolerance remain unchanged. Even though this process has the potential of being capable, it's performance is not because the process average is located too close to the LSL. The process has plenty of potential capability for producing the proper length connectors, but lacks acceptable performance capability due to improper centering.

Changing the Process Standard Deviation

Eventually the process output for connector length is centered at the middle of the tolerance, as inferred by the new estimate of 0 for the process average.

$$\hat{\mu} = \overline{X} = 0 \qquad\qquad \hat{\sigma}_{ST} = \frac{\overline{R}}{d_2} = 3$$

As a side effect of shifting the average, the standard deviation increased to 3, and now the $6\sigma_{ST}$ process spread of 18 (6 × 3) is too wide to fit within the tolerance of only 12, as Figure 6.15 discloses.

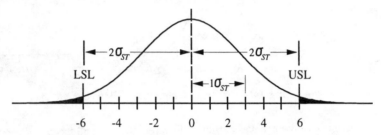

Fig. 6.15 An increase in process spread makes Z_{MIN} decrease.

When a symmetrical distribution is centered at the middle of the tolerance, all three Z values are equal.

$$\hat{Z}_{MIN,ST} = \text{Minimum} \left(\frac{\hat{\mu} - LSL}{\hat{\sigma}_{ST}}, \frac{USL - \hat{\mu}}{\hat{\sigma}_{ST}} \right)$$

$$= \text{Minimum} \left(\frac{0 - -6}{3}, \frac{6 - 0}{3} \right)$$

$$= \text{Minimum} (2, 2) = 2$$

The Z_{MIN} measure identifies this process as *not* meeting the minimum requirement for short-term performance capability, as the estimate of $Z_{MIN,ST}$ is quite a bit less than 3.

With this new estimate of σ_{ST}, $\hat{C}_{ST}$ is now just 4.00, while $\hat{C}_p$ is only .67.

$$\hat{C}_{ST} = \frac{\text{Tolerance}}{\hat{\sigma}_{ST}} = \frac{6 - -6}{3} = 4.00 \qquad\qquad \hat{C}_P = \frac{\text{Tolerance}}{6\hat{\sigma}_{ST}} = \frac{6 - -6}{6(3)} = .67$$

These measures of potential capability also indicate this process will have serious difficulty producing parts within tolerance, no matter where the process output is centered. Since the process lacks potential capability, it will lack performance capability. In fact, the C_{ST} index discloses the best this process can do is fit only a $4\sigma_{ST}$ spread into the tolerance. Therefore, the largest $Z_{MIN,ST}$ can possibly be is half of this, or 2.0. Since $Z_{MIN,ST}$ is already 2.0, moving the process average in either direction will make matters worse (a lower $Z_{MIN,ST}$ value) by decreasing performance capability. When both potential and performance capability are low, improvement efforts must initially concentrate on reducing common-cause variation.

Whenever the process average moves closer to the middle of the tolerance, Z_{MIN} increases, and the percentage of nonconforming product decreases (assuming σ remains constant). If σ decreases, and the process average remains constant, Z_{MIN} increases, while the percentage of nonconforming product decreases. When the process average moves toward the USL (assuming $M < \mu <$ USL), Z_{MIN} decreases and the percentage of nonconforming product increases. Likewise, if the process average moves toward the LSL (assuming LSL $< \mu < M$), Z_{MIN} declines, while the percentage of nonconforming product rises.

Relationship of Z_{MIN} to the Process Average

Because Z_{MIN} incorporates both σ and μ in its calculation, changes in the process average have a profound effect on its value. This is illustrated in Figure 6.16 by the steeply slanted lines for Z_{LSL} and Z_{USL}.

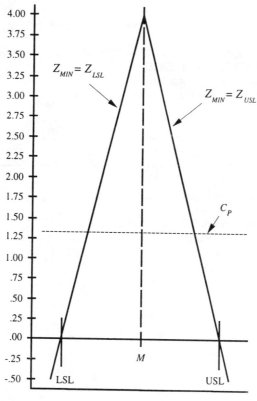

Process Average

Fig. 6.16 Z_{MIN} versus the process average.

Unlike measures of potential capability, Z_{MIN} varies as the process averages changes (assuming σ remains constant). Z_{MIN} equals Z_{LSL} when μ is less than the middle of the tolerance (M), while Z_{MIN} equals Z_{USL} when μ is greater than M. Z_{MIN} achieves its maximum value when the process average is located exactly at M. If the process average is ever centered at one of the specification limits, Z_{MIN} equals 0, and the percentage of nonconforming parts is at least 50 percent. For example, assume the process average equals the USL, then:

$$Z_{MIN} = \frac{USL - \mu}{\sigma} = \frac{USL - USL}{\sigma} = \frac{0}{\sigma} = 0$$

This process condition is displayed in Figure 6.17. With the average centered on the USL, half the process output is above the USL.

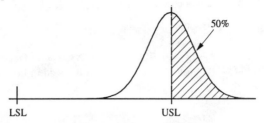

Fig. 6.17   When μ is equal to the USL, Z_{MIN} is 0.

However, a Z_{MIN} of 0 doesn't necessarily mean exactly 50 percent nonconforming parts. As shown in Figure 6.18, Z_{MIN} is 0, yet more than 50 percent of the output is outside of tolerance due to the additional 1 percent below the LSL. This is why *both* Z_{LSL} and Z_{USL} ought to be checked when estimating the total percentage of nonconforming parts.

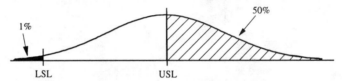

Fig. 6.18   There may be more than 50 percent nonconforming parts when Z_{MIN} is 0.

If the process average is located *outside* a print limit, Z_{MIN} is negative, indicating a serious capability problem since there is significantly more than 50 percent nonconforming parts. Suppose the process average is greater than the USL, as is the case in Figure 6.19. This forces Z_{USL} to be less than zero.

$$USL < \mu$$

$$USL - \mu < 0$$

$$\frac{USL - \mu}{\sigma} < \frac{0}{\sigma}$$

$$Z_{USL} < 0$$

Fig. 6.19   A negative Z_{MIN} implies there are more than 50% nonconforming parts.

In this situation, moving the process average to the middle of the tolerance will greatly improve performance capability, thereby dramatically reducing the percentage of nonconforming parts. Note that the opposite isn't necessarily true: there may be more than 50 percent nonconforming parts being produced, yet Z_{MIN} may be greater than zero. Figure 6.20 presents a process having almost 54 percent of its output outside of tolerance, but Z_{MIN} is .60, a positive number.

$$Z_{MIN} = \text{Minimum}\left(\frac{\mu - LSL}{\sigma}, \frac{USL - \mu}{\sigma}\right)$$

$$= \text{Minimum}\left(\frac{0--1}{1.667}, \frac{1-0}{1.667}\right)$$

$$= \text{Minimum}\,(.60, .60) = .60$$

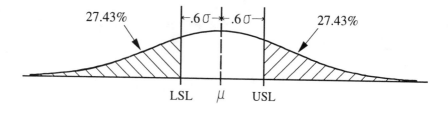

Fig. 6.20 A positive Z_{MIN} may be associated with more than 50% nonconforming parts.

Example 6.5 will provide more details on how Z values are converted into estimates of the percentage of nonconforming product.

Definition of Z_{MAX}

Occasionally, reference is made to Z_{MAX}, which is the larger of Z_{LSL} and Z_{USL}.

$$Z_{MAX} = \text{Maximum}\,(Z_{LSL}, Z_{USL})$$

Just as for Z_{MIN}, there is a $Z_{MAX.ST}$ and a $Z_{MAX.LT}$.

$$Z_{MAX.ST} = \text{Maximum}\,(Z_{LSL,ST}, Z_{USL,ST}) \qquad Z_{MAX.LT} = \text{Maximum}\,(Z_{LSL,LT}, Z_{USL,LT})$$

Although not as important as Z_{MIN}, Z_{MAX} provides a useful insight to process capability when interest is in the total percentage of nonconforming parts. As the last section just pointed out, Z_{MIN} focuses attention on only the specification with the highest percentage of nonconforming parts, however, the other specification might also be contributing a high percentage of non-conforming parts. In fact, it could be as much as for the specification that determines Z_{MIN}. The next example explains how to estimate the percentage of nonconforming parts for a given Z value. Summing the percentage of nonconforming parts corresponding to Z_{MIN} with that corresponding to Z_{MAX} will yield the total percentage nonconforming for the characteristic under study.

Example 6.5 Floppy Disks

A field problem for floppy disks is traced to variation in the diameter of the inner hole highlighted in Figure 6.21. If this hole is too small, it jams the delicate disk drive, causing a major warranty claim. If the hole is too big, the floppy disk fails to rotate properly and cannot store or retrieve information. In Example 5.9 on page 111, information collected from an $\overline{X}, R$ chart monitoring the hole-cutting process allowed computation of $\hat{\sigma}_{ST}$ (.898) and $\hat{\mu}$ (40.9).

FLOPCO, INC.
High Quality Diskettes

I.D.

Fig. 6.21 A floppy disk.

Given that the LSL is 39.5 and the USL is 44.5, the short-term Z measures are:

$$\hat{Z}_{LSL,ST} = \frac{\hat{\mu} - \text{LSL}}{\hat{\sigma}_{ST}} = \frac{40.9 - 39.5}{.898} = 1.56 \qquad \hat{Z}_{USL,ST} = \frac{\text{USL} - \hat{\mu}}{\hat{\sigma}_{ST}} = \frac{44.5 - 40.9}{.898} = 4.01$$

The potential measures of process capability had indicated a problem with hole size (recall that C_{ST} = 5.568 and C_P = .928), but did not reveal if the major problem involves holes that are oversized, undersized, or both. Observe how these performance measures immediately point out, that with a $Z_{USL,ST}$ of 4.01, there is little problem with oversize holes. However, a $Z_{LSL,ST}$ of only 1.56 indicates many floppies are produced with undersized holes.

Estimates of $Z_{MIN,ST}$ and $Z_{MAX,ST}$ are derived from these first two Z values.

$$\hat{Z}_{MIN,ST} = \text{Minimum}\,(\hat{Z}_{LSL,ST}, \hat{Z}_{USL,ST}) = \text{Minimum}\,(1.56, 4.01) = 1.56$$

$$\hat{Z}_{MAX,ST} = \text{Maximum}\,(\hat{Z}_{LSL,ST}, \hat{Z}_{USL,ST}) = \text{Maximum}\,(1.56, 4.01) = 4.01$$

An estimated $Z_{MIN,ST}$ less than 3 indicates a lack of performance capability. With an estimated C_{ST} of 5.568, the highest value $Z_{MIN,ST}$ can achieve by shifting the process average is only 2.784.

$$\text{Maximum } \hat{Z}_{MIN,ST} = \frac{\hat{C}_{ST}}{2} = \frac{5.568}{2} = 2.784$$

As this result is still less than 3.0, the process spread must be reduced if this process is ever to become capable. Beyond keeping the process in control, there is very little an operator can do to improve this process after the average is located at M. Management intervention is necessary for significant process improvements (changing one or more of the five Ms), or for alterations to the specifications. Typically, one of management's first requests for additional detailed process information to aid in their decision making is for an estimate of the percentage of nonconforming product.

Estimating the Percentage Nonconforming

The percentage of parts below the LSL is denoted by the symbol p'_{LSL}, whereas that above the USL is designated with p'_{USL}. The sum of these two represents the total percentage of nonconforming parts, and is labeled p'_{TOTAL}.

$$p'_{TOTAL} = p'_{LSL} + p'_{USL}$$

In practice, these process parameters are not known and must be estimated from the sample data. If their estimation is based on short-term variation, the following notation is employed.

$$\hat{p}'_{TOTAL,ST} = \hat{p}'_{LSL,ST} + \hat{p}'_{USL,ST}$$

In situations where long-term process variation is being studied, the notation becomes:

$$\hat{p}'_{TOTAL,LT} = \hat{p}'_{LSL,LT} + \hat{p}'_{USL,LT}$$

Z values for a process with a normal distribution are turned into estimates of the percentage nonconforming by referring to Appendix Table III, a portion of which is displayed in Figure 6.22. This table has Z values to one decimal place listed in the far left-hand column from 0.0 to 7.7. The column headings across the top of the page represent the second decimal place for a given Z value (.00 to .09). The body of the table contains the percentage nonconforming corresponding to these Z values.

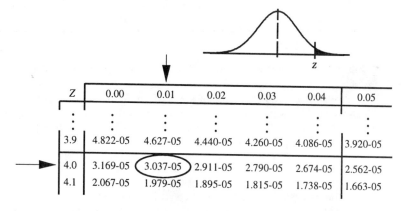

Z	0.00	0.01	0.02	0.03	0.04	0.05
⋮	⋮	⋮	⋮	⋮	⋮	⋮
3.9	4.822-05	4.627-05	4.440-05	4.260-05	4.086-05	3.920-05
4.0	3.169-05	3.037-05	2.911-05	2.790-05	2.674-05	2.562-05
4.1	2.067-05	1.979-05	1.895-05	1.815-05	1.738-05	1.663-05

Fig. 6.22 A portion of Appendix Table III for Z values.

For example, the estimate of $Z_{USL,ST}$ in the last example is 4.01. To discover what percentage of floppies have hole sizes larger than the USL, first find 4.0 in the left-hand column of Appendix Table III. The percentage nonconforming is somewhere in this 4.0 row. To determine where, locate the .01 column at the top of the table. Go down this column until the 4.0 row is reached, as is illustrated by the two arrows in Figure 6.22. At the intersection of the 4.0 row and the .01 column is 3.037-05, which is the estimated percentage of floppies with holes larger than the USL, *i.e.*, more than 4.01σ from the process average.

$$\hat{p}'_{USL,ST} = 3.037 - 05$$

To save space in this table, the percentage nonconforming values are expressed in scientific notation. Thus, 3.037-05 is actually 3.037×10^{-5}. Recall that multiplying by 10^{-5} is the same as dividing by 100,000.

$$10^{-5} = \frac{1}{10^5} = \frac{1}{100,000}$$

This translates into moving the decimal point 5 places to the left.

$$\hat{p}'_{USL,ST} = 3.037\text{-}05 = 3.037 \times 10^{-5} = .00003037$$

To convert this from decimal form to a percentage, multiply .00003037 by 100 to move the decimal point two places to the right. Therefore, the amount of floppies with holes above the USL is estimated as about .003 percent.

$$\hat{p}'_{USL,ST} \cong .003\%$$

Because the normal distribution is symmetrical, the percentage of parts below a given $Z_{LSL,ST}$ value is equal to the percentage above an identical $Z_{USL,ST}$ value, as is illustrated in Figure 6.23. Thus, Appendix Table III can help find the percentage of nonconforming parts for both $Z_{LSL,ST}$ and $Z_{USL,ST}$.

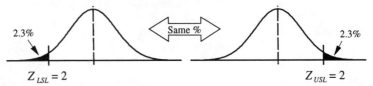

Fig. 6.23 The percentage below a $Z_{LSL,ST}$ of 2 equals the percentage above a $Z_{USL,ST}$ of 2.

In this example, the percentage of nonconforming holes corresponding to a $Z_{LSL,ST}$ value of 1.56 is found by going across the 1.5 row of Appendix Table III (see Figure 6.24) until the .06 column is reached. The value located here is 5.938×10^{-2}, which in decimal form is .05938, or equivalently, about 5.9 percent.

$$\hat{p}'_{LSL,ST} = .05938 \cong 5.9\%$$

Based on this estimate, almost 6 percent of the floppies produced with this process have undersized holes, which is most likely the cause of the recent customer complaints.

Z	0.00	0.01	0.02	0.03	0.04	0.05	0.06	0.07
⋮	⋮	⋮	⋮	⋮	⋮	⋮	⋮	⋮
1.4	8.076-02	7.927-02	7.780-02	7.636-02	7.493-02	7.353-02	7.214-02	7.078-02
1.5	6.681-02	6.552-02	6.426-02	6.301-02	6.178-02	6.057-02	5.938-02	5.821-02
1.6	5.480-02	5.370-02	5.262-02	5.155-02	5.050-02	4.947-02	4.846-02	4.746-02

Fig. 6.24 The portion of Table III needed for $Z_{LSL,ST}$ of 1.56.

By combining the percentage nonconforming from $Z_{LSL,ST}$ and $Z_{USL,ST}$, an estimate of the total percentage nonconforming is derived.

$$\hat{p}'_{TOTAL,ST} = \hat{p}'_{LSL,ST} + \hat{p}'_{USL,ST} = 5.938\% + .003\% = 5.941\%$$

If no alterations are made to this hole-forming process, a little over 94 percent (100 minus 5.9) of the floppies will have their hole sizes within print specification. This total is far short of the 99.73 percent required to meet the minimum requirement of performance capability for a characteristic with a bilateral specification.

Visual Interpretation of Capability

With all the information collected so far about both potential and performance capability concerning the hole-cutting process, what recommendations can be formulated about reducing the number of customer complaints regarding this critical characteristic:

> 1) in the next few days?
>
> 2) in the next few months?

At this point in a process capability study, it's a good idea to construct a histogram to gain additional insight about the current level of process capability and help determine what should to be done to improve it. Given that the process average is estimated as 40.9 and the short-term standard deviation as .898, a sketch of the process output is drawn by first estimating the upper and lower $3\sigma_{ST}$ tails.

$$\text{Upper } 3\hat{\sigma}_{ST} \text{ tail} = \hat{\mu} + 3\hat{\sigma}_{ST} = 40.9 + 3(.898) = 43.6$$

$$\text{Lower } 3\hat{\sigma}_{ST} \text{ tail} = \hat{\mu} - 3\hat{\sigma}_{ST} = 40.9 - 3(.898) = 38.2$$

Assuming a normal distribution with a process average of 40.9, a picture of the expected process output is created in Figure 6.25. As this is the estimated distribution for hole sizes on individual floppies, the specification limits may also be added to this histogram. Now it can be seen that the upper $3\sigma_{ST}$ tail of 43.6 is a fair distance below the USL of 44.5 (reflected by the $Z_{USL,ST}$ of 4.01), suggesting there should be few, if any, problems with over-sized holes. In fact, the estimate from the Z table was only .003 percent. Unfortunately, the lower $3\sigma_{ST}$ tail of 38.2 extends quite a ways *below* the LSL of 39.5, asserting that considerably more than .135 percent of these holes (about 5.9 percent) are smaller than this LSL.

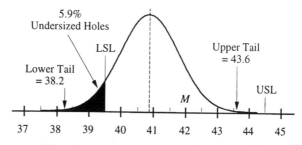

Fig. 6.25 Histogram of floppy disk hole sizes.

By studying this histogram, one apparent maneuver for reducing the number of floppies with undersized holes involves moving the process average higher. But how much higher? Centering the process average at the middle of the tolerance will help reduce the percentage of nonconforming holes, but will not completely eliminate the problem. Recall from Examples 5.19 and 5.26, that the potential measures of capability indicated a problem with capability (C_{ST} = 5.568, C_P = .928), even if the process is centered at M, as is illustrated in Figure 6.26. With a C_P of only .928, the largest $Z_{MIN,ST}$ can possibly be is 2.78.

$$\text{Maximum } \hat{Z}_{MIN.\,ST} = 3\,\hat{C}_P = 3(.928) = 2.78$$

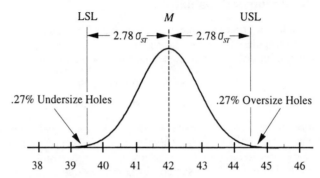

Fig. 6.26   Process centered at 42.0, the middle of the tolerance.

Even though the process is still not capable, the percentage of nonconforming holes is minimized with this shift in μ. By calculating Z values for $\hat{\mu}$ equal to 42.0, the percentage of holes outside both specifications may be estimated. As the process average is located at the middle of tolerance, $Z_{LSL,ST}$ is equal to $Z_{USL,ST}$, so only one needs to be calculated.

$$\hat{Z}_{MIN.\,ST} = \hat{Z}_{USL,ST} = \frac{USL - \hat{\mu}_{42}}{\hat{\sigma}_{ST}} = \frac{44.5 - 42.0}{.898} = 2.78$$

The amount of nonconforming holes above the USL corresponding to a $Z_{USL,ST}$ of 2.78 is .27 percent (2.718-03 from Appendix Table III). Due to the process being centered at the middle of the tolerance, this same percentage of holes is also below the LSL. Instead of a total of 5.9 percent nonconforming holes when the process was centered at 40.9, the total percentage out of specification decreases to just .54 when the average is located at 42.0, over a 90 percent reduction.

This represents an immense improvement, but there is another important consideration. For this proposed situation, the percentage above the USL (.27 percent) is identical to that below the LSL (.27 percent). However, the *severity* of the nonconformance is certainly different for a hole above the USL, where the customer simply returns the floppy for replacement, versus a floppy with an undersize hole that can damage the customer's expensive disk drive. It seems logical to move the process average even higher than 42.0 in order to bring the lower tail for hole size well above the LSL. But how much higher should it be shifted? Section 7.3 will explain a more systematic method (based on cost) of selecting the optimum process average, but assume a decision is made to have this process produce no more than one undersize hole out of every 100,000 disks. To determine where the process average must be centered to achieve this objective, look up this percentage in the body of Appendix Table III, and then find its corresponding $Z_{LSL,ST}$ value.

$$\frac{1}{100,000} = .00001 = 1.0 \times 10^{-5}$$

The $Z_{LSL,ST}$ value for this objective is discovered by searching for 1.000-05 in the body of Appendix Table III. The closest percentage without interpolation is 1.023×10^{-5}, which corresponds to a Z value of 4.26. Substituting this goal of 4.26 into the formula for $Z_{LSL,ST}$ allows solving for the required process average (μ_{GOAL}) to produce this desired percentage of nonconforming holes below the LSL.

$$\hat{Z}_{LSL,ST,GOAL} = \frac{\hat{\mu}_{GOAL} - LSL}{\hat{\sigma}_{ST}}$$

$$4.26 = \frac{\hat{\mu}_{GOAL} - 39.5}{.898}$$

$$39.5 + 4.26(.898) = \hat{\mu}_{GOAL}$$

$$43.33 = \hat{\mu}_{GOAL}$$

If the average hole size is raised to 43.33, there will be a distance of 4.26 standard deviations from the process average to the LSL of 39.5, meaning only 1 out of every 100,000 floppies will have an undersized hole. However, this change in μ greatly increases the number of oversized holes, as is discovered by calculating $Z_{USL,ST}$ for this new process average.

$$\hat{Z}_{USL,ST} = \frac{USL - \hat{\mu}_{GOAL}}{\hat{\sigma}_{ST}} = \frac{44.5 - 43.33}{.898} = 1.30$$

A $Z_{USL,ST}$ of 1.30 implies about 9.7 percent (9.680-02) of the floppies will have oversized holes if the average is shifted to 43.33 (Figure 6.27). Because oversize holes cannot be reworked, scrap costs will soar as well as the associated inspection costs. And because even 100 percent inspection won't catch all the oversize holes, some will still reach the customer, resulting in continued customer complaints.

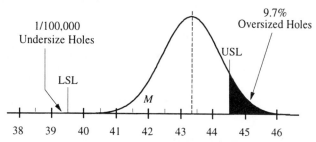

Fig. 6.27 Centering μ at 43.33 almost eliminates undersize holes, but greatly increases the percentage of oversize holes.

Many would set the goal for undersize holes at 0 percent, but theoretically, this means raising the process average to plus infinity, which is certainly not a practical solution. Besides, almost 100 percent of the holes would now be produced above the USL.

When performance and potential capability are both very low, the only viable solution for eliminating customer complaints is to reduce process variation. This means concentrating on common-cause items, such as tooling, materials, operator training, work methods, environmental concerns (temperature, humidity), and gaging. Lawson recommends applying statistical techniques such as analysis of variance to determine the major contributors of process variation. Once changes are implemented to this hole-cutting operation, they must be verified on the control chart by witnessing a run below the centerline of the range chart, like the one portrayed in Figure 6.28.

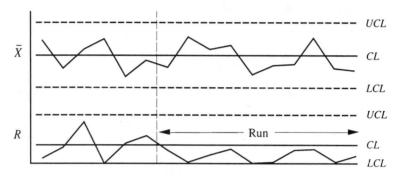

Fig. 6.28 A run below the centerline on a range chart implies reduced process variation.

When the reduction of variation is confirmed, the process should be centered at the middle of the tolerance to provide a sufficient safety margin on both sides, as is depicted in Figure 6.29. With a large enough reduction in process spread, practically all hole sizes should meet specification. This eliminates the cost of scrap, rework, and 100 percent inspection. Unfortunately, reducing variation typically requires process changes that are expensive or take considerable time to implement. For instance, weeks (if not months) may be needed to approve, order, and install new tooling.

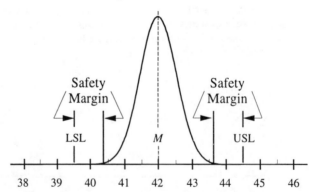

Fig. 6.29 Process centered at 42.0 and spread reduced.

Thus, the short-term solution is to adjust the process average to minimize losses. In the long-term, process changes must be implemented to reduce variation. When this is accomplished, the process can be centered at middle of the tolerance. Valuable insight concerning the performance of a process is gained from a careful and detailed analysis of both potential and performance capability measures, in conjunction with histograms of the current and proposed process outputs.

Example 6.6 Particle Contamination

An *IX & MR* chart is monitoring particle contamination of a cleaning solution for printed circuit boards after going through a wave soldering process. The following estimates of process parameters are made from the 25 most recent in-control readings.

$$\hat{\mu} = 127.32 \, ppm \qquad\qquad \hat{\sigma}_{ST} = 12.079 \, ppm$$

Estimates of potential capability were already made in Examples 5.21 and 5.28.

$$\hat{C'}_{PU} = 1.24 \qquad\qquad \hat{C'}_{STU} = 7.45$$

As contamination has just an USL of 165 *ppm*, only $Z_{USL,ST}$ can be estimated. By default, it also becomes the estimated $Z_{MIN,ST}$ value as well.

$$\hat{Z}_{MIN.\,ST} = \hat{Z}_{USL,ST} = \frac{USL - \hat{\mu}}{\hat{\sigma}_{ST}} = \frac{165 - 127.32}{12.079} = 3.12$$

Although this process meets the minimum requirement for performance capability, it does not meet a mandated performance capability goal of 4.0. With a unilateral specification, improvement in performance can always be brought about by shifting the process average (in this case, lower), or by reducing process spread. For example, if μ is moved down to 120 *ppm*, $Z_{MIN,ST}$ increases to 3.725.

$$\hat{Z}_{MIN.\,ST} = \hat{Z}_{USL,ST} = \frac{USL - \hat{\mu}}{\hat{\sigma}_{ST}} = \frac{165 - 120}{12.079} = 3.725$$

If μ is lowered to 110 *ppm*, $Z_{MIN,ST}$ increases to 4.55 and now exceeds the goal of 4.0.

$$\hat{Z}_{MIN.\,ST} = \hat{Z}_{USL,ST} = \frac{USL - \hat{\mu}}{\hat{\sigma}_{ST}} = \frac{165 - 110}{12.079} = 4.55$$

Assume 120 *ppm* is as low as the average can be decreased without extensive process modifications. In order to achieve a $Z_{USL,ST}$ goal of 4.0, σ_{ST} must be reduced to 11.25 *ppm*.

$$Z_{USL,ST,GOAL} = \frac{USL - \mu}{\sigma_{ST,GOAL}}$$

$$4.0 = \frac{165 \, ppm - 120 \, ppm}{\sigma_{ST,GOAL}}$$

$$\sigma_{ST,GOAL} = \frac{45 \, ppm}{4.0}$$

$$\sigma_{ST,GOAL} = 11.25 \, ppm$$

Example 6.7 Center Gap Size

In Example 3.3, estimates of μ (.3182 cm) and σ_{ST} (.003339 cm) were made from an $\overline{X}, R$ chart monitoring the center gap between the bottom flaps of corrugated boxes (Figure 6.30).

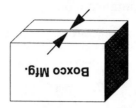

Fig. 6.30 Gap size between bottom flaps of a corrugated box.

The print specification for center gap size is .3200 cm $\pm$ .0160 cm, while the goal for this process is a $Z_{MIN,ST}$ of at least 4.00.

$$\hat{Z}_{MIN,\,ST} = \text{Minimum}\,(\hat{Z}_{LSL,ST}, \hat{Z}_{USL,ST})$$

$$= \text{Minimum}\left(\frac{\hat{\mu} - LSL}{\hat{\sigma}_{ST}}, \frac{USL - \hat{\mu}}{\hat{\sigma}_{ST}}\right)$$

$$= \text{Minimum}\left(\frac{.3182 - .3040}{.003339}, \frac{.3360 - .3182}{.003339}\right)$$

$$= \text{Minimum}\,(4.25, 5.33) = 4.25$$

This flap-folding operation is judged a very competent process as the estimated $Z_{MIN,ST}$ value of 4.25 exceeds the goal for short-term performance capability of 4.00. In fact, the percentage of nonconforming gaps is just a little over .001 percent. The Φ operator (pronounced "fie") means find the corresponding percentage for the Z value in parentheses in Appendix Table III. For example, $\Phi(3.0)$ equals .00135, while $\Phi(4.0)$ equals .00003169.

$$\hat{p}'_{TOTAL,ST} = \hat{p}'_{LSL,ST} + \hat{p}'_{USL,ST}$$

$$= \Phi(\hat{Z}_{LSL,ST}) + \Phi(\hat{Z}_{USL,ST})$$

$$= \Phi(4.25) + \Phi(5.33)$$

$$= .00001070 + .00000005$$

$$= .00001075 \cong .001\%$$

Performance capability could be maximized by adjusting the machinery to make the box flaps come slightly closer together, such that average gap size is equal to .3200 cm, the middle of the tolerance.

$$\text{Maximum } \hat{Z}_{MIN,\,ST} = \frac{\hat{Z}_{LSL,ST} + \hat{Z}_{LT,ST}}{2} = \frac{4.25 + 5.33}{2} = 4.79$$

This would reduce $\hat{p}'_{TOTAL,ST}$ to only .0002 percent.

$$\hat{p}'_{TOTAL,ST} = \Phi(\hat{Z}_{LSL,ST}) + \Phi(\hat{Z}_{USL,ST})$$

$$= \Phi(4.79) + \Phi(4.79)$$

$$= .00000084 + .00000084$$

$$= .00000168 \cong .0002\%$$

However, centering the process output at the middle of the tolerance makes sense only if M is the desired average gap size. If there is a different target for the process average, then a modified form of Z_{MIN} must be used to correctly assess capability (see Section 6.2).

Advantages and Disadvantages

This is the first estimate of process capability which incorporates both process average and standard deviation, as well as the LSL and USL, in its calculation. In addition, the percentage of parts above or below the specification limits can be estimated from Appendix Table III, which gives areas under the normal curve for various Z values. Although this is not direct, it can still be done, provided such a table is readily available.

The Z_{MIN} measure adapts easily to unilateral specifications, as there is only one Z value to calculate. If just an USL is given, the following formula applies:

$$Z_{MIN} = Z_{USL} = \frac{USL - \mu}{\sigma}$$

When only a LSL is given, switch to this formula:

$$Z_{MIN} = Z_{LSL} = \frac{\mu - LSL}{\sigma}$$

A different formula is not required, as was the case for most measures of potential capability. Neither does a target average need to be specified, nor a special rating scale established. Thus, the Z_{MIN} measure furnishes a straightforward quantitative assessment of performance capability for both bilateral and unilateral specifications.

A minor annoyance is the scale change to 3 as the minimum requirement for Z_{MIN} rather than 1, as with most other measures of capability mentioned so far. This makes comparison to the more commonly used potential capability measures (like C_p) rather awkward. However, a capability index introduced later in this chapter remedies this particular problem.

Major drawbacks are the assumption of normality and the need for an additional table to determine the percentage nonconforming. In addition, Z_{MIN} also assumes the middle of the tolerance is best for the process average. For many characteristics this is true, but there are some specifications (like 105 plus 5, minus 3) where it is desired to center the process output somewhere else. This disadvantage is addressed with a measure called Z^*_{MIN}.

6.2 Z^*_{MIN} for Cases Where $T \neq M$

Consider the situation where the average for a characteristic's output should be centered at a location other than the middle of the tolerance. Call this preferred location T, for "target" of the process output average. For example, in gold-plating silicon wafers, it is desired for economical reasons to keep the average plating thickness as low as possible, yet still meet the performance capability requirement. Thus, T would be set nearer to the LSL, as Figure 6.31 reveals.

Fig. 6.31 The target average is closer to the LSL than the USL.

To properly estimate process capability in this situation, create an artificial USL (called USL_T) that is the same distance above T as the LSL is below T (see Figure 6.32). This causes T to be positioned midway between the LSL and this newly created USL_T.

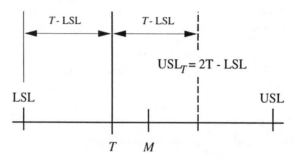

Fig. 6.32 Creating a surrogate upper specification limit, USL_T.

The formula for calculating USL_T is derived by adding the distance between T and the LSL (T minus LSL) to T. This positions USL_T at an identical distance above T as the LSL is below T. LSL_T in this situation is just equal to LSL.

$$\text{When } T < M, \quad USL_T = T + (T - LSL) = 2T - LSL \quad \text{and } LSL_T = LSL$$

Insert USL_T and LSL_T in place of the original USL and LSL in all formulas related to the Z_{MIN} method. Do *not* use them for anything else, like sorting parts or determining the amount of nonconforming parts. As mentioned in Chapter 5, the "*" notation denotes an altered version of the original capability measure for dimensions where T is not equal to M.

$$\hat{Z}^*_{LSL,ST} = \frac{\hat{\mu} - LSL_T}{\hat{\sigma}_{ST}} \qquad \hat{Z}^*_{USL,ST} = \frac{USL_T - \hat{\mu}}{\hat{\sigma}_{ST}}$$

Just as before, $Z^*_{MIN,ST}$ equals the smaller of these two Z values.

$$\hat{Z}^*_{MIN,ST} = \text{Minimum} \, (\hat{Z}^*_{LSL,ST}, \hat{Z}^*_{USL,ST})$$

When T is nearer the USL, create an artificial LSL_T to replace the original LSL in the capability formulas (see Figure 6.33). LSL_T is set to be as far below T as USL is above it. This causes T to be located midway between the LSL_T and the USL.

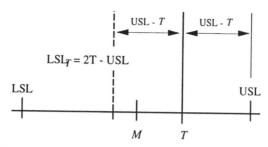

Fig. 6.33 Calculate LSL_T when T is closer to the USL than the LSL.

The formula for calculating LSL_T is derived by subtracting USL minus T from T. In this case, USL_T is left as the original USL.

$$\text{When } T > M, \quad LSL_T = T - (USL - T) = 2T - USL \quad \text{and } USL_T = USL$$

Employ this formula to estimate $\hat{Z}^*_{MIN.ST}$.

$$\hat{Z}^*_{MIN.ST} = \text{Minimum}\,(\hat{Z}^*_{LSL.ST}, \hat{Z}^*_{USL.ST}) = \text{Minimum}\left(\frac{\hat{\mu} - LSL_T}{\hat{\sigma}_{ST}}, \frac{USL_T - \hat{\mu}}{\hat{\sigma}_{ST}}\right)$$

Note that when T equals M, LSL_T is equal to the original LSL, and USL_T is equal to the original USL.

$$\text{If } T = M = \frac{USL + LSL}{2}$$

$$\text{then } LSL_T = 2T - USL = 2\left(\frac{USL + LSL}{2}\right) - USL = LSL$$

$$\text{and } USL_T = 2T - LSL = 2\left(\frac{USL + LSL}{2}\right) - LSL = USL$$

As there is no difference in the specifications employed in their calculation, Z^*_{MIN} now equals Z_{MIN}.

$$Z^*_{MIN.ST} = \text{Minimum}\left(\frac{\mu - LSL_T}{\sigma_{ST}}, \frac{USL_T - \mu}{\sigma_{ST}}\right)$$

$$= \text{Minimum}\left(\frac{\mu - LSL}{\sigma_{ST}}, \frac{USL - \mu}{\sigma_{ST}}\right) = Z_{MIN.ST}$$

Z^*_{MIN} is just a more generalized formula for Z_{MIN}, one which doesn't require T to equal the middle of the tolerance. In addition Z^*_{MIN} motivates shop floor personnel to center the process average at T, as will be demonstrated in this next example.

Example 6.8 Contact Closing Speed

A South Korean manufacturer checks circuit breakers for contact motion to verify they move at the proper opening and closing speed. Closing too quickly causes premature wearout of the contacts, while closing too slowly could burn or weld the contacts so the breaker could not be reopened. Their engineering department has developed the following window for closing speed to optimize the life of this product, with a target speed of .0170 seconds.

$$LSL = .0140 \text{ seconds} \qquad USL = .0250 \text{ seconds}$$

The middle of the tolerance is found to be .0195 seconds.

$$M = \frac{LSL + USL}{2} = \frac{.0140 \text{ seconds} + .0250 \text{ seconds}}{2} = .0195 \text{ seconds}$$

This information is portrayed below in Figure 6.34.

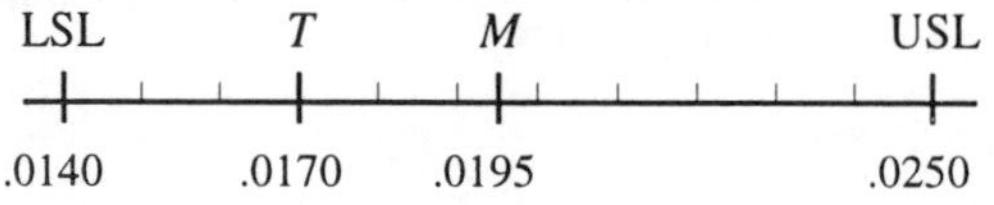

Fig. 6.34 Closing time requirements for a circuit breaker (in seconds).

From the control chart, σ_{ST} is estimated as .00129 seconds and μ as .0189. $Z_{MIN.ST}$ (which assumes M is best) is estimated as 3.80, which indicates good short-term performance capability.

$$\hat{Z}_{MIN.ST} = \text{Minimum}\left(\frac{\hat{\mu} - LSL}{\hat{\sigma}_{ST}}, \frac{USL - \hat{\mu}}{\hat{\sigma}_{ST}} \right)$$

$$= \text{Minimum}\left(\frac{.0189 - .0140}{.00129}, \frac{.0250 - .0189}{.00129} \right)$$

$$= \text{Minimum} (3.80, 4.73) = 3.80$$

As $Z_{USL.ST}$ is larger than $Z_{LSL.ST}$, performance capability can be increased by moving the process average higher. When it is eventually centered at the middle of the tolerance ($\mu = .0195$ seconds), $Z_{USL.ST}$ equals $Z_{LSL.ST}$ and $Z_{MIN.ST}$ is at its maximum of 4.26.

$$\text{Maximum } \hat{Z}_{MIN.ST} = \text{Minimum}\left(\frac{\hat{\mu} - LSL}{\hat{\sigma}_{ST}}, \frac{USL - \hat{\mu}}{\hat{\sigma}_{ST}} \right)$$

$$= \text{Minimum}\left(\frac{.0195 - .0140}{.00129}, \frac{.0250 - .0195}{.00129} \right)$$

$$= \text{Minimum} (4.26, 4.26) = 4.26$$

But this decision completely ignores engineering's desire for an average closing speed of .0170 seconds, which will maximize product life. The standard formula for $Z_{MIN.ST}$ misleads an operator into moving the average even further from the target. If $Z^*_{MIN.ST}$ is chosen to replace $Z_{MIN.ST}$ for measuring the performance capability of this process, the operator receives the correct signals for guiding his or her process to be centered at .0170 seconds. Because T is less than M in this example, the first step in applying this modified index is determining LSL_T and USL_T with these formulas:

$$LSL_T = LSL = .0140$$

$$USL_T = 2T - LSL = 2(.0170) - .0140 = .0200$$

The initial estimate for $Z^*_{MIN.ST}$ (with $\hat{\mu} = .0189$) is only .85 because the current average of .0189 seconds is too close to the USL_T of .0200, as indicated by the low $Z^*_{USL.ST}$ value.

$$\hat{Z}^*_{MIN.ST} = \text{Minimum} \left(\hat{Z}^*_{LSL.ST}, \hat{Z}^*_{USL.ST} \right)$$

$$= \text{Minimum} \left(\frac{\hat{\mu} - LSL_T}{\hat{\sigma}_{ST}}, \frac{USL_T - \hat{\mu}}{\hat{\sigma}_{ST}} \right)$$

$$= \text{Minimum} \left(\frac{.0189 - .0140}{.00129}, \frac{.0200 - .0189}{.00129} \right)$$

$$= \text{Minimum} (3.80, .85) = .85$$

The total percentage of breakers with nonconforming closing times is estimated by summing $\hat{p}'_{LSL.ST}$ and $\hat{p}'_{USL.ST}$. Because the percentage nonconforming is calculated with the actual specification limits (LSL and USL), the standard Z values of 3.80 and 4.73 replace $Z^*_{LSL.ST}$ and $Z^*_{USL.ST}$ in these next formulas.

$$\hat{p}'_{TOTAL.ST} = \Phi(\hat{Z}_{LSL.ST}) + \Phi(\hat{Z}_{USL.ST})$$

$$= \Phi(3.80) + \Phi(4.73)$$

$$= .0000725 + .0000011 = .0000736 \cong .0074\%$$

As $Z^*_{USL.ST}$ is much smaller than $Z^*_{LSL.ST}$, the operator realizes $Z^*_{MIN.ST}$ can be significantly increased by adjusting the process settings to *lower* the average closing speed. This response is exactly opposite the one prompted by the $Z_{MIN.ST}$ measure.

The largest $Z^*_{MIN.ST}$ can be without reducing σ_{ST} is 2.33, as is derived here.

$$\text{Maximum } \hat{Z}^*_{MIN.ST} = \frac{\hat{Z}_{LSL.ST} + \hat{Z}_{USL.ST}}{2} = \frac{3.80 + .85}{2} = 2.33$$

The required change in the process average to achieve this maximum value for $Z^*_{MIN.ST}$ is determined from the difference in estimated Z^* values.

$$\text{Shift in Process Average} = \frac{\hat{\sigma}_{ST}}{2}(\hat{Z}^*_{USL.ST} - \hat{Z}^*_{LSL.ST}) = \frac{.00129}{2}(.85 - 3.80) = -.0019$$

Suppose the average is successfully lowered by .0019 from its current value of .0189 to the target of .0170, as is displayed in Figure 6.35. With this new average, $Z^*_{MIN.ST}$ achieves its maximum possible value of 2.33.

$$\hat{Z}^*_{MIN.ST} = \text{Minimum} \left(\frac{\hat{\mu} - LSL_T}{\hat{\sigma}_{ST}}, \frac{USL_T - \hat{\mu}}{\hat{\sigma}_{ST}} \right)$$

$$= \text{Minimum} \left(\frac{.0170 - .0140}{.00129}, \frac{.0200 - .0170}{.00129} \right)$$

$$= \text{Minimum} \, (2.33, 2.33) = 2.33$$

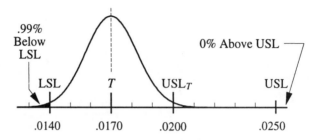

Fig. 6.35   Process average is shifted to the target of .0170 seconds.

Any movement in the process average higher, or lower, will cause $Z^*_{MIN.ST}$ to decrease, providing an incentive for the operator to leave the process centered at the target average of .0170 seconds. Unfortunately, the process fails to meet the minimum prerequisite for performance capability when centered here. Work must begin to reduce process variation. In fact, centering the process at T increases the total percentage of breakers with nonconforming closing times from .0074 percent to almost 1 percent. Remember, the standard Z values for μ equal to .0170 seconds must be used in this calculation.

$$\hat{p}'_{TOTAL,ST} = \Phi(\hat{Z}_{LSL,ST}) + \Phi(\hat{Z}_{USL,ST})$$

$$= \Phi(2.33) + \Phi(6.20)$$

$$= .0099030 + .0000000$$

$$= .0099030 \cong .99\%$$

The penalty for specifying a target different from M is an increased percentage of nonconforming parts. For a process having a normally distributed output, this percentage will always be minimized when the average is located exactly at the middle of the tolerance.

Relationship Between $Z^*_{MIN,ST}$ and μ When $T < M$

$Z^*_{MIN,ST}$ attains its largest value for a given σ_{ST} when μ equals T, as is conveyed by the graph in Figure 6.36. This property encourages those operating the process to keep the process average centered directly at T, the desired target value, because doing this maximizes performance capability as measured by $Z^*_{MIN,ST}$.

If μ is equal to either the LSL_T or USL_T, $Z^*_{MIN,ST}$ is equal to 0. When μ is above USL_T, then $Z^*_{MIN,ST}$ becomes negative. However, unlike $Z_{MIN,ST}$, there is *less* than 50 percent nonconforming parts produced when μ equals USL_T and $Z^*_{MIN,ST}$ becomes 0. The standard $Z_{USL,ST}$ value must be examined to determine the percentage of parts above the USL. Nevertheless, a strong signal is sent to move the process average back down toward the target.

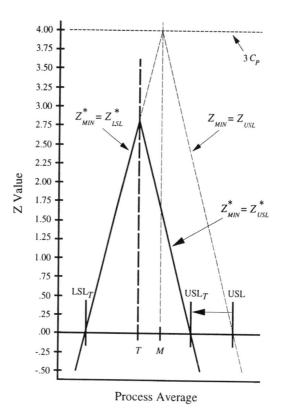

Fig. 6.36 Z^*_{MIN} versus the process average.

As a result of reducing the effective tolerance by the difference between USL_T and the original USL, the maximum possible value for $Z^*_{MIN,ST}$ (attained when μ equals T) is somewhat *less* than the maximum for $Z_{MIN,ST}$ (achieved when μ equals M). This reduction can be quantified in several ways, as is discussed in the next section.

Relationship of Maximum $Z^*_{MIN,ST}$ to C^*_P

C^*_P estimates short-term potential capability for contact closing speed, assuming μ is centered at T.

$$\hat{C}^*_P = \text{Minimum}\left(\frac{T-\text{LSL}}{3\hat{\sigma}_{ST}}, \frac{\text{USL}-T}{3\hat{\sigma}_{ST}}\right)$$

$$= \text{Minimum}\left(\frac{.0170-.0140}{3(.00129)}, \frac{.0250-.0170}{3(.00129)}\right)$$

$$= \text{Minimum}(.775, 2.067) = .775$$

The largest possible $Z^*_{MIN,ST}$ value for the contact closing speed example (assuming constant σ_{ST}) can be estimated by multiplying this C^*_P value by 3.

$$\text{Maximum } \hat{Z}^*_{MIN.\,ST} = 3\,\hat{C}^*_P = 3\,(.775) = 2.33$$

USL_T and LSL_T can also be used in the estimation of C^*_P by calculating the "effective" tolerance. This is the tolerance available to a symmetric output distribution centered at the target average.

$$\text{Effective Tolerance} = USL_T - LSL_T$$

In situations where T equals M, LSL_T equals LSL and USL_T equals USL. This makes the effective tolerance identical to the original tolerance.

$$\text{Effective Tolerance} = USL_T - LSL_T = USL - LSL = \text{Tolerance}$$

If T is not equal to M, LSL is less than or equal to LSL_T, while USL_T is less than or equal to USL.

$$LSL \le LSL_T$$

$$USL_T \le USL$$

This causes the effective tolerance to become less than the original tolerance. This is seen by adding the two above inequalities, then rearranging terms.

$$USL_T + LSL \le USL + LSL_T$$

$$USL_T - LSL_T \le USL - LSL$$

$$\text{Effective Tolerance} \le \text{Tolerance}$$

For contact closing speed, the full tolerance is .0110 (.0250 minus .0140), whereas the effective tolerance is only .0060 seconds.

$$\text{Effective Tolerance} = USL_T - LSL_T = .0200 \text{ seconds} - .0140 \text{ seconds} = .0060 \text{ seconds}$$

C^*_P may now be calculated in much the same manner as C_P, but replacing the full tolerance with the effective tolerance.

$$\hat{C}^*_P = \frac{\text{Effective Tolerance}}{6\hat{\sigma}_{ST}}$$

For contact closing time, C^*_P is estimated with this formula as .775.

$$\hat{C}^*_P = \frac{\text{Effective Tolerance}}{6\hat{\sigma}_{ST}} = \frac{.0060 \text{ seconds}}{6(.00129 \text{ seconds})} = .775$$

As this is identical to the result obtained from the original formula, either method may be selected for calculating C^*_P. Thus, the largest possible $Z^*_{MIN.ST}$ value for this contact closing time example may also be computed directly from the effective tolerance.

$$\text{Maximum } \hat{Z}^*_{MIN.\,ST} = 3\,\hat{C}^*_P$$

$$= \frac{3 \text{ Effective Tolerance}}{6\hat{\sigma}_{ST}}$$

$$= \frac{\text{Effective Tolerance}}{2\hat{\sigma}_{ST}}$$

$$= \frac{.0060 \text{ seconds}}{2(.00129 \text{ seconds})} = 2.33$$

Relationship Between Maximum $Z^*_{MIN,ST}$ and C_P

The following relationship between C_P and C^*_P was derived in Chapter 5 (page 143).

$$\hat{C}^*_P = \hat{C}_P (1 - k_T) \quad \text{where } k_T = \frac{|T - M|}{\text{Tolerance} / 2}$$

As the largest possible $Z^*_{MIN,ST}$ value is equal to 3 times C^*_P, then:

$$\text{Maximum } \hat{Z}^*_{MIN.\,ST} = 3\,\hat{C}^*_P = 3\,\hat{C}_P (1 - k_T)$$

In the contact closing speed example, C_P is estimated as 1.421, while k_T is .4545.

$$\hat{C}_P = \frac{\text{Tolerance}}{6\hat{\sigma}_{ST}} = \frac{.0250 - .0140}{6(.00129)} = 1.421$$

$$k_T = \frac{|T - M|}{\text{Tolerance} / 2} = \frac{|.0170 - .0195|}{(.0250 - .0140)/2} = .4545$$

This makes the largest possible $Z^*_{MIN,ST}$ value equal to 2.33

$$\text{Maximum } \hat{Z}^*_{MIN.\,ST} = 3\,\hat{C}_P (1 - k_T) = 3(1.421)(1 - .4545) = 2.33$$

k_T measures the loss of potential capability due to specifying a target not equal to M.

Relationship of Maximum $Z^*_{MIN,ST}$ to Maximum $Z_{MIN,ST}$

Because $Z_{MIN,ST}$ is equal to $3C_P$, then:

$$\text{Maximum } \hat{Z}^*_{MIN.\,ST} = 3\,\hat{C}_P (1 - k_T) = \text{Maximum } \hat{Z}_{MIN.\,ST} (1 - k_T)$$

In the case study for contact closing time:

$$\text{Maximum } \hat{Z}^*_{MIN.ST} = \text{Maximum } \hat{Z}_{MIN.ST} (1 - k_T) = 4.263(1 - .4545) = 2.33$$

Specifying a target different than M reduces the maximum obtainable performance capability. For this process, the maximum $Z^*_{MIN.ST}$ value is 2.33, compared to a possible 4.263 if the full tolerance was available. This represents a rather substantial reduction in capability of over 45 percent. Notice that when T equals M, k_T becomes 0, and the maximum $Z^*_{MIN.ST}$ value equals the maximum for $Z_{MIN.ST}$.

$$\text{Maximum } Z^*_{MIN,ST} = \text{Maximum } Z_{MIN,ST}(1-0) = \text{Maximum } Z_{MIN,ST}$$

Whenever T is not equal to M, k_T is positive, which makes 1 minus k_T less than 1. Under this condition, the maximum $Z^*_{MIN,ST}$ value is always less than the maximum for $Z_{MIN,ST}$.

$$1 - k_T < 1$$

$$\text{Maximum } Z_{MIN,ST}(1-k_T) < \text{Maximum } Z_{MIN,ST}$$

$$\text{Maximum } Z^*_{MIN,ST} < \text{Maximum } Z_{MIN,ST}$$

Example 6.9 Center Gap Size

In Examples 3.1 and 3.3, estimates of μ (.3182 cm) and σ_{ST} (.003339 cm) were made from an $\overline{X}, R$ chart monitoring the center gap between the two bottom flaps folded in a process making corrugated boxes. With a print specification of .3200 cm $\pm$ .0160 cm, the largest estimate possible for $Z_{MIN,ST}$ occurs when μ is located at .3200 cm, the middle of the tolerance.

$$\text{Maximum } \hat{Z}_{MIN,ST} = \text{Minimum} \left(\frac{\hat{\mu} - LSL}{\hat{\sigma}_{ST}}, \frac{USL - \hat{\mu}}{\hat{\sigma}_{ST}} \right)$$

$$= \text{Minimum} \left(\frac{.3200 - .3040}{.003339}, \frac{.3360 - .3200}{.003339} \right)$$

$$= \text{Minimum}\,(4.792, 4.792) = 4.792$$

This analysis assumes the middle of the tolerance is best for gap size. What if a target average of .3215 cm is specified instead? As T is greater than M, the following formulas apply for figuring LSL_T and USL_T.

$$LSL_T = 2T - USL = 2(.3215) - .3360 = .3070$$

$$USL_T = USL = .3360$$

The maximum estimate for $Z^*_{MIN,ST}$ occurs when the process average is centered at the target of .3215 cm.

$$\text{Maximum } \hat{Z}^*_{MIN,ST} = \text{Minimum }(\hat{Z}^*_{LSL,ST}, \hat{Z}^*_{USL,ST})$$

$$= \text{Minimum} \left(\frac{\hat{\mu} - LSL_T}{\hat{\sigma}_{ST}}, \frac{USL_T - \hat{\mu}}{\hat{\sigma}_{ST}} \right)$$

$$= \text{Minimum} \left(\frac{.3215 - .3070}{.003339}, \frac{.3360 - .3215}{.003339} \right)$$

$$= \text{Minimum }(4.343, 4.343) = 4.343$$

C^*_P, C_P, and k_T may also be calculated for this process.

$$\hat{C}_P = \frac{\text{Tolerance}}{6\hat{\sigma}_{ST}} = \frac{.3360 - .3040}{6(.003339)} = 1.597$$

$$\hat{C}_P^* = \frac{\text{Effective Tolerance}}{6\hat{\sigma}_{ST}} = \frac{.3360 - .3070}{6(.003339)} = 1.448$$

$$k_T = \frac{|T - M|}{\text{Tolerance}\,/\,2} = \frac{|.3215 - .3200|}{(.3360 - .3040)\,/2} = .0938$$

In addition to the approach taken above, the maximum possible $Z^*_{MIN.ST}$ estimate may be derived with any of the following three methods, since all three yield comparable results. First, from C^*_P:

$$\text{Maximum } \hat{Z}^*_{MIN.\,ST} = 3\,\hat{C}_P^* = 3\,(1.448) = 4.344$$

By using C_P and k_T:

$$\text{Maximum } \hat{Z}^*_{MIN.\,ST} = 3\,\hat{C}_P\,(1 - k_T) = 3\,(1.597)(1 - .0938) = 4.342$$

From the maximum possible estimate of $Z_{MIN.ST}$ and k_T:

$$\text{Maximum } \hat{Z}^*_{MIN.\,ST} = \text{Maximum } \hat{Z}_{MIN.\,ST}\,(1 - k_T) = 4.792\,(1 - .0938) = 4.343$$

By imposing the additional stipulation that μ must be centered at the target average of .3215 cm, the highest achievable performance capability (without reducing σ_{ST}) is 9.38 percent less than without this restriction.

Example 6.10 Solder Pull Test

Wire leads are soldered onto printed circuit boards like the one displayed in Figure 6.37. Each hour, a lead is checked by measuring the force required to pull it from the board, called a "pull test." Acceptable leads must be able to withstand a minimum pull force of 25 grams (LSL = 25 gm).

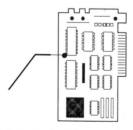

Fig. 6.37 Wire lead soldered to circuit board.

When this process is in control, the average pull test is estimated as 35 grams, with the short-term standard deviation estimated as 5 grams. A target of 40 grams is specified for the average pull test result. In Example 5.23 on page 148, potential capability was estimated with C'_{PL} as 1.00, and with C'_{STL} as 6.00. These modified versions of C_P and C_{ST} were applicable due to the unilateral specification for pull strength.

Because a target average is given for this process, many quality practitioners believe the

$Z^*_{MIN,ST}$ index should be chosen for measuring the performance capability of this soldering process. However, with only a LSL for pull strength, most improvement efforts will concentrate on making the average pull test as high as possible, even higher than the target. This objective differs greatly from that for characteristics with bilateral specifications, where the process average should be located exactly at the target, not higher or lower. This second situation is where $Z^*_{MIN,ST}$ applies, since this index reaches its maximum value only when the process average equals T.

To correctly measure performance capability for this unilateral situation, the standard $Z_{MIN,ST}$ index should be selected (with only a LSL given, $Z_{MIN,ST}$ equals $Z_{LSL,ST}$). If improvements to this soldering operation are successful in moving the average pull test higher, $Z_{MIN,ST}$ will increase, thus encouraging the stakeholders in this process to continue their efforts.

$$\hat{Z}_{MIN,ST} = \hat{Z}_{LSL,ST} = \frac{\hat{\mu} - LSL}{\hat{\sigma}_{ST}} = \frac{35 - 25}{5} = 2.00$$

If the average pull test can be increased to the target of 40 grams, $Z_{MIN,ST}$ jumps to 3.00.

$$\hat{Z}_{MIN,ST} = \frac{\hat{\mu} - LSL}{\hat{\sigma}_{ST}} = \frac{40 - 25}{5} = 3.00$$

Since μ is now in the target zone, $Z_{MIN,ST}$ becomes equal to half of C'_{STL}.

$$\hat{Z}_{LSL,ST} = \frac{\hat{C}'_{STL}}{2} = \frac{6.00}{2} = 3.00$$

When the average is moved to 45 grams, $Z_{MIN,ST}$ climbs to 4.00.

$$\hat{Z}_{MIN,ST} = \frac{45 - 25}{5} = 4.00$$

As μ is pushed higher, performance capability continues to increase, as is displayed in Figure 6.38.

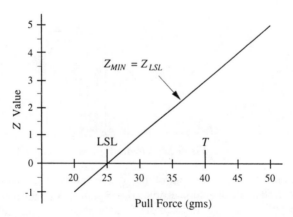

Fig. 6.38 Z_{MIN} continues to increase as μ becomes larger.

If, for some reason, the customer decides that μ should be located at T, and *not* higher, then $Z^*_{MIN,ST}$ becomes the appropriate capability measure. As there is no USL, T is closer to the LSL and these formulas apply for determining LSL_T and USL_T.

$$LSL_T = LSL = 25 \qquad USL_T = 2T - LSL = 2(40) - 25 = 55$$

When capability is assessed with this measure, moving the average to 45 grams decreases performance to 2.00, compared with 3.00 when centered at the target of 40 grams.

$$\hat{Z}^*_{MIN.ST} = \text{Minimum} \left(\hat{Z}^*_{LSL.ST}, \hat{Z}^*_{USL.ST} \right)$$

$$= \text{Minimum} \left(\frac{\hat{\mu} - LSL_T}{\hat{\sigma}_{ST}}, \frac{USL_T - \hat{\mu}}{\hat{\sigma}_{ST}} \right)$$

$$= \text{Minimum} \left(\frac{45 - 25}{5}, \frac{55 - 45}{5} \right)$$

$$= \text{Minimum} \left(4.00, 2.00 \right) = 2.00$$

A signal is sent to keep the average at the target of 40 grams, accomplishing the customer's wish. This mechanism is portrayed graphically in Figure 6.39.

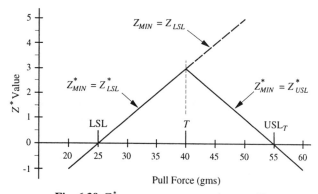

Fig. 6.39 Z^*_{MIN} encourages centering μ at T.

Remember, USL_T is merely a surrogate specification, created for the sole purpose of calculating Z^*_{MIN}, which helps keep the process centered at T. The portion of the output distribution extending above USL_T, but below USL, represents *conforming* parts. However, because they are so far from T, these parts are certainly less desirable to the customer than those between LSL_T and USL_T

Advantages and Disadvantages

These are similar to those stated for Z_{MIN}, with the exception that Z^*_{MIN} does not assume T is equal to M. This index of performance capability is related to the target average of the characteristic being assessed. As the process average moves closer to T, Z^*_{MIN} increases. When the average moves away from T, Z^*_{MIN} decreases. However, this more generalized measure requires additional calculations for LSL_T and USL_T, plus the determination of a target value (references for selecting appropriate target values are listed in Section 5.9).

A marginal drawback of this index is the unusual rating scale of 3, which is remedied with the performance capability measures introduced in the next section.

6.3 C_{PL}, C_{PU}, P_{PL}, and P_{PU} Indexes

One minor inconvenience associated with the Z_{MIN} method is its unusual rating scaling, where 3.0 represents attainment of the minimum performance capability requirement. For most other measures, 1.0 indicates meeting this prerequisite. The C_{PL}, C_{PU}, P_{PL}, and P_{PU} indexes address this problem by dividing their respective Z values by 3 (Kane, 1986). C_{PL} and P_{PL} are referred to as the lower performance capability indexes, whereas C_{PU} and P_{PU} are called the upper performance capability indexes.

$$C_{PL} = \frac{Z_{LSL,ST}}{3} = \frac{\mu - LSL}{3\sigma_{ST}} \qquad \hat{C}_{PL} = \frac{\hat{Z}_{LSL,ST}}{3} = \frac{\hat{\mu} - LSL}{3\hat{\sigma}_{ST}}$$

$$C_{PU} = \frac{Z_{USL,ST}}{3} = \frac{USL - \mu}{3\sigma_{ST}} \qquad \hat{C}_{PU} = \frac{\hat{Z}_{USL,ST}}{3} = \frac{USL - \hat{\mu}}{3\hat{\sigma}_{ST}}$$

$$P_{PL} = \frac{Z_{LSL,LT}}{3} = \frac{\mu - LSL}{3\sigma_{LT}} \qquad \hat{P}_{PL} = \frac{\hat{Z}_{LSL,LT}}{3} = \frac{\hat{\mu} - LSL}{3\hat{\sigma}_{LT}}$$

$$P_{PU} = \frac{Z_{USL,LT}}{3} = \frac{USL - \mu}{3\sigma_{LT}} \qquad \hat{P}_{PU} = \frac{\hat{Z}_{USL,LT}}{3} = \frac{USL - \hat{\mu}}{3\hat{\sigma}_{LT}}$$

By measuring how many 3σ spreads fit into the distance from the process average to a specification limit, these indexes have the more common rating scale where 1.0 denotes the minimum capability requirement.

< 1 discloses the process fails to meet the minimum condition for performance capability,

$= 1$ implies the process just meets the minimum requirement,

> 1 indicates that the process exceeds the minimum criterion.

A C_{PL} of 1.2 implies the process has short-term performance capability with regard to the LSL, as the process average is a distance of $3.6\sigma_{ST}$ (1.2×3) above the LSL.

$$C_{PL} = \frac{\mu - LSL}{3\sigma_{ST}} = 1.2 \quad \Rightarrow \quad \mu - LSL = 1.2(3\sigma_{ST}) = 3.6\sigma_{ST}$$

Another way of expressing this relationship is to say the length of one $3\sigma_{ST}$ spread is only 83 percent ($1/1.2 \times 100$) of the distance from the process average to the LSL.

$$1.2(3\sigma_{ST}) = \mu - LSL \quad \Rightarrow \quad 3\sigma_{ST} = \frac{1}{1.2}(\mu - LSL) = .83(\mu - LSL)$$

This percentage equals the C_{RL} ratio described in Section 5.6 on page 119. As mentioned there, due to the inclusion of the process average in the calculation of C_{RL}, it should really be considered a measure of performance capability.

$$3\sigma_{ST} = .83\,(\mu - LSL)$$

$$\frac{3\sigma_{ST}}{\mu - LSL} = .83$$

$$C_{RL} = .83$$

These two measures are related as follows:

$$C_{RL} = \frac{1}{C_{PL}} = \frac{1}{1.2} = .83$$

Likewise, the following measures are related as shown here:

$$C_{PU} = \frac{1}{C_{RU}} \qquad P_{PL} = \frac{1}{P_{RL}} \qquad P_{PU} = \frac{1}{P_{RU}}$$

Goals

A very common goal for these indexes is to be at least 1.33. This corresponds to a Z_{MIN} goal of 4 or more, as demonstrated below for the C_{PL} index.

$$C_{PL,GOAL} = \frac{Z_{LSL,ST,GOAL}}{3} = \frac{4}{3} = 1.33$$

Other often-used goals are 1.50 (equates to a Z value of 4.5), 1.67 (a Z value of 5), and 2.00 (identical to a Z value of 6).

$$C_{PU,GOAL} = \frac{Z_{LSL,ST,GOAL}}{3} = \frac{4.5}{3} = 1.50$$

$$P_{PL,GOAL} = \frac{Z_{LSL,LT,GOAL}}{3} = \frac{5}{3} = 1.67 \qquad\qquad P_{PU,GOAL} = \frac{Z_{USL,LT,GOAL}}{3} = \frac{6}{3} = 2.00$$

Example 6.11 Center Gap Size

Estimates of μ (.3182 cm) and σ_{ST} (.003339 cm) are made from an $\overline{X},R$ chart monitoring the center gap between the two bottom flaps folded in a process making corrugated boxes. The print specification for gap size is .3200 cm $\pm$ .0160 cm, while the goal for this process is to have both C_{PL} and C_{PU} be at least 1.33.

$$\hat{C}_{PL} = \frac{\hat{\mu} - LSL}{3\hat{\sigma}_{ST}} = \frac{.3182 - .3040}{3(.003339)} = 1.42 \qquad \hat{C}_{PU} = \frac{USL - \hat{\mu}}{3\hat{\sigma}_{ST}} = \frac{.3360 - .3182}{3(.003339)} = 1.78$$

Since both these measures exceed their goal of 1.33, this process has good short-term performance capability. Because C_{PL} is less than C_{PU}, performance capability can be increased by moving the process average lower until C_{PL} and C_{PU} are equal, which will occur when μ equals .3200 cm. At this point, both C_{PL} and C_{PU} equal C_P.

Example 6.12 Particle Contamination

An *IX & MR* chart is monitoring particle contamination (in *ppm*) of a cleaning solution for printed circuit boards after they go through wave soldering. The following estimates of process parameters are made from the 25 most recent in-control readings.

$$\hat{\mu} = 127.32 \, ppm \qquad \hat{\sigma}_{ST} = 12.079 \, ppm \qquad \hat{C}'_{PU} = 1.24$$

Because contamination has just an USL of 165 *ppm*, only C_{PU} can be estimated. This is why C_{PU} (and C_{PL}) is sometimes referred to as a "one-sided" performance capability index.

$$\hat{C}_{PU} = \frac{USL - \hat{\mu}}{3\hat{\sigma}_{ST}} = \frac{165 - 127.32}{3(12.079)} = 1.04$$

Although this process meets the minimum requirement for performance capability, it does not meet a performance capability goal of 1.67.

The formula for C_{PU} (C_{PL}) is identical to the formula for C'_{PU} (C'_{PL}) when no target average is specified, or if the average is located in the target zone. As mentioned in Section 5.10, and can be seen here, C_{PU} (C_{PL}) is actually a measure of performance capability because it is influenced by both process spread and centering.

$$C_{PU} = C'_{PU} = \frac{USL - \mu}{3\sigma_{ST}} \qquad\qquad C_{PL} = C'_{PL} = \frac{\mu - LSL}{3\sigma_{ST}}$$

Example 6.13 Solder Pull Test

Wire leads are soldered onto printed circuit boards. Each hour, a lead is checked by measuring the force required to pull it from the board, called a "pull test." Acceptable leads must be able to withstand a minimum pull force of 25 grams. Test results are plotted on an *IX & MR* chart to check the stability of this soldering process. When in control, the average pull test result is estimated as 35 grams, with the short-term standard deviation estimated as 5 grams. With a target average for pull test of 40 grams, the short-term potential capability of this process is estimated as 1.00 with the C'_{PL} index (review Example 5.23).

$$\hat{C}'_{PL} = \text{Maximum} \, (\, \hat{C}^*_{PL}, \, \hat{C}_{PL} \,)$$

$$= \text{Maximum} \left(\frac{T - LSL}{3\hat{\sigma}_{ST}}, \frac{\hat{\mu} - LSL}{3\hat{\sigma}_{ST}} \right)$$

$$= \text{Maximum} \left(\frac{40 - 25}{3(5)}, \frac{35 - 25}{3(5)} \right)$$

$$= \text{Maximum} \, (\, 1.00, .67 \,)$$

$$= 1.00$$

Because the average pull test result of 35 grams is less than the target of 40 grams, the performance capability of this process, as measured by C_{PL}, is much less than the potential of 1.00.

$$\hat{C}_{PL} = \frac{\hat{\mu} - \text{LSL}}{3\hat{\sigma}_{ST}} = \frac{35 - 25}{3(5)} = .67$$

If the process average is raised to 40 grams, performance capability will equal potential capability.

$$\hat{C}_{PL} = \frac{\hat{\mu} - \text{LSL}}{3\hat{\sigma}_{ST}} = \frac{40 - 25}{3(5)} = 1.00 = \hat{C}'_{PL}$$

When the process average is boosted above the target to 45 grams, both potential capability (as measured by C'_{PL}) and performance capability (as measured by C_{PL}) increase.

$$\hat{C}'_{PL} = \text{Maximum}\left(\frac{T - \text{LSL}}{3\hat{\sigma}_{ST}}, \frac{\hat{\mu} - \text{LSL}}{3\hat{\sigma}_{ST}} \right)$$

$$= \text{Maximum}\left(\frac{40 - 25}{3(5)}, \frac{45 - 25}{3(5)} \right) \qquad \hat{C}_{PL} = \frac{\hat{\mu} - \text{LSL}}{3\hat{\sigma}_{ST}} = \frac{45 - 25}{3(5)} = 1.33$$

$$= \text{Maximum}\,(\,1.00\,,1.33\,)$$

$$= 1.33$$

Relationship of C_{PL} and C_{PU} to the Process Average

Because C_{PL} and C_{PU} are calculated from both σ_{ST} and μ, changes in the process average have a major effect on their values, as is evidenced by their slanted lines in Figure 6.40. Observe how both measures vary when the process average changes, with one becoming larger as the other becomes smaller. They equal each other only when the process average is at M, the middle of the tolerance. If the process average equals the LSL (USL), then C_{PL} (C_{PU}) is 0. When the process average is outside a print limit, either C_{PL} or C_{PU} is negative, indicating a serious capability problem with more than a 50 percent nonconformance rate.

Notice how the scaling is now comparable to C_P, *i.e.*, a measure of 1.0 indicates attainment of the minimum capability requirement. When the Z values were plotted on a similar graph (Figure 6.36), a y-axis scale extending all the way to 4.0 was needed.

When the process average is equal to M, C_{PL} equals C_{PU}, which now also equals C_P. Under these circumstances, performance capability is identical to potential capability, which is the best performance possible (*i.e.*, the minimum percentage of nonconforming parts) for the current process. In addition, the distance from the process average to the LSL is identical to that between the average and the USL. Both of these distances are now equal to one-half of the tolerance.

$$\text{When } \mu = M, \ \mu - \text{LSL} = \text{USL} - \mu = 1/2 \text{ Tolerance}$$

For this special situation, the C_P formula may be rewritten as shown on the next page underneath Figure 6.40.

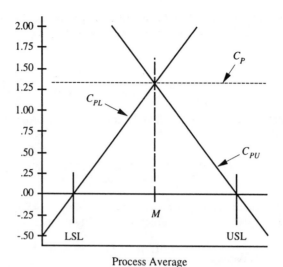

Fig. 6.40 C_{PL} and C_{PU} versus the process average.

$$C_P = \frac{\text{Tolerance}}{6\sigma_{ST}} = \frac{1/2\,\text{Tolerance}}{1/2\,(6\sigma_{ST})}$$

$$= \frac{\mu - \text{LSL}}{3\sigma_{ST}} = C_{PL}$$

$$= \frac{\text{USL} - \mu}{3\sigma_{ST}} = C_{PU}$$

Example 6.14 Floppy Disks

A field problem plaguing a manufacturer of floppy disks is traced to variation in the diameter of an inner hole. The following information is noted from an $\overline{X}, R$ chart monitoring the hole-cutting process, as well as the results of a capability study.

$$\text{LSL} = 39.5 \quad \text{USL} = 44.5 \quad \hat{C}_R = 1.078 \quad \hat{P}_R = 1.099 \quad \hat{C}_p = .928 \quad \hat{P}_p = .909$$

$$\hat{\mu} = 40.9 \quad \hat{\sigma}_{ST} = .898 \quad \hat{\sigma}_{LT} = .916 \quad \hat{Z}_{LSL,ST} = 1.559 \quad \hat{Z}_{USL,ST} = 4.009$$

The short-term performance capability measures are calculated from $\hat{\sigma}_{ST}$ and $\hat{\mu}$.

$$\hat{C}_{PL} = \frac{\hat{\mu} - \text{LSL}}{3\hat{\sigma}_{ST}} = \frac{40.9 - 39.5}{3(.898)} = .52 \qquad \hat{C}_{PU} = \frac{\text{USL} - \hat{\mu}}{3\hat{\sigma}_{ST}} = \frac{44.5 - 40.9}{3(.898)} = 1.34$$

These two capability measures can also be derived directly from their respective Z values.

$$\hat{C}_{PL} = \frac{\hat{Z}_{LSL,ST}}{3} = \frac{1.559}{3} = .52 \qquad \hat{C}_{PU} = \frac{\hat{Z}_{USL,ST}}{3} = \frac{4.009}{3} = 1.34$$

Because 1.0 is the minimum requirement for performance capability, the estimate of .52 for C_{PL} flags a serious capability problem with meeting the LSL of 39.5 for hole size. An estimate of 1.34 for C_{PU} shows the process has no trouble holding the USL of 44.5. Moving the process average closer to the USL will increase C_{PL}, but at the expense of C_{PU}. When the average is centered at 42.0, both measures will equal the C_p index of .928.

$$\hat{C}_{PL} = \frac{\hat{\mu} - LSL}{3\hat{\sigma}_{ST}} = \frac{42.0 - 39.5}{3(.898)} = .928 = \hat{C}_p$$

$$\hat{C}_{PU} = \frac{USL - \hat{\mu}}{3\hat{\sigma}_{ST}} = \frac{44.5 - 40.9}{3(.898)} = .928 = \hat{C}_p$$

The long-term performance capability measures are based on estimates of σ_{LT} and μ. As seen previously with the relationship between P_p and C_p, P_{PL} is normally less than C_{PL} (and P_{PU} less than C_{PU}), due to σ_{LT} being typically greater than σ_{ST}.

$$\hat{P}_{PL} = \frac{\hat{Z}_{LSL,LT}}{3} = \frac{\overline{\overline{X}} - LSL}{3\hat{\sigma}_{LT}} = \frac{40.9 - 39.5}{3(.916)} = .51$$

$$\hat{P}_{PU} = \frac{\hat{Z}_{USL,LT}}{3} = \frac{USL - \overline{\overline{X}}}{3\hat{\sigma}_{LT}} = \frac{44.5 - 40.9}{3(.916)} = 1.31$$

These new measures make comparing potential and performance capability measures much easier as they are now all on the same rating scale. C_p doesn't have to be multiplied by 3 so it can be compared to $Z_{MIN,ST}$. In this floppy disk example, the C_p index of .928 may be directly compared to the C_{PL} index of .52, revealing that C_{PL} can be increased to .928 by centering the process average at M.

The following formula determines the required shift in the process average to maximize C_{PL} and C_{PU}. In this example, an upward movement of 1.10 units is necessary.

$$\text{Shift in } \mu = M - \hat{\mu} = 42.0 - 40.9 = 1.10$$

Alternatively, the shift may be figured with this next formula to also be 1.10. Because $\hat{\mu}$ is currently 40.9, the process average should be increased to 42.0 (40.9 + 1.10).

$$\text{Shift in } \mu = 1.5\,\hat{\sigma}_{ST}(\hat{C}_{PU} - \hat{C}_{PL}) = 1.5(.898)(1.34 - .52) = 1.10$$

C_p (P_p) could also be directly calculated from C_{PL} and C_{PU} (P_{PL} and P_{PU}) when the process is normally distributed.

$$\hat{C}_p = \frac{\hat{C}_{PL} + \hat{C}_{PU}}{2} = \frac{.52 + 1.34}{2} = .93 \qquad \hat{P}_p = \frac{\hat{P}_{PL} + \hat{P}_{PU}}{2} = \frac{.51 + 1.31}{2} = .91$$

If the process average is less than M, C_p is always greater than C_{PL} and less than C_{PU} (review Figure 6.40), while P_p is always greater than P_{PL} and less than P_{PU}.

$$\text{When } \mu < M, \quad C_{PL} < C_p < C_{PU} \quad \text{and} \quad P_{PL} < P_p < P_{PU}$$

On the other hand, if the process average is greater than M, then C_P is less than C_{PL} and greater than C_{PU}, while P_P is always less than P_{PL} and larger than P_{PU}.

$$\text{When } \mu > M, \quad C_{PL} > C_P > C_{PU} \quad \text{and} \quad P_{PL} > P_P > P_{PU}$$

For the particular situation where the process average is centered at the middle of tolerance ($\mu = M$), all three are equal.

$$\text{When } \mu = M, \quad C_{PL} = C_P = C_{PU} \quad \text{and} \quad P_{PL} = P_P = P_{PU}$$

Advantages and Disadvantages

These indexes have the same advantages and disadvantages of the Z_{MIN} method with the exception of the scale change. They also work equally well with bilateral or unilateral specifications. The performance measures C_{PL} and C_{PU} can be directly compared to C_P, a measure of potential capability. Likewise, P_{PL} and P_{PU} may be compared to P_P, as they are all on the same rating scale.

Although C_{PL} and C_{PU} are easier to work with than their Z values as a result of being on the same rating scale as C_P, some people prefer having only *one* number to represent process capability instead of two. For them, the C_{PK} index covered in Section 6.5 provides a solution.

A minor disadvantage is that these indexes must first be multiplied by 3 in order to look up the corresponding percentage of nonconforming parts in Appendix Table III.

$$C_{PL} = \frac{Z_{LSL,ST}}{3} \quad \Rightarrow \quad 3\,C_{PL} = Z_{LSL,ST}$$

They also assume the middle of the tolerance is best for the process average ($T = M$). If this is not true, then these measures must be modified so they encourage centering the process average at the desired target.

6.4 C^*_{PL}, C^*_{PU}, P^*_{PL}, and P^*_{PU} for Cases Where $T \neq M$

The modification made to the Z measures for characteristics where T is not equal to M can also be applied to C_{PL}, C_{PU}, P_{PL}, and P_{PU}. This adjustment requires first determining LSL_T and USL_T with these formulas. If T is less than M:

$$LSL_T = LSL \qquad\qquad USL_T = 2T - LSL$$

When T is greater than M, find LSL_T and USL_T with this formula:

$$LSL_T = 2T - USL \qquad\qquad USL_T = USL$$

Once LSL_T and USL_T are derived, calculate the desired measure with one of these formulas:

$$C^*_{PL} = \frac{Z^*_{LSL,ST}}{3} = \frac{\mu - LSL_T}{3\sigma_{ST}} \qquad\qquad C^*_{PU} = \frac{Z^*_{USL,ST}}{3} = \frac{USL_T - \mu}{3\sigma_{ST}}$$

$$P^*_{PL} = \frac{Z^*_{LSL,LT}}{3} = \frac{\mu - LSL_T}{3\sigma_{LT}} \qquad\qquad P^*_{PU} = \frac{Z^*_{USL,LT}}{3} = \frac{USL_T - \mu}{3\sigma_{LT}}$$

Example 6.15 Contact Closing Speed

A South Korean manufacturer checks circuit breakers for contact motion to verify they move at the proper opening and closing speed. Closing too quickly causes premature wearout of the contacts, while closing too slowly could burn or weld the contacts so the breaker could not be reopened. Their engineering department has developed the following window for closing speed to optimize the life of this product, with a target speed of .0170 seconds.

$$\text{LSL} = .0140 \text{ seconds} \qquad \text{USL} = .0250 \text{ seconds} \qquad M = .0195 \text{ seconds}$$

From the control chart, σ_{ST} is estimated as .00129 seconds, and μ as .0189. To incorporate engineering's desire for an average closing speed of .0170 seconds in the measure of performance capability, C_{PL} and C_{PU} must be replaced with C^*_{PL} and C^*_{PU}. As the target average is less than M, the first step is determining USL_T and LSL_T with these formulas:

$$\text{LSL}_T = \text{LSL} = .0140 \qquad \text{USL}_T = 2T - \text{LSL} = 2(.0170) - .0140 = .0200$$

For the current average of .0189 seconds, C^*_{PL} is 1.27, but C^*_{PU} is estimated as only .28.

$$\hat{C}^*_{PL} = \frac{\hat{\mu} - \text{LSL}_T}{3\hat{\sigma}_{ST}} = \frac{.0189 - .0140}{3(.00129)} = 1.27 \qquad \hat{C}^*_{PU} = \frac{\text{USL}_T - \hat{\mu}}{3\hat{\sigma}_{ST}} = \frac{.0200 - .0189}{3(.00129)} = .28$$

Because C^*_{PL} is much larger than C^*_{PU}, the operator realizes that performance capability can be significantly increased by adjusting the process to lower the average closing speed. This will increase C^*_{PU} by the amount C^*_{PL} decreases. When the average is finally positioned at T, both these measures will equal .78, as well as equaling C^*_P.

$$\hat{C}^*_P = \text{Minimum}\left(\frac{T - \text{LSL}}{3\hat{\sigma}_{ST}}, \frac{\text{USL} - T}{3\hat{\sigma}_{ST}}\right)$$

$$= \text{Minimum}\left(\frac{.0170 - .0140}{3(.00129)}, \frac{.0250 - .0170}{3(.00129)}\right)$$

$$= \text{Minimum}\,(.78, 2.07) = .78$$

The following formula determines the required shift in the process average to maximize C^*_{PL} and C^*_{PU}. In this example, a downward shift of .0019 seconds is necessary.

$$\text{Shift in } \mu = T - \hat{\mu} = .0170 - .0189 = -.0019$$

Alternatively, the shift may be figured with this formula:

$$\text{Shift in } \mu = 1.5\,\hat{\sigma}_{ST}(\hat{C}^*_{PU} - \hat{C}^*_{PL}) = 1.5\,(.00129)\,(.28 - 1.27) = -.0019$$

$C^*_P (P^*_P)$ could also be directly calculated from C^*_{PL} and C^*_{PU} (P^*_{PL} and P^*_{PU}) when the process is normally distributed.

$$\hat{C}^*_P = \frac{\hat{C}^*_{PL} + \hat{C}^*_{PU}}{2} = \frac{1.27 + .28}{2} = .775$$

When the process average is less than T, C^*_P is always greater than C^*_{PL} and less than C^*_{PU}, while P^*_P is always greater than P^*_{PL} and less than P^*_{PU}.

$$\text{When } \mu < M, \quad C^*_{PL} < C^*_P < C^*_{PU} \quad \text{and} \quad P^*_{PL} < P^*_P < P^*_{PU}$$

On the other hand, if the process average is less than T, then C^*_P is less than C^*_{PL} and greater than C^*_{PU}, while P^*_P is always less than P^*_{PL} and larger than P^*_{PU}.

$$\text{When } \mu > M, \quad C^*_{PL} > C^*_P > C^*_{PU} \quad \text{and} \quad P^*_{PL} > P^*_P > P^*_{PU}$$

C^*_{PL} and C^*_{PU} can equal C^*_P only when the process average is located at the target of .0170.

$$\hat{C}^*_{PL} = \frac{\hat{\mu} - LSL_T}{3\hat{\sigma}_{ST}} = \frac{.0170 - .0140}{3(.00129)} = .78 = \hat{C}^*_P$$

$$\hat{C}^*_{PU} = \frac{USL_T - \hat{\mu}}{3\hat{\sigma}_{ST}} = \frac{.0200 - .0170}{3(.00129)} = .78 = \hat{C}^*_P$$

Any movement in the process average higher, or lower, will cause these measures to decrease, providing an incentive for the operator to keep the process centered at the target average of .0170 seconds. Unfortunately, even if this process is centered here, it lacks short-term performance capability. Work must now focus on making significant reductions in the amount of inherent process variation.

Note that when T equals M, LSL_T is just the original LSL, while USL_T equals the original USL. As there is no difference in the specifications included in their calculations, C^*_{PL} equals C_{PL}, and C^*_{PU} equals C_{PU}. Of course, a similar relationship holds for the long-term capability measures as well.

$$\text{When } T = M, \quad C^*_{PL} = C_{PL} \quad \text{and} \quad C^*_{PU} = C_{PU}$$

When T equals M *and* the process average equals M, the following relationships are true.

$$\text{When } \mu = T = M, \quad C^*_{PL} = C_{PL} = C^*_{PU} = C_{PU} = C^*_P = C_P = \frac{1}{C^*_R} = \frac{1}{C_R}$$

These modified measures are more generalized formulas for C_{PL} and C_{PU}, which don't require T to equal the middle of the tolerance.

Advantages and Disadvantages

These are similar to those listed for C_{PL}, C_{PU}, P_{PL}, and P_{PU}, with the exception that these modified measures do not assume T is equal to M. These indexes of performance capability are related to the target average of the characteristic being assessed. As the process average moves closer to T, they increase, and when the average moves away from T, they decrease.

Again, the difficulty for some managers is that *two* measures are required to be calculated and reported for each process. Some practitioners prefer to have process capability summarized in just one number, even though this can be dangerous. The next section explains how the C_{PK} (or P_{PK}) index incorporates both proximity to the target average and the amount of process variation into a single measure of performance capability.

6.5 C_{PK} and P_{PK} Indexes

The C_{PK} and P_{PK} indexes are closely related to C_{PL}, C_{PU}, P_{PL}, and P_{PU}, and are defined as follows:

$$C_{PK} = \text{Minimum}\,(\,C_{PL}\,,C_{PU}\,) \qquad\qquad P_{PK} = \text{Minimum}\,(\,P_{PL}\,,P_{PU}\,)$$

This relationship means C_{PK} and P_{PK} are also associated with Z_{MIN}. For C_{PK}:

$$\begin{aligned}
C_{PK} &= \text{Minimum}\,(\,C_{PL}\,,C_{PU}\,) \\[2mm]
&= \text{Minimum}\left(\frac{\mu - LSL}{3\sigma_{ST}}\,,\frac{USL - \mu}{3\sigma_{ST}}\right) \\[2mm]
&= \frac{1}{3}\text{Minimum}\left(\frac{\mu - LSL}{\sigma_{ST}}\,,\frac{USL - \mu}{\sigma_{ST}}\right) \\[2mm]
&= \frac{1}{3}\text{Minimum}\,(\,Z_{LSL,ST}\,,Z_{USL,ST}\,) \\[2mm]
&= \frac{Z_{MIN,ST}}{3}
\end{aligned}$$

More typically, they are determined directly from estimates of μ and σ.

$$\hat{C}_{PK} = \text{Minimum}\,(\,\hat{C}_{PL}\,,\hat{C}_{PU}\,) = \text{Minimum}\left(\frac{\hat{\mu} - LSL}{3\hat{\sigma}_{ST}}\,,\frac{USL - \hat{\mu}}{3\hat{\sigma}_{ST}}\right)$$

$$\hat{P}_{PK} = \text{Minimum}\,(\,\hat{P}_{PL}\,,\hat{P}_{PU}\,) = \text{Minimum}\left(\frac{\hat{\mu} - LSL}{3\hat{\sigma}_{LT}}\,,\frac{USL - \hat{\mu}}{3\hat{\sigma}_{LT}}\right)$$

When the process output just meets the minimum requirements for short-term performance capability, $Z_{MIN,ST}$ is 3.0 and the corresponding C_{PK} index is 1.0.

$$C_{PK} = \frac{Z_{MIN,ST}}{3} = \frac{3.0}{3} = 1.0$$

These indexes have the same rating scale as C_{PL}, C_{PU}, P_{PL}, and P_{PU}.

<1 reveals that the process fails to meet the minimum criterion for performance capability,

$=1$ indicates that the process just meets the minimum requirement,

>1 implies that the process exceeds the minimum prerequisite.

Like the Z_{MIN} method, these measures summarize performance capability into just one number. Whereas C_P and P_P are called "first-generation" capability measures because they are a function of only the process standard deviation, C_{PK} and P_{PK} are considered to be "second-generation" indexes as they are a function of both the process average *and* standard deviation (Pearn and Kotz).

Goals

If the process has good capability with $Z_{MIN,ST}$ equal to 4.0, then C_{PK} equals $4/3$ or 1.33, which is a very common goal for this index.

$$C_{PK} = \frac{Z_{MIN,ST}}{3} = \frac{4.0}{3} = 1.33$$

Other common goals are listed in Table 6.4 along with their corresponding Z_{MIN} values.

Table 6.4 Z_{MIN} versus C_{PK} and P_{PK}.

$Z_{MIN,ST}$	C_{PK}	$Z_{MIN,LT}$	P_{PK}
3	1.00	3	1.00
4	1.33	4	1.33
4.5	1.50	4.5	1.50
5	1.67	5	1.67
6	2.00	6	2.00

However, other goals have been proposed by several authors. Messina recommends separate goals for different types of products, as presented in Table 6.5.

Table 6.5 C_{PK} goals versus type of product.

Type of Product	C_{PK} Goal
Plastic	1.48
Machined	1.44
Sheet Metal	1.39
Casting	1.33

Montgomery assigns minimum capability goals based on the type of process and whether the characteristic has a bilateral or unilateral specification (see Table 6.6 on the facing page).

Denissoff writes about a manufacturer of capacitors who established a C_{PK} goal of 1.67 in 1988, and more recently upped it to 2.00. Also in the electronics industry, Skellie and Ngo recommend eliminating 100 percent inspection for electrical component parameters when their C_{PK} values exceed 1.40. The automotive industry (AIAG) in their QS-9000 standard has set a default C_{PK} goal of 1.33 if no other goal is specified. The Electronic Industries Association has also established guidelines for the application of C_{PK}.

When setting capability goals, consideration must be given to the "volatility" of the process. Recall from Figure 2.13 that a 1.5σ change in the process average has only a 50 percent chance of detection on an $\overline{X}$, R chart with n equal to 4. If the process under study is very sensitive and susceptible to frequent changes, a higher C_{PK} goal must be specified in order to provide a sufficient safety margin. Imagine a characteristic having only an USL, with the average for its output being $4\sigma_{ST}$ below this USL. With a C_{PK} of 1.333 (4 / 3), this process will produce only .0032 percent of its parts above the USL. But this is true only as long as the process average remains stable. An upward shift in μ of $1.5\sigma_{ST}$ causes C_{PK} to drop to just .833 (2.5 / 3), with .6210 percent of the process's output now above the USL. This percentage of nonconformance is 194 times greater than the original.

Table 6.6 C_{PK} goals versus type of process.

	C_{PK} Goal	
Type of Process	Bilateral	Unilateral
Existing	1.33	1.25
New	1.50	1.45
Safety related, existing	1.50	1.45
Safety related, new	1.67	1.60

Contrast this performance to a process output having $5\sigma_{ST}$ between its average and the USL, making C_{PK} equal to 1.667. A $1.5\sigma_{ST}$ shift in this process's average lowers C_{PK} to 1.167 (3.5 / 3), while increasing the percentage of parts above the USL from .00006 to .00233, an increase of 39 times. Despite a "shock" to its system, this process still easily meets the minimum requirement for performance capability.

If increasing performance capability is not feasible, the subgroup size should be increased and/or the time interval between subgroups decreased. Either action helps increase a control chart's ability to detect these types of changes sooner.

Maximum C_{PK} and Maximum P_{PK}

Similar to Z_{MIN}, the largest C_{PK} possible for a process, without reducing variation, can be ascertained in several ways. First, from the amount of potential capability:

$$\text{Maximum } C_{PK} = C_P = \frac{1}{C_R} = \frac{C_{ST}}{6}$$

It may also be found from the average of C_{PL} and C_{PU}.

$$\text{Maximum } C_{PK} = \frac{C_{PL} + C_{PU}}{2}$$

And, of course, from the largest possible Z_{MIN} value.

$$\text{Maximum } C_{PK} = \frac{1}{3}\text{Maximum } Z_{MIN.\,ST}$$

Likewise, the maximum P_{PK} possible is found as follows:

$$\text{Maximum } P_{PK} = P_P = \frac{1}{P_R} = \frac{P_{ST}}{6} = \frac{P_{PL} + P_{PU}}{2} = \frac{1}{3}\text{Maximum } Z_{MIN.\,LT}$$

Example 6.16 Keyway Width

From an $\overline{X}$, S chart monitoring the width of a keyway cut into a shaft on a milling machine, σ_{ST} is estimated from $\overline{S}$ as .01540 mm, while the average keyway width is estimated from $\overline{\overline{X}}$ as 4.99135 mm. With a LSL of 4.92500 mm and an USL of 5.05500 mm, the short-term performance capability can be measured with the C_{PK} index.

$$\hat{C}_{PK} = \text{Minimum}\,(\hat{C}_{PL}, \hat{C}_{PU})$$

$$= \text{Minimum}\left(\frac{\hat{\mu} - \text{LSL}}{3\hat{\sigma}_{ST}}, \frac{\text{USL} - \hat{\mu}}{3\hat{\sigma}_{ST}}\right)$$

$$= \text{Minimum}\left(\frac{4.99135 - 4.92500}{3(.01540)}, \frac{5.05500 - 4.99135}{3(.01540)}\right)$$

$$= \text{Minimum}\,(1.44, 1.38) = 1.38$$

As the goal for this characteristic is 1.33, this process has good capability. The amount of undersize and oversize keyways can be predicted from the estimates for C_{PL} and C_{PU} by calculating their corresponding Z values.

$$\hat{Z}_{LSL,ST} = 3\hat{C}_{PL} = 3(1.44) = 4.32 \qquad \hat{Z}_{USL,ST} = 3\hat{C}_{PU} = 3(1.38) = 4.14$$

Looking up these Z values in Appendix Table III, the predicted amount of undersize keyways is about .0008 percent, while the number oversize is forecast to be .0017 percent.

$$\hat{p}'_{TOTAL,ST} = \Phi(\hat{Z}_{LSL,ST}) + \Phi(\hat{Z}_{USL,ST})$$

$$= \Phi(4.32) + \Phi(4.14)$$

$$= .000008 + .000017$$

$$= .000025 \cong .0025\%$$

By adjusting the process so μ equals M (assuming σ_{ST} remains unchanged), C_{PK} can be increased to a maximum of 1.41, which is just the average of C_{PL} and C_{PU}.

$$\text{Maximum}\ \hat{C}_{PK} = \frac{\hat{C}_{PL} + \hat{C}_{PU}}{2} = \frac{1.44 + 1.38}{2} = 1.41$$

For this situation, there would be .0012 percent undersize as well as .0012 percent oversize, for a total of .0024 percent nonconforming keyways.

C_{PK} Matrix

Suppose in the last example, shafts made from various types of materials are run on the milling machine. As a result of this diversity in material, the process standard deviation and/or average for keyway width varies from part number to part number. This means the estimate of C_{PK} could be unique for each type of shaft. If their tolerances are similar, this information can be recorded in matrix form, as displayed in Table 6.7.

Table 6.7 C_{PK} versus type of material.

	Type of Material			
	Brass	Stainless	Copper	Plastic
C_{PK}	1.82	1.41	1.73	2.06

If there are different keyway width requirements, the matrix can be expanded to list C_{PK} values for each type of material versus width of cut. An example of this matrix is presented in Table 6.8. In addition to showing differences in capability between type of materials, this matrix reveals the process has greater difficulty producing wider cuts than narrow ones.

Table 6.8 C_{PK} versus width and type of material.

	C_{PK} by Type of Material			
Width	Brass	Stainless	Copper	Plastic
5.0	1.82	1.41	1.73	2.06
6.5	1.79	1.37	1.66	1.99
9.0	1.70	1.24	1.58	1.85

This type of capability matrix could be developed for an entire machining department, where capabilities of various machines are listed versus type of material, or type of operation, or even feeds and speeds. Such a graphic tool would help supervisors minimize scrap and rework by allowing them to easily schedule jobs to the most capable process. A more detailed example of this approach is provided by Juran *et al.* Similar matrices are also seen in companies doing plastic-injection molding, where process capability is usually cross-tabulated by part size and type of plastic, as is shown in Table 6.9.

Table 6.9 C_{PK} versus size and color of plastic.

	C_{PK} by Color of Plastic		
Part Size	White	Black	Beige
Small	1.69	1.46	1.51
Large	1.57	1.38	1.32

Example 6.17 Flange Thickness

The flange thickness of an aircraft part near a slot is considered a key characteristic. After two weeks of charting the grinding operation responsible for flange thickness, this process finally displays good control. From this chart, the average flange thickness is estimated as 9.087 mm, with a short-term standard deviation of .024 mm. With this information, the C_{PK} measure of performance capability may be estimated (LSL = 9.00 mm, USL = 9.20 mm).

$$\hat{C}_{PK} = \text{Minimum} (\hat{C}_{PL}, \hat{C}_{PU})$$

$$= \text{Minimum} \left(\frac{\hat{\mu} - \text{LSL}}{3\hat{\sigma}_{ST}}, \frac{\text{USL} - \hat{\mu}}{3\hat{\sigma}_{ST}} \right)$$

$$= \text{Minimum} \left(\frac{9.087 - 9.00}{3(.024)}, \frac{9.20 - 9.087}{3(.024)} \right)$$

$$= \text{Minimum} (1.21, 1.57) = 1.21$$

This grinding operation exceeds the minimum performance capability requirement for flange thickness of 1.00, but not its assigned goal of 1.33. To determine the percentage of flanges that are undersize, multiply the estimate of C_{PL} by 3 to obtain $\hat{Z}_{LSL.ST}$, then find the corresponding percentage nonconforming for this Z value in Appendix Table III.

$$\hat{Z}_{LSL,ST} = 3\hat{C}_{PL} = 3(1.21) = 3.63$$

As a Z value of 3.63 correlates to 1.4-04 in this table, the amount of undersize flanges is estimated to be about .014 percent, assuming a normal distribution for thickness. A picture of this process output is given in Figure 6.41.

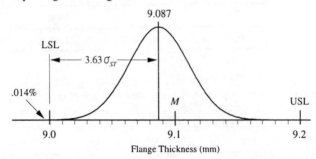

Fig. 6.41 The process output for flange thickness.

Assume efforts are successfully made to move the process average so it is now centered at 9.10 mm, the middle of the tolerance. Unfortunately, this shift causes the short-term standard deviation to increase slightly to .02755 mm. The combined effect of these changes on performance capability can be determined by estimating the new C_{PK} index.

$$\hat{C}_{PK} = \text{Minimum} \left(\frac{\hat{\mu} - \text{LSL}}{3\hat{\sigma}_{ST}}, \frac{\text{USL} - \hat{\mu}}{3\hat{\sigma}_{ST}} \right)$$

$$= \text{Minimum} \left(\frac{9.10 - 9.00}{3(.02755)}, \frac{9.20 - 9.10}{3(.02755)} \right)$$

$$= \text{Minimum} (1.21, 1.21) = 1.21$$

This result is identical to the original estimate of C_{PK}. The improvement in centering was countered by the increase in variation. The original and current process outputs have identical C_{PK} values, yet are definitely different, as Figure 6.42 plainly discloses.

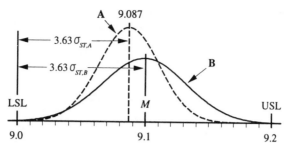

Fig. 6.42 The revised output for flange thickness (**B**) versus the original output (**A**).

Some quality practitioners falsely believe C_{PK} is an indicator of process centering: larger C_{PK} values mean the process average is closer to M. The above example demonstrates the fallacy of this perception - the process average moved closer to M, yet C_{PK} did *not* increase. In fact, as is illustrated next, it is possible for C_{PK} to increase significantly as the process average moves further *away* from M.

Additional adjustments are made to this grinding operation. After the process stabilizes, the output distribution for flange thickness looks like curve **C** in Figure 6.43. The average of 9.05 for **C** is quite a distance below the 9.10 average for B, which is centered right at M. However, the C_{PK} index for **C** is 2.00, which is much greater than the 1.21 rating for curve **B**. The dramatic increase in performance capability occurs due to the substantial decrease in σ_{ST} for **C**, from .024 to only .00833. This reduction in variation puts the average for **C** at a distance of $6\sigma_{ST,C}$ above the LSL, whereas **B**'s average is only 3.63 of its standard deviations above this specification.

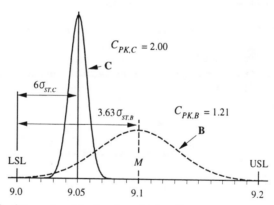

Fig. 6.43 **C** is centered farther from M than **B**, yet has a higher C_{PK} value.

The Dow Jones Industrial Average (DJIA) is an index reflecting the activity of a certain group of stocks traded on the New York Stock Exchange. When the DJIA becomes higher, it is not immediately clear which of the stocks included in the average are responsible. Some may have actually decreased, while others have significantly increased. Furthermore, the DJIA remaining unchanged from one day to the next does not imply the price of its component stocks has also remained steady. Increases in some stocks could be offset by decreases in others. The summarization into a single index obscures the detail concerning individual stocks. Like the DJIA, C_{PK} is a composite of several factors: μ, σ_{ST}, LSL, and USL. Changes in one can be countered, or accentuated, by changes in others. This is why it's always best to evaluate several measures, and a histogram, before formulating any final decision about process capability.

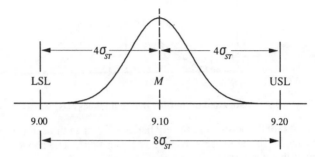

Fig. 6.44 Process centered at middle of tolerance.

If the process output is finally centered at the middle of the tolerance, as is illustrated above in Figure 6.44, then the following relationships hold.

$$C_{PK} = C_P = \frac{1}{C_R} \quad \text{and} \quad P_{PK} = P_P = \frac{1}{P_R}$$

When the process average moves in either direction, C_P (P_P) remains constant while C_{PK} (P_{PK}) decreases. This is because $Z_{MIN,ST}$ ($Z_{MIN,LT}$) becomes smaller as the average moves closer to either the LSL or the USL, assuming σ does not change. Always remember that performance capability can never exceed potential capability.

$$C_{PK} \leq C_P \qquad P_{PK} \leq P_P$$

The disparity between C_{PK} and C_P (or between P_{PK} and P_P) provides an indication of the amount the process average deviates from the middle of the tolerance. Quantifying the extent of this deviation exposes additional useful information about process performance, as will be demonstrated in the next section.

The *k* Factor Relating C_{PK} and C_P

C_{PK} and C_P (or P_{PK} and P_P) are related by the factor 1 minus k (Kane, 1989, p. 272).

$$C_{PK} = C_P(1-k) \quad \text{and} \quad P_{PK} = P_P(1-k) \quad \text{where } k = \frac{|M - \mu|}{(USL - LSL)/2}$$

Referred to as the "process location ratio" by De La Rosa, k is a dimensionless number comparing the distance the process average is off target, M minus μ, to one half of the tolerance, (USL minus LSL)/2. Unfortunately, there is some confusion with this notation, as the letter k is also used to represent the number of subgroups on a control chart. Hopefully, the context of how k is being applied will reveal which is meant. A full derivation of the formula for k is given in Section 6.15 (see Derivation 1 on page 338).

The quantity 1 minus k decreases C_{PK} (the performance capability) by the percentage the actual process average, μ, differs from the middle of the tolerance, M. For illustration, the process average is less than M in Figure 6.45. This difference is measured as M minus μ. To standardize this measurement, it is compared to half of the tolerance, which is equal to the distance from M to the LSL.

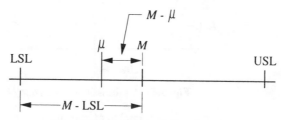

Fig. 6.45 Explanation of the k factor.

Thus, the quantity 1 minus k discloses how accurately the process is centered with respect to the middle of the tolerance. As μ moves farther away from M, k increases. The increase in k makes this measure of accuracy, 1 minus k, decrease, causing C_{PK} to decrease as well. Thus, 1 minus k expresses how much potential capability is lost due to poor centering. The absolute value of M minus μ is taken because this same analysis applies when the process average is above M.

To better understand this relationship between centering and performance capability, assume a process's output is centered at M, as is shown in Figure 6.46. In this particular situation, μ equals M, making k equal to 0.

$$ k = \frac{|M-\mu|}{(\text{USL}-\text{LSL})/2} = \frac{|M-M|}{(\text{USL}-\text{LSL})/2} = \frac{|0|}{(\text{USL}-\text{LSL})/2} = 0 $$

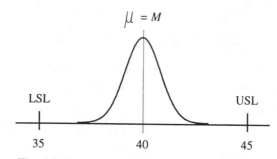

Fig. 6.46 Process is centered at middle of tolerance.

Because 1 minus 0 is 1, C_{PK} is identical to C_P. Performance equals potential capability and the process is performing at its best possible capability when the average is centered at M.

$$ C_{PK} = C_P(1-0) = C_P $$

In settings where μ and M differ, as is the case in Figure 6.47 where μ is greater than M, k reflects the distance between M and μ, as a percentage of one-half the tolerance.

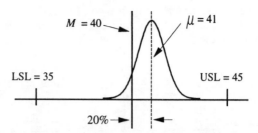

Fig. 6.47 A process not centered at *M*.

When the process average is not equal to *M*, the quantity $|M - \mu|$ is greater than zero.

$$|M - \mu| > 0$$

$$\frac{|M - \mu|}{(\text{USL - LSL})/2} > 0$$

$$k > 0$$

As *k* is now greater than zero, 1 minus *k* is less than 1, resulting in C_{PK} being less than C_P.

$$k > 0$$

$$-k < 0$$

$$1 - k < 1$$

$$C_P(1 - k) < C_P$$

$$C_{PK} < C_P$$

For the example presented in Figure 6.47, μ is one unit above *M*. As the upper half of the tolerance is 5 units (45 minus 40), *k* equals .20 (1/5).

$$k = \frac{|M - \mu|}{(\text{USL} - \text{LSL})/2} = \frac{|40 - 41|}{(45 - 35)/2} = \frac{1}{5} = .20$$

This makes 1 minus *k* equal to .80.

$$1 - k = 1 - .20 = .80$$

If C_P is 2.00, then C_{PK} is 80 percent of 2.00, or 1.60, meaning performance capability is only 80 percent of what it could be if the process were centered at *M*. This lack of accuracy in centering the process output results in losing 20 percent of the available potential capability.

$$C_{PK} = C_P(1 - k) = 2.00(1 - .20) = 2.00(.80) = 1.60$$

Had the process average been one unit *below M*, *k* would still be .80 because it is derived from the absolute value of the difference between *M* and μ. Thus, a shift in the process average above *M* reduces performance capability by the same amount as an equal shift below *M*. Bringing μ closer to *M* decreases *k*, causing 1 - *k* to increase, which in turn increases C_{PK}.

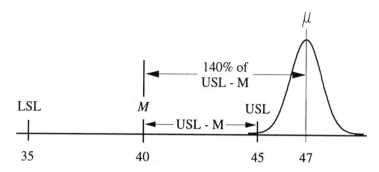

Fig. 6.48 Process centered at 47, which is above the USL.

When μ is located outside a specification limit, as is the case in Figure 6.48, $|M - \mu|$ is greater than the quantity (USL - LSL)/2. In this undesirable situation, k becomes greater than 1, resulting in 1 minus k being less than 0.

$$|M - \mu| > (USL - LSL)/2$$

$$\frac{|M - \mu|}{(USL - LSL)/2} > \frac{(USL - LSL)/2}{(USL - LSL)/2}$$

$$k > 1$$

$$0 > 1 - k$$

For the process output displayed in Figure 6.48, k turns out to be 1.40.

$$k = \frac{|40 - 47|}{(45 - 35)/2} = \frac{7}{5} = 1.40$$

This causes 1 minus k to equal -.40.

$$1 - k = 1 - 1.40 = -.40$$

A negative result causes C_{PK} to also be negative, revealing a serious performance capability problem where more than 50 percent of the output is nonconforming.

$$C_{PK} = C_p(1 - 1.40) = 2.00(-.40) = -.80$$

Due to the absolute value signs in its formula, k cannot be negative, forcing 1 minus k to always be less than, or equal to, 1.

$$0 \le k$$

$$1 \le 1 + k$$

$$1 - k \le 1$$

Because 1 minus k can never be greater than 1, C_{PK} can never be greater than C_p.

$$1 - k \leq 1$$

$$C_P(1 - k) \leq C_P$$

$$C_{PK} \leq C_P$$

Thus, performance capability cannot exceed potential capability. The only way to increase C_{PK} (performance) is to increase C_P (potential) or decrease k (improve centering). Doing both creates the largest improvement in performance capability.

$$C_{PK} = C_P (1 - k)$$

As a result of M being the middle of the tolerance, M minus the LSL equals one half of the tolerance.

$$1/2 \text{ Tolerance} = \frac{\text{USL - LSL}}{2} = M - \text{LSL}$$

Similarly, the USL minus M is equal to the other half of the tolerance.

$$\frac{\text{USL - LSL}}{2} = \text{USL} - M = M - \text{LSL}$$

These relationships allow alternative formulas for k to be derived.

$$k = \frac{|M - \mu|}{(\text{USL} - \text{LSL})/2} = \frac{|M - \mu|}{(M - \text{LSL})} = \frac{|M - \mu|}{(\text{USL} - M)}$$

Recall the earlier analogy of C_{PK} to the Dow Jones Industrial Average (DJIA). With the DJIA index, it is extremely difficult to discern how any one particular stock included in the average is performing. Summary statistics are good for an overall picture, but quite often more specific details are needed for decision making. If performance capability is unsatisfactory, it is hard to tell by looking at only the C_{PK} index whether the poor performance is due to an off-center average, a large standard deviation, or both. Expressing C_{PK} as C_P times the quantity 1 minus k separates performance capability into two components, precision and accuracy. C_P reveals information about the process standard deviation (precision), whereas the quantity 1 minus k quantifies the effect on performance capability due to process centering (accuracy). Note that these two components are completely independent of each other: C_P is based solely on σ_{ST}, whereas k is calculated from only μ. Figure 6.49 illustrates these two separate components of performance capability.

Fon refers to k as C_A, with the "A" subscript representing accuracy. As C_A is determined without either the short- or long-term standard deviation, it appears in the formulas for both C_{PK} and P_{PK}.

$$C_{PK} = C_P (1 - C_A) \qquad\qquad P_{PK} = P_P (1 - C_A)$$

Fon has even established a rating scale for different values of C_A, which is presented in Table 6.10.

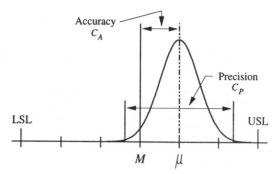

Fig. 6.49 Precision versus accuracy.

Table 6.10 Accuracy ratings for various C_A values.

C_A Range	Accuracy Rating
$C_A \leq .125$	Excellent
$.125 < C_A \leq .250$	Good
$.250 < C_A \leq .500$	Fair
$.500 < C_A \leq .750$	Poor
$.750 < C_A$	Terrible

By separating performance capability into these two components, each process parameter may be studied individually for its contribution to overall capability. A person conducting a capability study can infer if work should concentrate on centering by analyzing C_A (or k), or if variation reduction should become the focus of attention by examining C_P. A large 1 minus k factor (a small k) is desirable as well as a large C_P, since the product of these two will generate a large C_{PK} value.

The relationship between C_{PK}, C_A, and C_P (or C_R) is summarized in the graph displayed in Figure 6.50. The x axis of this graph is for accuracy, as measured by C_A (or k) on the bottom scale, and by 1 minus C_A (or 1 minus k) on the top scale. The y axis is for potential capability, measured by C_P on the left-hand side and by C_R on the right. The slanted lines drawn on the graph represent various C_{PK} values (.50, 1.00, 1.33, 1.67, 2.00, 3.00, and 5.00) and are labeled accordingly. These slanted lines reflect the necessary trade-offs between shifts in the process average and changes in potential capability to retain a given level of performance capability. When the process average is located at M, C_A equals 0 and C_{PK} becomes identical to C_P. When μ is centered at either the LSL or USL, C_A is 1.0 and C_{PK} is 0.

In order to have C_{PK} remain constant when the process average moves away from M (C_A is increasing), potential capability must increase (C_P becomes larger). This is why all these lines have positive slopes. For example, point A on Figure 6.50 shows that when C_P is 2.00 and k is 0, C_{PK} is 2.00.

$$C_{PK} = C_P(1 - k) = 2.00(1 - 0) = 2.00$$

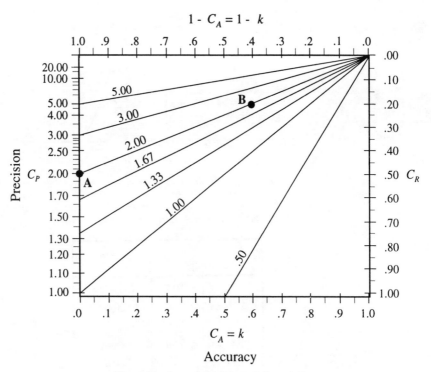

Fig. 6.50 C_{PK} as a function of C_A and C_P.

In order to keep C_{PK} at 2.00 when k changes to .6, C_P must increase to 5.00, as indicated by point B on the graph.

$$C_{PK} = C_P(1-k) = 5.00(1-.6) = 2.00$$

Phillips defines a "target ratio," labeled TR, which is equal to 3 times k when the target average is M.

$$TR = 3k = \frac{3|M-\mu|}{(USL - LSL)/2} = \frac{6|M-\mu|}{USL - LSL}$$

When the average is on target, k is zero, as is TR.

$$\text{When } \mu = M, \ k = 0, \text{ and } TR = 3(0) = 0$$

If the average is at one of the specification limits, k equals 1 and TR equals 3.

$$\text{When } \mu = USL, \ k = 1, \text{ and } TR = 3(1) = 3$$

Phillips recommends reporting TR along with C_P (or C_R) to identify improvement opportunities in either centering (high TR values) or in process spread (low C_P values). Gitlow *et al* derive another relationship between C_{PK} and C_P.

$$C_{PK} = C_P(1-k)$$

$$= C_P - C_P k$$

$$= C_P - \left(\frac{\text{USL - LSL}}{6\sigma_{ST}}\right)\frac{|M-\mu|}{(\text{USL - LSL})/2}$$

$$= C_P - \frac{|M-\mu|}{3\sigma_{ST}}$$

$$= C_P - \frac{|Z_{M,ST}|}{3} \qquad \text{where } Z_{M,ST} = \frac{M-\mu}{\sigma_{ST}}$$

A similar relationship exists between P_{PK} and P_P.

$$P_{PK} = P_P - \frac{|Z_{M,LT}|}{3} \quad \text{where } Z_{M,LT} = \frac{M-\mu}{\sigma_{LT}}$$

Z_M measures the distance between the process average and the middle of the tolerance in units of standard deviations. As this distance increases, the absolute value of Z_M becomes larger and more potential capability is lost. The exact relationship between $Z_{M,ST}$ and C_{PK} is graphed for several values of C_P in Figure 6.51, with $Z_{M,ST}$ on the x axis and C_{PK} on the y axis. For a given level of potential capability (C_P), when the absolute value of $Z_{M,ST}$ becomes larger (less accuracy), C_{PK} becomes lower. This is demonstrated by the slanted lines, each representing a particular level of potential capability, with C_P equal to 1.00, 1.50, and 2.00. These lines peak when $Z_{M,ST}$ equals 0, then decrease steadily as $Z_{M,ST}$ becomes either more positive or more negative. The shaded area of this graph identifies combinations of C_P and $Z_{M,ST}$ resulting in a lack of performance capability, *i.e.*, where C_{PK} is less than 1.0.

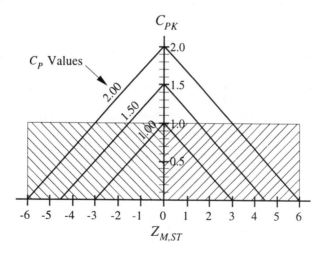

Fig. 6.51 C_{PK} as a function of $Z_{M,ST}$ for certain C_P levels.

The drawback to this particular relationship is that it fails to separate the effect on C_{PK} of changes in centering versus changes in process spread. As before, C_P measures the contribution to performance capability coming from the process spread (precision), as C_P depends on only σ_{ST}. However, $Z_{M,ST}$, which indicates shifts in process centering, is a function of both μ *and* σ_{ST}. Thus, changes in σ_{ST} affect both C_P and $Z_{M,ST}$, meaning these two indicators are not independent. Because the effect on C_{PK} due to changes in $Z_{M,ST}$ depends on the value of C_P, this relationship between C_{PK} and C_P is seldom analyzed in practice.

Example 6.18 Valve O.D. Size

On page 175, an estimate for $Z_{MIN,LT}$ of 3.472 was made for the O.D. size of an exhaust valve run on a grinding machine. From this result, an estimate of P_{PK} can be derived.

$$\hat{P}_{PK} = \frac{\hat{Z}_{MIN,LT}}{3} = \frac{3.472}{3} = 1.157$$

As the LSL is 9.9400 cm and the USL is 10.0600 cm, M is 10.0000 cm, resulting in an estimated k factor of 6.83 percent for this process. Note how the dimensions cancel, leaving k a dimensionless number. Because an estimate of μ is part of its calculation, the result is an point estimator of the true process k factor, which is why the $\wedge$ sign appears above k.

$$\mu = 10.0041 \text{ cm} \qquad M = 10.0000 \text{ cm} \qquad \hat{P}_p = 1.242$$

$$\hat{k} = \frac{|M - \hat{\mu}|}{(USL - LSL)/2} = \frac{|10.0000 \text{ cm} - 10.0041 \text{ cm}|}{(10.0600 \text{ cm} - 9.9400 \text{ cm})/2} = \frac{.0041 \text{ cm}}{.06 \text{ cm}} = .0683$$

The estimated k factor of .0683 means the current process average of 10.0041 cm is located above M (10.0000 cm) by 6.83 percent of the distance from M to the USL of 10.0600 cm. With an estimate for P_p of 1.242 calculated in Example 5.17 on page 135, P_{PK} may be estimated from this following relationship:

$$\hat{P}_{PK} = \hat{P}_p(1 - \hat{k}) = 1.242(1 - .0683) = 1.242(.9317) = 1.157$$

This agrees with the estimate for P_{PK} of 1.157 derived directly from the estimate of $Z_{MIN,LT}$ at the beginning of this example. The lack of aim (accuracy) in centering the process reduces long-term performance capability by 6.83 percent to only 93.17 percent of the process's long-term potential capability.

If working with a computer printout that doesn't provide k or the values needed to estimate it, k can also be estimated from this formula:

$$\hat{P}_{PK} = \hat{P}_p(1 - \hat{k})$$

$$\frac{\hat{P}_{PK}}{\hat{P}_p} = 1 - \hat{k}$$

$$\hat{k} = 1 - \frac{\hat{P}_{PK}}{\hat{P}_p}$$

In this example, the estimate of P_{PK} is 1.157, while that for P_p is 1.242. This makes the estimate for k equal to .0684, in close agreement with the original estimate of .0683.

$$\hat{k} = 1 - \frac{\hat{P}_{PK}}{\hat{P}_P} = 1 - \frac{1.157}{1.242} = .0684$$

The target ratio (TR) for this process is estimated as .2049.

$$\hat{TR} = 3\hat{k} = 3(.0683) = .2049$$

$Z_{M,LT}$ is estimated as a minus .2547, meaning the estimated process average is .2547 long-term standard deviations above M.

$$\hat{Z}_{M,LT} = \frac{M - \hat{\mu}}{\hat{\sigma}_{LT}} = \frac{10.0000 \text{ cm} - 10.0041 \text{ cm}}{.01610 \text{ cm}} = -.2547$$

This estimate of $Z_{M,LT}$, along with the estimate of P_P, helps determine P_{PK}.

$$\hat{P}_{PK} = \hat{P}_P - \frac{|\hat{Z}_{M,LT}|}{3} = 1.242 - \frac{|-.2547|}{3} = 1.157$$

This agrees with the original estimate of 1.157 for P_{PK} made at the beginning of this example.

Relationship of C_{PL}, C_{PU}, P_{PL}, and P_{PU} to C_P

In a manner similar to that for C_{PK}, a relationship between these indexes and C_P can be developed.

$$C_{PL} = C_P(1 - k_L) \quad C_{PU} = C_P(1 - k_U) \quad P_{PL} = P_P(1 - k_L) \quad P_{PU} = P_P(1 - k_U)$$

The k_L and k_U factors are defined in the formulas below. The absence of the absolute values signs means these factors may at times be negative.

$$k_L = \frac{M - \mu}{(USL - LSL)/2} = \frac{M - \mu}{M - LSL} \qquad k_U = \frac{\mu - M}{(USL - LSL)/2} = \frac{\mu - M}{USL - M}$$

k_L and k_U are always identical, except for their signs. One is positive and the other negative, except when μ equals M and both are 0. Learn how these alternative formulas work for estimating C_{PL} and C_{PU} in the floppy disk example presented next.

Example 6.19 Floppy Disks

A field problem plaguing a manufacturer of floppy disks is traced to variation in the diameter of an inner hole. The following information is noted from an $\overline{X}, R$ chart monitoring the hole-cutting process and the results of a recent capability study.

$$LSL = 39.5 \quad USL = 44.5 \quad M = 42.0 \quad \overline{\overline{X}} = 40.9 \quad \hat{\sigma}_{ST} = .898 \quad \hat{C}_P = .928$$

The short-term, upper and lower performance capability measures are calculated from estimates of σ_{ST} and μ.

$$\hat{C}_{PL} = \frac{\overline{\overline{X}} - LSL}{3\hat{\sigma}_{ST}} = \frac{40.9 - 39.5}{3(.898)} = .52 \qquad \hat{C}_{PU} = \frac{USL - \overline{\overline{X}}}{3\hat{\sigma}_{ST}} = \frac{44.5 - 40.9}{3(.898)} = 1.34$$

They may also be estimated from k_L and k_U.

$$\hat{k}_L = \frac{M - \hat{\mu}}{(\text{USL} - \text{LSL})/2} = \frac{42.0 - 40.9}{(44.5 - 39.5)/2} = .44$$

$$\hat{k}_U = \frac{\hat{\mu} - M}{(\text{USL} - \text{LSL})/2} = \frac{40.9 - 42.0}{(44.5 - 39.5)/2} = -.44$$

These estimates of the modified k factors, along with $\hat{C}_P$, are substituted into the following equation to estimate C_{PL}.

$$\hat{C}_{PL} = \hat{C}_P(1 - \hat{k}_L) = .928(1 - .44) = .52$$

Because the process average of 40.9 is less than M (42.0), the quantity 1 minus k_L is less than 1, making the estimate for C_{PL} less than the C_P estimate of .928. On the other hand, μ being less than M makes k_U negative, and 1 minus k_U *greater* than 1. As a result, the estimate of C_{PU} is greater than C_P.

$$\hat{C}_{PU} = \hat{C}_P(1 - \hat{k}_U) = .928(1 - -.44) = 1.34$$

These results agree with those made at the beginning of this example. Note that k is just the absolute value of either k_L and k_U.

$$\hat{k} = |\hat{k}_L| = |\hat{k}_U| = .44$$

C_{PK} can be estimated from $\hat{C}_{PL}$ and $\hat{C}_{PU}$, or from $\hat{C}_P$ and $\hat{k}$.

$$\hat{C}_{PK} = \text{Minimum}(\hat{C}_{PL}, \hat{C}_{PU}) = \text{Minimum}(.52, 1.34) = .52$$

$$\hat{C}_{PK} = \hat{C}_P(1 - \hat{k}) = .928(1 - .44) = .52$$

Increasing the process average so it becomes closer to M will decrease k_L and increase k_U. These changes cause C_{PL} to increase and C_{PU} to decrease. When the average is finally centered at 42.0, both k_L and k_U become zero.

$$\hat{k}_L = \frac{M - \hat{\mu}}{(\text{USL} - \text{LSL})/2} = \frac{42.0 - 42.0}{(44.5 - 39.5)/2} = 0$$

$$\hat{k}_U = \frac{\hat{\mu} - M}{(\text{USL} - \text{LSL})/2} = \frac{42.0 - 42.0}{(44.5 - 39.5)/2} = 0$$

Thus, both C_{PL} and C_{PU} will equal the C_P measure of .928 when μ is located at M.

$$\hat{C}_{PL} = \hat{C}_P(1 - \hat{k}_L) = .928(1 - 0) = .928 \qquad \hat{C}_{PU} = \hat{C}_P(1 - \hat{k}_U) = .928(1 - 0) = .928$$

As C_{PK} is the minimum of these two indexes, it also becomes equal to C_P, which is the maximum possible for C_{PK} without a reduction in σ_{ST}.

$$\hat{C}_{PK} = \text{Minimum}(\hat{C}_{PL}, \hat{C}_{PU}) = \text{Minimum}(.928, .928) = .928$$

As just seen, C_{PK} (and P_{PK}) can be estimated from k_L and k_U. However, this formula also requires an estimate of C_P. For this example, where hole size has a bilateral specification, estimating C_P is no problem. But for characteristics with unilateral specifications, C_P is not defined. Thus, an alternative measure of potential capability must be chosen, and k_L and k_U modified accordingly, before C_{PK} may be estimated.

k Factors for Unilateral Specifications

Just as C_P must be replaced with either C'_{PL} or C'_{PU} when the characteristic under study has a unilateral specification, k must be replaced with either k'_L or k'_U. When only a LSL is specified, the following relationship holds:

$$C_{PK} = C'_{PL}(1 - k'_L)$$

$$\text{where } k'_L = 0 \qquad \text{if } \mu \geq T \quad (\mu \text{ is inside the target zone})$$

$$= \frac{T - \mu}{T - \text{LSL}} \text{ if } \mu < T \quad (\mu \text{ is outside the target zone})$$

Recall the formula for C'_{PL} from Section 5.10.

$$C'_{PL} = \text{Maximum}(C^T_{PL}, C_{PL}) = \text{Maximum}\left(\frac{T - \text{LSL}}{3\sigma_{ST}}, \frac{\mu - \text{LSL}}{3\sigma_{ST}}\right)$$

When the process average is centered somewhere in the target zone ($\mu \geq T$), k'_L is zero by the above definition, and C_{PK} equals the full potential capability.

$$C_{PK} = C'_{PL}(1 - k'_L) = C'_{PL}(1 - 0) = C'_{PL}$$

On the other hand, suppose the process average is centered somewhere below the target zone ($\mu < T$), as is displayed in Figure 6.52. Now k'_L equals the ratio of the distance between T and μ ($T - \mu$) to that between T and the LSL (T - LSL).

$$k'_L = \frac{T - \mu}{T - \text{LSL}}$$

Whenever μ is close to its target, T minus μ is small, and the above ratio is also small. If μ moves further away from the target zone, T minus μ becomes larger, resulting in even a larger ratio, and a corresponding larger loss of potential capability. Thus, k'_L is a measure of accuracy (aim), as it reflects how far the process average is off target.

If μ is below T, k'_L is greater than zero, making 1 minus k'_L less than 1. This deficiency in proper centering forces performance capability, as measured by C_{PK}, to become less than the potential, as measured by C'_{PL}.

$$1 - k'_L < 1$$

$$C'_{PL}(1 - k'_L) < C'_{PL}$$

$$C_{PK} < C'_{PL}$$

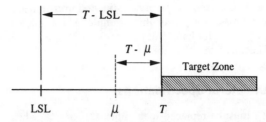

Fig. 6.52 Process average centered below the target zone.

When the process average is below the LSL, T minus μ is greater than T minus LSL, and k'_L becomes greater than 1.

$$\mu \ < \ \text{LSL}$$

$$-\mu \ > \ -\text{LSL}$$

$$T - \mu \ > \ T - \text{LSL}$$

$$\frac{T - \mu}{T - \text{LSL}} \ > \ \frac{T - \text{LSL}}{T - \text{LSL}}$$

$$k'_L \ > \ 1$$

With k'_L greater than 1, 1 minus k'_L is less than 0, making C_{PK} negative.

$$k'_L \ > \ 1$$

$$-k'_L \ < \ -1$$

$$1 - k'_L \ < \ 0$$

$$C'_{PL}(1 - k'_L) \ < \ 0$$

$$C_{PK} \ < \ 0$$

Thus, this relationship between C_{PK} and k'_L is similar to that between C_{PK} and k. Likewise, a similar association exists between C_{PK}, C'_{PU}, and k'_U.

$$C_{PK} \ = \ C'_{PU}(1 - k'_U)$$

$$\text{where } k'_U \ = \ 0 \qquad \text{if } \mu \le T \quad (\mu \text{ is inside target zone})$$

$$= \ \frac{\mu - T}{\text{USL} - T} \text{ if } \mu > T \quad (\mu \text{ is outside target zone})$$

From Section 5.10, C'_{PU} is defined as:

$$C'_{PU} \ = \ \text{Maximum}(C^T_{PU}, C_{PU}) \ = \ \text{Maximum}\left(\frac{\text{USL} - T}{3\sigma_{ST}}, \frac{\text{USL} - \mu}{3\sigma_{ST}}\right)$$

As k'_L and k'_U are independent of σ, they apply equally well to P'_{PL} and P'_{PU}.

$$P_{PK} = P'_{PL}(1 - k'_L) \qquad\qquad P_{PK} = P'_{PU}(1 - k'_U)$$

Example 6.20 Solder Pull Test

Wire leads are soldered onto printed circuit boards. Each hour, an inspector named Betsy checks a lead by measuring the force required to pull it from the board, called a "pull test." Acceptable leads must be able to withstand a minimum pull force of 25 grams. The average pull force is estimated as 35 grams, with the short-term standard deviation estimated as 5 grams. With a target average for pull force of 40 grams, the short-term potential capability of this process is estimated as 1.00.

$$\hat{C}'_{PL} = \text{Maximum}\left(\frac{T - \text{LSL}}{3\hat{\sigma}_{ST}}, \frac{\hat{\mu} - \text{LSL}}{3\hat{\sigma}_{ST}} \right) = \text{Maximum}\left(\frac{40 - 25}{3(5)}, \frac{35 - 25}{3(5)} \right) = 1.00$$

Due to the average pull force of 35 grams being less than the target of 40 grams, the following formula is chosen to estimate k'_L.

$$\hat{k}'_L = \frac{T - \hat{\mu}}{T - \text{LSL}} = \frac{40 - 35}{40 - 25} = .333$$

Because the estimate of k'_L is greater than zero, the performance capability of this process is less than the potential of 1.00.

$$\hat{C}_{PK} = \hat{C}'_{PL}(1 - \hat{k}'_L) = 1.00(1 - .333) = .67$$

Now it is known how performance capability is affected by lack of potential capability versus lack of accuracy in centering the process average in the target zone. The reason for the severe lack of performance capability is due both to low potential capability ($\hat{C}'_{PL}$ is only 1.00) and lack of proper centering (k'_L is .333).

Example 6.21 Particle Contamination

An *IX & MR* chart is monitoring particle contamination of a cleaning solution for printed circuit boards after passing through wave soldering. The following estimates of process parameters are made from the 25 most recent in-control readings.

$$\hat{\mu} = 127.32 \, ppm \qquad\qquad \hat{\sigma}_{ST} = 12.079 \, ppm$$

As contamination has just an USL of 165 *ppm* with a target average of 120, the potential capability of this process is estimated with C'_{PU}.

$$\hat{C}'_{PU} = \text{Maximum}\left(\frac{\text{USL} - T}{3\hat{\sigma}_{ST}}, \frac{\text{USL} - \hat{\mu}}{3\hat{\sigma}_{ST}} \right)$$

$$= \text{Maximum}\left(\frac{165 - 120}{3(12.079)}, \frac{165 - 127.32}{3(12.079)} \right)$$

$$= \text{Maximum}(1.24, 1.04) = 1.24$$

k'_U is estimated with the following formula:

$$\hat{k}'_U = \frac{\mu - T}{USL - T} = \frac{127.32 - 120}{165 - 120} = .1627$$

Due to the process average of 127.32 being less than the target of 120, performance capability will be 16.27 percent less than potential capability, as is seen when the C_{PK} index is estimated from these two components.

$$\hat{C}_{PK} = \hat{C}'_{PU}(1 - \hat{k}'_U) = 1.24(1 - .1627) = 1.04$$

This last equation quantifies the effect of potential capability versus centering on the performance capability for this process. Because more than 16 percent of the potential capability is lost due to the centering of the process output, an improved filtration system is installed in the cleaning bath in hopes of reducing the average contamination level. After the process regains stability, the new process average is 110 *ppm*, which is now less than the target of 120 *ppm*. Assuming no change in $\hat{\sigma}_{ST}$, the new C'_{PU} index for this improved process is estimated as follows (C'_{PU} must be re-estimated because μ changed):

$$\hat{C}'_{PU} = \text{Maximum}\left(\frac{USL - T}{3\hat{\sigma}_{ST}}, \frac{USL - \hat{\mu}}{3\hat{\sigma}_{ST}} \right)$$

$$= \text{Maximum}\left(\frac{165 - 120}{3(12.079)}, \frac{165 - 110}{3(12.079)} \right)$$

$$= \text{Maximum}(1.24, 1.52) = 1.52$$

As μ is now less than T, k'_U equals 0 and C_{PK} equals C'_{PU}.

$$\hat{C}_{PK} = \hat{C}'_{PU}(1 - \hat{k}'_U) = 1.52(1 - 0) = 1.52$$

Because the process average is in the target zone, there is no loss of potential capability due to improper centering (k'_U is 0). A graph of this relationship between performance and potential capability appears in Figure 6.53 on the facing page. When the process average is greater than the target of 120, performance capability is less than potential capability. This disparity is identified by the shaded area on this graph, which represents the loss in potential capability due to μ being located above the target zone. The farther μ is above T, the greater is the loss. Initially, when the process average for particle contamination was 127.32 *ppm*, the estimate for C_{PK} came out to only 1.04 (point A on this graph). After μ was decreased to only 110 *ppm*, the estimate of C_{PK} increased to 1.52 (point B).

As μ moves lower toward T, the difference between C'_{PU} and C_{PK} decreases, until they become identical when μ equals T. As k'_U is 0 whenever the process average is less than T, performance and potential capability are equal, with both increasing the farther μ is centered below the target.

Example 6.22 Acrylic Coating

A supplier to the aerospace industry produces integrated avionics systems for commercial aircraft. Many of the circuit boards in these systems have surface mount components requiring

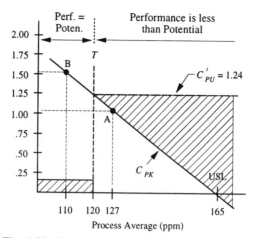

Fig. 6.53 C_{PK} versus μ for a unilateral specification.

an organic-acrylic coating to protect them from corrosion during their useful life. Boards are coated by them passing through an automated spraying process, with coating thickness being one of the key output characteristics.

Control of the spraying process has recently been established for a new board, with the average coating thickness estimated as 1.884 mm, and the long-term standard deviation as .028 mm. There is only a minimum thickness requirement of 1.774 mm, with a target of 1.900 mm. In Example 5.22, potential capability was estimated with the P'_{PL} index as 1.50, while the goal for P_{PK} is 2.00. Because the process average is less than T, the k'_L factor for this process is estimated with this formula:

$$\hat{k}'_L = \frac{T - \hat{\mu}}{T - \text{LSL}} = \frac{1.900 - 1.884}{1.900 - 1.774} = .127$$

This produces an estimate of 1.31 for performance capability.

$$\hat{P}_{PK} = \hat{P}'_{PL}(1 - \hat{k}'_L) = 1.50(1 - .127) = 1.31$$

After several changes to the spraying process, coating thickness is brought into control with an average of 1.909 mm and long-term standard deviation of .026 mm. With these new process parameters, the estimate of potential capability, P'_{PL}, turns out to be 1.73.

$$\hat{P}'_{PL} = \text{Maximum}\left(\frac{T - \text{LSL}}{3\hat{\sigma}_{LT}}, \frac{\hat{\mu} - \text{LSL}}{3\hat{\sigma}_{LT}}\right)$$

$$= \text{Maximum}\left(\frac{1.900 - 1.774}{3(.026)}, \frac{1.909 - 1.774}{3(.026)}\right)$$

$$= \text{Maximum}(1.62, 1.73) = 1.73$$

As the process average is now greater than T, the k'_L factor for this process is 0, making the estimate of performance capability, P_{PK}, also equal to 1.73.

$$\hat{P}_{PK} = \hat{P}'_{PL}(1 - \hat{k}'_L) = 1.73(1 - .0) = 1.73$$

Although this is a marked improvement over the original process, it is still short of the 2.00 goal. More work must be done on this process to either raise the average thickness, reduce variation, or both.

Special Values of C_{PK} and P_{PK}

Could either C_{PK} or P_{PK} ever be equal to 0? Because Z_{MIN} can be 0, this is entirely possible, although certainly not desirable, as Figure 6.54 illustrates. This figure presents a process that is in control, with its process average centered directly on the USL.

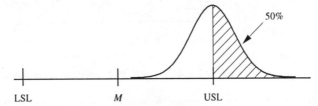

Fig. 6.54 Process centered at the USL.

When μ equals the USL, the quantity USL minus μ is 0, making C_{PK} equal to 0 as well.

$$\hat{C}_{PK} = \frac{\text{USL} - \hat{\mu}}{3\hat{\sigma}_{ST}} = \frac{0}{3\hat{\sigma}_{ST}} = 0$$

In this dire state, at least 50 percent of the process output is outside the print specification for this characteristic. This is a process in tremendous need of improvement.

Can an estimate of C_{PK} or P_{PK} ever be negative? Again, this is definitely possible, as is revealed in Figure 6.55.

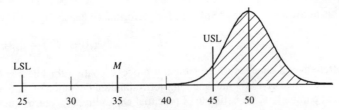

Fig. 6.55 Process centered above the USL.

With a process centered above the USL, the difference between the USL and the average is negative. This circumstance results in a negative C_{PK} index, as seen here.

$$\hat{C}_{PK} = \frac{\text{USL} - \overline{X}}{3\hat{\sigma}_{ST}} = \frac{45 - 50}{3(2)} = \frac{-5}{6} = -.833$$

Well over 50 percent nonconforming parts are being produced by this process. Note that for this particular situation, decreasing process spread will actually *increase* the percentage of nonconforming parts, as is illustrated in Figure 6.56.

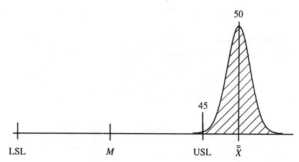

Fig. 6.56 A greater percentage of parts are above the USL.

If σ_{ST} is reduced from 2.0 to 1.667, C_{PK} is now more negative than originally, and has even a greater percentage of parts above the USL.

$$\hat{C}_{PK} = \frac{USL - \overline{\overline{X}}}{3\hat{\sigma}_{ST}} = \frac{45 - 50}{3(1.667)} = -1.00$$

When C_{PK} is exactly minus 1, μ is either $3\sigma_{ST}$ above the USL, or $3\sigma_{ST}$ below the LSL. For example, if the average is above the USL, then:

$$\frac{USL - \mu}{3\sigma_{ST}} = -1$$

$$USL - \mu = -3\sigma_{ST}$$

$$\mu - USL = 3\sigma_{ST}$$

$$\mu = USL + 3\sigma_{ST}$$

In either case, there are 99.865 percent nonconforming parts, assuming normality. Definitely an unpleasant situation in desperate need of efforts to center the process average nearer to *M*. Again, note that reducing variation will not help until the average is first located within the tolerance.

Example 6.23 Floppy Disks

Recall the field problem which is traced to variation in the diameter of an inner hole of the floppy disk. Some useful information has been collected so far and is listed here:

$$\overline{X} = \hat{\mu} = 40.9 \qquad \hat{C}_P = .928 \qquad \hat{P}_P = .909$$

$$\hat{C}_{PL} = .520 \qquad \hat{C}_{PU} = 1.336 \qquad \hat{P}_{PL} = .509 \qquad \hat{P}_{PU} = 1.310$$

Short-term performance capability for this process is measured by estimating C_{PK}, whereas long-term performance capability is estimated with P_{PK}.

$$\hat{C}_{PK} = \text{Minimum}(\hat{C}_{PL}, \hat{C}_{PU}) = \text{Minimum}(.520, 1.336) = .520$$

$$\hat{P}_{PK} = \text{Minimum} \, (\, \hat{P}_{PL}, \hat{P}_{PU}\,) = \text{Minimum} \, (\, .509, 1.310 \,) = .509$$

Both indexes indicate a severe lack of performance capability in view of their being well below 1.0. Comparing them to the potential capability measures (C_P of .928 and P_P of .909) reveals problems with both centering and process spread. The extent of the centering problem is quantified by estimating the k factor for this process. However, M must first be determined from the LSL of 39.5 and the USL of 44.5.

$$M = \frac{\text{LSL} + \text{USL}}{2} = \frac{39.5 + 44.5}{2} = 42.0$$

As the k formula considers only the difference between M and μ (no estimate of σ involved), k is the same for both C_{PK} and P_{PK}.

$$\hat{k} = \frac{|\, M - \hat{\mu} \,|}{(\,\text{USL} - \text{LSL} \,) / 2} = \frac{42.0 - 40.9}{(44.5 - 39.5)/2} = \frac{1.1}{2.5} = .44$$

The process average being off center by 44 percent exacts a tremendous toll on short-term performance capability. C_{PK} is just .520, which is only 56 percent of the potential capability of .928.

$$\hat{C}_{PK} = \hat{C}_P (\, 1 - \hat{k}\,) = .928 \, (\, 1 - .44 \,) = .928 \, (\, .56 \,) = .520$$

This same k factor is used to estimate P_{PK}.

$$\hat{P}_{PK} = \hat{P}_P (\, 1 - \hat{k}\,) = .909 \, (\, 1 - .44 \,) = .509$$

As discovered previously when the Z_{MIN} value was calculated in Example 6.5, one of the first attempts to improve this process should be positioning the average at the middle of the tolerance. Although it won't solve the entire problem (this process also lacks sufficient potential capability), better centering will certainly help. In addition, moving the process average is usually relatively easy, and inexpensive, compared to reducing process variation.

Tracking Changes in C_{PK} and P_{PK}

Changes in C_{PK} (or P_{PK}) over time may be tracked on a graph showing the relationship between the two components of performance capability: process precision, measured by either C_P or C_R; and process accuracy, measured by k_U.

$$C_P = \frac{\text{Tolerance}}{6\sigma_{ST}} = \frac{1}{C_R} \qquad k_U = \frac{\mu - M}{(\,\text{USL} - \text{LSL}\,) / 2}$$

The k_U factor reveals how much μ deviates from M, as well as the *direction*; k_U is positive when μ is above M, whereas a negative k_U means μ is below M.

$$k_U < 0 \implies \mu < M \qquad\qquad k_U > 0 \implies \mu > M$$

Recall that a k_U value of 0 occurs when μ is centered at M.

$$k_U = 0 \implies \mu = M$$

In addition, k_U equals minus 1 (plus 1) when μ is located at the LSL (USL).

$$k_U = -1 \quad \Rightarrow \quad \mu = \text{LSL} \qquad\qquad k_U = +1 \quad \Rightarrow \quad \mu = \text{USL}$$

The k_U factor and a measure of potential capability form the x and y coordinates, respectively, for plotting C_{PK} values on the A/P graph displayed in Figure 6.57 ("A" is for accuracy, "P" for precision). The scale on the left-hand y axis is for C_P, while the one on the right is for C_R. The slanted lines on this graph portray various combinations of C_P and k_U required to maintain a constant level of performance capability, as measured by C_{PK}. Lines are drawn for C_{PK} values ranging from 1.00 to 5.00.

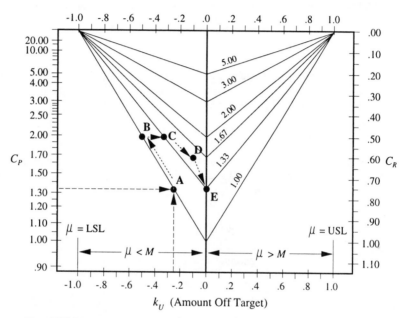

Fig. 6.57 Plotting C_{PK} values as a function of C_P and k_U on the A/P graph.

When interpreting this graph, note that points closer to the middle vertical line imply better centering, while points higher on the A/P graph mean better potential capability. For instance, point **A** represents a process that initially has a C_P of 1.33 (C_R of .75) and a k_U factor of minus .25, which means μ is below M by 25 percent of half of the tolerance. This combination generates a C_{PK} value of 1.00. Remember that the absolute value of k_U, which equals k, is used when calculating C_{PK}.

$$C_{PK,A} = C_P(1 - |k_U|) = 1.33(1 - |-.25|) = 1.00$$

The slanted line labeled "1.00" on the A/P graph represents all combinations of these two components that result in a C_{PK} of 1.00. For example, assume a change in tooling causes the process average to move even farther below M, such that k_U is now -.5. In order for C_{PK} (the performance capability) to remain at 1.00, the C_P index (potential capability) must increase to 2.00 in order to compensate for the decrease in accuracy, as is indicated by point **B**.

$$C_{PK,B} = C_P(1 - |k_U|) = 2.00(1 - |-.5|) = 1.00$$

An improvement in fixturing causes the process average to move close enough to M that k_U decreases to minus .333. Assuming C_P remains at 2.00, this mix of centering and potential capability increases C_{PK} to 1.33 (point **C**). Notice that negative k_U factors signal the process average must be shifted higher in order to be centered at M.

$$C_{PK,C} = C_P(1 - |k_U|) = 2.00(1 - |-.333|) = 1.33$$

Suppose the next adjustment to bring μ closer to the middle of the tolerance results in k_U declining to only minus .10, but inadvertently causes C_P to fall to 1.67. This new combination of precision and accuracy generates a C_{PK} of 1.50, the best performance capability so far (point **D**). Notice how point **D** is approximately half way between the lines for C_{PK} values of 1.33 and 1.67.

$$C_{PK,D} = C_P(1 - |k_U|) = 1.67(1 - |-.10|) = 1.50$$

When the process average is finally centered at M, k_U becomes 0. Unfortunately, this change makes C_P end up somewhat lower, at just 1.33. However, being centered at M means C_{PK} now equals the full potential capability of 1.33 (point **E**).

$$C_{PK,E} = C_P(1 - |k_U|) = 1.33(1 - |0|) = 1.33$$

This type of graph can be employed to track progress in performance capability over time. For instance, estimates of C_{PK} for a fabrication operation have fluctuated over the past six months, as is listed in Table 6.11.

Table 6.11 Monthly capability results.

Month	C_P	k_U	C_{PK}
1	1.19	.38	.74
2	1.36	.05	1.29
3	1.56	.21	1.23
4	1.56	-.05	1.48
5	1.49	-.13	1.30
6	1.92	-.04	1.84

The capability information for this six-month period may be better understood by plotting it on a A/P graph, which was done in Figure 6.58. The overall tendency of the plot points to move closer to the middle vertical line bears witness that the efforts expended to better center this process were successful. The points are also moving in a general upward direction, implying the process spread was steadily reduced during these six months. The combination of these two positive movements has resulted in a tremendous improvement in performance capability, with C_{PK} increasing from .74 all the way up to 1.84.

As the units of measurement on both the x and y axis are dimensionless, this type of graph is also very helpful for summarizing the capability results of multiple characteristics (see Section 12.4). Buckfelder and Powell describe a similar, but inverted, graph, called a "C_{PK} dart chart," while Gabel presents a comparable graph, called a "process performance chart."

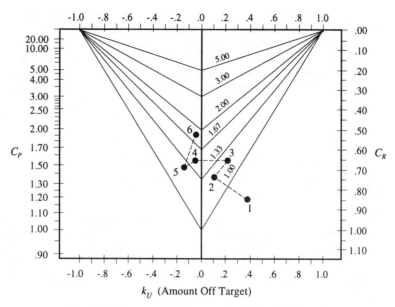

Fig. 6.58 Changes in C_{PK} are tracked on an A/P graph.

Singhal (1991-1992, 1992) has developed a plotting procedure, referred to as a "multi-parameter analysis chart," which accomplishes the same objectives as the A/P graph. A blank copy of the A/P graph is provided in Section 6.16 (page 343).

Alternative Formulas for C_{PK} and P_{PK}

The C_{PK} formula may be written in several ways, many of which are listed below. Choose whichever one seems easiest, as all generate identical estimates of short-term performance capability.

$$C_{PK} = \text{Minimum}\,(\,C_{PL}, C_{PU}\,)$$

$$= \text{Minimum}\left(\frac{Z_{LSL,ST}}{3}, \frac{Z_{USL,ST}}{3}\right)$$

$$= \frac{1}{3}\,\text{Minimum}\,(\,Z_{LSL,ST}, Z_{USL,ST}\,)$$

$$= \frac{1}{3}\,\text{Minimum}\left(\frac{\mu - LSL}{\sigma_{ST}}, \frac{USL - \mu}{\sigma_{ST}}\right)$$

$$= \text{Minimum}\left(\frac{\mu - LSL}{3\sigma_{ST}}, \frac{USL - \mu}{3\sigma_{ST}}\right)$$

$$= \frac{1}{3\sigma_{ST}}\,\text{Minimum}\,(\,\mu - LSL\,, USL - \mu\,)$$

Likewise, formulas for P_{PK} can be expressed in any of the above forms by simply replacing σ_{ST} with σ_{LT}.

Sometimes the formula for C_{PK} is presented as given below. The abbreviation "N.S." is for "nearest specification" to the process average (Gunter), although Finley refers to N.S. as "CS" for "closest specification." As this variation of the C_{PK} formula is seen quite often (Shunta, Smith), it is apparently chosen to save space in writing out the full formula.

$$C_{PK} = \frac{|\,\text{N.S.} - \overline{\overline{X}}\,|}{3\sigma}$$

However, there are several problems with this formula. First, it doesn't specify which standard deviation should be used, short-term or long-term. Second, it assumes all control charts are $\overline{X}$ charts, where the estimate of the process average is $\overline{\overline{X}}$. Many companies utilize *IX* charts, where $\overline{X}$ is the estimate of μ. These first two concerns are not very serious, but computing the absolute value of the difference between the "nearest specification" and the process average is a major problem, as this is incorrect. Previously, it was demonstrated that C_{PK} becomes negative when the process average is centered outside a specification limit. This proposed formula would make the result of *every* C_{PK} calculation positive (due to the absolute value signs involved), leading to incorrect decision making, as the following example clearly demonstrates.

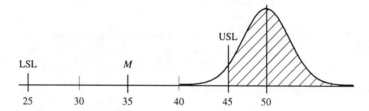

Fig. 6.59 Process centered above the USL.

For the distribution displayed in Figure 6.59, the process average is centered above the USL, and the true C_{PK} value should be a minus .833 ($\hat{\mu} = \overline{\overline{X}} = 50$).

$$\hat{C}_{PK} = \frac{\text{USL} - \hat{\mu}}{3\hat{\sigma}_{ST}} = \frac{45 - 50}{3\,(2)} = \frac{-5}{6} = -.833$$

However, the formula with the absolute value of N.S. minus the average makes the C_{PK} for this process equal to a *positive* .833.

$$\hat{C}_{PK} = \frac{|\,\text{N.S.} - \overline{\overline{X}}\,|}{3\hat{\sigma}} = \frac{|\,45 - 50\,|}{3(2)} = \frac{|-5|}{6} = \frac{5}{6} = .833$$

By moving the process average even further above the USL so it is centered at 65, the true C_{PK} becomes even more negative (-3.333), with almost 100 percent nonconforming parts being produced. This significant decrease in C_{PK} would send a strong signal to lower the process average ($\hat{\mu}$ is now 65).

$$\hat{C}_{PK} = \frac{\text{USL} - \hat{\mu}}{3\hat{\sigma}_{ST}} = \frac{45 - 65}{3\,(2)} = \frac{-20}{6} = -3.333$$

Because the absolute value signs make N.S. minus the average a positive number, the upward shift in μ to 65 has the appearance of *increasing* the performance capability of this process from .833 to 3.333. This suggests the average should be moved even *higher*, which would be a terrible mistake.

$$\hat{C}_{PK} = \frac{|\,\text{N.S.} - \overline{X}\,|}{3\hat{\sigma}} = \frac{|\,45 - 65\,|}{3(2)} = \frac{20}{6} = 3.333$$

As this formula leads to erroneous conclusions concerning process capability, it should never be considered for analysis.

Advantages and Disadvantages

C_{PK} and P_{PK} are similar to Z_{MIN} in that they are dimensionless indexes which summarize performance capability into one number. However, they have the additional benefit of using the more familiar rating scale where 1.0 indicates attainment of the minimum requirement for performance capability. This scale allows a direct comparison of performance capability (C_{PK} or P_{PK}) to potential capability (C_P or P_P). Note that because C_{PK} and P_{PK} are derived from different types of process variation, they should seldom be compared to each other. C_{PK} may be compared to C_P, as both are based on the short-term standard deviation, while P_{PK} may be compared to P_P, as these two are based on the long-term standard deviation.

C_{PK} may be separated into a component representing precision (C_P), and a second reflecting accuracy, called k. Likewise, P_{PK} may be divided into two portions: P_P for precision, and k for accuracy. An analysis of these two components identifies which facet of process improvement (reducing spread or shifting average) will produce the greatest increase in performance capability.

In order to derive reasonably accurate estimates of process capability, this index, as with all others mentioned so far, assumes that the process output is normally distributed. If not, assessment errors may occur, as the following example for an $\overline{X}, S$ chart illustrates. When the process is in control, its output has the skewed distribution displayed in Figure 6.60. Notice that both lower and upper tails are well within the specification limits, meaning very few nonconforming parts are being produced.

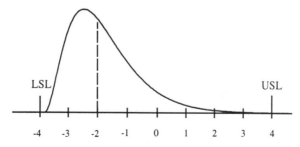

Fig. 6.60 Process output is non-normally distributed.

Based on the control chart centerlines, process parameters are estimated as follows:

$$\hat{\mu} = \overline{X} = -2 \qquad \hat{\sigma}_{ST} = \hat{\sigma}_{\overline{s}} = \frac{\overline{S}}{c_4} = 1$$

From these process parameters, estimates of short-term potential and performance process capability are computed, *without* first checking the normality assumption (LSL is -4 and USL is +4).

$$\hat{C}_P = \frac{\text{USL} - \text{LSL}}{6\hat{\sigma}_{ST}} = \frac{4 - (-4)}{6 \times 1} = 1.33$$

The estimated C_P index indicates the process has good potential capability. Unfortunately, performance capability, as measured by C_{PK}, leaves much to be desired.

$$\hat{C}_{PK} = \frac{\hat{\mu} - \text{LSL}}{3\hat{\sigma}_{ST}} = \frac{-2 - (-4)}{3 \times 1} = .67$$

With a C_{PK} well below 1.0, the process is (falsely) assessed as definitely lacking performance capability, when, in fact, essentially all parts are within tolerance. The large difference between potential and performance measures implies there is a sizable centering problem. Calculating k quantifies of how much of the performance problem is due to poor centering.

$$\hat{k} = \frac{|M - \hat{\mu}|}{(\text{USL} - \text{LSL})/2} = \frac{|0 - -2|}{(4 - -4)/2} = .50$$

A k of .50 means a full half of the potential capability is lost because the process average is not centered at M. In view of these results, the naturally logical decision is to shift the process average upward to the middle of the tolerance, as Figure 6.61 illustrates. But unknown to the personnel involved with this decision, this action pushes a considerable portion of the distribution's upper tail above the USL.

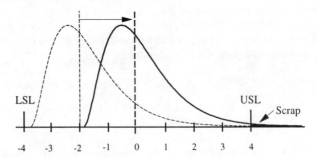

Fig. 6.61 Centering process at M increases scrap.

After the process is brought into control at this new level, estimates of the parameters for this new output distribution are calculated.

$$\hat{\mu} = \overline{X} = 0 \qquad \hat{\sigma}_{ST} = \hat{\sigma}_{\overline{s}} = \frac{\overline{S}}{c_4} = 1$$

Capability measures are estimated for this revised process, again without checking the normality assumption.

$$\hat{C}_P = \frac{\text{USL} - \text{LSL}}{6\hat{\sigma}_{ST}} = \frac{4 - (-4)}{6 \times 1} = 1.33 \qquad \hat{C}_{PK} = \frac{\hat{\mu} - \text{LSL}}{3\hat{\sigma}_{ST}} = \frac{0 - (-4)}{3 \times 1} = 1.33$$

With the process average equal to M, k is 0. Now potential and performance capability are equal.

$$\hat{k} = \frac{|M - \hat{\mu}|}{(\text{USL} - \text{LSL})/2} = \frac{|0 - 0|}{(4 - -4)/2} = 0 \qquad \hat{C}_{PK} = \hat{C}_P(1 - k) = 1.33(1 - 0) = 1.33$$

Moving the process average to the middle of the tolerance seems like the correct adjustment based on the revised C_{PK} estimate of 1.33, which now equals the measure of potential capability. However, all these capability calculations assume the process output is *normally* distributed. Neglecting to check this important assumption may lead to faulty decision making. In this example, shifting the process average upward actually causes scrap to be produced above the USL when none was initially being produced. Fortunately, the next chapter shows how to check this important normality assumption so these costly mistakes in interpretation are avoided.

There are some additional drawbacks to the C_{PK} and P_{PK} measures of capability. Because these measures are based on Z_{MIN}, they look at only *half* of the problem. What about Z_{MAX}? Is it equal to Z_{MIN}, a little greater, or is Z_{MAX} substantially greater? This makes a significant difference in the *total* amount of nonconforming parts being produced by the process, as this could be anywhere from equal to the percentage for Z_{MIN}, to double that amount. Yet C_{PK} is identical for both scenarios since it always equals just $Z_{MIN,ST}$.

To illustrate this point, consider two suppliers (A and B) who are bidding for a customer's business. Both are similar in cost, percentage of on-time deliveries, etc. The final decision will be based on which supplier provides the "best" quality. A fairly common practice in many industries is to associate the best quality with the highest C_{PK} index. The process outputs for suppliers A and B are displayed in Figure 6.62. Supplier A's process has half the variation as B's, but is centered halfway between the middle of the tolerance and the USL, with .1 percent above the USL. Supplier B, on the other hand, is centered exactly at the middle of the tolerance, with .1 percent of its output outside each of the two specifications. Which supplier should be awarded the contract? Visually, most people prefer A over B, but whatever their preference, agree the two are quite different.

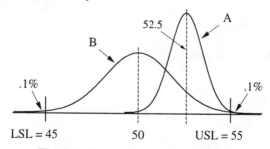

Fig. 6.62 Outputs for suppliers A and B.

To base this decision on a capability measure, the respective process averages and standard deviations are estimated for these two suppliers.

$$\hat{\mu}_A \ = \ 52.5 \qquad \hat{\sigma}_{ST,A} \ = \ .809 \qquad\qquad \hat{\mu}_B \ = \ 50.0 \qquad \hat{\sigma}_{ST,B} \ = \ 1.618$$

These estimates are substituted into the C_{PK} formula to determine this measure for each supplier. For supplier A:

$$\hat{C}_{PK,A} \ = \ \text{Minimum} \left(\frac{\hat{\mu}_A - LSL}{3\hat{\sigma}_{ST,A}} , \frac{USL - \hat{\mu}_A}{3\hat{\sigma}_{ST,A}} \right)$$

$$= \ \text{Minimum} \left(\frac{52.5 - 45}{3(.809)} , \frac{55 - 52.5}{3(.809)} \right)$$

$$= \ \text{Minimum} \ (\ 3.09 , 1.03 \ ) \ = \ 1.03$$

For supplier B:

$$\hat{C}_{PK,B} \ = \ \text{Minimum} \left(\frac{\hat{\mu}_B - LSL}{3\hat{\sigma}_{ST,B}} , \frac{USL - \hat{\mu}_B}{3\hat{\sigma}_{ST,B}} \right)$$

$$= \ \text{Minimum} \left(\frac{50 - 45}{3(1.618)} , \frac{55 - 50}{3(1.618)} \right)$$

$$= \ \text{Minimum} \ (\ 1.03 , 1.03 \ ) \ = \ 1.03$$

Based on these C_{PK} results, both suppliers have *equal* quality levels, yet B is producing *twice* the percentage of nonconforming parts as A! The C_{PK} index by itself does not provide a direct indication of how far the process average deviates from the target or how wide the process spread is. Even though A and B have obvious differences in their averages, standard deviations, *and* percentages nonconforming, the C_{PK} index rates them as equal. Later in this book, other measures of process capability (ppm_{TOTAL}, C_{PM}, C_{PMK}, and Product C_{PK}) provide solutions to this problem.

As mentioned several times before, it's always a good idea to consider more than a single measure of capability. If the C_P measure is also calculated for each supplier, a definite distinction between the two is uncovered.

$$\hat{C}_{P,A} \ = \ \frac{\text{Tolerance}}{6\hat{\sigma}_{ST,A}} \ = \ \frac{55 - 45}{6(.809)} \ = \ 2.06 \qquad\qquad \hat{C}_{P,B} \ = \ \frac{\text{Tolerance}}{6\hat{\sigma}_{ST,B}} \ = \ \frac{55 - 45}{6(1.618)} \ = \ 1.03$$

This measure reveals that supplier A has twice the potential capability of B. Their C_{PK} values are equal only because of a difference in centering. The k factor can be estimated for each supplier to help quantify this disparity in centering.

$$\hat{k}_A \ = \ \frac{| M - \hat{\mu}_A |}{(USL - LSL)/2} \ = \ \frac{| 50 - 52.5 |}{(55 - 45)/2} \ = \ .50 \qquad \hat{k}_B \ = \ \frac{| M - \hat{\mu}_B |}{(USL - LSL)/2} \ = \ \frac{| 50 - 50 |}{(55 - 45)/2} \ = \ 0$$

Although their C_{PK} values are identical, supplier A needs to work on properly centering the process average ($k_A = .50$), while supplier B must concentrate on reducing common-cause variation ($C_{P.A} = 1.03$). Supplier A has good precision but poor accuracy. On the other hand, B has excellent accuracy, but lacks precision. If only C_{PK} is considered, management would have a difficult time choosing the correct course of action to pursue. Always be wary when someone attempts to quantify capability with just a solitary measure, as most manufacturing processes are not that simple to describe.

A quick look at their $Z_{MAX,ST}$ values would also prove beneficial. Even though their C_{PK} values are identical, supplier B has a lower $Z_{MAX,ST}$ than A (3.09 versus 9.27), disclosing it has the poorer performance capability.

$$\hat{Z}_{MAX.\ ST.\ A} = \text{Maximum} \left(\frac{\hat{\mu}_A - LSL}{\hat{\sigma}_{ST,A}}, \frac{USL - \hat{\mu}_A}{\hat{\sigma}_{ST,A}} \right)$$

$$= \text{Maximum} \left(\frac{52.5 - 45}{.809}, \frac{55 - 52.5}{.809} \right)$$

$$= \text{Maximum} (9.27, 3.09) = 9.27$$

$$\hat{Z}_{MAX.\ ST.\ B} = \text{Maximum} \left(\frac{\hat{\mu}_B - LSL}{\hat{\sigma}_{ST,B}}, \frac{USL - \hat{\mu}_B}{\hat{\sigma}_{ST,B}} \right)$$

$$= \text{Maximum} \left(\frac{50 - 45}{1.618}, \frac{55 - 50}{1.618} \right)$$

$$= \text{Maximum} (3.09, 3.09) = 3.09$$

Another common misconception involving this index is believing a higher C_{PK} value implies a lower nonconforming rate. As revealed in Figure 6.63, this is not always true. Process output distribution A has an average of 3, with a standard deviation of 1.143.

$$Z_{LSL,ST} = \frac{\mu - LSL}{\sigma_{ST}} = \frac{3 - -5}{1.143} = 7.00 \qquad\qquad Z_{USL,ST} = \frac{USL - \mu}{\sigma_{ST}} = \frac{5 - 3}{1.143} = 1.75$$

This results in a C_{PK} rating of .58 for A.

$$C_{PK,A} = \frac{1}{3} \text{Minimum} (Z_{LSL,ST}, Z_{USL,ST}) = \frac{1}{3} \text{Minimum} (1.75, 7.00) = .58$$

The total percentage nonconforming is:

$$p'_{TOTAL,ST,A} = p'_{LSL,ST} + p'_{USL,ST} = .00 + .04 = .04$$

Now consider process output distribution B, which has an average of 0 and a standard deviation of 2.660. The Z values for this process are:

$$Z_{LSL,ST} = \frac{\mu - LSL}{\sigma_{ST}} = \frac{0 - -5}{2.660} = 1.88 \qquad\qquad Z_{USL,ST} = \frac{USL - \mu}{\sigma_{ST}} = \frac{5 - 0}{2.660} = 1.88$$

This produces a C_{PK} of .63 for B.

$$C_{PK,B} = \frac{1}{3} \text{Minimum} (Z_{LSL,ST}, Z_{USL,ST}) = \frac{1}{3} \text{Minimum} (1.88, 1.88) = .63$$

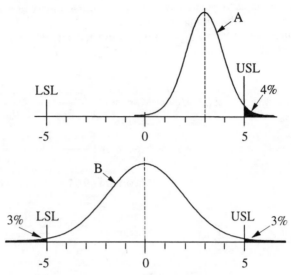

Fig. 6.63 A's process output compared to B's.

B has a higher C_{PK} than A (.63 vs .58), yet its percentage nonconforming is greater than A, 6 percent versus 4 percent.

$$p'_{TOTAL,ST,B} = p'_{LSL,ST} + p'_{USL,ST} = .03 + .03 = .06$$

The C_{PK} and P_{PK} measures have one other disadvantage in that they are "biased" measures of performance capability. This means if estimates of C_{PK} are made for a process from different samples, the average of all these estimates would not quite equal the true C_{PK} value for the process (Dovich, Price and Price). This issue is addressed further during the discussion on confidence bounds in Section 11.18.

Relationship Between C_{PK} and the Process Average

Because C_{PK} includes μ in its measure of process capability, movements in μ cause this index to change. This association is portrayed graphically in Figure 6.64 on the next page. Note how C_{PK} is equal to C_{PL} when μ is less than M and is equal to C_{PU} when μ is above M. C_{PK} attains its maximum value of C_P when the process average is located precisely at the middle of the tolerance. This is also where the lines for C_{PL} and C_{PU} intersect, meaning they are equal. Because both lines have identical slopes (except for sign), moving μ a certain distance below M reduces C_{PK} by the same amount as moving μ an equal distance above M. As noted earlier, when μ is equal to one of the specification limits, C_{PK} equals 0. C_{PK} is negative if μ is ever located outside the tolerance.

Since C_{PK} achieves its highest value when the process is centered at M, operators will make efforts to locate the process average here. If M is in fact the desired center of the process output, then this leads to maximizing customer satisfaction. However, there are some processes which should be centered closer to one specification than the other. For these situations, C_{PK} and P_{PK} should not be chosen to measure process capability. Instead, either C^*_{PK} or P^*_{PK} would be a much better choice.

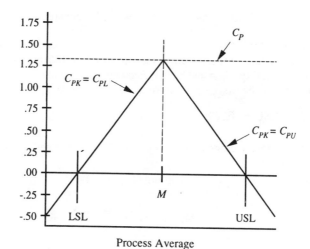

Fig. 6.64 C_{PK} as a function of the process average.

6.6 C^*_{PK} and P^*_{PK} for Cases Where $T \neq M$

The modification made to the C_{PL}, C_{PU}, P_{PL}, and P_{PU} measures for characteristics where T is not equal to M also applies to the C_{PK} and P_{PK} indexes.

$$C^*_{PK} = \text{Minimum}\,(C^*_{PL}, C^*_{PU}) = \text{Minimum}\left(\frac{\mu - \text{LSL}_T}{3\sigma_{ST}}, \frac{\text{USL}_T - \mu}{3\sigma_{ST}}\right)$$

$$P^*_{PK} = \text{Minimum}\,(P^*_{PL}, P^*_{PU}) = \text{Minimum}\left(\frac{\mu - \text{LSL}_T}{3\sigma_{LT}}, \frac{\text{USL}_T - \mu}{3\sigma_{LT}}\right)$$

As before, LSL_T and USL_T are determined by these formulas when T is less than M:

$$\text{LSL}_T = \text{LSL} \qquad\qquad \text{USL}_T = 2T - \text{LSL}$$

If T is greater than M, compute LSL_T and USL_T with this set of formulas:

$$\text{LSL}_T = 2T - \text{USL} \qquad\qquad \text{USL}_T = \text{USL}$$

LSL_T and USL_T are then used to estimate C^*_{PK} and P^*_{PK}, as is demonstrated in the next example (see also Chen, Vannman).

Example 6.24 Contact Closing Speed

Circuit breakers are checked for contact motion to verify movement at the proper opening and closing speed. Closing too quickly causes premature wearout of the contacts, while closing too slowly could burn or weld the contacts so the breaker could not be reopened. The engineering department has developed the following window for closing speed to optimize the life of this product, with a target average speed of .0170 seconds.

$$\text{LSL} = .0140 \text{ seconds} \qquad \text{USL} = .0250 \text{ seconds} \qquad M = .0195 \text{ seconds}$$

From the control chart, σ_{ST} is estimated as .00129 seconds and μ as .0189. To incorporate engineering's desire for an average closing speed of .0170 seconds in the measure of performance capability, C_{PK} must be replaced with C^*_{PK}. Since the target average (.0170) is less than M (.0195), the first step is determining LSL_T and USL_T with this set of formulas.

$$\text{LSL}_T = \text{LSL} = .0140 \qquad\qquad \text{USL}_T = 2T - \text{LSL} = 2(.0170) - .0140 = .0200$$

For the current average of .0189 seconds, C^*_{PK} is estimated as a meager .28.

$$\hat{C}^*_{PK} = \text{Minimum}\,(\hat{C}^*_{PL}, \hat{C}^*_{PU})$$

$$= \text{Minimum}\left(\frac{\hat{\mu} - \text{LSL}_T}{3\hat{\sigma}_{ST}}, \frac{\text{USL}_T - \hat{\mu}}{3\hat{\sigma}_{ST}} \right)$$

$$= \text{Minimum}\left(\frac{.0189 - .0140}{3(.00129)}, \frac{.0200 - .0189}{3(.00129)} \right)$$

$$= \text{Minimum}\,(1.27, .28) = .28$$

Because $\hat{C}^*_{PL}$ is greater than $\hat{C}^*_{PU}$, $\hat{C}^*_{PK}$ can be increased by shifting the process average lower. If moving μ does not cause σ_{ST} to change, the maximum $\hat{C}^*_{PK}$ possible without reducing σ_{ST} is found by averaging $\hat{C}^*_{PU}$ and $\hat{C}^*_{PK}$.

$$\text{Maximum } \hat{C}^*_{PK} = \frac{\hat{C}^*_{PL} + \hat{C}^*_{PU}}{2} = \frac{1.27 + .28}{2} = .78$$

This maximum can also be determined by estimating C^*_P.

$$\text{Maximum } \hat{C}^*_{PK} = \hat{C}^*_P = \frac{\text{Effective Tolerance}}{6\hat{\sigma}_{ST}}$$

$$= \frac{.0200 - .0140}{6(.00129)}$$

$$= .78$$

Achieving this maximum requires positioning μ halfway between LSL_T and USL_T. Of course, these modified specifications were calculated so this halfway point is exactly the target average of .0170 seconds.

$$\frac{\text{LSL}_T + \text{USL}_T}{2} = \frac{.0140 + .0200}{2} = .0170$$

Centering the process average at .0170 seconds will maximize C^*_{PK}.

$$\hat{C}_{PK}^* = \text{Minimum}\left(\frac{\hat{\mu} - \text{LSL}_T}{3\hat{\sigma}_{ST}}, \frac{\text{USL}_T - \hat{\mu}}{3\hat{\sigma}_{ST}}\right)$$

$$= \text{Minimum}\left(\frac{.0170 - .0140}{3(.00129)}, \frac{.0200 - .0170}{3(.00129)}\right)$$

$$= \text{Minimum}\ (.78, .78) = .78$$

Any movement in the average closing speed away from .0170 seconds will cause C^*_{PK} to decrease. This provides an incentive for the operator to keep the process centered at T. Unfortunately, the process still lacks performance capability even when left on target. Now work must begin to reduce process variation (see also Abraham).

If the process average equals T, the following relationships are true:

$$C_{PK}^* = C_{PL}^* = C_{PU}^* = C_P^*$$

When T equals M, LSL_T equals the original LSL, and USL_T equals the original USL. As there is no difference in the specifications used for their calculations, C^*_{PK} equals C_{PK}.

$$C_{PK}^* = C_{PK} \quad \text{and} \quad C_P^* = C_P$$

Whenever T equals M, *and* the process average is equal to M, these relationships are true:

$$C_{PK}^* = C_{PL}^* = C_{PU}^* = C_{PK} = C_{PL} = C_{PU} = C_P^* = C_P = \frac{1}{C_R^*} = \frac{1}{C_R}$$

Example 6.25 Gold Plating

The LSL for the thickness of gold being plated on silicon wafers is 50 angstroms and the USL is 100 angstroms. An economic decision is made to target the average thickness at 65 angstroms.

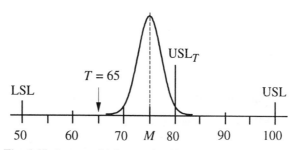

Fig. 6.65 Average thickness of gold plating is 75 angstroms.

If $\hat{\sigma}_{LT}$ is 2.5 and $\overline{\overline{X}}$ is originally equal to 75 (as portrayed in Figure 6.65), P_{PK} is estimated as 3.33. Even though this indicates very high performance capability, the process is currently plating *no* wafers at the desired thickness of 65.

$$\hat{P}_{PK} = \text{Minimum}\left(\frac{\overline{\overline{X}} - \text{LSL}}{3\hat{\sigma}_{LT}}, \frac{\text{USL} - \overline{\overline{X}}}{3\hat{\sigma}_{LT}} \right)$$

$$= \text{Minimum}\left(\frac{75 - 50}{3 \times 2.5}, \frac{100 - 75}{3 \times 2.5} \right)$$

$$= \text{Minimum}\,(\,3.33\,,\,3.33\,) = 3.33$$

To estimate P^*_{PK}, USL_T and LSL_T must be derived first. As T (65) is less than M (75), these modified limits are determined by this set of formulas:

$$\text{LSL}_T = \text{LSL} = 50 \qquad\qquad \text{USL}_T = 2T - \text{LSL} = 2(65) - 50 = 80$$

With these modified specification limits, P^*_{PK} is estimated to a disappointing .67.

$$\hat{P}^*_{PK} = \text{Minimum}\left(\frac{\overline{\overline{X}} - \text{LSL}_T}{3\hat{\sigma}_{LT}}, \frac{\text{USL}_T - \overline{\overline{X}}}{3\hat{\sigma}_{LT}} \right)$$

$$= \text{Minimum}\left(\frac{75 - 50}{3 \times 2.5}, \frac{80 - 75}{3 \times 2.5} \right) = \text{Minimum}\,(\,3.33\,,\,.67\,) = .67$$

Because P^*_{PL} is much larger than P^*_{PU}, P^*_{PK} can be substantially increased by lowering the average plating thickness. Following this advice, adjustments are made that shift the process average from 75 to 70 angstroms ($\overline{X} = 70$), as indicated in Figure 6.66. T is still 65, while $\hat{\sigma}_{LT}$ remains at 2.5 angstroms after this shift in μ.

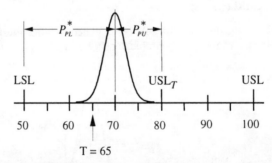

Fig. 6.66 Average thickness of gold plating is shifted to 70 angstroms.

The two performance capability indexes are recalculated with this new process average.

$$\hat{P}_{PK} = \text{Minimum}\left(\frac{\overline{\overline{X}} - \text{LSL}}{3\hat{\sigma}_{LT}}, \frac{\text{USL} - \overline{\overline{X}}}{3\hat{\sigma}_{LT}} \right)$$

$$= \text{Minimum}\left(\frac{70 - 50}{3 \times 2.5}, \frac{100 - 70}{3 \times 2.5} \right)$$

$$= \text{Minimum}\,(\,2.67\,,\,4.00\,) = 2.67$$

$$\hat{P}^*_{PK} = \text{Minimum} \left(\frac{\overline{\overline{X}} - LSL_T}{3\hat{\sigma}_{LT}}, \frac{USL_T - \overline{\overline{X}}}{3\hat{\sigma}_{LT}} \right)$$

$$= \text{Minimum} \left(\frac{70 - 50}{3 \times 2.5}, \frac{80 - 70}{3 \times 2.5} \right)$$

$$= \text{Minimum} (2.67, 1.33) = 1.33$$

The estimate for P_{PK} has decreased from 3.33 to 2.67, indicating this change in process average was a bad idea and the average should be shifted higher as soon as possible. On the other hand, the P^*_{PK} index tells a completely different story. Increasing from .67 to 1.33, it signals this was a very *good* move and, if anything, the average should be moved even *lower*. These diametrically opposed conclusions demonstrate why it is so crucial to pick the correct measure for assessing process capability. If M is "best" for the process average, choose P_{PK}. When T is "best," switch to P^*_{PK}.

As P^*_{PL} is still larger than P^*_{PU} (2.67 versus 1.33), a decision is made to lower μ even further, from 70 to 65 angstroms, as is portrayed in Figure 6.67.

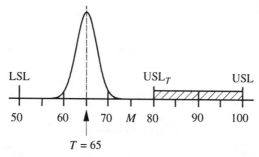

Fig. 6.67 Average is shifted to the target of 65 angstroms.

Recalculating estimates of these two capability measures with $\overline{X}$ equal to the target average of 65 angstroms yields the following results:

$$\hat{P}_{PK} = \text{Minimum} \left(\frac{\overline{\overline{X}} - LSL}{3\hat{\sigma}_{LT}}, \frac{USL - \overline{\overline{X}}}{3\hat{\sigma}_{LT}} \right)$$

$$= \text{Minimum} \left(\frac{65 - 50}{3 \times 2.5}, \frac{100 - 65}{3 \times 2.5} \right)$$

$$= \text{Minimum} (2.00, 4.67) = 2.00$$

$$\hat{P}^*_{PK} = \text{Minimum} \left(\frac{\overline{\overline{X}} - LSL_T}{3\hat{\sigma}_{LT}}, \frac{USL_T - \overline{\overline{X}}}{3\hat{\sigma}_{LT}} \right)$$

$$= \text{Minimum} \left(\frac{65 - 50}{3 \times 2.5}, \frac{80 - 65}{3 \times 2.5} \right)$$

$$= \text{Minimum} (2.00, 2.00) = 2.00$$

Again, P_{PK} has been lowered while P^*_{PK} has increased. The first index complains this is another "bad" move; the second says it's another "good" move. Because P^*_{PL} now equals P^*_{PU}, no further shifts in the average plating thickness are necessary. But what if a decision is made to again lower μ, this time to 60 angstroms, as has happened in Figure 6.68?

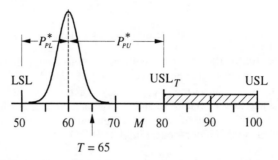

Fig. 6.68 Average plating thickness is lowered to only 60 angstroms.

Lowering the average thickness to 60 puts μ below the target of 65, causing both capability indexes to decrease.

$$\hat{P}_{PK} = \text{Minimum}\left(\frac{\overline{\overline{X}} - \text{LSL}}{3\hat{\sigma}_{LT}}, \frac{\text{USL} - \overline{\overline{X}}}{3\hat{\sigma}_{LT}}\right)$$

$$= \text{Minimum}\left(\frac{60 - 50}{3 \times 2.5}, \frac{100 - 60}{3 \times 2.5}\right)$$

$$= \text{Minimum}(1.33, 5.33) = 1.33$$

$$\hat{P}^*_{PK} = \text{Minimum}\left(\frac{\overline{\overline{X}} - \text{LSL}_T}{3\hat{\sigma}_{LT}}, \frac{\text{USL}_T - \overline{\overline{X}}}{3\hat{\sigma}_{LT}}\right)$$

$$= \text{Minimum}\left(\frac{60 - 50}{3 \times 2.5}, \frac{80 - 60}{3 \times 2.5}\right)$$

$$= \text{Minimum}(1.33, 2.67) = 1.33$$

Both measures decline to 1.33, and now agree that shifting the process average to 60 is a mistake. When the process average is less than T, these two indexes are always equal because both are now based on the relationship of the average to the LSL, and thus make the same recommendation: raise the average.

When the process average is greater than T, and less than the middle of the tolerance, P_{PK} signals for the process average to be raised, while P^*_{PK} indicates it should be lowered (review Figure 6.69). An operator attempting to maximize capability for this process by reacting to P_{PK} will obviously respond differently than one reacting to P^*_{PK}. It's extremely important to fully understand a process and its related goals before deciding which measures to employ in a capability study. In the same plant, diverse processes will usually require different capability measures. Very seldom will the same measure correctly apply to all situations in a single company, much less an entire industry.

Relationship Between C^*_{PK} and μ

C^*_{PK} attains its maximum value when μ equals T, as is shown in Figure 6.69. This property encourages operators to keep the process average centered at T, the desired target value.

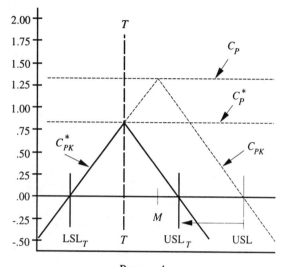

Process Average

Fig. 6.69 C^*_{PK} versus the process average.

USL_T is always less than, or equal to, the USL, while LSL_T is always greater than, or equal to, the LSL. Example 5.20 on page 140 disclosed how this relationship causes the effective tolerance to be less than, or equal to, the engineering tolerance. As a result, C^*_P is *less* than, or equal to, C_P. C^*_P equals C_P only when T equals M.

$$\text{Effective Tolerance} \leq \text{Tolerance}$$

$$\text{Effective Tolerance} \leq USL - LSL$$

$$\frac{\text{Effective Tolerance}}{6\sigma_{ST}} \leq \frac{USL - LSL}{6\sigma_{ST}}$$

$$C^*_P \leq C_P$$

Since the largest C^*_{PK} can ever be is equal to C^*_P (attained when μ equals T), C^*_{PK} is always less than, or equal to, C_P as well. C^*_{PK} equals C_P only when T equals M *and* μ equals M.

$$C^*_{PK} \leq C^*_P \leq C_P$$

The k^* Factor Relating C^*_{PK} and C^*_P

C^*_{PK} is related to C^*_P by this next formula (Kane, 1986; Grant and Leavenworth). k^* discloses how far the process average is from T, as a percentage of half the effective tolerance.

$$C_{PK}^* = C_P^*(1 - k^*)$$

$$\text{where } C_P^* = \text{Minimum}\left(\frac{T - \text{LSL}}{3\sigma_{ST}}, \frac{\text{USL} - T}{3\sigma_{ST}}\right) = \frac{\text{Effective Tolerance}}{6\sigma_{ST}} = \frac{\text{USL}_T - \text{LSL}_T}{6\sigma_{ST}}$$

$$\text{and } k^* = \frac{|T - \mu|}{\text{Minimum}(T - \text{LSL}, \text{USL} - T)} = \frac{|T - \mu|}{(\text{Effective Tolerance})/2} = \frac{|T - \mu|}{(\text{USL}_T - \text{LSL}_T)/2}$$

P_{PK}^* is related to P_P^* by this same k^* factor.

$$P_{PK}^* = P_P^*(1 - k^*)$$

A TR^* factor can be defined as 3 times k^*.

$$\text{TR}^* = 3k^* = \frac{3|T - \mu|}{(\text{USL}_T - \text{LSL}_T)/2} = \frac{6|T - \mu|}{\text{USL}_T - \text{LSL}_T}$$

In a similar fashion, k_L^* and k_U^* are defined in the following manner:

$$C_{PL}^* = C_P^*(1 - k_L^*) \qquad\qquad C_{PU}^* = C_P^*(1 - k_U^*)$$

$$P_{PL}^* = P_P^*(1 - k_L^*) \qquad\qquad P_{PU}^* = P_P^*(1 - k_U^*)$$

$$\text{where } k_L^* = \frac{T - \mu}{T - \text{LSL}_T} \text{ and } k_U^* = \frac{\mu - T}{\text{USL}_T - T}$$

These results can then be utilized to find C_{PK}^* and P_{PK}^*.

$$C_{PK}^* = \text{Minimum}(C_{PL}^*, C_{PU}^*) \qquad\qquad P_{PK}^* = \text{Minimum}(P_{PL}^*, P_{PU}^*)$$

Example 6.26 Contact Closing Speed

In the earlier case study involving contact closing speed (Example 6.24, page 255), the process average is .0189 seconds, σ_{ST} equals .00129, T is .0170, LSL_T equals .0140 and USL_T is .0200. C_P^* and k^* are derived from this information by first finding the effective tolerance.

$$\text{Effective Tolerance} = \text{USL}_T - \text{LSL}_T = .0200 - .0140 = .0060$$

This value is needed to estimate both C_P^* and k^*.

$$\hat{C}_P^* = \frac{\text{Effective Tolerance}}{6\hat{\sigma}_{ST}} = \frac{.0060}{6(.00129)} = .775$$

$$\hat{k}^* = \frac{|T - \hat{\mu}|}{(\text{Effective Tolerance})/2} = \frac{|.0170 - .0189|}{(.0060)/2} = .633$$

These two are combined to produce an estimate of .284 for C^*_{PK}, an outcome duplicating that obtained in Example 6.24.

$$\hat{C}^*_{PK} = \hat{C}^*_P(1 - \hat{k}^*) = .775(1 - .633) = .284$$

Observe that when μ is equal to either LSL_T or USL_T, the quantity $|T - \mu|$ equals half the effective tolerance, resulting in k^* equaling 1.

$$|T - \mu| = (\text{Effective Tolerance})/2$$

$$\frac{|T - \mu|}{(\text{Effective Tolerance})/2} = 1$$

$$k^* = 1$$

With a k^* factor of 1, C^*_{PK} equals 0.

$$C^*_{PK} = C^*_P(1 - k^*) = C^*_P(1 - 1) = 0$$

If the average closing speed ever becomes less than the LSL, the distance between T and μ would be larger than half the effective tolerance, resulting in a k^* factor greater than 1.

$$|T - \mu| \geq (\text{Effective Tolerance})/2$$

$$\frac{|T - \mu|}{(\text{Effective Tolerance})/2} \geq 1$$

$$k^* \geq 1$$

With this k^* factor, C^*_{PK} is negative, implying that more than 50 percent of the circuit breakers would have a contact closing time less than the LSL of .0140 seconds.

$$k^* > 1$$

$$-k^* < -1$$

$$1 - k^* < 0$$

$$C^*_P(1 - k^*) < 0$$

$$C^*_{PK} < 0$$

Furthermore, if μ is above the USL_T, k^* will also become greater than 1, which again results in a negative C^*_{PK} value. However, if μ is below the USL, there is *less* than 50 percent nonconforming parts being produced, as Figure 6.70 clearly shows. The negative C^*_{PK} value associated with this illustration means 50 percent of the closing times are above the USL_T of .0200, but they may not be above the *real* USL of .0250 seconds. Only those closing times above .0250 are truly nonconforming. To determine this percentage, estimate $Z_{USL,ST}$.

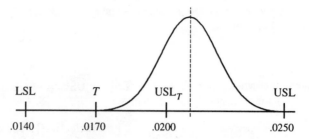

Fig. 6.70 C^*_{PK} is negative when μ is above USL_T.

Although less than 50 percent of closing times are nonconforming, a negative C^*_{PK} value still sends a strong signal to move the process average closer to the target, which is the primary purpose of this measure.

Relationship Between C^*_{PK} and C_P

In Section 5.9 (page 139), C^*_P was shown to be related to C_P by the quantity 1 minus k_T.

$$C^*_P = C_P(1 - k_T) \qquad \text{where } k_T = \frac{|T - M|}{(\text{Tolerance})/2}$$

If this result is substituted into the following equation, a direct relationship is formed between C^*_{PK} and C_P.

$$C^*_{PK} = C^*_P(1 - k^*) = C_P(1 - k_T)(1 - k^*)$$

C_P represents the maximum potential capability available to this process. k_T is the percentage of potential capability lost by having a target average different than M, whereas k^* is the percentage of potential capability lost due to the process average not being centered at T. For the contact closing speed example, C_P is found to be 1.421 and k_T is .455.

$$\hat{C}_P = \frac{\text{Tolerance}}{6\hat{\sigma}_{ST}} = \frac{.0250 - .0140}{6(.00129)} = 1.421$$

$$M = \frac{\text{Tolerance}}{2} = \frac{.0250 + .0140}{2} = .0195$$

$$\hat{k}_T = \frac{|T - M|}{(\text{Tolerance})/2} = \frac{|.0170 - .0195|}{(.0250 - .0140)/2} = .455$$

With $\hat{k}^*$ computed previously as .633, C^*_{PK} is estimated as .284, the same result obtained at the beginning of Example 6.26 on page 262.

$$\hat{C}^*_{PK} = \hat{C}_P(1 - \hat{k}_T)(1 - \hat{k}^*) = 1.421(1 - .455)(1 - .633) = .284$$

Performance capability is first decreased by 45.5 percent from its maximum potential of 1.421 due to *T* being different than *M*. Then, this remainder is reduced by another 63.3 percent because μ is not centered at the target average. The design engineers are responsible for the first loss in capability, whereas the manufacturing engineers (or operators) are responsible for the second.

Advantages and Disadvantages

These are similar to the ones given for C_{PK} and P_{PK}, with the exception that these modified measures do not assume *T* is equal to *M*. These indexes of performance capability are related to the target average of the characteristic being assessed. As the process average moves closer to *T*, they increase, and when the average moves away from *T*, they decrease. A more complicated alternative to C^*_{PK} is proposed by Littig and Lam for situations where parts produced in the smaller "half" of a non-symmetrical tolerance are considered less desirable than those in the larger "half."

As mentioned often in this book, it's always prudent to assess capability with more than just one type of index. Some of the more popular supplemental measures reflect process yield in some manner, quite often as the number of nonconforming parts per million pieces produced.

6.7 *ppm* Method

One disadvantage with the capability measures covered so far is the difficulty in comprehending the exact degree of improvement when positive changes occur in the process. Consider a process extruding vinyl sheets (Figure 6.71) which is finally brought into a good state of control for minimum thickness.

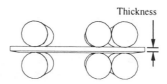

Fig. 6.71

Using measurement data from the *IX & MR* chart monitoring thickness, the performance capability for this critical characteristic is estimated with the C_{PK} index.

$$\hat{C}_{PK,ORIGINAL} = 1.24$$

Because the capability goal for thickness is a C_{PK} of at least 1.33, money is invested in new rollers in order to help reduce common-source variation. After these rollers are installed, the chart is brought into control at a lower level of variation. Measurement data from this improved process output produces the following updated estimate of C_{PK}:

$$\hat{C}_{PK,IMPROVED} = 1.30$$

Although this capability estimate is still short of the goal, the variation-reduction team is happy with the amount of improvement. However, their top management wants to know exactly how much of an improvement in capability resulted from their substantial investment in expensive new rollers. Is it 6 percent (1.30 minus 1.24)? More than 6 percent? Less? It is very difficult to judge the amount of improvement with C_P or C_{PK} values.

The *ppm* (parts per million) method easily answers this question by considering the total amount of nonconforming parts produced by a process (Fisher). *ppm* is just the percentage of nonconforming parts (*p'*) times 10 to the 6th power.

$$ppm = p' \times 10^6$$

For example, here's how 1 percent (.01) is converted to 10,000 parts per million.

$$ppm = p' \times 10^6 = .01 \times 10^6 = 10,000$$

There is a connection between *ppm* and several of the other capability measures mentioned so far, as revealed by Table 6.12 on the facing page. For example, a C_{PK} of 1.33 means there are 4 standard deviations from the process average to the nearest specification ($Z_{MIN,ST}$ is 4.00) and there are 32 *ppm* outside of this specification, assuming a normal distribution. If the characteristic under study has a unilateral specification, this is the total *ppm*. However, if it has a bilateral specification, an additional 32 *ppm* is outside the other specification if the process output is centered at the middle of the tolerance. When µ is centered anywhere else, less than 32 *ppm* will exceed this other specification limit. Thus, the *ppm* for a C_{PK} of 1.33 is 32 for a characteristic with a unilateral specification, and somewhere between a minimum of 32 and a maximum of 63 for a characteristic having a bilateral specification.

A C_{PK} index of 1.50 arises when there are 4.5 standard deviations from the process average to the nearest specification, and thus, 3 *ppm* outside this specification. For bilateral specifications, up to an additional 3 *ppm* could be outside the other specification, depending on where the process average is centered, making the total *ppm* somewhere between 3 and 6.

From Table 6.12, the original extrusion process with a C_{PK} of 1.24 had 100 *ppm* below the LSL for thickness, whereas the improved process with its C_{PK} of 1.30 has only 50 *ppm* below.

$$\hat{C}_{PK,ORIGINAL} = 1.24 \implies ppm = 100 \qquad \hat{C}_{PK,IMPROVED} = 1.30 \implies ppm = 50$$

Installing the new rollers caused a 50 percent reduction in the amount of vinyl sheets below the minimum thickness specification. It is very difficult to sense, and therefore appreciate, the magnitude of this improvement from the small change in C_{PK} values (1.24 vs. 1.30). However, cutting the number of nonconforming sheets in half (100 *ppm* to 50 *ppm*) is a very substantial achievement. This change in process performance becomes quite obvious with *ppm* values, especially to shop floor personnel, as well as top management. Most people, including customers, find it much easier to understand the amount of improvement in capability when expressed in *ppm* (Carr), and some believe all capability indexes should be direct measures of process yield (Pyzdek).

Calculating *ppm* from Z Values

For processes having normal distributions, *ppm* values are calculated by first looking up the percentage (in decimal form) outside print specification for both Z_{LSL} and Z_{USL} in Appendix Table III (Section 8.13 explains how to estimate this percentage for non-normal distributions). Each percentage is then multiplied by 10 to the sixth power. The parts per million below the LSL are denoted as ppm_{LSL}, those above the USL as ppm_{USL}. The operator "$\Phi(\quad)$" means find the corresponding percentage for the Z value inside the parentheses from Appendix Table III (Φ is pronounced "fie"). For example, $\Phi(3)$ equals .00135 and $\Phi(4)$ equals .000032.

$$ppm_{LSL} = \Phi(Z_{LSL}) \times 10^6 \qquad\qquad ppm_{USL} = \Phi(Z_{USL}) \times 10^6$$

The total *ppm* is found by adding the *ppm* below the LSL to the *ppm* above the USL.

$$ppm_{TOTAL} = ppm_{LSL} + ppm_{USL} = \Phi(Z_{LSL}) \times 10^6 + \Phi(Z_{USL}) \times 10^6 = [\Phi(Z_{LSL}) + \Phi(Z_{USL})] \times 10^6$$

Table 6.12 C_{PK} values versus *ppm*.

C_{PK}	$Z_{MIN,ST}$	Number of Nonconforming Parts per Million (*ppm*)	
		One Limit	Both Limits*
.33	1.00	158,700	317,400
.43	1.28	100,000	200,000
.50	1.50	66,800	133,600
.52	1.56	59,380	118,760
.55	1.64	50,000	100,000
.67	2.00	22,800	45,600
.78	2.33	10,000	20,000
.83	2.50	6,200	12,400
.86	2.58	5,000	10,000
1.00	3.00	1,350	2,700
1.03	3.09	1,000	2,000
1.10	3.29	500	1,000
1.13	3.40	337	674
1.20	3.60	159	318
1.24	3.72	100	200
1.27	3.80	72	145
1.30	3.89	50	100
1.33	4.00	32	63
1.40	4.20	13	27
1.42	4.26	10	20
1.47	4.42	5	10
1.50	4.50	3	7
1.58	4.75	1	2
1.63	4.89	.5	1
1.67	5.00	.3	.6
1.73	5.20	.1	.2
1.78	5.33	.05	.1
1.87	5.61	.01	.02
1.91	5.73	.005	.01
2.00	6.00	.001	.002
2.04	6.11	.0005	.001

* Assumes process is centered at middle of tolerance and has a normal distribution.

Short-term or long-term *ppm* values are designated by the additional subscripts "*ST*" and "*LT*." Because of the association between C_{PL} and C_{PU} with their respective Z values, short-term *ppm* values are computed with these two formulas:

$$ppm_{LSL,ST} = \Phi(Z_{LSL,ST}) \times 10^6 = \Phi(3\,C_{PL}) \times 10^6$$

$$ppm_{USL,ST} = \Phi(Z_{USL,ST}) \times 10^6 = \Phi(3\,C_{PU}) \times 10^6$$

The total *ppm* is then found as follows for characteristics with bilateral specifications.

$$\begin{aligned}
ppm_{TOTAL,ST} &= ppm_{LSL,ST} + ppm_{USL,ST} \\
&= \Phi(3\,C_{PL}) \times 10^6 + \Phi(3\,C_{PU}) \times 10^6 \\
&= [\Phi(3\,C_{PL}) + \Phi(3\,C_{PU})] \times 10^6 \\
&= [\Phi(Z_{LSL,ST}) + \Phi(Z_{USL,ST})] \times 10^6
\end{aligned}$$

As illustrated in Table 6.12, a one-to-one relationship does not exist between ppm_{TOTAL} and Z_{MIN} for characteristics with bilateral specifications. However, ppm_{TOTAL} is always bracketed by the following two values:

$$\Phi(Z_{MIN}) \times 10^6 \le ppm_{TOTAL} \le 2\,\Phi(Z_{MIN}) \times 10^6$$

For characteristics with unilateral specifications, C_{PK} equals C_{PL} if only a LSL is involved and equals C_{PU} if just an USL is given. In this case, ppm_{TOTAL} is directly related to Z_{MIN}.

$$ppm_{TOTAL,ST} = \Phi(3\,C_{PK}) \times 10^6 = \Phi(Z_{MIN,ST}) \times 10^6$$

Likewise, the long-term *ppm* values for characteristics with bilateral specifications are derived from these formulas:

$$ppm_{LSL,LT} = \Phi(3\,P_{PL}) \times 10^6 \qquad\qquad ppm_{USL,LT} = \Phi(3\,P_{PU}) \times 10^6$$

$ppm_{TOTAL,LT}$ is found by summing these two values.

$$\begin{aligned}
ppm_{TOTAL,LT} &= ppm_{LSL,LT} + ppm_{USL,LT} \\
&= \Phi(3\,P_{PL}) \times 10^6 + \Phi(3\,P_{PU}) \times 10^6 \\
&= [\Phi(3\,P_{PL}) + \Phi(3\,P_{PU})] \times 10^6 \\
&= [\Phi(Z_{LSL,LT}) + \Phi(Z_{USL,LT})] \times 10^6
\end{aligned}$$

Another useful measure is $ppm_{MAX,LT}$, which is the larger of $ppm_{LSL,LT}$ and $ppm_{USL,LT}$:

$$ppm_{MAX,LT} = \text{Maximum}(ppm_{LSL,LT}, ppm_{USL,LT})$$

For characteristics with unilateral specifications, switch to this formula:

$$\begin{aligned}
ppm_{TOTAL,LT} &= \Phi(3\,P_{PK}) \times 10^6 \\
&= \Phi(Z_{MIN,LT}) \times 10^6
\end{aligned}$$

The minimum performance capability requirement for characteristics with bilateral specifications is to have both ppm_{LSL} and ppm_{USL} less than 1,350, which means ppm_{MAX} (the larger of the two) must be less than 1,350. This guarantees ppm_{TOTAL} will be less than 2,700, thus meeting the performance capability requirement of at least 99.73 percent conforming parts. In view of this stipulation, the rating scale for bilateral specifications becomes:

$ppm_{MAX} < 1,350$ indicates that the process exceeds the minimum performance capability requirement

$ppm_{MAX} = 1,350$ implies that the process just meets the minimum performance capability requirement

$ppm_{MAX} > 1,350$ means that the process lacks performance capability

For those characteristics having a unilateral specification, ppm_{TOTAL} must be less than 1,350 to meet the minimum performance capability requirement of at least 99.865 percent conforming parts. Thus, the rating scale is:

$ppm_{TOTAL} < 1,350$ reveals that the process exceeds the minimum performance capability requirement

$ppm_{TOTAL} = 1,350$ means that the process just meets the minimum performance capability requirement

$ppm_{TOTAL} > 1,350$ discloses that the process lacks performance capability

Because many companies desire to have an additional safety margin for capability, higher goals for performance capability are typically mandated in practice.

Goals

A common goal for characteristics with bilateral specifications is to have ppm_{MAX} less than 32. This is similar to a C_{PK} goal of 1.33, or a Z_{MIN} goal of 4.0. Occasionally, some companies establish a goal of having ppm_{MAX} less than 3.4, which is equivalent to a C_{PK} goal of 1.50, which is identical to a Z_{MIN} goal of 4.5. A few companies establish the ppm_{MAX} goal as being less than .3, which is equivalent to a C_{PK} goal of 1.67 and a Z_{MIN} goal of 5.0. Litsikas reports Universal Instruments Corporation has established goals for its suppliers varying from 100 to 500 *ppm*, whereas Keenan (1996*a*) predicts single-digit *ppm* goals for the automotive industry in the year 2000, with Magna International Corporation's Decoma Exterior Group already reporting 0 *ppm* for some critical characteristics (Keenan, 1996*b*). Vaucher *et al* state a goal of 100 *ppm* for the manufacture and assembly of printed wired circuit boards. Donnelly Corporation (automotive mirrors and windows) is striving for a goal of 10 ppm.

For unilateral specifications, the goals are just half of those for bilateral specifications. Thus, a ppm_{TOTAL} goal of 32 is now comparable to a C_{PK} goal of 1.33 or a Z_{MIN} goal of 4.0. Note that the C_{PK} and Z_{MIN} goals usually do not distinguish between bilateral and unilateral specifications (Montgomery is the exception, see Table 6.6), whereas the goal for the *ppm* index does.

Some managers faced with a process having a C_{PK} of 1.00 are interested in improving its performance capability. They set as a goal for this process to have a C_{PK} of 1.10 by the end of the next quarter and publicize this desire. They feel this is a reasonable goal as it represents only a 10 percent increase in capability. However, for a characteristic with a unilateral specification, this goal represents a *63 percent* reduction in ppm_{TOTAL}, from 1,350 to 500. This will present a formidable challenge to those assigned this task, especially if they are given no additional resources.

Example 6.27 Floppy Disks

Recall the field problem that is traced to variation in the diameter of a floppy disk's inner hole. The following estimates have been calculated during the course of a capability study.

$$\hat{Z}_{LSL,ST} = 1.56 \qquad \hat{Z}_{USL,ST} = 4.01 \qquad \hat{P}_{PL} = .51 \qquad \hat{P}_{PU} = 1.31$$

The short-term *ppm* estimates are based on $\hat{Z}_{LSL,ST}$ and $\hat{Z}_{USL,ST}$.

$$\hat{p}pm_{LSL,ST} = \Phi(\hat{Z}_{LSL,ST}) \times 10^6 = \Phi(1.56) \times 10^6 = (5.938 \times 10^{-2}) \times 10^6 = 59,380$$

$$\hat{p}pm_{USL,ST} = \Phi(\hat{Z}_{USL,ST}) \times 10^6 = \Phi(4.01) \times 10^6 = (3.037 \times 10^{-5}) \times 10^6 = 30$$

Summing these two produces an estimate for the total *ppm*.

$$\hat{p}pm_{TOTAL,ST} = \hat{p}pm_{LSL,ST} + \hat{p}pm_{USL,ST} = 59,380 + 30 = 59,410$$

$$\hat{p}pm_{MAX,ST} = \text{Maximum}(\hat{p}pm_{LSL,ST}, \hat{p}pm_{USL,ST}) = \text{Maximum}(59,380, 30) = 59,380$$

As $ppm_{MAX,ST}$ is considerably more than 1,350, this process definitely lacks short-term performance capability. Because $ppm_{LSL,ST}$ is so much greater than $ppm_{USL,ST}$, moving the process average higher would improve performance capability. However, unlike $Z_{LSL,ST}$ and $Z_{USL,ST}$, these estimates for $ppm_{LSL,ST}$ and $ppm_{USL,ST}$ cannot be averaged to determine what the minimum $ppm_{TOTAL,ST}$ will be when μ is centered at M. Rather, the estimates of $Z_{LSL,ST}$ and $Z_{USL,ST}$ must first be averaged to find the largest possible $Z_{MIN,ST}$ value that may be obtained by changing only the process average.

$$\text{Maximum } \hat{Z}_{MIN,ST} = \frac{\hat{Z}_{LSL,ST} + \hat{Z}_{USL,ST}}{2} = \frac{1.56 + 4.01}{2} = 2.78$$

When μ is centered at M, $Z_{LSL,ST}$ will equal $Z_{USL,ST}$, which will in turn equal this maximum $Z_{MIN,ST}$ value.

$$\hat{Z}_{LSL,ST} = \hat{Z}_{USL,ST} = \text{Maximum } \hat{Z}_{MIN,ST} = 2.78$$

The percentage nonconforming for these two Z values can be converted into *ppm*, then summed to derive the minimum possible $ppm_{TOTAL,ST}$.

$$\hat{p}pm_{LSL,ST} = \Phi(\hat{Z}_{LSL,ST}) \times 10^6 = \Phi(2.78) \times 10^6 = (2.718 \times 10^{-3}) \times 10^6 = 2,718$$

$$\hat{p}pm_{USL,ST} = \Phi(\hat{Z}_{USL,ST}) \times 10^6 = \Phi(2.78) \times 10^6 = (2.718 \times 10^{-3}) \times 10^6 = 2,718$$

$$\text{Minimum } \hat{p}pm_{TOTAL,ST} = \hat{p}pm_{LSL,ST} + \hat{p}pm_{USL,ST} = 2,718 + 2,718 = 5,436$$

Even if the average hole size could be centered at M, the process would still lack short-term performance capability since $ppm_{MAX,ST}$ would still be greater than 1,350.

The long-term *ppm* measures may be estimated from $\hat{P}_{PL}$ and $\hat{P}_{PU}$.

$$\hat{p}pm_{LSL,LT} = \Phi(3\hat{P}_{PL}) \times 10^6 = \Phi(3 \times .51) \times 10^6 = \Phi(1.53) \times 10^6 = 60,010$$

$$\hat{p}pm_{USL,LT} = \Phi(3\hat{P}_{PU}) \times 10^6 = \Phi(3 \times 1.31) \times 10^6 = \Phi(3.93) \times 10^6 = 43$$

$ppm_{TOTAL,LT}$ is predicted by summing these estimates for $ppm_{LSL,LT}$ and $ppm_{USL,LT}$.

$$\hat{p}pm_{TOTAL,LT} = \hat{p}pm_{LSL,LT} + \hat{p}pm_{USL,LT} = 63,010 + 43 = 63,053$$

Although the $ppm_{USL,LT}$ estimate indicates little problem with holes being oversize, the $ppm_{LSL,LT}$ estimate of over 63,000 discloses a serious capability problem associated with meeting the LSL of 39.5.

Example 6.28 Bolt Torque

Cylinder head bolts for medium-duty diesel engines are tightened with an air-powered nut runner. Being a critical fastener, an $\overline{X}$, R control chart is kept on breakaway torque. The top-center bolt is checked on five consecutive heads every half hour, with the subgroup average and range plotted on the chart. When stability is demonstrated, σ_{ST} is estimated as 11.83 newton-meters and μ as 191.59 newton-meters. The estimated process output distribution is exhibited in Figure 6.72. The LSL is 155 newton-meters, while the goal for this process is to have no more than 1,000 *ppm* outside this specification.

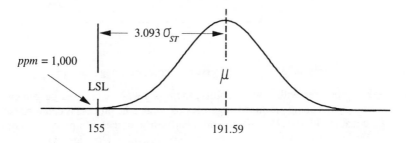

Fig. 6.72 Distribution of breakaway torque (newton-meters).

Assuming a normal distribution for breakaway torque, $ppm_{LSL,ST}$ is estimated by first computing $Z_{LSL,ST}$.

$$\hat{Z}_{LSL,ST} = \frac{\hat{\mu} - LSL}{\hat{\sigma}_{ST}} = \frac{191.59 - 155}{11.83} = 3.093$$

The percentage below this Z value is found in Appendix Table III, then used to estimate $ppm_{LSL,ST}$.

$$\hat{p}pm_{LSL,ST} = \Phi(\hat{Z}_{LSL,ST}) \times 10^6 = \Phi(3.093) \times 10^6 = (1.000 \times 10^{-3}) \times 10^6 = 1,000$$

This process just demonstrates attainment of its capability goal of 1,000 *ppm*.

Relationship Between *ppm*_TOTAL and the Process Average

Just as are all other measures of performance capability, ppm_{TOTAL} is also a function of the process average. Figure 6.73 reveals how this measure changes as μ moves between the specification limits. Six processes with different levels of capability are displayed, with the label on the curves reflecting the C_P value for that process. ppm_{TOTAL} is always at its lowest when the process is centered at the middle of the tolerance. As the average shifts toward a specification limit, ppm_{TOTAL} begins to increase, and reaches 500,000 *ppm* (50 percent) when μ equals the specification limit.

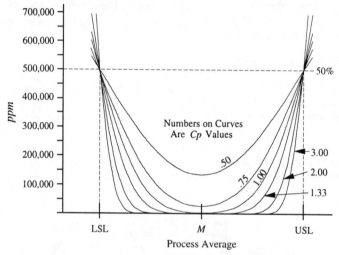

Fig. 6.73 ppm_{TOTAL} versus the process average for various C_P values.

For processes with low capability, there is always a fairly large percentage of noncon-forming parts being produced, regardless of where the process is centered (see the curves for C_P equal to .50 and .75). As potential capability increases, a process becomes more robust to changes in its average. For the curve representing a process with a C_P index of 2.00, μ may move away from *M* by more than half the distance from *M* to the USL, yet this process will still be producing only an insignificant number of nonconforming parts.

When the process average moves above the USL (or below the LSL), the percentage of nonconforming parts increases at a faster rate for those processes having a higher potential capability. Although not intuitive at first, this phenomenon occurs because high capability means most of the output is grouped around the process average. Thus, moving the average from the USL, to just slightly above the USL, pulls a greater percentage of the process output above the USL than a process having a very wide spread.

If a target average different than *M* is specified, there is no change in these curves because LSL and USL, *not* LSL_T and USL_T, determine ppm_{TOTAL}.

Advantages and Disadvantages

A majority of authors agree that ease of understanding is perhaps the principal advantage for the *ppm* measure of process capability (Dovich; Electronic Industries Association (1990); Gitlow *et al*; M. Johnson; Kaminsky *et al*; Pearn *et al*; Pyzdek). There are no confusing or arbitrary scales as the number of nonconforming parts is the measure. The current level of capability and any improvements in capability are readily comprehended by everyone. Thus, this measure is consistent with Albert Einstein's philosophy to "Make things as simple as possible, and no simpler."

Unlike C_{PK} or P_{PK}, ppm_{TOTAL} expresses the total percentage of nonconforming parts, not just those outside one of the specifications. Thus, ppm_{TOTAL} is the only capability measure where the risk (probability) of producing a nonconforming part is readily apparent.

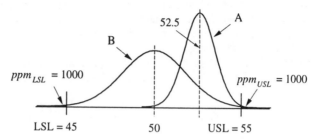

Fig. 6.74   Supplier A versus supplier B.

Recall that supplier A and B (see Figure 6.74) were rated as having the same performance capability by the C_{PK} method. For supplier A:

$$\hat{C}_{PK,A} = \frac{USL - \overline{\overline{X}}_A}{3\hat{\sigma}_{ST,A}} = \frac{55 - 52.5}{3(.809)} = 1.03$$

For supplier B:

$$\hat{C}_{PK,B} = \frac{USL - \overline{\overline{X}}_B}{3\hat{\sigma}_{ST,B}} = \frac{55 - 50}{3(1.618)} = 1.03$$

Notice how the *ppm* method distinguishes between the two, clearly identifying supplier A as having better performance capability, because it has only half the nonconforming parts that B does (1,000 *ppm* versus 2,000 *ppm*).

$$ppm_{TOTAL,ST,A} = ppm_{LSL,ST,A} + ppm_{USL,ST,A} = 0 + 1,000 = 1,000$$

$$ppm_{TOTAL,ST,B} = ppm_{LSL,ST,B} + ppm_{USL,ST,B} = 1,000 + 1,000 = 2,000$$

In addition, Section 8.14 details how this measure can assess capability for processes having non-normally distributed outputs, while Section 9.2 explains how the *ppm* method also applies to processes generating attribute-type data. Thus, this one measure works with every variety of industrial data.

A ppm_{TOTAL} representing potential capability can be computed by assuming the process average is centered at M, then calculating the resulting Z scores and their associated *ppm* values. The sum of these is the lowest possible ppm_{TOTAL} for this process with the current standard deviation. Comparing this sum to the actual ppm_{TOTAL} reveals how much improvement is possible if the process average can be centered at the middle of the tolerance.

Disadvantages include the extra work involved to calculate this measure by having to look up the Z value, as well as confusion over where the decimal point goes when converting from percent to *ppm*. As a note of caution: *ppm* values less than 1,000 are usually not very reliable estimates. Accurately determining the exact shape of a process output distribution this far from its average is extremely difficult unless a substantial number of measurements are included in the capability study (Pearn *et al*). Because the distribution's shape plays such an important role in estimating *ppm* values, some authors strongly advise against converting capability indexes to *ppm* (Wheeler and Chambers, p. 129). In fact, Porter and Oakland have discovered that even the choice of subgroup size can affect the estimated *ppm* level.

ppm_{TOTAL} is minimized when the process average is centered at the middle of the tolerance. This causes a problem if the desired target for the process average is located somewhere else ($T \neq M$). ppm_{TOTAL} should *not* be applied in these situations as it discourages operators from running at the targeted average.

Even when T equals M, problems with interpretation can arise. Suppose the average for distribution A in Figure 6.74 is raised from 52.50 to 52.67. This makes $Z_{USL,ST}$ equal to 2.88.

$$\hat{Z}_{USL,ST} = \frac{USL - \overline{\overline{X}}_A}{\hat{\sigma}_{ST}} = \frac{55 - 52.67}{.809} = 2.88$$

The *ppm* for this Z value is 2,000. Because $ppm_{LSL,ST}$ is essentially 0, $ppm_{TOTAL,ST}$ for A is now 2,000, the same as for distribution B, as is revealed in Figure 6.75.

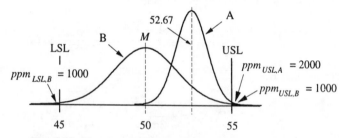

Fig. 6.75 The average for distribution A is moved to 52.67.

Distributions A and B have identical $ppm_{TOTAL,ST}$ values, yet B is producing significantly more parts near the target average than is A. If parts produced at M perform better than those produced elsewhere, customers will be much happier with parts made by B. As one major reason for assessing process capability is to gauge customer satisfaction, a good capability measure should be able to distinguish between these two distributions. The next set of capability measures accomplish this objective by considering the size of the gap between the process and target averages.

6.8 C_{PM} and P_{PM} Indexes

A relatively recent concept for measuring process capability is called the C_{PM} index (Chan *et al*; Taguchi). Because it considers variation between the process average and the target average, as well as the process standard deviation, Boyles notes that this measure is preferred by many Taguchi followers. The formula for C_{PM} is given below for cases where the target average equals M, the middle of the tolerance. Section 6.9 covers situations where this assumption is not true. The Greek letter τ is pronounced "tau" (see also Wright).

$$C_{PM} = \frac{\text{Tolerance}}{6\tau_{ST}} = \frac{USL - LSL}{6\tau_{ST}}$$

τ_{ST} is a function of both the process standard deviation and the difference between the process average and M. Thus, it measures both the precision and accuracy of a process.

$$\tau_{ST} = \sqrt{\sigma_{ST}^2 + (\mu - M)^2}$$

To better understand this formula, notice that when μ and M are equal, as is depicted in Figure 6.76, the difference between them is 0 (μ - M = 40 - 40 = 0).

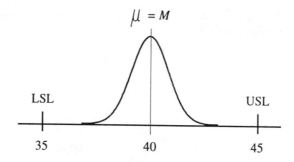

Fig. 6.76 Process average centered at M.

For this special situation where the process average is centered right at the middle of the tolerance, τ_{ST} is equal to σ_{ST}.

$$\tau_{ST} = \sqrt{\sigma_{ST}^2 + (\mu - M)^2} = \sqrt{\sigma_{ST}^2 + 0^2} = \sigma_{ST}$$

As τ_{ST} equals σ_{ST}, C_{PM} is now identical to the C_P measure.

$$C_{PM} = \frac{\text{Tolerance}}{6\tau_{ST}} = \frac{\text{Tolerance}}{6\sigma_{ST}} = C_P$$

Assume μ is not equal to M, as is the case in Figure 6.77.

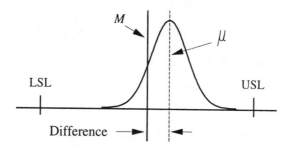

Fig. 6.77 Process average is centered higher than M.

Now $(\mu - M)^2$ is greater than zero, which causes τ_{ST} to be greater than σ_{ST}.

$$(\mu - M)^2 > 0$$

$$\sigma_{ST}^2 + (\mu - M)^2 > \sigma_{ST}^2 + 0$$

$$\sqrt{\sigma_{ST}^2 + (\mu - M)^2} > \sqrt{\sigma_{ST}^2}$$

$$\tau_{ST} > \sigma_{ST}$$

This means C_{PM} is *less* than C_P for any process whose average is not equal to M, making C_P the highest C_{PM} can ever be.

$$\tau_{ST} > \sigma_{ST}$$

$$\frac{1}{\tau_{ST}} < \frac{1}{\sigma_{ST}}$$

$$\frac{\text{Tolerance}}{6\tau_{ST}} < \frac{\text{Tolerance}}{6\sigma_{ST}}$$

$$C_{PM} < C_P$$

Because the process parameters μ and σ_{ST} are seldom known when conducting a capability study, they are estimated from measurement data collected on the control chart monitoring this process. These estimates in turn help predict C_{PM}, as demonstrated in these next equations.

$$\hat{C}_{PM} = \frac{\text{Tolerance}}{6\hat{\tau}_{ST}} \qquad \text{where } \hat{\tau}_{ST} = \sqrt{\hat{\sigma}_{ST}^2 + (\hat{\mu} - M)^2}$$

The P_{PM} measure is similar to the C_{PM} index, except that the long-term standard deviation replaces σ_{ST}.

$$P_{PM} = \frac{\text{Tolerance}}{6\tau_{LT}} = \frac{\text{USL - LSL}}{6\tau_{LT}}$$

$$\text{where } \tau_{LT} = \sqrt{\sigma_{LT}^2 + (\mu - M)^2}$$

P_{PM} is estimated with these formulas:

$$\hat{P}_{PM} = \frac{\text{Tolerance}}{6\hat{\tau}_{LT}} \qquad \text{where } \hat{\tau}_{LT} = \sqrt{\hat{\sigma}_{LT}^2 + (\hat{\mu} - M)^2}$$

Boyles recommends estimating τ_{LT} via this formula.

$$\hat{\tau}_{LT} = \sqrt{\frac{\sum_{i=1}^{kn} (X_i - M)^2}{kn - 1}}$$

This last formula discloses that τ_{LT} is based on the deviation of the individual measurements from the desired process average, in this case, the middle of the tolerance. The farther measurements are from M, the greater is τ_{LT}. Compare this to the long-term process standard deviation used in calculating P_{PK}, which is based on the deviation of individual measurements from the *actual* process average.

$$\hat{\sigma}_{LT} = \frac{1}{c_4} \sqrt{\frac{\sum_{i=1}^{kn} (X_i - \hat{\mu})^2}{kn - 1}}$$

When the process average is centered at M, τ_{LT} equals σ_{LT}, resulting in P_{PM} equaling P_p.

$$\text{When } \mu = M, \qquad \tau_{LT} = \sigma_{LT}$$

$$\frac{\text{Tolerance}}{\tau_{LT}} = \frac{\text{Tolerance}}{\sigma_{LT}}$$

$$P_{PM} = P_p$$

If the process average is located anywhere else, τ_{LT} is greater than σ_{LT}, causing P_{PM} to be less than P_p.

$$\text{When } \mu \neq M, \quad P_{PM} < P_p$$

Because C_{PM} and P_{PM} incorporate both process spread and centering in their measure of capability, they are called "second-generation" capability measures. Recall that C_{PK} and P_{PK} are also second-generation measures, whereas C_p and P_p are considered "first-generation" measures. Read through the next example to see how these measures of capability are estimated from actual manufacturing data.

Example 6.29 Valve O.D. Size

An estimate of potential capability ($\hat{P}_p = 1.24$) was made for the O.D. size of an exhaust valve run on a grinding machine (Figure 6.78). M is 10.0000 cm, σ_{LT} is .01610 cm, and $\hat{\mu}$ is 10.0041 cm. From this information, τ_{LT} may be estimated.

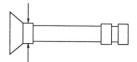

Fig. 6.78

$$\hat{\tau}_{LT} = \sqrt{\hat{\sigma}_{LT}^2 + (\hat{\mu} - M)^2} = \sqrt{(.01610 \text{ cm})^2 + (10.0041 \text{ cm} - 10.0000 \text{ cm})^2} = .01661 \text{ cm}$$

$\hat{\tau}_{LT}$ is somewhat greater than the $\hat{\sigma}_{LT}$ of .01610 cm because the process is not centered at 10.0000 cm. With this estimate of τ_{LT}, and the print specifications for O.D. size (LSL is 9.9400 cm, USL is 10.0600 cm), P_{PM} is predicted to be 1.20. Notice how the units of measurement cancel out, leaving a dimensionless index.

$$\hat{P}_{PM} = \frac{\text{USL} - \text{LSL}}{6\hat{\tau}_{LT}} = \frac{10.0600 \text{ cm} - 9.9400 \text{ cm}}{6(.01661 \text{ cm})} = 1.20$$

The behavior of P_{PM} is similar to P_{PK} in that the actual performance capability is less than the potential, 1.20 versus 1.24, due to the process average not being centered at M. However, the rate of decrease in P_{PM} as the process average deviates from the target is substantially different than that for P_{PK}. To understand why, the concept of an economic loss function must first be introduced.

Economic Loss Functions

The motivation behind analyzing the measures of process capability covered so far has been to help produce parts within print specifications. However, none of these past measures care what the actual measurement of a conforming part is, only that it is somewhere within tolerance. This line of thinking, which classifies parts as either conforming or nonconforming, is sometimes referred to as the "goalpost" philosophy of quality.

The Goalpost Loss Function

Under this philosophy, any part within tolerance is considered conforming. Parts with measurements of A, B, or C in Figure 6.79 are all considered equally "good" parts, while parts with measurements D and E are regarded as equally "bad," *i.e.*, out of specification. The

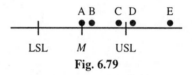

Fig. 6.79

loss in terms of scrap, rework, and customer satisfaction for producing parts D and E is identical, while no loss is associated with A, B, or C because they are within specification.

An economic loss function plots the loss associated with parts produced at various dimensions. The x axis represents the part measurement and the y axis reflects the loss. The shape of the loss function associated with producing parts under this first philosophy is high for parts below the LSL, zero for parts produced within specification, and high again for parts above the USL. This pattern creates a step function resembling a goalpost, as is seen below.

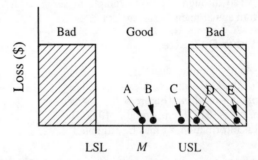

Fig. 6.80 The economic loss function for the goalpost philosophy.

This pattern is responsible for the label "goalpost" due to the similarity of this loss function's shape to the goalpost found in football (Figure 6.81). If the ball is kicked between the "uprights" (the specification limits), the scoring attempt is considered "good," with no distinction based on where it went through. If the ball is kicked outside of the uprights, the attempt is classified "no good," no matter how close it passes to the uprights. There are no partial points awarded. If the ball hits one of the uprights and bounces through, it's still considered completely good, somewhat analogous to a material review board's decision to accept parts with marginal quality.

Fig. 6.81 Football goalpost analogy.

However, assuming the print specification accurately reflects a customer requirement, a part produced at the middle of the tolerance, like A, performs somewhat better for the customer than a part slightly larger, like B or C. Typically, design engineers specify a target value which provides optimum performance for the product, then establish some tolerance around this target. Parts B and C may both provide *acceptable* performance to the customer, but not as good as part A. On the other hand, the product performance of C and D will be quite similar, but C is considered completely "good" while D is completely "bad." In fact, the difference in performance between A and C will likely be much greater than the difference between C and D.

This philosophy also forms the basis for the "zero defects" concept. If a process output looks like the one displayed in Figure 6.82, there is a considerable number of nonconforming parts being produced and a sizable loss being incurred. There is a definite monetary incentive for reducing inherent process variation to improve quality and thereby decrease scrap, rework, and warranty costs.

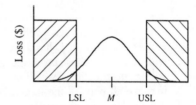

Fig. 6.82

Once process variation is reduced to that of curve 2 in Figure 6.83, there are almost no nonconforming parts being made and no loss incurred. A state of "zero defects" is achieved.

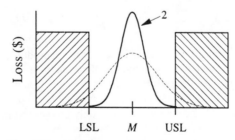

Fig. 6.83 A process producing "zero defects."

At this point, any incentive to further improve the process vanishes because there is nothing more to gain as no parts are produced in the loss areas. Because additional improvement would only cost money, the process is left as is, since it has reached its optimum "economical" level of quality.

The Taguchi Loss Function

Even though the goalpost principle is widespread, there are other loss function models. The most widely known alternative is the "Taguchi" loss function, named for Genichi Taguchi who promoted its use (Taguchi). Instead of the step function loss of the goalpost model, the Taguchi loss function assumes parts produced at the target average of M are the best, and therefore, have zero loss associated with them. As parts deviate farther and farther from the target average, their ability to satisfy customers decreases, and the average part loss increases. Although the exact shape of this loss function for any given characteristic is generally unknown, it usually is assumed to take the shape of a continuous parabola (Taguchi and Clausing), as is exhibited in Figure 6.84.

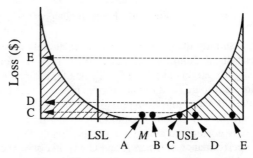

Fig. 6.84 The Taguchi loss function.

Most parts *within* the print specifications incur some loss under this model. In fact, the print specifications have little significance here. With this parabolic loss function, E has a greater loss than D. D has a greater loss than C, and C a greater loss than B. The only parts with no loss are those produced at the target, like A. This model distinguishes between parts within specification, as well between parts in or out of specification. The economic goal is to make all parts at the target, the design intent, where there is no loss because customer satisfaction is maximized.

Tunner has shown the average loss per part ($\overline{L}$) for the output of a process with a parabolic loss function is related to the amount of off-center product. The derivation of his formula assumes zero loss for a part produced at M.

$$\overline{L}_{ST} = \frac{c}{d^2}[\sigma_{ST}^2 + (\mu - M)^2]$$

In this equation, c is the loss associated with a unit produced at a specification limit (Phadke), while d is the distance from M to the specification limit. As the process spread (σ) increases, so does the loss. As the process average deviates farther from M, the loss also increases. Minimum loss for a given σ_{ST} occurs when all parts are produced exactly on target ($\mu = M$). Once centered at M, the average loss can only be reduced further by decreasing σ_{ST}.

$$\overline{L}_{ST} = \frac{c}{d^2}[\sigma_{ST}^2 + 0^2] = \frac{c\,\sigma_{ST}^2}{d^2}$$

Caldwell has developed a detailed worksheet for estimating $\overline{L}$, illustrated with an example taken from the cable industry.

The Taguchi loss function is the driving force behind continuous process improvement (Jessup). Recall how the goalpost model encourages quality improvement only as long as parts are produced outside of the print specification. With the Taguchi loss function, it makes economic sense to continue reducing process variation until *all* parts are produced at the target value. Thus, curve 3 in Figure 6.85 has less total loss than curve 2, which has less loss than that associated with curve 1. Ideally, all parts would be produced exactly at *M*, with absolutely no variation, and no loss.

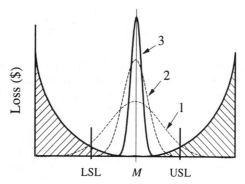

Fig. 6.85 The Taguchi loss function promotes continuous improvement.

Companies adopting the goalpost philosophy spend considerable time arguing about specifications. Those companies embracing the Taguchi loss function model have little time to worry about specifications because they are too busy reducing process variation around the target average (Sullivan). C_{PM} is the first index developed to incorporate this loss function concept in its measure of process capability (T. Johnson). C_{PM} and $\overline{L}_{ST}$ are linked by τ_{ST}, which measures variation from the target in terms of its two components: process spread, σ_{ST}; and process centering, the difference between μ and M.

$$\tau_{ST} = \sqrt{\sigma_{ST}^2 + (\mu - M)^2} \;\; \Rightarrow \;\; \tau_{ST}^2 = \sigma_{ST}^2 + (\mu - M)^2$$

The average loss is directly proportional to τ_{ST}^2. If τ_{ST} increases, the average loss per part also increases, as is shown here (Tsui).

$$\overline{L}_{ST} = \frac{c}{d^2}[\,\sigma_{ST}^2 + (\mu - M)^2\,] = \frac{c}{d^2}\tau_{ST}^2 = c\left(\frac{\tau_{ST}}{d}\right)^2$$

This last equation can be rearranged to make τ_{ST} a function of the average loss.

$$\frac{\overline{L}_{ST}}{c} = \left(\frac{\tau_{ST}}{d}\right)^2$$

$$\sqrt{\frac{\overline{L}_{ST}}{c}} = \frac{\tau_{ST}}{d}$$

$$d\sqrt{\frac{\overline{L}_{ST}}{c}} = \tau_{ST}$$

This relationship for τ_{ST} can be substituted into the formula for C_{PM}.

$$C_{PM} = \frac{\text{Tolerance}}{6\tau_{ST}} = \frac{\text{Tolerance}}{6d\sqrt{\overline{L}_{ST}/c}} = \frac{\text{Tolerance}}{6d}\sqrt{\frac{c}{\overline{L}_{ST}}}$$

As d is the distance from M to a specification limit, it equals one half of the tolerance.

$$C_{PM} = \frac{\text{Tolerance}}{6(\text{Tolerance}/2)}\sqrt{\frac{c}{\overline{L}_{ST}}} = \frac{1}{3}\sqrt{\frac{c}{\overline{L}_{ST}}}$$

Thus, C_{PM} is inversely related to $\overline{L}_{ST}$. When the process spread decreases and/or the process average moves closer to its target, average loss per part decreases, which causes process capability as measured by C_{PM} to increase. Conversely, when the loss increases, C_{PM} decreases. The average loss may also be expressed as a function of C_{PM}.

$$C_{PM} = \frac{1}{3}\sqrt{\frac{c}{\overline{L}_{ST}}}$$

$$C_{PM}^2 = \frac{1}{9}\frac{c}{\overline{L}_{ST}}$$

$$\overline{L}_{ST} = \frac{c}{9C_{PM}^2}$$

Comparing C_{PM} to C_{PK}

Although also including an estimate of μ in its calculation, C_{PK} quantifies performance capability in a slightly different manner than C_{PM}, as this next example demonstrates. Three similar machines (A, B, and C) are producing the identical characteristic of the same part number. The target process average for this characteristic is the middle of the tolerance ($M = 0$), with the LSL equal to -8, and the USL equal to +8. In Figure 6.86, the output from machine A is compared to that for machine B. Visually, it appears machine B is doing a better job of producing parts to specification. When estimates of capability are calculated, C_{PM} and C_{PK} both agree the capability of machine B (2.67) is better than that for A (1.33).

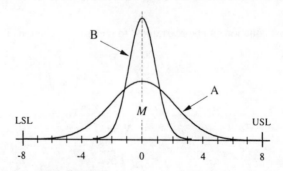

Fig. 6.86 The output for machine A compared to machine B.

$$\begin{array}{ll} \underline{\text{For Machine B}} & \qquad\qquad \underline{\text{For Machine A}} \end{array}$$

$$\hat{\mu}_B = 0 \qquad \hat{\sigma}_{ST,B} = 1 \qquad\qquad \hat{\mu}_A = 0 \qquad \hat{\sigma}_{ST,A} = 2$$

$$\hat{\tau}_{ST,B} = \sqrt{1^2 + (0-0)^2} = 1 \qquad\qquad \hat{\tau}_{ST,A} = \sqrt{2^2 + (0-0)^2} = 2$$

$$\hat{C}_{PK,B} = \frac{8-0}{3\,(1)} = 2.67 \qquad\qquad \hat{C}_{PK,A} = \frac{8-0}{3\,(2)} = 1.33$$

$$\hat{C}_{PM,B} = \frac{8--8}{6\,(1)} = 2.67 \qquad\qquad \hat{C}_{PM,A} = \frac{8--8}{6\,(2)} = 1.33$$

In Figure 6.87, the process output of machine B is contrasted to that of machine C. From this comparison, B's output looks much better than C's, so it's not surprising that both C_{PM} and C_{PK} estimate the capability of machine B to be better than that for C. However, the C_{PM} measure for machine C is much lower than its C_{PK} index (.84 versus 1.67) due to the large distance between its average and M. This difference makes $\tau_{ST,C}$ much greater than $\sigma_{ST,C}$.

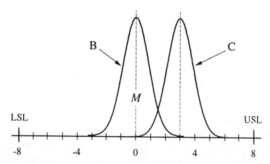

Fig. 6.87 The output of machine B compared to machine C.

$$\begin{array}{ll} \underline{\text{For Machine B}} & \qquad\qquad \underline{\text{For Machine C}} \end{array}$$

$$\hat{\mu}_B = 0 \qquad \hat{\sigma}_{ST,B} = 1 \qquad\qquad \hat{\mu}_C = 3 \qquad \hat{\sigma}_{ST,C} = 1$$

$$\hat{\tau}_{ST,B} = \sqrt{1^2 + (0-0)^2} = 1 \qquad\qquad \hat{\tau}_{ST,C} = \sqrt{1^2 + (3-0)^2} = 3.162$$

$$\hat{C}_{PK,B} = \frac{8-0}{3\,(1)} = 2.67 \qquad\qquad \hat{C}_{PK,C} = \frac{8-3}{3\,(1)} = 1.67$$

$$\hat{C}_{PM,B} = \frac{8--8}{6\,(1)} = 2.67 \qquad\qquad \hat{C}_{PM,C} = \frac{8--8}{6\,(3.162)} = .84$$

Finally, a comparison between the outputs for machine A and C is displayed in Figure 6.88. Visually, most people prefer the output of machine C to that of A. The C_{PK} index agrees with this evaluation by rating the capability of machine C much higher than machine A (1.67 versus 1.33). However, the C_{PM} measure disagrees and ranks C's capability considerably *lower* than A's (.84 versus 1.33).

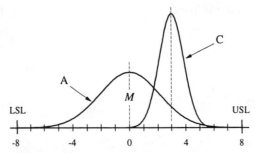

Fig. 6.88 The output of machine A compared to machine C.

For Machine A

$$\hat{\mu}_A = 0 \qquad \hat{\sigma}_{ST,A} = 2$$

$$\hat{\tau}_{ST,A} = \sqrt{2^2 + (0 - 0)^2} = 2$$

$$\hat{C}_{PK,A} = \frac{8 - 0}{3\,(2)} = 1.33$$

$$\hat{C}_{PM,A} = \frac{8 - -8}{6\,(2)} = 1.33$$

For Machine C

$$\hat{\mu}_C = 3 \qquad \hat{\sigma}_{ST,C} = 1$$

$$\hat{\tau}_{ST,C} = \sqrt{1^2 + (3 - 0)^2} = 3.162$$

$$\hat{C}_{PK,C} = \frac{8 - 3}{3\,(1)} = 1.67$$

$$\hat{C}_{PM,C} = \frac{8 - -8}{6\,(3.162)} = .84$$

Why this difference in analysis? C_{PM} considers how well the process is minimizing the total loss of parts produced both within and outside of tolerance. C_{PM} enlarges its estimate of process variation (τ_{ST}) by the square of the deviation of the estimated process average from the target average.

$$\tau_{ST} = \sqrt{\sigma_{ST}^2 + (\mu - M)^2}$$

Even though all its parts are well within specification, C is producing practically every part in a loss area with almost *no parts at the desired target average* of 0, as is shown in Figure 6.89. Parts produced on target are the only ones which have zero loss under the Taguchi loss function philosophy.

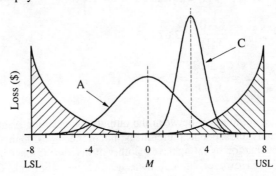

Fig. 6.89 The loss of machine A compared to machine C.

Assuming a part produced at M has a loss of \$0, while one made at the USL creates a loss of \$20, the average loss for parts produced on machine C is calculated as \$3.12. The distance (d) between M and the USL is 8 (8 minus 0).

$$\overline{L}_{ST,C} = c\left(\frac{\hat{\tau}_{ST,C}}{d}\right)^2 = \$20\left(\frac{3.162}{8}\right)^2 = \$3.12$$

Although A manufactures some parts in loss areas, the majority of its parts are right at the target value, and therefore, have zero loss. In fact, the average loss for A is only \$1.25, which is just 40 percent of the average loss for parts produced on machine C (\$3.12 × .40 = \$1.25).

$$\overline{L}_{ST,A} = c\left(\frac{\hat{\tau}_{ST,A}}{d}\right)^2 = \$20\left(\frac{2}{8}\right)^2 = \$1.25$$

Thus, while the goalpost philosophy prefers C over A, the Taguchi loss function philosophy nominates A as the better machine for meeting customer requirements by minimizing total loss, and thereby maximizing customer satisfaction. If A and C were different suppliers, many companies with highly automated factories would prefer A's output. Even though well within specification, the extra adjustment costs and problems caused by supplier C's off-centered parts would eliminate that supplier from contention.

When a company decides to measure capability with the C_{PM} index, measures previously covered in this book should be excluded from consideration, as they may generate inconsistent results, leading to confusion. Likewise, if a company chooses to base capability decisions on the C_{PK} index, or any previously mentioned measure, it should not consider C_{PM} in any capability assessment. Because these measures are founded on different value systems (whether to give customers parts that are good enough, or give customers the best parts possible), top management must first pick a quality philosophy to follow. Once this selection is made, capability measures consistent with this philosophy can be correctly chosen and properly analyzed.

Example 6.30 Roller Width

After molding and curing, hard rubber rollers (Figure 6.90) for guiding and driving paper sheets through copying machines are ground to a final diameter. Rollers are then sliced with a rotating knife to the proper width. Because excessive variation in roller width results in paper jams, this dimension must be held to within plus or minus .0127 cm of the nominal length of 31.2250 cm. This criterion makes the tolerance equal to .0254 cm.

Fig. 6.90 A hard rubber roller for copying machines.

When a good state of control is demonstrated, the short-term standard deviation is estimated as .005314 cm and the average as 31.2306 cm. Potential capability is often measured with the C_P index.

$$\hat{C}_P = \frac{\text{Tolerance}}{6\hat{\sigma}_{ST}} = \frac{.0254}{6(.005314)} = .797$$

Many companies adopting the goalpost philosophy evaluate performance capability with the C_{PK} index.

$$\hat{C}_{PK} = \text{Minimum}\left(\frac{\hat{\mu} - \text{LSL}}{3\hat{\sigma}_{ST}}, \frac{\text{USL} - \hat{\mu}}{3\hat{\sigma}_{ST}}\right)$$

$$= \text{Minimum}\left(\frac{31.2306 - 31.2123}{3(.005314)}, \frac{31.2377 - 31.2306}{3(.005314)}\right)$$

$$= \text{Minimum}\,(\,1.148\,,.445\,) = .445$$

Those companies preferring the Taguchi loss function quite often employ C_{PM} to assess process capability.

$$\hat{\tau}_{ST} = \sqrt{\hat{\sigma}_{ST}^2 + (\hat{\mu} - M)^2} = \sqrt{(.005314)^2 + (31.2306 - 31.2250)^2} = .00772$$

$$\hat{C}_{PM} = \frac{\text{Tolerance}}{6\hat{\tau}_{ST}} = \frac{.0254}{6(.00772)} = .548$$

Sometimes the average loss per part is estimated as well as C_{PM}. Assuming a loss of $15.79 for a roller produced at one of the specification limits, $\overline{L}_{ST}$ is estimated as $5.83.

$$\overline{L}_{ST} = c\left(\frac{\hat{\tau}_{ST}}{d}\right)^2 = \$15.79\left(\frac{.00772}{.0127}\right)^2 = \$5.83$$

As a first step in improving the performance of this cutting operation, adjustments to the process are made so the average width is now centered at 31.2250 cm, the middle of the tolerance. If no change in the estimate of σ_{ST} occurs, C_{PK} increases to equal C_P.

$$\hat{C}_{PK} = \text{Minimum}\left(\frac{\hat{\mu} - \text{LSL}}{3\hat{\sigma}_{ST}}, \frac{\text{USL} - \hat{\mu}}{3\hat{\sigma}_{ST}}\right)$$

$$= \text{Minimum}\left(\frac{31.2250 - 31.2123}{3(.005314)}, \frac{31.2377 - 31.2250}{3(.005314)}\right)$$

$$= \text{Minimum}\,(.797\,,.797\,) = .797$$

With this new process average, C_{PM} also equals C_P, as well as C_{PK}.

$$\hat{\tau}_{ST} = \sqrt{\hat{\sigma}_{ST}^2 + (\hat{\mu} - M)^2} = \sqrt{(.005314)^2 + (31.2250 - 31.2250)^2} = .005314$$

$$\hat{C}_{PM} = \frac{\text{Tolerance}}{6\hat{\tau}_{ST}} = \frac{.0254}{6(.005314)} = .797$$

This change in average also reduces the average loss to $2.76.

$$\overline{L}_{ST} = c\left(\frac{\hat{\tau}_{ST}}{d}\right)^2 = \$15.79\left(\frac{.005314}{.0127}\right)^2 = \$2.76$$

This shift in the process average represents a substantial improvement in performance capability, as well as a major reduction in the average loss per part. With the process correctly centered, future improvements in capability (or decreases in loss) must come by reducing process variation. Suppose the introduction of thicker knives to this cutting process shrinks the estimate of σ_{ST} to only .00316. Assuming average roller width still equals M, estimates of the above capability measures change as follows:

$$\hat{C}_{PM} = \hat{C}_{PK} = \hat{C}_P = \frac{\text{Tolerance}}{6\hat{\sigma}_{ST}} = \frac{.0254}{6(.00309)} = 1.370$$

The new average loss is now only $ .93.

$$\overline{L}_{ST} = c\left(\frac{\hat{\tau}_{ST}}{d}\right)^2 = \$15.79\left(\frac{.00309}{.0127}\right)^2 = \$.93$$

This reduction in loss can be compared to the investment needed to install the new equipment. Anticipated annual volume is forecasted to be 4,000 rollers. As the thicker knives reduce the average loss per roller by $1.83 ($2.76 minus $ .93), $7,320 is saved yearly.

$$(\$1.83 / \text{roller}) \times (4,000 \text{ rollers} / \text{year}) = \$7,320 / \text{year}$$

If the cost of installing the new knives is $11,750, it will take 1.6 years to recoup this investment.

$$\frac{\$11,750}{\$7,320 / \text{year}} = 1.6 \text{ years}$$

This result can be compared to the required payback period to help management make an economic decision about implementing this process improvement.

The m_{ST} Factor Relating C_{PM} and C_P

When the process is centered at the middle of the tolerance, and M is also the target for the process average ($T = M$), then C_{PM} equals C_P. This happens because μ minus M is 0, resulting in τ_{ST} becoming identical to σ_{ST}.

$$\tau_{ST} = \sqrt{\sigma_{ST} + (\mu - M)^2} = \sqrt{\sigma_{ST} + 0^2} = \sigma_{ST}$$

$$C_{PM} = \frac{\text{USL - LSL}}{6\tau_{ST}} = \frac{\text{USL - LSL}}{6\sigma_{ST}} = C_P$$

For those times when the process average is centered elsewhere, C_{PM} is related to C_P as is shown here:

$$C_{PM} = C_P m_{ST} \quad \text{where} \quad m_{ST} = \frac{\sigma_{ST}}{\sqrt{\sigma_{ST}^2 + (\mu - M)^2}} = \frac{\sigma_{ST}}{\tau_{ST}}$$

Although somewhat similar to the k factor relating C_{PK} to C_P in that it is a function of the difference between μ and M, m_{ST} is independent of the tolerance whereas k is not. On the other hand, k is independent of σ_{ST}, whereas m_{ST} is not. The subscript "ST" appearing with m_{ST} is needed to distinguish it from m_{LT}, which relates P_{PM} to P_P.

$$P_{PM} = P_P \, m_{LT} \qquad \text{where} \quad m_{LT} = \frac{\sigma_{LT}}{\sqrt{\sigma_{LT}^{\,2} + (\mu - M)^2}} = \frac{\sigma_{LT}}{\tau_{LT}}$$

As seen previously, when μ equals M, τ_{ST} equals σ_{ST}, and m_{ST} equals 1.

$$m_{ST} = \frac{\sigma_{ST}}{\sqrt{\sigma_{ST}^{\,2} + 0}} = \frac{\sigma_{ST}}{\sigma_{ST}} = 1$$

This condition causes C_{PM} to equal C_P.

$$C_{PM} = C_P \, m_{ST} = C_P(1) = C_P$$

Because C_P also equals C_{PK} when μ equals M, C_{PM} is also equal to C_{PK}.

$$C_{PM} = C_P = C_{PK}$$

If μ differs from M, $(\mu - M)^2$ is greater than 0, resulting in m_{ST} being less than 1.

$$(\mu - M)^2 > 0$$

$$\sigma_{ST}^2 + (\mu - M)^2 > \sigma_{ST}^2$$

$$\sqrt{\sigma_{ST}^2 + (\mu - M)^2} > \sigma_{ST}$$

$$1 > \frac{\sigma_{ST}}{\sqrt{\sigma_{ST}^2 + (\mu - M)^2}}$$

$$1 > m_{ST}$$

An m_{ST} factor less than 1 makes C_{PM} less than C_P.

$$m_{ST} < 1$$

$$C_P \, m_{ST} < C_P$$

$$C_{PM} < C_P$$

Because both σ_{ST} and τ_{ST} are positive, m_{ST} is always positive (> 0).

$$m_{ST} = \frac{\sigma_{ST}}{\tau_{ST}} > 0$$

As m_{ST} cannot be greater than 1, it is always between 0 and 1.

$$0 < m_{ST} \leq 1$$

Due to C_P always being positive, C_{PM} can never be less than zero, as it is the product of C_P and m_{ST}, two positive numbers.

$$0 < m_{ST} \leq 1$$

$$C_P(0) < C_P m_{ST} \leq C_P(1)$$

$$0 < C_{PM} \leq C_P$$

The relationship between C_{PM} and C_P is occasionally written as:

$$C_{PM} = \frac{\text{Tolerance}}{6\sqrt{\sigma_{ST}^2 + (\mu - M)^2}}$$

$$= \frac{\text{Tolerance}}{6\sigma_{ST}\sqrt{1 + (\mu - M)^2/\sigma_{ST}^2}}$$

$$= \frac{C_P}{\sqrt{1 + (\mu - M)^2/\sigma_{ST}^2}}$$

$$= \frac{C_P}{\sqrt{1 + [(\mu - M)/\sigma_{ST}]^2}}$$

$$= \frac{C_P}{\sqrt{1 + Z_{M,ST}^2}} \quad \text{where } Z_{M,ST} = \frac{M - \mu}{\sigma_{ST}}$$

$Z_{M,ST}$ expresses how far the process average is from the middle of the tolerance in units of short-term standard deviations. When the average is equal to M, $Z_{M,ST}$ is 0 and C_{PM} equals C_P. If the average is 1 standard deviation from M, then $Z_{M,ST}$ is 1. This makes C_{PM} equal to .707 C_P.

$$C_{PM} = \frac{C_P}{\sqrt{1 + Z_{M,ST}^2}} = \frac{C_P}{\sqrt{1 + 1^2}} = .707\, C_P$$

Likewise, the relationship between C_{PM} and C_P for $Z_{M,ST}$ of 2, 3, and 4 is as follows:

$$C_{PM} = \frac{C_P}{\sqrt{1 + Z_{M,ST}^2}} = \frac{C_P}{\sqrt{1 + 2^2}} = .447\, C_P$$

$$C_{PM} = \frac{C_P}{\sqrt{1 + Z_{M,ST}^2}} = \frac{C_P}{\sqrt{1 + 3^2}} = .316\, C_P$$

$$C_{PM} = \frac{C_P}{\sqrt{1 + Z_{M,ST}^2}} = \frac{C_P}{\sqrt{1 + 4^2}} = .243\, C_P$$

Figure 6.91 illustrates how C_{PM} equals a continually dwindling portion of the available potential capability (x axis) as the distance between the process average and M increases. This distance is measured in units of standard deviations on the y axis, which is exactly what $Z_{M,ST}$ represents.

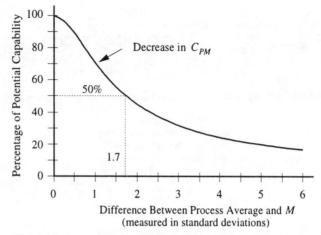

Fig. 6.91 Percentage loss of potential capability versus $Z_{M,ST}$.

When the average is centered at M, C_{PM} equals 100 percent of the potential capability (m_{ST} = 1.0). As the average moves away from M, the reduction in performance capability is initially quite dramatic. At a difference of about 1.7 standard deviations, C_{PM} is only half of C_P (m_{ST} = .50). By the time the average is more than 4 standard deviations from M, the ratio of performance to potential capability levels out at about 20 percent ($m_{ST} \cong .20$).

P_{PM} is related to P_P in a similar fashion.

$$P_{PM} = \frac{P_P}{\sqrt{1 + Z_{M,LT}^2}} \quad \text{where } Z_{M,LT} = \frac{M - \mu}{\sigma_{LT}}$$

Example 6.31 Roller Width

In Example 6.30, rollers for copying machines were cut to the proper width. The original short-term standard deviation was estimated as .005314 cm and the average as 31.2306 cm. With a tolerance of .0127 cm, this information generated an estimate of .797 for C_P and .00772 cm for τ_{ST}. C_{PM} can be estimated by first finding m_{ST}. Observe how the units of measurement cancel out, leaving m_{ST} a dimensionless number.

$$\hat{m}_{ST} = \frac{\hat{\sigma}_{ST}}{\hat{\tau}_{ST}} = \frac{.005314 \text{ cm}}{.00772 \text{ cm}} = .6883$$

Multiplying C_P by m_{ST} provides an estimate of C_{PM}. Because m_{ST} is .6883, over 31 percent of potential capability is lost to this process because of poor centering.

$$\hat{C}_{PM} = \hat{C}_P \, \hat{m}_{ST} = .797(.6883) = .548$$

This equals the result from the original formula for C_{PM}.

$$\hat{C}_{PM} = \frac{\text{Tolerance}}{6\hat{\tau}_{ST}} = \frac{.0254}{6(.00772)} = .548$$

Notice that m_{ST} is related to $Z_{M,ST}$, the number of standard deviations between the process average and M. The nominal roller length of 31.2250 cm is specified as M.

$$\hat{Z}_{M,ST} = \frac{M - \hat{\mu}}{\hat{\sigma}_{ST}} = \frac{31.2250 \text{ cm} - 31.2306 \text{ cm}}{.005314 \text{ cm}} = -1.054$$

$$\hat{m}_{ST} = \frac{1}{\sqrt{1 + \hat{Z}_{M,ST}^2}} = \frac{1}{\sqrt{1 + (-1.054)^2}} = .6883$$

As the absolute value of $Z_{M,ST}$ increases, m_{ST} decreases. For example, if the process average shifts such that $|Z_{M,ST}|$ grows to 2.0, m_{ST} drops from .6883 to .4472, a reduction of about 35 percent (review Figure 6.91).

$$\hat{m}_{ST} = \frac{1}{\sqrt{1 + \hat{Z}_{M,ST}^2}} = \frac{1}{\sqrt{1 + 2^2}} = .4472$$

This shift means an additional 35 percent of the potential capability is lost to C_{PM}, making it equal to only .356, down from the initial .548.

$$\hat{C}_{PM} = \hat{C}_p \, \hat{m}_{ST} = .797(.4472) = .356$$

Relationship Between C_{PM} and C_{PK}

C_P is related to C_{PK} as follows:

$$C_{PK} = C_P(1 - k)$$

Because C_{PM} is also related to C_P, it is related to C_{PK}, as is demonstrated in the following equations (Kotz and Johnson, 1999; Mittag and Rinne).

$$C_{PM} = C_P \, m_{ST} \quad \Rightarrow \quad C_P = \frac{C_{PM}}{m_{ST}}$$

$$C_{PK} = C_{PM}\left(\frac{1 - k}{m_{ST}}\right) = C_{PM}(1 - k)\sqrt{1 + Z_{M,ST}^2}$$

This last equation may be rearranged to solve for C_{PM}, although it is better to apply the above equation in practice because 1 minus k could be 0 for some situations.

$$C_{PM} = C_{PK}\left(\frac{m_{ST}}{1 - k}\right)$$

Through a similar derivation, P_{PM} can be related to P_{PK}.

$$P_{PM} = P_{PK}\left(\frac{m_{LT}}{1-k}\right) \qquad \text{where } m_{LT} = \frac{\sigma_{LT}}{\tau_{LT}}$$

If the process average is centered at the middle of the tolerance ($\mu = M$), then k equals 0, and m_{ST} is 1.

$$k = \frac{|\mu - M|}{(\text{USL - LSL})/2} = \frac{0}{(\text{USL - LSL})/2} = 0$$

$$m_{ST} = \frac{\sigma_{ST}}{\sqrt{\sigma_{ST}^2 + (\mu - M)^2}} = \frac{\sigma_{ST}}{\sqrt{\sigma_{ST}^2 + 0}} = \frac{\sigma_{ST}}{\sigma_{ST}} = 1$$

This means the quantity $m_{ST}/(1 - k)$ equals 1, resulting in C_{PM} equaling C_{PK}.

$$C_{PM} = C_{PK}\left(\frac{m_{ST}}{1-k}\right) = C_{PK}\left(\frac{1}{1-0}\right) = C_{PK}(1) = C_{PK}$$

In fact, for a fixed standard deviation, as the process average nears M, C_{PM} and C_{PK} both approach C_P.

$$\text{As } \mu \to M, \ C_{PM} \to \frac{\text{Tolerance}}{6\sigma_{ST}} \quad \text{and} \quad C_{PK} \to \frac{\text{Tolerance}}{6\sigma_{ST}}$$

However, for a fixed process average, as the standard deviation goes to zero, C_{PM} approaches a fixed value (depending on how far the process average is from M), while C_{PK} becomes infinitely large.

$$\text{As } \sigma_{ST} \to 0, \ C_{PM} \to \frac{\text{Tolerance}}{6|\mu - M|} \quad \text{and} \quad C_{PK} \to +\infty$$

When μ differs from M and the standard deviation is not near zero, the relationship between C_{PM} and C_{PK} becomes somewhat complex. For very large differences, C_{PM} is greater than C_{PK}. However, for smaller differences, whether C_{PM} is greater or less than C_{PK} depends on the relationship between σ_{ST} and the tolerance. This complicated association between C_{PK} and C_{PM} is discussed further in the next two sections.

Relationship of C_P, C_{PK}, and C_{PM} to μ

The set of diagrams in Figures 6.92a through 6.92e display the relationship between several measures of capability and μ when σ_{ST} is 1, M is 0, and the specifications are ± 6. Recall that any measure of potential capability, such as C_P, remains constant no matter where the process average is located.

$$C_P = \frac{\text{Tolerance}}{6\sigma_{ST}} = \frac{12}{6} = 2.00$$

On the other hand, measures of performance capability, such as C_{PK} and C_{PM}, change as the difference between μ and M increases. As revealed in these figures, the rate of change in C_{PK} and C_{PM} is distinctly different. Initially C_{PM} is less than C_{PK}, but after μ passes 4.77, C_{PM} becomes greater than C_{PK}.

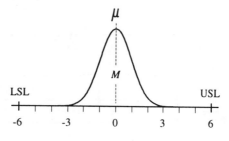

Fig. 6.92a

μ	C_P	$1-k$	m_{ST}	C_{PK}	C_{PM}
0	2.00	.00	1.00	2.00	2.00

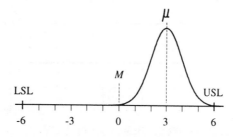

Fig. 6.92b

μ	C_P	$1-k$	m_{ST}	C_{PK}	C_{PM}
3	2.00	.50	.315	1.00	.63

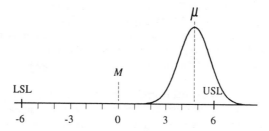

Fig. 6.92c

μ	C_P	$1-k$	m_{ST}	C_{PK}	C_{PM}
4.77	2.00	.205	.205	.41	.41

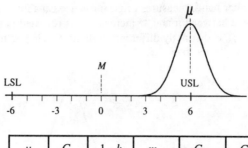

Fig. 6.92d

μ	C_P	1 - k	m_{ST}	C_{PK}	C_{PM}
6	2.00	.00	.164	.00	.33

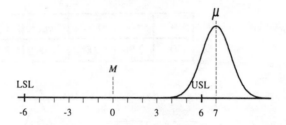

Fig. 6.92e

μ	C_P	1 - k	m_{ST}	C_{PK}	C_{PM}
7	2.00	-.16	.141	-.33	.28

When the process average is located outside one of the specification limits, C_{PK} is negative while C_{PM} remains positive. This relationship between μ and C_{PM} is displayed graphically in the next section (see also Spiring, 1997).

Relationship Between C_{PM} and μ

Due to C_{PM} being calculated with τ_{ST}, a nonlinear relationship develops between it and μ, as illustrated in Figure 6.93. When there is a very small deviation between μ and *M*, C_{PM} is usually less than C_{PK}. As the process average moves farther from *M*, C_{PM} becomes greater than C_{PK} and remains so no matter how large the difference between μ and *M*. Unlike C_{PK}, C_{PM} is always positive, even when the process average is located well outside the print specifications.

C_{PM} will always be greater than C_{PK} when the process average is outside a print specification. However, when the process average is between the specification limits, the connection between C_{PM} and C_{PK} is more complex and depends on the ratio of σ_{ST} to the tolerance. If this ratio is more than .146, C_{PM} is greater than C_{PK} for all values of μ.

$$\text{If } \frac{\sigma_{ST}}{\text{Tolerance}} \geq .146, \text{ then } C_{PM} \geq C_{PK}$$

In the example shown in Figures 6.92a through 6.92e, the short-term standard deviation is 1.0 and the tolerance is 12 (6 minus -6), making the ratio between these two .0833, which is less than the above critical value of .146.

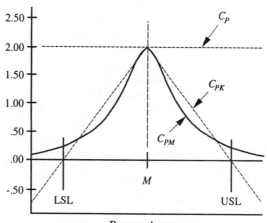

Fig. 6.93 C_{PM} versus μ.

$$\frac{\sigma_{ST}}{\text{Tolerance}} = \frac{1.0}{12} = .0833 \le .146$$

As this ratio is less than .146, C_{PM} is less than C_{PK} for a certain range of process averages (-4.77 to +4.77). Outside of this range, C_{PM} is greater than C_{PK}. There are some processes for which C_{PM} is always greater than C_{PK}, as will be discovered in Example 6.32. Because C_R is related to σ_{ST} divided by the tolerance, the above condition can be restated as follows:

$$\frac{\sigma_{ST}}{\text{Tolerance}} \ge .146$$

$$\frac{6\sigma_{ST}}{\text{Tolerance}} \ge 6(.146)$$

$$C_R \ge .876$$

Thus, C_{PM} is always greater than C_{PK} for processes where C_R is greater than .876, or equivalently, where C_P is less than 1.142 (recall that C_P is just the inverse of C_R).

$$C_R \ge .876$$

$$\frac{1}{C_R} \le \frac{1}{.876}$$

$$C_P \le 1.142$$

So whenever C_P is under 1.142, C_{PM} will be greater than C_{PK} for all values of μ. Conversely, if C_P is *over* 1.142, C_{PM} is greater than C_{PK} for only certain values of μ, while being *less* than C_{PK} for others. For the example covered by Figures 6.92a through e, C_P equaled 2.00, which is greater than 1.142. In Figures 6.92a and b, C_{PM} is seen to be less than C_{PK}, whereas in Figures 6.92d and e, C_{PM} is greater than C_{PK}.

Rating Scale

Unlike all the capability measures presented so far in this book, C_{PM} and P_{PM} do not have any particular rating scale. Under the Taguchi loss function philosophy, emphasis is on reducing variation around the target average because this lowers total loss. As perpetual cost reduction is a goal of every organization, efforts must concentrate on continuously increasing the C_{PM} value for every process, no matter what its current value. C_{PM} does, however, describe a symmetrical range around M where the process average must be located. This is seen by first rearranging the formula for C_{PM} to solve for τ_{ST}.

$$C_{PM} = \frac{\text{Tolerance}}{6\tau_{ST}} \quad \Rightarrow \quad \tau_{ST} = \frac{\text{Tolerance}}{6C_{PM}} \quad \text{where } \tau_{ST} = \sqrt{\sigma_{ST} + (\mu - M)^2}$$

Assuming that σ_{ST} is greater than 0, then:

$$\sigma_{ST} > 0$$

$$\sigma_{ST} + (\mu - M)^2 > (\mu - M)^2$$

$$\sqrt{\sigma_{ST} + (\mu - M)^2} > \sqrt{(\mu - M)^2}$$

$$\tau_{ST} > \sqrt{(\mu - M)^2}$$

$$\tau_{ST} > |\mu - M|$$

This inequality for τ_{ST} is now substituted into the relationship between τ_{ST} and C_{PM}.

$$\tau_{ST} = \frac{\text{Tolerance}}{6C_{PM}}$$

$$|\mu - M| < \frac{\text{Tolerance}}{6C_{PM}}$$

$$|\mu - M| < \frac{1}{6C_{PM}}\text{Tolerance}$$

Assume C_{PM} is 1.0. The above relationship means the process average cannot be more than 1/6th of the tolerance away from M in either direction.

$$|\mu - M| < \frac{1}{6(1.0)}\text{Tolerance} = \frac{1}{6}\text{Tolerance}$$

As the average could range from 1/6th of the tolerance above M to 1/6th of the tolerance below M, it must fall within the middle third (1/6 plus 1/6) of the tolerance, as is illustrated in Figure 6.94. If C_{PM} is 1.33, the process average must fall within the middle fourth of the tolerance (1/8 below to 1/8 above M).

$$|\mu - M| < \frac{1}{6(1.33)}\text{Tolerance} = \frac{1}{8}\text{Tolerance}$$

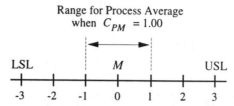

Fig. 6.94 $C_{PM} = 1$ means μ is in the middle third of the tolerance.

Table 6.13 lists the middle segment of the tolerance wherein the process average must lie for several C_{PM} values. Notice that when C_{PM} is greater than .33, the process average must be located within the tolerance. If C_{PM} is less than .33, it is possible for the average to be centered outside the tolerance.

Table 6.13 Relationship between C_{PM} and location of process average.

C_{PM}	Middle Portion of Tolerance
.25	4/3 = 1.33
.33	3/3 = 1.00
.50	2/3 = .667
.60	5/18 = .556
.67	1/2 = .500
.75	4/9 = .444
1.00	1/3 = .333
1.33	1/4 = .250
1.50	2/9 = .222
1.67	1/5 = .200
2.00	1/6 = .167
2.33	1/7 = .143
2.67	1/8 = .125
3.00	1/9 = .111

Compared to C_{PM}, C_{PK} provides very little information as to where the process average is located (Parlar and Wesolowsky). When C_{PK} is less than 0, the process average is known to lie outside of the tolerance. But when C_{PK} is greater than 0, all that is known is that the process average must be somewhere within the tolerance. This is demonstrated with the following logic: when C_{PK} is greater than 0, 1 minus k must be greater than 0, meaning k is less than 1.

$$0 < 1 - k$$

$$k < 1$$

Replacing k with its formula yields:

$$\frac{|\mu - M|}{\text{Tolerance}/2} < 1$$

$$|\mu - M| < \frac{1}{2}\text{Tolerance}$$

Thus, whenever C_{PK} is positive, the process average cannot be more than half the tolerance less than M, or half the tolerance greater than M. Because M is the middle of the tolerance, this simply means μ is somewhere within the tolerance. A small C_{PK} value could arise from a process where the average is centered right on M, or located just below the USL. Likewise, a process with a large C_{PK} could have its average centered at M, or have it located just below the USL. Knowledge of C_{PK} alone is not enough to identify a specific region of the tolerance where the process average must be centered.

Figure 6.95 displays a graph of the values tabulated in Table 6.13. As seen on this chart, a C_{PM} of 1.00 means the process average must be located within the middle third of the tolerance. As C_{PM} increases, the maximum possible difference between μ and M decreases. This restriction is unlike the C_{PK} index, where it is possible for μ to be *anywhere* within the tolerance for a positive C_{PK} value, no matter how large (or small) C_{PK} becomes.

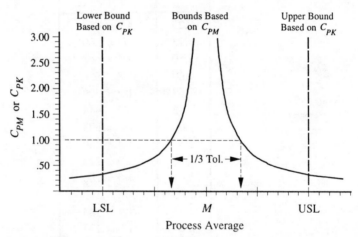

Fig. 6.95 Boundaries on C_{PM} and C_{PK} for differing values of μ.

Tracking Changes in C_{PM}

Changes in C_{PM} for a single process over time can be tracked on a chart similar to that presented in Figure 6.96 (Holmes). After the k_U factor and C_P (or C_R) are calculated for a process, a point is plotted on this graph using k_U (accuracy) as the x coordinate and C_P (precision) as the y coordinate. The association between these three (C_{PM}, C_P, and k_U) can be derived by first reviewing the formula for k_U.

$$k_U = \frac{\mu - M}{\text{Tolerance}/2}$$

$$k_U(\text{Tolerance}/2) = \mu - M$$

The formula for C_{PM} can be rewritten as a function of k_U and C_P. This relationship is derived by first substituting k_U (Tolerance/2) for μ minus M into the formula for C_{PM}.

$$C_{PM} = \frac{\text{Tolerance}}{6\tau_{ST}} = \frac{\text{Tolerance}}{6\sqrt{\sigma_{ST}^2 + (\mu - M)^2}} = \frac{\text{Tolerance}}{6\sqrt{\sigma_{ST}^2 + (k_U \text{ Tolerance}/2)^2}}$$

The formula for C_P may be rearranged as follows:

$$C_P = \frac{\text{Tolerance}}{6\sigma_{ST}}$$

$$6\sigma_{ST} C_P = \text{Tolerance}$$

$$3\sigma_{ST} C_P = \text{Tolerance}/2$$

Substituting $3\sigma_{ST}C_P$ in place of tolerance/2 into the above formula for C_{PM} provides this next relationship:

$$C_{PM} = \frac{6\sigma_{ST} C_P}{6\sqrt{\sigma_{ST}^2 + (k_U \, 3\sigma_{ST} C_P)^2}}$$

$$= \frac{\sigma_{ST} C_P}{\sqrt{\sigma_{ST}^2 + \sigma_{ST}^2(k_U \, 3 \, C_P)^2}}$$

$$= \frac{\sigma_{ST} C_P}{\sigma_{ST} \sqrt{1 + (3k_U \, C_P)^2}}$$

$$= \frac{C_P}{\sqrt{1 + (3k_U \, C_P)^2}}$$

$$= \frac{(1/C_P) C_P}{(1/C_P)\sqrt{1 + (3k_U \, C_P)^2}}$$

$$= \frac{1}{\sqrt{1/C_P^2 + (3k_U)^2}}$$

$$= \frac{1}{\sqrt{1/C_P^2 + 9k_U^2}}$$

Recognize that increasing C_P (potential capability) increases C_{PM}, whereas increasing k_U (decreasing accuracy) reduces C_{PM}. As C_R is the inverse of C_P, this alternative formula may be used as well.

$$C_{PM} = \frac{1}{\sqrt{1/C_P^2 + 9k_U^2}} = \frac{1}{\sqrt{C_R^2 + 9k_U^2}}$$

Figure 6.96 reveals the trade-offs that must be made between k_U (plotted on the x axis) and C_P (on the left-hand y axis) or C_R (on the right-hand y axis) to maintain a constant performance capability as measured by C_{PM}. For example, the semicircle labeled 1.00 shows how rapidly C_P must increase, and C_R decrease, as the process average drifts farther from M. Combinations of k_U and C_P values required for C_{PM} values ranging from 1.20 to 5.00 are also given. A blank copy of this graph is provided in Section 6.16 on page 342.

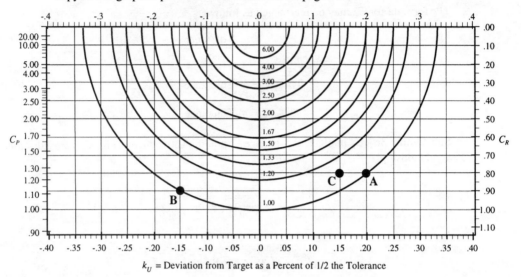

k_U = Deviation from Target as a Percent of 1/2 the Tolerance

Fig. 6.96 C_{PM} versus k_U and C_P.

Plotted points are compared to the semicircles drawn on the graph to estimate C_{PM} for that process. For example, the process represented by point A has a k_U of .20 and a C_P of 1.25, a combination resulting in a C_{PM} of 1.00.

$$C_{PM} = \frac{1}{\sqrt{1/C_P^2 + 9k_U^2}} = \frac{1}{\sqrt{1/1.25^2 + 9(.20)^2}} = 1.00$$

Each semicircle drawn on the graph identifies all pairs of k_U and C_P which generate the C_{PM} value of that semicircle. Note that point B also falls on the semicircle for processes having C_{PM} values equal to 1.00. This is because B's k_U of minus .15 and C_P of 1.12 also produce a C_{PM} of 1.00.

$$C_{PM} = \frac{1}{\sqrt{1/C_P^2 + 9k_U^2}} = \frac{1}{\sqrt{1/1.12^2 + 9(-.15)^2}} = 1.00$$

A process with k_U equal to .15 and C_P equal to 1.25 would have a point plotted at C. As this point is about halfway between the semicircles for 1.00 and 1.20, the C_{PM} for this process is about 1.10. Points moving toward the "0" vertical line, or moving toward the top of the chart, represent increasing C_{PM} values.

$$C_{PM} = \frac{1}{\sqrt{1/1.25^2 + 9(.15)^2}} = 1.09$$

Notice that the semicircle for C_{PM} equaling 1.00 ends at +.33 and -.33 on the top of the graph. This means it is impossible for a process to have a C_{PM} greater than 1.0 and also have its average be more than one-third of half the tolerance above M (or below M), even with a standard deviation of 0 ($C_R = 0$). Likewise, a process with a C_{PM} of 1.67 must have its average within the middle 20 percent of the tolerance (review Table 6.13). Contrast this to the line for C_{PK} equaling 1.00 in Figure 6.57. This line, as do all the others, ends at +1 and -1, meaning μ could be anywhere within the full tolerance and yet still have C_{PK} equal to 1.00.

By plotting the C_{PM} value after each modification to a process, the change in this measure can be tracked over time. In addition, as both x and y axes have dimensionless scales, C_{PM} values from different processes, even with dissimilar units of measurement, can be compared on this one special graph.

Goals

Even though the goal is continuous improvement, some practitioners like having a reference point for process capability. In these cases, a common goal for C_{PM} (P_{PM}) is to equal the goal set for C_P (P_P). In order for this to occur, μ must be centered at M, the middle of the tolerance.

$$C_{PM,GOAL} = C_{P,GOAL} \qquad P_{PM,GOAL} = P_{P,GOAL}$$

Many supporters of these indexes stress the importance of never-ending process improvement through reducing the process spread and centering the process average on the target. Their objective is to bring about continuous increases in C_{PM} rather than just achieving some stationary goal (Sullivan).

Example 6.32 Floppy Disks

A field problem is traced to variation in the diameter of a floppy disk's inner hole. The LSL for hole size is 39.5, while the USL is 44.5, making the tolerance 5.0, and M equal to 42.0. In earlier examples, the following estimates were made:

$$\overline{X} = \hat{\mu} = 40.9 \quad \hat{\sigma}_{LT} = .916 \quad \hat{k} = .44 \quad \hat{P}_R = 1.099 \quad \hat{P}_P = .909 \quad \hat{P}_{PK} = .509$$

τ_{LT} is estimated as follows:

$$\hat{\tau}_{LT} = \sqrt{\hat{\sigma}_{LT}^2 + (\hat{\mu} - M)^2}$$

$$= \sqrt{.916^2 + (40.9 - 42.0)^2} = 1.431$$

From this estimate of τ_{LT}, P_{PM} can be estimated.

$$\hat{P}_{PM} = \frac{USL - LSL}{6\hat{\tau}_{LT}} = \frac{44.5 - 39.5}{6(1.431)} = .582$$

A comparison of the P_{PM} estimate (.582) to the P_P estimate (.909) indicates the process average is centered some distance from M. Interpolating in Table 6.13 for a P_{PM} value of .582 indicates μ must be located within the middle 60 percent of the tolerance ($M \pm .30$ tolerance). With M equal to 42.0 and a tolerance of 5.0, this restriction means the average hole size must be somewhere between 40.5 and 43.5. Note that the estimated average hole size of 40.9 is indeed within this interval.

$$M - .30(\text{Tolerance}) \le \mu \le M + .30(\text{Tolerance})$$

$$42.0 - .30(5.0) \le \mu \le 42.0 + .30(5.0)$$

$$40.5 \le \mu \le 43.5$$

The connection between P_{PM} and P_P is also quantified by the m_{LT} factor.

$$\hat{m}_{LT} = \frac{\hat{\sigma}_{LT}}{\sqrt{\hat{\sigma}_{LT}^2 + (\hat{\mu} - M)^2}} = \frac{\hat{\sigma}_{LT}}{\hat{\tau}_{LT}} = \frac{.916}{1.431} = .640$$

A full 36 percent of the potential long-term process capability for hole size is lost to P_{PM} as a result of the process being currently centered at 40.9 instead of 42.0.

$$\hat{P}_{PM} = \hat{P}_P (\hat{m}_{LT}) = .909 (.640) = .582$$

Note that 44 percent of the potential capability is lost to the P_{PK} index because of this centering problem.

$$\hat{P}_{PK} = \hat{P}_P (1 - \hat{k}) = .909 (1 - .44) = .509$$

By raising the average hole size to 42.0, both these measures of performance capability will increase to their maximums, which is the P_P estimate of .909. Now consider the ratio between σ_{LT} and the tolerance, which is .1832.

$$\frac{\hat{\sigma}_{LT}}{\text{Tolerance}} = \frac{.916}{5.0} = .1832 \ge .146$$

Because this ratio is greater than critical value of .146, P_{PM} is always greater than P_{PK}, no matter where the process average is located, as the graph in Figure 6.97 verifies (review the discussion concerning Figure 6.93 on page 294).

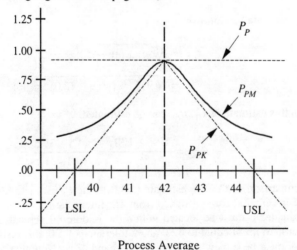

Fig. 6.97 P_{PM} versus P_{PK} for different process averages.

For this particular relationship between σ_{LT} and the tolerance, P_{PK} is more sensitive to changes in the process average than P_{PM}. If the process average moves from its current location of 40.9 to the target of 42.0, P_{PM} increases by 56 percent (.582 to .909), whereas P_{PK} grows by 79 percent (from .509 to .909). Thus, the "Taguchi" capability measure is not always as sensitive to centering problems as the "goalpost" measure.

The target ratio (TR = 3k) proposed by Phillips can be used in conjunction with C_R (or P_R) to derive C_{PM} (or P_{PM}). A complete derivation is provided in Section 6.15 on page 340.

$$C_{PM} = \frac{1}{\sqrt{C_R^2 + TR^2}} \qquad P_{PM} = \frac{1}{\sqrt{P_R^2 + TR^2}}$$

If process precision is increased, as reflected by a smaller P_R index, P_{PM} will increase. P_{PM} also increases if accuracy is improved, which is revealed by smaller TR values. In this floppy disk example, TR is estimated as 1.320.

$$\hat{TR} = \frac{6|M - \hat{\mu}|}{USL - LSL} = \frac{6|42.0 - 40.9|}{44.5 - 39.5} = 1.320$$

This result, along with the estimate for P_R of 1.099, are combined to estimate P_{PM} as .582, the same answer encountered at the beginning of this example.

$$\hat{P}_{PM} = \frac{1}{\sqrt{\hat{P}_R^2 + \hat{TR}^2}} = \frac{1}{\sqrt{1.099^2 + 1.320^2}} = .582$$

One advantage of separating P_{PM} into these two components is the disclosure of whether process spread (P_R) or centering (TR) is more responsible for the observed low performance capability. In this example, P_R is 1.099 and TR is 1.320. Although both indicate serious problems, the lack of proper centering should be addressed first, as TR is 30 percent greater than P_R.

Improvement Ratios

The floppy disk capability team has a choice of improving process variation by 20 percent, or improving accuracy by 10 percent. Both improvements cost the same, but unfortunately, the team can afford to implement only one. To discover which has the largest effect on C_{PM}, the following improvement ratio is derived. Let the subscript "I" denote the improved condition, and the subscript "O" the original condition.

$$\hat{C}_{PM,I} = \frac{1}{\sqrt{\hat{C}_{R,I}^2 + \hat{TR}_I^2}} \qquad \hat{C}_{PM,O} = \frac{1}{\sqrt{\hat{C}_{R,O}^2 + \hat{TR}_O^2}}$$

The ratio between these two is the percentage of improvement.

$$C_{PM} \text{ Improvement Ratio} = \frac{\hat{C}_{PM,I}}{\hat{C}_{PM,O}} = \frac{\sqrt{\hat{C}_{R,O}^2 + \hat{TR}_O^2}}{\sqrt{\hat{C}_{R,I}^2 + \hat{TR}_I^2}} = \sqrt{\frac{\hat{C}_{R,O}^2 + \hat{TR}_O^2}{\hat{C}_{R,I}^2 + \hat{TR}_I^2}}$$

Of course, a similar improvement ratio can be developed for P_{PM}.

$$P_{PM} \text{ Improvement Ratio} = \sqrt{\frac{\hat{P}_{R,O}^2 + \hat{TR}_O^2}{\hat{P}_{R,I}^2 + \hat{TR}_I^2}}$$

These improvement ratios disclose the expected percentage increase in C_{PM}, or P_{PM}, for reducing variation and/or improving aim. As a first scenario, assume process spread is decreased by 20 percent in the floppy disk example, with no change in centering. Because the original P_R is 1.099, the P_R ratio for the improved process would be .879 (20 percent less), while the improved TR remains unchanged from the original.

$$\hat{P}_{R,I} = .80\,\hat{P}_{R,O} = .80(1.099) = .879 \qquad \hat{TR}_I = \hat{TR}_0 = 1.320$$

Substituting these results into the improvement ratio formula predicts the expected P_{PM} index for the improved process should be 8.3 percent larger.

$$P_{PM} \text{ Improvement Ratio} = \sqrt{\frac{\hat{P}_{R,O}^2 + \hat{TR}_O^2}{\hat{P}_{R,I}^2 + \hat{TR}_I^2}} = \sqrt{\frac{1.099^2 + 1.320^2}{.879^2 + 1.320^2}} = 1.083$$

On the other hand, if process aim is improved by 10 percent, the improved TR value is 90 percent of the original, or 1.188.

$$\hat{TR}_I = .90\,\hat{TR}_O = .90(1.320) = 1.188$$

Assuming no change in variation, the improved P_R ratio is identical to the original.

$$\hat{P}_{R,I} = \hat{P}_{R,O} = 1.099$$

Substituting these changes into the improvement ratio formula forecasts an overall 6.1 percent increase in P_{PM}.

$$P_{PM} \text{ Improvement Ratio} = \sqrt{\frac{\hat{P}_{R,O}^2 + \hat{TR}_O^2}{\hat{P}_{R,I}^2 + \hat{TR}_I^2}} = \sqrt{\frac{1.099^2 + 1.320^2}{1.099^2 + 1.188^2}} = 1.061$$

Based on this analysis, the team should choose to reduce variation, (assuming equal costs) as this course of action provides the largest improvement in performance capability.

A similar improvement ratio formula can be worked out for C_{PK}.

$$\hat{C}_{PK,I} = \hat{C}_{P,I}(1 - \hat{k}_I) = \frac{1 - \hat{k}_I}{\hat{C}_{R,I}} \qquad \hat{C}_{PK,O} = \hat{C}_{P,O}(1 - \hat{k}_O) = \frac{1 - \hat{k}_O}{\hat{C}_{R,O}}$$

The ratio between these becomes:

$$C_{PK} \text{ Improvement Ratio} = \frac{\hat{C}_{PK,I}}{\hat{C}_{PK,O}} = \frac{(1 - \hat{k}_I)/\hat{C}_{R,I}}{(1 - \hat{k}_O)/\hat{C}_{R,O}} = \frac{\hat{C}_{R,O}}{\hat{C}_{R,I}}\left(\frac{1 - \hat{k}_I}{1 - \hat{k}_O}\right)$$

Example 6.33 Comparison of Improvement Ratios

A process output for a certain characteristic originally has an estimated average of 2 and a short-term standard deviation of 1. With a LSL of minus 8 and an USL of plus 8, the $C_{R,O}$ ratio, the k_O factor, and TR_O are estimated as follows:

$$\hat{C}_{R,O} = \frac{6\hat{\sigma}_{ST}}{\text{USL - LSL}} = \frac{6(1)}{8--8} = .375$$

$$\hat{k}_O = \frac{2|M-\hat{\mu}|}{\text{USL - LSL}} = \frac{2|0-2|}{8--8} = .25$$

$$\hat{\text{TR}}_O = 3\hat{k}_O = 3(.25) = .75$$

From these, $C_{PM,O}$ and $C_{PK,O}$ are estimated as 1.19 and 2.00, respectively.

$$\hat{C}_{PM,O} = \frac{1}{\sqrt{\hat{C}_{R,O}^2 + \hat{\text{TR}}_O^2}} = \frac{1}{\sqrt{.375^2 + .75^2}} = 1.19$$

$$\hat{C}_{PK,O} = \frac{1-\hat{k}_O}{\hat{C}_{R,O}} = \frac{1-.25}{.375} = 2.00$$

A 20 percent reduction in process spread makes $C_{R,I}$ equal to .30.

$$\hat{C}_{R,I} = \hat{C}_{R,O}(1-.20) = .375(.80) = .30$$

With no change in accuracy, the expected improvements in C_{PM} and C_{PK} are estimated as shown here:

$$C_{PM} \text{ Improvement Ratio } = \sqrt{\frac{\hat{C}_{R,O}^2 + \hat{\text{TR}}_O^2}{\hat{C}_{R,I}^2 + \hat{\text{TR}}_I^2}} = \sqrt{\frac{.375^2 + .75^2}{.30^2 + .75^2}} = 1.038$$

$$C_{PK} \text{ Improvement Ratio } = \frac{\hat{C}_{R,O}}{\hat{C}_{R,I}}\left(\frac{1-\hat{k}_I}{1-\hat{k}_O}\right) = \frac{.375}{.30}\left(\frac{1-.25}{1-.25}\right) = 1.250$$

The 20 percent reduction in variation would cause C_{PM} to edge up by 3.8 percent, while pushing C_{PK} up by 25 percent. Instead of this 20 percent decrease in variation, suppose a 20 percent improvement in centering the process average is implemented. This modification makes k_I equal to .20 and TR_I equal to .60.

$$\hat{k}_I = \hat{k}_O(1-.20) = .25(.80) = .20 \qquad \hat{\text{TR}}_I = 3\hat{k}_I = 3(.20) = .60$$

With this option, C_{PM} is predicted to rise by 18.5 percent, while C_{PK} would increase by only 6.7 percent.

$$C_{PM} \text{ Improvement Ratio } = \sqrt{\frac{\hat{C}_{R,O}^2 + \hat{\text{TR}}_O^2}{\hat{C}_{R,I}^2 + \hat{\text{TR}}_I^2}} = \sqrt{\frac{.375^2 + .75^2}{.375^2 + .60^2}} = 1.185$$

$$C_{PK} \text{ Improvement Ratio } = \frac{\hat{C}_{R,O}}{\hat{C}_{R,I}}\left(\frac{1-\hat{k}_I}{1-\hat{k}_O}\right) = \frac{.375}{.375}\left(\frac{1-.20}{1-.25}\right) = 1.067$$

C_{PM} increases by 18.5 percent when the average is shifted, versus 3.8 percent for the reduction in variation. If only one option can be chosen, this analysis strongly recommends shifting the process average. However, C_{PK} increases by just 6.7 percent when the average is shifted, compared to 25 percent when the variation is reduced. This second analysis reaches the *opposite* conclusion, *i.e.*, choose the variation-reduction option. As this example points out once again, taking sufficient time to carefully consider which capability measure will be employed to assess a process is extremely crucial. If centering the output at the target is of utmost importance, then go with C_{PM}. When variation reduction is valued more, choose C_{PK}. If process yield is stressed, select ppm_{TOTAL} to measure capability.

Advantages and Disadvantages

This measure considers both process centering and process variation in computing process capability. It incorporates the Taguchi loss function which vigorously encourages the centering of the process output at the middle of the tolerance. However, for some processes, M is not the desired location for the process average, making the C_{PM} index inappropriate for measuring performance capability. In addition, C_{PM} cannot handle unilateral specifications where the middle of the tolerance is undefined (Pillet *et al*) nor non-normal distributions (Spiring *et al*).

The shape of the loss function isn't always known or is difficult to determine. The C_{PM} formula assumes a parabolic loss function centered at the middle of the tolerance. The symmetry of this loss function model implies the loss associated with a part located a certain distance below M is identical to that for a part located an equal distance above M. But, being too far below M may mean scrapping a part versus reworking a part too far above the target. A customer may have more difficulty with parts produced on one side of M versus the other, as in the floppy disk example. These factors affect the loss unequally and are not modeled correctly with a symmetrical loss function. The asymmetric loss function presented in Figure 6.98 is more appropriate when the loss related to an undersize part is greater than that for an oversize one.

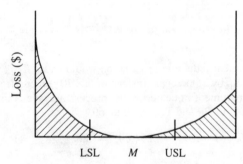

Fig. 6.98 A non-symmetric loss function.

Johnson and Kotz offer a measure of performance capability to handle cases where the loss function is not symmetrical. Referred as the C_{JKP} index (Kotz and Lovelace), it is estimated with this next formula. Because of its complexity, this index is seen infrequently in practice.

$$\hat{C}_{JKP} = \frac{1}{3\sqrt{2}} \text{Minimum} \left(\frac{M - \text{LSL}}{\sqrt{S_- / kn}}, \frac{\text{USL} - M}{\sqrt{S_+ / kn}} \right)$$

$$\text{where } S_- = \sum_{X_i < M} (X_i - M)^2 \quad \text{and} \quad S_+ = \sum_{X_i > M} (X_i - M)^2$$

Either symmetric or asymmetric, the parabolic loss function model has no maximum loss. As parts move farther from the target average, their corresponding loss continues to increase without limit. Because this is unrealistic, Spring (1991*b*) has proposed modifying the parabolic model to that of an inverse normal curve, as displayed in Figure 6.99.

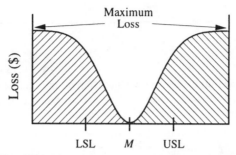

Fig. 6.99 An inverse-normal loss function model.

With this model, losses still increase as parts move away from the target average, but the rate of increase in loss diminishes and eventually levels off at some maximum loss. To handle the situation of unequal losses on one side of the tolerance versus the other, Spring also proposed the loss function presented in Figure 6.100. The upper portion of the curve applies to part measurements above *M*, while the lower curve applies to measurements below *M*. Choi and Owen have experimented with a capability measure for this particular type of loss function.

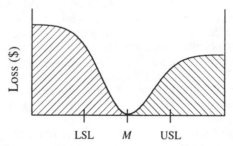

Fig. 6.100 An asymmetrical, inverse-normal loss function.

The percentage of nonconforming parts is extremely difficult to estimate from the C_{PM} index. However, followers of the Taguchi loss function philosophy argue this is not a prime consideration as more emphasis is placed on reducing variation around the target value than on the percentage of nonconforming parts.

C_{PM} is touted as a better performance capability measure than C_{PK} because C_{PK} does not always represent how far the process average is located from the target. The two distributions displayed in Figure 6.101 are the process outputs from two different spindles of the same boring machine. Which spindle has better capability? Distribution A has an average of 0, with a standard deviation of 5. B has an average of 4, with a standard deviation of only 3. Even with these big differences in process parameters, C_{PK} turns out to be .67 for both.

$$C_{PK,A} = \text{Minimum}\left(\frac{0--10}{3(5)}, \frac{10-0}{3(5)}\right) = .67 \quad C_{PK,B} = \text{Minimum}\left(\frac{4--10}{3(3)}, \frac{10-4}{3(3)}\right) = .67$$

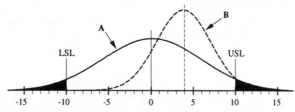

Fig. 6.101 Two different process outputs.

Although the outputs are quite a bit different, the C_{PK} measure rates the spindles as equal in terms of performance capability. How does the C_{PM} index appraise them? First the τ_{ST} values must be calculated for each distribution.

$$\tau_{ST,A} = \sqrt{\sigma_{ST,A}^2 + (\mu_A - M)^2} = \sqrt{5^2 + (0 - 0)^2} = 5$$

$$\tau_{ST,B} = \sqrt{\sigma_{ST,B}^2 + (\mu_B - M)^2} = \sqrt{3^2 + (4 - 0)^2} = 5$$

Now the C_{PM} value for each spindle can be determined.

$$C_{PM,A} = \frac{\text{Tolerance}}{6\tau_{ST,A}} = \frac{10--10}{6(5)} = .67 \qquad C_{PM,B} = \frac{\text{Tolerance}}{6\tau_{ST,B}} = \frac{10--10}{6(5)} = .67$$

As demonstrated here, C_{PM} is no better than C_{PK} at detecting the difference in performance capability, as C_{PM} also rates the spindles as being equal. In this example, neither C_{PK} nor C_{PM} identifies the difference in where the outputs are centered. In addition to either C_{PM} or C_{PK}, other measures, as well as histograms, must be considered before reaching any final conclusion concerning process capability (Mirabella). In this example, C_P for A is just .67, whereas C_P for B is 1.11. Even though their C_{PM} indexes are identical, $ppm_{TOTAL,ST}$ for A is 45,500, while only 22,750 for B. The k factor is 0 for A, and .40 for B. These supplemental capability measures point out important differences between these two spindles not revealed by either the C_{PK} or C_{PM} indexes.

The measuring scale for C_{PM} does not directly relate to any other measure of process capability. For example, what does a C_{PM} of 1.00 mean? Table 6.13 shows that it indicates the estimate of μ falls somewhere within the middle third of the tolerance. But how does this compare to a C_{PK} of 1.00? And what about $ppm_{TOTAL,ST}$ or k? As C_{PM} is always positive, one cannot tell immediately if the process average is located outside, or inside, the tolerance. In addition, this measure does not work well with unilateral specifications.

There are times when this measure of capability is not as sensitive to changes in the process average as others, such as C_{PK}. This problem is remedied by switching to a modified form of the C_{PM} measure, called C_{PMK}, which is covered in Section 6.12. The problem with requiring the target average to be the middle of the tolerance is eliminated with the C^*_{PM} measure, which is introduced next.

6.9 C^*_{PM} and P^*_{PM} Indexes for Cases Where $T \neq M$

One major assumption associated with the C_{PM} and P_{PM} indexes is that the target average equals M, the middle of the tolerance, *i.e.*, the specification limits are symmetrical about the target average. There are obviously circumstances where this is not the case, such as the gold-plating example previously covered in Example 6.25. With a slight change in its formula, the C_{PM} measure can be modified to handle these special situations (Chan *et al*). The resulting measure of capability is called C^{*}_{PM} (Spiring, 1991a) and is calculated as follows, where T replaces M as the target average, while τ^{*}_{ST} replaces τ_{ST}.

$$C^{*}_{PM} = \text{Minimum}\left(\frac{T-\text{LSL}}{3\tau^{*}_{ST}}, \frac{\text{USL}-T}{3\tau^{*}_{ST}}\right) = \frac{1}{3\tau^{*}_{ST}}\text{Minimum}\,(\,T-\text{LSL},\,\text{USL}-T\,)$$

$$\text{where}\quad \tau^{*}_{ST} = \sqrt{\sigma^{2}_{ST}+(\mu-T\,)^{2}}$$

Compare these to the original formulas for C_{PM} and τ_{ST}.

$$C_{PM} = \frac{\text{USL}-\text{LSL}}{6\tau_{ST}} \qquad \text{where}\quad \tau_{ST} = \sqrt{\sigma^{2}_{ST}+(\mu-M\,)^{2}}$$

C^{*}_{PM} and C_{PM} have different perceptions of the allowable tolerance. C_{PM} compares $6\tau_{ST}$ to the full tolerance, USL minus LSL. C^{*}_{PM} compares $3\tau^{*}_{ST}$ to the smaller "half" of the tolerance, based on where T is located. If T equals the middle of the tolerance, then τ^{*}_{ST} equals τ_{ST}, and C^{*}_{PM} equals C_{PM}.

$$\text{When}\quad T = M, \quad C^{*}_{PM} = C_{PM}$$

If T is located elsewhere, then either USL minus T or T minus LSL is less than one-half the tolerance, and C^{*}_{PM} is less than C_{PM}.

$$\text{When}\quad T \neq M, \quad C^{*}_{PM} \leq C_{PM}$$

As before, an alternative formula for C^{*}_{PM} can be derived from the modified specification limits LSL_{T} and USL_{T}. If T is less than M, choose these formulas to define the effective tolerance.

$$\text{USL}_{T} = 2T-\text{LSL} \qquad\qquad \text{LSL}_{T} = \text{LSL}$$

$$\text{Effective Tolerance} = \text{USL}_{T}-\text{LSL}_{T} = (\,2T-\text{LSL}\,)-\text{LSL} = 2(\,T-\text{LSL}\,)$$

Because T is less than M, the distance between T and the LSL is less than that between M and the LSL, causing the effective tolerance to be less than the full tolerance.

$$T-\text{LSL} < M-\text{LSL}$$

$$2(\,T-\text{LSL}\,) < 2(\,M-\text{LSL}\,)$$

$$\text{Effective Tolerance} < \text{Tolerance}$$

When T is greater than M, these formulas apply for defining the effective tolerance.

$$\text{USL}_T = \text{USL} \qquad\qquad \text{LSL}_T = 2T - \text{USL}$$

$$\text{Effective Tolerance} = \text{USL}_T - \text{LSL}_T = \text{USL} - (2T - \text{USL}) = 2(\text{USL} - T)$$

The effective tolerance is then used to calculate C_{PM}^*.

$$C_{PM}^* = \frac{\text{Effective Tolerance}}{6\tau_{ST}^*} = \frac{\text{USL}_T - \text{LSL}_T}{6\tau_{ST}^*}$$

If T is less than M, the above becomes:

$$C_{PM}^* = \frac{2(T - \text{LSL})}{6\tau_{ST}^*} = \frac{T - \text{LSL}}{3\tau_{ST}^*}$$

When T is greater than M, C_{PM}^* can be estimated with this formula:

$$C_{PM}^* = \frac{2(\text{USL} - T)}{6\tau_{ST}^*} = \frac{\text{USL} - T}{3\tau_{ST}^*}$$

In most capability studies, μ and σ_{ST} are not known and must be estimated from the measurement data collected on a control chart. Under these circumstances, C_{PM}^* is estimated with this formula:

$$\hat{C}_{PM}^* = \text{Minimum}\left(\frac{T - \text{LSL}}{3\hat{\tau}_{ST}^*}, \frac{\text{USL} - T}{3\hat{\tau}_{ST}^*}\right) = \frac{1}{3\hat{\tau}_{ST}^*}\text{Minimum}(T - \text{LSL}, \text{USL} - T)$$

$$\text{where} \quad \hat{\tau}_{ST}^* = \sqrt{\hat{\sigma}_{ST}^2 + (\hat{\mu} - T)^2}$$

If using modified specification limits, estimate C_{PM}^* with this formula:

$$\hat{C}_{PM}^* = \frac{\text{USL}_T - \text{LSL}_T}{6\hat{\tau}_{ST}^*}$$

In a similar manner, P_{PM}^* is a modification of P_{PM} that allows the target for the process average to be different than M.

$$\hat{P}_{PM}^* = \text{Minimum}\left(\frac{T - \text{LSL}}{3\hat{\tau}_{LT}^*}, \frac{\text{USL} - T}{3\hat{\tau}_{LT}^*}\right) = \frac{1}{3\hat{\tau}_{LT}^*}\text{Minimum}(T - \text{LSL}, \text{USL} - T)$$

$$\text{where} \quad \hat{\tau}_{LT}^* = \sqrt{\hat{\sigma}_{LT}^2 + (\hat{\mu} - T)^2}$$

Or, if the modified specifications are preferred, the formula for P_{PM}^* becomes like that shown on the top of the facing page.

$$\hat{P}^{*}_{PM} = \frac{USL_T - LSL_T}{6\hat{\tau}^{*}_{LT}}$$

Just as for C_{PM} and P_{PM}, there is no specific rating scale for these modified measures. Working to continuously increase these indexes is the main objective.

Goals

The maximum C^{*}_{PM} possible for a process (without reducing σ_{ST}) occurs when μ equals T. In this situation, τ^{*}_{ST} becomes equal to σ_{ST}.

$$\tau^{*}_{ST} = \sqrt{\sigma^2_{ST} + (\mu - T)^2} = \sqrt{\sigma^2_{ST} + (0)^2} = \sqrt{\sigma^2_{ST}} = \sigma_{ST}$$

For this value of τ^{*}_{ST}, C^{*}_{PM} is identical to C^{*}_{P}.

$$C^{*}_{PM} = \frac{\text{Effective Tolerance}}{6\tau^{*}_{ST}} = \frac{\text{Effective Tolerance}}{6\sigma_{ST}} = C^{*}_{P}$$

Therefore, C^{*}_{P} is the largest C^{*}_{PM} can be.

$$\text{Maximum } C^{*}_{PM} = C^{*}_{P}$$

Likewise, the maximum possible for P^{*}_{PM} is the P^{*}_{P} index for the process.

$$\text{Maximum } P^{*}_{PM} = P^{*}_{P}$$

Thus, if a goal is established for C^{*}_{P} (P^{*}_{P}), these become interim goals for C^{*}_{PM} (P^{*}_{PM}). Although, as mentioned in the discussion about C_{PM}, the real goal is reducing variation around the target average. Shifting the process average closer to the target and/or reducing the process standard deviation will cause C^{*}_{PM} and P^{*}_{PM} to increase.

Example 6.34 Hole Size

When holes are drilled in a substrate used for mounting electronic components, it is preferred to have each hole slightly oversize to prevent bent leads. With a LSL of 1.40 mm and an USL of 1.60 mm, the desired hole size (T) is 1.54 mm. If $\overline{X}$ equals 1.53 and $\hat{\sigma}_{LT}$ is .014, $\hat{\tau}^{*}_{LT}$ is estimated as follows:

$$\hat{\tau}^{*}_{LT} = \sqrt{\hat{\sigma}^2_{LT} + (\hat{\mu} - T)^2} = \sqrt{.014^2 + (1.53 - 1.54)^2} = .0172$$

P^{*}_{PM} is now estimated from $\hat{\tau}^{*}_{LT}$.

$$\hat{P}^{*}_{PM} = \frac{1}{3\hat{\tau}^{*}_{LT}} \text{Minimum } (T - LSL, USL - T)$$

$$= \frac{1}{3(.0172)} \text{Minimum } (1.54 - 1.40, 1.60 - 1.54)$$

$$= \frac{1}{.0516} \text{Minimum} (.14, .06) = 1.16$$

As with the P_{PM} index, this measure becomes a baseline for gauging future improvement efforts. To determine how much potential capability is lost due to the process average not being centered at the target of 1.54 mm, the performance measure P^*_{PM} is compared to P^*_P.

$$\hat{P}^*_P = \frac{1}{3\hat{\sigma}_{LT}} \text{Minimum} (T - LSL, USL - T)$$

$$= \frac{1}{3(.014)} \text{Minimum} (1.54 - 1.40, 1.60 - 1.54)$$

$$= \frac{1}{.042} \text{Minimum} (.14, .06) = 1.43$$

If centered properly, this process could achieve a performance capability for P^*_{PM} of 1.43 instead of the current 1.16, a difference of 19 percent. Efforts must focus on moving the process average to the target of 1.54 mm.

Relationship of C^*_{PM} to C^*_P, and P^*_{PM} to P^*_P

Just as C_{PM} and C_P are related by m_{ST}, C^*_{PM} and C^*_P are related via m^*_{ST}. This factor is derived as follows:

$$C^*_{PM} = C^*_P m^*_{ST}$$

$$\frac{1}{3\tau^*_{ST}} \text{Min} (T - LSL, USL - T) = \frac{1}{3\sigma_{ST}} \text{Min} (T - LSL, USL - T) m^*_{ST}$$

$$\frac{1}{3\tau^*_{ST}} = \frac{1}{3\sigma_{ST}} m^*_{ST}$$

$$\frac{\sigma_{ST}}{\tau^*_{ST}} = m^*_{ST}$$

In a similar manner, P^*_{PM} and P^*_P are related by m^*_{LT}.

$$P^*_{PM} = P^*_P m^*_{LT} \qquad \text{where} \quad m^*_{LT} = \frac{\sigma_{LT}}{\tau^*_{LT}}$$

For the hole size example, P^*_{PM} was estimated as 1.16. This capability measure can also be estimated from the combination of P^*_P and m^*_{LT}.

$$\hat{\tau}^*_{LT} = .0172 \qquad \hat{\sigma}_{LT} = .014 \qquad \hat{P}^*_P = 1.43$$

$$\hat{P}^*_{PM} = \hat{P}^*_P \hat{m}^*_{LT} = 1.43 \left(\frac{.014}{.0172} \right) = 1.16$$

The target ratio proposed by Phillips can also be incorporated with the C^*_{PM} and P^*_{PM} measures by setting TR^* equal to $3k^*$.

$$TR^* = 3k^* = \frac{3|\mu - T|}{\text{Minimum}(T - LSL, USL - T)}$$

$$C^*_{PM} = \frac{1}{\sqrt{C^{*2}_R + TR^{*2}}} \qquad\qquad P^*_{PM} = \frac{1}{\sqrt{P^{*2}_R + TR^{*2}}}$$

In this example, TR^* and P^*_R are estimated as .50 and .70, respectively.

$$\hat{TR}^* = \frac{3|\hat{\mu} - T|}{\text{Minimum}(T - LSL, USL - T)} = \frac{3|1.53 - 1.54|}{\text{Minimum}(1.54 - 1.40, 1.60 - 1.54)} = .50$$

$$\hat{P}^*_R = \frac{1}{\hat{P}^*_P} = \frac{1}{1.43} = .70$$

Substituting these estimates into the formula for P^*_{PM} yields the following estimate:

$$\hat{P}^*_{PM} = \frac{1}{\sqrt{\hat{P}^{*2}_R + \hat{TR}^{*2}}} = \frac{1}{\sqrt{.70^2 + .50^2}} = 1.16$$

As P^*_R is larger than TR^*, reducing process spread offers the better opportunity for improving P^*_{PM}. However, decision makers must also consider the associated cost of quality improvements. Reducing process spread typically demands a greater expenditure in resources than required for shifting the process average closer to the target.

Relationship of P_P, P^*_P, P_{PK}, P^*_{PK}, P_{PM}, and P^*_{PM} to μ

The series of diagrams in Figure 6.102 display the relationship between several measures of capability and μ, assuming σ_{LT} equals 1, T is 2, the LSL is -6, and the USL is +6. These conditions make LSL_T equal to -2 and USL_T equal to +6.

$$LSL_T = 2T - USL = 2(2) - 6 = -2 \qquad\qquad USL_T = USL = +6$$

P_P and P^*_P are measures of potential capability and do not change with μ.

$$P_P = \frac{USL - LSL}{6\sigma_{LT}} = \frac{6 - -6}{6(1)} = 2.00$$

$$P^*_P = \text{Minimum}\left(\frac{T - LSL}{3\sigma_{LT}}, \frac{USL - T}{3\sigma_{LT}}\right)$$

$$= \text{Minimum}\left(\frac{2 - -6}{3(1)}, \frac{+6 - 2}{3(1)}\right)$$

$$= \text{Minimum}(2.67, 1.33) = 1.33$$

P_{PK} and P_{PM} achieve their maximum value of 2.00 when the process is centered at 0, while P^*_{PK} and P^*_{PM} both hit their peak of 1.33 when the process is centered at 2.

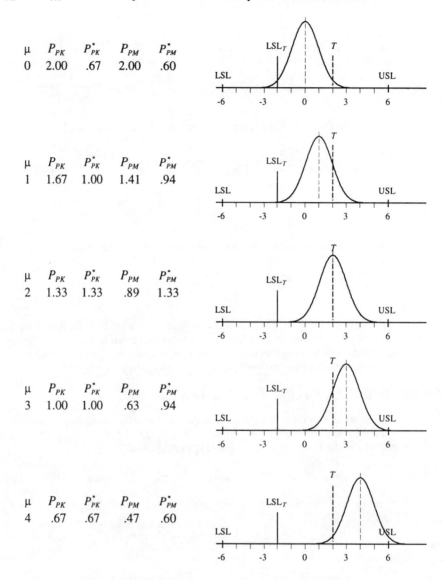

μ	P_{PK}	P^*_{PK}	P_{PM}	P^*_{PM}
0	2.00	.67	2.00	.60

μ	P_{PK}	P^*_{PK}	P_{PM}	P^*_{PM}
1	1.67	1.00	1.41	.94

μ	P_{PK}	P^*_{PK}	P_{PM}	P^*_{PM}
2	1.33	1.33	.89	1.33

μ	P_{PK}	P^*_{PK}	P_{PM}	P^*_{PM}
3	1.00	1.00	.63	.94

μ	P_{PK}	P^*_{PK}	P_{PM}	P^*_{PM}
4	.67	.67	.47	.60

Fig. 6.102 Comparison of several capability measures.

P_{PK} continues to decrease as the process average moves away from 0, and eventually becomes negative when μ is above the USL. P^*_{PK} increases as μ moves above 0 until reaching the target average of 2, where it achieves its highest value of 1.33 (which equals P^*_P). As μ continues moving beyond this target, P^*_{PK} begins decreasing and becomes identical to P_{PK}.

In a similar manner, P_{PM} decreases from its high of 2.00 (which is equal to P_P) as the process average shifts away from 0, although it always remains positive, even when μ is located above the USL. P^*_{PM} increases as μ moves above 0 until 2 is reached, at which point P^*_{PM} achieves its highest value of 1.33, which is equal to P^*_P. As μ continues above 2, P^*_{PM} begins decreasing, but does *not* become identical to P_{PM}.

P^*_{PM} is similar to the P^*_{PK} index in many ways. Both are applied when the target average for the process output is not located at the middle of the tolerance. Both attain their maximum value when μ equals T. As μ deviates from T, they both decrease, but at different rates. However, P^*_{PM} is always positive, whereas P^*_{PK} may be zero, or even negative when μ is outside a specification limit. Being calculated from τ^*_{LT}, P^*_{PM} has a nonlinear relationship with μ, which is responsible for the curved appearance of the P^*_{PM} line in Figure 6.103.

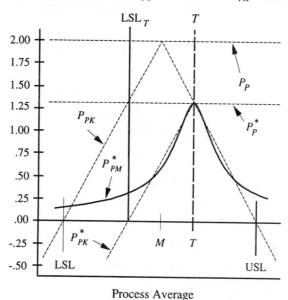

Fig. 6.103 P^*_{PM} versus changes in the process average.

P^*_{PM} and P^*_{PK} are related as follows:

$$P^*_{PK} = \frac{P^*_{PM}}{m^*_{LT}}(1 - k^*)$$

When the process average equals T, P^*_{PM} equals P^*_{PK}, which equals P^*_P.

$$\text{When } \mu = T, \quad P^*_{PM} = P^*_{PK} = P^*_P$$

If the target average is the same as the middle of the tolerance, P^*_{PM} equals P_{PM}.

$$\text{When } T = M, \quad P^*_{PM} = P_{PM} \quad \text{and} \quad P^*_{PK} = P_{PK} \quad \text{and} \quad P^*_P = P_P$$

In the special case where the process average is centered at the target, *and* the target equals M, the following relationship holds.

$$P^*_{PM} = P_{PM} = P^*_{PK} = P_{PK} = P^*_P = P_P = \frac{1}{P^*_R} = \frac{1}{P_R}$$

Similar relationships hold for C^*_{PM}.

Example 6.35 Gold Plating

Even though the LSL for the thickness of gold plating on a silicon wafer is 50 angstroms and the USL is 100 angstroms, it is decided to target the average thickness at 65 ($T = 65$). Currently, $\hat{\mu}$ is 70 angstroms and $\hat{\sigma}_{LT}$ is 2.5, as is shown in Figure 6.104.

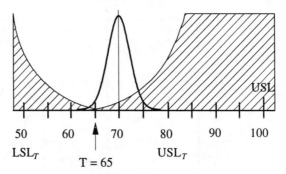

Fig. 6.104 Process centered at 70 angstroms.

From this process information, P^*_{PM} is estimated to be .89, as demonstrated below. Compare this measure of process capability to the P^*_{PK} of 1.33 calculated previously for this example.

$$\hat{\tau}^*_{LT} = \sqrt{\hat{\sigma}^2_{LT} + (\hat{\mu} - T)^2} = \sqrt{2.5^2 + (70 - 65)^2} = 5.59$$

$$\hat{P}^*_{PM} = \frac{1}{3\hat{\tau}^*_{LT}} \text{Minimum} (T - LSL, USL - T)$$

$$= \frac{1}{3(5.59)} \text{Minimum} (65 - 50, 100 - 65)$$

$$= \frac{1}{16.77} \text{Minimum} (15, 35) = .89$$

When centered at 70 angstroms, a majority of wafers are plated with a thickness in the loss area, while only a few are plated at the target value of 65, which has the minimum loss. Moving the average plating thickness lower should help minimize the total loss, and thus, improve performance capability. Assume the process is now adjusted so $\hat{\mu}$ is 65 angstroms, as is exhibited in Figure 6.105. This shift in the process average decreases the estimate of τ^*_{LT} to only 2.5.

$$\hat{\tau}^*_{LT} = \sqrt{\hat{\sigma}^2_{LT} + (\hat{\mu} - T)^2} = \sqrt{2.5^2 + (65 - 65)^2} = \sqrt{2.5^2} = 2.5$$

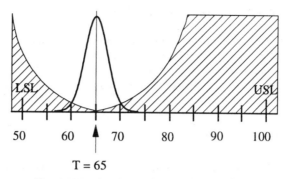

Fig. 6.105 Process centered at 65 angstroms.

This reduction in $\hat{\tau}^*_{LT}$ increases the estimate of P^*_{PM} to 2.00.

$$\hat{P}^*_{PM} = \frac{1}{3\hat{\tau}^*_{LT}} \text{Minimum} \, (T - \text{LSL}, \text{USL} - T)$$

$$= \frac{1}{3(2.5)} \text{Minimum} \, (65 - 50, 100 - 65)$$

$$= \frac{1}{7.5} \text{Minimum} \, (15, 35) = 2.00$$

When this process is centered at 65 angstroms, the majority of wafers are plated with the desired thickness and therefore have minimum loss, which translates into a maximum P^*_{PM} value of 2.00 for the given estimate of σ_{LT}. If the average thickness is lowered, $\hat{\tau}^*_{LT}$ increases, causing P^*_{PM} to decrease. Thus, an operator monitoring this plating process is signaled to leave the process centered at the desired average thickness of 65 angstroms.

Remember that T is established to minimize *total* loss, which includes losses to customers as well as internal manufacturing losses. If the thinner gold plating obtained when the process average is at 65 angstroms is not best for wafer performance, the target average must be increased to the thickness that maximizes customer satisfaction.

Advantages and Disadvantages

These measures share the same advantages as C_{PM} and P_{PM}. In line with the Taguchi loss function philosophy, centering the process output at the target average is encouraged, no matter where T is located. Despite this advantage, Kotz and Johnson strongly recommend choosing this index to measure process capability only when T equals M. Assuming they have equal standard deviations, the C^*_{PM} index for a process whose average is at T plus a is identical to that for a process whose average is at T minus a. However, if T does not equal M, the percentage of nonconforming parts could be significantly different for these two processes.

Although some authors state C^*_{PM} will work with unilateral specifications (Chan *et al*, p. 170), this is not always true due to the nature of the loss function. Suppose there is only a LSL of 200 kg for tensile strength and the target average is specified as being at least 250 kg. The loss function would look like the one displayed in Figure 6.106.

As the process average is raised to 250 kg, C^*_{PM} increases until the target is reached. When the average is raised above 250 kg, which is even a more desirable condition as stronger tensile strengths are desired, τ^*_{ST} *increases* because the difference between the process average and T is increasing.

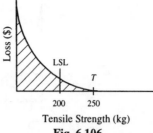

$$\tau^*_{ST} = \sqrt{\sigma^2_{ST} + (\mu - T)^2}$$

Fig. 6.106

This increase in τ^*_{ST} causes C^*_{PM} to *decrease*. In order to maintain C^*_{PM} at its maximum value, efforts are now made to *lower* the process average, returning it to T. This reaction results in reduced tensile strength and less satisfaction to the customer.

C^*_{PM} may be meaningfully applied in the special case where there is only an USL, the lower physical bound is zero, and the target average is zero. An example is the surface finish of a roller bearing where the smoothest a bearing can be is zero. The loss function associated with this characteristic is given in Figure 6.107. As the process average approaches T, C^*_{PM} increases, attaining its maximum value only when μ equals zero. Because the average surface finish cannot move below zero, there is an incentive to keep the process centered on target.

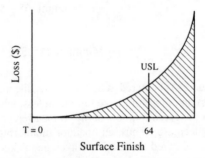

Fig. 6.107 The loss function for roller bearing surface finish.

6.10 C_{PG} and P_{PG} Ratios

Under the Taguchi loss function model, loss increases as the square of the distance between the process average and the target average. C_{PM} inversely reflects this loss: it decreases when the loss increases, and increases as the loss decreases. However, Marcucci and Beazley point out that the change in C_{PM} is not directly proportional to the change in loss: when the loss doubles, C_{PM} decreases, but not by a factor of 2 (see page 282). These authors propose a measure of capability, referred to as C_{PG}, which is equal to the inverse of the square of C_{PM}.

$$C_{PG} = \frac{1}{C^2_{PM}}$$

By expanding this formula, its relationship to the average loss becomes more apparent.

$$C_{PG} = \frac{1}{C^2_{PM}} = \frac{(6\tau_{ST})^2}{\text{Tolerance}^2} = \frac{36 \, [\, \sigma^2_{ST} + (\mu - M)^2 \,]}{\text{Tolerance}^2} = \frac{36}{\text{Tolerance}^2} [\, \sigma^2_{ST} + (\mu - M)^2 \,]$$

Recall the formula for the average loss.

$$\bar{L}_{ST} = \frac{c}{d^2}[\sigma_{ST}^2 + (\mu - M)^2] \quad \Rightarrow \quad \frac{\bar{L}d^2}{c} = \sigma_{ST}^2 + (\mu - M)^2$$

Substituting this last result into the previous equation for C_{PG} yields the following:

$$C_{PG} = \frac{36}{\text{Tolerance}^2}[\sigma_{ST}^2 + (\mu - M)^2] = \frac{36}{\text{Tolerance}^2}\frac{\bar{L}_{ST}d^2}{c} = \frac{36\,d^2}{c\,\text{Tolerance}^2}\bar{L}_{ST}$$

Due to d being equal to one-half of the tolerance (the distance from M to a specification limit), the above can be somewhat simplified (see also Quesenberry).

$$C_{PG} = \frac{36\,(1/2\,\text{Tolerance})^2}{c\,\text{Tolerance}^2}\bar{L}_{ST} = \frac{9}{c}\bar{L}_{ST}$$

Because c is the cost of a part produced at one of the specification limits, which would be a constant value, this new capability measure is directly proportional to the average short-term loss, $\bar{L}_{ST}$. When the loss doubles, C_{PG} also doubles; when the loss decreases by a third, C_{PG} decreases by a third.

Notice that *lower* values of C_{PG} indicate better capability, which is similar to the C_R ratio. Improving C_{PG} is achieved by reducing σ_{ST}, or decreasing the distance between the process average and M. Remember, any combination of changes that causes the quantity σ_{ST} plus $(\mu - M)^2$ to decrease by x percent, also causes C_{PG} to drop by an identical x percent.

If μ equals M, this measure equals the square of C_R.

$$\text{When } \mu = M, C_{PG} = \frac{1}{C_{PM}^2} = \frac{1}{C_P^2} = C_R^2$$

The formula for P_{PG} is similar, but with τ_{LT} replacing τ_{ST}, and $\bar{L}_{LT}$ replacing $\bar{L}_{ST}$.

$$P_{PG} = \frac{1}{P_{PM}^2} = \frac{(6\tau_{LT})^2}{\text{Tolerance}^2} = \frac{9}{c}\bar{L}_{LT}$$

Example 6.36 Roller Width

In Example 6.31 on page 290, rollers for copying machines were cut to the proper width. The estimate of C_{PM} for the original process was .548, with an average loss per part of $5.83. C_{PG} is the inverse of C_{PM} squared.

$$\hat{C}_{PG} = \frac{1}{\hat{C}_{PM}^2} = \frac{1}{(.548)^2} = 3.330$$

After improvements in centering the process average were made, the estimate of C_{PM} jumped to .797, a change of 45 percent. The estimate of C_{PG} for this revised process decreases by 52.7 percent to 1.574.

$$\hat{C}_{PG} = \frac{1}{\hat{C}_{PM}^2} = \frac{1}{(.797)^2} = 1.574$$

Notice how the average loss also drops by 52.7 percent, from \$5.83 to \$2.76.

$$\$5.83\,(1 - .527) = \$5.83\,(.473) = \$2.76$$

Following the introduction of thicker knives to this cutting process, the process variation decreased, causing the estimate of C_{PM} to increase to 1.370 (a change of 72 percent), with the average loss shrinking to \$ .93. These same process changes lower the estimate of C_{PG} to .533, which is a reduction of 66.2 percent.

$$\hat{C}_{PG} = \frac{1}{\hat{C}_{PM}^2} = \frac{1}{(1.370)^2} = .533 = 1.574\,(1 - .662)$$

Again, the average loss decreases by 66.2 percent, just as the estimate for C_{PG} did.

$$\$2.76\,(1 - .662) = \$2.76\,(.338) = \$\,.93$$

Goals

These are similar in concept to the goals for C_{PM} and P_{PM}, where the objective is really continuous improvement. Just remember, with C_{PG} and P_{PG}, *smaller* is better. For a given process standard deviation, the smallest C_{PM} (P_{PM}) can be is the square of C_R (P_R).

$$\text{Minimum } C_{PG} = C_R^2 \qquad \text{Minimum } P_{PG} = P_R^2$$

The next example demonstrates how this last relationship works in practice.

Example 6.37 Valve O.D. Size

A capability study is conducted for the O.D. size of an exhaust valve run on a grinding machine. M is 10.0000 cm, $\hat{\sigma}_{LT}$ is .01610 cm, and $\hat{\mu}$ is 10.0041 cm. Given a LSL of 9.9400 and an USL of 10.0600, P_{PM} is predicted to be 1.20.

$$\hat{P}_{PM} = \frac{USL - LSL}{6\hat{\tau}_{LT}} = \frac{10.0600 - 9.9400}{6\sqrt{(.01610)^2 + (10.0041 - 10.0000)^2}} = 1.20$$

P_{PG} is estimated directly from its formula as .69.

$$\hat{P}_{PG} = \frac{36}{\text{Tolerance}^2}[\,\hat{\sigma}_{LT}^2 + (\hat{\mu} - M)^2\,]$$

$$= \frac{36}{(10.0600 - 9.9400)^2}[\,(.01610)^2 + (10.0041 - 10.0000)^2\,] = .69$$

This estimate of P_{PG} is also equal to 1 over P_{PM} squared.

$$\hat{P}_{PG} = \frac{1}{\hat{P}_{PM}^2} = \frac{1}{1.20^2} = .69$$

Recall that P_{PM} can be split into separate components for precision, P_R, and accuracy, TR.

$$P_{PM} = \frac{1}{\sqrt{P_R^2 + TR^2}}$$

$$P_{PM}^2 = \frac{1}{P_R^2 + TR^2}$$

$$\frac{1}{P_{PM}^2} = P_R^2 + TR^2$$

As P_{PG} is related to P_{PM}, it can also be expressed as a function of these two components.

$$P_{PG} = \frac{1}{P_{PM}^2} = P_R^2 + TR^2$$

For this example, P_R is estimated as .805, while TR is estimated as .205.

$$\hat{P}_R = \frac{6\hat{\sigma}_{LT}}{USL - LSL} = \frac{6(.01610)}{10.0600 - 9.9400} = .805$$

$$\hat{TR} = 3\hat{k} = \frac{3|M - \hat{\mu}|}{(USL - LSL)/2} = \frac{3|10.0000 - 10.0041|}{(10.0600 - 9.9400)/2} = .205$$

These can now be substituted into the formula for P_{PG}.

$$\hat{P}_{PG} = \hat{P}_R^2 + \hat{TR}^2 = .805^2 + .205^2 = .69$$

As P_R is approximately four times larger than TR, this separation into precision and accuracy discloses that the most significant improvements in P_{PG} will come about through reductions in process spread. In fact, the smallest P_{PG} possible by just shifting the process average occurs when μ equals 10.0000 cm, the middle of the tolerance. At this location, TR is zero, and P_{PG} equals the square of P_R.

$$\hat{P}_{PG} = \hat{P}_R^2 + \hat{TR}^2 = \hat{P}_R^2 + 0^2 = \hat{P}_R^2 = .805^2 = .648$$

To reduce P_{PG} any further below .648, σ_{LT} must be decreased.

Relationship of C_{PG} and P_{PG} to Other Capability Measures

Because C_{PM} is related to C_P via the m_{ST} factor, C_{PG} can also be expressed as:

$$C_{PG} = \frac{1}{C_{PM}^2} = \frac{1}{(C_P m_{ST})^2} = \left(\frac{C_R}{m_{ST}}\right)^2$$

Since C_{PM} is related to C_{PK}, C_{PG} is also related to C_{PK}.

$$C_{PG} = \frac{1}{C_{PM}^2} = \frac{1}{[C_{PK} m_{ST}/(1-k)]^2} = \left(\frac{1-k}{C_{PK} m_{ST}}\right)^2$$

Likewise, P_{PG} is related to P_P and P_{PK}.

$$P_{PG} = \frac{1}{P_{PM}^2} = \frac{1}{(P_P m_{LT})^2} = \left(\frac{P_R}{m_{LT}}\right)^2$$

$$P_{PG} = \frac{1}{P_{PM}^2} = \frac{1}{[P_{PK} m_{LT}/(1-k)]^2} = \left(\frac{1-k}{P_{PK} m_{LT}}\right)^2$$

Advantages and Disadvantages

These are similar to those for C_{PM} and P_{PM}, with the exception that this measure is directly proportional to the average loss per part. Another advantage of this ratio is that, unlike most measures of performance capability, it is an unbiased estimator of C_{PG}. Unfortunately, this measure is not very well known and is seldom applied in practice. In addition, the loss function must be symmetrical around M, which also must be the target average for the process output. For dimensions where T is not equal to M, C_{PG} ought to be replaced with the C_{PG}^* ratio.

6.11 C_{PG}^* and P_{PG}^* Ratios for Cases Where $T \neq M$

In a modification similar to converting C_{PM} to C_{PM}^*, C_{PG} (P_{PG}) can be easily transformed to C_{PG}^* (P_{PG}^*) when T differs from M.

$$C_{PG}^* = \frac{1}{C_{PM}^{*\,2}} = \frac{(6\tau_{ST}^*)^2}{(\text{Effective Tolerance})^2}$$

τ_{ST}^* and the effective tolerance are the same as previously defined.

$$\tau_{ST}^* = \sqrt{\sigma_{ST}^2 + (\mu - T)^2} \qquad\qquad \text{Effective Tolerance} = \text{USL}_T - \text{LSL}_T$$

Thus, C_{PG}^* can also be expressed as follows:

$$C_{PG}^* = \frac{36\,[\sigma_{ST}^2 + (\mu - T)^2]}{(\text{USL}_T - \text{LSL}_T)^2}$$

P_{PG}^* has a similar formula, but with τ_{LT} and σ_{LT} replacing τ_{ST} and σ_{ST}.

$$P_{PG}^* = \frac{1}{P_{PM}^{*\,2}} = \frac{(6\tau_{LT}^*)^2}{(\text{Effective Tolerance})^2} = \frac{36\,[\sigma_{LT}^2 + (\mu - T)^2]}{(\text{USL}_T - \text{LSL}_T)^2}$$

When the process average equals T, $(\mu - T)^2$ equals 0, and C_{PG}^* equals C_R^{*2}.

$$C_{PG}^* = \frac{36\,[\sigma_{ST}^2 + (\mu - T)^2]}{(\text{USL}_T - \text{LSL}_T)^2} = \frac{36\sigma_{ST}^2}{(\text{Effective Tolerance})^2} = \left(\frac{6\sigma_{ST}}{\text{Effective Tolerance}}\right)^2 = C_R^{*2}$$

In addition to equaling C_R^{*2}, C_{PG}^* is also equal to the following when μ equals T:

$$C_{PG}^* = \frac{1}{C_{PM}^{*\,2}} = \frac{1}{C_{PK}^{*\,2}} = \frac{1}{C_P^{*2}}$$

If the process average is not equal to T, then $(\mu - T)^2$ is greater than 0, and C_{PG}^* (performance capability) is less than C_R^{*2} (potential capability).

For the unique case where T equals M, the following relationships are true.

$$C_{PG}^* = C_{PG} = \frac{1}{C_{PM}^{*\,2}} = \frac{1}{C_{PM}^{\,2}} = \frac{1}{C_{PK}^{*\,2}} = \frac{1}{C_{PK}^{\,2}} \qquad \text{while} \quad C_R^* = C_R = \frac{1}{C_P^*} = \frac{1}{C_P}$$

The next set of relationships hold in the special situation where the process average equals T, and T is the same as M.

$$C_{PG}^* = C_{PG} = \frac{1}{C_{PM}^{*\,2}} = \frac{1}{C_{PM}^{\,2}} = \frac{1}{C_{PK}^{*\,2}} = \frac{1}{C_{PK}^{\,2}} = C_R^{*2} = C_R^2 = \frac{1}{C_P^2} = \frac{1}{C_P^{*2}}$$

Example 6.38 Hole Size

When holes are drilled in a substrate used for mounting electronic components, it is preferred to have each hole slightly oversize to prevent bent leads. With a LSL of 1.40 mm and an USL of 1.60 mm, the desired hole size (T) is 1.54 mm. This makes the LSL_T equal to 1.48 mm, with the USL_T equal to 1.60 mm.

$$LSL_T = 2T - USL = 2(1.54\,\text{mm}) + 1.60\,\text{mm} = 1.48\,\text{mm}$$

$$USL_T = USL = 1.60\,\text{mm}$$

Thus, the effective tolerance is .12 mm.

$$\text{Effective Tolerance} = USL_T - LSL_T = 1.60\,\text{mm} - 1.48\,\text{mm} = .12\,\text{mm}$$

If $\hat{\mu}$ equals 1.53 mm and $\hat{\sigma}_{LT}$ is .014 mm, P_{PG}^* is equal to .74. Note how the units of measurement cancel out.

$$\hat{P}_{PG}^* = \frac{36[\hat{\sigma}_{LT}^2 + (\hat{\mu} - T)^2]}{(\text{Effective Tolerance})^2} = \frac{36[(.014\,\text{mm})^2 + (1.53\,\text{mm} - 1.54\,\text{mm})^2]}{(.12\,\text{mm})^2} = .74$$

Assuming the cost of a hole produced .06 mm away from the target average is $3.42, the average loss per hole is $ .2812.

$$\overline{L}_{LT} = \frac{c}{d^2}[\hat{\sigma}_{LT}^2 + (\hat{\mu} - T)^2] = \frac{\$3.42}{(.06\,\text{mm})^2}[(.014\,\text{mm})^2 + (1.53\,\text{mm} - 1.54\,\text{mm})^2] = \$.2812$$

This measure becomes a baseline for benchmarking future improvement efforts in terms of their impact on the total loss. For example, a switch is made to a new brand of drills, prompting $\hat{\mu}$ to increase to 1.55 mm, and $\hat{\sigma}_{LT}$ to decrease to .012 mm. Although these new drills make the process average slightly higher than the target of 1.54 mm, they shrink the standard deviation to .012. The combined effect of these changes reduce the estimate of P_{PG}^* to .61 for this revised process.

$$\hat{P}^*_{PG} = \frac{36[\hat{\sigma}^2_{LT} + (\hat{\mu} - T)^2]}{(\text{Effective Tolerance})^2} = \frac{36[.012^2 + (1.55 - 1.54)^2]}{.12^2} = .61$$

Thus, the combination of these two shifts (one "good," the other "bad"), resulted in an overall improvement in performance capability. P^*_{PG} has decreased from .74 to .61, a reduction of 17.6 percent. As expected, the average loss has also decreased by 17.6 percent.

$$\bar{L}_{LT} = \frac{c}{d^2}[\hat{\sigma}^2_{LT} + (\hat{\mu} - T)^2] = \frac{\$3.42}{.06^2}[.012^2 + (1.55 - 1.54)^2] = \$.2318$$

In the same manner done for P_{PG}, P^*_{PG} may also be split into separate components for precision (P^*_R) and accuracy (TR*).

$$P^*_{PG} = P^{*2}_R + \text{TR}^{*2}$$

For the last set of operating conditions in this example, P^*_R and TR* are estimated as .60 and .50, respectively.

$$\hat{P}^*_R = \frac{6\hat{\sigma}_{LT}}{\text{Effective Tolerance}} = \frac{6(.012)}{.12} = .60$$

$$\hat{\text{TR}}^* = 3\hat{k}^* = \frac{6|T - \hat{\mu}|}{\text{Effective Tolerance}} = \frac{6|1.54 - 1.55|}{.12} = .50$$

P^*_{PG} may be estimated from these two components, yielding a result of .61, which checks with its previous estimate.

$$\hat{P}^*_{PG} = \hat{P}^{*2}_R + \hat{\text{TR}}^{*2} = .60^2 + .50^2 = .61$$

As P^*_R is slightly larger than TR*, reducing process spread offers the better opportunity (assuming costs are similar) for increasing the performance capability of this drilling process.

Advantages and Disadvantages

These are similar to the ones for C_{PG} and P_{PG}, except that the target average need not be the middle of the tolerance. The output distribution must be normally distributed and the loss function must be symmetrical about T.

6.12 C_{PMK} and P_{PMK} Indexes

A variation of C_{PM} was proposed by Pearn *et al*, which is considered to be a "third-generation" capability index. Recall that C_P is referred to as a "first-generation" index, while C_{PK} and C_{PM} are called "second-generation." They labeled this index C_{PMK} (Choi and Owen refer to it as C_{PN}), as it is a combination of both the C_{PM} and C_{PK} measures (Pearn and Chen).

$$C_{PMK} = \text{Minimum}\left(\frac{\mu - \text{LSL}}{3\tau_{ST}}, \frac{\text{USL} - \mu}{3\tau_{ST}}\right) \qquad \text{where } \tau_{ST} = \sqrt{\sigma^2_{ST} + (\mu - M)^2}$$

As usual, there is an equivalent formula for long-term performance capability. Called P_{PMK}, it is a composite of P_{PM} and P_{PK}.

$$P_{PMK} = \text{Minimum}\left(\frac{\mu - LSL}{3\tau_{LT}}, \frac{USL - \mu}{3\tau_{LT}}\right) = \frac{1}{3\tau_{LT}}\text{Minimum}\,(\mu - LSL\,,\,USL - \mu\,)$$

$$\text{where } \tau_{LT} = \sqrt{\sigma_{LT}^2 + (\mu - M)^2}$$

Notice how these measures of customer satisfaction are "penalized" *twice* when the process average is off target. The first penalty is the reduction in effective tolerance created by taking the minimum difference between μ and either specification limit. The second occurs when the difference between μ and M becomes part of the calculation of τ. For situations where the process average equals M, τ_{ST} is equal to σ_{ST} and the following relationships hold.

$$\text{When } \mu = M, \quad C_{PMK} = C_{PM} = C_{PK} = C_P = \frac{1}{C_R}$$

If estimates of the process parameters are all that's available, these two new measures are estimated with these formulas:

$$\hat{C}_{PMK} = \text{Minimum}\left(\frac{\hat{\mu} - LSL}{3\hat{\tau}_{ST}}, \frac{USL - \hat{\mu}}{3\hat{\tau}_{ST}}\right) \quad \text{where } \hat{\tau}_{ST} = \sqrt{\hat{\sigma}_{ST}^2 + (\hat{\mu} - M)^2}$$

$$\hat{P}_{PMK} = \text{Minimum}\left(\frac{\hat{\mu} - LSL}{3\hat{\tau}_{LT}}, \frac{USL - \hat{\mu}}{3\hat{\tau}_{LT}}\right) \quad \text{where } \hat{\tau}_{LT} = \sqrt{\hat{\sigma}_{LT}^2 + (\hat{\mu} - M)^2}$$

An example applying these formulas follows.

Example 6.39 Valve O.D. Size

An estimate of potential capability ($\hat{P}_P = 1.24$) was made for the O.D. size of an exhaust valve run on a grinding machine. M is 10.0000 cm, $\hat{\sigma}_{LT}$ equals .01610 cm, LSL is 9.9400 cm, USL is 10.0600 cm, and $\hat{\mu}$ equals 10.0041 cm. P_{PK} is estimated from this information as follows:

$$\hat{P}_{PK} = \text{Minimum}\left(\frac{\hat{\mu} - LSL}{3\hat{\sigma}_{LT}}, \frac{USL - \hat{\mu}}{3\hat{\sigma}_{LT}}\right)$$

$$= \text{Minimum}\left(\frac{10.0041 - 9.9400}{3(.01610)}, \frac{10.0600 - 10.0041}{3(.01610)}\right) = 1.157$$

τ_{LT} and P_{PM} are estimated as:

$$\hat{\tau}_{LT} = \sqrt{\hat{\sigma}_{LT}^2 + (\hat{\mu} - M)^2} = \sqrt{(.01610)^2 + (10.0041 - 10.0000)^2} = .01661$$

$$\hat{P}_{PM} = \frac{USL - LSL}{6\hat{\tau}_{LT}} = \frac{10.0600 - 9.9400}{6(.01661)} = 1.204$$

P_{PMK} may also be estimated for this grinding operation. Notice how it and P_{PM} are calculated from the same estimate of τ_{LT}.

$$\hat{P}_{PMK} = \text{Minimum} \left(\frac{\hat{\mu} - \text{LSL}}{3\hat{\tau}_{LT}}, \frac{\text{USL} - \hat{\mu}}{3\hat{\tau}_{LT}} \right)$$

$$= \text{Minimum} \left(\frac{10.0041 - 9.9400}{3(.01661)}, \frac{10.0600 - 10.0041}{3(.01661)} \right) = 1.122$$

P_{PMK} turns out to be the lowest of these three estimates of performance capability. In addition, slight changes in the process average have a more profound effect on this measure than the other two. For instance, assume the average is shifted from 10.0041 to 10.0010 cm, with the estimate of σ_{LT} remaining at .01610 cm. P_{PK} increases from 1.157 to 1.222, a change of 5.5 percent.

$$\hat{P}_{PK} = \text{Minimum} \left(\frac{\hat{\mu} - \text{LSL}}{3\hat{\sigma}_{LT}}, \frac{\text{USL} - \hat{\mu}}{3\hat{\sigma}_{LT}} \right)$$

$$= \text{Minimum} \left(\frac{10.0010 - 9.9400}{3(.01610)}, \frac{10.0600 - 10.0010}{3(.01610)} \right) = 1.222$$

This shift in average reduces τ_{LT} from .01661 to .01613.

$$\hat{\tau}_{LT} = \sqrt{\hat{\sigma}_{LT}^2 + (\hat{\mu} - M)^2} = \sqrt{(.01610)^2 + (10.0010 - 10.0000)^2} = .01613$$

The reduction in τ_{LT} increases P_{PM} from 1.204 to 1.240, an increase of 3.0 percent.

$$\hat{P}_{PM} = \frac{\text{USL} - \text{LSL}}{6\hat{\tau}_{LT}} = \frac{10.0600 - 9.9400}{6(.01613)} = 1.240$$

By lowering μ to 10.0010 cm, P_{PMK} increases by 8.6 percent.

$$\hat{P}_{PMK} = \text{Minimum} \left(\frac{\hat{\mu} - \text{LSL}}{3\hat{\tau}_{LT}}, \frac{\text{USL} - \hat{\mu}}{3\hat{\tau}_{LT}} \right)$$

$$= \text{Minimum} \left(\frac{10.0010 - 9.9400}{3(.01613)}, \frac{10.0600 - 10.0010}{3(.01613)} \right) = 1.219$$

This improvement in process centering translates into a 5.5 percent rise for P_{PK}, a 3.0 percent enlargement in P_{PM}, and over an 8.6 percent increase for P_{PMK}. Being the most sensitive to shifts in the process average, P_{PMK} will always register the greatest percentage change.

Relationships to C_p and P_p

Just as C_{PK} and C_{PM} are related to C_P, C_{PMK} is also related to C_P.

$$C_{PMK} = C_P m_{ST} (1 - k)$$

$$\text{where } m_{ST} = \frac{\sigma_{ST}}{\sqrt{\sigma_{ST}^2 + (\mu - M)^2}} = \frac{\sigma_{ST}}{\tau_{ST}} \text{ and } k = \frac{|M - \mu|}{(\text{USL} - \text{LSL})/2}$$

In a similar fashion, P_{PMK} is also related to P_P.

$$P_{PMK} = P_P(m_{LT})(1-k) \quad \text{where } m_{LT} = \frac{\sigma_{LT}}{\sqrt{\sigma_{LT}^2 + (\mu - M)^2}} = \frac{\sigma_{LT}}{\tau_{LT}}$$

Because these two new performance measures receive a double penalty when μ is not centered at M, they become much more sensitive to movements in the process average than either C_{PM} or C_{PK} (P_{PM} or P_{PK}), as was pointed out in the last example.

Since k is always greater than, or equal to, 0, the quantity 1 minus k must always be equal to, or less than 1.

$$1 - k \leq 1$$

As m_{ST} is always positive, both sides of the above inequality can be multiplied by m_{ST} without changing the direction of the inequality sign.

$$m_{ST}(1-k) \leq m_{ST}$$

m_{ST} is also always less than, or equal to, 1. This means the product of $m_{ST}(1-k)$ is always less than, or equal to, 1.

$$m_{ST}(1-k) \leq m_{ST} \leq 1 \quad \Rightarrow \quad m_{ST}(1-k) \leq 1$$

So C_{PMK} (or P_{PMK}) is at best equal to C_P (or P_P), and usually quite a bit less as the process average is seldom centered at M.

$$C_P m_{ST}(1-k) \leq C_P$$

$$C_{PMK} \leq C_P$$

Because 1 minus k is never greater than 1, and C_P and m_{ST} are always positive, C_{PMK} can never exceed C_{PM}.

$$1 - k \leq 1$$

$$C_P m_{ST}(1-k) \leq C_P m_{ST}$$

$$C_{PMK} \leq C_{PM}$$

Due to m_{ST} always being less than, or equal to, 1, C_{PMK} is always less than C_{PK} when 1 minus k is nonnegative (*i.e.*, 1 minus $k \geq 0$). Recall this condition occurs when the process average is centered within the tolerance.

$$m_{ST} \leq 1$$

$$C_P m_{ST}(1-k) \leq C_P(1-k)$$

$$C_{PMK} \leq C_{PK}$$

When μ is located outside the tolerance, 1 minus k is negative (1 minus $k \leq 0$) and C_{PMK} is always *greater* than C_{PK}. Remember that multiplying an inequality by a negative number reverses the inequality sign.

$$m_{ST} \leq 1$$

$$m_{ST}(1 - k) \geq (1 - k)$$

$$C_P m_{ST}(1 - k) \geq C_P(1 - k)$$

$$C_{PMK} \geq C_{PK}$$

C_{PMK} is also related to C_{PM}, as well as C_{PK}, as is seen in these next two sets of equations.

$$C_{PMK} = C_P m_{ST}(1 - k) = C_{PM}(1 - k)$$

Because 1 minus k may be less than 0, C_{PMK} can also be negative, as is seen in Figure 6.108.

$$C_{PMK} = C_P m_{ST}(1 - k) = C_P(1 - k)m_{ST} = C_{PK}m_{ST}$$

These four indexes are also related as follows:

$$C_{PMK} = \frac{C_{PM} C_{PK}}{C_P}$$

Relationship of P_{PMK} to μ

Due to P_{PMK} being calculated with τ_{LT}, it has a nonlinear relationship with μ, as is displayed in Figure 6.108. Although similar in shape to the P_{PM} curve, P_{PMK} is always less than P_{PM}.

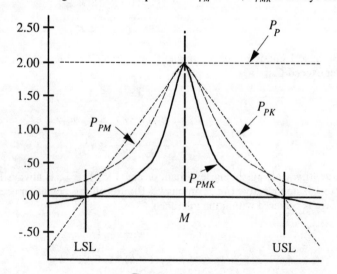

Fig. 6.108 P_{PMK} versus μ.

However, unlike P_{PM}, P_{PMK} is always less P_{PK} when the process average is within the tolerance. Like P_{PK}, the P_{PMK} index is also 0 when μ is centered at one of the specification limits. When the process average is located beyond a specification limit, P_{PMK} and P_{PK} are both negative, with P_{PMK} now always being greater (less negative) than P_{PK}.

Goals

A very common goal for C_{PMK} (P_{PMK}) is to equal to the goal set for C_P (P_P). In order for this to occur, μ must be centered at M.

$$C_{PMK,GOAL} = C_{P,GOAL} \qquad\qquad P_{PMK,GOAL} = P_{P,GOAL}$$

As with C_{PM}, the major objective is minimizing variation around the target value, either by centering the process output at T and/or reducing the process standard deviation.

Advantages and Disadvantages

C_{PMK} and P_{PMK} consider both process centering and process variation in computing process capability. They incorporate the Taguchi loss function, which vigorously encourages the centering of the process output at the target for the process average. However, the shape of this loss function isn't always known for every characteristic, nor does it always have a parabolic, or even symmetrical, shape (see pages 306 and 307). Also because of the loss function, this index doesn't always apply to characteristics with unilateral specifications.

The percentage of nonconforming parts is extremely difficult to estimate from these indexes. Although followers of the Taguchi loss function concept would argue that this is not a major consideration because more emphasis is placed on continuously reducing variation around the target value.

The measuring scale for C_{PMK} and P_{PMK} is more difficult to comprehend, for instance, what does a C_{PMK} of 1 mean? In a manner similar to the C_{PM} index, the value of C_{PMK} brackets a specific region of the tolerance wherein the process average must lie. It is shown in Section 6.15 (Derivation 3 on page 341) that the following relationship holds when the process average is within the tolerance ($C_{PMK} \geq 0$).

$$\text{When LSL} < \mu < \text{USL}, \quad |\mu - M| < \frac{1}{6C_{PMK} + 2} \text{Tolerance}$$

For example, if C_{PMK} equals 1.0, then the process average must be no farther away from M than one-eighth of the tolerance.

$$|\mu - M| < \frac{1}{6C_{PMK} + 2} \text{Tolerance} = \frac{1}{6(1) + 2} \text{Tolerance} = \frac{1}{8} \text{Tolerance}$$

Thus, it must be within the middle two-eighths, or one-fourth, of the tolerance.

$$M - \frac{1}{8} \text{Tolerance} < \mu < M + \frac{1}{8} \text{Tolerance}$$

In general, the process average must be within these two bounds for a given value of C_{PMK}.

$$M - \frac{1}{6C_{PMK} + 2} \text{Tolerance} < \mu < M + \frac{1}{6C_{PMK} + 2} \text{Tolerance}$$

Table 6.14 lists the middle segment of the tolerance wherein the process average must lie for several additional values of C_{PMK}.

Table 6.14 C_{PMK} versus location of μ within the tolerance.

C_{PMK}	Middle Portion of Tolerance
.00	$1/1 = 1.00$
.17	$2/3 = .667$
.33	$1/2 = .500$
.50	$2/5 = .400$
.67	$1/3 = .333$
1.00	$1/4 = .250$
1.33	$1/5 = .200$
1.67	$1/6 = .167$
2.00	$1/7 = .143$
2.33	$1/8 = .125$
2.67	$1/9 = .111$
3.00	$1/10 = .100$

Just as was the case for C_{PK}, when C_{PMK} is positive, the process average must be located within the tolerance. Figure 6.109 displays a graph of the values in Table 6.14 overlaid on top of those for the C_{PM} index which were listed in Table 6.13 on page 297. Observe how the two curves for C_{PMK} are always between those for C_{PM}.

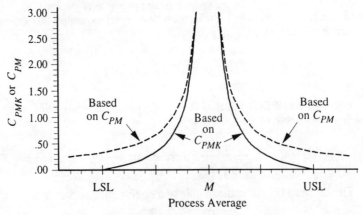

Fig. 6.109 Boundaries for the location of μ based on C_{PMK} versus those based on C_{PM}.

For example, a C_{PM} of 1.00 means μ must be located within the middle third of the tolerance, whereas a C_{PMK} of 1.00 means μ must be within the middle fourth of the tolerance. The C_{PM} index must be as large as 1.33 in order to know for sure that μ is within the middle fifth of the tolerance.

Example 6.40 Floppy Disks

Customer complaints are traced to variation in the diameter of the inner hole of a floppy disk, with the following information collected during the course of a capability study.

$$\hat{\mu} = 40.9 \quad \hat{\sigma}_{ST} = .898 \quad \hat{\sigma}_{LT} = .916 \quad \text{LSL} = 39.5 \quad \text{USL} = 44.5 \quad \text{Tol.} = 5.0 \quad M = 42.0$$

$$\hat{C}_P = .928 \quad \hat{C}_{PK} = .520 \quad \hat{C}_{PM} = .587 \quad \hat{P}_P = .909 \quad \hat{P}_{PK} = .509 \quad \hat{P}_{PM} = .582$$

τ_{ST} and C_{PMK} are estimated as follows (assume M is 42.0):

$$\hat{\tau}_{ST} = \sqrt{\hat{\sigma}_{ST}^2 + (\hat{\mu} - M)^2} = \sqrt{.898^2 + (40.9 - 42.0)^2} = 1.420$$

$$\hat{C}_{PMK} = \text{Minimum}\left(\frac{\hat{\mu} - \text{LSL}}{3\hat{\tau}_{ST}}, \frac{\text{USL} - \hat{\mu}}{3\hat{\tau}_{ST}}\right)$$

$$= \text{Minimum}\left(\frac{40.9 - 39.5}{3(1.420)}, \frac{44.5 - 40.9}{3(1.420)}\right)$$

$$= \text{Minimum}(.329, .845) = .329$$

Relating this result to the maximum possible of .928 (the C_P estimate) for the given short-term standard deviation shows there is room for much improvement in performance capability. The C_{PMK} value of .329 compares to an estimated C_{PK} of .520 and a C_{PM} estimated as .587. As noted before, C_{PMK} will always be the lowest of these three measures of performance capability when μ is centered within the tolerance. A C_{PMK} of .329 means μ must be within the middle 50.4 percent (2 × .252) of the tolerance.

$$|\mu - M| < \frac{1}{6C_{PMK} + 2}\text{Tolerance} = \frac{1}{6(.329) + 2}\text{Tolerance} = .252\,\text{Tolerance}$$

Because M is 42.0 and the tolerance is 5.0, the above result means μ must fall between 40.74 and 43.26, which, at 40.9, it does.

$$M - .252\,\text{Tolerance} < \mu < M + .252\,\text{Tolerance}$$

$$42.0 - .252(5.0) < \mu < 42.0 + .252(5.0)$$

$$40.74 < \mu < 43.26$$

A C_{PM} value of .587 means μ must be between 40.58 and 43.42 (the middle 56.8 percent of the tolerance), a slightly wider interval than that generated by the C_{PMK} index.

$$M - \frac{1}{6C_{PM}}\text{Tolerance} < \mu < M + \frac{1}{6C_{PM}}\text{Tolerance}$$

$$42.0 - \frac{1}{6(.587)}5.0 < \mu < 42.0 + \frac{1}{6(.587)}5.0$$

$$40.58 < \mu < 43.42$$

The only information revealed by a C_{PK} value of .520 about process centering is that, because it is positive, μ is somewhere between the LSL and USL.

$$39.50 < \mu < 44.50$$

When a measure of long-term performance capability is desired, τ_{LT} and P_{PMK} are estimated in a similar manner to that done for τ_{ST} and C_{PMK}.

$$\hat{\tau}_{LT} = \sqrt{\hat{\sigma}_{LT}^2 + (\hat{\mu} - M)^2} = \sqrt{.916^2 + (40.9 - 42.0)^2} = 1.431$$

$$\hat{P}_{PMK} = \text{Minimum} \left(\frac{\hat{\mu} - \text{LSL}}{3\hat{\tau}_{LT}}, \frac{\text{USL} - \hat{\mu}}{3\hat{\tau}_{LT}} \right)$$

$$= \text{Minimum} \left(\frac{40.9 - 39.5}{3(1.431)}, \frac{44.5 - 40.9}{3(1.431)} \right)$$

$$= \text{Minimum} (.326, .839) = .326$$

A comparison of the P_{PMK} estimate of .326 to the P_P estimate of .909 for potential capability indicates the process average is centered some distance from M. Moving the process average closer to 42.0 will greatly increase this measure of performance capability. This P_{PMK} estimate of .326 is quite a bit less than the estimated P_{PM} value of .582 due to the double penalty for poor centering.

$$\hat{m}_{LT} = \frac{\hat{\sigma}_{LT}}{\sqrt{\hat{\sigma}_{LT}^2 + (\hat{\mu} - M)^2}} = \frac{\hat{\sigma}_{LT}}{\hat{\tau}_{LT}} = \frac{.916}{1.431} = .64$$

The potential capability measured by P_P is reduced by this m_{LT} factor to determine P_{PM}.

$$\hat{P}_{PM} = \hat{P}_P \hat{m}_{LT} = .909(.64) = .582$$

However, P_P is reduced by *both* m_{LT} and k when figuring P_{PMK}.

$$\hat{k} = \frac{|\hat{\mu} - M|}{(\text{USL} - \text{LSL})/2} = \frac{|40.9 - 42.0|}{(44.5 - 39.5)/2} = \frac{1.1}{2.5} = .44$$

Only 64 percent of the long-term potential capability is actually available for performance capability as measured by P_{PMK} due to the m factor. Of this remainder, only 56 percent (1 minus $k = .56$) is available for P_{PMK} as a result of the k factor. The combination of these two penalties is responsible for a total loss in potential capability of 64.16 percent (1 minus .3584).

$$\hat{P}_{PMK} = \hat{P}_P \hat{m}_{LT} (1 - \hat{k}) = .909(.64)(1 - .44) = .909(.3584) = .326$$

Due to this double penalty, P_{PMK} emphasizes proper centering even more than P_{PM}. P_{PMK} also increases at a greater rate than either P_{PM} or P_{PK} when μ is moved closer to M. Suppose the process average is shifted to 41.5 with no change in the estimate of σ_{LT}. Estimates of m_{LT} and k would change as computed here:

$$\hat{\tau}_{LT} = \sqrt{\hat{\sigma}_{LT}^2 + (\hat{\mu} - M)^2} = \sqrt{.916^2 + (41.5 - 42.0)^2} = 1.0436$$

$$\hat{m}_{LT} = \frac{\hat{\sigma}_{LT}}{\hat{\tau}_{LT}} = \frac{.916}{1.0436} = .878$$

$$\hat{k} = \frac{|\hat{\mu} - M|}{(USL - LSL)/2} = \frac{|41.5 - 42.0|}{(44.5 - 39.5)/2} = \frac{.5}{2.5} = .20$$

The changes affect these three measures of performance capability as follows:

$$\hat{P}_{PK} = \hat{P}_P(1 - \hat{k}) = .909(1 - .20) = .727 \qquad \hat{P}_{PM} = \hat{P}_P \hat{m}_{LT} = .909(.878) = .798$$

$$\hat{P}_{PMK} = \hat{P}_P \hat{m}_{LT}(1 - \hat{k}) = .909(.878)(1 - .20) = .638$$

The effect of this shift in process average is summarized in Table 6.15. Although both P_{PK} and P_{PM} have substantial increases of around 40 percent, P_{PMK} more than doubles, going from .326 to .683.

Table 6.15 Increases in capability when μ is centered at *M*.

Capability Index	μ = 40.9	μ = 41.5	Percentage Increase
P_P	.909	.909	0
P_{PK}	.509	.727	43
P_{PM}	.582	.798	37
P_{PMK}	.326	.683	109

A graph displaying the relationship between these four measures is given in Figure 6.110. Note how the P_{PMK} index increases the fastest as the process average moves toward the middle of the tolerance.

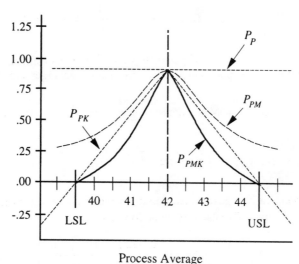

Process Average

Fig. 6.110 P_{PMK} versus μ for the floppy disk example.

6.13 C^*_{PMK} and P^*_{PMK} Indexes for Cases Where $T \neq M$

When the target for the process average (T) is not the middle of the tolerance (M), the formulas for C^*_{PMK} and P^*_{PMK} need to be modified in a manner similar to that done for C^*_{PK}. First, the specification limits need to be modified to LSL_T and USL_T. When T is less than M:

$$LSL_T = LSL \qquad\qquad USL_T = 2T - LSL$$

For cases where T is greater than M, switch to these formulas:

$$LSL_T = 2T - USL \qquad\qquad USL_T = USL$$

These special specification limits are then substituted for the standard ones in the formula for C_{PMK}, while T replaces M in the formula for τ^*_{ST}.

$$C^*_{PMK} = \text{Minimum}\left(\frac{\mu - LSL_T}{3\tau^*_{ST}}, \frac{USL_T - \mu}{3\tau^*_{ST}}\right) \qquad \text{where } \tau^*_{ST} = \sqrt{\sigma^2_{ST} + (\mu - T)^2}$$

Note that when T equals M, LSL_T equals LSL, USL_T equals USL, and the following relationships are true.

$$\text{When } T = M, \quad C^*_{PMK} = C_{PMK} \quad \text{and} \quad C^*_{PM} = C_{PM} \quad \text{and} \quad C^*_{PK} = C_{PK}$$

When the process average equals T, these measure are all identical:

$$\text{When } \mu = T, \quad C^*_{PMK} = C^*_{PM} = \frac{1}{\sqrt{C^*_{PG}}} = C^*_{PK} = C^*_P = \frac{1}{C^*_R}$$

And finally, when the process average equals T, and T equals M, then:

$$C^*_{PMK} = C_{PMK} = C^*_{PM} = C_{PM} = \frac{1}{\sqrt{C^*_{PG}}} = \frac{1}{\sqrt{C_{PG}}} = C^*_{PK} = C_{PK} = C^*_P = C_P = \frac{1}{C^*_R} = \frac{1}{C_R}$$

As a result of the process parameters typically being estimated from control chart data, these formulas are applied in practice (see also Jessenberger and Weihs).

$$\hat{C}^*_{PMK} = \text{Minimum}\left(\frac{\hat{\mu} - LSL_T}{3\hat{\tau}^*_{ST}}, \frac{USL_T - \hat{\mu}}{3\hat{\tau}^*_{ST}}\right) = \hat{C}^*_P \hat{m}^*_{ST}(1 - \hat{k}^*)$$

$$\text{where } \hat{\tau}^*_{ST} = \sqrt{\hat{\sigma}^2_{ST} + (\hat{\mu} - T)^2}, \quad \hat{m}^*_{ST} = \frac{\hat{\sigma}_{ST}}{\hat{\tau}^*_{ST}}, \text{and } \hat{k}^* = \frac{2|T - \hat{\mu}|}{USL_T - LSL_T}$$

The identical changes are incorporated into the P_{PMK} formula to transform it into the formula for P^*_{PMK}. Again, T replaces M in the formula for τ^*_{LT}.

$$P^*_{PMK} = \text{Minimum}\left(\frac{\mu - LSL_T}{3\tau^*_{LT}}, \frac{USL_T - \mu}{3\tau^*_{LT}}\right) = P^*_P m^*_{LT}(1 - k^*)$$

$$\text{where} \quad \tau_{LT}^* = \sqrt{\sigma_{LT}^2 + (\mu - T)^2} \quad \text{and} \quad m_{LT}^* = \frac{\sigma_{LT}}{\tau_{LT}^*}$$

Example 6.41 Gold Plating

The LSL for the thickness of gold plating on a silicon wafer is 50 angstroms, while the USL is 100 angstroms. It is decided to target the average thickness at 65. If $\hat{\sigma}_{LT}$ is 2.5 and $\hat{\mu}$ equals 70 as is shown in Figure 6.111, estimates for P_{PK}^*, P_{PM}^*, and P_{PMK}^* are found by first estimating τ_{LT}^*.

$$\hat{\tau}_{LT}^* = \sqrt{\hat{\sigma}_{LT}^2 + (\hat{\mu} - T)^2} = \sqrt{2.5^2 + (70 - 65)^2} = 5.59$$

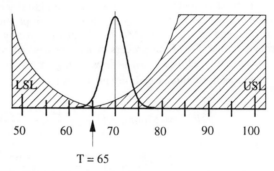

Fig. 6.111 Process is centered at 70 angstroms, whereas the target is 65.

Next, LSL_T and USL_T need to be determined. As T (65) is less than M (75), these two values become:

$$\text{LSL}_T = \text{LSL} = 50 \qquad \text{USL}_T = 2T - \text{LSL} = 2(65) - 50 = 80$$

Now these three indexes of performance capability may be estimated, beginning with P_{PK}^*.

$$\hat{P}_{PK}^* = \text{Minimum}\left(\frac{\hat{\mu} - \text{LSL}_T}{3\hat{\sigma}_{LT}}, \frac{\text{USL}_T - \hat{\mu}}{3\hat{\sigma}_{LT}}\right)$$

$$= \text{Minimum}\left(\frac{70 - 50}{3(2.5)}, \frac{80 - 70}{3(2.5)}\right)$$

$$= \text{Minimum}(2.67, 1.33) = 1.33$$

Next to be computed is P_{PM}^*.

$$\hat{P}_{PM}^* = \frac{\text{USL}_T - \text{LSL}_T}{6\hat{\tau}_{LT}^*} = \frac{80 - 50}{6(5.59)} = .89$$

Finally, P_{PMK}^* is estimated.

$$\hat{P}^*_{PMK} = \text{Minimum}\left(\frac{\hat{\mu} - \text{LSL}_T}{3\hat{\tau}^*_{LT}}, \frac{\text{USL}_T - \hat{\mu}}{3\hat{\tau}^*_{LT}}\right)$$

$$= \text{Minimum}\left(\frac{70 - 50}{3(5.59)}, \frac{80 - 70}{3(5.59)}\right)$$

$$= \text{Minimum}\,(1.19, .60) = .60$$

Note that this last measure could have been estimated with this formula:

$$\hat{P}^*_{PMK} = \hat{P}^*_P \hat{m}^*_{LT}(1 - \hat{k}^*) = 2.00(.447)(1 - .333) = .60$$

When the process is centered at 70, it is plating a majority of parts in the loss area with only a few at the target value. P^*_{PMK} is most sensitive to this problem as it is only .60, compared to a P^*_{PK} of 1.33, and a P^*_{PM} of .89. Moving the process average lower should help improve the performance capability of this process to plate the desired thickness of gold. Assume the process is successfully adjusted so $\hat{\mu}$ is now 65 angstroms, as is depicted in Figure 6.112. This shift increases these three capability measures as follows, beginning with P^*_{PK}.

$$\hat{P}^*_{PK} = \text{Minimum}\left(\frac{\hat{\mu} - \text{LSL}_T}{3\hat{\sigma}_{LT}}, \frac{\text{USL}_T - \hat{\mu}}{3\hat{\sigma}_{LT}}\right)$$

$$= \text{Minimum}\left(\frac{65 - 50}{3(2.5)}, \frac{80 - 65}{3(2.5)}\right)$$

$$= \text{Minimum}\,(2.00, 2.00) = 2.00$$

Next to be calculated are $\hat{\tau}^*_{LT}$ and P^*_{PM}:

$$\hat{\tau}^*_{LT} = \sqrt{\hat{\sigma}^2_{LT} + (\hat{\mu} - M)^2} = \sqrt{2.5^2 + (65 - 65)^2} = \sqrt{2.5^2} = 2.5$$

$$\hat{P}^*_{PM} = \frac{\text{USL}_T - \text{LSL}_T}{6\hat{\tau}^*_{LT}} = \frac{80 - 50}{6(2.5)} = 2.00$$

And finally, P^*_{PMK} is estimated:

$$\hat{P}^*_{PMK} = \text{Minimum}\left(\frac{\hat{\mu} - \text{LSL}_T}{3\hat{\tau}^*_{LT}}, \frac{\text{USL}_T - \hat{\mu}}{3\hat{\tau}^*_{LT}}\right)$$

$$= \text{Minimum}\left(\frac{65 - 50}{3(2.5)}, \frac{80 - 65}{3(2.5)}\right)$$

$$= \text{Minimum}\,(2.00, 2.00) = 2.00$$

Notice how each of these three performance measures respond when the process average moves from 70 to 65: P^*_{PK} improves from 1.33 to 2.00, an increase of 50 percent; P^*_{PM} goes up from .89 to 2.00, a rise of 125 percent; while P^*_{PMK} undergoes the most dramatic improvement, rising from .60 to 2.00, a growth of 233 percent.

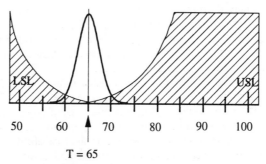

Fig. 6.112 Plating thickness centered at 65 angstroms.

When this process is centered at 65, the majority of wafers are plated at the desired target thickness and, thus, at a minimum loss. The attainment of this objective is reflected by all three measures, as they are now at their highest possible values, equal to P^*_P. Moving the process average either higher or lower will force all three to decrease, with P^*_{PMK} experiencing the greatest reduction. An operator monitoring this process with anyone of these measures is signaled to leave the process centered at the target of 65, because this is where performance capability is maximized. However, if very small deviations in μ from T are crucial, the P^*_{PMK} index should receive more consideration, as it is the most sensitive of these three for detecting shifts in the process average.

Relationships to Other Measures

Just as with the C_{PMK} index, C^*_{PMK} is related to the C^*_P, C^*_{PK}, C^*_{PM}, C^*_{PG}, and C_P indexes.

$$C^*_{PMK} = C^*_P m^*_{ST}(1-k^*) = C^*_{PK} m^*_{ST} = C^*_{PM}(1-k^*) = \frac{1-k^*}{\sqrt{C^*_{PG}}} = C_P(1-k_T)(1-k^*)m^*_{ST}$$

Likewise, P^*_{PMK} is a function of P^*_P, P^*_{PK}, P^*_{PM}, P^*_{PG}, and P_P.

$$P^*_{PMK} = P^*_P m^*_{LT}(1-k^*) = P^*_{PK} m^*_{LT} = P^*_{PM}(1-k^*) = \frac{1-k^*}{\sqrt{P^*_{PG}}} = P_P(1-k_T)(1-k^*)m^*_{LT}$$

6.14 Summary

Performance measures are more revealing indicators of actual process capability than potential measures because they consider both centering and process spread. As seen in this chapter, there are quite a few indexes available to assess performance capability. Some measures, such as Z_{MIN}, C_{PL}, C_{PU}, C_{PK}, P_{PL}, P_{PU}, P_{PK}, C_{PM}, P_{PM}, C_{PG}, P_{PG}, C_{PMK}, and P_{PMK}, assume centering the process output at the middle of the tolerance is best. Other indexes maximize their measure of performance capability when the process average is centered at some target that is different from M. These include: Z^*_{MIN}, C^*_{PK}, P^*_{PK}, C^*_{PM}, P^*_{PM}, C^*_{PG}, P^*_{PG}, C^*_{PMK}, and P^*_{PMK}. Others just reveal the process yield, ppm_{LSL}, ppm_{USL}, and ppm_{TOTAL}, regardless of where the process should be centered.

Many measures of performance capability are founded on the goalpost philosophy of quality, where making parts to print is the major concern. In contrast to these, the C_{PM}, P_{PM}, C_{PG}, P_{PG}, C^*_{PM}, P^*_{PM}, C^*_{PG}, and P^*_{PG} measures are based on the Taguchi loss function, where emphasis is placed on producing parts on target. If the loss per piece is known for parts produced at a given distance from the target, the average loss per piece may be estimated.

The C_{PMK}, P_{PMK}, C^*_{PMK}, and P^*_{PMK} indexes are a hybrid of the measures for the goalpost and Taguchi philosophies. Being penalized twice whenever μ is off target, they are extremely sensitive to changes in the process average.

To help identify if process spread or process centering is the cause of low performance capability, many of the above measures can be split into two independent components with the k, k^*, TR, or TR* factors. Whereas measures of potential capability from the previous chapter convey the effect of process spread on performance capability, these additional factors disclose the effect of process aim on performance capability.

Although this chapter explains measures for both short-term and long-term performance capability, a company should select one type or the other, and avoid reporting both kinds. Comparing the C_{PK} index of one process to the P_{PK} index of a second is not very meaningful, and will only lead to confusion. Don't attempt to calculate all performance measures for every process. Choose only the vital few which add value to your decision-making process.

These last two chapters have described in great detail how to evaluate potential and performance capability for processes generating variable-type data. The capability measures introduced provide a methodology for quantifying how well a process satisfies customer demands. They also help management make better operating decisions, inform shop floor personnel on how to schedule equipment more effectively, alert personnel to the presence of potential manufacturing problems, provide useful product information to design engineers, assist in prioritizing quality concerns, and, finally, help allocated scarce resources more efficiently. However, one major assumption for both potential and performance measures is that the process output has a normal distribution, or one very close to it. English and Taylor have published two articles investigating the adverse effect of severe non-normality on estimates of capability. The next chapter explains several techniques that can help verify the normality assumption, while Chapter 8 describes numerous alternate measures for assessing capability when this crucial assumption is not met.

6.15 Derivations

Derivation 1: The *k* Factor

The k factor is derived as follows:

$$C_{PK} = C_P(1 - k)$$

$$\text{Minimum}\left(\frac{\mu - \text{LSL}}{3\sigma_{ST}}, \frac{\text{USL} - \mu}{3\sigma_{ST}}\right) = \frac{\text{USL - LSL}}{6\sigma_{ST}}(1 - k)$$

$$\frac{1}{3\sigma_{ST}}\text{Minimum}\left(\mu - \text{LSL}, \text{USL} - \mu\right) = \frac{\text{USL - LSL}}{6\sigma_{ST}}(1 - k)$$

$$\text{Minimum}\left(\mu - \text{LSL}, \text{USL} - \mu\right) = \frac{\text{USL - LSL}}{2}(1 - k)$$

First, assume that μ is less than M. This assumption makes the quantity μ minus LSL less than the quantity USL minus μ.

$$\text{Minimum}\,(\mu - \text{LSL}\,,\text{USL} - \mu\,) = \frac{\text{USL - LSL}}{2}(1 - k)$$

$$\mu - \text{LSL} = \frac{\text{USL - LSL}}{2}(1 - k)$$

$$\frac{2(\mu - \text{LSL})}{\text{USL - LSL}} = 1 - k$$

$$k = 1 - \frac{2(\mu - \text{LSL})}{\text{USL - LSL}}$$

$$= \frac{\text{USL - LSL} - 2(\mu - \text{LSL})}{\text{USL - LSL}}$$

$$= \frac{\text{USL + LSL} - 2\mu}{\text{USL - LSL}}$$

As M is equal to $(\text{USL} + \text{LSL})/2$, the quantity $(\text{USL} + \text{LSL})$ can be replaced by $2M$.

$$k = \frac{2M - 2\mu}{\text{USL - LSL}} = \frac{2(M - \mu)}{\text{USL - LSL}} = \frac{M - \mu}{(\text{USL - LSL})/2}$$

Because μ is less than M, M minus μ is always positive and, therefore, equals $|\mu - M|$.

$$k = \frac{M - \mu}{(\text{USL - LSL})/2} = \frac{|\mu - M|}{(\text{USL - LSL})/2}$$

Now assume μ is greater than M. This makes the quantity $(\text{USL} - \mu)$ less than $(\mu - \text{LSL})$.

$$\text{Minimum}\,(\mu - \text{LSL}\,,\text{USL} - \mu\,) = \frac{\text{USL - LSL}}{2}(1 - k)$$

$$\text{USL} - \mu = \frac{\text{USL - LSL}}{2}(1 - k)$$

$$\frac{2(\text{USL} - \mu)}{\text{USL - LSL}} = 1 - k$$

The above equation can be rearranged to solve for k.

$$k = 1 - \frac{2(\text{USL} - \mu)}{\text{USL - LSL}}$$

$$= \frac{\text{USL - LSL} - 2(\text{USL} - \mu)}{\text{USL - LSL}}$$

$$= \frac{\text{-USL - LSL} + 2\mu}{\text{USL - LSL}}$$

$$= \frac{2\mu - (\text{USL + LSL})}{\text{USL - LSL}}$$

As M is equal to $(USL + LSL)/2$, $2M$ equals $(USL + LSL)$.

$$k = \frac{2\mu - 2M}{USL - LSL} = \frac{2(\mu - M)}{USL - LSL} = \frac{\mu - M}{(USL - LSL)/2}$$

Because μ is greater than M, μ minus M is always positive and, therefore, equals $|\mu - M|$.

$$k = \frac{|\mu - M|}{(USL - LSL)/2}$$

Thus, no matter if μ is greater than M or less than M, k is always equal to:

$$k = \frac{|\mu - M|}{(USL - LSL)/2}$$

Derivation 2: The Relationship Between C_{PM} and TR

The relationship between C_{PM} and TR is derived as follows:

$$C_{PM} = \frac{\text{Tolerance}}{6\tau_{ST}}$$

$$\frac{6\tau_{ST}}{\text{Tolerance}} = \frac{1}{C_{PM}}$$

$$\frac{6}{\text{Tolerance}}\sqrt{\sigma_{ST}^2 + (\mu - M)^2} = \frac{1}{C_{PM}}$$

The quantity $(6/\text{Tolerance})$ can be brought inside the radical sign as follows:

$$\sqrt{\frac{36}{\text{Tolerance}^2}}\sqrt{\sigma_{ST}^2 + (\mu - M)^2} = \frac{1}{C_{PM}}$$

$$\sqrt{\frac{36\sigma_{ST}^2}{\text{Tolerance}^2} + \frac{36(\mu - M)^2}{\text{Tolerance}^2}} = \frac{1}{C_{PM}}$$

$$\sqrt{\frac{36\sigma_{ST}^2}{\text{Tolerance}^2} + \frac{9(\mu - M)^2}{(\text{Tolerance}/2)^2}} = \frac{1}{C_{PM}}$$

$$\sqrt{\left(\frac{6\sigma_{ST}}{\text{Tolerance}}\right)^2 + \left(\frac{3(\mu - M)}{\text{Tolerance}/2}\right)^2} = \frac{1}{C_{PM}}$$

$$\sqrt{C_R^2 + (3k)^2} = \frac{1}{C_{PM}}$$

Recall that TR equals 3 times k.

$$\sqrt{C_R^2 + TR^2} = \frac{1}{C_{PM}}$$

$$C_{PM} = \frac{1}{\sqrt{C_R^2 + TR^2}}$$

Derivation 3: Maximum Deviation Between μ and *M* as a Function of C_{PMK}

The formula to determine the maximum possible deviation between the process average and *M* for a given value of C_{PMK} is derived as follows:

$$C_{PMK} = C_p \, m_{ST}(1-k) = \frac{\text{Tolerance}}{6\sigma_{ST}}\left(\frac{\sigma_{ST}}{\tau_{ST}}\right)(1-k) = \frac{\text{Tolerance}}{6\tau_{ST}}(1-k)$$

Substituting the formula for τ_{ST} into the above yields:

$$C_{PMK} = \frac{\text{Tolerance}}{6\sqrt{\sigma_{ST}^2 + (\mu - M)^2}}(1-k)$$

$$C_{PMK}\sqrt{\sigma_{ST}^2 + (\mu - M)^2} = \frac{\text{Tolerance}}{6}(1-k)$$

An assumption that σ_{ST} is greater than 0 is made for this next step, which makes these following inequalities true.

$$\sqrt{(\mu - M)^2} < \sqrt{\sigma_{ST}^2 + (\mu - M)^2}$$

$$C_{PMK}\sqrt{(\mu - M)^2} < C_{PMK}\sqrt{\sigma_{ST}^2 + (\mu - M)^2}$$

Substituting the previous equation for the right-hand side of the above inequality yields:

$$C_{PMK}\sqrt{(\mu - M)^2} < \frac{\text{Tolerance}}{6}(1-k)$$

For this next step, C_{PMK} must be positive, as division by a negative number would reverse the inequality sign. This means the final relationship applies only for situations where the process average is located between the specification limits, as this is where C_{PMK} is always positive.

$$\sqrt{(\mu - M)^2} < \frac{\text{Tolerance}}{6\,C_{PMK}}(1-k)$$

$$|\mu - M| < \frac{\text{Tolerance}}{6\,C_{PMK}}(1-k)$$

Now the expression for *k* is substituted into the above equation.

$$|\mu - M| < \frac{\text{Tolerance}}{6C_{PMK}}\left(1 - \frac{|\mu - M|}{\text{Tolerance}/2}\right)$$

$$|\mu - M| < \frac{\text{Tolerance}}{6C_{PMK}} - \frac{|\mu - M|}{3C_{PMK}}$$

$$|\mu - M| + \frac{|\mu - M|}{3C_{PMK}} < \frac{\text{Tolerance}}{6C_{PMK}}$$

$$|\mu - M|\left(1 + \frac{1}{3C_{PMK}}\right) < \frac{\text{Tolerance}}{6C_{PMK}}$$

$$|\mu - M|\left(\frac{3C_{PMK} + 1}{3C_{PMK}}\right) < \frac{\text{Tolerance}}{6C_{PMK}}$$

$$|\mu - M|(3C_{PMK} + 1) < \frac{\text{Tolerance}}{2}$$

$$|\mu - M| < \frac{1}{6C_{PMK} + 2}\text{Tolerance}$$

6.16 Blank Graphs

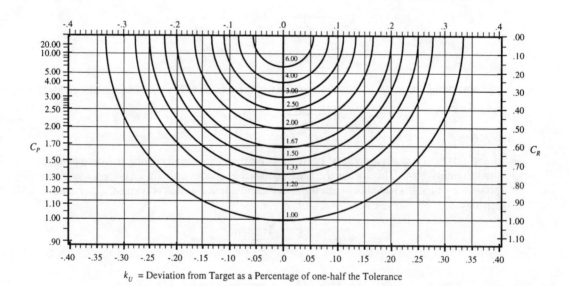

k_U = Deviation from Target as a Percentage of one-half the Tolerance

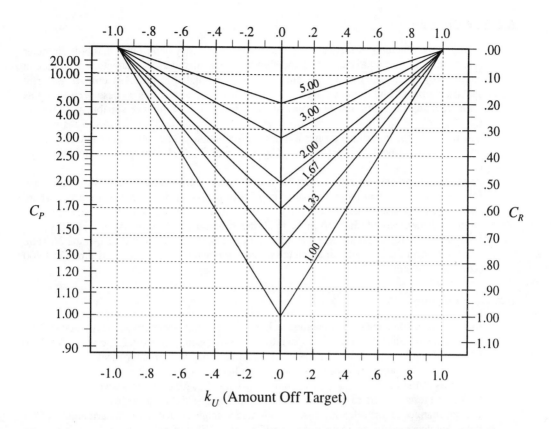

6.17 Exercises

6.1 Connecting Force

The force required to join the male and female parts of an electrical connector should be between 1.30 and 1.50 kilograms. A subgroup of three consecutive connectors are checked every hour, with the average and range of each subgroup plotted on an $\overline{X}, R$ chart. After eliminating several assignable sources of variation, the most recent 26 subgroups show no out-of-control conditions. The short-term process standard deviation is estimated as .02999 kg, while the average is estimated to be 1.41615 kg. Assuming joining force has a normal distribution, estimate the short-term performance capability with these measures: $Z_{LSL.ST}$, $Z_{USL.ST}$, $Z_{MIN.ST}$, $Z_{MAX.ST}$, C_{PL}, C_{PU}, C_{PK}, C_{PM}, C_{PG}, C_{PMK}, $ppm_{LSL.ST}$, $ppm_{USL.ST}$, and $ppm_{TOTAL.ST}$. With the current process standard deviation, what is the maximum $Z_{MIN.ST}$ possible for this process? The maximum C_{PK}? The maximum C_{PM}? The minimum $ppm_{TOTAL.ST}$? Where must the process average be located in order to achieve these values?

Compare these performance measures to the potential measures estimated in Exercise 5.1 on page 161. Estimate k and m_{ST}. How well does the output of this process meet the customer's requirement? What are your recommendations for improving performance capability?

6.2 Bolt Diameter

A major manufacturer of fasteners for the aerospace industry determines that bolt diameter is extremely important to customers and should be considered a key characteristic. Once in control, σ_{ST} is estimated as .01300 cm, while μ is estimated to be 2.191 cm. If the print specifications for bolt diameter are 2.200 cm $\pm$ .050 cm, how is this process performing as far as meeting the minimum requirement for short-term performance capability? Use the $C_{PL}, C_{PU}, C_{PK}, C_{PM}, C_{PG}, C_{PMK}, ppm_{LSL.ST}, ppm_{USL.ST}$, and $ppm_{TOTAL.ST}$ measures to answer this question (assume bolt diameter has a normal distribution). The following capability goals have been established: 1.33 for $C_{PL}, C_{PU}, C_{PK}, C_{PM}$, and C_{PMK}; 31.5 for $ppm_{LSL.ST}$ and $ppm_{USL.ST}$; 63 for $ppm_{TOTAL.ST}$.

Assuming the process average cannot be shifted, how much of a reduction in σ_{ST} is necessary for this process to meet its performance capability goals? If σ_{ST} cannot be reduced, where should the process average be centered to maximize performance capability? How much of an improvement would this shift make in all these measures of performance capability?

If the loss is $0 for a part produced at 2.200 cm, and is $1.52 for a part measuring 2.250 cm, what is the average loss per part when the process average is centered at 2.191 cm? At 2.200 cm? At 2.209 cm?

6.3 Relay Voltage

In Exercise 3.3 on page 48, the average voltage to lift a relay armature was estimated as 29.6 volts, while the short-term standard deviation (σ_{ST}) was estimated to be 2.136 volts. With an USL of 40 volts and a target average of 32 volts, does this process meet the minimum requirement for short-term performance capability for a unilateral specification (assume normality)? Estimate the $Z_{USL.ST}, C_{PU}, ppm_{USL.ST}$, and C_{PMK} measures to answer this question. Compare these answers to the results obtained in Exercise 5.3 on page 161.

Design changes to this relay increase the average voltage needed to lift the armature to 33.4 volts, but decrease the estimate of σ_{ST} to 1.975 volts. What effect do these changes have on the above estimates of performance capability?

6.4 Piston Ring Diameter

In Exercise 3.4 on page 48, the process average and long-term standard deviation were estimated as 9.02 cm and .02170 cm, respectively, for the inside diameter of a piston ring. With a LSL of 8.9250 cm and an USL of 9.0750 cm, estimate the $Z_{LSL.LT}, Z_{USL.LT}, Z_{MIN.LT}, P_{PL}, P_{PU}, P_{PK}, P_{PM}, P_{PG}, P_{PMK}, ppm_{LSL.LT}, ppm_{USL.LT}$, and $ppm_{TOTAL.LT}$ measures of long-term performance capability (assume normality). Does this process meet the minimum requirement for performance capability? Does it meet a goal of 1.33 for P_{PK}? Estimate k, k_L, k_U, and m_{LT}. Review the measures of potential capability calculated in Exercise 5.4 on page 161. What are your suggestions for improvement?

Assume a target for the process average is established at 9.0250 cm. How does this affect the assessment of performance capability? What must be done to increase performance capability? Use the $Z^*_{MIN.LT}, P^*_{PK}, P^*_{PM}, P^*_{PG}, P^*_{PMK}$, and $ppm_{TOTAL.LT}$ measures to answer this question.

Estimate k^*, m^*_{LT}, and TR^*. What is the maximum P^*_{PK} possible with the current estimate of the process standard deviation?

6.5 Shaft Length

In Exercise 3.5 on page 49, estimates of the process average and short-term standard deviation were made for the length of a water pump shaft. Assuming a normal distribution, how does this process output compare to the customer specifications of ± 3 mm? Use the C_{PK} and C_{PM} measures to answer this question. Does this process meet the minimum requirement for short-term performance capability? Does the process meet a capability goal for C_{PK} of 1.67? What percentage of shaft lengths are out of specification?

A target for the process average of minus .1 mm is specified. How does this off-center target change the determination of performance capability (use C^*_{PK} and C^*_{PM})?

Suppose the introduction of new tooling decreases the estimate of σ_{ST} by 68 percent. What effect does this have on these estimates of capability?

6.6 Rib Width

In Exercise 3.6 on page 49, μ, σ_{ST}, and σ_{LT} were estimated from an *IX & MR* chart for the width of "ribs" used in constructing wings for military aircraft. Given a specification of 1.580 $\pm$ .016 cm, estimate C_{PK}, C_{PM}, and C_{PMK}. Then estimate the long-term process spread, P_{PK}, P_{PM}, and P_{PMK}. Why the differences? Compare these results to the potential capability measures estimated in Exercise 5.6 (page 162) and comment.

Estimate the k_U factor. Then plot k_U and C_{PK} on the graph shown in Figure 6.57, and k_U and C_{PM} on the graph displayed in Figure 6.96. Compute estimates of $Z_{LSL.LT}$ and $Z_{USL.LT}$, then use these to estimate $ppm_{LSL.LT}$, $ppm_{USL.LT}$, and $ppm_{TOTAL.LT}$.

6.7 Connector Length

Electrical connectors are cut to the specified length on a shearing operation. From the length measurements given in Exercise 4.7 on page 78, the process average is estimated as 67.085, while the long-term standard deviation is estimated as .0883. With a LSL for connector length of 66.700 and an USL of 67.300, estimate P_{PK}, P_{PM}, and P_{PMK}.

Assume a target of 67.100 for connector length is desired. How does this additional customer requirement alter this capability analysis (answer in part by estimating P^*_{PK}, P^*_{PM}, and P^*_{PMK})? Estimate k^* and m^*_{LT}. Combine these two estimates with that for P^*_P from Exercise 5.7 (page 162) to again estimate P^*_{PK}, P^*_{PM}, and P^*_{PMK}. These answers should agree with those obtained at the beginning of this paragraph.

6.8 Cam Roller Waviness

Use the estimates of μ and σ_{LT} calculated in Exercise 4.8 for the waviness of cam rollers to estimate $Z_{USL.LT}$, P_{PU}, P_{PK}, and $ppm_{TOTAL.LT}$. The USL for waviness is 75 microinches. The histogram of these measurements implies the process output for waviness may not be normally distributed. What effect would this have on your estimates of performance capability?

6.9 Fan Test

Electric fans for cooling personal computers are assembled and tested. In Exercise 4.9 on page 80, oversize armature shafts were identified as a major reason why fans fail on final test. In response, a team is formed to conduct a formal capability study for the overall length of this part. After achieving process stability, average shaft length is estimated as 2.554 cm, while the short-term standard deviation is estimated as .00279 cm. With a tolerance of 2.550 $\pm$ .012 cm for shaft length, estimate C_{PK} and compare it to the goal of 1.33. In addition, predict $ppm_{LSL.ST}$, $ppm_{USL.ST}$, and $ppm_{TOTAL.ST}$.

Estimate the k factor. With the estimate of C_P derived in Exercise 5.9 on page 163, estimate C_{PK} with k. On what should this team focus its process-improvement activities, increasing precision or accuracy? Compute estimates of TR and C_R, then use these to estimate C_{PG}.

6.10 Compressive Strength

Layers of woven glass-fiber fabric are laminated together with a special type of heat-resistant resin. After curing, tests are performed to measure the compressive strength of this material, which must meet a minimum requirement of 50 ksi and has a target average of 62 ksi. The average and short-term standard deviation for compressive strength are estimated as 59.3 and 2.40, respectively. Estimate performance capability with $Z_{LSL,ST}$, C_{PL}, $ppm_{LSL,ST}$, and C_{PK}. Does the process meet these capability goals: 4.50 for $Z_{LSL,ST}$; 3.4 for $ppm_{LSL,ST}$; 1.50 for C_{PK}? What is the value of k'_L? Use the estimate of C'_{PL} computed in Exercise 5.10 on page 163 to estimate C_{PK}. How does this result compare to the estimate of C_{PL}?

The introduction of a new reformulated resin increases the average compressive strength to 63.8 ksi, but also increases the estimate of σ_{ST} to 2.76. What effect does this second resin have on the ability of this process to satisfy the customer's specification? Which resin would you recommend, assuming equal costs?

6.11 Pellet Weight

A process produces a certain size of ferrite pellets. Each pellet must weigh between a minimum of 20.5 grams and a maximum of 24.5 grams. After stability is established, the process average is estimated as 21.9 grams and the long-term standard deviation as .43 grams. With a target average of 23.1 grams, estimate the P^*_{PK}, P^*_{PM}, P^*_{PG}, and P^*_{PMK} measures (assume normality). If it is possible to move the actual process average to 22.6 grams (with no change in σ_{LT}), how do these capability measures respond? If the process average can be shifted to 23.3 grams, how do these measures react? Contrast the changes in these performance measures to the changes observed in Exercise 5.11 for the potential capability measures.

Estimate the k^* and m^*_{LT} factors. Use them with the P^*_P index from Exercise 5.11 (page 163) to estimate P^*_{PK}, P^*_{PM}, and P^*_{PMK}.

6.12 Clevis Width

Cockpit levers for controlling the landing gear and wing flaps on commercial aircraft are machined in a small Mexican job shop. During one critical operation, a clevis is cut into the connecting end of the lever, with the minimum clevis width being a key characteristic. After process stability is established via a control chart, the average width is estimated as 1.396 cm, while the long-term standard deviation is estimated as .0095 cm. The LSL for clevis width is 1.365 cm, with a target width of 1.405 cm. Estimate P'_{PL}, then P_{PL}, which by default becomes P_{PK}. Does this process meet a P_{PK} goal of 1.33? Assuming clevis width has a normal distribution, estimate $ppm_{LSL,LT}$, which is also $ppm_{TOTAL,LT}$.

Estimate k'_L, then multiply P'_{PL} by 1 minus this estimate to see if the product equals your prior estimate of P_{PL}.

The introduction of new tooling raises the estimate of μ to 1.408, but unfortunately increases the estimate of σ_{LT} to .0132. Estimate P'_{PL}, P_{PK}, and $ppm_{TOTAL,LT}$. Why is there no increase in performance capability?

6.18 References

Abraham, G., *Quality Improvement Through Statistical Methods*, Birkhauser, Boston, MA, 1998

AIAG, *Quality System Requirements: QS-9000*, 2nd edition, Chrysler Corp., Ford Motor Co., General Motors Corp., Automotive Industry Action Group, 1995, p. 30

Boyles, Russel, A., "The Taguchi Capability Index", *Journal of Quality Technology*, Vol. 23, No. 1, January 1991, pp. 17-26

Buckfelder, John, and Powell, Jon, "C_{PK} Dart Charts: An Easy Way to Track Continuous Improvement," *Quality Progress*, Vol. 26, No. 2, February 1993, p. 120

Burns, Clarence R., "SPC Training and Decision Making with OC Curves," *46th ASQC Annual Quality Congress Transactions*, Nashville, TN, May 1992, pp. 513-517

Caldwell, Andrew, "Communicating Process Improvement Measurements," *Quality*, Vol. 33, No. 3, March 1994, p. 54

Carr, Wendall E., "A New Capability Index: Parts Per Million," *Quality Progress*, Vol. 24, No. 8, August 1991, p. 152

Chan, Lai K., Cheng, Smiley W., and Spiring, Frederick A., "A New Measure of Process Capability: C_{PM}", *Journal of Quality Technology*, Vol. 20, No. 3, July 1988, pp. 162-175

Chen, K.S., "Incapability Index With Asymmetric Tolerances," *Stat. Sinica*, 8, 1998, pp. 253-262

Choi, Byoung-Chul, and Owen, Donald B., "A Study of a New Process Capability Index," *Communications in Statistics - Theory and Methods*, Vol. 19, No. 4, 1990, pp. 1231-1245

De La Rosa, Joseph, "Process vs. Specification," *Quality*, 24, 4, April 1985, pp. 57-58

Denissoff, Basile A., "War With Defects and Peace With Quality," *Quality Progress*, Vol. 26, No. 9, September 1993, pp. 97-101

Donnelly Corp., "Facing the Harsh Reality," *Ward's Auto World*, July 1997, p. 53

Dovich, Robert, "Say No to Cpk," *Cutting Tool Engineering*, April 1994, pp. 85-86

Electronic Industries Association, *EIA Interim Standard IS-32: Assessment of Quality Levels in ppm Using Variables Test Data*, 1990, Electronic Industries Association, Washington, DC

Electronic Industries Association, *EIA Bulletin QB6: Guidelines on the Use and Application of Cpk*, 1991, Electronic Industries Association, Washington, DC

English, J.R., and Taylor, G.D., "Process Capability Analysis in Departures from Normality," *Proceedings of the 1st IE Research Conference*, 1992, Chicago, IL, pp. 335-338

English, J.R., and Taylor, G.D., "Process Capability Analysis - A Robustness Study," *International Journal of Production Research*, Vol. 31, No. 7, July 1993, pp. 1621-1635

Finley, John C., "What is Capability? Or What is Cp and Cpk?," *46th ASQC Annual Quality Congress Transactions*, Nashville, TN, May 1992, pp. 186-192

Fisher, Lawrence D., "Estimating Component Quality in Parts Per Million (PPM)," *11th Annual Rocky Mountain Quality Conference Transactions*, Denver, CO, June 1987, p. 15

Fon, K. C., *Control Charts*, 1980, Chinese Society for Quality Control, Taipei, Taiwan, (in Chinese)

Gabel, Stanley H., "Process Performance Chart," *44th ASQC Annual Quality Congress Transactions*, San Francisco CA, May 1990, pp. 683-688

Gitlow, Howard, Gitlow, Shelly, Oppenheim, Alan, and Oppenheim, Rosa, 1989, *Tools and Methods for the Improvement of Quality*, Irwin, Homewood, IL, p. 455-457

Grant, Eugene L., and Leavenworth, Richard S., *Statistical Quality Control*, 6th edition, 1988, McGraw-Hill, New York, NY, p. 172

Gunter, Bert, "The Use and Abuse of Cpk," *Quality Progress*, 22, 1, Jan. 1989, pp. 72-73

Holmes, Donald, "A Quality Portfolio Management Chart," *Quality*, 25, 12, Dec. 1986, p. 67

Jessenberger, Jutta and Weihs, Claus, "A Note of the Behavior of C_{PMK} With Asymmetric Specification Limits," *Journal of Quality Technology*, 32, 4, Oct. 2000, pp. 440-443

Jessup, P., "The Value of Continuing Improvement," *Proceedings of the International Communications Conf.*, Institute of Electrical and Electronics Engineers, 1985, Chicago, IL

Johnson, Michael, "Statistics Simplified," *Quality Progress*, 25, 1, Jan. 1992, pp. 10-11

Johnson, Norman L., and Kotz, Samuel, "Flexible Process Capability Indices," 1992, *The Institute of Statistics and Department of Statistics Mimeo Series*, #2072, University of North Carolina, Chapel Hill, NC

Johnson, Thomas, "The Relationship of C_{PM} to Squared Error Loss," *Journal of Quality Technology*, Vol. 24, No. 4, October 1992, pp. 211-215

Juran, Joseph M., Gryna, Frank M., Jr., and Bingham, R. S., Jr., *Quality Control Handbook*, 3rd edition, 1979, McGraw-Hill, New York, NY, p. 9-8

Kaminsky, F.C., Dovich, R.A., Burke, R.J., "Process Capability Indices: Now and in the Future," *Quality Engineering*, Vol. 10, No. 3, 1998, pp. 445-453

Kane, Victor E., "Process Capability Indices," *Journal of Quality Technology*, Vol. 18, No. 1, January 1986, pp. 45-46

Kane, Victor E., *Defect Prevention: Use of Simple Statistical Tools*, 1989, Marcel Dekker, Inc., New York, NY

Keenan, Tim, "At What Price *ppm*?," *Ward's Auto World*, Vol. 32, No. 4, April 1996*a*, p. 55

Keenan, Tim, "Quality Questions," *Ward's Auto World*, Vol. 32, No. 8, August 1996*b*, p. 28

Kotz, Samuel and Lovelace, Cynthia R., *Process Capability Indices in Theory and Practice*, 1998, Arnold, London, U.K., pp. 101-107

Kotz, Samuel, and Johnson, Norman L., *Process Capability Indices*, 1993, Chapman & Hall, London, England, p. 91

Kotz, Samuel, and Johnson, Norman L., "Delicate Relations Among the Basic Process Capability Indices C_P, C_{PK}, and C_{PM}, and Their Modifications," *Communications in Statistics - Theory and Methods*, Vol. 26, 1999, pp. 849-861

Lawson, John S., "A Case Study of Effective Use of Statistical Experimental Design in a Smoke Stack Industry," *Journal of Quality Technology*, 20, 1, Jan. 1988, pp. 51-55

Litsikas, Mary, "Universal Instuments Reaches Out to Customers, Suppliers," *Quality*, Vol. 34, No. 12, December 1995, pp. 48-50

Littig, Steven J., and Lam, Theresa C., "Case Studies in Process Capability Measurement," *47th ASQC Annual Quality Congress Transactions*, Boston, MA, May 1993, pp. 572-574

Marcucci, Mark O., and Beazley, Charles C., "Capability Indices: Process Performance Measures," *42nd ASQC Annual Quality Congress Trans.*, Dallas, TX, 1988, pp. 516-523

Meagher, Jack, "Process Capability: Understanding the Concept," *Quality Progress*, Vol. 33, No. 1, January 2000, p. 136

Messina, William S., 1987, *Statistical Quality Control for Manufacturing Managers*, John Wiley & Sons, New York, NY

Mirabella, Jim, "Determining Which Capability Index to Use," *Quality Progress*, Vol. 24, No. 8, August 1991, pp. 8-10

Mittag, H.J., and Rinne, H., *Statistical Methods of Quality Assurance*, 1993, Chapman & Hall, London, England, p. 454

Montgomery, Douglas C., *Introduction to Statistical Quality Control*, 2nd edition, 1991, John Wiley & Sons, New York, NY, p. 373

Parlar, Mahmut and Wesolowsky, George O., "Specification Limits, Capability Ratios, and Process Centering in Assembly Manufacture," *Journal of Quality Technology*, 31, 3, July 1999, pp. 317-325

Paulk, Mark, Weber, Charles, Garcia, Suzanne, Chrissis, Marybeth, and Bush, Marilyn, *Key Practices of the Capability Maturity Model*, Version 1.1, CMU/SEI-93-TR-25, 1993, Software Engineering Institute, Carnegie Mellon University, Pittsburgh, PA, pp. L4-1

Pearn, W.L., Kotz, S., and Johnson, N. L., "Distributional and Inferential Properties of Process Capability Indices," *Journal of Quality Technology*, Vol. 24, No. 4, 1992, pp. 216-231

Pearn, W.L. and Kotz, Samuel, "Application of Clement's Method for Calculating Second- and Third-Generation Process Capability Indices for Non-normal Pearsonian Populations," *Quality Engineering*, Vol. 7, No. 1, 1994-1995, pp. 139-145

Pearn, W.L. and Chen, K.S., "New Generalization of the Process Capability Index C_{PK}," Journal of Applied Statistics, Vol. 25, 1998, pp. 801-810

Phadke, M.S., *Quality Engineering Using Robust Design*, 1989, Prentice-Hall, Englewood Cliffs, NJ, pp. 18-19

Phillips, Gary P., "Target Ratio Simplifies Capability Index System, Makes It Easy to Use Cpm," *Quality Engineering*, Vol. 7, No. 2, 1994-1995, pp. 299-313

Pillet, Maurice, Rochon, Sylvain, and Duclos, Emmanuel, "SPC - Generalization of Capability Index C_{PM}: Case of Unilateral Tolerances," *Quality Engineering*, 10, 1, 1997-1998, pp. 171-176

Pitt, Hy, *SPC for the Rest of Us: A Personal Path to Statistical Process Control*, 1994, Addison-Wesley Publishing Co., Reading, MA, p. 336

Porter, L.J., and Oakland, J.S., "Measuring Process Capability Using Indices - Some New Considerations," *Quality and Reliability Engineering International*, Vol. 6, 1990, pp. 19-27

Price, Barbara, and Price, Kelly, "Estimating Cpk Accuracy," *46th ASQC Annual Quality Congress Transactions*, Nashville, TN, May 1992, pp. 9-15

Pyzdek, Thomas, "Process Capability Analysis Using Personal Computers," *Quality Engineering*, Vol. 4, No. 3, 1992, p. 432

Quesenberry, Charles P., *SPC Methods for Quality Improvement*, 1997, John Wiley & Sons, New York, NY, pp. 496-500

Shunta, Joseph P., *Achieving World Class Manufacturing through Process Control*, 1995, Prentice Hall PTR, Englewood Cliffs, NJ, p. 41

Singhal, Subhash C., "Multi-Process Performance Analysis Chart (MPPAC) With Capability Zones," *Quality Engineering*, Vol. 4, No. 1, 1991-1992, pp. 75-81

Singhal, Subhash C., "Multi-Parameter Analysis Chart (MPAC)," *46th ASQC Annual Quality Congress Transactions*, Nashville, TN, May 1992, pp. 524-530

Skellie, Joel, and Ngo, Phung, "Use Statistical Process Monitoring Instead of Inspection," *Quality Progress*, Vol. 23, No. 6, June 1990, pp. 99-100

Smith, Gerald, *Statistical Process Control and Quality Improvement*, 1991, Macmillian Publishing Co., New York, NY, p. 111

Spiring, Fred, "The Cpm Index," *Quality Progress*, Vol. 24, No. 2, February 1991*a*, pp. 57-61

Spiring, Fred, "An Alternative to Taguchi's Loss Function," *45th ASQC Annual Quality Congress Transactions*, Milwaukee, WI, May 1991*b*, pp. 660-665

Spiring, Fred, "A Unifying Approach to Process Capability Indices," *Journal of Quality Technology*, Vol. 29, No. 1, January 1997, pp. 49-58

Spiring, Fred, Yeung, Anthony, and Leung, P.K., "The Robustness of C_{PM} to Departures from Normality," *Proceedings of the Section on Quality and Productivity*, American Statistical Association, 1997, pp. 95-100

Sullivan, Larry P., "Reducing Variability: A New Approach," *Quality Progress*, Vol. 17, No. 7, July 1984, pp. 15-21

Taguchi, Genichi, "A Tutorial on Quality Control and Assurance - The Taguchi Methods," American Statistical Association Annual Meeting, Las Vegas, NV, 1985

Taguchi, Genichi, and Wu, Yuan, *Introduction to Off-Line Quality Control*, 1980, Central Japan Quality Control Association, Nagoya, Japan

Taguchi, Genichi, and Clausing, D., "Robust Design," *Harvard Business Review*, January-February 1990, pp. 65-75

Tsui, Kowk-Leung, "Interpretation of Process Capability Indices and Some Alternatives," *Quality Engineering*, Vol. 9, No. 4, 1997, pp. 587-596

Tunner, Joseph, "Is an Out-of-Spec Product Really Out of Spec?," *Quality Progress*, Vol. 23, No. 12, December 1990, pp. 57-59

Vannman, K., "Distribution and Moments in Simplified Form for a General Class of Capability Indices," *Communications in Statistics - Theory and Methods*, Vol. 26, 1997, pp. 2049-2072

Vaucher, C. L., Balme, L. J., and Benali, A., "The ppm Myth in Board Assembly," *Quality Engineering*, Vol. 8, No. 4, 1996, pp. 615-621

Wheeler, Donald J., and Chambers, David S., *Understanding Statistical Process Control*, 2nd edition, 1992, SPC Press, Knoxville, TN

Wright, P.A., "The Cumulative Distribution of the Process Capability Index C_{PM}," *Statistics and Probability Letters*, Vol. 47, 2000, pp. 249-251

Chapter 7

Checking the Normality Assumption

A common assumption for all measures of capability presented so far has been that the process output for the characteristic under study is normally distributed. If not, Tosch warns that the standard measures of capability may be quite misleading. In theory, the normal distribution is continuous, with its upper tail going to plus infinity and the lower tail to minus infinity, as is shown in Figure 7.1.

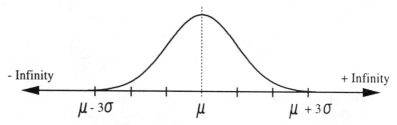

Fig. 7.1 The normal distribution.

However, a distribution such as this could *never* occur in manufacturing (Pitt, p. 153) as there are no "infinitely small" or "infinitely large" parts. Every characteristic has some real, physical boundary that cannot be exceeded, *e.g.*, it is impossible to make an infinitely long shaft, or even one that's a mile long. In addition, there is no truly continuous data in practice. Depending on the level of gage resolution, a part can be measured to only a certain number of decimal places. Every measurement is truncated at the last significant decimal place, with the remainder of the measurement dropped.

Because an actual process output can never be truly normally distributed, the normality requirement means having a process output which is very similar to a normal distribution, so that this distribution provides a reasonably good model of the actual process output (Pyzdek, 1995). But how similar must the actual output be to a normal curve, and how is this similarity quantified so a decision concerning the normality assumption can be made? To further complicate this issue, the decision must be made based on just one *sample* of data, rather than the entire process output.

Most quality practitioners rely on one of three often-used procedures for answering the normality question: histograms, normal probability paper, and goodness-of-fit tests. All three of these popular methods are discussed in this chapter, beginning with histograms. Less common techniques such as hypothesis testing for skewness and kurtosis can be found in Duncan, while Lilliefor's test for normality is covered in Iman and Conover.

351

7.1 Histograms

Histograms are a valuable tool for visually displaying a set of measurements and are quite helpful in capability studies. However, due to random sampling variation, a histogram of sample data will seldom form a "perfect" normal distribution, even if the data come from a process known to be normally distributed. This predicament is illustrated in Figure 7.2, where the solid line represents a histogram of the sample data, while the dashed line depicts a normal distribution having the same average and standard deviation as the sample data. The histogram is quite close to the normal curve for some intervals, while for others, there is a fair discrepancy between the two. How far can one interval deviate from the ideal before concluding the process output is not close to a normal distribution? Because it is so hard to tell from just this diagram, Gunter (1994) stresses that superimposing a normal distribution curve on top of a histogram is not a very reliable method for assessing the shape of the process output.

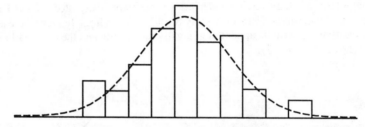

Fig. 7.2 The normal distribution curve superimposed on a histogram of sample data.

To further demonstrate the problems with this first method, a computer program was written to randomly select 75 measurements from a known normal distribution, then display a histogram of these measurements on the monitor screen for a few seconds. The program would then clear the screen, select another random sample of 75 from the same normal distribution, make a new histogram, and display it on the screen. In this fashion, many histograms of sample data taken from this normal distribution could be viewed over a short period of time.

As might be expected, there was variation among the shapes of the displayed histograms. Some appeared very close to having the "bell-shaped" curve associated with a normal distribution. However, the appearance of a few histograms were not at all like the anticipated normal curve. Figures 7.3a and 7.3b display two such histograms, each based on a different random sample of 75 measurements taken from a process output known to be normally distributed. By looking at histogram A in Figure 7.3a, one is led to believe the sample measurements making up this histogram perhaps originate from a process with a bimodal distribution. On the other hand, histogram B in Figure 7.3b implies the process output has a distribution that is skewed to the right. However, it is known that both come from the same process, which definitely has a normal distribution.

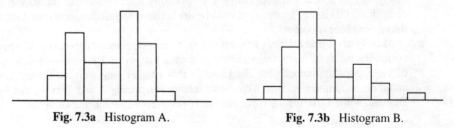

Fig. 7.3a Histogram A.   **Fig. 7.3b** Histogram B.

When the normality assumption is checked with this method in practice, only *one* sample of 75 measurements is taken and plotted as a histogram. A person unlucky enough to get either one of the above histograms would most likely *falsely* conclude the process is non-normally distributed.

In addition to these difficulties, simply changing the class interval size can significantly alter a histogram's shape and influence a decision concerning normality. The class size of the histogram displayed in Figure 7.3a is doubled, and another histogram constructed using the identical set of measurements. This new histogram (Figure 7.4) appears to substantiate the assumption of normality much better than the first, even though both histograms are derived from the same group of sample data.

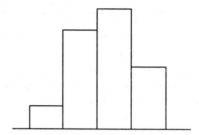

Fig. 7.4 Histogram A with interval size doubled.

As a rule, there should seldom be fewer than 6 classes, nor more than 15 (Freund and Williams). Sturge's formula given below provides a good "rule of thumb" for determining the number of classes. For a more complex statistical algorithm for determining class intervals, see Terrell and Scott.

$$\text{Number of Classes} = 1 + 3.3 \log kn$$

With k being the number of subgroups and n the subgroup size, kn becomes the total number of individual measurements being assembled into the histogram. For example, if kn is 125, the approximate number of classes should be around 8.

$$\text{Number of Classes} = 1 + 3.3 \log 125 = 7.9$$

Furthermore, simply lowering the beginning value of the first class, while keeping the same class width, could cause a considerable change in the histogram's appearance by shifting measurements from one class to another.

Example 7.1 Shaft Length

Data for constructing a histogram typically comes from a control chart displaying good stability. A process cutting the length of a tapered steel shaft (Figure 7.5) is in a rea-sonably good state of control, as was demonstrated by the chart constructed in Exercise 2.5 back on page 24.

Fig. 7.5

The last 10 subgroups ($n = 5$) are taken from the data collection portion of the chart (see Table 7.1) and used to build a histogram. With 50 (10 × 5) length measurements, Sturge's formula recommends around seven classes. In order to have each interval be a convenient width of 1.0, eight classes are actually chosen.

$$\text{Number of Classes} = 1 + 3.3 \log kn = 1 + 3.3 \log 50 = 6.61 \cong 7$$

Table 7.1 50 shaft length measurements.

x_i	Subgroup Number									
	1	2	3	4	5	6	7	8	9	10
x_1	-1	1	0	-1	-4	-1	2	-3	0	0
x_2	0	-2	0	0	1	1	0	-1	3	-2
x_3	0	-1	-3	3	-1	1	-2	0	0	1
x_4	-2	0	1	0	0	0	-2	2	-1	-1
x_5	2	1	-2	2	-1	0	0	1	0	-1

A histogram of these measurements is exhibited in Figure 7.6, which reveals these sample data to be "somewhat" normally distributed. Unfortunately, interest is not in the distribution of this particular group of *sample* measurements, but in the distribution of the *process* from which these measurements were taken.

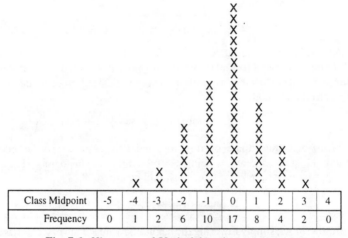

Class Midpoint	-5	-4	-3	-2	-1	0	1	2	3	4
Frequency	0	1	2	6	10	17	8	4	2	0

Fig. 7.6 Histogram of 50 shaft length measurements.

As the histogram consists of individual measurements, the specification limits of plus and minus 3 may be drawn on this diagram. 2 percent (1/50) of the sample measurements are below the LSL of -3 while 0 percent are above the USL of +3. Again, these results are just for the pieces in this particular sample of 50. What about the results of a different group of 50? Even more importantly, what about the *entire* process, which consists of thousands, if not millions, of pieces? This is really the distribution of interest for capability studies and the one that must be checked for normality as well as the percentage of nonconforming shafts.

Advantages and Disadvantages

Histograms are fairly easy to construct and present a rough picture of the process output. As they represent individual measurements, specification limits may be added and the actual percentage of out-of-specification measurements determined for that particular sample. Hopefully, all sample measurements are within specification.

When a process output is highly non-normal, a histogram of sample measurements will probably reveal it as such. For example, Figure 7.7 displays a histogram of data collected from a bimodal process output, while Figure 7.8 shows a histogram from a process with a highly skewed (asymmetric) output.

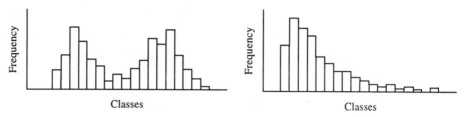

Fig. 7.7 A histogram from a
bimodal distribution.

Fig. 7.8 A histogram from a
skewed distribution.

Looking at these extreme examples, almost everyone would reject the possibility these histograms come from a process with a normal distribution. But for the vast majority of cases where the underlying distributions are not so extreme, histograms can provide only a very rough idea about the true shape of the process output.

The shape of a histogram is subject to the size and number of class intervals chosen, which is somewhat of an arbitrary decision. The selection of the lower class boundary for the first interval (again, a somewhat arbitrary choice) may also affect the histogram's appearance. Thus, two quality practitioners analyzing the same set of data could construct two dissimilar histograms and, perhaps, arrive at opposite conclusions concerning the issue of normality. Although relatively easy to apply, and useful for many other applications, a histogram is not a very dependable technique for verifying the normality assumption.

In addition, it is very difficult to obtain estimates of the process average, process standard deviation, percentage nonconforming in the process, or any measure of capability directly from the histogram. Fortunately, other procedures exist for checking the normality assumption based on sample data, with one of the best being normal probability paper.

7.2 Normal Probability Paper

A second method for checking the normality assumption involves a special type of graph paper called normal probability paper (abbreviated NOPP). By plotting the *cumulative* percentage distribution of the sample data on this paper, and properly analyzing the results, a decision can quickly be made about the likely shape of the process output distribution. The concept of probability paper is very similar to that of log paper. For example, suppose the temperature of a cup of hot coffee placed on a table is monitored over a period of time. The coffee's temperature decreases rapidly at first, but as time passes, the rate of decrease slows, as is depicted in Figure 7.9. This curved line can be turned into a straight line by simply converting the scale of the y axis to a log scale, as was done in Figure 7.10. Notice the unequal spacing of intervals for the temperature scale on this graph.

Although the information contained on both graphs is identical (the points 2, 40 and 5, 20 are on both lines), most people prefer to work with the straight line in Figure 7.10. First of all, fitting a line through the points on this second graph is certainly much easier compared to the first. As the points now lie in a fairly straight line, simply lay a straightedge on the graph and draw a line through the points. Second, it's quite simple to extend this line beyond the last plotted point to predict coffee temperature at some future time, say at ten minutes.

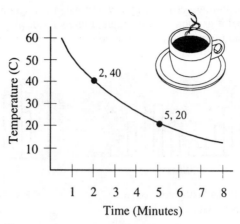

Fig. 7.9 Coffee temperature versus time, plotted on regular graph paper.

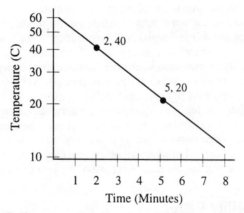

Fig. 7.10 Coffee temperature versus time, plotted on log graph paper.

Similar to log paper, the y axis of NOPP is scaled in such a way so that the cumulative percentage distribution of a process having a normal distribution will plot as a straight line, as is shown in Figure 7.11. Again, notice the unequal spacing of the intervals on the y-axis scale. In addition to explaining how this special type of graph paper can check for normality, this chapter also discloses how NOPP can estimate the process average, process standard deviation, and the percentage of nonconforming parts (or *ppm*).

But before all this can be done, the cumulative percentage distribution for the process must be estimated from the sample data. This special distribution reveals what percentage of all measurements are equal to, or less than, a particular measurement. It is calculated by accumulating the area under the distribution of individuals, moving from the lowest to highest values. For a normal distribution, the cumulative curve equals 0 percent at minus infinity. Because .135 percent of the process output lies from minus infinity to μ minus 3σ, the cumulative distribution equals .135 percent at μ minus 3σ. At the process average, it equals 50 percent, while at μ plus 3σ, it equals 99.865 percent, as is shown in Figure 7.12. At plus infinity, the cumulative percentage distribution equals 100 percent.

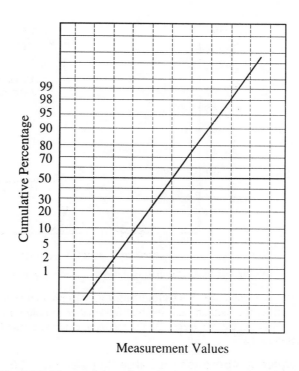

Fig. 7.11   The special *y*-axis scale of normal probability paper.

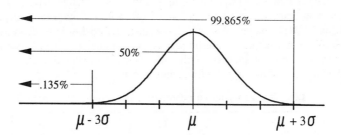

Fig. 7.12   Areas under the normal distribution.

When plotted on regular graph paper, the cumulative distribution for a normal distribution appears as the curve shown in Figure 7.13. As the area under a normal curve is accumulated from minus infinity to plus infinity, an "S-shaped" curve develops due to changes in the rate of accumulation. Initially, this rate is quite low due to the small area contained in the extreme lower tail. However, the accumulation rate accelerates rapidly as the middle of the distribution is reached, since this is where the majority of the process output lies. As the accumulation continues after the average, which is the inflection point for the S-shaped curve, the accumulation rate decreases gradually at first, then tapers off rather quickly as the relatively small area of the upper tail is being included. Eventually, all the area under the normal distribution is accumulated, and the curve levels off at 100 percent.

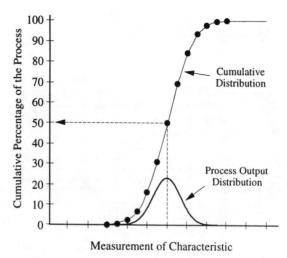

Fig. 7.13 Accumulating the area under a normal distribution.

There are two popular methods for estimating the cumulative percentage distribution of a process output: the median rank method for smaller sample sizes; the grouped frequency approach for larger sample sizes. As the second method is applied more often in practice, it is discussed first.

Grouped Frequency Approach for Large Samples ($kn \geq 50$)

A common method for estimating the cumulative percentage distribution for large samples is the grouped frequency approach given by Charbonneau and Webster. Using data from the shaft length example in the histogram section (Table 7.1 on page 354), the class midpoints of the histogram are written in column 1 of Table 7.2, with their respective frequencies in column 2. The frequency for each class is then shifted down one class, as was done in column 3 (Shift). For each class, the frequency in column 2 (Freq.) is added to that in column 3 (Shift), and their sum placed in column 4 (Sum). Note that the total of this "Sum" column is 100, twice that of the original frequency of 50.

Table 7.2 Procedure for estimating the cumulative percentage distribution.

(1) Class Midpoint	(2) Freq.	(3) Shift	(4) Sum	(5) Accum. Freq.	(6) Cum. Percentage
-4	1	0	1	1	1
-3	2	1	3	4	4
-2	6	2	8	12	12
-1	10	6	16	28	28
0	17	10	27	55	55
1	8	17	25	80	80
2	4	8	12	92	92
3	2	4	6	98	98
4	0	2	2	100	100
Total	50	50	100		

The accumulated frequency for each class is calculated and entered into column 5 (Accum. Freq.). The accumulated frequency for a given class is the frequency of that class, plus the frequencies of all prior classes. For the first class in this example (whose midpoint equals -4), the accumulated frequency is just 1, as there are no prior classes. For the second class, it is 4 (1 + 3). The 4 is derived from the 3 measurements in the second class, plus the 1 in the first class. The accumulated frequency for the third class is 12 (1 + 3 + 8).

Finally, the cumulative percentage (last column) for the process output is estimated by dividing the accumulated frequency for each class by twice the original sample size. The cumulative percentage for the first class is 1 percent (1/100). For the second, it is 4 percent (4/100). These cumulative percentages are then plotted against the class midpoints and a curve drawn through the points, as illustrated in Figure 7.14. This S-shaped curve estimates the cumulative percentage distribution of the process and is often referred to as an "ogive," as it resembles the shape of a pointed Gothic arch.

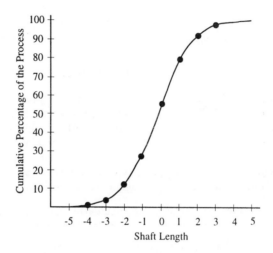

Fig. 7.14 The estimated cumulative distribution of the process output for shaft length.

For a process with a normal output distribution, drawing a horizontal line from the 50 percent mark on the *y* axis over to the cumulative curve, then down to the *x* axis, provides an estimate of the process average (see Figure 7.15 at the top of the next page). In this case, μ is estimated to be about minus .2.

Because this is a curve for the process output, the specification limits of ±3 may be drawn on the graph. Based on the cumulative percentage curve displayed in Figure 7.15, 4 percent of the *process* output is estimated to be below the LSL and 98 percent below the USL. This second percentage implies there is 2 percent above the USL (100 - 98). Notice these process estimates are significantly different than the percentage nonconforming calculated for the *sample* measurements used to construct the histogram of this same data (review Figure 7.6).

Unfortunately, it is rather difficult to fit a curved line through these points (a French curve is needed), or to extend this curve beyond the last plot point to estimate what percentage of shafts will have a length greater than 4. Working with a straight line would certainly be much easier, as will be discovered in the next section, where this cumulative percentage curve is plotted on normal probability paper.

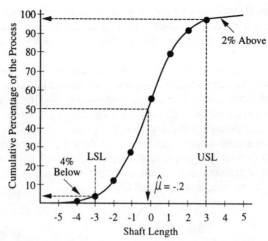

Fig. 7.15 Estimating the process average and percentage of nonconforming shafts.

Plotting the Cumulative Percentage Distribution on NOPP

There are many different varieties of NOPP, with the one shown in this book called the 6σ Capability Analysis Worksheet™, which is sold by the International Quality Institute of Sacramento, CA (916-933-2318). All types are fairly similar in construction, usage and interpretation, and are filled out by following the general instructions given next.

First, collect measurement data from a control chart and tally the frequency for each class as if a histogram were being prepared (see Table 7.3). Measurements from the shaft length example (Table 7.1 on page 354) are again used to illustrate the various steps.

Table 7.3 Frequency distribution of shaft length measurements.

Class Midpoint	Freq.	Tally
-5	0	
-4	1	X
-3	2	XX
-2	6	XXXXXX
-1	10	XXXXXXXXXX
0	17	XXXXXXXXXXXXXXXXX
1	8	XXXXXXXX
2	4	XXXX
3	2	XX
4	0	

Transcribe this data into the boxes marked "**Class Midpoints**" and "**Frequency**" located at the bottom of the 6σ Capability Analysis Worksheet™, as was done in Figure 7.16a. To help center the curve on this worksheet, put the middle class of this frequency distribution in the box above the solid triangle. There should be a minimum of 6 classes with data in order to have a good idea of how to fit a line through the plot points.

Estimated Accumulated Frequency (EAF)				0	1	4	12	28	55	80	92	98	100	
Shift and Sum														
Frequency				0	1	2	6	10	17	8	4	2	0	0
kn = Sample Size of Study Class Midpoints				-5	-4	-3	-2	-1	0	1	2	3	4	5

Fig. 7.16a Entering data on the NOPP.

Calculate the estimated accumulated frequency (EAF) for each class by following the sequence of arrows and equal signs drawn on the NOPP (Figure 7.16b). For example, the frequency of 1 for the -4 class is brought up to the line above the arrow. To calculate the EAF for the next class, follow the arrow from this value of 1 down to where it points to the frequency of 1 for the -4 class. Add this frequency of 1 to the original 1, making a total of 2. Then continue following the arrow as it points to the frequency of 2 in the -3 class. Add this frequency to the total, now making it 4 (2 + 2). Finally, write this amount in the EAF box above the -3 class. Then repeat this procedure to determine the EAF for the -2 class. Add the EAF of 4 to the frequency of 2 for the -3 class, for a subtotal of 6. Now add the frequency of 6 for the -2 class to make a final total of 12, and write this result in the EAF box for the -2 class.

% Plot Points = (EAF / 2kn) X 100				-	1	4	12	28	55	80	92	98	-	
Estimated Accumulated Frequency (EAF)				0	1	4	12	28	55	80	92	98	100	
Shift and Sum														
Frequency				0	1	2	6	10	17	8	4	2	0	0
kn = Sample Size of Study Class Midpoints				-5	-4	-3	-2	-1	0	1	2	3	4	5

Fig. 7.16b Calculating the EAFs.

Beginning with this total of 12, add the frequency of 6 for the -2 class, then the frequency of 10 for the -1 class to arrive at a total of 28. This becomes the EAF for the -1 class. Continue these calculations through to the +4 class. To check your math, the last EAF should equal **twice** the number of original measurements, which in this example is 100 (2 × 50). This sequence combines the calculations involved with shifting the frequencies down one category, summing the two columns and then accumulating the frequencies, as done back in Table 7.2 on page 358. This can be verified by comparing the results listed in Figure 7.16b with those for the original ogive calculations displayed in Table 7.2.

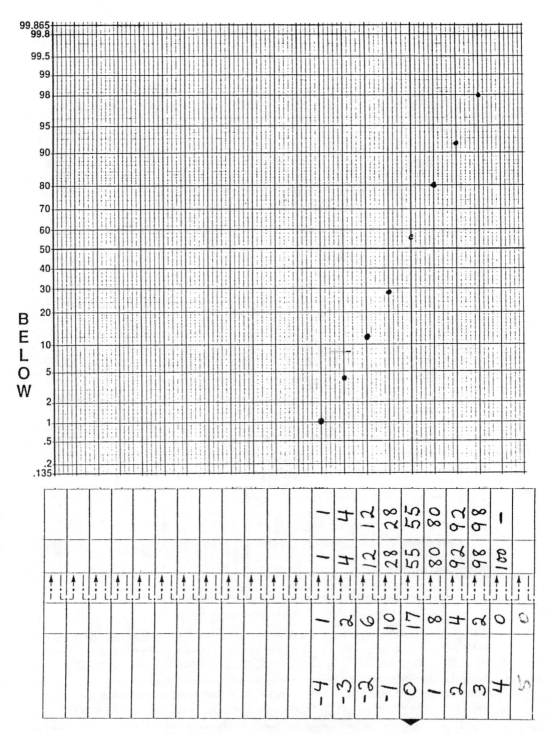

Fig. 7.17   Plotting the cumulative plot points on NOPP.

Compute the NOPP plot points by dividing the EAF for each class by two times the number of measurements. In this example it is fairly easy because each EAF is divided by 100 (2 × 50). This duplicates the calculations done in Table 7.2 to estimate the cumulative percentage for each class and produces identical results. This cumulative percentage becomes the plot point for its class midpoint on the NOPP. For example, the cumulative percentage of 1 percent will be plotted against the class midpoint of -4. Some authors, like Bissell (p. 246), recommend plotting this point at the upper class boundary as this is a cumulative curve, while Kimball offers several additional strategies. Whichever method is adopted, one must be consistent in how the points are plotted and interpreted.

Locate and plot the remaining cumulative percentage plot points on the NOPP above their respective class midpoints, as is displayed in Figure 7.17. Note that the 0 and 100 percent points are *not* plotted. Theoretically, 0 percent would be accumulated at minus infinity, while 100 percent is accumulated at plus infinity.

In order to present more detail, Figure 7.17 shows the graph from only .135 percent (-3σ) to 99.865 percent ($+3\sigma$). This particular brand of NOPP actually extends from .00000012 percent (-6σ) to 99.99999988 percent ($+6\sigma$) and is 17 inches by 22 inches (43 cm by 56 cm) in size. Although less intuitive for most users, some probability papers reverse the axes, with the cumulative percentage on the x axis and the characteristic measurements on the y axis.

After all points are plotted, draw a line of best fit through the points, as is illustrated in Figure 7.18. Try to get approximately half the points on either side of the fitted line (a transparent straightedge often helps with this task). Give more weight to the points between 10 percent and 90 percent when drawing the line as the majority of data is located here. Avoid fitting the line by considering only the highest and lowest percentage plot points because these two are determined by the least amount of data. When a line is finally drawn, count the number of points falling above the line, as well as the number of points falling below (ignore points falling directly on the line). If the difference between these counts exceeds two, the line is not a good fit and should be redrawn (King, 1988). A second test for fit can be done by drawing a line from the lowest plot point to the highest. If all plot points in between fall on one side of this line, the process output is most likely not normally distributed. A later step describes several likely shapes of process output distributions when the best-fit line is not a straight one.

NOPP is designed so that if a straight line fits these plot points fairly well, the output of this process can be assumed to follow a normal distribution, or something very close to it. As this appears to be true for this shaft length example, the assumption of normality is verified, and the standard measures may be applied to assess capability.

To estimate the percentage of shafts with lengths outside the print specifications, draw a vertical line from each specification limit (LSL is -3, USL is +3) up to the fitted line. At the intersection involving the LSL line, draw a horizontal line to the y axis on the **left**-hand side of the worksheet and read off the percentage below the LSL. Figure 7.18 shows this to be approximately 4 percent.

The percentage below the USL is read from the y axis in the same manner as for the LSL. This percentage must then be subtracted from 100 to derive the percentage of shaft lengths above the USL. An alternative method is to read the percentage from the y axis located on the **right**-hand side of the NOPP, as is illustrated in Figure 7.19. The scale on this side is 100 percent minus the corresponding percentage on the left-hand side, thus representing the percentage of shafts having a length measurement *above* a given x value.

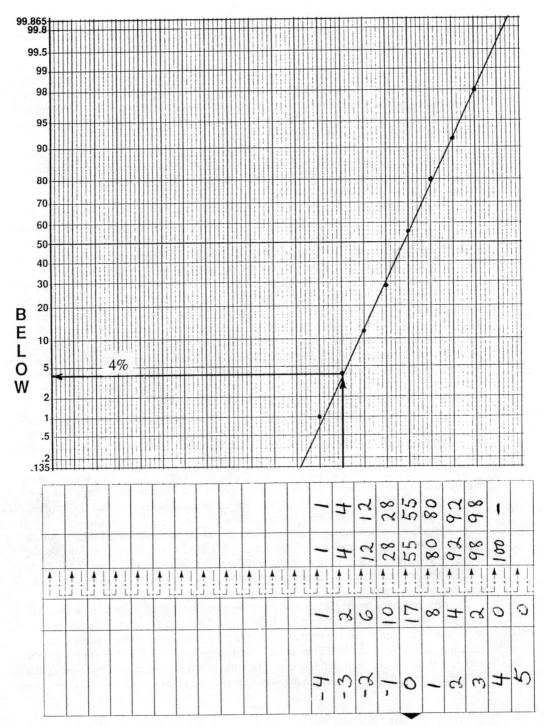

Fig. 7.18   Fitting a line through the plot points.

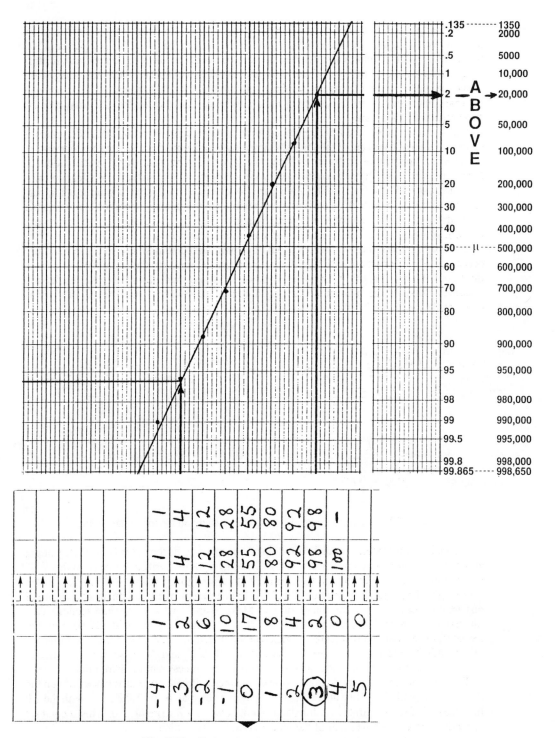

Fig. 7.19 Estimating the percentage nonconforming from NOPP.

For example, the estimated percentage of shafts with lengths *shorter* than the USL is 98 percent, which implies that the estimated percentage *longer* than the USL is 2 percent (100 minus 98). These estimates are identical to those derived previously from the ogive for this sample data displayed in Figure 7.15. On this particular worksheet, these percentages are also expressed as a *ppm* (parts per million) value in the column adjacent to the percentage column. For instance, the 2 percent of nonconforming shafts above the USL may also be expressed as 20,000 *ppm*.

The 4 percent below the LSL combined with the 2 percent above the USL add up to an estimated total of 6 percent of the shafts having their length outside of tolerance. Conversely, this means 94 percent of shafts are estimated to have their length cut within specification, as long as this process remains in control. Because a process with a bilateral specification must have at least 99.73 percent conforming parts to meet the minimum requirement for performance capability (recall the definition of capability), this sawing operation is *not* capable for shaft length. Improvement efforts need to begin as quickly as possible to center the process average at the middle of the tolerance, and then to reduce process spread.

Information about these estimates should be summarized somewhere so it can be easily accessed. For this particular NOPP, these results are entered into the appropriate spaces at the top of the 6σ Capability Analysis Worksheet™ to permanently record the results and document some of the more important conditions under which this capability study was conducted (see Figure 7.20).

PART NUMBER   *123456*
PART NAME   *XYZ-9 Shaft*
SAMPLE SIZE FOR STUDY   *50*

CHARACTERISTIC   *Length*
UPPER SPEC. LIMIT (USL)   *+3*
LOWER SPEC. LIMIT (LSL)   *-3*

MACHINE NAME / NUMBER   *B-1301*
OPERATION NUMBER   *150*
GAGE NUMBER   *DE-1176*

% (or ppm) ABOVE USL   *2%*
% (or ppm) BELOW LSL   *4%*
CAPABILITY MEASURE

Fig. 7.20   Entering results of the capability study on the top of the NOPP.

A sketch of the process output distribution can be made at the top of the NOPP. Beginning at the upper right-hand corner of the NOPP, draw a very light horizontal line from the "3.0" listed in the "Z Score" column to the fitted line, then straight up to the "3.0" line in the box at the top of the graph. The height of this "3.0" line represents the *y* coordinate of the normal distribution curve at an *x* coordinate of 3σ. Make a small mark at this intersection, as is illustrated in Figure 7.21. Return to the "Z Score" column. Beginning at a Z value of 2.5, repeat this procedure so that a mark is made at the matching "2.5" line in the box at the top of the NOPP. Then do the same for the 2.0, 1.5, 1.0, .5, and .0 (μ) scores. The marks corresponding to these values define the upper half the process output distribution.

The negative Z scores are done in a similar manner, with only one difference. Match the Z score of -.5 with the .5 line at the top of the page. Then match the -1.0 Z score with the 1.0 line, the -1.5 Z score with the 1.5 line, and so on, down to a Z score of -3.0. This second set of marks portray the lower half of the process output distribution.

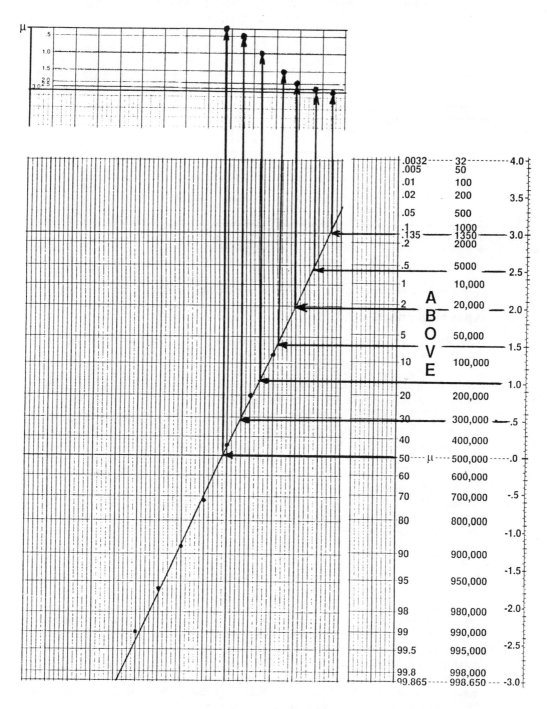

Fig. 7.21 Using NOPP to sketch the estimated process output distribution.

When finished, draw a smooth curve through all marks made at the top of the NOPP, as is exhibited in Figure 7.22. This curve represents the estimated distribution for the process output. For normal distributions, the 6σ spread extends from the lower 3σ tail to the upper 3σ tail. Draw vertical lines from the LSL and the USL up to this curve to discover how well the 6σ spread of the process output fits within the tolerance. If both tails are within the specification limits, there is good potential and performance capability. If one tail is outside a specification limit, the process lacks performance capability, although it may still have acceptable potential capability. If both tails are out, the process lacks both potential and performance capability.

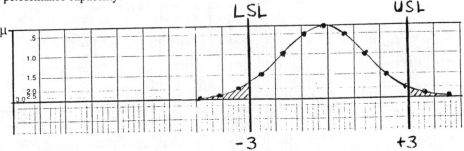

Fig. 7.22   Sketching the estimated process output distribution at the top of the NOPP.

Other Types of Distributions

The shape of the line fitting the cumulative percentage plot points provides an indication of what the process output distribution looks like. With NOPP, a straight line implies the process output has a normal distribution. However, there are times when the fitted line is not straight, but appears more like one of those exhibited in Figures 7.23a and 7.23b. Both of these curves indicate the process output distribution is skewed.

The term "skewness" describes the symmetry, or more precisely, the *lack* of symmetry of a distribution. Symmetrical distributions, like the normal distribution, have "zero" skewness. Non-symmetrical distributions usually have either "positive" or "negative" skewness depending on the shape of the distribution. Although it is possible to have a non-symmetrical distribution with zero skewness, these types of distributions are not likely to appear often in manufacturing situations.

A distribution with a long tail to the left (like the one in Figure 7.23a) is said to have negative skewness and is occasionally referred to as being skewed left. On the other hand, a distribution with a long tail to the right (see Figure 7.23b) has positive skewness and is sometimes called skewed to the right. When the area under a skewed-right distribution is accumulated, the rate of accumulation is very rapid at first, resulting in a relatively steep slope for the start of the curve drawn on NOPP. After reaching the median, the rate of accumulation decreases, and the slope of the accumulation curve lessens. This change in rate is responsible for the shape of the curve seen at the top of Figure 7.23b.

Skewed distributions commonly occur when there is some upper or lower physical bound in the process. Manufacturing characteristics such as tensile strength, material hardness, and weld strength have an *upper* physical bound that cannot be exceeded and will tend to have a skewed-left distribution, which is one having a longer tail on the left side. Many other characteristics like taper, flatness, waviness, concentricity, and squareness have a *lower* measurement bound of zero, and thus display a skewed right distribution (longer tail extending to the right) when the process average is close to zero.

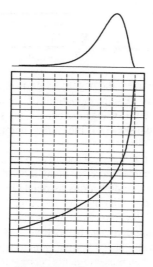

Fig. 7.23a Skewed left.   **Fig. 7.23b** Skewed right.

For instance, the process output distribution for surface finish may be fairly close to normal when the average of the distribution is far from zero, as is the case for distribution 1 of Figure 7.24. When this process is improved, its average is pushed closer and closer to zero (see distributions 2 through 5). Because the lower tail of each surface finish distribution cannot go below zero, these distributions start to become more and more "lopsided," or asymmetrical. Parts with very high surface finish are still possible, but no part can have a surface finish measurement that is less than zero. Note that a process with a non-normal distribution can be in control. The $\overline{X}$s of subgroups sampled from a process with a non-normal distribution will still tend to have a normal distribution, courtesy of the central limit theorem.

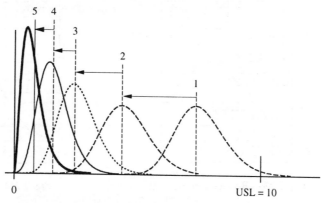

Fig. 7.24 As the average is moved lower, the distribution for surface finish becomes skewed.

Just as there are formulas for estimating the process average and standard deviation, there is a formula for estimating skewness, labeled a_3 (Wheeler and Chambers, p. 322). In this formula, note the difference in the calculation of S_n compared to the more common S.

$$a_3 = \frac{\sum_{i=1}^{kn} (X_i - \overline{X})^3}{nS_n^3} \qquad \text{where } S_n = \sqrt{\frac{\sum_{i=1}^{kn} (X_i - \overline{X})^2}{n}}$$

Remember that when a negative value is cubed, the negative sign is retained ($-2 \times -2 \times -2$ $= -8$). Because a symmetrical distribution has the same deviations above the average (which result in positive cubes) as below (which end up as negative cubes), the sum of all cubes is zero. Dividing zero by nS_n^3 equals zero, so the skewness measure generated by this formula is zero for all symmetrical distributions.

Now consider a distribution with a long tail to the right (positive skew). A few individual values are much greater than the average, whereas there are many individual values slightly below the average. However, because the large deviations associated with measurements above the average are cubed, their sum of cubes will outweigh the sum of the negative cubes coming from the smaller deviations associated with measurements below the average. As a result, the sum of all cubes is positive, causing a_3 to also be positive.

Bounded distributions (like the one shown in Figure 7.25a) may materialize when a process has automatic sensors that stop production when measurements become too high or too low. Occasionally referred to as a truncated curve, this type of distribution may also occur due to sorting, which is likely if the curve is bounded at the LSL and the USL. For example, plastic pellets are sized by first passing them over a screen with small holes to allow undersize pellets to drop out. The remaining pellets then pass over a second screen with slightly larger holes to allow the desired sized pellets to fall through, but not the oversized ones.

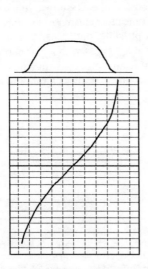

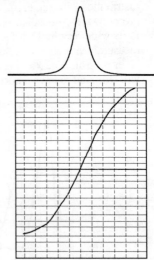

Fig. 7.25a Bounded (truncated).   **Fig. 7.25b** Unbounded.

Unbounded curves with greatly extended tails, like the one presented in Figure 7.25b, seldom happen in manufacturing operations. Sometimes such a curve may appear due to a lack of gage discrimination. This condition occurs when process variation is so small that most measurements fall within just one increment on the gage. Now and then, a measurement may fall in the next higher or lower gage increment, but there are very few of these, as Figure 7.26 illustrates.

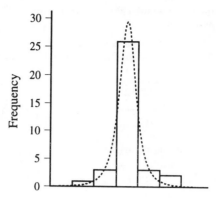

Fig. 7.26 Unbounded distribution caused by lack of gage discrimination.

Inadequate gage discrimination can often lead to incorrectly-calculated control limits. Because the majority of measurements are identical, many subgroup ranges will be zero. This causes the average range to be artificially small, which results in "tighter" control limits around the centerline of the $\overline{X}$ chart, as is displayed in Figure 7.27. These artificially tight limits make many $\overline{X}$ points appear as if they were out of control.

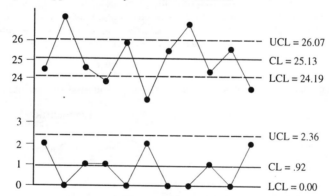

Fig. 7.27 Lack of gage discrimination causes many ranges to be zero.

One method for identifying lack of gage discrimination is to count how many different values a subgroup range may take on and still fall within the control limits of the range chart. Five or fewer possible values indicate the measurement system suffers from a lack of gage discrimination (Wheeler and Chambers, p. 213). For the range chart in Figure 7.27, there are only three possible values for an in-control range (0, 1, and 2), indicating a serious problem with gage discrimination.

An obvious solution (although an expensive one) is buying a gage with better discrimination. If option this is not possible, or is too costly, increase the subgroup size. By including more measurements in a subgroup, a greater likelihood exists of including at least one different value, which would produce a non-zero subgroup range. If measuring additional pieces is too expensive or time consuming, switch to an *IX & MR* chart with a long interval between measurements. By spreading the measurements out over time, more variation will be seen between readings, resulting in larger moving range values. Unfortunately, this procedure lessens the sensitivity of the chart for detecting changes in the process average.

The relationship of how much area of a distribution is in its tails (how "bounded" it is) to that near the center is measured by a term called "kurtosis." Just as there is a formula for quantifying skewness, there is a formula for estimating kurtosis, denoted by the symbol a_4 (Wheeler and Chambers, p. 322). Again, note the difference in the calculation of S_n.

$$a_4 = \frac{\sum\limits_{i=1}^{kn} (X_i - \overline{X})^4}{nS_n^4} \qquad \text{where } S_n = \sqrt{\frac{\sum\limits_{i=1}^{kn} (X_i - \overline{X})^2}{n}}$$

To better understand this kurtosis formula, recall that a negative value taken to the fourth power gives a positive result ($-2^4 = -2 \times -2 \times -2 \times -2 = +16$). As a positive value to the fourth power is also positive, the sum in the numerator of the kurtosis formula is always positive (or zero). Because n and S_n are also always positive, a_4 can never be negative.

Taking the fourth power of deviations of the individuals from the average also means observations near the center of a distribution contribute very little to the overall sum. Observations in the tails have large deviations from the average, and thus have a much greater influence on the total. Hence, saying kurtosis measures the "peakedness" of a distribution isn't very accurate, because observations near the "peak" have very little weight in determining the actual degree of kurtosis.

Three distributions with differing amounts of kurtosis are displayed in Figure 7.28. The one drawn with a solid line is a normal distribution, which has a kurtosis value of 3. Curves similar to this shape are said to be "mesokurtic," which means a medium amount of kurtosis.

A distribution with a relatively large number of observations near its center and long heavy tails has a high degree of kurtosis, defined as being greater than 3 (see distribution designated by the long-dashed line in Figure 7.28). These types of distributions are called "leptokurtic" and are commonly referred to as having a large degree of "peakedness."

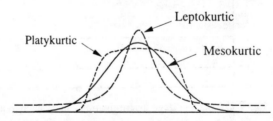

Fig. 7.28 Three distributions with varying degrees of kurtosis.

The distribution identified with the short-dashed line has less area right at its average, with very few observations in its tails. Most observations are in the "shoulders" of the distribution rather than near the middle. Distributions with this shape have a low amount of kurtosis (less than 3) and are referred to as being "platykurtic" by Bissell (p. 271).

Sometimes kurtosis is defined in a manner slightly different than above, with the following formula occasionally seen in practice.

$$a'_4 = \frac{\sum\limits_{i=1}^{n} (X_i - \overline{X})^4}{nS_n^4} - 3$$

The amount of kurtosis for a normal distribution computed with the original formula is 3. With this second formula, kurtosis for a normal distribution is 0, positive for a leptokurtic distribution, and negative for a mesokurtic distribution. Some practitioners prefer this revised rating scale as skewness is also zero for a normal distribution. With this second formula, bounded curves have negative kurtosis, whereas unbounded curves have positive kurtosis.

Whichever rating scale is chosen, skewness and kurtosis provide some assistance in visualizing the shape of a distribution. Figure 7.29 exhibits the shapes of distributions with various combinations of skewness and kurtosis.

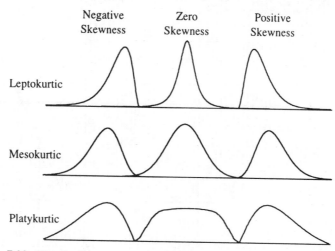

Fig. 7.29 Distributions with various combinations of skewness and kurtosis.

One parting word about these two shape parameters: because they deal with the tail areas of a distribution, a fair amount of data is needed to get accurate estimates of skewness and kurtosis. Most authors recommend a minimum of at least 200 observations, with some suggesting 400 or more (Wheeler and Chambers, p. 326). Of course, all data included in these calculations must originate from a stable process in order to produce reliable estimates of these two additional process parameters.

Estimating Process Parameters from NOPP

For a normal distribution, the process average is estimated from NOPP by drawing a horizontal line from the 50 percent mark on the y axis over to the cumulative percentage distribution curve, then down to the x axis, as is exhibited in Figure 7.30. This x coordinate, which is about -.2 in this example, is called the 50th percentile, since 50 percent of the process output lies below it. In general, the yth percentile is the x coordinate for which y percent of the distribution lies below. For example, 10 percent of the process output lies below the 10th percentile, whereas 80 percent lies below the 80th percentile.

The 50th percentile is commonly referred to as the median and is written as x_{50}, or occasionally as $\tilde{x}$. For normal distributions, and all other symmetrical distributions as well, the process average and the median are equal. In this shaft length example, the following relationship holds true.

$$\hat{\mu} = \hat{x}_{50} = \hat{\tilde{x}} = -.2$$

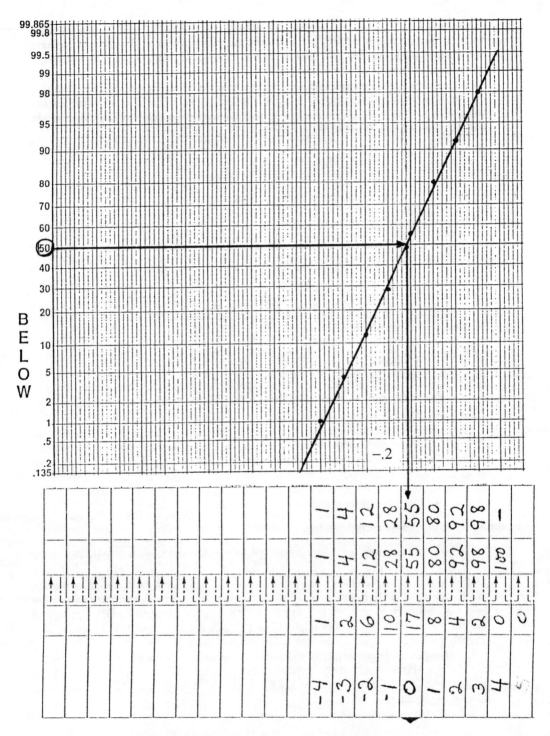

Fig. 7.30 Estimating the process average from NOPP.

This estimate of -.2 for average shaft length generated by the NOPP is quite reasonable. When the $\overline{X}$, R chart for this process was created in Exercise 2.5 (p. 24), $\overline{\overline{X}}$, which is a good estimate of μ, was -.192. Rounded to one decimal place, this becomes -.2.

The long-term process standard deviation, σ_{LT}, is estimated by first finding the difference between the 99.865 percentile and the .135 percentile. For a normal distribution, .135 percent of the process output lies above the upper $3\sigma_{LT}$ tail, which means 99.865 percent is below it, as is revealed in Figure 7.31. According to the percentile definition, this makes μ plus $3\sigma_{LT}$ the 99.865 percentile, labeled accordingly as $x_{99.865}$.

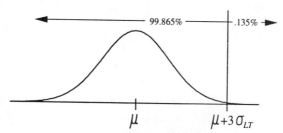

Fig. 7.31 The 99.865 percentile.

This particular percentile is found on NOPP by drawing a horizontal line from the 99.865 percent mark on the left-hand y axis over to the fitted line, then down to the x axis. In the shaft length example, the x coordinate is about 4.5 (halfway between 4 and 5), as can be seen in Figure 7.32. In addition to being an estimate of $x_{99.865}$, this length value is also an estimate of the location for the upper $3\sigma_{LT}$ tail of this distribution.

Likewise, because .135 percent of the process output lies below it, the x coordinate for the lower $3\sigma_{LT}$ tail is the .135 percentile, $x_{.135}$. The .135 percentile is found in the same manner as $x_{99.865}$, only beginning at the .135 percent mark on the y axis. For the shaft length example, $x_{.135}$ is located half an interval above -5.0, at approximately -4.9. The difference between these two of 9.4 (4.5 minus -4.9) represents the $6\sigma_{LT}$ spread of the process output, which is one estimate of long-term potential process capability. As this estimate of the $6\sigma_{LT}$ spread is larger than the tolerance of 6 (+3 minus -3) for length, this process lacks potential capability.

$$6\hat{\sigma}_{LT} \text{ Spread} = \hat{x}_{99.865} - \hat{x}_{.135} = 4.5 - (-4.9) = 9.4$$

Dividing both sides of the above equation by 6 provides an estimate of σ_{LT}. This result represents an estimate of *long-term* variation as measurements within and between subgroups are mixed on the NOPP (Charbonneau and Webster, p. 120; Podolski, p. 161). In Exercise 3.5 (p. 49), an estimate of 1.53 was generated for the long-term standard deviation of this process, which is almost identical to the estimate of 1.57 derived from the NOPP.

$$\hat{\sigma}_{LT} = \frac{6\hat{\sigma}_{LT}}{6} = \frac{\hat{x}_{99.865} - \hat{x}_{.135}}{6} = \frac{4.5 - (-4.9)}{6} = \frac{9.4}{6} = 1.57$$

In summary: from NOPP, this process output was determined to be normally distributed, with an average of -.2 and a long-term standard deviation of 1.57. As the $6\sigma_{LT}$ spread is estimated to be larger than the tolerance, this process lacks long-term potential capability. In addition, NOPP estimates this process produces 2 percent of the shaft lengths below the LSL and 4 percent above the USL, for a total of 6 percent nonconforming, information which is not readily available from a control chart. On account of having only 94 percent of its output

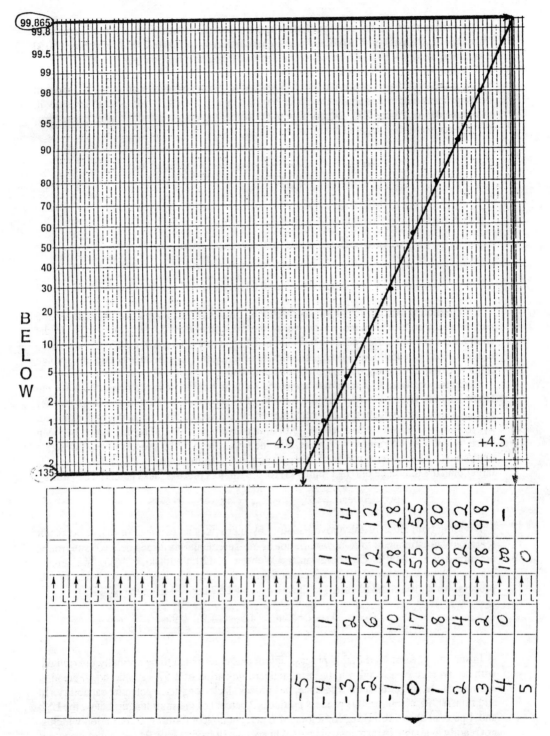

Fig. 7.32 Estimating the process standard deviation from NOPP.

within tolerance, this process does *not* meet the minimum requirement for long-term performance capability of producing at least 99.73 percent conforming parts. Quite a bit of information from just one piece of graph paper!

Example 7.2 Electrical Connector Length

Electrical connectors are cut to the specified length on a shearing operation. Because connector length is critical to the proper functioning of the finished plug, an $\overline{X}, R$ control chart ($n = 4$) is kept on this characteristic to help stabilize the process. After control is finally established, measurements from the last 25 in-control subgroups are arranged into the frequency tally displayed in Table 7.4, which is similar to Table 4.1 (p. 78). With 25 subgroups of size 4 each, there is a total of 100 measurements. When these measurements are entered on the bottom of a piece of NOPP, the cumulative percentage plot points can be calculated and plotted, as was done in Figure 7.33. By centering the classes near the middle of the NOPP, there is plenty of room for the fitted line to be extended beyond the first and last class.

Table 7.4 Frequency tally for 100 connector length measurements.

Class Midpoint	Freq.		EAF	Plot Points
66.80	0		0	-
66.85	1		1	.5
66.90	2		4	2.0
66.95	4		10	5.0
67.00	16		30	15.0
67.05	21		67	33.5
67.10	30		118	59.0
67.15	12		160	80.0
67.20	7		179	89.5
67.25	5		191	95.5
67.30	2		198	99.0
67.35	0		200	-

The process output is close to being normally distributed, as a straight line fits reasonably well through these plotted points. For a process with a normal distribution, the process average can be estimated as the 50th percentile. Drawing a horizontal line from the 50 percent mark on the y axis over to the fitted line, then from this intersection down to the x axis, provides an estimate of 67.087 for x_{50}.

$$\hat{\mu} = \hat{x}_{50} = 67.087$$

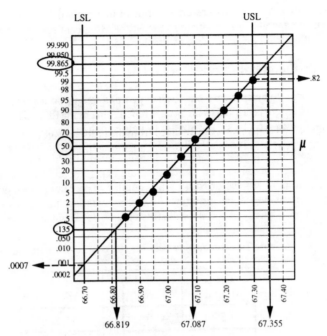

Fig. 7.33 NOPP for connector length.

The $6\sigma_{LT}$ spread is estimated as the difference between $x_{.135}$ and $x_{99.865}$. As this spread is less than the tolerance of .600 (67.300 minus 66.700), this process meets the minimum requirement for long-term potential capability.

$$6\hat{\sigma}_{LT} \text{ Spread } = \hat{x}_{99.865} - \hat{x}_{.135} = 67.355 - 66.819 = .536$$

σ_{LT} is estimated by dividing the $6\hat{\sigma}_{LT}$ spread by 6.

$$\hat{\sigma}_{LT} = \frac{\hat{x}_{99.865} - \hat{x}_{.135}}{6} = \frac{67.355 - 66.819}{6} = \frac{.536}{6} = .0893$$

The specification limits for connector length may be added to the NOPP. Based on the fitted line, the amount of connectors shorter than the LSL of 66.700 is predicted to be about .0007 percent, or 7 *ppm*, and the amount longer than the USL of 67.300 as .82 percent, or 8,200 *ppm*. The maximum of these two becomes $ppm_{MAX.LT}$, which is one estimate of long-term performance capability.

$$\hat{p}pm_{MAX.LT} = \text{Maximum}(\hat{p}pm_{LSL.LT}, \hat{p}pm_{USL.LT}) = \text{Maximum}(7, 8,200) = 8,200$$

Because the $ppm_{MAX.LT}$ value of 8,200 exceeds 1,350 *ppm*, which is the maximum allowed to meet the minimum capability requirement for a bilateral specification, this process is classified as lacking long-term performance capability. This conclusion is verified by substituting the estimates of μ and σ_{LT} into the formula for P_{PK}. As the estimate for P_{PK} is less than 1.0, this shearing operation is classified not capable.

$$\hat{P}_{PK} = \text{Minimum}\left(\frac{\hat{\mu} - \text{LSL}}{3\hat{\sigma}_{LT}}, \frac{\text{USL} - \hat{\mu}}{3\hat{\sigma}_{LT}}\right)$$

$$= \text{Minimum}\left(\frac{67.087 - 66.700}{3(.0893)}, \frac{67.300 - 67.087}{3(.0893)}\right)$$

$$= \text{Minimum}(1.44, .80) = .80$$

Figure 7.34 shows how the distance from the process average to the lower $3\sigma_{LT}$ tail is equal to x_{50} minus $x_{.135}$. Likewise, the distance between the upper $3\sigma_{LT}$ tail and μ is equal to $x_{99.865}$ minus x_{50}.

$$3\sigma_{LT} = \mu - (\mu - 3\sigma_{LT}) = x_{50} - x_{.135} \qquad 3\sigma_{LT} = (\mu + 3\sigma_{LT}) - \mu = x_{99.865} - x_{50}$$

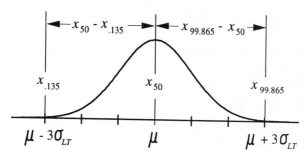

Fig. 7.34 Distances to the $3\sigma_{LT}$ tails.

With these two relationships, an estimate of the long-term performance capability could have also been obtained directly from the percentile values (Lai *et al*). As seen here, both formulas provide the same answer.

$$\hat{P}_{PK} = \text{Minimum}\left(\frac{\hat{\mu} - \text{LSL}}{3\hat{\sigma}_{LT}}, \frac{\text{USL} - \hat{\mu}}{3\hat{\sigma}_{LT}}\right)$$

$$= \text{Minimum}\left(\frac{\hat{x}_{50} - \text{LSL}}{\hat{x}_{50} - \hat{x}_{.135}}, \frac{\text{USL} - \hat{x}_{50}}{\hat{x}_{99.865} - \hat{x}_{50}}\right)$$

$$= \text{Minimum}\left(\frac{67.087 - 66.700}{67.087 - 66.819}, \frac{67.300 - 67.087}{67.355 - 67.087}\right)$$

$$= \text{Minimum}(1.44, .80) = .80$$

With an estimate of σ_{LT} for the connector length output distribution, estimates can also be made for these two popular measures of long-term potential capability.

$$\hat{P}_R = \frac{6\hat{\sigma}_{LT}}{\text{Tolerance}} = \frac{6(.0893)}{67.300 - 66.700} = .89 \qquad \hat{P}_P = \frac{\text{Tolerance}}{6\hat{\sigma}_{LT}} = \frac{67.300 - 66.700}{6(.0893)} = 1.12$$

This process has good long-term potential capability as it would require only 89 percent of the tolerance to produce 99.73 percent of the connector lengths when the process is centered at M. The above measures of potential capability may also be estimated with this next set of formulas, where the difference between the 99.865 and .135 percentiles replaces $6\hat{\sigma}_{LT}$.

$$\hat{P}_R = \frac{6\hat{\sigma}_{LT}}{\text{Tolerance}} = \frac{\hat{x}_{99.865} - \hat{x}_{.135}}{\text{Tolerance}} = \frac{67.355 - 66.819}{67.30 - 66.70} = .89$$

$$\hat{P}_P = \frac{\text{Tolerance}}{6\hat{\sigma}_{LT}} = \frac{\text{Tolerance}}{\hat{x}_{99.865} - \hat{x}_{.135}} = \frac{67.30 - 66.70}{67.355 - 66.819} = 1.12$$

On account of having potential capability which exceeds the minimum requirement, this process must lack long-term performance capability due to improper centering. The k factor reveals performance capability is 29 percent less than potential capability.

$$\hat{k} = \frac{|\hat{\mu} - M|}{(\text{USL} - \text{LSL})/2} = \frac{|67.087 - 67.000|}{(67.300 - 66.700)/2} = .29$$

If the average can be lowered to 67.000 (the middle of the tolerance), P_{PK} will increase from .80 to 1.12, while $ppm_{MAX,LT}$ will decrease from 8,200 to about 390, representing a Z value of 3.36 (3 × 1.12).

Example 7.3 Metal Block Length

Table 7.5 displays a frequency tally for the lengths of 100 metal blocks (Figure 7.35). These measurements were taken from the last 20 in-control subgroups of an $\overline{X}$, S chart with a subgroup size of 5. The LSL for block length is 3.275, while the USL is 3.625.

Fig. 7.35

Table 7.5   Metal block length measurements.

Class Midpoint	Freq.		EAF	Plot Points
3.250	0		0	-
3.300	3		3	1.5
3.350	3		9	4.5
3.400	9		21	10.5
3.450	32		62	31.0
3.500	38		132	66.0
3.550	10		180	90.0
3.600	3		193	96.5
3.650	1		197	98.5
3.700	1		199	99.5
3.750	0		200	-

The cumulative percentage plot points are calculated from this data and plotted on the sheet of NOPP displayed in Figure 7.36. The line of best fit is a curved line, indicating the process output is *not* normally distributed. The elongated S-shaped curve fitted through these plot points suggests this process has an unbounded distribution similar to the one exhibited in Figure 7.25b on page 370, with about 1 percent of the block lengths below the LSL and approximately 2 percent above the USL. The total of 3 percent nonconforming (only 97 percent conforming) offers strong evidence this process lacks long-term performance capability for machining the length of these blocks.

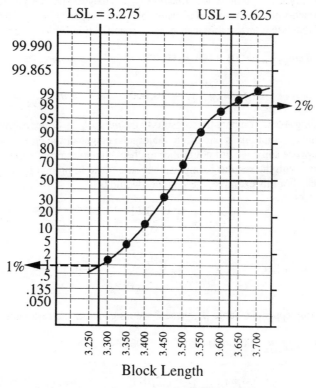

Fig. 7.36 NOPP for metal block length.

Due to the process output not being normally distributed, estimates of the process average and standard deviation cannot be made from NOPP. Instead, they must be calculated directly from the 100 individual length measurements, as is shown here.

$$\hat{\mu} = \frac{\sum\limits_{i=1}^{kn} X_i}{kn} = \frac{347.60}{100} = 3.476$$

$$\hat{\sigma}_{LT} = \frac{1}{c_4} \sqrt{\frac{\sum\limits_{i=1}^{kn} (X_i - \hat{\mu})^2}{kn - 1}} = \frac{1}{.9975}(.0649319) = .065$$

The median may be estimated from NOPP for any type of distribution by drawing a line from the 50 percent mark on the y axis to the fitted curve, then down to the x axis. However, for non-symmetrical distributions, x_{50} is usually not the process average. Half of the block lengths are estimated to be below this x coordinate of 3.474 ($x_{50} = 3.474$). This result is close to the estimated average of 3.476, as an unbounded distribution is fairly symmetrical.

Estimates of process capability, such as P_P or P_{PK}, cannot be calculated from their traditional formulas because the process output is not normally distributed. The next chapter presents two alternative methods of estimating these capability measures, which do not rely on the normality assumption. Notice that $ppm_{TOTAL, LT}$ may still be estimated from NOPP, even though the process has a non-normal distribution. $ppm_{LSL, LT}$ and $ppm_{USL, LT}$ are the y-axis values where the fitted line intersects their respective specification limits. From Figure 7.36:

$$\hat{p}pm_{LSL,LT} = .01 \times 10^6 = 10,000 \qquad \hat{p}pm_{USL,LT} = .02 \times 10^6 = 20,000$$

$$\hat{p}pm_{TOTAL,LT} = 30,000 \qquad \hat{p}pm_{MAX. LT} = 20,000$$

Being able to compare these results to the $ppm_{TOTAL, LT}$ values estimated for processes having normal distributions is one of the major advantages of the ppm capability measure.

Suppose someone did not bother to check the normality assumption, but just estimated μ and σ_{LT} with the regular formulas, then attempted to predict ppm by computing the usual Z values. For this example, the Z values would be:

$$\hat{Z}_{LSL,LT} = \frac{\hat{\mu} - LSL}{\hat{\sigma}_{LT}} = \frac{3.476 - 3.275}{.0651} = 3.09$$

$$\hat{Z}_{USL,LT} = \frac{USL - \hat{\mu}}{\hat{\sigma}_{LT}} = \frac{3.625 - 3.476}{.0651} = 2.29$$

From Appendix Table III, these Z values generate the following ppm estimates.

$$\hat{p}pm_{LSL,LT} = 1.001 \times 10^{-3} \times 10^6 = 1,001 \qquad \hat{p}pm_{USL,LT} = 1.101 \times 10^{-2} \times 10^6 = 11,010$$

$$\hat{p}pm_{TOTAL,LT} = 12,011 \qquad \hat{p}pm_{MAX. LT} = 11,010$$

In this situation, assuming a normal distribution generates ppm estimates that are quite a bit less than the actual values read off the NOPP for this process output. The reason for this discrepancy can be seen in Figure 7.37, where the straight line of a normal distribution is drawn on top of the unbounded curve for this process. The unbounded curve is leptokurtic, meaning there is relatively more area in its tails than in those for a normal distribution, which is why the ppm values estimated under the assumption of normality are so much less than the actual ppm values. As Figure 7.37 discloses, the LSL limit intersects the line for the normal distribution at a much lower cumulative percentage (.1 percent) than that for the unbounded curve (1 percent). Likewise, the USL intersects the normal curve at 1.1 percent, while intersecting the unbounded curve at 2 percent. It's always important to verify all assumptions concerning capability measures. If any are violated, serious errors may arise that could lead to an inaccurate assessment of process capability (Gunter, 1989).

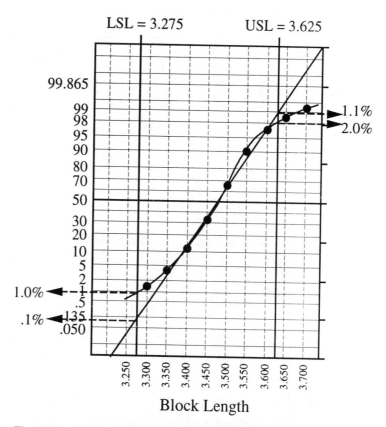

LSL = 3.275　　　USL = 3.625

99.865

99
98
95
90
80
70
50
30
20
10
5
2

1.0% ◄── 1

.5

.1% ◄── .135
.050

1.1%
2.0%

3.250　3.300　3.350　3.400　3.450　3.500　3.550　3.600　3.650　3.700

Block Length

Fig. 7.37 Comparison of unbounded and normal distributions on NOPP.

Example 7.4 Waviness of Cam Rollers

An $\overline{X}, R$ chart ($n = 4$) is monitoring a critical characteristic of an automotive machining operation, waviness of the surface finish for the cam roller shown in Figure 7.38. After three weeks of observing the process and devising several improvements, the chart finally exhibits good control. Before estimating process capability, Chad, the department supervisor, wishes to determine if the process output for waviness is normally distributed or not.

Fig. 7.38 Cam roller.

To find out, he takes the last 31 in-control subgroups from the chart and plots the 124 (31 × 4) measurements in the form of a frequency distribution, as is displayed in Figure 7.39 on the next page. Still uncertain about the output's shape, he decides to calculate the cumulative percentage distribution for this process (results given in Table 7.6) and plot this curve on the NOPP shown in Figure 7.40 on page 385. From the line fitted through these points, a decision can be made about the normality issue. In addition, a prediction can be made concerning the percentage of cam rollers having their waviness above the USL of 75.

Class Midpoint	Freq.	Tally
5	0	X
10	3	XXX
15	16	XXXXXXXXXXXXXXXX
20	27	XXXXXXXXXXXXXXXXXXXXXXXXXXX
25	22	XXXXXXXXXXXXXXXXXXXXXX
30	20	XXXXXXXXXXXXXXXXXXXX
35	14	XXXXXXXXXXXXXX
40	8	XXXXXXXX
45	6	XXXXXX
50	3	XXX
55	2	XX
60	1	X
65	0	
70	1	X
75	0	
80	0	
85	1	X

Fig. 7.39   The frequency distribution for cam roller waviness.

Table 7.6   NOPP plot point calculations for roller waviness.

Class Midpoint	Freq.		EAF	Plot Points
5	0		0	-
10	3		3	1.2
15	16		22	8.9
20	27		65	26.2
25	22		114	46.0
30	20		156	62.9
35	14		190	76.6
40	8		212	85.5
45	6		226	91.1
50	3		235	94.8
55	2		240	96.8
60	1		243	98.0
65	0		244	98.4
70	1		245	98.8
75	0		246	99.2
80	0		246	99.2
85	1		247	99.6
90	0		248	-

Based on the curve drawn through the points, Chad confirms his belief that the process output is most likely *not* normally distributed. Instead, the output appears to be heavily skewed to the right, with about .8 percent (8,000 *ppm*) above the USL of 75. It's always a good idea to check the predicted percentage nonconforming from the NOPP with that estimated from the sample measurements. In this case, there is one roller in the sample of 124 with a waviness reading more than 75, for an estimated .81 percent nonconforming (8,100 *ppm*), which is almost identical to that predicted by the NOPP.

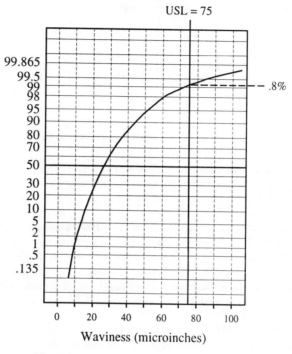

Fig. 7.40 NOPP for waviness of cam rollers.

In view of the output distribution being non-normal, estimates of process parameters cannot be made from this piece of NOPP. Thus, traditional measures of capability, other than *ppm*, are inappropriate for this process. Nonetheless, a broad judgment concerning capability can be made as follows. There must be at least 99.865 percent conforming parts to meet the minimum requirement of performance capability for a unilateral specification. Because this process is producing only 99.2 percent (100 minus .8) conforming rollers, it must be considered *not* capable for the waviness characteristic.

Estimating the Cumulative Percentage Distribution from Small Samples (*kn* < 50)

Assume only 10 measurements collected during the course of a capability study. This small number may be the result of a study conducted on prototype parts, a small lot size run in a job shop, or a characteristic evaluated through destructive (or expensive) testing. The first step in this method for determining the cumulative percentage distribution is to rank these ten measurements from lowest to highest, as was done in Table 7.7. The first column in this table is for the rank number, *i*, where *i* goes from 1 to *kn*, which is 10 in this example. The second column has the ranked measurements, from smallest to largest.

Table 7.7 The ranked measurements and their cumulative percentages.

Rank # (*i*)	Sample Meas.	Cum. Freq.	Sample Cum. %
1	2.1	1	10
2	2.7	2	20
3	3.0	3	30
4	3.3	4	40
5	3.5	5	50
6	3.5	6	60
7	3.7	7	70
8	4.1	8	80
9	4.3	9	90
10	4.8	10	100

The third column (Cum. Freq) accumulates the frequency from the lowest measurement to the highest. In this case, there is one measurement (2.1) at, or below, 2.1, while there are two measurements (2.1 and 2.7) at, or below 2.7. 3 measurements at, or below 3.0, whereas 4 are at, or below, 3.3, etc.

The cumulative percentage recorded in the last column is found by dividing the cumulative frequency by the total number of sample measurements, which in this example is 10. The cumulative frequency of 1 for the 2.1 measurement is divided by 10, giving a cumulative percentage of 10 percent. This means 10 percent of the sample measurements are at, or below 2.1. For the 2.7 measurement, 2 divided by 10 is 20 percent, and so on until finally, for 4.8, 10 divided by 10 is a cumulative percentage of 100 percent. All 10 measurements in this sample are at, or below, 4.8.

In general, the cumulative percentage for the *i*th ranked sample measurement is found from the following formula, where *kn* is the total number of measurements in the sample.

$$\text{Cumulative Percentage for } i \text{ th Ranked Measurement} = \frac{i}{kn} \times 100$$

Unfortunately, this cumulative percentage distribution is for only this one particular *sample*, not for the entire *process*. It's rather likely there are (or will be) some measurements produced by this process that are greater than 4.8.

Median Ranks

One method for estimating the cumulative percentage distribution for a process based on sample measurements is through the application of median ranks (Lipson and Sheth). Instead of using *i*/*kn* to calculate the *sample* cumulative percentage, the next formula is chosen to estimate the approximate *process* cumulative percentage.

$$\text{Median Rank for the } i \text{ th Ranked Measurement} = \frac{i - .3}{kn + .4} \times 100$$

In this example, the median rank for the first sample measurement (2.1) is 6.7 percent.

$$\text{Median Rank for the 1st Ranked Measurement} = \frac{1 - .3}{10 + .4} \times 100 = \frac{.7}{10.4} \times 100 = 6.7$$

For the second, it is 16.3 percent.

$$\text{Median Rank for the 2nd Ranked Measurement} = \frac{2 - .3}{10 + .4} \times 100 = \frac{1.7}{10.4} \times 100 = 16.3$$

Table 7.8 displays median ranks for the remaining eight sample measurements. Notice the last measurement (4.8) has a cumulative percentage of 93.3, which implies the existence of process measurements greater than 4.8. This is quite plausible, as one would not expect to see the largest process measurement in a sample of only ten. Appendix Table V beginning page 850 contains median ranks calculated for sample sizes of 10 through 50. If there are more than 50 measurements, go with the grouped frequency method previously described.

Table 7.8   The ranked sample measurements and their median ranks.

Rank # (i)	Sample Meas.	Median Rank
1	2.1	6.7
2	2.7	16.3
3	3.0	26.0
4	3.3	35.6
5	3.5	45.2
6	3.5	54.8
7	3.7	64.4
8	4.1	74.0
9	4.3	83.7
10	4.8	93.3

There are other formulas for estimating median ranks. Bissell (p. 246) recommends the following formula to simplify the plot point calculations.

$$\text{Median Rank for the } i \text{ th Ranked Measurement} = \frac{i - .5}{kn} \times 100$$

However, because this formula tends to underestimate the true process standard deviation (the fitted line will have a steeper slope), King (1980) prefers this formula:

$$\text{Median Rank for the } i \text{ th Ranked Measurement} = \frac{i - .375}{kn + .25} \times 100$$

Table 7.9 shows that all three formulas yield similar results, especially for measurements ranked near the middle.

Table 7.9 Comparison of median rank formulas.

Rank # (i)	Sample Meas.	$\dfrac{i - .3}{kn + .4}$	$\dfrac{i - .5}{kn}$	$\dfrac{i - .375}{kn + .25}$
1	2.1	6.7	5.0	6.1
2	2.7	16.3	15.0	15.9
3	3.0	26.0	25.0	25.6
4	3.3	35.6	35.0	35.4
5	3.5	45.2	45.0	45.1
6	3.5	54.8	55.0	54.9
7	3.7	64.4	65.0	64.6
8	4.1	74.0	75.0	74.4
9	4.3	83.7	85.0	84.1
10	4.8	93.3	95.0	93.9

The estimated cumulative percentage distribution for this process can be drawn by plotting the median ranks against their respective measurements on NOPP, as illustrated in Figure 7.41. The smooth curve fitted through the plotted points represents the predicted process cumulative percentage distribution. A straight line can reasonably be fitted through the points, thus supporting the assumption the process output is close to a normal distribution.

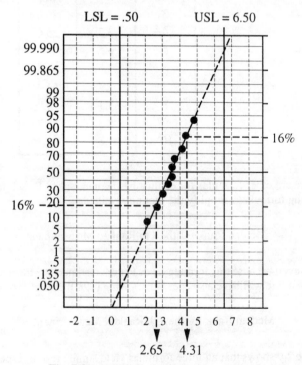

Fig. 7.41 NOPP for median rank example.

Assuming normality, the process average may be estimated by drawing a horizontal line from the 50 percent mark on the y axis over to the fitted line, then down to the x axis. For this example, the estimate of 3.48 is very close to the sample average of 3.50 for these ten measurements, which is also an estimate of the process average.

$$\hat{\mu} = \hat{x}_{50} = 3.48 \qquad\qquad \hat{\mu} = \frac{\sum_{i=1}^{10} X_i}{10} = 3.50$$

The process standard deviation may also be estimated from this NOPP worksheet. However, due to the limited amount of data, finding the .135 and 99.865 percentiles (the $\pm3\sigma$ tails) would involve quite an extended extrapolation of the fitted line. So instead of the 3σ tails, the 1σ tails are used. As there is about 68 percent of the normal distribution between $\mu \pm 1\sigma$, there is 32 percent outside these tails. Half of this, or 16 percent, is below the lower 1σ tail, with a like amount above the upper 1σ tail. Thus, the lower 1σ tail is the 16th percentile, while the upper 1σ tail represents the 84th percentile. Drawing a line from the 16 percent mark on the left-hand y axis to the fitted line, then down to the x axis yields an estimated x_{16} value of 2.66. Drawing a second line from the 16 percent mark on the *right*-hand y axis to the fitted line, then down to the x axis provides an estimate of 4.34 for x_{84}. The difference between these two percentiles, divided by 2, is an estimate of σ_{LT}.

$$2\hat{\sigma}_{LT} = \hat{x}_{84} - \hat{x}_{16} \;\Rightarrow\; \hat{\sigma}_{LT} = \frac{\hat{x}_{84} - \hat{x}_{16}}{2} = \frac{4.31 - 2.65}{2} = .83$$

S_{TOT} can be calculated for the ten measurements and divided by the appropriate c_4 factor to obtain an estimate of σ_{LT}. This result agrees closely with the NOPP estimate of .83.

$$\hat{\sigma}_{LT} = \frac{S_{TOT}}{c_4} = \frac{.790}{.9727} = .81$$

Given a LSL of .50 and an USL of 6.50, process capability may now be assessed from the estimates of μ and σ_{LT} derived from the NOPP.

$$\hat{P}_P = \frac{\text{Tolerance}}{6\hat{\sigma}_{LT}} = \frac{6.50 - .50}{6(.83)} = 1.205$$

$$\hat{P}_{PK} = \text{Minimum}\left(\frac{\hat{\mu} - LSL}{3\hat{\sigma}_{LT}}, \frac{USL - \hat{\mu}}{3\hat{\sigma}_{LT}}\right)$$

$$= \text{Minimum}\left(\frac{3.48 - .50}{3(.83)}, \frac{6.50 - 3.48}{3(.83)}\right)$$

$$= \text{Minimum}(1.197, 1.213) = 1.197$$

Although this process exceeds the minimum requirement for both potential and performance capability, it does not demonstrate attainment of a 1.33 goal. The process appears to be properly centered as the estimates of P_P and P_{PK} are almost equal. In fact, the k factor discloses that less than 1 percent of potential capability is lost due to off-target centering. Therefore, improvement efforts must focus on reducing common-cause variation.

$$\hat{k} = \frac{|\hat{\mu} - M|}{(USL - LSL)/2} = \frac{|3.48 - 3.50|}{(6.50 - .50)/2} = .0067$$

Take care not to rely too heavily on the results of small samples, as another sample of ten may look considerably different than this first one. Chapter 11 explains how to construct confidence bounds for process parameters and capability measures in order to provide some idea of how "soft" these numbers may be.

Also note the problem with this method when there are duplicate measurements. The two median ranks plotted for 3.5 make fitting a line a bit tricky. If there are many duplicate measurements, drawing a line becomes fairly confusing. Other techniques that might be considered for checking the normality assumption with small sample sizes are given by Gunter, Lin and Mudholkar, L.S. Nelson, Ramsey and Ramsey, as well as Zylstra.

Example 7.5 Gap Size

During the development phase of a new automobile program, 16 prototype vehicles are run down a simulated assembly line to check the fit of doors, hood, trunk, and windows. Gap size measurements taken at one particular location of the windshield (windscreen) are plotted on an *IX & MR* chart. Although there is very limited data, the chart indicates the assembly process appeared to operate in a state of statistical control during this trial run. Before estimating any measures of capability, the development engineers decide to check the normality assumption with the median rank method of estimating the cumulative percentage distribution for this process. The 16 gap size measurements are first ranked from lowest to highest, as was done in Table 7.10. Then the corresponding median ranks are found in Appendix Table V and written next to these measurements.

Table 7.10 The ranked gap measurements and their median ranks.

Rank (*i*)	Gap Size	Median Rank
1	.1	4.3
2	.2	10.4
3	.3	16.5
4	.3	22.6
5	.6	28.7
6	.6	34.8
7	.7	40.9
8	.9	47.0
9	1.0	53.0
10	1.2	59.1
11	1.5	65.2
12	1.8	71.3
13	2.0	77.4
14	2.4	83.5
15	3.0	89.6
16	3.9	95.7

When the estimated cumulative percentage distribution is plotted on the NOPP in Figure 7.42, the line of best fit indicates the process most likely has a skewed-right output distribution. Thus, of all the capability measures covered so far in this book, only the *ppm* measure is applicable.

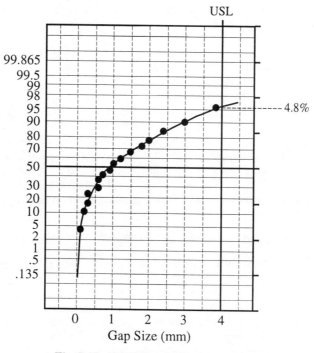

Fig. 7.42 NOPP for windshield gap size.

The development team originally considered setting an USL of 4.0 mm for gap size. If these prototypes accurately represent what will be seen under normal production conditions (which could be a big assumption), approximately 4.8 percent of the gaps will be too large. This percentage translates into 48,000 *ppm*. As this is most likely unacceptable, the specification must be revised, or work done to improve the assembly process, before regular production begins.

$$\hat{p}\,pm_{TOTAL,LT} = .048 \times 10^6 = 48,000$$

Assume no check of the normality assumption is made, and the team calculates the sample average and standard deviation of these 16 measurements.

$$\overline{X} = \frac{\sum\limits_{i=1}^{kn} X_i}{kn} = \frac{20.7}{16} = 1.29375 \qquad S_{TOT} = \sqrt{\frac{\sum\limits_{i=1}^{kn} (X_i - \overline{X})^2}{kn-1}} = \sqrt{\frac{18.969376}{15}} = 1.12456$$

These two statistics would most likely be used to estimate the process average and long-term standard deviation for gap size.

$$\hat{\mu} = \overline{X} = 1.29375 \qquad\qquad \hat{\sigma}_{LT} = \frac{S_{TOT}}{c_4} = \frac{1.12456}{.9835} = 1.14342$$

To discover what percentage of the gaps are expected to be larger than 4.0 mm, the team would calculate a $Z_{USL,\,LT}$ value, then find the corresponding percentage of nonconforming gaps in Appendix Table III.

$$\hat{Z}_{USL,LT} = \frac{USL - \hat{\mu}}{\hat{\sigma}_{LT}} = \frac{4.0 - 1.29375}{1.14342} = 2.37$$

A Z value of 2.37 implies only about .9 percent of the gap measurements can be expected to exceed the proposed USL of 4.0 mm. However, the NOPP analysis discloses that almost 5 percent of the gaps will be above 4.0 mm. Why the large difference? The estimate of .9 percent is based on the process output for gap size having a normal distribution. The curved line on NOPP clearly demonstrates the normality assumption is not valid for this process. Due to its skewed-right distribution, this process output has a much larger percentage of gaps above the USL than that predicted for a normal distribution with the same average and standard deviation. It's crucial to verify *all* assumptions before estimating any capability measure or percentage of nonconforming product.

Advantages and Disadvantages of NOPP

NOPP is a relatively easy graphical method for checking the normality assumption associated with many measures of capability. If a straight line is the best fit for the cumulative percentage plot points, the process output is probably very close to a normal distribution. Madden claims this method is vastly superior to histograms, because if a straight line doesn't fit the points very well, NOPP reveals what type of distribution is most likely present.

For normal distributions, μ may be estimated as the x coordinate where the fitted line intersects the 50 percent line of the y axis. This value is called the 50 percentile and is labeled x_{50} or $\tilde{x}$. For both normal and non-normal distributions, this x coordinate is known as the median, or middle, of the distribution.

The long-term standard deviation is estimated for normal distributions by dividing the difference between the 99.865 percentile and the .135 percentile by 6.

$$\hat{\sigma}_{LT} = \frac{\hat{x}_{99.865} - \hat{x}_{.135}}{6}$$

For both normal and non-normal distributions, the percentage (or *ppm*) of product outside specification limits is readily estimated from NOPP for either unilateral or bilateral specifications. Cumulative percentage distribution curves from different processes can be overlaid on one sheet of NOPP for ease of comparison. Madden explains how realistic tolerances can be determined for critical process parameters, while the next section describes how proposed changes to the process can quickly be simulated and effects analyzed *before* actual adjustments are implemented. Section 7.4 explains how NOPP can establish the centerline and control limits for an *IX & MR* chart when the individual measurements have a non-normal distribution.

Not only can NOPP generate estimates of the P_P and P_{PK} indexes, a cursory glance at this graph reveals how well the process is doing with regard to both centering *and* variability. This valuable insight helps determine what action will be most effective for enhancing process performance. In addition, estimates of process capability may be derived for non-normal distributions, as will be explained in the next chapter.

Unfortunately, there is no easy way with NOPP to estimate μ or σ for non-normal distributions. These process parameters must be calculated with their traditional formulas.

Being a graphical method, different people may fit slightly different curves through the plotted points, especially for non-normal distributions. These differences could affect the estimated percentage of nonconforming parts, as this estimate typically involves the tails of the cumulative distribution where there are fewer plotted points to help guide the fitting of the curve. The application of linear regression eliminates this problem as it fits an identical line through the points no matter who calculates it. Read Crocker or Montgomery for an excellent explanation of this technique.

7.3 Simulating Effects of Process Changes on Process Capability

NOPP is very helpful in simulating how changes in the process parameters affect the process output (Brewer). As shown in this section, NOPP can also help determine the economical machine setting for operating a process that is not capable. Ideally, a curve plotted on NOPP would look like the one presented in Figure 7.43. This curve implies fairly good capability as both 4σ tails are well within the specification limits, indicating this process far exceeds the minimum capability requirement of 99.73 percent conforming parts.

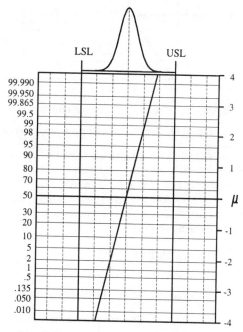

Fig. 7.43 A process with good capability.

Unfortunately, the curve for a sawing operation cutting the length of shafts looks like the one in Figure 7.44. The process output is centered at the middle of the print specifications (at μ_0), but 1.6 percent of the parts are outside either specification limit due to excessive process spread. Part containment and 100 percent inspection must be initiated at once, then work begun to reduce inherent process variation. But until this work is completed, where should the process average be centered to minimize operating costs?

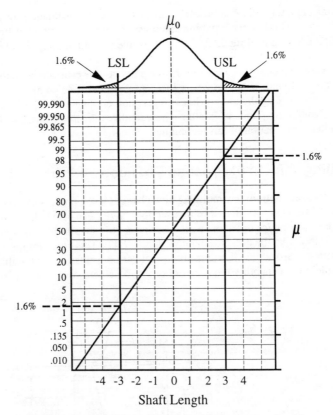

Fig. 7.44 A process with poor capability centered at *M*.

Assume an oversize shaft (length greater than the USL of +3) can be reworked for $1.00, whereas an undersize shaft (length less than the LSL of -3) must be scrapped at a cost of $4.00. With the process centered at 0, the rework and scrap cost for cutting 1,000 shafts is $16 for rework ($1.6\% \times 1,000 \times \$1$) and $64 for scrap ($1.6\% \times 1,000 \times \$4$). This is a total of $80 ($16 + $64) per 1,000 shafts. This loss may also be expressed as an estimated cost per part by using the following formula.

$$\text{Cost per Part} = C_{LSL}(\hat{p}'_{LSL,LT}) + C_{LSL}(\hat{p}'_{USL,LT})$$

where C_{LSL} = Cost for producing a part below the LSL

C_{USL} = Cost for producing a part above the USL

$\hat{p}'_{LSL,LT}$ = Percentage nonconforming below the LSL

$\hat{p}'_{USL,LT}$ = Percentage nonconforming above the USL

In this example, the average cost for scrap and rework is $ 0.080 per shaft.

$$\text{Cost per Shaft} = \$4.00(.016) + \$1.00(.016) = \$0.080$$

Because an undersize shaft is four times as costly as an oversize one, it makes sense to move the process average somewhat higher than 0 to help reduce the total cost of scrap and rework. But how much higher should it be moved, and how much will this move change these costs?

Changing the Process Average

To simulate the effect of moving the process average 1 unit higher, draw a line parallel to the fitted line but located 1 unit to the right of the fitted line, as was done in Figure 7.45a. This second line represents what the cumulative percentage distribution for this process output would look like if it were centered at 1 (μ_1). The percentage of parts below the LSL is now only .25 percent, which should reduce scrap costs by an equal amount. However, the percentage of parts above the USL increases to about 8.0 percent. This 500 percent increase (1.6 to 8.0) in rework is much more than the 84 percent reduction (1.6 to .25) in scrap.

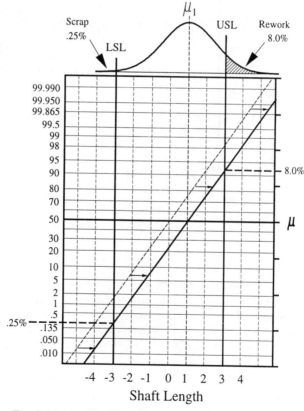

Fig. 7.45a Shifting the process average one unit above *M*.

Cost per 1,000 parts will decrease to just $10 for scrap (.25% × 1,000 × $4), but increase to $80 for rework (8.0% × 1,000 × $1). This new total of $90 per 1000 shafts is *greater* than the initial total of $80, indicating that moving the process average 1 unit higher would be too large of an upward adjustment. Instead, a smaller shift of only 1/2 a unit above the middle of the tolerance ($\mu_{.5} = .5$) should be tried if it is possible to adjust the machine this precisely.

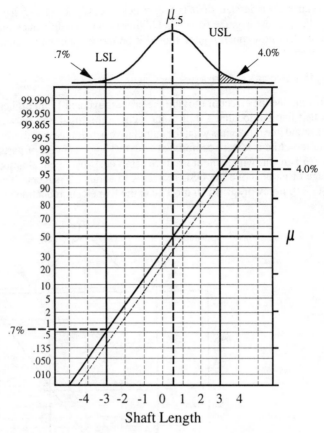

Fig. 7.45b   Shifting the process average one-half unit above *M*.

The effect of this .5 shift is illustrated in Figure 7.45b, where the cost of scrap becomes $28 (1,000 × .7% × $4), while rework is $40 (1,000 × 4.0% × $1). The total cost for running these shafts at an average length of .5 is $68 per 1000, which is less than the $80 when the average was μ_0, as well as the $90 cost when centered at μ_1. These three findings are summarized in Table 7.11. By simply shifting the fitted line on NOPP, costs are easily estimated for various settings of the process average. This paper simulation permits a determination of the optimal setting for μ *before* any parts are actually produced.

Table 7.11   Simulation results for scrap and rework costs.

| | Cost ($) per 1,000 Shafts | | |
Average	Scrap	Rework	Total
μ_0	64	16	80
$\mu_{.5}$	28	40	68
μ_1	10	80	90

An identical analysis can also be performed by first estimating σ_{LT}, then calculating a Z value for both LSL and USL for each simulated setting of the process average. The average cost per shaft is computed for each set of Z values with this formula.

$$\text{Cost per Part} = C_{LSL} \, \Phi(Z_{LSL,LT}) + C_{LSL} \, \Phi(Z_{USL,LT}) = C_{LSL}(p'_{LSL,LT}) + C_{LSL}(p'_{USL,LT})$$

$$\text{where } Z_{LSL,LT} = \text{Simulated Z value for the LSL}$$

$$\text{and } Z_{USL,LT} = \text{Simulated Z value for the USL}$$

However, this approach necessitates frequent trips to a Z table and requires quite a few more calculations than the NOPP method. But if a software program can be written for these formulas, the simulation will be completed in far less time when run on a personal computer.

Be aware that this procedure of shifting of the process average to minimize only internal production costs is not consistent with the Taguchi loss function concept. Under this philosophy, great effort is expended to keep the process centered at the target average in order to minimize *total* costs, both internal and external.

Changes in performance capability are determined by calculating P_{PK} for the various process averages. The potential capability measures should remain constant, assuming the standard deviation does not change when the average is shifted. However, before a P_{PK} value may be computed, an estimate of the long-term standard deviation is needed. This estimate is obtained by dividing the difference between the $x_{.135}$ and $x_{99.865}$ percentiles by 6.

$$\hat{\sigma}_{LT} = \frac{\hat{x}_{99.865} - \hat{x}_{.135}}{6} = \frac{4.2 - -4.2}{6} = 1.4$$

Now P_{PK} can be estimated for this process when centered at 0.

$$\hat{P}_{PK,0} = \frac{1}{3}\text{Minimum}\left(\frac{\hat{\mu} - \text{LSL}}{\hat{\sigma}_{LT}}, \frac{\text{USL} - \hat{\mu}}{\hat{\sigma}_{LT}}\right)$$

$$= \frac{1}{3}\text{Minimum}\left(\frac{0 - -3}{1.4}, \frac{+3 - 0}{1.4}\right)$$

$$= \frac{1}{3}\text{Minimum}(2.14, 2.14) = .71$$

From these Z values of 2.14, $ppm_{TOTAL,\,LT}$ may also be predicted for this process.

$$\hat{p}pm_{TOTAL,LT,0} = \hat{p}pm_{LSL,LT,0} + \hat{p}pm_{USL,LT,0} = 16,180 + 16,180 = 32,360$$

These two measures of performance capability may also be estimated for this process when centered at .5, assuming σ_{LT} doesn't change.

$$\hat{P}_{PK,.5} = \frac{1}{3}\text{Minimum}\left(\frac{\hat{\mu} - \text{LSL}}{\hat{\sigma}_{LT}}, \frac{\text{USL} - \hat{\mu}}{\hat{\sigma}_{LT}}\right)$$

$$= \frac{1}{3}\text{Minimum}\left(\frac{.5 - -3}{1.4}, \frac{+3 - .5}{1.4}\right)$$

$$= \frac{1}{3}\text{Minimum} \, (\, 2.50 \, , \, 1.79 \,) \, = \, .60$$

$$\hat{p}pm_{TOTAL,LT,.5} \; = \; \hat{p}pm_{LSL,LT,.5} + \hat{p}pm_{USL,LT,.5} \; = \; 6,210 + 36,730 \; = \; 42,940$$

Process capability, as measured by P_{PK}, decreases from .71 to .60 when the process average shifts from 0 to .5. This translates into an increase from 32,360 *ppm* to 42,940 *ppm*, both of which are unacceptable. The only way to significantly improve this process, and minimize production costs, is to reduce the amount of process variation. Fortunately, NOPP is also able to predict the effect of changes in the process standard deviation, as will be explained in the next section.

Changing the Process Spread

As just demonstrated, moving the curve **parallel** to its original position simulates the effect of shifting the process average. Because the slope of the fitted line is a measure of the standard deviation, **tilting** the curve about the average, as is shown in Figure 7.46, simulates the effect of decreasing the existing amount of process variation. A 40 percent reduction is simulated by moving the 4σ tails in by 40 percent. In this example, the upper tail is brought in from 5 to 3, and the lower tail from -5 to -3, representing an overall reduction of 40 percent (2/5) in process variation.

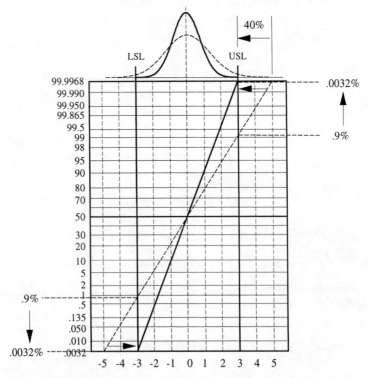

Fig. 7.46 Tilting the fitted line simulates a reduction in process variation.

Operating costs can be estimated for both curves as was done in the previous section when simulating changes in the process average. The difference between these two costs reflects the amount of savings for this reduction in variation, which can then be compared to the cost of the required process improvement. For example, new tooling and fixtures are expected to reduce variation by 40 percent for a process boring the center hole in a flywheel casting, but would cost $15,290 to implement. If the center hole is oversized, the flywheel is scrapped out at a cost of $157. If undersized, the flywheel can be reworked at a cost of $63. The anticipated volume for flywheels is 10,000 units per year.

Based on the NOPP displayed in Figure 7.46, .9 percent of the bores produced with the current tooling will be undersized, while another .9 percent will be oversized. For 10,000 units produced in a year, 90 units ($10,000 \times .009$) would be scrapped out due to oversize holes at a cost of $14,130 ($90 \times \$157$). An additional 90 flywheels would be reworked at a cost of $5,670 ($90 \times \$63$). Thus, the sum of expected yearly scrap and rework costs is $19,800 ($14,130 plus $5,670) for the current tooling.

Installation of the new tooling will reduce scrap to just 32 *ppm* by bringing the $4\sigma_{LT}$ tail from -4 to the LSL of -3, thus reducing yearly scrap costs to just $50 ($10,000 \times .000032 \times \$157$). As a result of the process being centered at the middle of the tolerance, rework also drops to 32 *ppm* when the upper $4\sigma_{LT}$ tail shrinks from +4 to +3, making this cost a meager $20 ($10,000 \times .000032 \times \$63$). The expected total yearly cost for scrap and rework produced by the new tooling is only $70 ($50 plus $20).

A cost comparison of old and new tooling reveals a projected annual savings of $19,730 ($19,800 minus $70). Since this amount is more than the $15,290 required for purchasing the new tooling, buying it makes economic sense as the payback period is less than one year.

$$\text{Payback Period} \; = \; \frac{\text{New Tooling Cost}}{\text{Anticipated Savings}} = \frac{\$15,290}{\$19,730 / \text{Year}} = .77 \text{ Years}$$

Once variation is reduced, additional simulations may be run in an attempt to lower production costs even further through changes in the process average. The procedure for shifting the new fitted line in a parallel manner was explained in the last section. Process capability can also be estimated for any of these simulations as previously demonstrated. For the interested reader, more formal mathematical treatments concerning this type of decision making are proposed by R. Clements, Dodson, Roth, and Schmidt and Pfeifer. Hunter and Kartha suggest applying evolutionary operation (EVOP) for determining the optimal operating conditions of a process.

One note of caution: Be careful about scale changes on the *x* axis of NOPP. These can make the curve appear steeper or flatter, which will affect the perception of process variability. For proper interpretation when comparing the variation of one process to that of another on a different sheet of NOPP, make sure the scales of the *x* axis are identical.

Changing Specification Limits

NOPP can also predict the effect on capability resulting from changes to the specification limits. Suppose the specification limits on the NOPP displayed in Figure 7.47 are to be reduced by 50 percent, from 75 ± 15 to 75 ± 7.5. How much of an effect is there on potential and performance capability? To find out, the process average and standard deviation must first be estimated from NOPP.

$$\hat{\mu} = \hat{x}_{50} = 75 \qquad\qquad \hat{\sigma}_{LT} = \frac{\hat{x}_{99.865} - \hat{x}_{.135}}{6} = \frac{86 - 64}{6} = 3.67$$

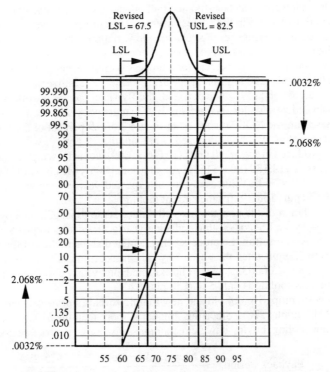

Fig. 7.47 Simulating a 50 percent reduction in the tolerance.

With these, the P_{PK} index can be estimated for this process with the original specifications. As the process is centered at the middle of the tolerance, potential and performance capability are equal.

$$\hat{P}_{P,ORIGINAL} = \hat{P}_{PK,ORIGINAL} = \text{Minimum}\left(\frac{75-60}{3(3.67)}, \frac{90-75}{3(3.67)}\right) = \text{Minimum}\,(1.36, 1.36) = 1.36$$

Redraw the LSL at 67.5 (instead of 60) and the USL at 82.5 (instead of 90). Now estimate P_{PK} with these revised specifications.

$$\hat{P}_{PK,REVISED} = \text{Minimum}\left(\frac{75-67.5}{3(3.67)}, \frac{82.5-75}{3(3.67)}\right) = \text{Minimum}\,(.68, .68) = .68$$

The process capability goes from very good (1.36) to very unacceptable (.68) when the tolerance is cut in half. In fact, the estimated amount of nonconforming parts leaps from 46 ppm to 41,360 ppm.

$$\hat{p}\,pm_{TOTAL,LT,ORIGINAL} = \hat{p}\,pm_{LSL=60,LT} + \hat{p}\,pm_{USL=90,LT} = 23 + 23 = 46$$

$$\hat{p}\,pm_{TOTAL,LT,REVISED} = \hat{p}\,pm_{LSL=67.5,LT} + \hat{p}\,pm_{USL=82.5,LT} = 20,680 + 20,680 = 41,360$$

When this proposed reduction in tolerance becomes a reality, process variation must be decreased by at least 50 percent in order to restore the original high level of process capability and low number of nonconforming parts. If this is not possible, the increased costs associated with producing the expected 41,360 *ppm* should be estimated and reported so management is well aware of the full ramifications of this decision.

Sometimes management falsely believes it can force higher quality parts to be produced by simply tightening tolerances. As just witnessed, reducing a tolerance without a matching process improvement increases the amount of nonconforming parts, which leads to substantial rises in scrap, rework, and inspection costs, and a corresponding drop in productivity. The best approach for increasing quality levels is to implement process improvements, verify their effectiveness by the changes detected on the control chart monitoring the process, then measure the amount of improvement by recalculating process capability. Only *after* a process has been sufficiently improved should consideration be given to reducing the tolerance.

7.4 Using NOPP to Set Control Limits on *IX & MR* Charts

Formulas for the control limits of an *IX* chart assume the process output has a normal distribution. The *UCL* is set at a distance of 3σ above the centerline, while the *LCL* is 3σ below. The 3σ distance is estimated by 2.66 times the average moving range, as is seen in the formulas for the *IX* control limits.

$$LCL_{IX} = \overline{X} - 2.66\overline{MR} \qquad\qquad UCL_{IX} = \overline{X} + 2.66\overline{MR}$$

When a process with a normally distributed output is in control, 99.73 percent of all individual measurements will fall between these control limits. There is only a .135 percent chance an *IX* plot point will fall above the *UCL* and a .135 percent chance of falling below the *LCL* when the process is, in fact, in control. This is called the α risk (Figure 7.48).

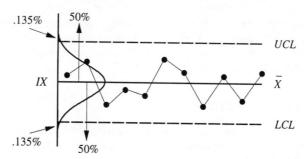

Fig. 7.48 Control limits for the *IX* chart when the process output is normally distributed.

In addition, $\overline{X}$ divides the plot points such that 50 percent of them fall above this centerline and 50 percent fall below, a property used to develop rules for identifying significantly long runs. For example, a shift in the process average has most likely occurred when seven consecutive *IX* plot points fall on the same side of $\overline{X}$. This decision is based on the small probability of seeing seven points in a row on one side of the centerline if the process is still in control.

$$P \ (7 \text{ in a row}) = (.50)^7 = .0078$$

In view of this probability being so small, it is assumed the only way to see a run of seven points is when the process average has shifted. Thus, a run of seven signals an out-of-control condition that should be addressed by the person monitoring the chart.

As a result of individual measurements being plotted on the *IX* chart, the normality assumption must be verified if the centerline and control limits are to function properly for interpretation. NOPP can verify normality and, if it appears the distribution is non-normal, can also help establish control limits for the *IX* chart. For non-normal distributions, estimate $x_{.135}$, x_{50}, and $x_{99.865}$ from NOPP. On the *IX* chart, x_{50} replaces $\overline{X}$ as the centerline (*CL*), $x_{.135}$ becomes the *LCL* and $x_{99.865}$ the *UCL*, as is illustrated in Figure 7.49. This modified chart is sometimes referred to as a Johnson control chart (Pyzdek, 1991, 1992).

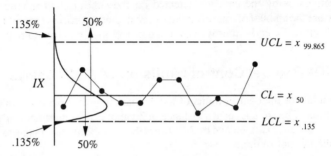

Fig. 7.49 Control limits for the *IX* chart when the process output is non-normally distributed.

An *IX* chart with these control limits has the same false signal rate ($\alpha = .27$ percent) as one for a normally distributed process having conventionally calculated control limits. The rules for testing "runs" is identical as well, because half the *IX* plot points should fall on either side of the median for the *IX* values when the process is stable. However, these "percentile" limits are based on long-term process variation whereas conventional limits are derived from short-term process variation.

Unless very severe, skewness is not usually a problem for calculating control limits on $\overline{X}$ charts due to the central limit theorem. In those few cases where acute skewness does exist, Bai and Choi have devised a procedure for constructing asymmetric control limits based on the amount and direction of skewness.

7.5 Goodness-of-Fit Tests

A more statistically sophisticated method for checking the normality assumption is conducting a goodness-of-fit test. This kind of test calculates how well the distribution of a set of sample data fits a particular theoretical distribution. If there is a "good" fit, then the sample measurements are assumed to come from a process having that type of theoretical distribution. When the fit is poor, a conclusion is made that the process generating the sample measurements does *not* have the proposed theoretical distribution.

Ideally, a histogram of sample measurements taken from a process with a normally distributed output would form a perfect, normal distribution. In reality, random sampling variation causes sample results to differ from the true population, as was seen in previous sections. Thus, a "small" disparity between the distribution of the sample measurements and a normal distribution is expected, as is seen in Figure 7.50. But how much difference can be observed between these two before suspicions arise that the sample data do not come from a process having a normal distribution? This is the question a goodness-of-fit test answers.

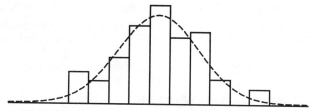

Fig. 7.50 A normal curve superimposed on a histogram.

These types of statistical tests compute a difference between the distribution of sample data and a normal distribution for each class interval. Then this deviation between the actual results and the theoretical is compared to a critical value. If the computed difference is less than the critical value, the sample data are assumed to come from a process having a normal distribution, or one very close to it. On the other hand, when the calculated difference is *greater* than the critical value, the sample measurements are assumed to have originated from a process *not* having a normal distribution.

Although any theoretical distribution can be tested, this book will always apply goodness-of-fit tests to check for normality. Two different tests are discussed, the χ^2 test for the grouped frequency data of large samples, and the Anderson-Darling test for the individual measurements in small samples.

The χ^2 Goodness-of-fit Test

Perhaps the best-known, although definitely not the best statistically according to D'Agnostino and Stephens, is the χ^2 goodness-of-fit test. In applying this test for normality, the actual frequencies in each class of the histogram of sample data are compared to theoretical frequencies expected for a normal distribution. This theoretical normal distribution is based on the process average and standard deviation estimated from the sample data. Of course, to be meaningful, the sample data must be collected from a stable process. For a control chart with k subgroups of size n each, μ and σ are estimated with these formulas:

$$\hat{\mu} = \frac{\sum\limits_{i=1}^{kn} X_i}{kn} \qquad \hat{\sigma}_{LT} = \frac{1}{c_4} \sqrt{\frac{\sum\limits_{i=1}^{kn} (X_i - \hat{\mu})^2}{kn - 1}}$$

Once parameters for the theoretical normal curve are estimated, differences between this distribution and the histogram of sample data are computed. Exactly how this is done is fully explained in the next section. From these differences, a χ^2 test statistic is calculated and then compared to a critical χ^2 value found in Appendix Table VI, which begins on page 854. This critical value is determined by the desired level of confidence (1 minus α), the degrees of freedom, ν ($\nu = j - 3$, where j is the number of class intervals in the histogram), and the assumption the measurements come from a normal distribution.

If the calculated χ^2 statistic exceeds the critical χ^2 value, the sample data most likely did *not* come from a process having a normal output distribution. This conclusion is drawn because the observed differences between theoretical and actual frequencies are too large to be explained by sampling variation alone.

If $\chi^2_{ACTUAL} \geq \chi^2_{CRITICAL}$, assume a non-normal distribution.

If the critical χ^2 value is not exceeded, the process generating the sample data is assumed to have a normal distribution, or one very close to a normal distribution, as the observed differences are small enough to be completely explained by random sampling variation.

$$\text{If } \chi^2_{ACTUAL} < \chi^2_{CRITICAL}, \text{ assume a normal distribution.}$$

Unfortunately, the calculations involved in computing the actual χ^2 test statistic are quite lengthy, as the examples in the next section painfully attest. Fortunately, many SPC software programs now include some type of goodness-of-fit test.

Specific Steps for χ^2 Goodness-of-Fit Test

Follow these ten steps for applying the χ^2 goodness-of-fit test to check for normality. To help better understand this procedure, look at the calculations for the shaft length example given in Table 7.12 as these instructions are read.

1. List the upper (UB) and lower (LB) class boundaries for each class. For example, the fourth class has a midpoint of -1. As the width of each class is 1.0, half of this width (.5) is below the midpoint and half above. Thus, the LB is -1.5 (-1.0 minus .5), while the UB is -.5 (-1.0 plus .5) for this particular class.

2. Calculate the Z value for each upper class boundary (UB) using the estimates of μ and σ_{LT} made from the sample data.

$$Z \text{ value for UB of } i \text{ th class } = \frac{UB_i - \hat{\mu}}{\hat{\sigma}_{LT}}$$

For the example in Table 7.12, μ is estimated as -.2 and σ_{LT} as 1.57. The Z value for the UB of the fourth class (which extends from -1.5 to -.5) is -.19.

$$Z \text{ value for UB of 4th class } = \frac{UB_4 - \hat{\mu}}{\hat{\sigma}_{LT}} = \frac{-.5 - (-.2)}{1.57} = -.19$$

3. The cumulative percentage for each Z value must be found. Because the process output is assumed to have a normal distribution, the expected cumulative percentages for these Z values can be found in Appendix Table IV on pages 848 and 849. For instance, the cumulative percentage associated with the Z value of -.19 for the UB of the fourth class is .4247 (42.47%), meaning 42.47 percent of the shaft lengths lie below -.5.

4. Take successive differences of these cumulative percentages to derive the expected percentage for each class interval under the assumption of normality. For example, the expected cumulative percentage up to the UB of the fourth class is .4247, while the expected cumulative percentage up to the UB of the third class is .2033. Therefore, the expected percentage falling in *just* the fourth class (from -1.5 to -.5) is the difference of .2214 (.4247 minus .2033) between these two cumulative percentages. This result is written in the fifth column of Table 7.12.

Table 7.12 Calculations for the χ^2 goodness-of-fit test.

Class Midpoint	Class LB to UB	Z for UB	Expected Cum. %	% in Class	Expected Freq. (f_E)	Combined f_E
-4	$-\infty$ - -3.5	-2.10	.0179	.0179	.90	-
-3	-3.5 - -2.5	-1.46	.0721	.0542	2.71	3.61
-2	-2.5 - -1.5	-.83	.2033	.1312	6.56	6.56
-1	-1.5 - -.5	-.19	.4247	.2214	11.07	11.07
0	-.5 - .5	.46	.6772	.2525	12.62	12.62
1	.5 - 1.5	1.08	.8599	.1827	9.13	9.13
2	1.5 - 2.5	1.72	.9573	.0974	4.87	7.01
3	2.5 - $+\infty$	$+\infty$	1.0000	.0427	2.14	-
					50.00	50.00

5. Multiply these class percentages by *kn* to calculate the expected frequency, which is denoted by f_E. This is the number of sample observations predicted to be in each class assuming the process has a perfect normal distribution. In this shaft length example, there are 10 subgroups of size 5 each, so *kn* equals 50. The expected frequency for the fourth class is then .2214 times 50, or 11.07.

6. Because this test requires a minimum expected frequency of 3 in each class, several classes at either the high or low end may have to be combined (Cochran). In the shaft length example, the first class has an expected frequency of only .90, so it is combined with the expected frequency of the second class (2.71) for a total of 3.61, which now exceeds the recommended minimum of 3. This combined class now extends from minus infinity to -2.5. The seventh and eighth classes are merged to get an expected frequency of 7.01 (see last column of Table 7.12).

7. Table 7.13 illustrates how the actual frequencies (f_A) are compared to their expected frequencies (f_E) by subtracting the actual frequency for each class from its expected frequency. Each difference is squared, then divided by the expected frequency for that class. If there is a good match between actual and expected frequencies, this result will be rather small. When the actual frequency is significantly different than expected, this result will be large. This is how the "goodness of fit" is measured between theoretical and actual frequencies.

$$\text{The "goodness-of-fit" measure for the } i \text{ th class} = \frac{(f_{E_i} - f_{A_i})^2}{f_{E_i}}$$

For the fourth class in the shaft length example, the expected frequency is 11.07, while the actual frequency is 10.

$$\frac{(f_{E_4} - f_{A_4})^2}{f_{E_4}} = \frac{(11.07 - 10)^2}{11.07} = .1034$$

Table 7.13 Comparing actual to expected frequencies.

Class Midpoint	i	Expected Freq. (f_E)	Actual Freq. (f_A)	$f_{E_i} - f_{A_i}$	$(f_{E_i} - f_{A_i})^2$	$\dfrac{(f_{E_i} - f_{A_i})^2}{f_{E_i}}$
-4	-	-	-	-	-	-
-3	1	3.61	3	.61	.3721	.1031
-2	2	6.56	6	.56	.3136	.0478
-1	3	11.07	10	1.07	1.1449	.1034
0	4	12.62	17	-4.38	19.1844	1.5202
1	5	9.13	8	1.13	1.2769	.1399
2	6	7.01	6	1.01	1.0201	.1455
3	-	-	-	-	-	-
Total		50.00	50	0.00		2.0599

8. The test statistic, χ^2_{ACTUAL}, is the sum of the goodness-of-fit measures calculated in step 7. If these measures are small for most classes, this sum will also be small, implying the sample measurements appear to have originated from a process with a normal distribution. On the other hand, if there are large differences between the actual and expected frequencies for many classes, this sum will be large, suggesting the sample data came from a process with a non-normal distribution. In the shaft length example, this sum is 2.0599.

$$\chi^2_{ACTUAL} = \sum_{i=1}^{j} \frac{(f_{E_i} - f_{A_i})^2}{f_{E_i}} = 2.0599$$

9. How does one know if χ^2_{ACTUAL} is "small" or "large"? This judgment is made by comparing the test statistic to a critical value, labeled $\chi^2_{v,\alpha}$, which is found in Appendix Table VI. v (pronounced "new") is called the degrees of freedom and equals j minus 3 for this particular test. j is the final number of class intervals, which equals the beginning number of intervals minus those which were combined in step 6. In this shaft example, j is 6 (8 minus 2). Additional insight about the number of degrees of freedom is provided in Chapter 11 on page 604.

   With all statistical testing procedures, there is always a chance of selecting an oddball sample that doesn't really reflect the actual process. A goodness-of-fit test run on this unusual sample could lead to a false conclusion. α is the level of this risk associated with concluding the sample measurements came from a non-normal output distribution, when in fact, they really did come from a normal output distribution. Called the significance level of the test, α is typically chosen as .05 (5 percent).

$$\alpha = .05$$

For this shaft length example, there are six categories with expected frequencies, making j equal to 6. With this information, v is calculated as 3.

$$\nu = j - 3 = 6 - 3 = 3$$

From Appendix Table VI, the critical $\chi^2_{\nu,\alpha}$ value for ν equal to 3 and α equal to .05 is found to be 7.8147, as is illustrated in Figure 7.51.

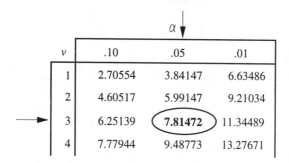

ν	.10	.05	.01
1	2.70554	3.84147	6.63486
2	4.60517	5.99147	9.21034
3	6.25139	7.81472	11.34489
4	7.77944	9.48773	13.27671

Fig. 7.51 Locating $\chi^2_{3,.05}$ in Appendix Table VI.

10. The last step is to make a decision. If χ^2_{ACTUAL} is less than the critical $\chi^2_{\nu,\alpha}$ value, the observed differences in frequencies could be explained by sampling variation, and there is no reason to believe the sample data came from other than a normal distribution. On the other hand, if χ^2_{ACTUAL} is greater than, or equal to the critical $\chi^2_{\nu,\alpha}$ value, the observed differences are too large to be explained by random sampling variation, and it is more probable the sample data came from a process with a non-normal distribution. In the shaft length example, the χ^2_{ACTUAL} of 2.0599 is substantially less than the critical $\chi^2_{\nu,\alpha}$ of 7.8147.

$$\chi^2_{ACTUAL} = 2.0599 < \text{Critical } \chi^2_{\nu,\alpha} = 7.8147$$

Thus, a decision is made that the sample data came from a process output very close to a normal distribution (at the .05 significance level). This is the same conclusion for this sample data as reached with NOPP at the beginning of this chapter (review Figure 7.18 on page 364).

Example 7.7 Waviness of Cam Roller

In Figure 7.40 on page 385, the cumulative percentage distribution for cam roller waviness was plotted on NOPP. Visual interpretation of the NOPP indicated the 124 sample measurements appear to have come from a process with a non-normal output distribution (most likely skewed right). This same set of data can be evaluated with the χ^2 goodness-of-fit test to statistically check for normality at the α equal .05 significance level. From the 124 sample measurements, μ and σ_{LT} are estimated as follows:

$$\hat{\mu} = 35 \qquad \hat{\sigma}_{LT} = 16 \qquad kn = 124$$

These estimates are used to perform the necessary calculations of the χ^2 goodness-of-fit test, as is done in Tables 7.14a and b on the next page.

Table 7.14a Calculations for the χ^2 goodness-of-fit test, part 1.

Class Midpoint	Class LB to UB	Z for UB	Expected Cum. %	% in Class	Expected Freq. (f_E)	Combined f_E
5	- ∞ - 7.5	-1.72	.0427	.0427	5.30	5.30
10	7.5 - 12.5	-1.41	.0793	.0366	4.54	4.54
15	12.5 - 17.5	-1.09	.1379	.0586	7.27	7.27
20	17.5 - 22.5	-.78	.2177	.0798	9.90	9.90
25	22.5 - 27.5	-.47	.3192	.1015	12.59	12.59
30	27.5 - 32.5	-.16	.4364	.1172	14.53	14.53
35	32.5 - 37.5	.16	.5636	.1272	15.77	15.77
40	37.5 - 42.5	.47	.6808	.1172	14.53	14.53
45	42.5 - 47.5	.78	.7823	.1015	12.59	12.59
50	47.5 - 52.5	1.09	.8621	.0798	9.90	9.90
55	52.5 - 57.5	1.41	.9207	.0586	7.27	7.27
60	57.5 - 62.5	1.72	.9573	.0366	4.54	4.54
65	62.5 - 67.5	2.03	.9788	.0215	2.67	5.30
70	67.5 - 72.5	2.34	.9904	.0116	1.44	-
75	72.5 - 77.5	2.66	.9961	.0057	.71	-
80	77.5 - 82.5	2.97	.9985	.0024	.30	-
85	82.5 - + ∞	+∞	1.0000	.0015	.18	-

Table 7.14b Calculations for the χ^2 goodness-of-fit test, part 2.

Class Midpoint	i	Expected Freq. (f_E)	Actual Freq. (f_A)	$f_{E_i} - f_{A_i}$	$(f_{E_i} - f_{A_i})^2$	$\dfrac{(f_{E_i} - f_{A_i})^2}{f_{E_i}}$
5	1	5.30	0	5.30	28.09	5.30
10	2	4.54	3	1.54	2.37	.52
15	3	7.27	16	-8.73	76.21	10.48
20	4	9.90	27	-17.10	292.41	29.54
25	5	12.59	22	-9.41	88.55	7.03
30	6	14.53	20	-5.47	29.92	2.06
35	7	15.77	14	1.77	3.13	.20
40	8	14.53	8	6.53	42.64	2.93
45	9	12.59	6	6.59	43.43	3.45
50	10	9.90	3	6.90	47.61	4.81
55	11	7.27	2	5.27	27.77	3.82
60	12	4.54	1	3.54	12.53	2.76
65	13	5.30	2	3.30	10.89	2.05
					Total	74.95

When these calculations are completed, j and v are determined as 13 and 10, respectively.

$$v = j - 3 = 13 - 3 = 10$$

With an α risk of .05, the critical $\chi^2_{v,\alpha}$ value for 10 degrees of freedom is located in Appendix Table VI.

$$\text{Critical } \chi^2_{v,\alpha} = \chi^2_{10,.05} = 18.31$$

The χ^2_{ACTUAL} is just the total of the last column in Table 7.14b.

$$\chi^2_{ACTUAL} = \sum_{i=1}^{13} \frac{(f_{E_i} - f_{A_i})^2}{f_{E_i}} = 74.95 > \text{Critical } \chi^2_{v,\alpha} = 18.31$$

Because χ^2_{ACTUAL} is much greater than the critical $\chi^2_{v,\alpha}$ value, the difference observed between the sample data and a normal distribution is too great to be attributed to only random sampling variation. The sample data most likely (at the .05 significance level) come from a process having a non-normal distribution.

Note that the χ^2 test does not indicate what other type of distribution the data may have come from, only that it isn't a normal distribution. More work must be done to identify this other type of distribution. In addition, there is no information as to the percentage of non-conforming rollers, nor any way to estimate process capability. With the NOPP method, as soon as a line was fitted through the plot points, it became apparent the output distribution for waviness was skewed right. In addition, the percentage of nonconforming rollers was readily available. The next chapter demonstrates how NOPP also provides estimates of capability for non-normal distributions. There are far fewer calculations associated with NOPP, yet much more useful information is divulged about the process and its capability.

Example 7.8 Metal Block Length

In Figure 7.36, length measurements from a process producing metal blocks were plotted on NOPP. Visually, it appeared these sample measurements came from a process having a definite non-normal distribution, most likely a distribution with fairly extended tails (unbounded). Therefore, estimates of the process average and long-term standard deviation must be derived from the individual length measurements, which was done in Example 7.3.

$$\hat{\mu} = 3.476 \qquad \hat{\sigma}_{LT} = .065 \qquad kn = 100$$

The χ^2 goodness-of-fit test can also be applied to check the normality assumption at the α equal .10 significance level, as is computed in Tables 7.15a and b on the next page. With 6 classes, the number of degrees of freedom is 3.

$$v = j - 3 = 6 - 3 = 3$$

With an α level of .10 and 3 degrees of freedom, the critical $\chi^2_{v,\alpha}$ is 6.2514.

$$\text{Critical } \chi^2_{v,\alpha} = \chi^2_{3,.10} = 6.2514$$

Table 7.15a   Calculations for metal block length, part 1.

Class Midpoint	Class LB to UB	Z for UB	Expected Cum. %	% in Class	Expected Freq. (f_E)	Combined f_E
3.300	-∞ - 3.325	-2.32	.0102	.0102	1.02	-
3.350	3.325 - 3.375	-1.55	.0606	.0504	5.04	6.06
3.400	3.375 - 3.425	-.78	.2177	.1571	15.71	15.71
3.450	3.425 - 3.475	-.02	.4920	.2743	27.43	27.43
3.500	3.475 - 3.525	.75	.7734	.2814	28.14	28.14
3.550	3.525 - 3.575	1.52	.9357	.1623	16.23	16.23
3.600	3.575 - 3.625	2.29	.9889	.0532	5.32	6.43
3.650	3.625 - 3.675	3.06	.9988	.0099	.99	-
3.700	3.675 - +∞	+∞	1.0000	.0012	.12	-

Table 7.15b   Calculations for metal block length, part 2.

Class Midpoint	i	Expected Freq. (f_E)	Actual Freq. (f_A)	$f_{E_i} - f_{A_i}$	$(f_{E_i} - f_{A_i})^2$	$\dfrac{(f_{E_i} - f_{A_i})^2}{f_{E_i}}$
3.350	1	6.06	6	.06	.0036	.0006
3.400	2	15.71	9	6.71	45.0241	2.8660
3.450	3	27.43	32	-4.57	20.8849	.7614
3.500	4	28.14	38	-9.86	97.2196	3.4549
3.550	5	16.23	10	6.23	38.8129	2.3914
3.600	6	6.43	5	1.43	2.0449	.3180
					Total	9.7923

χ^2_{ACTUAL} is the total of the last column in Table 7.15b.

$$\chi^2_{ACTUAL} = \sum_{i=1}^{6} \frac{(f_{E_i} - f_{A_i})^2}{f_{E_i}} = 9.7923$$

Because χ^2_{ACTUAL} is greater than the critical $\chi^2_{v,\alpha}$ value, it doesn't seem likely the sample measurements of block length came from a process output having a normal distribution. This judgment is in agreement with the decision made previously when analyzing this same set of measurements with NOPP.

Unfortunately, the critical χ^2 value, and thus, the test's outcome, is influenced by the size of the class intervals, as well as the starting location for the first interval. Changing one of these will most likely influence the number of measurements grouped into each interval. As both choices are left up to the individual conducting the test, different people could possibly reach opposite conclusions from the same set of data.

The Anderson-Darling Goodness-of-Fit Test

If a capability study is conducted on a process monitored by an *IX & MR* chart, there are typically only 20 to 30 sample measurements available. This is too few to construct a meaningful frequency distribution (W. Nelson), resulting in the χ^2 goodness-of-fit test becoming fairly insensitive due to v being so very small. In these situations, an alternative is to apply the Anderson-Darling test, which performs well with smaller sample sizes by working directly with the individual measurements. Because this test is done on these individual measurements, it has an advantage over the χ^2 method, which is sensitive to how the measurements are grouped into classes.

The six steps for computing the Anderson-Darling test statistic and comparing it to the appropriate critical value are listed below. To better understand this procedure, follow the example displayed in Tables 7.16 and 7.17 as these steps are read. The set of 12 sample measurements is taken from Shapiro.

1. Calculate the average and standard deviation of the sample data with these formulas. In this example, *kn* is 12, $\overline{X}$ is 12.38, and *S* is .37.

$$\overline{X} = \frac{\sum\limits_{i=1}^{kn} X_i}{kn} \qquad S = \sqrt{\frac{\sum\limits_{i=1}^{kn} (X_i - \overline{X})^2}{kn - 1}}$$

2. Rank the *kn* measurements from smallest to largest.

$$X_1 \leq X_2 \leq X_3 \leq \ldots \leq X_{kn}$$

The measurements in Table 7.16 are ranked in this order (*kn* = 12).

Table 7.16 Example for Anderson-Darling calculations.

i	Ranked Order	Z_i	CP_i
1	12.01	-1.00	.159
2	12.07	-0.84	.201
3	12.11	-0.73	.233
4	12.13	-0.68	.248
5	12.14	-0.65	.258
6	12.18	-0.54	.295
7	12.22	-0.43	.334
8	12.34	-0.11	.456
9	12.47	0.24	.595
10	12.88	1.35	.912
11	12.97	1.59	.944
12	13.02	1.73	.958

3. Calculate a Z value for each measurement with the $\overline{X}$ and S computed in step 1.

$$Z \text{ value for } X_i = Z_i = \frac{X_i - \overline{X}}{S}$$

For this example, the Z value for the first ranked measurement is -1.00.

$$Z_1 = \frac{X_1 - \overline{X}}{S} = \frac{12.01 - 12.38}{.37} = -1.00$$

4. Convert these Z values into their respective cumulative percentages (CP) for a normal distribution by using Appendix Table IV on pages 848 and 849.

$$CP_i = P(x \le Z_i)$$

For Z_1, this cumulative percentage is .159.

$$CP_1 = P(x \le Z_1) = P(x \le -1.00) = .159$$

5. Calculate A^2, then use it to determine the A^{2*} test statistic with these formulas, where "ln" is the natural log:

$$A^{2*} = A^2 \left(1 + \frac{.75}{kn} + \frac{2.25}{kn^2} \right)$$

$$\text{where } A^2 = -\frac{\left\{ \sum_{i=1}^{kn} (2i - 1)[\ln(CP_i) + \ln(1 - CP_{kn+1-i})] \right\}}{kn} - kn$$

Determining A^2 is the most difficult calculation of this procedure, as can be deduced from Table 7.17 on the next page. Fortunately, many SPC software programs include this goodness-of-fit test.
The A^2 test statistic is derived from the sum of the last column in Table 7.17 (-155.97396) as follows:

$$A^2 = -\frac{\left\{ \sum_{i=1}^{kn} (2i - 1)[\ln(CP_i) + \ln(1 - CP_{kn+1-i})] \right\}}{kn} - kn = \frac{155.97396}{12} - 12 = .99783$$

A^{2*} is calculated from A^2 and kn.

$$A^{2*} = A^2 \left(1 + \frac{.75}{kn} + \frac{2.25}{kn^2} \right) = .99783 \left(1 + \frac{.75}{12} + \frac{2.25}{12^2} \right) = 1.076$$

Table 7.17 Calculations for Anderson-Darling goodness-of-fit test.

i	$2i-1$	CP_i	$\ln CP_i$	$1 - CP_{12+1-i}$	$\ln(1 - CP_{12+1-i})$	Product
1	1	.159	-1.83885	.042	-3.17009	-5.00894
2	3	.201	-1.60445	.056	-2.88240	-13.46056
3	5	.233	-1.45672	.088	-2.43042	-19.43568
4	7	.248	-1.39433	.405	-0.90387	-16.08736
5	9	.258	-1.35480	.544	-0.60881	-17.67242
6	11	.295	-1.22078	.666	-0.40647	-17.89970
7	13	.334	-1.09661	.705	-0.34956	-18.80023
8	15	.456	-0.78526	.742	-0.29841	-16.25503
9	17	.595	-0.51919	.752	-0.28502	-13.67162
10	19	.912	-0.09212	.767	-0.26527	-6.79029
11	21	.944	-0.05763	.799	-0.22439	-5.92249
12	23	.958	-0.04291	.841	-0.17316	-4.96964
						-155.97396

6. Compare the A^{2*} test statistic to the critical values corresponding to the desired significance level (α) listed in Table 7.18, then make a decision concerning the normality of the process output. If A^{2*} exceeds the critical value for the selected α significance level, then there is 1 minus α confidence that the distribution under study is not normally distributed.

Table 7.18 Critical values for Anderson-Darling test.

	α		
	.10	.05	.01
Critical Value	.631	.752	1.035

In this example, an α of .05 is chosen. Because the calculated A^{2*} value of 1.076 is greater than the critical value of .752, it is very unlikely that these measurements came from a process with a normally distributed output. The process under study cannot be assumed to have a normal distribution, and additional work must be done to find a suitable model for the process output.

Due to the complexity of the calculations required, the Anderson-Darling goodness-of-fit test is hardly ever applied in practice, unless the appropriate software is available.

Other Goodness-of-Fit Tests

There are many other goodness-of-fit tests available, as revealed by the partial listing below. Most of these are much better tests than the χ^2 test, but the calculations associated with them are usually more difficult. Of these, Rodriguez recommends the Kolmogorov-Smirnov, Cramer-von Mises W^2, and Anderson-Darling tests. An excellent reference for learning more about these tests is provided by D'Agnostino and Stephens.

1. Pearson's method of moments
2. Kolmogorov-Smirnov
3. Kuiper V
4. Pyke's C
5. Brunk's B
6. Durbin's B
7. Durbin's M
8. Watson's U^2
9. Fisher's π
10. Hartley-Pfaffenberger S^2
11. Cramer-von Mises W^2
12. Geary's Z
13. Shapiro-Wilks W
14. D'Agnostino's D

7.6 Curve Fitting Non-Normal Distributions

If a process appears to have a non-normal distribution, the next step in a capability study is to determine what type of distribution it does have. Three major methods are listed below for accomplishing this task. However, one must always keep in mind the difference between modeling and curve fitting. Modeling is based on engineering judgment and understanding of the process. For example, tensile strength of copper wire is expected to have a negatively skewed output distribution, like the one presented in Figure 7.52. The maximum material strength forms an upper bound on tensile measurements, but due to slight imperfections in material, some wires could have a much lower strength. The choice of a negatively skewed distribution as a model for the process is based on these theoretical expectations.

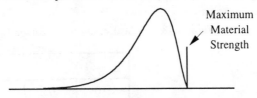

Actual Tensile Strength

Fig. 7.52 A skewed-left distribution.

The three methods presented in this section select a curve based on only how well it fits the actual data, regardless of whether this curve makes theoretical sense or not. So in addition to checking the appropriateness of the curve by comparing the actual percentage of non-conforming product to what the fitted curve predicts, one should also make sure the chosen curve makes sense theoretically, as well as practically.

Normal Probability Paper

The first method for fitting a distribution curve to non-normally distributed data is with NOPP. Fit a line through the points as best as possible by sight. A French curve is often helpful for this task. This curve is an estimate of the cumulative percentage distribution of the process generating the sample data. The percentage of parts outside specification limits can be estimated from this line as well as the median and other percentiles. These estimates play important roles in determining process capability for processes with non-normal distributions, as explained in the next chapter. For the interested reader, Zwick describes a hybrid procedure for curve fitting that utilizes a combination of NOPP and the Kolmogrov-Smirnov goodness-of-fit test to select the best curve.

Data Transformations

Another method is to first "transform" every measurement, then plot them on NOPP (Hoaglin *et al*). For instance, if the sample data are suspected to come from a skewed right (log-normal) distribution, compute the logarithms of the individual measurements. A log transformation compresses the scale for large measurements more than it does for smaller ones, thus shortening the elongated right tail of a positively-skewed distribution (Crow and Shimizu). Plot the cumulative percentage plot points on NOPP using these transformed measurements as the x coordinates. If the plot points of the transformed data follow a reasonably straight line, there is a good indication the sample measurements come from a process having a log-normal distribution. For example, recall the cam roller waviness data from Example 7.4 that plotted as a skewed-right distribution on NOPP. As the individual measurements are grouped into classes, logs of the class midpoints are calculated and written in the second column of Table 7.19.

Table 7.19 NOPP plot point calculations for roller waviness.

Class Midpoint	Log of Midpoint	Freq.		EAF	Plot Points
5	.699	0		0	-
10	1.000	3		3	1.2
15	1.176	16		22	8.9
20	1.301	27		65	26.2
25	1.398	22		114	46.0
30	1.477	20		156	62.9
35	1.544	14		190	76.6
40	1.602	8		212	85.5
45	1.653	6		226	91.1
50	1.699	3		235	94.8
55	1.740	2		240	96.8
60	1.778	1		243	98.0
65	1.813	0		244	98.4
70	1.845	1		245	98.8
75	1.875	0		246	99.2
80	1.903	0		246	99.2
85	1.929	1		247	99.6
90	1.954	0		248	-

When the cumulative percentages are plotted on NOPP against the logs of the midpoints, a fairly straight line fits through the plot points, as is seen in Figure 7.53. This strongly suggests the measurements follow a log-normal distribution.

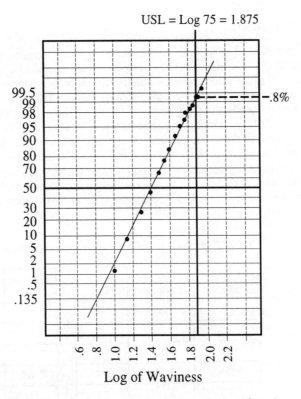

Fig. 7.53 Cumulative percentage versus the log of waviness.

The percentage of nonconforming parts can be estimated by obtaining the logs of the specification limits and drawing them on the NOPP. For the cam roller example, the USL is 75, with the log of 75 being 1.875. This transformed specification is drawn on the NOPP exhibited in Figure 7.53 and indicates that approximately .8 percent of the rollers are expected to have their waviness above the USL. This prediction agrees exactly with the percentage estimated from NOPP in Figure 7.40 on page 385.

Also from Figure 7.40, the 50 percentile for waviness is read as about 26 microinches. From the NOPP in Figure 7.53, the 50 percentile is seen to be about 1.4. The antilog of 1.42 is 25.1, in good agreement with the first estimate.

From the NOPP in Figure 7.53 above, $x_{.135}$ and $x_{99.865}$ are found as .8 and 2.0, respectively. The long-term standard deviation of the log measurements is estimated from these two percentiles to be .2.

$$\hat{\sigma}_{LT} = \frac{x_{99.865} - x_{.135}}{6} = \frac{2.0 - .8}{6} = .2$$

Combining this result with the estimated average of the logs (1.4), a Z value may be computed for the USL.

$$\hat{Z}_{USL,LT} = \frac{USL - \hat{\mu}}{\hat{\sigma}_{LT}} = \frac{1.875 - 1.4}{.2} = 2.38$$

From this Z value, the percentage of rollers with a waviness reading above the USL is estimated as .86 percent, in close agreement to that obtained directly from the NOPP.

What if the logs of the data do not form a straight line on NOPP? Then the log-normal model is not appropriate and another transformation must be tried. If the sample measurements appear skewed right (long tail to the right) with a lower bound of 0, try in addition to log x, ln x, $\sqrt{x}$, $\sqrt[3]{x}$, $-1/\sqrt{x}$, or $-1/x$. When the lower bound is not 0, but say b, then try log $(x - b)$, $\sqrt{x - b}$, or $\sqrt[3]{x - b}$. Note that b must be less than the minimum measurement in the sample.

For a skewed-left distribution, try x^2 or x^3. These transformations compress the scale for smaller measurements more than they do for larger ones, thereby shortening the left tail. If the output is suspected to be bounded, try these following transformations: log $[x / (1 - x)]$, ln $[x / (1 - x)]$, or even arcsin$\sqrt{x}$. All of these expand the scale for both very small and very large measurement values, thus stretching out both tails.

The obvious drawback of this method is having to first guess the type of distribution, transform all the measurements, plot on NOPP, and then judge if a straight line fits best. If not, another transformation must be tried. This could result in quite a bit a work if you're not a good guesser. In addition, the transformed scale is less familiar, and therefore more difficult for some to understand. For additional information on transformations, read Berry, Box and Cox, Carroll and Ruppert, Gaudard and Karson, as well as Lin and Vonesh.

To make the transformation task easier, other probability papers have been developed for many types of distributions: Weibull, logarithmic, exponential, extreme value. With these special graph papers, which are available from TEAM (Tamworth, NH, 603-323-8843), the measurements no longer need to be manually transformed. The actual measurements become the x coordinates for the plot points, while the x- and y-axis scales have been appropriately transformed instead. If a straight line develops on a particular type of graph paper, then there exists strong evidence that the process output has that kind of distribution. For an interesting example of using Weibull probability paper to estimate process spread, see the paper by Goodson.

The cumulative percentage distribution for the cam roller waviness measurements is plotted on the log-normal probability paper shown in Figure 7.54 on the next page. Note the log scale on the x axis where the actual waviness measurements are listed.

Mathematical Curve Fitting

Perhaps the most sophisticated method is to employ mathematical curve-fitting procedures that statistically determine a curve for the process distribution based on the sample measurements. Two of the more popular techniques are the Johnson family of curves (Chou and Polansky, Elderton and Johnson) and Pearson's method of moments (Mirkhani *et al*). A lesser-known, and rarely-used, method has been proposed by Flaig.

Based on the first four moments of the sample measurements, the Pearson curve-fitting procedure select an appropriate distribution from the following: normal, Student's t, log-normal, exponential, beta, gamma, J-shaped, U-shaped, uniform (Pearn and Kotz). These same four sample moments are then used to estimate the parameters of the selected distribution (Gruska *et al*). The first sample moment (m_1) is the average of the sample measurements, while the second sample moment (m_2) is the average of the squared deviations from the sample average.

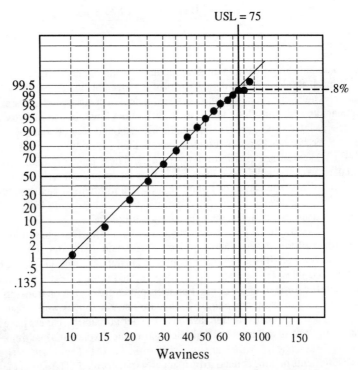

Fig. 7.54 Cumulative percentage versus waviness on log-normal probability paper.

$$m_1 = \frac{1}{n} \sum_{i=1}^{kn} X_i \qquad\qquad m_2 = \frac{1}{n} \sum_{i=1}^{kn} (X_i - m_1)^2$$

The third moment (m_3) is related to the skewness (a_3) of the sample measurements, whereas the fourth moment (m_4) is related to their kurtosis (a_4).

$$a_3 = \frac{m_3}{m_2^3} \quad \text{where } m_3 = \frac{1}{n} \sum_{i=1}^{kn} (X_i - m_1)^3 \qquad a_4 = \frac{m_4}{m_2^2} \quad \text{where } m_4 = \frac{1}{n} \sum_{i=1}^{kn} (X_i - m_1)^4$$

The type of Pearson curve best fitting the data is based on a_3 and a_4. Unfortunately, the mathematics for fitting this distribution to the data is quite lengthy as each type of curve requires solving a different set of complicated equations (Hill *et al*). The concept for this procedure is explained in detail by Kendall and Stuart, while the formulas needed for each curve type are given by Elderton. A much briefer introduction is provided by J. Clements as well as Farnum. Due to the difficult computations, this method is practical only when access to software programs providing this procedure is available.

The Johnson family of curves has three members (Johnson):

    1. Johnson S_L, which represents skewed, or log-normal distributions,

    2. Johnson S_B, for bounded distributions,

    3. Johnson S_U, for unbounded distributions.

Figures 7.23a, 7.23b (p. 369), 7.25a, and 7.25b (p. 370) show typical manufacturing examples of these curves. Like the Pearson method, the family of curve appropriate for a set of data is also determined by a_3 and a_4. For an excellent explanation of this second method, see Chapter 6 of Hahn and Shapiro, or Chapter 16 of Durand. Unfortunately, the Johnson method is similar to the Pearson method in that the required mathematics are also fairly complex, even with the simplified procedures proposed by Slifker and Shapiro, and definitely require a special software program, as well as the help of a qualified statistician.

7.7 Summary

Many measures of capability assume the process output is close to being normally distributed. If not, English and Taylor point out that serious mistakes may arise when assessing capability. There are three popular methods of verifying this important assumption; histograms, NOPP and goodness-of-fit tests. Histograms are easy to construct, but are not very definitive. NOPP is somewhat more involved, but provides a wealth of desirable information about the process, in addition to answering the question concerning normality. Goodness-of-fit tests are the most difficult, but allow checking the normality assumption at a predetermined level of significance, which is why they are preferred by most statisticians.

Simulation of changes in the process average, standard deviation, or specification limits can be conducted with NOPP. The effect of these proposed changes on measures of capability, percentage of nonconforming parts, and trade offs between scrap and rework costs can be quickly evaluated.

Traditional control limits for the *IX* chart assume the plot points are normally distributed. If they aren't, appropriate control limits can be determined from NOPP by using x_{50} as the centerline, $x_{.135}$ as the lower control limit, and $x_{99.865}$ as the upper. Non-normally distributed measurements are analyzed with these probability limits in the same manner (and with similar risks) as normally distributed measurements are with the standard limits.

If a distribution is judged to be non-normal, an appropriate curve must be chosen for the distribution. This selection can be assisted by analyzing NOPP, applying data transformations, or utilizing mathematical curve-fitting techniques, such as the Pearson and Johnson methods. The next chapter explains how this fitted curve helps estimate capability measures for processes with non-normally distributed outputs.

7.8 Exercises

7.1 Slot Width

A 10 cm by 3 cm slot is cut near the center of a control panel for the front of a popular brand of baking oven. During final assembly, a digital timing clock is installed into this slot. To assure there is neither interference nor too large of a gap between these two, the width of the slot must be carefully controlled. A subgroup of five slots is checked every 15 minutes, with the average and range plotted on an $\overline{X}, R$ chart. When the process displays good control, the individual slot widths from the most recent 35 in-control subgroups are organized by size and frequency in Table 7.20. Make a histogram of these measurements, then plot their cumulative distribution on NOPP. What conclusion can be made about the probable shape of the output distribution for this process?

Table 7.20 Frequency tally for control panel slot width.

Class Midpoint	Freq.
2.92	2
2.94	6
2.96	21
2.98	37
3.00	51
3.02	30
3.04	17
3.06	8
3.08	3

Estimate the process average and long-term standard deviation for slot width. Given a specification of $3.00 \pm .08$ cm for slot width, estimate the $6\sigma_{LT}$ spread, as well as the P_P and P_{PK} indexes. Finally, estimate $ppm_{TOTAL, LT}$. Does this process meet the minimum requirement for potential capability? Does it satisfy the minimum requirement for performance capability? How would these estimates change if the specification limits were reduced to $3.00 \pm .06$ cm?

7.2 Powdered Metal Size

Molten metal flows through a small hole, then falls 8 meters inside a tower containing only the inert gas argon. As this stream of metal descends, it forms into drops, which harden by the time they reach the bottom of the tower, creating powdered metal. After cooling, the powder must be sorted into the proper size by passing over a series of vibrating screens. Samples are collected periodically, measured for particle size, and then charted on an $\overline{X}$, R chart, with n equal to 3. When the process becomes reasonably stable, the 123 individual size measurements from the last 41 in-control subgroups are tallied by frequency in Table 7.21 on the facing page. Construct a histogram from this data and evaluate its shape. Analyze the data on NOPP to determine if the process output for size satisfies the normality assumption. If not, what type of distribution seems likely?

With the specifications for particle size being a minimum of 17.5 and a maximum of 23.5, estimate $ppm_{LSL, LT}$, $ppm_{USL, LT}$, and $ppm_{TOTAL, LT}$. What is the median particle size? Where should the median be located to minimize $ppm_{TOTAL, LT}$?

7.3 Thermostat Opening Time

A Brazilian manufacturing company checks thermostats on a special fixture at the end of the assembly line. A certain amount of current is applied, and the time required by the thermostat to open is recorded. After process stability is achieved, individual times from the last 30 in-control subgroups ($n = 5$) are tabulated in Table 7.22 on the facing page. Make a histogram of this data and surmise the shape of the process output distribution, then plot these measurements on NOPP and analyze. The LSL is 75 seconds, while the USL is 105 seconds.

Table 7.21 Frequency tally for powdered metal particle size.

Class Midpoint	Freq.
17.5	1
18.0	13
18.5	11
19.0	16
19.5	17
20.0	15
20.5	18
21.0	16
21.5	15
22.0	1

Table 7.22 Frequency tally for thermostat opening times.

Class Midpoint	Freq.
70	4
75	5
80	7
85	18
90	25
95	41
100	35
105	10
110	5

What percentage of opening times are below the LSL? What percentage is above the USL? Does this process meet the minimum requirement for performance capability (at least 99.73 percent conforming parts)? What is the estimated process median? Between what two opening times does the middle 99.73 percent of this process output lie? How does this span compare to the tolerance of 30 seconds?

7.4 *IX & MR* Chart for Opening Time

A decision is made to switch to an *IX & MR* chart for the opening times of Exercise 7.3. As the output distribution for closing time appears to be non-normal, use the curve on the NOPP to derive control limits for this chart.

7.5 Tab Locator Length

The length of a tab locator on a new model of catalytic converter is considered a critical characteristic. There is only enough money in the budget to produce 24 converters during a pilot run prior to releasing this model for full production. The tab locator lengths of these 24 are plotted on an *IX & MR* chart, with only one point out of control. The remaining 23 measurements are ranked from smallest to largest in Table 7.23. Using the median rank method, plot the cumulative distribution for this process on NOPP. Does the process output for tab locator length appear to have a normal distribution? Find the average of these 23 measurements. Compare this to the median read off the NOPP. Estimate the long-term standard deviation with the S_{TOT} formula. If these readings come from a process with a normal distribution, estimate the long-term standard deviation by computing one sixth of the difference between the .135 and 99.865 percentiles. Compare these results and comment.

Table 7.23 Ranking of tab locator lengths (cm).

Rank	Length	Rank	Length
1	1.458	13	1.501
2	1.467	14	1.502
3	1.471	15	1.506
4	1.478	16	1.509
5	1.481	17	1.511
6	1.485	18	1.514
7	1.488	19	1.516
8	1.490	20	1.520
9	1.492	21	1.524
10	1.494	22	1.531
11	1.497	23	1.539
12	1.499		

If the proposed specification for tab locator length is $1.50 \pm .05$ cm, what percentage of the locators will be too short, and what percentage will be too long? Does this meet the minimum requirement for performance capability? Estimate P_P, P_{PK}, and $ppm_{TOTAL, LT}$.

How much of an increase in the tolerance is necessary so that this process can produce a minimum of 99.73 percent conforming lengths? If engineering refuses to "loosen" the specifications, how much of a reduction in the standard deviation is necessary so this process just meets the minimum requirement for performance capability?

7.6 Wafer Strength

Silicon wafers are checked for strength after a sizing operation, and must be able to withstand a minimum force of 40 psi before breaking. Because of the high cost associated with this destructive testing procedure, only 14 tests are performed. When plotted on an *IX & MR* chart, all 14 measurements appear to be in control. Before estimating any measures of process capability, the normality assumption must be verified. Due to the small number of strength readings, use the median rank approach to accomplish this task. These 14 readings are presented in ranked order in Table 7.24.

Table 7.24 Ranking of 14 wafer strength measurements (psi).

Rank	Strength	Rank	Strength
1	54.4	8	63.4
2	58.1	9	63.4
3	59.8	10	64.0
4	60.7	11	64.5
5	61.4	12	64.6
6	62.3	13	65.3
7	62.6	14	65.5

Does the process output for wafer strength appear to have a normal distribution? Estimate the process average by summing the measurements and dividing by 14. Compare this to the estimate of the process median from the NOPP. Why the difference? What percentage of the wafers are predicted to break when subjected to a force less than the LSL of 40 psi? Does this meet the minimum requirement for performance capability?

As the process output appears to be non-normal, the centerline and control limits of the *IX & MR* chart should not be calculated from the standard formulas. Use the NOPP method to derive modified limits for this chart and compare them to those computed from the standard control limit formulas.

Suppose the normality assumption had not been checked, and both μ and σ_{LT} are estimated directly from the 14 strength measurements. Calculate the resulting $Z_{LSL,\ LT}$ value, find the corresponding percentage nonconforming, and then compare it to that read from the NOPP. Why the difference? Which estimate is better?

7.7 Slot Width Simulation

Using the NOPP of Exercise 7.1, simulate a 20 percent reduction in the process standard deviation. With the original specification of $3.00 \pm .08$ cm, estimate the $6\sigma_{LT}$ spread, the P_P and P_{PK} indexes, and $ppm_{TOTAL,\ LT}$.

Control panels with either an oversized or undersized slot must be scrapped at a cost of $1.89 per panel. If the process improvements required to reduce the standard deviation by 20 percent cost $14,530, how long would it take to recoup this investment. Assume annual production is 24,000 units.

7.8 Floppy Disks

In the floppy disk case study on page 111, the average hole size was computed as 40.9, with the long-term standard deviation estimated at .916. Assuming hole size has a normal distribution, a straight line representing the cumulative distribution for this process can be drawn on NOPP. This line crosses the 50 percent line at the average hole size of 40.9. It will also cross the 99.865 percent line at the average plus $3\sigma_{LT}$, which in this case is 40.9 plus 3(.916), or 43.65. A straight line can now be drawn through these two points on the NOPP.

A floppy with an undersize hole costs the company \$ .39, while one with an oversize hole costs \$ .11. Simulate where the process average should be centered to minimize the total coat of scrap and rework (LSL = 39.5, USL = 44.5). If the standard deviation is reduced to .793, where should the process average be centered to minimize scrap and rework costs?

7.9 Slot Width Goodness-of-Fit Test

Perform a χ^2 goodness-of-fit test to check the normality assumption for the slot width data presented in Table 7.20. Test at the α equals .05 level. Compare the conclusion about normality made from this test to those drawn from studying the histogram and NOPP in Exercise 7.1.

7.10 Powdered Metal Size

Conduct a χ^2 goodness-of-fit test to check the normality assumption for the particle size data presented in Table 7.21. Test at the α equals .10 level. Compare the conclusion about normality made from this test to those drawn from studying the histogram and NOPP in Exercise 7.2.

7.11 Thermostat Opening Times

Run a χ^2 goodness-of-fit test to check the normality assumption for the opening time data presented in Table 7.22. Test at the α equals .01 level. Compare the conclusion about normality made from this test to those drawn from studying the histogram and NOPP in Exercise 7.3.

7.12 Tab Locator Length

Perform the Anderson-Darling goodness-of-fit test on the tab locator length measurements of Exercise 7.5. Use a .05 significance level. Compare the conclusion about normality made from this test to that drawn from analyzing the NOPP in Exercise 7.5.

7.13 Wafer Strength

Conduct an Anderson-Darling goodness-of-fit test on the wafer strength measurements of Exercise 7.6. Use a .10 significance level. Compare the conclusion about normality made from this test to that drawn from analyzing the NOPP in Exercise 7.6.

7.14 Median Ranks

According to the Anderson-Darling goodness-of-fit test, the 12 measurements listed in Table 7.16 were believed to come from a non-normal distribution. Using median ranks, plot these measurements on NOPP and determine the most likely shape of the process output distribution.

7.15 Wafer Strength

Transform the wafer strength data given in Exercise 7.6 by cubing the measurements, then plotting them on NOPP to see if a straight line appears. If this doesn't work, try other data transformations.

7.16 Gap Size

16 gap measurements of windshields are recorded in Table 7.10 on page 390. When plotted on the NOPP in Figure 7.42 (p. 391), the process output distribution appears to be definitely non-normal. The resulting curved line on the probability paper makes extrapolation to a proposed USL of 6.0 mm rather difficult. Apply the cube root transformation on this set of data, then re-plot on NOPP. If a reasonably straight line can be drawn through the points, predict what percentage of the gap sizes will be greater than 6.0 mm. Remember to transform the proposed specification limit as well as the data.

7.17 Rib Width

Check the normality assumption for the wing rib width data in Exercise 2.6 (p. 25) with the median rank method. Find the .135 and 99.865 percentiles, then use them to estimate the $6\sigma_{LT}$ spread, as well as P_P. Compare these results to those computed in Exercise 5.6 (p. 162). Estimate $ppm_{LSL,\ LT}$, $ppm_{USL,\ LT}$, and $ppm_{TOTAL,\ LT}$ from the NOPP. Compare these answers to those obtained in Exercise 6.6 (p. 345) and explain any differences.

7.9 References ▬▬▬▬▬▬▬▬▬▬▬▬▬▬▬▬▬▬

Bai, D.S., and Choi, I.S., "$\overline{X}$ and R Control Charts for Skewed Populations," *Journal of Quality Technology*, Vol. 27, No. 2, April 1995, pp. 120-131

Berry, D.A., "Logarithmic Transformations in ANOVA," *Biometrics*, 43, 1987, pp. 439-456

Bissell, Derek, *Statistical Methods for SPC and TQM*, 1994, Chapman & Hall, London, England

Box, G.E.P., and Cox, D.R., "An Analysis of Transformations (with Discussion)," *Journal of the Royal Statistical Society B*, Vol. 26, 1964, pp. 211-252

Brewer, Robert F., "Use of Probability Plots in Analyzing Experiments," *46th ASQC Annual Quality Congress Transactions*, Nashville, TN, May 1992, p. 1218

Carroll, R.J., and Ruppert, D., "Power Transformations When Fitting Theoretical Models to Data," *Journal of the American Statistical Association*, Vol. 79, 1984, pp. 321-328

Charbonneau, Harvey C., and Webster, Gordon L., *Industrial Quality Control*, 1978, Prentice-Hall, Inc., Englewood Cliffs, NJ, pp. 31-40

Chou, Youn-Min and Polansky, Alan, "Fitting SPC Data Using a Sample Quantile Ratio," *Proceedings of the Section on Quality and Productivity*, American Statistical Association, August 1996, pp. 9-16

Clements, John A., "Process Capability Calculations for Non-Normal Distributions," *Quality Progress*, Vol. 22, No. 9, September 1989, pp. 95-100

Clements, Richard B., *Handbook of Statistical Methods in Manufacturing*, 1991, Prentice-Hall, Inc., Englewood Cliffs, NJ, pp. 60-62

Cochran, W.G., "Some Methods for Strengthening the Common χ^2 Tests," *Biometrics*, Vol. 10, 1954, pp. 47-52

Crocker, Douglas C., *Volume 9: How to Use Regression Analysis in Quality Control*, Quality Press, Milwaukee, WI, 1985

Crow, E.L., and Shimizu, K., eds., *Lognormal Distributions, Theory and Applications*, 1988, Marcel Dekker, Inc. New York, NY

D'Agnostino, Ralph B., and Stephens, Michael A., eds., *Goodness-of-fit Techniques*, 1986, Marcel Dekker, Inc., New York, NY

Dodson, B.L., "Determining the Optimal Target for a Process with Upper and Lower Specifications," *Quality Engineering*, Vol. 5, No. 3, 1993, pp. 393-402

Durand, David, *Stable Chaos*, 1971, General Learning Corp., Morristown, NJ

Duncan, Acheson, J., *Quality Control and Industrial Statistics*, 5th edition, 1986, Irwin, Inc., Homewood, IL, pp. 643-644

Elderton, W.P., *Frequency Curves and Correlation*, 4th ed., 1953, Cambridge Univ. Press

Elderton, W.P., and Johnson, Normal L., *Systems of Frequency Curves*, 1969, Cambridge University Press, New York, NY

English, J.R., and Taylor, G.D., "Process Capability Analysis in Departures from Normality", *Proceedings of the 1st IE Research Conference*, 1992, Chicago, IL, pp. 335-338

Farnum, Nicholas R., "Using Johnson Curves to Describe Non-normal Process Data," *Quality Engineering*, Vol. 9, No. 2, 1996-97, pp. 329-336

Flaig, John J., "A New Approach to Process Capability Analysis," *Quality Engineering*, Vol. 9, No. 2, 1996-97, pp. 205-211

Freund, John E., and Williams, Frank J., *Modern Business Statistics*, 1958, Prentice-Hall, Inc., Englewood Cliffs, NJ, pp. 20-21

Gaudard, Marie, and Karson, Marvin, "Transforming Data to Normality," *Proceedings of the Section on Quality and Productivity*, American Statistical Association, August 1996, pp. 138-143

Goodson, Richard M., "A Simple Method for Estimating Process Spread," *46th ASQC Annual Quality Congress Transactions*, Nashville, TN, May 1992, pp. 1030-1036

Gruska, G. F., Mirkhani, K., and Lamberson, L. R., *Non Normal Data Analysis*, 1989, Applied Computer Solutions, Inc., St. Clair Shores, MI

Gunter, Bert, "The Use and Abuse of Cpk, Part 2" *Quality Progress*, Vol. 22, No. 3, March 1989, pp. 109-109

Gunter, Bert, "Q-Q Plots," *Quality Progress*, Vol. 27, No. 2, February 1994, pp. 81-86

Hahn, Gerald J., and Shapiro, Samuel S., *Statistical Models in Engineering*, 1967, John Wiley & Sons, New York, NY

Hill, I.D., Hill, R., and Holder, R.L., "Fitting Johnson Curves by Moments," *Applied Statistician*, 25, 1976, pp. 180-189

Hoaglin, D.C., *et al*, eds., *Understanding Robust and Exploratory Data Analysis*, 1983, John Wiley, New York, NY

Hunter, William G., and Kartha, C.P., "Determining the Most Profitable Target Value for a Production Product," *Journal of Quality Technology*, Vol. 9, No. 4, October 1977, pp. 176-181

Iman, Ronald, L., and Conover, W.J., *A Modern Approach to Statistics*, 1983, John Wiley & Sons, New York, NY, pp. 153-155

Johnson, Norman L., "System of Frequency Curves Generated by Multiple Methods of Translation," *Biometrika*, 36, 1949, pp. 149-176

Kendal, M.G., and Stuart, A., *The Advanced Theory of Statistics*, Vol. 1, 1958, Hafner Publishing Co., New York, NY

Kimball, B.F., "On the Choice of Plotting Positions in Probability Paper," *Journal of the American Statistical Association*, Vol. 55, 1960, pp. 546-560

King, James R., *Frugal Sampling Schemes*, 1980, Technical and Engineering Aids for Management, Tamworth, NH, p. 22

King, James R., "Elementary Probability Plotting for Statistical Data Analysis," *Quality Progress*, Vol. 21, No. 4, April 1988, p. 62

Lai, K. Chan, Cheng, Smiley W., and Spiring, Fred A., "A Graphical Technique for Process Capability," *42nd ASQC Annual Quality Congress Trans.*, Dallas TX, May 1988, pp. 269-272

Lin, C.C., and Mudholkar, G.S., "A Simple Test for Normality Against Asymmetrical Alternatives," *Biometrika*, 1980, pp. 455-461

Lin, L.I., and Vonesh, E.F., "Empirical Nonlinear Data-Fitting Approach for Transforming Data to Normality," *The American Statistician*, Vol. 43, 1989, pp. 237-243

Lipson, Charles, and Sheth, Narendra J., *Statistical Design and Analysis of Engineering Experiments*, 1973, McGraw-Hill Book Co., New York, NY, pp. 17-18

Madden, Dale A., "Development of Tolerance Limits for an Industry - Ethylene Oxide Sterilization of Medical Devices," *46th ASQC Annual Quality Congress Transactions*, Nashville, TN, May 1992, pp. 741-747

Mirkhani, Kazem, Chitra, Surya, and Lazur, John, "Mixed Moments Estimation for Non-normal Analysis," *41st ASQC Annual Quality Congress Transactions*, Minneapolis, MN, May 1987, pp. 539-545

Montgomery, Douglas, *Introduction to Linear Regression Analysis*, 1982, John Wiley & Sons, New York, NY

Nelson, L.S., "A Simple Test for Normality," *Journal of Quality Technology*, Vol. 13, 1981, pp. 76-77

Nelson, Wayne, *Volume 1: How to Analyze Data with Simple Plots*, 1979, ASQC, Milwaukee, WI, p. 5

Pearn, W.L., and Kotz, S., "Application of Clement's Method for Calculating Second- and Third-Generation Process Capability Indices for Non-Normal Pearsonian Populations," *Quality Engineering*, Vol. 7, No. 1, 1994, pp. 139-145

Pitt, Hy, *SPC for the Rest of Us: A Personal Path to Statistical Process Control*, 1994, Addison-Wesley Publishing Co., Readings, MA, p. 153

Podolski, Garry, "Standard Deviation: Root Mean Square versus Range Conversion," *Quality Engineering*, Vol. 2, No. 2, 1989-1990, p. 161

Pyzdek, Thomas, "Johnson Control Charts," *Quality*, Vol. 30, No. 2, February 1991, p. 41

Pyzdek, Thomas, "Process Capability Analysis Using Personal Computers," *Quality Engineering*, Vol. 4, No. 3, 1992, p. 431

Pyzdek, Thomas, "Why Normal Distributions Aren't [All That Normal]," *Quality Engineering*, Vol. 7, No. 4, 1995, pp. 769-777

Ramsey, Patricia P., and Ramsey, Philip H., "Simple Tests of Normality in Small Samples," *Journal of Quality Technology*, Vol. 22, No. 4, October 1990, pp. 299-309

Rodriguez, Robert N., "Recent Developments in Process Capability Analysis," *Journal of Quality Technology*, Vol. 24, No. 4, October 1992, p. 183

Roth, Gregory, "A Predictive Failure Cost Tool," *Quality*, 32, 10, October 1993, p. 58

Schmidt, Robert L., and Pfeifer, Phillip E., "An Economic Evaluation of Improvements in Process Capability for a Single-Level Canning Problem," *Journal of Quality Technology*, Vol. 21, No. 1, January 1989, pp. 16-19

Shapiro, S.S., *Volume 3: How to Test for Normality and Other Distributional Assumptions*, 1986, ASQC Quality Press, Milwaukee, WI, p. 41

Slifker, J.F., and Shapiro, S.S., "The Johnson System: Selection and Parameter Estimation," *Technometrics*, Vol. 22, No. 2, May 1980, pp. 238-246

Sturge, H. A., "The Choice of Class Interval," *Journal of the American Statistical Association*, Vol. 21, 1926

Terrell, G.R., and Scott, D.W., "Oversmoothed Non-Parametric Density Estimation," *Journal of American Statistical Association*, Vol. 80, 1985, pp. 209-214

Tosch, Thomas J., "Translating Process Capability into Good Products," *1993 Proceedings of the Section on Quality and Productivity*, American Statistical Association, San Francisco, CA, August 1993, p. 198

Wheeler, Donald J., and Chambers, David S., *Understanding Statistical Process Control*, 2nd edition, 1992, SPC Press, Knoxville, TN

Zwick, D., "A Hybrid Method for Fitting Distributions to Data and Its Use in Computing Process Capability Indices," *Quality Engineering*, Vol. 7, No. 3, 1995, pp. 601-613.

Zylstra, R.R., "Normality Tests for Small Sample Sizes," *Quality Engineering*, Vol. 7, No. 1, 1994-1995, pp. 45-58

Chapter 8

Measuring Capability for Non-Normal Variable Data

If the output of a process has a non-normal distribution, the capability measures covered in Chapters 5 and 6 are *not* valid (Farnum; Somerville and Montgomery). Thus, measures not relying on the normality assumption must be developed, as there are many manufacturing processes that typically generate non-normally distributed outputs, even when in control. Below is a list of twenty such characteristics:

1. Taper	11. Squareness
2. Flatness	12. Weld or Bond Strength
3. Surface Finish	13. Tensile Strength
4. Concentricity	14. Casting Hardness
5. Eccentricity	15. Particle Contamination
6. Perpendicularity	16. Hole Location
7. Angularity	17. Shrinkage
8. Roundness	18. Dynamic Imbalance
9. Warpage	19. Insertion Depth
10. Straightness	20. Parallelism

These characteristics share a common trait, in that they are bounded by some physical limit: surface finish cannot be less than zero; weld strength cannot exceed the strength of the material being welded; particle contamination cannot be less than zero. When a process producing this type of characteristic is improved, the average of the output distribution is usually pushed closer and closer to this bound. For example, efforts to reduce surface finish will focus on shifting the process average down to zero. Moving the average closer to this physical limit tends to make the output distribution skewed, as is illustrated in Figure 8.1 with curves 1 through 5. Because the lower tail cannot move below zero, but the upper tail has no upper limit, the distribution of surface finish becomes more and more asymmetric as the average approaches zero. Note that this change in shape from normal to non-normal is desirable as more parts are near the target of zero with curve 5 than with curve 1.

There are two recognized methods for estimating a measure of capability for non-normally distributed variable data:

1. using the .135 and 99.865 percentiles,

2. using the percentage nonconforming.

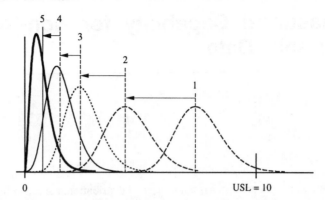

Fig. 8.1 The output distribution for surface finish is often skewed.

Both approaches are explained in this chapter, beginning with the percentile method, which is the more popular of the two.

8.1 Measuring Capability Based on the .135 and 99.865 Percentiles

This first method of measuring process capability for a process with a non-normally distributed output involves the use of certain percentiles. With this procedure, the following capability measures may be estimated:

1. Equivalent $6\sigma_{LT}$ spread
2. Equivalent P_P and P_R
3. Equivalent $P'_{RL}, P'_{RU}, P'_{PL},$ and P'_{PU}
4. Equivalent P^*_P and P^*_R
5. Equivalent P_{ST} and P^*_{ST}

6. Equivalent P'_{STL} and P'_{STU}
7. Equivalent $P_{PL}, P_{PU},$ and P_{PK}
8. Equivalent $P^*_{PL}, P^*_{PU},$ and P^*_{PK}
9. Equivalent $Z_{LSL, LT}, Z_{USL, LT},$ and $Z_{MIN, LT}$
10. Equivalent $Z^*_{LSL, LT}, Z^*_{USL, LT},$ and $Z^*_{MIN, LT}$

The percentile values required for calculating these special measures are typically estimated from the cumulative distribution drawn on NOPP for the process output. Thus, the "*P*" and "*LT*" labels representing long-term process capability are appropriate, as the points plotted on NOPP are computed by using *all* individual measurements from in-control subgroups. As explained in Chapter 7, this technique provides an estimate of the long-term process standard deviation, σ_{LT}, as it combines variation originating from both within and between subgroups.

The term "equivalent" distinguishes these specially calculated measures from the standard capability indexes. "Equivalent" was chosen because these new measures are based on a distance that is equivalent to the $6\sigma_{LT}$ spread of a normal distribution.

Example 8.1 Hole Size

Control is finally established for an operation drilling a hole in a bearing support bracket for a missile's inertial guidance system (see Figure 8.2). An estimate of σ_{LT} for the inner diameter of this hole is usually calculated directly from the individual measurements with the following formulas:

Fig. 8.2

$$\hat{\sigma}_{LT} = \frac{S_{TOT}}{c_4} \quad \text{where} \quad S_{TOT} = \sqrt{\frac{\sum\limits_{i=1}^{kn}(x_i - \hat{\mu})^2}{kn - 1}}$$

For normal distributions, $6\hat{\sigma}_{LT}$ predicts the span required to produce the middle 99.73 percent of the process output and is one measure of potential capability.

$$6\hat{\sigma}_{LT} \text{ Spread} = 6\hat{\sigma}_{LT}$$

But for this $6\hat{\sigma}_{LT}$ spread to accurately represent the middle 99.73 percent of the process output requires the output to have a normal distribution. Thus, before making any decisions based on this measure of capability (in fact, before even estimating it), the normality assumption must be verified. To check, the hole size measurements are plotted on the NOPP exhibited in Figure 8.3. The curve drawn through the plotted points indicates the process output distribution for hole size is non-normal and appears to be skewed right. Hence, the standard $6\hat{\sigma}_{LT}$ spread is inappropriate for measuring the potential capability of this hole-drilling operation, and a different technique is needed for estimating the width of the middle 99.73 percent spread. The measure presented in the next section accomplishes this task by using the .135 and 99.865 percentiles of the process output distribution.

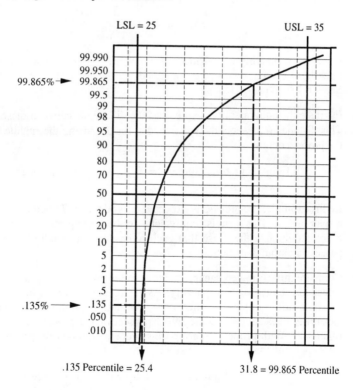

Fig. 8.3 $x_{.135}$ and $x_{99.865}$ are estimated from NOPP.

8.2 Equivalent $6\sigma_{LT}$ Spread

After a curve is fitted to the points plotted on a piece of NOPP, estimate the equivalent $6\sigma_{LT}$ spread (one which captures the middle 99.73 percent of the process output) by following these four steps. To illustrate the procedure, these steps use information derived from the NOPP displayed in Figure 8.3 for the inner diameter of a hole drilled in a support bracket.

1. Draw a horizontal line from the 99.865 percent mark on the y axis to the curve. At the intersection, draw a vertical line down to the x axis and read the 99.865 percentile, which is labeled $x_{99.865}$. For the support bracket example, $x_{99.865}$ is estimated as 31.8.

$$99.865 \text{ Percentile} = \hat{x}_{99.865} = 31.8$$

Because 99.865 percent of the process output lies below this percentile, .135 percent is *above* 31.8.

2. Draw a horizontal line from the .135 percent mark on the y axis to the curve on the NOPP drawn through the plot points. At the intersection, draw a vertical line down to the x axis and read the .135 percentile, or $x_{.135}$. For the support bracket example, $x_{.135}$ is estimated as 25.4, meaning .135 percent of the holes sizes are *below* this value.

$$.135 \text{ Percentile} = \hat{x}_{.135} = 25.4$$

Instead of estimating $x_{99.865}$ and $x_{.135}$ from NOPP, Farnum (as well as Pyzdek) explains how these percentiles can be computed via the Johnson system of curves (review Section 7.6 on page 418).

3. Estimate the equivalent $6\sigma_{LT}$ spread as the difference between these two percentiles. This distance represents the span needed for producing the middle 99.73 percent of the process output.

$$\text{Equivalent } 6\hat{\sigma}_{LT} \text{ Spread} = \hat{x}_{99.865} - \hat{x}_{.135} = 31.8 - 25.4 = 6.4$$

In this example, the process is expected to produce 99.73 percent of the hole sizes in a span of 6.4. This span is "equivalent" to the $6\sigma_{LT}$ spread of a normally distributed process output in the sense that each represents the middle 99.73 percent of their process's output distribution.

4. Compare this estimated capability index to the goal for this characteristic, then make a decision about process capability. As the tolerance for hole size is 10 (35 minus 25), this hole-drilling operation meets the minimum requirement for long-term potential capability because the middle 99.73 percent of the process output is less than the tolerance. With proper centering, this process could easily produce a minimum of 99.73 percent conforming hole sizes.

$$\text{Equivalent } 6\hat{\sigma}_{LT} \text{ Spread} \leq \text{Tolerance}$$

$$6.4 \leq 35 - 25$$

$$\leq 10$$

Suppose the long-term potential capability goal for hole size is specified as having the equivalent $6\sigma_{LT}$ spread less than 75 percent of the tolerance. With a tolerance of 10, and a spread of only 6.4 for the middle 99.73 percent of its output, this process also surpasses this capability goal.

$$\text{Equivalent } 6\hat{\sigma}_{LT} \text{ Spread} \leq \text{Capability Goal}$$

$$6.4 \leq .75\,(\text{Tolerance})$$

$$\leq .75(35 - 25) = 7.5$$

Rating Scale

Conveniently, this measure of capability for non-normal distributions has the identical rating scale as the $6\sigma_{LT}$ spread does for normal distributions (review page 88).

Equivalent $6\sigma_{LT}$ Spread < Tolerance   means the minimum requirement for long-term potential capability is exceeded: the process has the ability to produce more than 99.73 percent conforming parts with proper centering.

Equivalent $6\sigma_{LT}$ Spread = Tolerance   indicates the minimum prerequisite for long-term potential capability is just met: the process has the ability to produce at most 99.73 percent conforming parts with proper centering.

Equivalent $6\sigma_{LT}$ Spread > Tolerance   discloses the minimum requirement for long-term potential capability is *not* satisfied: the process lacks the ability to produce 99.73 percent conforming parts, even with proper centering.

A similar rating scale is an attractive advantage of this measure, as it allows a direct comparison of capability for processes with non-normal output distributions to those with normal output distributions.

Goals

As the rating scale for the equivalent $6\sigma_{LT}$ spread is similar to the standard $6\sigma_{LT}$ spread, the goals are set in a similar manner as well. A very common long-term potential capability goal is to have the equivalent $6\sigma_{LT}$ spread be less than 75 percent of the tolerance.

$$\text{Goal for Equivalent } 6\sigma_{LT} \text{ Spread} = .75 \times \text{Tolerance}$$

Two other often-used goals are 60 percent of the tolerance and 50 percent of the tolerance.

Concept of an Equivalent Spread

The label "equivalent" for this capability measure is chosen because the percentage between these two percentiles is equivalent to the percentage contained in the $6\sigma_{LT}$ spread of a normal distribution, which is the middle 99.73 percent. To better understand this concept, consider a process with a normally distributed output, like the one displayed in Figure 8.4.

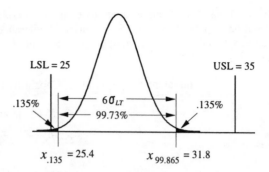

Fig. 8.4 Calculating the $6\sigma_{LT}$ spread for a normal distribution with percentile values.

The $6\sigma_{LT}$ spread represents the width required to produce the middle 99.73 percent of the process output. For normal distributions, this measure of potential capability is computed by multiplying σ_{LT} by 6. In Figure 8.4, σ_{LT} is 1.0667, resulting in a $6\sigma_{LT}$ spread of 6.4.

$$6\sigma_{LT}\text{ Spread } = 6\sigma_{LT} = 6\,(1.0667) = 6.4$$

For a process with a normal distribution, the length of the $6\sigma_{LT}$ spread is also equal to the difference between the 99.865 and .135 percentiles. If the individual measurements from this process are plotted on NOPP, these two percentiles are readily available. Their difference also measures the span required to produce the middle 99.73 percent of the process output.

$$6\sigma_{LT}\text{ Spread } = x_{99.865} - x_{.135} = 31.8 - 25.4 = 6.4$$

For a normal distribution, the spread of 6.4 for the middle 99.73 percent can be determined with either method.

Now consider a process with a non-normal distribution, such as the skewed process output distribution for bearing hole size from the previous example, which is shown in Figure 8.5.

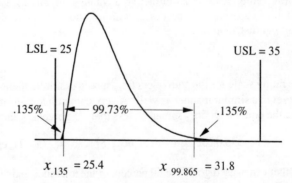

Fig. 8.5 Equivalent $6\sigma_{LT}$ spread calculated with percentiles for a skewed distribution.

If the difference between $x_{99.865}$ and $x_{.135}$ for this process is 6.4, then it has the same potential capability as the process displayed in Figure 8.4, because both require an identical span to produce the middle 99.73 percent of their output distributions.

$$\text{Equivalent } 6\sigma_{LT} \text{ Spread } = x_{99.865} - x_{.135} = 31.8 - 25.4 = 6.4$$

This "equivalent" $6\sigma_{LT}$ spread measure for non-normal outputs has the same meaning as the $6\sigma_{LT}$ spread measure for normal distributions, as both are derived from the difference between the 99.865 and .135 percentiles. The basic definition of potential capability considers whether or not a process has the ability to produce at least 99.73 percent conforming parts. This particular percentage was originally selected because it represents the area underneath a normal curve between its upper $3\sigma_{LT}$ tail ($\mu + 3\sigma_{LT}$) and its lower $3\sigma_{LT}$ tail ($\mu - 3\sigma_{LT}$).

$$6\sigma_{LT} \text{ Spread } = (\mu + 3\sigma_{LT}) - (\mu - 3\sigma_{LT}) = 6\sigma_{LT}$$

The $6\sigma_{LT}$ spread is simply a measure of the distance between these two tails. The smaller this distance, the better the long-term potential capability. For a process with a normally distributed output, this distance may be figured in one of two ways: the difference between the $3\sigma_{LT}$ tails, as was done above; or the difference between $x_{99.865}$ and $x_{.135}$, as is shown below.

$$6\sigma_{LT} \text{ Spread } = x_{99.865} - x_{.135}$$

However, for non-normal distributions, this distance can be found in only one way: the difference between the 99.865 and .135 percentiles.

$$\text{Equivalent } 6\sigma_{LT} \text{ Spread } = x_{99.865} - x_{.135}$$

As this is identical to the second method for normal distributions, the equivalent $6\sigma_{LT}$ formula is more general in that it can be applied to all types of distributions, both normal and non-normal.

Example 8.2 Forging Hardness

Lynette is a manufacturing engineer undertaking a capability study on an operation responsible for the hardness of an aluminum forging. Once control of this process is demonstrated, she plots the cumulative percentage distribution for hardness on the piece of NOPP displayed in Figure 8.6. After examining the NOPP, Lynette decides the process output most likely has a bounded distribution.

Realizing the standard $6\sigma_{LT}$ spread capability measure is inappropriate, this engineer chooses to estimate potential process capability with the equivalent $6\sigma_{LT}$ spread measure. Once the .135 and 99.865 percentiles are estimated from the NOPP, she can determine the range of hardness required to produce the middle 99.73 percent of the process output.

$$\text{Equivalent } 6\hat{\sigma}_{LT} \text{ Spread } = \hat{x}_{99.865} - \hat{x}_{.135} = 182 - 132 = 50$$

An estimated equivalent $6\sigma_{LT}$ spread of 50 implies the process meets the minimum requirement for capability, because at least one middle 99.73 percent spread could fit within the tolerance of 60 (190 minus 130). However, if the potential capability goal is to have the equivalent $6\sigma_{LT}$ spread be no more than 75 percent of the tolerance, this forging process falls somewhat short.

$$\text{Equivalent } 6\hat{\sigma}_{LT} \text{ Spread } \geq .75 \times \text{Tolerance}$$

$$50 \geq .75(190 - 130) = 45$$

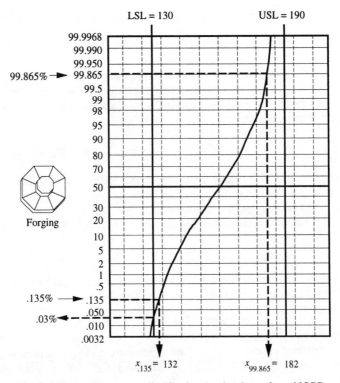

Fig. 8.6 Estimating percentiles for forging hardness from NOPP.

In order to meet this goal, work must begin on reducing process spread.

Advantages and Disadvantages

This first measure of capability for non-normal distributions is fairly easy to understand as it is just the span required to produce the middle 99.73 percent of the process output. This definition is similar in concept to the standard $6\sigma_{LT}$ spread and can be computed in an identical fashion: as the difference between $x_{99.865}$ and $x_{.135}$. Because of this similarity, the equivalent $6\sigma_{LT}$ spread has the same rating scale and, thus, may be compared directly to the $6\sigma_{LT}$ spread for normal distributions. In addition, their goals are similar as well.

The equivalent $6\sigma_{LT}$ spread formula works for both unilateral and bilateral tolerances, without any special modifications. However, on account of the units of measurements associated with this measure, the equivalent $6\sigma_{LT}$ spread for one type of process cannot easily be compared to the equivalent $6\sigma_{LT}$ spread of another.

Another disadvantage is that measurement data must be plotted on NOPP to determine the percentiles, which does require some time, although NOPP is needed anyway to check the normality assumption. Precisely reading percentiles off the x-axis scale of the NOPP can also be difficult. Yet, once the equivalent $6\sigma_{LT}$ spread is determined, several other familiar measures of potential capability may also be estimated for characteristics with non-normally distributed outputs.

8.3 Equivalent P_P and Equivalent P_R

After a curve is fitted to the points plotted on a piece of NOPP, and the equivalent $6\sigma_{LT}$ spread determined, the equivalent P_P index is estimated with this modified formula incorporating $x_{99.865}$ and $x_{.135}$ (Bissell, p. 260):

$$\text{Equivalent } \hat{P}_P = \frac{\text{Tolerance}}{\text{Equivalent } 6\hat{\sigma}_{LT} \text{ Spread}} = \frac{\text{USL - LSL}}{\hat{x}_{99.865} - \hat{x}_{.135}}$$

As before, the label "equivalent" is chosen because the percentage of the process output distribution between these two percentiles is identical to the percentage in the $6\sigma_{LT}$ spread of a normal distribution. This equivalent $6\sigma_{LT}$ spread, representing the *middle* 99.73 percent of the output distribution, is then used to calculate equivalent P_P, just as was done for processes having normally distributed data (review Section 5.8 on page 130). Similar to P_P, this index expresses the number of equivalent $6\sigma_{LT}$ spreads that will fit within the tolerance. Obviously, at least one equivalent $6\sigma_{LT}$ spread must be able to fit in the tolerance if the process is to be judged as meeting the minimum requirement for potential capability.

Rating Scale

With the above definition, the rating scale for equivalent P_P becomes identical to that for the standard P_P index given back on page 131.

Equivalent $P_P > 1$ indicates that the minimum requirement for long-term potential capability is exceeded. The process has the ability to produce at least 99.73 percent conforming parts with proper centering, as is shown in Figure 8.7a.

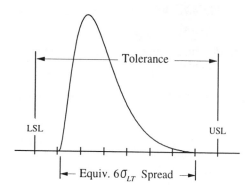

Fig. 8.7a A process displaying good potential capability.

Equivalent $P_P = 1$ means that the minimum prerequisite for long-term potential capability is just met, *i.e.*, process has the ability to produce at most 99.73 percent conforming parts with proper centering. This means only one equivalent $6\sigma_{LT}$ spread could fit within the tolerance, as is illustrated in Figure 8.7b on the next page.

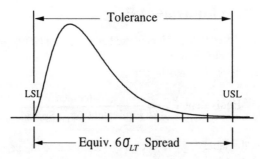

Fig. 8.7b A process displaying marginal potential capability.

Equivalent $P_P < 1$ implies that the minimum stipulation for long-term potential capability is *not* met. The process lacks the ability to produce 99.73 percent conforming parts, even with proper centering, as is depicted below in Figure 8.7c.

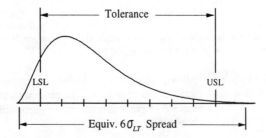

Fig. 8.7c A process displaying poor potential capability.

The equivalent P_R ratio is defined with the following formula, which is just the inverse of the formula for the equivalent P_P index:

$$\text{Equivalent } P_R = \frac{\text{Equivalent } 6\sigma_{LT} \text{ Spread}}{\text{Tolerance}} = \frac{x_{99.865} - x_{.135}}{\text{USL - LSL}} = \frac{1}{\text{Equivalent } P_P}$$

This ratio represents the percentage of the tolerance required to produce one equivalent $6\sigma_{LT}$ spread. In order to meet the minimum criterion for potential capability, the equivalent $6\sigma_{LT}$ spread must be less than the tolerance. When this is true, equivalent P_R is less than 1.0, as is seen here:

$$\text{Equivalent } 6\sigma_{LT} \text{ Spread } < \text{ Tolerance}$$

$$x_{99.865} - x_{.135} < \text{ Tolerance}$$

$$\frac{x_{99.865} - x_{.135}}{\text{Tolerance}} < 1$$

$$\text{Equivalent } P_R < 1$$

If the equivalent $6\sigma_{LT}$ spread is greater than the tolerance, equivalent P_R is greater than 1.0, and the process lacks potential capability.

$$\text{Equivalent } 6\sigma_{LT} \text{ Spread } > \text{ Tolerance}$$

$$x_{99.865} - x_{.135} > \text{ Tolerance}$$

$$\frac{x_{99.865} - x_{.135}}{\text{Tolerance}} > 1$$

$$\text{Equivalent } P_R > 1$$

Rating Scale for Equivalent P_R

These relationships establish a rating scale for equivalent P_R that mirrors the one for the standard P_R ratio.

Equivalent $P_R < 1$ means that the minimum prerequisite for long-term potential capability is exceeded: process has the ability to produce at least 99.73 percent conforming parts with proper centering.

Equivalent $P_R = 1$ shows that the minimum requirement for long-term potential capability is just met: process has the ability to produce at most 99.73 percent conforming parts with proper centering.

Equivalent $P_R > 1$ reveals that the minimum requirement for long-term potential capability is *not* met: process cannot produce 99.73 percent conforming parts, even with proper centering.

Both of these equivalent measures for non-normally distributed process outputs may be directly compared to the respective P_P and P_R measures calculated for normally distributed outputs. However, to realize full potential capability, the process average for a non-normal distribution may have to be centered somewhere other than the middle of the tolerance. More details on this subject are presented in a later section (see Example 8.9 on page 463).

Goals

Goals for equivalent P_P are comparable to those for the standard P_P index, with a common goal being 1.33. Other often-seen goals are 1.50, 1.67, and 2.00. Typical goals for equivalent P_R are .75, .67, .60, and .50, identical to those for the standard P_R ratio.

Example 8.3 Bearing Support Hole Size

Measurements for the inner diameter of a hole drilled in a bearing support bracket for a missile's inertial guidance system are analyzed on NOPP. The curve drawn through the plotted points in Figure 8.3 indicates the process output distribution for this characteristic appears to be skewed right. An estimate of the 99.865 percentile is read off the x axis of the NOPP as 31.8. Likewise, the .135 percentile is found to be 25.4. Given a tolerance for hole size of 10 (35 minus 25), equivalent P_P is estimated as follows:

$$\text{Equivalent } \hat{P}_P = \frac{\text{Tolerance}}{\hat{x}_{99.865} - \hat{x}_{.135}} = \frac{35 - 25}{31.8 - 25.4} = 1.56$$

An equivalent P_P of 1.56 means it is possible to fit the middle 99.73 percent of this process output into the tolerance 1.56 times. Because this exceeds the minimum capability requirement of being able to fit at least one such spread into the tolerance, this process is considered capable. If the long-term potential capability goal is 1.33, this drilling operation also easily meets this additional demand.

Equivalent P_R for hole size is estimated with this formula:

$$\text{Equivalent } \hat{P}_R = \frac{\hat{x}_{99.865} - \hat{x}_{.135}}{\text{Tolerance}} = \frac{31.8 - 25.4}{35 - 25} = .64$$

This second measure may also be determined by computing the inverse of the equivalent P_P index.

$$\text{Equivalent } \hat{P}_R = \frac{1}{\text{Equivalent } \hat{P}_P} = \frac{1}{1.56} = .64$$

An equivalent P_R ratio of .64 implies the span for the middle 99.73 percent of this process's output equals 64 percent of the tolerance. If properly centered, less than .27 percent non-conforming hole sizes would be produced. Consequently, this process meets the minimum requirement for potential capability. If the goal is to have equivalent P_R less than .75, the process also meets this additional stipulation. Unfortunately, just as with the standard measures, the actual percentage of nonconforming holes cannot be derived from either the equivalent P_P or equivalent P_R measures. Instead, NOPP must be used for estimating $p'_{LSL.LT}$ and $p'_{USL.LT}$.

Concept Behind Equivalent P_P

Consider a process with a normally distributed output like the one exhibited in Figure 8.8a.

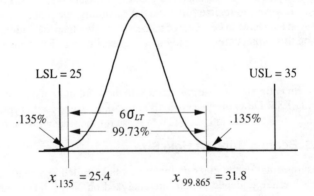

LSL = 25   USL = 35

.135% $6\sigma_{LT}$ .135%

                                        99.73%

$x_{.135} = 25.4$ $x_{99.865} = 31.8$

Fig. 8.8a Calculating P_P for a normal distribution.

For a process output with a normal distribution, P_P reveals how many $6\sigma_{LT}$ spreads could fit into the tolerance. With σ_{LT} equal to 1.067, P_P is calculated as follows:

$$P_P = \frac{\text{Tolerance}}{6\sigma_{LT}} = \frac{35 - 25}{6(1.067)} = \frac{10}{6.4} = 1.56$$

Note that the length of this $6\sigma_{LT}$ spread is identical to the difference between the 99.865 and .135 percentiles. If $x_{.135}$ is 25.4 and $x_{99.865}$ is 31.8, P_p could also be computed for this process as follows (Chan *et al*, 1988):

$$P_p = \frac{\text{Tolerance}}{x_{99.865} - x_{.135}} = \frac{35 - 25}{31.8 - 25.4} = \frac{10}{6.4} = 1.56$$

Now consider a process with a non-normal distribution, such as the skewed distribution for hole size from Example 8.1, which is shown in Figure 8.8b. For non-normal distributions, $6\sigma_{LT}$ doesn't necessarily represent the middle 99.73 percent of the process output and, thus, cannot be used to determine potential capability under the traditional definition.

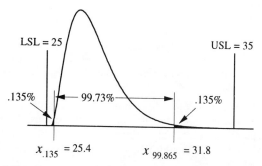

Fig. 8.8b Calculating equivalent P_p for a skewed distribution.

However, if the distance between the 99.865 and .135 percentiles equals 6.4, this process has the same potential capability as the prior example for a normal distribution, as the middle 99.73 percent of its output distribution will fit into the tolerance an equal number of times.

$$\text{Equivalent } P_p = \frac{\text{Tolerance}}{x_{99.865} - x_{.135}} = \frac{35 - 25}{31.8 - 25.4} = \frac{10}{6.4} = 1.56$$

This equivalent P_p formula for non-normal output distributions is similar to the P_p index for normally distributed outputs, based on the difference between the 99.865 and .135 percentiles. This new index represents the maximum number of equivalent $6\sigma_{LT}$ spreads that could fit into the tolerance when the process is properly centered. Note that for non-normal distributions, this "proper center" may not be the middle of the tolerance. The percentile formula is more general in that it may be applied to *every* type of distribution, normal or non-normal, as this next example illustrates.

Example 8.4 Forging Hardness

After examining the NOPP in Figure 8.6, Lynette decided the process output for the hardness of an aluminum forging has a non-normal distribution. In order to derive a measure of potential process capability, she chooses the percentile method to estimate both equivalent P_p and equivalent P_R. The LSL for hardness is 130, the USL is 190, the equivalent P_p goal is 1.33, while the goal for equivalent P_R is .75.

$$\text{Equivalent } \hat{P}_p = \frac{\text{Tolerance}}{\hat{x}_{99.865} - \hat{x}_{.135}} = \frac{190 - 130}{182 - 132} = \frac{60}{50} = 1.20$$

An equivalent P_P of 1.20 means at least one equivalent $6\sigma_{LT}$ spread will easily fit within the tolerance. In fact, up to 1.20 equivalent $6\sigma_{LT}$ spreads could fit, as is seen in Figure 8.9.

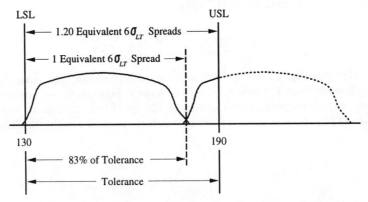

Fig. 8.9 1.20 equivalent $6\,\sigma_{LT}$ spreads could fit within the tolerance for forging hardness.

Whether or not the output distribution actually is within the tolerance depends on where the process is centered. Just as with normal distributions, this type of measure reveals only *potential* capability, not actual *performance*. As before, a histogram and a measure of performance capability should always be analyzed in conjunction with any measure of potential capability.

$$\text{Equivalent } \hat{P}_R = \frac{1}{\text{Equivalent } \hat{P}_P} = \frac{1}{1.20} = .83$$

Equivalent P_R equaling .83 indicates the middle 99.73 percent of the hardness output distribution would take up only 83 percent of the tolerance. Thus, both measures imply the minimum requirement for long-term potential capability is satisfied. But unfortunately, this process does not meet its capability goal for forging hardness, as the estimate of 1.20 for equivalent P_P is less than 1.33, while the estimate of .83 for equivalent P_R is greater than .75.

Advantages and Disadvantages

Equivalent P_R and equivalent P_P are fairly easy to calculate once the cumulative percentage distribution for the process is plotted on NOPP. Just read off $x_{.135}$ and $x_{99.865}$, then compare their difference to the tolerance. Equivalent P_R expresses this difference as a percentage of the tolerance, whereas equivalent P_P conveys the maximum number of these middle 99.73 percent spreads that could fit within the tolerance. Neither measure requires the process output to be normally distributed.

One disadvantage is having to extend the curved line on NOPP out far enough to obtain estimates of the .135 and 99.865 percentiles. This task is especially difficult for highly capable processes and/or those studies with small amounts of data. In addition, if the target for the process median is not equal to M, a modification must be made to the formulas for equivalent P_R and equivalent P_P. These two measures, called equivalent P^*_R and equivalent P^*_P, are covered in Section 8.5 on page 450. And finally, alternate formulas must be used to estimate potential capability when a characteristic has a unilateral specification. These modified measures constitute the topic of the next section.

8.4 Equivalent P'_{PL}, P'_{PU}, P'_{RL}, and P'_{RU} for Unilateral Specifications

Recall that for a normally distributed process, the P_P measure is calculated in a slightly different manner when dealing with a unilateral specification (review Section 5.10 on page 144). When only an USL is present, P'_{PU} is computed from $3\sigma_{LT}$, rather than $6\sigma_{LT}$.

$$P'_{PU} = \text{Maximum}\,(P^T_{PU}, P_{PU}) = \text{Maximum}\left(\frac{USL - T}{3\sigma_{LT}}, \frac{USL - \mu}{3\sigma_{LT}}\right)$$

For a normally distributed process output, the distance from the upper $3\sigma_{LT}$ tail to the process average equals half of the $6\sigma_{LT}$ spread, or $3\sigma_{LT}$.

$$(\mu + 3\sigma_{LT}) - \mu = 3\sigma_{LT}$$

$3\sigma_{LT}$ represents the upper half of the middle 99.73 percent span, which is 49.865 percent of the process output, as is seen in Figure 8.10a.

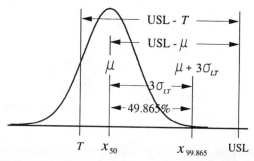

Fig. 8.10a The upper half of the middle 99.73 percent spread.

This upper half of the middle 99.73 percent may also be determined from the percentiles of a cumulative percentage distribution plotted on NOPP. For normal distributions, 50 percent of the process output lies below the process average, thus making μ the 50th percentile (x_{50}). For any type of a distribution, 99.865 percent of the output lies below the 99.865 percentile. Therefore, the area under a normal distribution between these two percentiles represents the same 49.865 percent of the process output as the upper half of the $6\sigma_{LT}$ spread. Thus, the distance from the upper $3\sigma_{LT}$ tail to μ equals the distance from $x_{99.865}$ to x_{50}.

$$3\sigma_{LT} = (\mu + 3\sigma_{LT}) - \mu = x_{99.865} - x_{50}$$

Substituting these percentile values into the original formula for P'_{PU} transforms it as displayed below. Either version of the P'_{PU} formula may be chosen to estimate potential capability for a process having a normally distributed output.

$$P'_{PU} = \text{Maximum}\left(\frac{USL - T}{3\sigma_{LT}}, \frac{USL - \mu}{3\sigma_{LT}}\right) = \text{Maximum}\left(\frac{USL - T}{x_{99.865} - x_{50}}, \frac{USL - x_{50}}{x_{99.865} - x_{50}}\right)$$

Due to the symmetry of the normal distribution, $3\sigma_{LT}$ is also equal to the lower half of the middle 99.73 percent spread, which equals the distance from x_{50} to $x_{.135}$.

$$3\sigma_{LT} = \mu - (\mu - 3\sigma_{LT}) = x_{50} - x_{.135}$$

Thus, for characteristics having only a LSL, P'_{PL} may be calculated from the above percentiles with the following modified formula:

$$P'_{PL} = \text{Maximum}\,(P^T_{PL}, P_{PL}) = \text{Maximum}\left(\frac{T - \text{LSL}}{3\sigma_{LT}}, \frac{\mu - \text{LSL}}{3\sigma_{LT}}\right)$$

$$= \text{Maximum}\left(\frac{T - \text{LSL}}{x_{50} - x_{.135}}, \frac{x_{50} - \text{LSL}}{x_{50} - x_{.135}}\right)$$

The percentile version of this formula may also be applied to a characteristic having a unilateral specification *and* a non-normal distribution. For instance, consider the skewed-right output distribution for taper displayed in Figure 8.10b.

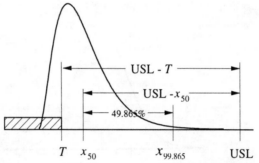

Fig. 8.10b Calculating potential capability for a non-normal distribution with a unilateral specification.

The long-term potential capability for this process is calculated with the percentile-version formula of P'_{PU}, called equivalent P'_{PU}.

$$\text{Equivalent } P'_{PU} = \text{Maximum}\,(\text{Equivalent } P^T_{PU}, \text{Equivalent } P_{PU})$$

$$= \text{Maximum}\left(\frac{\text{USL} - T}{x_{99.865} - x_{50}}, \frac{\text{USL} - x_{50}}{x_{99.865} - x_{50}}\right)$$

Equivalent P^T_{PU} represents how many times the upper half of the middle 99.73 percent spread could fit into the distance from the USL to the target. Equivalent P_{PU} measures how many upper halves of the middle 99.73 percent spread could fit into the distance from the USL to the process median. The larger of these two determines the potential capability of this process. For characteristics having only a LSL, switch to this formula:

$$\text{Equivalent } P'_{PL} = \text{Maximum}\,(\text{Equivalent } P^T_{PL}, \text{Equivalent } P_{PL})$$

$$= \text{Maximum}\left(\frac{T - \text{LSL}}{x_{50} - x_{.135}}, \frac{x_{50} - \text{LSL}}{x_{50} - x_{.135}}\right)$$

These equations for non-normal distributions are identical to those for normal distributions which incorporate the percentile values in place of μ and $3\sigma_{LT}$. Just as with the bilateral equivalent capability measures, those for non-normal distributions are more general and may be applied to all types of distributions.

As usual, the equivalent P'_{RU} value is just the inverse of the equivalent P'_{PU} index.

$$\text{Equivalent } P'_{RU} = \text{Minimum} \left(\text{Equivalent } P^T_{RU} , \text{Equivalent } P_{RU} \right)$$

$$= \text{Minimum} \left(\frac{x_{99.865} - x_{50}}{\text{USL} - T} , \frac{x_{99.865} - x_{50}}{\text{USL} - x_{50}} \right) = \frac{1}{\text{Equivalent } P'_{PU}}$$

Likewise, equivalent P'_{RL} is the inverse of equivalent P'_{PL}.

$$\text{Equivalent } P'_{RL} = \text{Minimum} \left(\text{Equivalent } P^T_{RL} , \text{Equivalent } P_{RL} \right)$$

$$= \text{Minimum} \left(\frac{x_{50} - x_{.135}}{T - \text{LSL}} , \frac{x_{50} - x_{.135}}{x_{50} - \text{LSL}} \right) = \frac{1}{\text{Equivalent } P'_{PL}}$$

The rating scales for these equivalent measures are identical to those for the standard P'_{RL}, P'_{RU} (see page 119), P'_{PL}, and P'_{PU} (see page 145) measures.

Example 8.5 Cam Roller Waviness

In Exercise 7.4 on page 383, the process output for the waviness of a cam roller was discovered to be non-normally distributed when its cumulative distribution was plotted on NOPP (Figure 8.11). Because waviness has only an USL of 75, equivalent P'_{PU} or equivalent P'_{RU} must be applied to correctly assess potential capability.

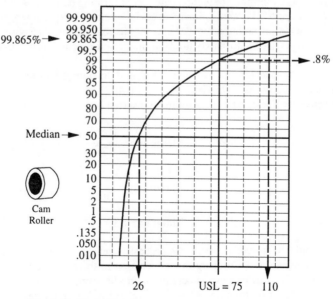

Fig. 8.11 Estimating the median and $x_{99.865}$ from NOPP.

x_{50} and $x_{99.865}$ are read off the NOPP as 26 and 110, respectively, whereas the target for centering the process is given as 30.

$$\text{Equivalent } \hat{P}'_{PU} = \text{Maximum} \left(\frac{USL - T}{\hat{x}_{99.865} - \hat{x}_{50}}, \frac{USL - \hat{x}_{50}}{\hat{x}_{99.865} - \hat{x}_{50}} \right)$$

$$= \text{Maximum} \left(\frac{75 - 30}{110 - 26}, \frac{75 - 26}{110 - 26} \right)$$

$$= \text{Maximum} (.536, .583) = .583$$

The estimated equivalent P'_{PU} index of .583 implies only 58.3 percent of the upper half of the process's middle 99.73 percent can fit within the distance from the median to the USL. In order to meet the minimum requirement of potential capability, at least 100 percent of this upper half must be able to fit into this distance. Because a maximum of only 58.3 could fit, the process responsible for waviness lacks long-term potential capability. Work must begin on lowering the average waviness or on reducing the process spread, as either action will lower the 99.865 percentile.

$$\text{Equivalent } \hat{P}'_{RU} = \frac{1}{\text{Equivalent } \hat{P}'_{PU}} = \frac{1}{.583} = 1.715$$

The estimated equivalent P'_{RU} ratio of 1.715 means the width required to produce the upper 49.865 percent of the process output is 71.5 percent greater than the distance from the median to the USL. Even though the median is centered well within the target zone (26 is less than 30), the 99.865 percentile for this process is located above the USL, causing more than .135 percent of the rollers to be nonconforming for waviness. Thus, this process lacks the potential of producing a minimum of 99.865 percent conforming parts. In fact, from the sheet of NOPP shown in Figure 8.11, about .8 percent of the rollers are expected to have a waviness reading greater than 75, meaning only 99.2 percent are conforming. Because at least 99.865 percent conforming is needed to satisfy the minimum requirement, this process is definitely deficient in potential capability.

Suppose no check for normality is done, and the output distribution for waviness is falsely assumed to be normal. In Example 7.7 (p. 407), σ_{LT} and μ were estimated as 16 and 35, respectively. These estimates would be inserted into the formula for P'_{PU}.

$$\hat{P}'_{PU} = \text{Maximum} \left(\frac{USL - T}{3\hat{\sigma}_{LT}}, \frac{USL - \hat{\mu}}{3\hat{\sigma}_{LT}} \right)$$

$$= \text{Maximum} \left(\frac{75 - 30}{3(16)}, \frac{75 - 35}{3(16)} \right)$$

$$= \text{Maximum} (.94, .83) = .94$$

This index incorrectly implies the process is nearly meeting the minimum prerequisite for potential capability. Compare this conclusion to the estimate of .583 for equivalent P'_{PU}, which revealed this process as actually being far from capable (see also Chan *et al*, 1986). Somerville and Montgomery provide several examples which clearly demonstrate the sizable errors involved when capability measures based on the normality assumption are estimated for processes whose output distributions are non-normal. For additional examples of these problems, see English and Taylor (1992, 1993), and Gunter.

Example 8.6 Tensile Strength

The tensile strength of a certain gauge of steel wire must be at least 450 kilograms, with the target being 600 kg. A NOPP analysis concludes the process output for tensile strength is skewed to the left, with the median estimated as 547 kg and the .135 percentile as 481 kg. Because a measure of long-term potential capability is desired, the equivalent P'_{PL} index is appropriately chosen.

$$\text{Equivalent } \hat{P}'_{PL} = \text{Maximum}\left(\text{Equivalent } \hat{P}^T_{PL}, \text{Equivalent } \hat{P}_{PL}\right)$$

$$= \text{Maximum}\left(\frac{T - \text{LSL}}{\hat{x}_{50} - \hat{x}_{.135}}, \frac{\hat{x}_{50} - \text{LSL}}{\hat{x}_{50} - \hat{x}_{.135}}\right)$$

$$= \text{Maximum}\left(\frac{600 - 450}{547 - 481}, \frac{547 - 450}{547 - 481}\right)$$

$$= \text{Maximum}\left(2.27, 1.47\right) = 2.27$$

As the goal for equivalent P'_{PL} is 1.67, the wire-forming process easily surpasses this additional requirement for potential capability.

If the median can be increased by 53 kg so it just equals the target of 600 kg, equivalent P^T_{PL} will equal equivalent P_{PL} (this analysis assumes $x_{.135}$ also shifts up by 53 kg to 534 kg).

$$\text{Equivalent } \hat{P}'_{PL} = \text{Maximum}\left(\text{Equivalent } \hat{P}^T_{PL}, \text{Equivalent } \hat{P}_{PL}\right)$$

$$= \text{Maximum}\left(\frac{T - \text{LSL}}{\hat{x}_{50} - \hat{x}_{.135}}, \frac{\hat{x}_{50} - \text{LSL}}{\hat{x}_{50} - \hat{x}_{.135}}\right)$$

$$= \text{Maximum}\left(\frac{600 - 450}{600 - 534}, \frac{600 - 450}{600 - 534}\right)$$

$$= \text{Maximum}\left(2.27, 2.27\right) = 2.27$$

If the median tensile strength increases above the target, equivalent P'_{PL} begins measuring performance capability, as it now equals equivalent P_{PL}, which is a function of process spread *and* centering (see Section 8.8 on page 456). This relationship is graphically portrayed in Figure 8.12, which is similar to the graph for the standard P'_{PL} displayed in Figure 5.56 (p. 148). Note how the x axis now represents x_{50} instead of μ.

This graph reveals that the formula for equivalent P'_{PL} may be written in another manner. When x_{50} is in the target zone:

$$\text{Equivalent } \hat{P}'_{PL} = \text{Equivalent } \hat{P}_{PL} = \frac{\hat{x}_{50} - \text{LSL}}{\hat{x}_{50} - \hat{x}_{.135}} \quad \text{if } x_{50} \geq T$$

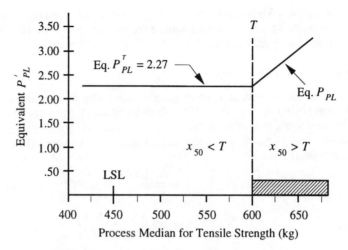

Fig. 8.12 Equivalent P'_{PL} versus the process median.

When x_{50} is below the target zone:

$$\text{Equivalent } \hat{P}'_{PL} = \text{Equivalent } \hat{P}^T_{PL} = \frac{T - LSL}{\hat{x}_{50} - \hat{x}_{.135}} \quad \text{if } x_{50} < T$$

Advantages and Disadvantages

A major advantage of these new measures is their similarity to the standard P'_{RL}, P'_{RU}, P'_{PL}, and P'_{PU} measures. These special measures have the same rating scale as the standard measures, with goals established and evaluated in an identical manner. However, they assume the process median can be shifted to the target without altering the distribution's shape, *i.e.*, the distance between x_{50} and $x_{.135}$, as well as between $x_{99.865}$ and x_{50}, does not change.

Unfortunately, extra time is required to plot the measurements on NOPP in order to derive the necessary percentiles, but NOPP (or some other test) is needed anyway to check the normality assumption. In addition, precise percentile values may be difficult to read off of NOPP. Sometimes, when a fitted curve is drawn by hand, different people may derive slightly different percentile estimates. Thus, estimates of potential capability could vary to some degree for the same set of data.

8.5 Equivalent P^*_P and Equivalent P^*_R

If a characteristic with a bilateral specification has a target designated for the process center, P_P and P_R become inappropriate for measuring the true potential capability. As operators will make efforts to center the process at this target, the potential capability implied by these two measures will never be realized. Under these circumstances, the quality practitioner should switch to the P^*_P and P^*_R measures to gain a better indication of the available potential capability under this added restriction. Formulas for these two indexes were developed in Sections 5.5 (p. 113) and 5.9 (p. 139), and are repeated here:

$$P_P^* = \text{Minimum}\,(P_{PL}^T, P_{PU}^T) = \text{Minimum}\left(\frac{T - \text{LSL}}{3\sigma_{LT}}, \frac{\text{USL} - T}{3\sigma_{LT}}\right)$$

$$P_R^* = \text{Maximum}\,(P_{RL}^T, P_{RU}^T) = \text{Maximum}\left(\frac{3\sigma_{LT}}{T - \text{LSL}}, \frac{3\sigma_{LT}}{\text{USL} - T}\right)$$

Both of these assume normality, so if the process output is known to be non-normal, these measures are inappropriate. However, they can be modified to handle non-normal distributions by replacing μ with x_{50}, the lower 3σ tail with x_{50} minus $x_{.135}$, and the upper 3σ tail with $x_{99.865}$ minus $x_{.50}$. Note that T now becomes the target for the process median.

$$\text{Equivalent } P_P^* = \text{Minimum}\,(\text{Equivalent } P_{PL}^T, \text{Equivalent } P_{PU}^T)$$

$$= \text{Minimum}\left(\frac{T - \text{LSL}}{x_{50} - x_{.135}}, \frac{\text{USL} - T}{x_{99.865} - x_{50}}\right)$$

$$\text{Equivalent } P_R^* = \text{Maximum}\,(\text{Equivalent } P_{RL}^T, \text{Equivalent } P_{RU}^T)$$

$$= \text{Maximum}\left(\frac{x_{50} - x_{.135}}{T - \text{LSL}}, \frac{x_{99.865} - x_{50}}{\text{USL} - T}\right)$$

As before, equivalent P_R^* is also equal to the inverse of equivalent P_P^*.

$$\text{Equivalent } P_R^* = \frac{1}{\text{Equivalent } P_P^*}$$

As Figure 8.13 reveals, equivalent P_{PL}^T represents how many lower "halves" of the middle 99.73 percent spread could possibly fit into the distance from the LSL to the target, assuming the process median is centered at T.

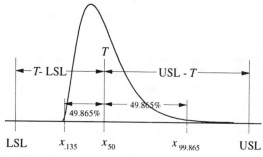

Fig. 8.13 Calculating potential capability when a target is specified.

If this process is to meet the minimum requirement for potential capability when centered on the target median, then T minus LSL must be greater than x_{50} minus $x_{.135}$.

$$T - \text{LSL} > x_{50} - x_{.135}$$

This relationship means equivalent $P^T{}_{PL}$ must be greater than 1.

$$\frac{T - \text{LSL}}{x_{50} - x_{.135}} > 1$$

$$\text{Equivalent } P^T_{PL} > 1$$

Likewise, the distance from the USL to T must be greater than that from $x_{99.865}$ to x_{50}, meaning equivalent $P^T{}_{PU}$ must also be greater than 1.

$$\text{USL} - T > x_{99.865} - x_{50}$$

$$\frac{\text{USL} - T}{x_{99.865} - x_{50}} > 1$$

$$\text{Equivalent } P^T_{PU} > 1$$

The potential capability of a process with a specified target is limited by the *smaller* of these two, which is exactly what equivalent P^*_P reflects.

$$\text{Equivalent } P^*_P = \text{Minimum} \left(\text{Equivalent } P^T_{PL}, \text{Equivalent } P^T_{PU} \right)$$

$$= \text{Minimum} \left(\frac{T - \text{LSL}}{x_{50} - x_{.135}}, \frac{\text{USL} - T}{x_{99.865} - x_{50}} \right)$$

The respective rating scales for these modified measures are the same as those for equivalent P_P and equivalent P_R, while their goals are established in a similar manner.

Example 8.7 Flange Thickness

A process fabricating composite aircraft components is finally brought into a good state of statistical control for the thickness of a critical flange. When the thickness measurements are plotted on NOPP by Alex (a consultant), a J-shaped curve appears, indicating the process output has a distribution which is skewed left (asymmetric). Due to this being a non-normal distribution, he realized the standard potential capability measures do not apply and should not be calculated. The equivalent P_P measure would be an acceptable alternative, but there is a target of 12.660 mm (which is not equal to M) specified for flange thickness. Thus, Alex decides to measure potential capability with the equivalent P^*_P index, which assumes the process median will eventually be located at T. With a LSL of 12.640 mm and an USL of 12.670 mm, a goal is established to have an equivalent P^*_P index of at least 1.75.

The following percentiles are estimated from the NOPP: $x_{.135}$ equals 12.647; x_{50} equals 12.658; and $x_{99.865}$ equals 12.664. Substituting these into the formula for equivalent P^*_P yields the following result:

$$\text{Equivalent } \hat{P}^*_P = \text{Minimum} \left(\text{Equivalent } \hat{P}^T_{PL}, \text{Equivalent } \hat{P}^T_{PU} \right)$$

$$= \text{Minimum} \left(\frac{T - \text{LSL}}{\hat{x}_{50} - \hat{x}_{.135}}, \frac{\text{USL} - T}{\hat{x}_{99.865} - \hat{x}_{50}} \right)$$

$$= \text{Minimum} \left(\frac{12.660 - 12.640}{12.658 - 12.647}, \frac{12.670 - 12.660}{12.664 - 12.658} \right)$$

$$= \text{Minimum} \, (1.82, 1.67) = 1.67$$

As the estimate for equivalent P^*_P is well over 1.00, this process meets the minimum requirement for long-term potential capability, *i.e.*, at least 99.73 percent of the flanges will have their thickness within specification when x_{50} equals T. However, it does not quite meet the goal of 1.75. If there were no restriction on where the process median must be centered, potential capability would be somewhat greater, as revealed by the equivalent P_P index.

$$\text{Equivalent } \hat{P}_P = \frac{\text{Tolerance}}{\hat{x}_{99.865} - \hat{x}_{.135}} = \frac{12.670 - 12.640}{12.664 - 12.647} = 1.7647$$

If the process median is free to move to a location different than the target of 12.660, potential capability could be increased to 1.7647, just surpassing the goal. This ideal location (called x'_{50}) is determined by setting the number of equivalent upper $3\sigma_{LT}$ spreads that will fit into the distance from the USL to this location equal to 1.7647, and then solving for x'_{50}. This analysis assumes the distance between the 99.865 percentile and the median (12.664 - 12.658 = .006) does not change when the median is shifted to the new location.

$$\text{Maximum Equivalent } P^*_P \implies 1.7647 = \frac{\text{USL} - x'_{50}}{x_{99.865} - x_{50}}$$

$$1.7647 = \frac{12.670 - x'_{50}}{.006}$$

$$1.7647(.006) = 12.670 - x'_{50}$$

$$x'_{50} = 12.670 - 1.7647(.006)$$

$$x'_{50} = 12.65941$$

If the current target for flange thickness is lowered to this ideal location of 12.65941, potential capability is maximized as equivalent P^*_P and equivalent P_P will both equal 1.7647. Again, the distances between x_{50} and $x_{.135}$ (12.658 - 12.647 = .011) and between $x_{99.865}$ and x_{50} (.006) are assumed to remain unchanged if the median is shifted.

$$\text{Equivalent } \hat{P}^*_P = \text{Minimum} \left(\frac{T - \text{LSL}}{\hat{x}_{50} - \hat{x}_{.135}}, \frac{\text{USL} - T}{\hat{x}_{99.865} - \hat{x}_{50}} \right)$$

$$= \text{Minimum} \left(\frac{12.65941 - 12.640}{.011}, \frac{12.670 - 12.65941}{.006} \right)$$

$$= \text{Minimum} \, (1.7645, 1.7645) = 1.7645$$

Note that x'_{50} is *not* the middle of the tolerance, which is 12.655. Specifying any target other than 12.65941 causes a reduction in potential capability.

Advantages and Disadvantages

Equivalent P^*_P and P^*_R have the same rating scale and interpretation as the standard P^*_P and P^*_R measures. Goals are also established and evaluated in a similar manner. However, they involve formulas that are slightly more complicated than other measures of potential capability for non-normal distributions.

Note that the concept of creating an USL_T and LSL_T at equal distances from the target won't work for non-normal distributions which are not symmetrical. A modified method for determining the appropriate USL_T and LSL_T values for asymmetrical distributions is presented later in this chapter (see Section 8.9 on page 483).

8.6 Equivalent P_{ST} and Equivalent P^*_{ST}

For normal distributions, the P_{ST} index was defined in Section 5.12 (p. 151) as:

$$P_{ST} = \frac{\text{Tolerance}}{\sigma_{LT}} = 6P_P$$

A P_{ST} value exceeding 6 indicates good potential capability, while a value less than 6 implies poor potential capability. Common goals are 8 and 10.

For non-normal distributions, this index must be modified to replace σ_{LT} with the proper percentiles. As the equivalent $6\sigma_{LT}$ spread equals the distance between $x_{99.865}$ and $x_{.135}$, one equivalent σ_{LT} spread equals:

$$\text{Equivalent } 6\sigma_{LT} \text{ Spread } = x_{99.865} - x_{.135} \implies \text{Equivalent } 1\sigma_{LT} \text{ Spread } = \frac{x_{99.865} - x_{.135}}{6}$$

Equivalent P_{ST} can now be defined as:

$$\text{Equivalent } P_{ST} = \frac{\text{Tolerance}}{\text{Equivalent } 1\sigma_{LT} \text{ Spread}}$$

$$= \frac{6\,\text{Tolerance}}{x_{99.865} - x_{.135}}$$

$$= 6\frac{\text{Tolerance}}{x_{99.865} - x_{.135}} = 6\,\text{Equivalent } P_P$$

With this definition, the rating scale and goals are similar to those for the standard P_{ST} index. If a target value for the process center is specified, the P^*_{ST} index should be chosen in place of P_{ST} (review Section 5.13 on page 152).

$$P^*_{ST} = 2\,\text{Minimum}\,(P^T_{STL}, P^T_{STU}) = 2\,\text{Minimum}\left(\frac{T - LSL}{\sigma_{LT}}, \frac{USL - T}{\sigma_{LT}}\right) = 6P^*_P.$$

The above formula is correct when the process output has a normal distribution. If not, the equivalent upper and lower $1\sigma_{LT}$ spreads must be substituted for σ_{LT}.

$$\text{Equiv. Lower } 3\sigma_{LT} \text{ Spread } = x_{50} - x_{.135} \quad \Rightarrow \quad \text{Equiv. Lower } 1\sigma_{LT} \text{ Spread } = \frac{x_{50} - x_{.135}}{3}$$

$$\text{Equiv. Upper } 3\sigma_{LT} \text{ Spread } = x_{99.865} - x_{50} \quad \Rightarrow \quad \text{Equiv. Upper } 1\sigma_{LT} \text{ Spread } = \frac{x_{99.865} - x_{50}}{3}$$

The equivalent P^*_{ST} index can now be defined as:

$$\text{Equivalent } P^*_{ST} = 2\,\text{Minimum}\,(\,\text{Equivalent } P^T_{STL}, \text{Equivalent } P^T_{STU}\,)$$

$$= 2\,\text{Min.}\left(\frac{T - LSL}{\text{Equiv. Lower } 1\sigma_{LT} \text{ Spread}}, \frac{USL - T}{\text{Equiv. Upper } 1\sigma_{LT} \text{ Spread}}\right)$$

$$= 2\,\text{Minimum}\left(\frac{3(T - LSL)}{x_{50} - x_{.135}}, \frac{3(USL - T)}{x_{99.865} - x_{50}}\right)$$

$$= 6\,\text{Minimum}\left(\frac{T - LSL}{x_{50} - x_{.135}}, \frac{USL - T}{x_{99.865} - x_{50}}\right)$$

$$= 6\,\text{Equivalent } P^*_P$$

This measure has the same goals and rating system as P^*_{ST}. Being directly related to equivalent P^*_P, the application and interpretation of the equivalent P^*_{ST} index are very similar, as are its advantages and disadvantages.

8.7 Equivalent P'_{STL} and Equivalent P'_{STU} for Unilateral Specifications

P'_{STL} and P'_{STU} were defined in Section 5.14 (p. 154) for process outputs that are normally distributed and have unilateral specifications. An indication of good potential capability exists when these indexes exceed 6, while poor potential capability is implied when they are less than 6. A common goal is 8 or 10.

$$P'_{STL} = 2\,\text{Maximum}\,(P^T_{STL}, P_{STL}) = 2\,\text{Maximum}\left(\frac{T - LSL}{\sigma_{LT}}, \frac{\mu - LSL}{\sigma_{LT}}\right) = 6P'_{PL}$$

$$P'_{STU} = 2\,\text{Maximum}\,(P^T_{STU}, P_{STU}) = 2\,\text{Maximum}\left(\frac{USL - T}{\sigma_{LT}}, \frac{USL - \mu}{\sigma_{LT}}\right) = 6P'_{PU}$$

For non-normal distributions, these two measures must be altered to incorporate percentiles for estimating σ_{LT}. In a manner similar to that for equivalent P^*_{ST}, the equivalent lower and upper σ_{LT} are replaced with the corresponding percentile values to create the following equivalent indexes:

Equivalent P'_{STL} $= 2\,\text{Maximum}\,(\,\text{Equivalent}\,P^T_{STL}\,,\text{Equivalent}\,P_{STL}\,)$

$$= 2\,\text{Max.}\left(\frac{T - \text{LSL}}{\text{Equiv. Lower }1\sigma_{LT}\text{ Spread}}\,,\frac{\mu - \text{LSL}}{\text{Equiv. Lower }1\sigma_{LT}\text{ Spread}}\right)$$

$$= 2\,\text{Maximum}\left(\frac{3(T - \text{LSL})}{x_{50} - x_{.135}}\,,\frac{3(\mu - \text{LSL})}{x_{50} - x_{.135}}\right)$$

$$= 6\,\text{Maximum}\left(\frac{T - \text{LSL}}{x_{50} - x_{.135}}\,,\frac{\mu - \text{LSL}}{x_{50} - x_{.135}}\right) = 6\,\text{Equivalent}\,P'_{PL}$$

Equivalent P'_{STU} $= 2\,\text{Maximum}\,(\,\text{Equivalent}\,P^T_{STU}\,,\text{Equivalent}\,P_{STU}\,)$

$$= 2\,\text{Max.}\left(\frac{\text{USL} - T}{\text{Equiv. Upper }1\sigma_{LT}\text{ Spread}}\,,\frac{\text{USL} - \mu}{\text{Equiv. Upper }1\sigma_{LT}\text{ Spread}}\right)$$

$$= 2\,\text{Maximum}\left(\frac{3(\text{USL} - T)}{x_{99.865} - x_{50}}\,,\frac{3(\text{USL} - \mu)}{x_{99.865} - x_{50}}\right)$$

$$= 6\,\text{Maximum}\left(\frac{\text{USL} - T}{x_{99.865} - x_{50}}\,,\frac{\text{USL} - \mu}{x_{99.865} - x_{50}}\right) = 6\,\text{Equivalent}\,P'_{PU}$$

With these formulas, the rating scale and goals for these indexes are identical to those for P'_{STL} and P'_{STU}. As these measures are directly related to the corresponding equivalent P_{PL} and P_{PU} measures, their interpretation is similar as well.

One major disadvantage of all potential capability measures is that they do not disclose any information about how well this process is actually meeting the print specifications. As Chapter 6 revealed, it's entirely possible to have terrific potential capability, but terrible performance capability. The next several indexes measure actual performance by including process centering in their assessment of capability. However, the process center is represented by the median, rather than the average, which is the center of only symmetrical distributions.

8.8 Equivalent P_{PL}, Equivalent P_{PU}, and Equivalent P_{PK}

For process outputs that are normally distributed, the P_{PL} and P_{PU} indexes are defined as follows, with P_{PK} equal to the smaller of these two.

$$P_{PL} = \frac{\mu - \text{LSL}}{3\sigma_{LT}} \qquad P_{PU} = \frac{\text{USL} - \mu}{3\sigma_{LT}} \qquad P_{PK} = \text{Minimum}\,(P_{PL}, P_{PU})$$

When the process output is non-normally distributed like the one presented in Figure 8.14, these formulas must be modified to incorporate the percentile values for both the process center and spread.

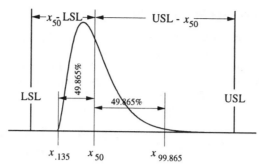

Fig. 8.14 Calculating equivalent P_{PL} and P_{PU} for a skewed distribution.

The distance from the median to the .135 percentile represents the lower half of the middle 99.73 percent of the process output, whereas the difference between the 99.865 percentile and the median equals the upper half of the middle 99.73 percent, as is shown in Figure 8.14.

Lower half of the middle 99.73% $= x_{50} - x_{.135}$ Upper half of the middle 99.73% $= x_{99.865} - x_{50}$

The number of "lower halves" that fit into the distance from the process median to the LSL is determined with the equivalent P_{PL} index, which is analogous to the regular P_{PL} index for normal distributions. In fact, this is the same equivalent P_{PL} index mentioned in Section 8.4 (p. 445) for figuring equivalent P'_{PL}. To meet the minimum performance capability criterion, at least one lower "half" (x_{50} minus $x_{.135}$) of the middle 99.73 percent spread must be within the distance from the median to the LSL.

$$\text{Equivalent } P_{PL} = \frac{x_{50} - \text{LSL}}{x_{50} - x_{.135}}$$

The number of upper "halves" ($x_{99.865}$ minus x_{50}) fitting into the distance from the USL to the median is determined with the equivalent P_{PU} index, which is just the percentile version of the standard P_{PU} index. Again, this equivalent P_{PU} index was explained previously in Section 8.4 when equivalent P'_{PU} was defined.

$$\text{Equivalent } P_{PU} = \frac{\text{USL} - x_{50}}{x_{99.865} - x_{50}}$$

Clements defines equivalent P_{PK} as the smaller of these two indexes, which represents the worst case for performance capability.

$$\text{Equivalent } P_{PK} = \text{Minimum} \left(\text{Equivalent } P_{PL}, \text{Equivalent } P_{PU} \right)$$

$$= \text{Minimum} \left(\frac{x_{50} - \text{LSL}}{x_{50} - x_{.135}}, \frac{\text{USL} - x_{50}}{x_{99.865} - x_{50}} \right)$$

The ISO Technical Committee 69 (ISO/TC69) on applications for statistical methods recommends this method for measuring capability when the process output is non-normally distributed. It was proposed in document N13 by Working Group 6 of subcommittee 4. Note that this formula also applies when the process output is normally distributed. For other approaches, see Bittanti *et al*, Deleryd, McCormack *et al*, Mukherjee and Singh, Polansky (1998, 2000), Polansky *et al*, Wu, and Yeh and Bhattacharya.

Example 8.8 Bearing Support Hole Size

Measurements for the inner diameter of a hole drilled in a bearing support bracket for a missile's inertial guidance system are analyzed on NOPP. The curve drawn through the plotted points indicates the process output distribution for this characteristic appears to be skewed right (recall Figure 8.3, p. 433). An estimate of the 99.865 percentile was read off the x axis of the NOPP as 31.8 in Example 8.2, while the .135 percentile was found to be 25.4. Given a tolerance for hole size of 10 (35 - 25), equivalent P_P was estimated as 1.56.

$$\text{Equivalent } \hat{P}_P \ = \ \frac{\text{Tolerance}}{\hat{x}_{99.865} - \hat{x}_{.135}} = \frac{35 - 25}{31.8 - 25.4} = 1.56$$

From the NOPP, the median is determined to be 26.7. This last bit of information allows equivalent P_{PK} to be estimated for this hole drilling operation.

$$\text{Equivalent } \hat{P}_{PK} \ = \ \text{Minimum} \ (\text{Equivalent } \hat{P}_{PL} , \text{Equivalent } \hat{P}_{PU})$$

$$= \ \text{Minimum} \left(\frac{\hat{x}_{50} - \text{LSL}}{\hat{x}_{50} - \hat{x}_{.135}} , \frac{\text{USL} - \hat{x}_{50}}{\hat{x}_{99.865} - \hat{x}_{50}} \right)$$

$$= \ \text{Minimum} \left(\frac{26.7 - 25.0}{26.7 - 25.4} , \frac{35.0 - 26.7}{31.8 - 26.7} \right)$$

$$= \ \text{Minimum} \ (1.31 , 1.63) = 1.31$$

As equivalent P_P is 1.56 and equivalent P_{PK} only 1.31, performance capability can be enhanced by shifting the median. Because equivalent P_{PL} (1.31) is less than equivalent P_{PU} (1.63), equivalent P_{PK} may be increased to 1.56 by moving the median hole size higher, assuming the process spread remains constant. This analysis is similar to that done in Chapter 6 for the standard capability measures.

Rating Scale

The rating scale for this measure is identical to that for the standard P_{PK} index.

Equivalent $P_{PK} > 1$ reveals that the minimum requirement for long-term performance capability is exceeded, and the process is producing more than 99.73 percent conforming parts.

Equivalent $P_{PK} = 1$ implies that the minimum condition for long-term performance capability is just met, *i.e.*, no more than .135 percent nonconforming parts outside either specification.

Equivalent $P_{PK} < 1$ discloses the minimum requirement for long-term performance capability is *not* satisfied because more than .135 percent of the process output is outside at least one of the specification limits.

Goals

Attainment of performance capability goals is evaluated in the same manner as done with the P_{PK} measure for normal distributions. Common goals for equivalent P_{PK} are 1.33, 1.50, 1.67, and 2.00. Note that attainment of a goal of 1.33 doesn't necessarily imply $ppm_{MAX, LT}$ equals 32, as is the case for a P_{PK} index of 1.33 for a process output having a normal distribution. The area in the tails of a non-normal distribution may be markedly different than that for a normal distribution. As an example, two machines (A and B) are producing the identical characteristic of the same product. After control is achieved for both, sample measurements from each process are plotted as separate lines on the same sheet of NOPP. The output from machine A appears to follow a normal distribution, while the one for B appears to be unbounded, as is displayed below in Figure 8.15.

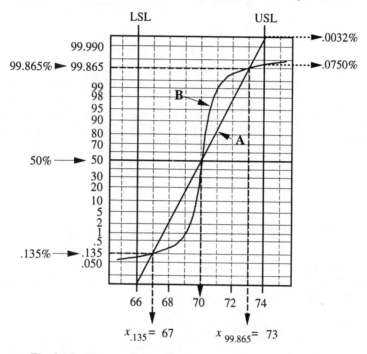

Fig. 8.15 The cumulative distributions for machines A and B.

Notice that curves A and B have the same .135 percentile (67), median (70) and 99.865 percentile (73). As they are machining the same characteristic, the specifications are also identical (LSL = 66, USL = 74). To establish the capability of machine A, these percentiles and specification limits are inserted into the formula for P_{PK}.

$$\hat{P}_{PK,A} = \text{Minimum} \left(\hat{P}_{PL,A}, \hat{P}_{PU,A} \right)$$

$$= \text{Minimum} \left(\frac{\hat{x}_{50} - \text{LSL}}{\hat{x}_{50} - \hat{x}_{.135}}, \frac{\text{USL} - \hat{x}_{50}}{\hat{x}_{99.865} - \hat{x}_{50}} \right)$$

$$= \text{Minimum} \left(\frac{70 - 66}{70 - 67}, \frac{74 - 70}{73 - 70} \right)$$

$$= \text{Minimum} \left(\frac{4}{3}, \frac{4}{3} \right) = 1.33$$

Machine B's percentiles are used to determine its equivalent P_{PK} index.

$$\text{Equivalent } \hat{P}_{PK,B} = \text{Minimum} \left(\text{Equivalent } \hat{P}_{PL,B}, \text{Equivalent } \hat{P}_{PU,B} \right)$$

$$= \text{Minimum} \left(\frac{\hat{x}_{50} - \text{LSL}}{\hat{x}_{50} - \hat{x}_{.135}}, \frac{\text{USL} - \hat{x}_{50}}{\hat{x}_{99.865} - \hat{x}_{50}} \right)$$

$$= \text{Minimum} \left(\frac{70 - 66}{70 - 67}, \frac{74 - 70}{73 - 70} \right)$$

$$= \text{Minimum} \left(\frac{4}{3}, \frac{4}{3} \right) = 1.33$$

Both machines have performance capability ratings of 1.33. However, the amount of nonconforming parts above the USL for A is 32 *ppm*, while for B it is 750 *ppm*. B has a much higher $ppm_{USL,\,LT}$ than A (and a higher $ppm_{LSL,\,LT}$), even though their performance capability indexes are identical. Even calculating their potential capability measures won't help in this situation, since they are equal as well.

$$\hat{P}_{P,A} = \frac{\text{Tolerance}}{\hat{x}_{99.865} - \hat{x}_{.135}} = \frac{74 - 66}{73 - 67} = 1.33$$

$$\text{Equivalent } \hat{P}_{P,B} = \frac{\text{Tolerance}}{\hat{x}_{99.865} - \hat{x}_{.135}} = \frac{74 - 66}{73 - 67} = 1.33$$

Let the specification limits be changed to a LSL of 67 and an USL of 73, as is the case in Figure 8.16. Substituting these revised limits into the capability formulas, both P_{PK} for A and equivalent P_{PK} for B equal 1.00.

$$\hat{P}_{PK,A} = \text{Minimum} \left(\frac{\hat{x}_{50} - \text{LSL}}{\hat{x}_{50} - \hat{x}_{.135}}, \frac{\text{USL} - \hat{x}_{50}}{\hat{x}_{99.865} - \hat{x}_{50}} \right)$$

$$= \text{Minimum} \left(\frac{70 - 67}{70 - 67}, \frac{73 - 70}{73 - 70} \right)$$

$$= \text{Minimum} \left(1.00, 1.00 \right) = 1.00$$

$$\text{Equivalent } \hat{P}_{PK,B} = \text{Minimum} \left(\frac{\hat{x}_{50} - \text{LSL}}{\hat{x}_{50} - \hat{x}_{.135}}, \frac{\text{USL} - \hat{x}_{50}}{\hat{x}_{99.865} - \hat{x}_{50}} \right)$$

$$= \text{Minimum} \left(\frac{70 - 67}{70 - 67}, \frac{73 - 70}{73 - 70} \right)$$

$$= \text{Minimum} \left(1.00, 1.00 \right) = 1.00$$

With these new specification limits, the amount of nonconforming parts above the USL for A increases to 1,350 *ppm*, while that below the new LSL is also 1,350 *ppm*. Notice that now machine B also has 1,350 *ppm* above this USL of 73, as well as below this revised LSL of 67. Under these circumstances, A and B again have identical ratings for both performance and potential capability, but now their *ppm* values are equal as well.

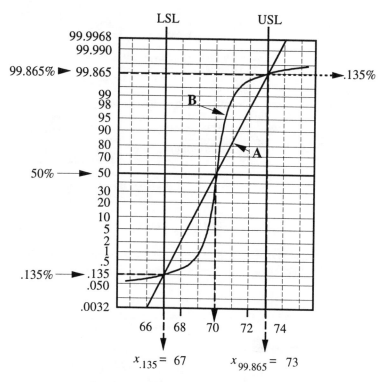

Fig. 8.16 The LSL is 67 and the USL is 73.

Nevertheless, even though all their capability measures are identical, most quality practitioners would prefer buying parts from B, as a greater percentage of its output is near the middle of the tolerance (see Figure 8.17).

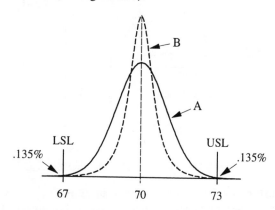

Fig. 8.17 B's output is clustered closer around *M*.

Finally, assume the LSL is moved to 68 and the USL to 72, as is shown in Figure 8.18. With these reduced limits, P_{PK} and equivalent P_{PK} both turn out to equal .67.

$$\hat{P}_{PK,A} = \text{Minimum}\left(\frac{\hat{x}_{50} - \text{LSL}}{\hat{x}_{50} - \hat{x}_{.135}}, \frac{\text{USL} - \hat{x}_{50}}{\hat{x}_{99.865} - \hat{x}_{50}} \right)$$

$$= \text{Minimum}\left(\frac{70 - 68}{70 - 67}, \frac{72 - 70}{73 - 70} \right)$$

$$= \text{Minimum}\,(\,.67\,,\,.67\,) = .67$$

$$\text{Equivalent } \hat{P}_{PK,B} = \text{Minimum}\left(\frac{\hat{x}_{50} - \text{LSL}}{\hat{x}_{50} - \hat{x}_{.135}}, \frac{\text{USL} - \hat{x}_{50}}{\hat{x}_{99.865} - \hat{x}_{50}} \right)$$

$$= \text{Minimum}\left(\frac{70 - 68}{70 - 67}, \frac{72 - 70}{73 - 70} \right)$$

$$= \text{Minimum}\,(\,.67\,,\,.67\,) = .67$$

Despite having identical capability ratings, the amount of nonconforming parts above the USL for A has jumped to 22,750 *ppm*, while that for B has increased to only 3,000 *ppm*.

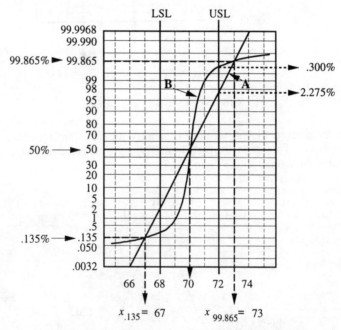

Fig. 8.18 The LSL is 68 and the USL is 72.

For normal distributions, the Z table is used to estimate *ppm*. The best method for estimating the percentage of nonconforming parts for non-normal distributions is directly from the y axis of NOPP, provided the sample size is large enough to accurately fit a curve extending to the specifications. See Flaig for a new approach to estimating the percentage of nonconforming parts.

Example 8.9 Metal Block Length

Performance capability for the metal block length discussed in Example 7.3 (p. 380) can be determined by first finding $\hat{\mu}$ and $\hat{\sigma}_{LT}$ for the 100 length measurements.

$$\hat{\mu} = \frac{\sum\limits_{i=1}^{100} X_i}{100} = 3.476 \qquad\qquad \hat{\sigma}_{LT} = \frac{S_{TOT}}{c_4} = .065$$

These results are then used to estimate P_{PK}.

$$\hat{P}_{PK} = \text{Minimum}\left(\frac{\hat{\mu} - LSL}{3\hat{\sigma}_{LT}}, \frac{USL - \hat{\mu}}{3\hat{\sigma}_{LT}}\right)$$

$$= \text{Minimum}\left(\frac{3.476 - 3.275}{3(.065)}, \frac{3.625 - 3.476}{3(.065)}\right)$$

$$= \text{Minimum}(1.03, .76) = .76$$

According to this index, the process is not capable. However, P_{PK} is valid only for processes which are normally distributed. When this assumption is checked with NOPP, the S-shaped curve on the NOPP shown in Figure 8.19 implies the process output for length is not normally distributed, but most likely has an unbounded distribution.

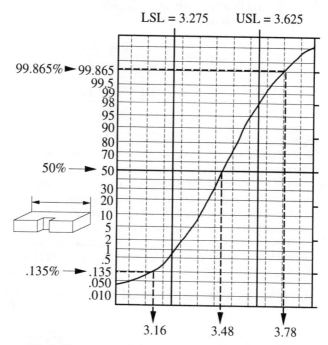

Fig. 8.19 The NOPP for the metal block length example.

Thus, the standard formula for P_{PK} is not applicable, but the percentile method is. The three essential percentiles are estimated from the NOPP as 3.16 for $x_{.135}$, 3.48 for x_{50}, and 3.78 for $x_{99.865}$. Recalling that the LSL is 3.275 and the USL is 3.625, the equivalent P_{PK} is estimated by substituting these values into the following formula:

$$\text{Equivalent } \hat{P}_{PK} = \text{Minimum } (\text{Equivalent } \hat{P}_{PL}, \text{Equivalent } \hat{P}_{PU})$$

$$= \text{Minimum} \left(\frac{\hat{x}_{50} - \text{LSL}}{\hat{x}_{50} - \hat{x}_{.135}}, \frac{\text{USL} - \hat{x}_{50}}{\hat{x}_{99.865} - \hat{x}_{50}} \right)$$

$$= \text{Minimum} \left(\frac{3.48 - 3.275}{3.48 - 3.16}, \frac{3.625 - 3.48}{3.78 - 3.48} \right)$$

$$= \text{Minimum } (.64, .48) = .48$$

This process does not meet the minimum requirement for long-term performance capability as only 48 percent of the equivalent upper $3\sigma_{LT}$ tail, which is the upper half of the middle 99.73 percent, will fit into the distance from the median to the USL. In addition, only 64 percent of the lower equivalent tail will fit between the median and the LSL, implying *more* than .135 percent of the lengths are below the LSL. The standard P_{PL} index of 1.03, which is invalid due to the non-normal output distribution, incorrectly indicates that *less* than .135 percent are below the LSL.

As the equivalent P_{PL} index of .64 is larger than the equivalent P_{PU} index of .48, moving the median lower will increase equivalent P_{PK} somewhat, but not much because the long-term potential capability is only about .56.

$$\text{Equivalent } \hat{P}_P = \frac{\text{Tolerance}}{\hat{x}_{99.865} - \hat{x}_{.135}}$$

$$= \frac{3.625 - 3.275}{3.78 - 3.16} = .56452$$

The highest equivalent P_{PK} can possibly be with the current amount of process variation is .56452. Part containment, accompanied with 100 percent inspection, must continue while work is done to reduce common-cause variation.

Where should the median be located to realize this maximum possible performance capability of .56452? Because non-normal distributions are quite often asymmetrical, centering the process output at the middle of the tolerance does not always guarantee maximum performance capability. Assuming the distance from the median to the 99.865 percentile remains constant at .30 (3.78 minus 3.48) when the median is shifted, the following formula can be solved for x'_{50}, the location for the median that maximizes equivalent P_{PK}.

$$\text{Maximum Performance Capability } \Rightarrow .56452 = \frac{\text{USL} - x'_{50}}{x_{99.865} - x_{50}}$$

$$.56452 = \frac{3.625 - x'_{50}}{.30}$$

$$.30(.56452) = 3.625 - x'_{50}$$

$$x'_{50} = 3.625 - .30(.56452)$$

$$x'_{50} = 3.45564$$

$$\cong 3.456$$

If the median is lowered .024 from its current location of 3.480 down to 3.456, equivalent P_{PK} will increase to its maximum of .5645. This analysis assumes the .135 percentile is also lowered by .024 from its current value of 3.160 to 3.136 while the 99.865 percentile moves down .024 from 3.780 to 3.756. These changes are displayed in Figure 8.20, with the original curve represented by a dashed line and the proposed curve with a solid line. Substituting these new percentiles into the equivalent P_{PK} formula yields the desired result.

$$\text{Equivalent } \hat{P}_{PK} = \text{Minimum}\left(\frac{x'_{50} - \text{LSL}}{x'_{50} - \hat{x}_{.135}}, \frac{\text{USL} - x'_{50}}{\hat{x}_{99.865} - x'_{50}} \right)$$

$$= \text{Minimum}\left(\frac{3.45564 - 3.275}{3.45564 - 3.136}, \frac{3.625 - 3.45564}{3.756 - 3.45564} \right)$$

$$= \text{Minimum}(.5645, .5645) = .5645$$

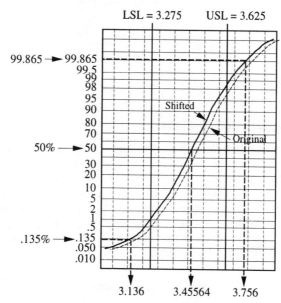

Fig. 8.20 The original curve is shifted down by .024 units.

Whenever the process median equals x'_{50}, equivalent P_{PK} equals equivalent P_P, and the process has achieved its maximum capability with the current amount of process spread. Notice that the desired location for the median of 3.45564 is *not* the middle of the tolerance, which is 3.450. Centering the process output at the middle of the tolerance to maximize performance capability works only when the distance from the median to the .135 percentile is identical to that from the median to the 99.865 percentile. For the block length example, the first distance (lower tail) is .32 and the second (upper tail) is .30. On account of this difference, the median must be centered slightly *higher* than the middle of the tolerance to accommodate the longer lower tail. In general, x'_{50} is found with this formula, whose full derivation is given at the end of this chapter (see Derivation 1 on page 509):

$$x'_{50} = \frac{(x_{50} - x_{.135})\,\text{USL} + (x_{99.865} - x_{50})\,\text{LSL}}{x_{99.865} - x_{.135}}$$

For the metal block example, the .135, 50, and 99.865 percentiles are 3.16, 3.48, and 3.78, respectively. With a LSL of 3.275 and an USL of 3.625, x'_{50} is estimated as 3.45564.

$$\hat{x}'_{50} = \frac{(3.48 - 3.16)3.625 + (3.78 - 3.48)3.275}{3.78 - 3.16} = 3.45564$$

In the case of symmetrical distributions, like the normal distribution, the following relationships are true:

$$x_{50} - x_{.135} = x_{99.865} - x_{50} = 1/2(x_{99.865} - x_{.135})$$

For these distributions, x'_{50} equals M, the middle of the tolerance.

$$x'_{50} = \frac{1/2(x_{99.865} - x_{.135})\,\text{USL} + 1/2(x_{99.865} - x_{.135})\,\text{LSL}}{x_{99.865} - x_{.135}}$$

$$= \frac{1/2(x_{99.865} - x_{.135})\,(\text{USL} + \text{LSL})}{x_{99.865} - x_{.135}}$$

$$= \frac{\text{USL} + \text{LSL}}{2} = M$$

Example 8.10 Forging Hardness

An estimate of long-term performance capability is desired for forging hardness to determine if it is meeting a customer-imposed goal of 1.67 for P_{PK}. After establishing control, the cumulative distribution for hardness is plotted on NOPP by members of a capability study team (Figure 8.6). The curved line that appears indicates the process output has a bounded distribution, thus the standard formula for P_{PK} does not apply. Instead, the team decides to utilize the percentile formula for estimating equivalent P_{PK}. After reading off the .135, 50, and 99.865 percentiles from the NOPP, long-term performance capability is estimated as follows (LSL is 130, USL is 190):

$$\text{Equivalent } \hat{P}_{PK} = \text{Minimum}\left(\frac{\hat{x}_{50} - \text{LSL}}{\hat{x}_{50} - \hat{x}_{.135}}, \frac{\text{USL} - \hat{x}_{50}}{\hat{x}_{99.865} - \hat{x}_{50}}\right)$$

$$= \text{Minimum}\left(\frac{162 - 130}{162 - 132}, \frac{190 - 162}{182 - 162}\right)$$

$$= \text{Minimum}\left(\frac{32}{30}, \frac{28}{20}\right)$$

$$= \text{Minimum}\,(1.07, 1.40) = 1.07$$

This forging process fulfills the minimum condition for performance capability, as slightly more than 99.73 percent conforming parts are being produced. Unfortunately, it falls far short of the desired goal of 1.67. As equivalent P_{PU} (1.40) is greater than equivalent P_{PL} (1.07), moving the process median higher will increase performance capability, but only slightly, as potential capability is just 1.20.

$$\text{Equivalent } \hat{P}_P = \frac{\text{Tolerance}}{\hat{x}_{99.865} - \hat{x}_{.135}} = \frac{190 - 130}{182 - 132} = 1.20$$

With a normal (or any symmetrical) distribution, the 1.20 could also be obtained by averaging P_{PL} and P_{PU}, as was done in Example 6.14 on page 215.

$$\hat{P}_P = \frac{\hat{P}_{PL} + \hat{P}_{PU}}{2}$$

For non-symmetrical distributions, this relationship is not true. Instead, the following formula applies.

$$\text{Equivalent } \hat{P}_P = \frac{(x_{99.865} - x_{50})\text{Equivalent } \hat{P}_{PL} + (x_{50} - x_{.135})\text{Equivalent } \hat{P}_{PU}}{x_{99.865} - x_{.135}}$$

In this forging hardness example, equivalent P_P is found to be 1.20 with this formula.

$$\text{Equivalent } \hat{P}_P = \frac{(182 - 162)1.07 + (162 - 132)1.40}{182 - 132} = \frac{60}{50} = 1.20$$

This result is the same as previously generated by the original formula for equivalent P_P. But where should the process median be centered to achieve this potential capability? With the formula introduced in the metal block example on page 466, x'_{50} is found to be 166.

$$\hat{x}'_{50} = \frac{(\hat{x}_{50} - \hat{x}_{.135})\text{USL} + (\hat{x}_{99.865} - \hat{x}_{50})\text{LSL}}{\hat{x}_{99.865} - \hat{x}_{.135}} = \frac{(162 - 132)190 + (182 - 162)130}{182 - 132} = 166$$

If the median is shifted upward 4 units from its current location of 162 to 166, equivalent P_{PK} will become 1.20. Again, this assumes the .135 percentile also shifts by 4 from its current value of 132 to 136, while the 99.865 percentile moves up 4 from 182 to 186. The net effect of all these proposed changes is presented as the solid curve in Figure 8.21. Substituting these new percentiles into the equivalent P_{PK} formula generates the maximum possible performance capability of 1.20.

$$\text{Equivalent } \hat{P}_{PK} = \text{Minimum} \left(\text{Equivalent } \hat{P}_{PL}, \text{Equivalent } \hat{P}_{PU} \right)$$

$$= \text{Minimum} \left(\frac{x'_{50} - \text{LSL}}{x'_{50} - \hat{x}_{.135}}, \frac{\text{USL} - x'_{50}}{\hat{x}_{99.865} - x'_{50}} \right)$$

$$= \text{Minimum} \left(\frac{166 - 130}{166 - 136}, \frac{190 - 166}{186 - 166} \right)$$

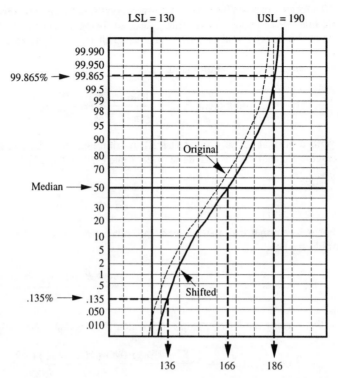

Fig. 8.21 The hardness curve is shifted 4 units to the right.

$$= \text{Minimum} \left(\frac{36}{30}, \frac{24}{20} \right)$$

$$= \text{Minimum} \, (\, 1.20 \, , 1.20 \,) \; = \; 1.20$$

For this process, the desired location of 166 for the median is not the middle of the tolerance, which is 160. Centering the process output at M to maximize performance capability is correct only when x_{50} minus $x_{.135}$ equals $x_{99.865}$ minus x_{50}, which is true for symmetrical distributions. In this forging hardness example, the first difference is 30 (166 minus 136), while the second is 20 (186 minus 166). Because the first difference is larger, the median must be centered *higher* than the middle of the tolerance to balance equivalent P_{PL} and equivalent P_{PU}.

Equivalent P_{PK} versus x_{50}

As the graph in Figure 8.22 for the forging example discloses, a shift in the process median above x'_{50} creates a greater loss in performance capability than an equal move below x'_{50}. This is a different response than that occurring for normal distributions, where an upward movement in the process median decreases capability by the same amount an equal movement lower does. Unlike situation for a normal distribution where the slope of the P_{PU} line equals that of the P_{PL} line (see Figure 6.64 on page 255), the line for equivalent P_{PU} has a much steeper slope than the line for equivalent P_{PL}.

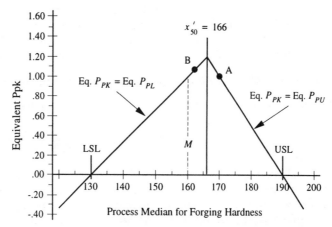

Fig. 8.22 Equivalent P_{PK} versus the process median.

For example, if the median is moved from 166 to 170, which is 4 units above x'_{50}, the estimate for equivalent P_{PK} becomes 1.00 (see point A in Figure 8.22).

$$\text{Equivalent } \hat{P}_{PK} = \text{Minimum} \left(\frac{\hat{x}_{50} - \text{LSL}}{\hat{x}_{50} - \hat{x}_{.135}}, \frac{\text{USL} - \hat{x}_{50}}{\hat{x}_{99.865} - \hat{x}_{50}} \right)$$

$$= \text{Minimum} \left(\frac{170 - 130}{170 - 140}, \frac{190 - 170}{190 - 170} \right)$$

$$= \text{Minimum} \left(1.33, 1.00 \right) = 1.00$$

This upward move of 4 units causes this index to drop about 17 percent, from 1.20 to 1.00. Compare this outcome to that obtained by moving the median to 162, 4 units *below* x'_{50} (represented by point B). This equal-sized downward move in the median creates a decrease of only 11 percent in equivalent P_{PK}, from 1.20 to 1.07. Thus, when x_{50} is above x'_{50}, upward shifts in the median are more harmful to the performance of this process than downward ones.

Just like the standard P_{PK} index, equivalent P_{PK} is 0 when the median is centered at either the LSL or the USL. For instance, when x_{50} equals the LSL:

$$\text{Equivalent } \hat{P}_{PK} = \text{Minimum} \left(\frac{\text{LSL} - \text{LSL}}{\hat{x}_{50} - \hat{x}_{.135}}, \frac{\text{USL} - \text{LSL}}{\hat{x}_{99.865} - \hat{x}_{50}} \right)$$

$$= \text{Minimum} \left(\frac{0}{\hat{x}_{50} - \hat{x}_{.135}}, \frac{\text{USL} - \text{LSL}}{\hat{x}_{99.865} - \hat{x}_{50}} \right) = 0$$

Whenever the median is below the LSL (or above the USL), equivalent P_{PK} is negative, again mimicking the standard P_{PK} index.

Maximum performance capability for normal distributions occurs when the process is centered at M. This is seldom true for non-normal distributions, where maximum performance capability is achieved when the median equals x'_{50}, which is seldom equal to M, especially for asymmetrical distributions.

The performance capability goal for forging hardness was set at having P_{PK} equal 1.67. Usually this goal is based on a desired maximum *ppm* level for each specification, in this case .3. It is entirely possible for equivalent P_{PK} to exceed 1.67, yet have more than .3 *ppm* outside one of the specifications. If the *ppm* level is of utmost concern, do not use this measure, but switch to equivalent P^*_{PK}, which will be covered in Section 8.15 (p. 499). When this alternative index equals 1.67, there is no more than .3 ppm outside either the LSL or USL.

The Equivalent *k* Factor Relating Equivalent P_{PK} and Equivalent P_P

P_{PK} and P_P are related by the k factor according to this formula:

$$P_{PK} = P_P(1 - k) \qquad \text{where } k = \frac{|\mu - M|}{(\text{USL - LSL})/2}$$

This relationship split performance capability into its two important components: precision, σ_{LT}, and accuracy, μ. An analysis of these two indicated where process improvement efforts should be focused, either on reducing variation, shifting the average, or both.

In a similar manner, equivalent P_{PK} and equivalent P_P are related through an equivalent k factor, whose derivation is provided at the end of this chapter (see Derivation 2 on page 509). Recall that x'_{50} represents the median value for which equivalent P_{PK} is maximized.

$$\text{Equivalent } P_{PK} = \text{Equivalent } P_P(1 - \text{Equivalent } k)$$

$$\text{where Equivalent } k = \text{Maximum}\left(\frac{x_{50} - x'_{50}}{\text{LSL} - x'_{50}}, \frac{x_{50} - x'_{50}}{\text{USL} - x'_{50}} \right)$$

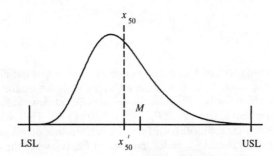

Fig. 8.23 Process median centered at x'_{50}.

Figure 8.23 shows a process output where x_{50} equals x'_{50}. This means x_{50} minus x'_{50} is zero, making equivalent k equal to zero.

$$\text{When } x_{50} = x'_{50}, \text{ Equivalent } k = \text{Maximum}\left(\frac{x_{50} - x'_{50}}{\text{LSL} - x'_{50}}, \frac{x_{50} - x'_{50}}{\text{USL} - x'_{50}} \right)$$

$$= \text{Maximum}\left(\frac{0}{\text{LSL} - x'_{50}}, \frac{0}{\text{USL} - x'_{50}} \right) = 0$$

When equivalent k equals zero, equivalent P_{PK} equals equivalent P_P, and maximum performance capability is achieved, *i.e.*, performance equals potential.

$$\text{Equivalent } P_{PK} = \text{Equivalent } P_P(1 - \text{Equivalent } k) = \text{Equivalent } P_P(1 - 0) = \text{Equivalent } P_P$$

Now assume x_{50} equals the LSL, as is displayed in Figure 8.24.

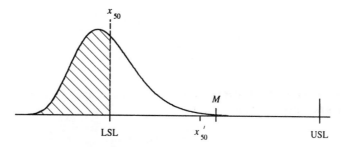

Fig. 8.24 Process median centered at LSL.

$$\text{Equivalent } k = \text{Maximum}\left(\frac{\text{LSL} - x'_{50}}{\text{LSL} - x'_{50}}, \frac{\text{LSL} - x'_{50}}{\text{USL} - x'_{50}}\right) = \text{Maximum}\left(1, \frac{\text{LSL} - x'_{50}}{\text{USL} - x'_{50}}\right)$$

Because x'_{50} is always between the LSL and USL, the difference between the LSL and x'_{50} is negative, while the difference between the USL and x'_{50} is positive. These circumstances makes their ratio negative.

$$\frac{\text{LSL} - x'_{50}}{\text{USL} - x'_{50}} < 0$$

This means equivalent k equals 1 when x'_{50} equals the LSL.

$$\text{Equivalent } k = \text{Maximum}\,(1, <0) = 1$$

Substituting an equivalent k of 1 into the formula for equivalent P_{PK} produces:

$$\text{Equivalent } P_{PK} = \text{Equivalent } P_P(1 - \text{Equivalent } k) = \text{Equivalent } P_P(1 - 1) = 0$$

Just as for normal distributions, when x_{50} equals the LSL (or USL), performance capability is 0, and at least 50 percent of the parts are nonconforming.

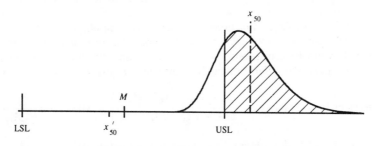

Fig. 8.25 Process median centered above USL.

Consider the case where x_{50} is outside the tolerance, say above the USL, as is illustrated in Figure 8.25. This makes x'_{50} greater than x_{50}.

$$x_{50} > \text{USL}$$

$$x_{50} - x'_{50} > \text{USL} - x'_{50}$$

$$\frac{x_{50} - x'_{50}}{\text{USL} - x'_{50}} > 1$$

As the difference between x_{50} and x'_{50} is positive, and the difference between the LSL and x'_{50} is negative, their ratio is negative.

$$\frac{x_{50} - x'_{50}}{\text{LSL} - x'_{50}} < 0$$

Under these conditions, equivalent k is always greater than 1.

$$\text{Equivalent } k = \text{Maximum}\left(\frac{x_{50} - x'_{50}}{\text{LSL} - x'_{50}}, \frac{x_{50} - x'_{50}}{\text{USL} - x'_{50}}\right) = \text{Maximum} \left(<0, >1\right) > 1$$

Consequently, 1 minus equivalent k is less than 0.

$$\text{Equivalent } k > 1$$

$$-\text{Equivalent } k < -1$$

$$1 - \text{Equivalent } k < 1 - 1 = 0$$

Multiplying both sides by equivalent P_P (which is always positive) reveals that equivalent P_{PK} will be less than 0 when x_{50} is located above the USL, with much more than 50 percent nonconforming parts.

$$1 - \text{Equivalent } k < 0$$

$$\text{Equivalent } P_P \left(1 - \text{Equivalent } k\right) < 0$$

$$\text{Equivalent } P_{PK} < 0$$

Again, equivalent k functions in a manner very similar to the standard k factor. In fact, when the process output under study has a normal distribution, equivalent k is identical to k. For all symmetrical distributions, x_{50} equals μ, and x'_{50} becomes M, the middle of the tolerance (review page 466). Substituting these into the formula for equivalent k yields:

$$\text{Equivalent } k = \text{Maximum}\left(\frac{x_{50} - x'_{50}}{\text{LSL} - x'_{50}}, \frac{x_{50} - x'_{50}}{\text{USL} - x'_{50}}\right) = \text{Maximum}\left(\frac{\mu - M}{\text{LSL} - M}, \frac{\mu - M}{\text{USL} - M}\right)$$

As M is the middle of the tolerance, the distance between the USL and M is equal to one-half of the tolerance.

$$\text{USL} - M = 1/2 \text{ Tolerance}$$

Likewise, the distance between M and the LSL is also equal to one-half the tolerance.

$$M - LSL \;=\; 1/2\,\text{Tolerance} \quad \Rightarrow \quad LSL - M \;=\; -1/2\,\text{Tolerance}$$

Making these additional substitutions into the equivalent k formula results in:

$$\text{Equivalent } k \;=\; \text{Maximum}\left(\frac{\mu - M}{LSL - M}, \frac{\mu - M}{USL - M}\right)$$

$$=\; \text{Maximum}\left(\frac{\mu - M}{-1/2\,\text{Tolerance}}, \frac{\mu - M}{1/2\,\text{Tolerance}}\right)$$

$$=\; \text{Maximum}\left(\frac{M - \mu}{1/2\,\text{Tolerance}}, \frac{\mu - M}{1/2\,\text{Tolerance}}\right)$$

$$=\; \frac{1}{1/2\,\text{Tolerance}}\,\text{Maximum}\,(M - \mu, \mu - M)$$

The quantities M minus μ and μ minus M are identical except for their sign. One will be positive while the other is negative, depending on whether μ is above M or below it. Because the larger of these two is always taken, either can be chosen as the maximum by ignoring its sign. This selection procedure functions in the same manner as the absolute value operator.

$$\text{Maximum}\,(M - \mu, \mu - M) \;=\; |\mu - M|$$

Thus, equivalent k is identical to the standard k value for all symmetrical distributions.

$$\text{Equivalent } k \;=\; \frac{1}{1/2\,\text{Tolerance}}\,\text{Maximum}\,(M - \mu, \mu - M) \;=\; \frac{|\mu - M|}{1/2\,\text{Tolerance}} \;=\; k$$

Unfortunately, equivalent k and equivalent P_P are not always independent. If process spread is reduced, equivalent P_P will change because it is a function of the difference between $x_{99.865}$ and $x_{.135}$. However, equivalent k is also determined from this same difference, via x'_{50}. Therefore, equivalent k is best used for assessing the current situation to make a decision on whether to shift the process median or to reduce process spread.

For those occasions where decreasing process spread reduces x_{50} minus $x_{.135}$ and $x_{99.865}$ minus x_{50} by an equal percentage, x'_{50} will not vary with changes in process spread. Suppose the spread of process output distribution is decreased uniformly by 10 percent. This means x_{50} minus $x_{.135}$, $x_{99.865}$ minus x_{50}, and $x_{99.865}$ minus $x_{.135}$ are all now only 90 percent of their old distances. With these changes in spread, the new x'_{50} equals the original x'_{50}.

$$\text{New } x'_{50} \;=\; \frac{.90(x_{50} - x_{.135})\,USL + .90(x_{99.865} - x_{50})\,LSL}{.90(x_{99.865} - x_{.135})}$$

$$=\; \frac{.90[(x_{50} - x_{.135})\,USL + (x_{99.865} - x_{50})\,LSL]}{.90(x_{99.865} - x_{.135})}$$

$$=\; \frac{(x_{50} - x_{.135})\,USL + (x_{99.865} - x_{50})\,LSL}{x_{99.865} - x_{.135}} \;=\; \text{Original } x'_{50}$$

Under this special condition, equivalent k and equivalent P_P are independent, and their analysis becomes identical to that for the standard k and P_P.

Recall the k_L and k_U factors relating P_P to P_{PL} and P_{PU}.

$$P_{PL} = P_P(1 - k_L) \quad \text{where } k_L = \frac{M - \mu}{\text{Tolerance}/2}$$

$$P_{PU} = P_P(1 - k_U) \quad \text{where } k_U = \frac{\mu - M}{\text{Tolerance}/2}$$

Equivalent k_L and equivalent k_U are defined as follows:

$$\text{Equivalent } P_{PL} = \text{Equivalent } P_P(1 - \text{Equivalent } k_L) \quad \text{where Equivalent } k_L = \frac{x'_{50} - x_{50}}{x'_{50} - \text{LSL}}$$

$$\text{Equivalent } P_{PU} = \text{Equivalent } P_P(1 - \text{Equivalent } k_U) \quad \text{where Equivalent } k_U = \frac{x_{50} - x'_{50}}{\text{USL} - x'_{50}}$$

For symmetrical distributions, x'_{50} equals M, x_{50} equals μ, and equivalent k_L becomes k_L.

$$\text{Equivalent } k_L = \frac{x'_{50} - x_{50}}{x'_{50} - \text{LSL}} = \frac{M - \mu}{M - \text{LSL}} = \frac{M - \mu}{\text{Tolerance}/2} = k_L$$

Likewise, equivalent k_U becomes k_U in this situation.

Example 8.11 Forging Hardness

In Example 8.4, an estimate of long-term potential capability was desired for forging hardness (LSL = 130, USL = 190). Having ascertained from NOPP that the output distribution for hardness is non-normal, equivalent P_P was selected to assess potential capability.

$$\text{Equivalent } \hat{P}_P = \frac{\text{Tolerance}}{\hat{x}_{99.865} - \hat{x}_{.135}} = \frac{190 - 130}{182 - 132} = \frac{60}{50} = 1.20$$

In Example 8.10 (p. 466), x'_{50} was estimated as 166.

$$\hat{x}'_{50} = \frac{(\hat{x}_{50} - \hat{x}_{.135})\,\text{USL} + (\hat{x}_{99.865} - \hat{x}_{50})\,\text{LSL}}{\hat{x}_{99.865} - \hat{x}_{.135}} = \frac{(162 - 132)190 + (182 - 162)130}{182 - 132} = 166$$

With a median hardness of 162, equivalent k is estimated as .1111, meaning a little over 11 percent of potential capability is wasted due to improper centering.

$$\text{Equivalent } \hat{k} = \text{Maximum}\left(\frac{\hat{x}_{50} - \hat{x}'_{50}}{\text{LSL} - \hat{x}'_{50}}, \frac{\hat{x}_{50} - \hat{x}'_{50}}{\text{USL} - \hat{x}'_{50}} \right)$$

$$= \text{Maximum}\left(\frac{162 - 166}{130 - 166}, \frac{162 - 166}{190 - 166} \right)$$

$$= \text{Maximum}(.1111, -.1667) = .1111$$

From these estimates of equivalent P_P and equivalent k, equivalent P_{PK} is estimated as 1.07, which is the same result obtained when it was estimated directly in Example 8.10.

$$\text{Equivalent } \hat{P}_{PK} = \text{Equivalent } \hat{P}_P (1 - \text{Equivalent } \hat{k}) = 1.20(1 - .1111) = 1.07$$

As an equivalent P_{PK} of 1.07 is far short of the 1.67 goal, process improvements must be made. An analysis of the equivalent P_P and equivalent k values suggests that, although shifting the median will help somewhat, efforts to reduce process spread should take priority (locating the median at 166 will raise equivalent P_{PK} to only 1.20, still far short of the 1.67 goal).

Assume the process spread can be uniformly reduced by 15 percent around the current median, meaning all differences between percentiles are only 85 percent of their former values. For example, the difference between $x_{99.865}$ and $x_{.135}$ is now 42.5 ($.85 \times 50$). This reduction in spread increases equivalent P_P to 1.412.

$$\text{Equivalent } \hat{P}_P = \frac{\text{Tolerance}}{\hat{x}_{99.865} - \hat{x}_{.135}} = \frac{190 - 130}{42.5} = 1.412$$

Because the median was not shifted, and x'_{50} does not change with a uniform reduction in spread, the equivalent k factor remains at .1111. Therefore, the above increase in precision causes equivalent P_{PK} to grow from 1.07 to 1.255.

$$\text{Equivalent } \hat{P}_{PK} = \text{Equivalent } \hat{P}_P (1 - \text{Equivalent } \hat{k}) = 1.412(1 - .1111) = 1.255$$

Unfortunately, this is not enough of an improvement to reach the goal of 1.67. To achieve this goal without changing the median, the process spread must be decreased until equivalent P_P equals 1.879.

$$1.67 = \text{Equivalent } \hat{P}_P (1 - .1111)$$

$$\frac{1.67}{1 - .1111} = \text{Equivalent } \hat{P}_P$$

$$1.879 = \text{Equivalent } \hat{P}_P$$

In order to have equivalent P_P equal 1.879, the difference between $x_{99.865}$ and $x_{.135}$ must be no more than 31.93.

$$\text{Equivalent } \hat{P}_P = \frac{\text{Tolerance}}{\hat{x}_{99.865} - \hat{x}_{.135}}$$

$$1.879 = \frac{190 - 130}{\hat{x}_{99.865} - \hat{x}_{.135}}$$

$$\hat{x}_{99.865} - \hat{x}_{.135} = \frac{60}{1.879}$$

$$\hat{x}_{99.865} - \hat{x}_{.135} = 31.93$$

This requires almost an additional 25 percent reduction in process spread.

$$42.5 \times (1 - \text{Reduction}) = 31.93$$

$$1 - \text{Reduction} = \frac{31.93}{42.5}$$

$$\text{Reduction} = 1 - .751$$

$$= .249$$

Example 8.12 Bearing Support Hole Size

Measurements for the inner diameter of a hole drilled in a bearing support bracket are analyzed on NOPP. The curve drawn through the plotted points indicates the process output distribution for this characteristic appears to be skewed right (Figure 8.3, p. 433). Estimates of the .135 percentile, median, and 99.865 percentile are read off the x axis of the NOPP as 25.4, 26.7, and 31.8, respectively. The tolerance for hole size is 10 (35 minus 25), while the goal for equivalent P_{PK} is 1.50. From this information, equivalent P_P and equivalent P_{PK} may be estimated to analyze the performance capability of this drilling process.

$$\text{Equivalent } \hat{P}_P = \frac{\text{Tolerance}}{\hat{x}_{99.865} - \hat{x}_{.135}} = \frac{35 - 25}{31.8 - 25.4} = 1.5625$$

$$\text{Equivalent } \hat{P}_{PK} = \text{Minimum} \left(\frac{\hat{x}_{50} - \text{LSL}}{\hat{x}_{50} - \hat{x}_{.135}}, \frac{\text{USL} - \hat{x}_{50}}{\hat{x}_{99.865} - \hat{x}_{50}} \right)$$

$$= \text{Minimum} \left(\frac{26.7 - 25.0}{26.7 - 25.4}, \frac{35.0 - 26.7}{31.8 - 26.7} \right)$$

$$= \text{Minimum} (1.3077 , 1.6275) = 1.3077$$

As equivalent P_{PK} is less than equivalent P_P, performance capability may be enhanced by shifting the median to x'_{50}. Because equivalent P_{PL} is less than equivalent P_{PU}, x'_{50} will be higher than the current median of 26.7.

$$\hat{x}'_{50} = \frac{(\hat{x}_{50} - \hat{x}_{.135}) \text{USL} + (\hat{x}_{99.865} - \hat{x}_{50}) \text{LSL}}{\hat{x}_{99.865} - \hat{x}_{.135}} = \frac{(26.7 - 25.4)35 + (31.8 - 26.7)25}{31.8 - 25.4} = 27.03125$$

The above estimate of x'_{50} allows equivalent k to be estimated.

$$\text{Equivalent } \hat{k} = \text{Maximum} \left(\frac{\hat{x}_{50} - \hat{x}'_{50}}{\text{LSL} - \hat{x}'_{50}}, \frac{\hat{x}_{50} - \hat{x}'_{50}}{\text{USL} - \hat{x}'_{50}} \right)$$

$$= \text{Maximum} \left(\frac{26.7 - 27.03125}{25.0 - 27.03125}, \frac{26.7 - 27.03125}{35.0 - 27.03125} \right)$$

$$= \text{Maximum} (.16308 , -.04157) = .16308$$

Over 16 percent of potential capability is not available for performance capability due to improper centering. Because equivalent P_P is over 1.56, and reducing spread is usually more costly, emphasis should be placed on moving the median closer to x'_{50} in order to reach the goal of 1.50. If the median can be centered at x'_{50}, equivalent k becomes zero and equivalent P_{PK} would equal equivalent P_P, which at 1.5625, is already greater than the 1.50 goal.

For practice, the equivalent k_L and equivalent k_U factors can be estimated for this example.

$$\text{Equivalent } \hat{k}_L = \frac{\hat{x}'_{50} - \hat{x}_{50}}{\hat{x}'_{50} - \text{LSL}} = \frac{27.03125 - 26.7}{27.03125 - 25} = .16308$$

$$\text{Equivalent } \hat{k}_U = \frac{\hat{x}_{50} - \hat{x}'_{50}}{\text{USL} - \hat{x}'_{50}} = \frac{26.7 - 27.03125}{35 - 27.03125} = -.04157$$

From these results, equivalent P_{PL} and equivalent P_{PU} may be estimated.

$$\begin{aligned}
\text{Equivalent } P_{PL} &= \text{Equivalent } P_P \, (\, 1 - \text{Equivalent } k_L \,) \\
&= 1.5625 \, (\, 1 - .16308 \,) \\
&= 1.3077
\end{aligned}$$

Thanks to the negative k_U factor, equivalent P_{PU} picks up a "bonus" due to the median being centered below x'_{50}.

$$\begin{aligned}
\text{Equivalent } P_{PU} &= \text{Equivalent } P_P \, (\, 1 - \text{Equivalent } k_U \,) \\
&= 1.5625 \, (\, 1 - -.04157 \,) \\
&= 1.6275
\end{aligned}$$

The smaller of these two is the estimate for equivalent P_{PK}.

$$\begin{aligned}
\text{Equivalent } \hat{P}_{PK} &= \text{Minimum} \, (\, \text{Equivalent } \hat{P}_{PL} \, , \text{Equivalent } \hat{P}_{PU} \,) \\
&= \text{Minimum} \, (\, 1.3077 , 1.6275 \,) = 1.3077
\end{aligned}$$

Example 8.13 Hole Location

A turret press punches holes in sheet metal that eventually will form a computer housing. There is some concern about hole location along the x axis, as this feature is critical to the assembly operation. To gain additional insight, Taylor (the production supervisor) starts monitoring this characteristic with an $\overline{X}, R$ chart. After several days of charting and working to eliminate several assignable causes of variation, the chart finally provides evidence the process is in a reasonably good state of control for this feature.

When measurements for hole location along the x axis are plotted on NOPP to check the normality assumption, he discovers the process output has a skewed-right distribution, as is displayed in Figure 8.26. Realizing the standard capability measures are inappropriate for this process, Taylor decides to assess potential capability via the equivalent P'_{PU} index. He derives estimates from the NOPP for the median (.0058) and for $x_{99.865}$ (.0087). The USL for hole location is given as .0080, with a target of .0035.

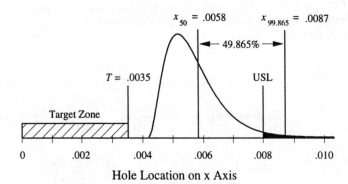

Fig. 8.26 The skewed output distribution for hole location.

$$\text{Equivalent } \hat{P}'_{PU} = \text{Maximum} \left(\text{Equivalent } \hat{P}^T_{PU}, \text{Equivalent } \hat{P}_{PU} \right)$$

$$= \text{Maximum} \left(\frac{\text{USL} - T}{\hat{x}_{99.865} - \hat{x}_{50}}, \frac{\text{USL} - \hat{x}_{50}}{\hat{x}_{99.865} - \hat{x}_{50}} \right)$$

$$= \text{Maximum} \left(\frac{.0080 - .0035}{.0087 - .0058}, \frac{.0080 - .0058}{.0087 - .0058} \right)$$

$$= \text{Maximum} \left(1.55, .76 \right) = 1.55$$

Happy to see the potential capability of this process to locate holes is so high, this supervisor decides to also estimate the performance capability with the equivalent P_{PK} index. Because hole location has only an USL, the formula for equivalent P_{PK} reduces to just the formula for equivalent P_{PU}.

$$\text{Equivalent } \hat{P}_{PK} = \text{Equivalent } \hat{P}_{PU} = \frac{\text{USL} - \hat{x}_{50}}{\hat{x}_{99.865} - \hat{x}_{50}} = \frac{.0080 - .0058}{.0087 - .0058} = .76$$

Because the process median is located quite a distance above the target of .0035, the process does not meet even the minimum requirement for performance capability of 1.00, much less the goal of 1.50 for this characteristic. As potential capability is 1.55, efforts must concentrate on lowering the process median so it is closer to the target value.

What if the median can be lowered by .0023 units so it is now positioned directly at the target of .0035? If the distribution's spread is not altered by this change in x_{50}, meaning $x_{99.865}$ is also lowered by .0023 units to .0064, the potential capability of this process will not change.

$$\text{Equivalent } \hat{P}'_{PU} = \text{Maximum} \left(\frac{\text{USL} - T}{\hat{x}_{99.865} - \hat{x}_{50}}, \frac{\text{USL} - \hat{x}_{50}}{\hat{x}_{99.865} - \hat{x}_{50}} \right)$$

$$= \text{Maximum} \left(\frac{.0080 - .0035}{.0064 - .0035}, \frac{.0080 - .0035}{.0064 - .0035} \right)$$

$$= \text{Maximum} \left(1.55, 1.55 \right) = 1.55$$

However, there is a substantial increase in performance capability. In fact, performance capability now equals potential capability, and exceeds its goal of 1.50.

$$\text{Equivalent } \hat{P}_{PK} = \frac{\text{USL} - \hat{x}_{50}}{\hat{x}_{99.865} - \hat{x}_{50}} = \frac{.0080 - .0035}{.0064 - .0035} = 1.55$$

Encouraged by this increase in performance capability, Taylor continues to introduce additional process improvements until the median is finally lowered to .0030. However, due the lower physical bound of zero for hole location, the process output distribution changes to that displayed in Figure 8.27. Although the median shifted by .0005 (.0035 to .0030), $x_{99.865}$ moved down by only .0003, from .0064 to .0061.

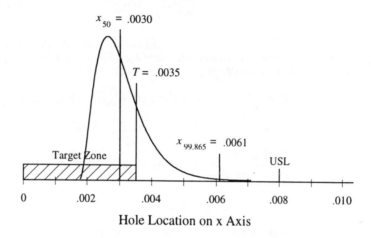

Fig. 8.27 The output distribution for hole location with $x_{50} = .0030$.

As the median is now in the target zone, potential and performance capability are identical.

$$\text{Equivalent } \hat{P}'_{PU} = \text{Maximum} \left(\frac{\text{USL} - T}{\hat{x}_{99.865} - \hat{x}_{50}}, \frac{\text{USL} - \hat{x}_{50}}{\hat{x}_{99.865} - \hat{x}_{50}} \right)$$

$$= \text{Maximum} \left(\frac{.0080 - .0035}{.0061 - .0030}, \frac{.0080 - .0030}{.0061 - .0030} \right)$$

$$= \text{Maximum} \, (\, 1.45 \, , 1.61 \,) \, = \, 1.61$$

$$\text{Equivalent } \hat{P}_{PK} = \frac{\text{USL} - \hat{x}_{50}}{\hat{x}_{99.865} - \hat{x}_{50}} = \frac{.0080 - .0030}{.0061 - .0030} = 1.61$$

This hole-punching process now surpasses its performance capability goal of 1.50 with room to spare.

Equivalent *k* Factors for Unilateral Specifications

Equivalent k'_L and equivalent k'_U can be developed to separate the effects of precision and accuracy on equivalent P_{PK} for characteristics having unilateral specifications. These are analogous to the k'_L and k'_U factors derived in Section 6.5 for use with P'_{PL} and P'_{PU} (read the discussion about Figure 6.52 on page 237). For equivalent P'_{PU}:

$$\text{Equivalent } P_{PK} = \text{Equivalent } P'_{PU} \, (\, 1 - \text{Equivalent } k'_U \,)$$

$$\text{where Equivalent } k'_U = 0 \qquad \text{if } x_{50} \leq T \quad (\, x_{50} \text{ inside target zone })$$

$$= \frac{x_{50} - T}{\text{USL} - T} \quad \text{if } x_{50} > T \quad (\, x_{50} \text{ outside target zone })$$

With only an USL, equivalent P_{PK} always equals equivalent P_{PU}. Whenever x_{50} is in the target zone, equivalent P'_{PU} equals equivalent P_{PU}. These two relationships mean equivalent k'_U must be 0, as is seen here.

$$\text{Equivalent } P_{PK} = \text{Equivalent } P'_{PU} \, (\, 1 - \text{Equivalent } k'_U \,)$$

$$\text{Equivalent } P_{PU} = \text{Equivalent } P_{PU} \, (\, 1 - \text{Equivalent } k'_U \,)$$

$$\frac{\text{Equivalent } P_{PU}}{\text{Equivalent } P_{PU}} = (\, 1 - \text{Equivalent } k'_U \,)$$

$$1 = 1 - \text{Equivalent } k'_U$$

$$\text{Equivalent } k'_U = 1 - 1$$

$$= 0$$

For those times when x_{50} is greater than T (above the target zone), equivalent P'_{PU} equals equivalent P^T_{PU}, and the following relationship holds true:

$$\text{Equivalent } P_{PK} = \text{Equivalent } P'_{PU} \, (\, 1 - \text{Equivalent } k'_U \,)$$

$$\text{Equivalent } P_{PU} = \text{Equivalent } P^T_{PU} \, (\, 1 - \text{Equivalent } k'_U \,)$$

$$\frac{\text{USL} - x_{50}}{x_{99.865} - x_{50}} = \frac{\text{USL} - T}{x_{99.865} - x_{50}} (\, 1 - \text{Equivalent } k'_U \,)$$

$$\frac{\text{USL} - x_{50}}{\text{USL} - T} = 1 - \text{Equivalent } k'_U$$

By rearranging terms, this last equation can be solved for equivalent k'_U.

$$\text{Equivalent } k'_U = 1 - \frac{\text{USL} - x_{50}}{\text{USL} - T} = \frac{(\text{USL} - T) - (\text{USL} - x_{50})}{\text{USL} - T} = \frac{x_{50} - T}{\text{USL} - T}$$

Under this condition where the median is outside the target zone, equivalent k'_U represents the ratio of the distance between x_{50} and T, to the distance between the USL and T.

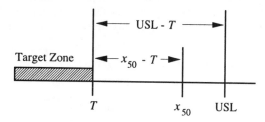

Fig. 8.28 Process centered above the target zone.

As Figure 8.28 illustrates, equivalent k'_U is a measure of accuracy, as it discloses how far away the median is from the target zone. Whenever x_{50} moves higher, meaning process aim is becoming worse, the difference between x_{50} and T becomes larger, causing equivalent k'_U to become larger as well. This increase in equivalent k'_U triggers a corresponding decrease in equivalent P_{PK}. In a similar manner, equivalent k'_L is defined as:

$$\text{Equivalent } P_{PK} = \text{Equivalent } P'_{PL}(1 - \text{Equivalent } k'_L)$$

$$\text{where Equivalent } k'_L = 0 \qquad \text{if } x_{50} \geq T$$

$$= \frac{T - x_{50}}{T - \text{LSL}} \quad \text{if } x_{50} < T$$

Unlike equivalent k, equivalent k_L, or equivalent k_U, both equivalent k'_L and equivalent k'_U are independent of process spread.

Example 8.14 Hole Location

In Example 8.13 (p. 477), a capability study was undertaken for hole location. As NOPP indicated that the output distribution was skewed right, the equivalent measures were correctly chosen to assess capability. With the median estimated as .0058, $x_{99.865}$ as .0087, and given an USL of .0080 with a target hole location of .0035, equivalent P'_{PU} was estimated as 1.55. Because the median is greater than the target, equivalent k'_U is estimated with this formula:

$$\text{Equivalent } \hat{k}'_U = \frac{\hat{x}_{50} - T}{\text{USL} - T} = \frac{.0058 - .0035}{.0080 - .0035} = .5111$$

Equivalent P_{PK} can now be estimated from equivalent P'_{PU} and equivalent k'_U.

$$\text{Equivalent } \hat{P}_{PK} = \text{Equivalent } \hat{P}'_{PU}(1 - \text{Equivalent } \hat{k}'_U) = (1.55)(1 - .5111) = .76$$

This is the same result obtained earlier in Example 8.13. However, now it can be quickly seen that over half (51.1 percent) of the potential capability is lost due to improper centering. Correcting the process aim should take priority over all other improvement efforts.

Example 8.15 Tensile Strength

In Example 8.6 (p. 449), the tensile strength of steel wire had to meet a LSL of 450 kilograms, with a target of 600 kg. From NOPP, the median and .135 percentile were estimated as 547 kg and 481 kg, respectively. This allowed equivalent P'_{PL} to be estimated as 2.2727.

$$\text{Equivalent } \hat{P}'_{PL} = \text{Maximum} \left(\frac{T - \text{LSL}}{\hat{x}_{50} - \hat{x}_{.135}}, \frac{\hat{x}_{50} - \text{LSL}}{\hat{x}_{50} - \hat{x}_{.135}} \right)$$

$$= \text{Maximum} \left(\frac{600 - 450}{547 - 481}, \frac{547 - 450}{547 - 481} \right)$$

$$= \text{Maximum} \left(2.2727, 1.4697 \right) = 2.2727$$

Equivalent $\hat{k}'_L$ can be calculated to check process aim.

$$\text{Equivalent } \hat{k}'_L = \frac{T - \hat{x}_{50}}{T - \text{USL}} = \frac{600 - 547}{600 - 450} = .3533$$

This large equivalent $\hat{k}'_L$ factor greatly reduces the performance capability of the process.

$$\text{Equivalent } \hat{P}_{PK} = \text{Equivalent } \hat{P}'_{PL} (1 - \text{Equivalent } \hat{k}'_L) = 2.2727 (1 - .3533) = 1.4697$$

The relationship between equivalent P_{PK} and equivalent P'_{PL} is portrayed in Figure 8.29. The shaded area represents potential capability lost due to x_{50} being located below the target zone. When x_{50} is at, or above, the target of 600 kg, there is no loss of potential capability, and equivalent P'_{PL} equals equivalent P_{PK}.

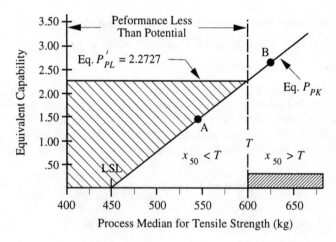

Fig. 8.29 Equivalent P_{PK} versus x_{50} for a unilateral specification.

If the median is increased from 547 to 625, and $x_{.135}$ from 481 to 559, equivalent P'_{PL} climbs to 2.6515 (see change from point A to point B in Figure 8.29).

$$\text{Equivalent } \hat{P}'_{PL} = \text{Maximum} \left(\frac{600-450}{625-559}, \frac{625-450}{625-559} \right)$$

$$= \text{Maximum} (2.2727, 2.6515) = 2.6515$$

Because the median is now located inside the target zone, equivalent k'_L is zero, causing equivalent P_{PK} to jump to 2.6515.

$$\text{Equivalent } \hat{P}_{PK} = \text{Equivalent } \hat{P}'_{PL}(1 - \text{Equivalent } \hat{k}'_L) = 2.6515(1-0) = 2.6515$$

If the median continues to increase, so will equivalent P'_{PL} and equivalent P_{PK}.

Advantages and Disadvantages

The generalized formula developed in this section for measuring equivalent P_{PK} applies to any type of output distribution, normal or non-normal. Because its estimate of performance capability is consistent with that derived from the standard P_{PK} formula, their rating scales and goals are similar as well. In addition, comparisons between equivalent potential and performance capability parallel those made with the standard capability formulas. However, P_{PK} is maximized when the process median is at M, whereas equivalent P_{PK} achieves its maximum when the median is located at x'_{50}. As an added benefit for companies involved in ISO 9000 or QS-9000 registration, this measure is recommended in the associated standards for estimating capability when non-normal distributions are involved.

The major disadvantage involves the formula for equivalent P_{PK}. It is more complex and somewhat less intuitive than the standard P_{PK} formula. And unfortunately, the percentage of nonconforming parts associated with a given equivalent P_{PK} value for a non-normal distribution is usually different than that for a normally distributed process output with an identical P_{PK} value.

Equivalent P_{PL}, equivalent P_{PU}, and equivalent P_{PK} all encourage centering the process median at x'_{50}. If a target other than x'_{50} is designated, including cases where T equals M, these measures must be modified, as is demonstrated in the next section.

8.9 Equivalent P^*_{PL}, P^*_{PU}, and P^*_{PK} for Cases Where $T \neq x'_{50}$

Equivalent P_{PL}, equivalent P_{PU}, and equivalent P_{PK} are maximized when the process median is located at x'_{50}. Thus, they are not appropriate when a target other than x'_{50} is specified for the center of the process output. In order to encourage centering the process at the designated target, an equivalent index must be developed, which achieves its maximum value when the median is located at T. This objective is accomplished in a manner similar to that done in Section 6.4 and 6.6 when the standard P^*_{PL}, P^*_{PU}, and P^*_{PK} measures were developed. This modification involved creating LSL_T and USL_T, which then replaced LSL and USL in the capability calculations. The formulas given here for LSL_T and USL_T were derived in Section 6.2 (p. 197) for situations where T is less than M:

$$\text{LSL}_T = \text{LSL} \qquad \text{USL}_T = T + (T - \text{LSL})$$

This establishes USL_T at the same distance above T as LSL is below T, as is depicted in Figure 8.30.

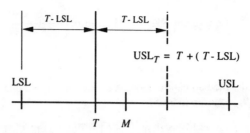

Fig. 8.30 USL_T is the same distance above T as LSL is below it.

With this formula, T becomes the middle of the effective tolerance.

$$\text{Middle of Effective Tolerance} = \frac{USL_T + LSL}{2} = \frac{T + (T - LSL) + LSL}{2} = \frac{2T}{2} = T$$

On the other hand, these formulas are appropriate when T is greater than M:

$$LSL_T = T - (USL - T) \qquad USL_T = USL$$

LSL_T is set the same distance below T as the USL is above T. Just like USL_T, this formula makes T the center of the effective tolerance. USL_T and LSL_T were then inserted into the formulas for P^*_{PL}, P^*_{PU}, and P^*_{PK}, as shown below. When the process average is centered at T, P^*_{PL} equals P^*_{PU}, and P^*_{PK} is at its maximum (review Figure 6.69 on page 261).

$$P^*_{PK} = \text{Minimum}(P^*_{PL}, P^*_{PU}) = \text{Minimum}\left(\frac{\mu - LSL_T}{3\sigma_{LT}}, \frac{USL_T - \mu}{3\sigma_{LT}}\right)$$

However, these formulas for LSL_T and USL_T were developed under the assumption the process output is normally distributed. By locating μ at the middle of the effective tolerance (which is T), P^*_{PK} is maximized. But as was just seen in the last section, centering the output at the middle of the tolerance will not maximize equivalent P_{PK} for non-symmetrical distributions. Likewise, centering the median at the middle of the effective tolerance defined by the difference between USL_T and LSL_T will not maximize equivalent P^*_{PK}.

Whenever the distance from the median to the .135 percentile is not equal to that between the median and the 99.865 percentile, equivalent P^*_{PK} is maximized when the median equals the x'_{50} value for the *effective* tolerance, called $x'_{50, T}$. Therefore, the standard formulas for LSL_T and USL_T must be adjusted so the specified target ends up being equal to $x'_{50, T}$. These adjusted formulas are presented below, with a full derivation available at the end of this chapter (see Derivation 3, page 511). Whenever T is less than the x'_{50} value for the full tolerance, let LSL_T and USL_T be defined as follows:

$$LSL_T = LSL \qquad USL_T = T + \frac{x_{99.865} - x_{50}}{x_{50} - x_{.135}}(T - LSL)$$

For symmetrical distributions, $x_{99.865}$ minus x_{50} equals x_{50} minus $x_{.135}$, making their ratio 1.

$$x_{99.865} - x_{50} = x_{50} - x_{.135} \quad \Rightarrow \quad \frac{x_{99.865} - x_{50}}{x_{50} - x_{.135}} = 1$$

This relationship makes the above formula for USL_T identical to the one developed previously for normal distributions.

$$USL_T = T + \frac{x_{99.865} - x_{50}}{x_{50} - x_{.135}}(T - LSL) = T + (T - LSL)$$

The generalized formula is more useful in that it applies to both normal and non-normal distributions. In cases where T is above the x'_{50} value for the full tolerance:

$$LSL_T = T - \frac{x_{50} - x_{.135}}{x_{99.865} - x_{50}}(USL - T) \qquad USL_T = USL$$

These values for USL_T and LSL_T are then inserted into the formulas for equivalent P^*_{PL}, equivalent P^*_{PU}, and equivalent P^*_{PK}, as is done here.

$$\text{Equivalent } P^*_{PK} = \text{Minimum} (\text{Equivalent } P^*_{PL}, \text{Equivalent } P^*_{PU})$$

$$= \text{Minimum} \left(\frac{x_{50} - LSL_T}{x_{50} - x_{.135}}, \frac{USL_T - x_{50}}{x_{99.865} - x_{50}} \right)$$

The next example demonstrates how the equivalent P^*_{PK} index is applied in practice.

Example 8.16 Flange Thickness

A process fabricating composite aircraft components is finally brought into a good state of statistical control for the thickness of a critical flange. When the thickness measurements are plotted on NOPP, a J-shaped curve appears, indicating the process output has a distribution which is skewed left. Due to this being a non-normal distribution, the standard performance capability measures do not apply. In addition, the target of 12.660 mm for flange thickness is not the middle of the tolerance, which is 12.655 mm. Because proper centering is extremely important for this characteristic, a decision is made to track performance capability with the equivalent P^*_{PK} index. With a LSL of 12.640 mm and an USL of 12.670 mm, a goal is established to have equivalent P^*_{PK} be at least 1.75.

The following percentiles are estimated from the NOPP: $x_{.135}$ equals 12.647; x_{50} equals 12.658; and $x_{99.865}$ equals 12.664. From these percentiles, x'_{50} can be determined for the full tolerance.

$$\hat{x}'_{50} = \frac{(\hat{x}_{50} - \hat{x}_{.135})USL + (\hat{x}_{99.865} - \hat{x}_{50})LSL}{\hat{x}_{99.865} - \hat{x}_{.135}}$$

$$= \frac{(12.658 - 12.647)12.670 + (12.664 - 12.658)12.640}{12.664 - 12.647} = 12.659$$

Because the target of 12.660 mm is greater than this estimate of 12.659 for x'_{50}, the following set of formulas is chosen for deriving USL_T and LSL_T:

$$USL_T = USL = 12.670$$

$$\text{LSL}_T = T - \frac{x_{50} - x_{.135}}{x_{99.865} - x_{50}}(\text{USL} - T)$$

$$= 12.660 - \frac{12.658 - 12.647}{12.664 - 12.658}(12.670 - 12.660) = 12.64167$$

Now equivalent P_{PK}^* may be estimated.

$$\text{Equivalent } \hat{P}_{PK}^* = \text{Minimum } (\text{Equivalent } \hat{P}_{PL}^*, \text{Equivalent } \hat{P}_{PU}^*)$$

$$= \text{Minimum }\left(\frac{\hat{x}_{50} - \text{LSL}_T}{\hat{x}_{50} - \hat{x}_{.135}}, \frac{\text{USL}_T - \hat{x}_{50}}{\hat{x}_{99.865} - \hat{x}_{50}}\right)$$

$$= \text{Minimum }\left(\frac{12.658 - 12.64167}{12.658 - 12.647}, \frac{12.670 - 12.658}{12.664 - 12.658}\right)$$

$$= \text{Minimum }(1.48, 2.00) = 1.48$$

This performance capability is currently quite a bit less than the goal of 1.75. Moving the process median up by .002 mm so it equals the target of 12.660 will help bring equivalent P_{PK}^* up to 1.67. Notice how equivalent P_{PL}^* equals equivalent P_{PU}^* when the median is centered at the target.

$$\text{Equivalent } \hat{P}_{PK}^* = \text{Minimum }\left(\frac{\hat{x}_{50} - \text{LSL}_T}{\hat{x}_{50} - \hat{x}_{.135}}, \frac{\text{USL}_T - \hat{x}_{50}}{\hat{x}_{99.865} - \hat{x}_{50}}\right)$$

$$= \text{Minimum }\left(\frac{12.660 - 12.64167}{12.660 - 12.649}, \frac{12.670 - 12.660}{12.666 - 12.660}\right)$$

$$= \text{Minimum }(1.67, 1.67) = 1.67$$

As this still does not meet the required capability goal of 1.75, work must now concentrate on reducing process variation.

When the process median is centered at T, performance capability, as measured by equivalent P_{PK}^*, equals potential capability, as measured by equivalent P_P^*.

$$\text{Equivalent } \hat{P}_P^* = \text{Minimum }\left(\frac{T - \text{LSL}}{\hat{x}_{50} - \hat{x}_{.135}}, \frac{\text{USL} - T}{\hat{x}_{99.865} - \hat{x}_{50}}\right)$$

$$= \text{Minimum }\left(\frac{12.660 - 12.640}{12.660 - 12.649}, \frac{12.670 - 12.660}{12.666 - 12.660}\right)$$

$$= \text{Minimum }(1.82, 1.67) = 1.67$$

The formulas for equivalent P_P^* and equivalent P_R^* may also be expressed as a function of LSL_T and USL_T. The derivation of these formulas are provided at the end of this chapter in Section 8.17 (Derivation 4, page 512).

$$\text{Equivalent } P_P^* = \frac{\text{Effective Tolerance}}{x_{99.865} - x_{.135}} = \frac{USL_T - LSL_T}{x_{99.865} - x_{.135}}$$

$$\text{Equivalent } P_R^* = \frac{x_{99.865} - x_{.135}}{\text{Effective Tolerance}} = \frac{x_{99.865} - x_{.135}}{USL_T - LSL_T}$$

For the above flange thickness example, equivalent P_P^* is estimated with this formula to be 1.67, agreeing with the original calculation.

$$\text{Equivalent } \hat{P}_P^* = \frac{USL_T - LSL_T}{\hat{x}_{99.865} - \hat{x}_{.135}} = \frac{12.670 - 12.64167}{12.664 - 12.647} = 1.67$$

Notice how moving the process median above the target results in a *lower* value for equivalent P_{PK}^*. If the median is increased by another .002 mm to 12.662 mm, equivalent P_{PK}^* drops to 1.33. As before, this analysis assumes the .135 and 99.865 percentiles are also increased by .002 mm.

$$\text{Equivalent } \hat{P}_{PK}^* = \text{Minimum} \left(\frac{\hat{x}_{50} - LSL_T}{\hat{x}_{50} - \hat{x}_{.135}}, \frac{USL_T - \hat{x}_{50}}{\hat{x}_{99.865} - \hat{x}_{50}} \right)$$

$$= \text{Minimum} \left(\frac{12.662 - 12.64167}{12.662 - 12.651}, \frac{12.670 - 12.662}{12.668 - 12.662} \right)$$

$$= \text{Minimum} (1.85, 1.33) = 1.33$$

Operators running this process will react to this decrease in performance capability by making adjustments to lower the median back to the target of 12.660 mm. This is the mechanism by which equivalent P_{PK}^* helps keep the process output centered at the desired target value.

Just as for the standard measures, when the median equals T, then:

$$\text{Equivalent } P_{PK}^* = \text{Equivalent } P_P^* = \frac{1}{\text{Equivalent } P_R^*}$$

When T equals x'_{50}, then:

$$\text{Equivalent } P_P^* = \text{Equivalent } P_P = \frac{1}{\text{Equivalent } P_R^*} = \frac{1}{\text{Equivalent } P_R}$$

And finally, when the median equals T, and T equals x'_{50}, then:

$$\text{Equivalent } P_{PK}^* = \text{Equivalent } P_{PK} = \text{Equivalent } P_P^* = \text{Equivalent } P_P$$

$$= \frac{1}{\text{Equivalent } P_R^*} = \frac{1}{\text{Equivalent } P_P}$$

Relationship Between Equivalent P^{*}_{PK} and x_{50}

In the flange thickness example just covered, the target is located above x'_{50}. Because the process output is non-symmetric, any shift in the process median above T decreases equivalent P^{*}_{PK} more than a comparable shift below T, as revealed in Figure 8.31. When the median was located .002 mm below T at 12.658 mm, equivalent P^{*}_{PK} was 1.48, which is represented by point A on the graph. After centering the median at the target of 12.660 mm, equivalent P^{*}_{PK} rose to 1.67. Raising the median further to 12.662, which is .002 mm above T, makes equivalent P^{*}_{PK} fall to 1.33 (point B), a value considerably less than the 1.48 for point A. This phenomenon occurs because the slope for the equivalent P^{*}_{PL} line is flatter than the one for equivalent P^{*}_{PU}, due to T not being the middle of the effective tolerance.

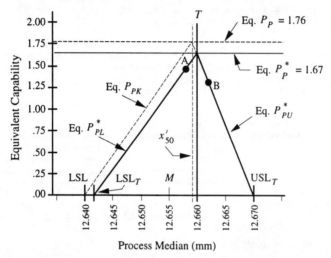

Fig. 8.31 Equivalent P^{*}_{PK} versus the process median.

By selecting a target that is not equal to x'_{50}, equivalent P^{*}_{P} will be less than equivalent P_{P}, while equivalent P^{*}_{PK} will be less than equivalent P_{PK}. This critical fact should be taken into consideration during the design phase when the tolerance and target are established for a characteristic suspected of having a non-normal output distribution.

Example 8.17 Ferrite Powder

Ferrite powder is mixed in 450 kilogram batches. A sample weighing .1 kg is taken from each batch, pressed into the shape of a torus, fired, and then checked for inductance. The specification is 56, plus 4, minus 6 kilohertz.

Once process stability is achieved, an analysis with NOPP concludes that the distribution for inductance is skewed right, with $x_{.135}$ equal to 52, x_{50} equal to 55, and $x_{99.865}$ equal to 59. From these percentiles, x'_{50} is found to be 54.2857 kHz.

$$\hat{x}'_{50} = \frac{(\hat{x}_{50} - \hat{x}_{.135})\,\text{USL} + (\hat{x}_{99.865} - \hat{x}_{50})\,\text{LSL}}{\hat{x}_{99.865} - \hat{x}_{.135}} = \frac{(55 - 52)60 + (59 - 55)50}{59 - 52} = 54.2857$$

As the target of 56 is greater than this estimate of x'_{50}, the following formulas are selected to determine LSL_T and USL_T:

$$\text{USL}_T = \text{USL} = 60$$

$$\text{LSL}_T = T - \frac{x_{50} - x_{.135}}{x_{99.865} - x_{50}}(\text{USL} - T) = 56 - \frac{55 - 52}{59 - 55}(60 - 56) = 53$$

Potential capability may be estimated with equivalent P_P^*.

$$\text{Equivalent } \hat{P}_P^* = \frac{\text{USL}_T - \text{LSL}_T}{\hat{x}_{99.865} - \hat{x}_{.135}} = \frac{60 - 53}{59 - 52} = 1.00$$

Performance capability is estimated with equivalent P_{PK}^*.

$$\text{Equivalent } \hat{P}_{PK}^* = \text{Minimum}\,(\,\text{Equivalent } \hat{P}_{PL}^*,\, \text{Equivalent } \hat{P}_{PU}^*\,)$$

$$= \text{Minimum}\left(\frac{\hat{x}_{50} - \text{LSL}_T}{\hat{x}_{50} - \hat{x}_{.135}}, \frac{\text{USL}_T - \hat{x}_{50}}{\hat{x}_{99.865} - \hat{x}_{50}}\right)$$

$$= \text{Minimum}\left(\frac{55 - 53}{55 - 52}, \frac{60 - 55}{59 - 55}\right)$$

$$= \text{Minimum}\,(\,.67\,,\,1.25\,) = .67$$

In view of equivalent P_{PL}^* being significantly less than equivalent P_{PU}^*, equivalent P_{PK}^* can be increased by moving x_{50} higher. If the process median for inductance is shifted 1 kHz higher to the target of 56, performance capability will equal potential capability, assuming $x_{.135}$ and $x_{99.865}$ also increase by 1 kHz.

$$\text{Equivalent } \hat{P}_{PK}^* = \text{Minimum}\left(\frac{\hat{x}_{50} - \text{LSL}_T}{\hat{x}_{50} - \hat{x}_{.135}}, \frac{\text{USL}_T - \hat{x}_{50}}{\hat{x}_{99.865} - \hat{x}_{50}}\right)$$

$$= \text{Minimum}\left(\frac{56 - 53}{56 - 53}, \frac{60 - 56}{60 - 56}\right)$$

$$= \text{Minimum}\,(\,1.00\,,\,1.00\,) = 1.00$$

At this point, .135 percent of the batches would have their inductance reading above the USL of 60 kHz. Remember that the LSL_T cannot be used to figure the percentage of batches with low inductance readings. Rather, read this percentage below the actual LSL of 50 kHz directly from the sheet of NOPP.

Had no target had been assigned, and the process median could be centered at x'_{50}, potential capability, as measured by equivalent P_P, would have been 1.43.

$$\text{Equivalent } \hat{P}_P = \frac{\text{Tolerance}}{\hat{x}_{99.865} - \hat{x}_{.135}} = \frac{60 - 50}{59 - 52} = 1.43$$

This disparity in capability evaluation is summarized in Figure 8.32.

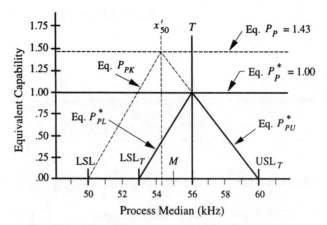

Fig. 8.32 Capability with and without a specified target.

Because the process output distribution is skewed right, specifying a target for the process median closer to USL than to the LSL results in a large loss of potential capability, as is depicted in Figure 8.33. This loss is brought about by attempting to squeeze the longer tail of the output distribution $(x_{99.865} - x_{50})$ into a small portion of the tolerance $(USL_T - T)$.

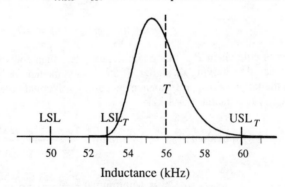

Fig. 8.33 A mismatch between the process output distribution and design requirements.

This clash between design objectives and the shape of the process output distribution inflates manufacturing costs and increases the chances of producing nonconforming product. To effectively establish specifications and target values, product designers must fully understand the nature of the expected process output, including its shape. They can best gain this vital information by conducting a formal process capability study.

The reduction in potential capability due to specifying a target different from x'_{50} is equal to 1 minus the ratio of the effective tolerance to the actual tolerance.

$$\text{Equivalent } P_P^* = \text{Equivalent } P_P \frac{\text{Effective Tolerance}}{\text{Tolerance}} = \text{Equivalent } P_P \frac{USL_T - LSL_T}{USL - LSL}.$$

In this last example, equivalent P_P was estimated as 1.43, the effective tolerance was computed as 7 (60 minus 53), and the actual tolerance was 10 (60 minus 50). Specifying a target causes a 30 percent (1 minus .70) loss in potential capability.

$$\text{Equivalent } \hat{P}^{*}_{P} = \text{Equivalent } \hat{P}_{P} \frac{\text{Effective Tolerance}}{\text{Tolerance}}$$

$$= 1.43 \frac{60 - 53}{60 - 50} = 1.43\,(.70) = 1.00$$

Equivalent k^{*}_{L}, Equivalent k^{*}_{U}, and Equivalent k^{*}

Similar to the formulas for k^{*}_{L} and k^{*}_{U} developed in Section 6.6 (p. 262), formulas for the equivalent version of these factors are presented here:

$$\text{Equivalent } k^{*}_{L} = \frac{T - x_{50}}{T - \text{LSL}_{T}} \qquad \text{Equivalent } k^{*}_{U} = \frac{x_{50} - T}{\text{USL}_{T} - T}$$

These aid in computing equivalent P^{*}_{PL} and P^{*}_{PU}.

$$\text{Equivalent } P^{*}_{PL} = \text{Equivalent } P^{*}_{P}\,(1 - \text{Equivalent } k^{*}_{L})$$

$$\text{Equivalent } P^{*}_{PU} = \text{Equivalent } P^{*}_{P}\,(1 - \text{Equivalent } k^{*}_{U})$$

Equivalent P^{*}_{PK} may also be derived from these values.

$$
\begin{aligned}
\text{Equiv. } P^{*}_{PK} &= \text{Minimum}\,(\text{Equivalent } P^{*}_{PL}, \text{Equivalent } P^{*}_{PU}) \\[4pt]
&= \text{Min.}\,[\,\text{Equivalent } P^{*}_{P}\,(1 - \text{Equivalent } k^{*}_{L}), \text{Equivalent } P^{*}_{P}\,(1 - \text{Equivalent } k^{*}_{U})\,] \\[4pt]
&= \text{Equivalent } P^{*}_{P}\,[\,\text{Minimum}\,(1 - \text{Equivalent } k^{*}_{L}, 1 - \text{Equivalent } k^{*}_{U})\,]
\end{aligned}
$$

The minimum of the quantities 1 minus equivalent k^{*}_{L} and 1 minus equivalent k^{*}_{U} is just the *maximum* of equivalent k^{*}_{L} and equivalent k^{*}_{U}.

$$\text{Equivalent } P^{*}_{PK} = \text{Equivalent } P^{*}_{P}\,[1 - \text{Maximum}\,(\text{Equivalent } k^{*}_{L}, \text{Equivalent } k^{*}_{U})\,]$$

The values for equivalent k^{*}_{L} and equivalent k^{*}_{U} may be substituted into this last formula to determine equivalent k^{*}.

$$
\begin{aligned}
\text{Equivalent } P^{*}_{PK} &= \text{Equivalent } P^{*}_{P}\left[\,1 - \text{Maximum}\!\left(\frac{T - x_{50}}{T - \text{LSL}_{T}}, \frac{x_{50} - T}{\text{USL}_{T} - T}\right)\right] \\[4pt]
&= \text{Equivalent } P^{*}_{P}\,(1 - \text{Equivalent } k^{*})
\end{aligned}
$$

Example 8.18 Ferrite Powder

Ferrite powder is tested for its inductance properties, with a specification of 56, plus 4, minus 6 kilohertz. Once process stability is achieved, an analysis with NOPP concludes the distribution for inductance is skewed right. In Example 8.17 on page 488, x'_{50} was found to be 54.2857. As the target of 56 is greater than this, LSL_{T} was determined as 53 and USL_{T} as 60. With these, equivalent k^{*} may be estimated.

$$\text{Equivalent } \hat{k}^* = \text{Maximum} \left(\frac{T - \hat{x}_{50}}{T - \text{LSL}_T}, \frac{\hat{x}_{50} - T}{\text{USL}_T - T} \right)$$

$$= \text{Maximum} \left(\frac{56 - 55}{56 - 53}, \frac{55 - 56}{60 - 56} \right)$$

$$= \text{Maximum} (\, .3333 \,, -.2500 \,) = .3333$$

Due to poor accuracy in centering, over a third of the potential capability is lost to performance capability.

Process precision was estimated with equivalent P_P^* to be 1.00. With this additional information, an estimate may be made for equivalent P_{PK}^*.

$$\text{Equivalent } \hat{P}_{PK}^* = \text{Equivalent } \hat{P}_P^* (1 - \text{Equivalent } \hat{k}^*) = 1.00 (1 - .3333) = .67$$

This answer is consistent with that obtained in Example 8.17. The benefit of assessing capability in this second manner is having the overall measure split into its two components. This separation allows one to determine if the reason for poor performance capability is a lack of precision, inadequate accuracy, or both, as it is in this example.

8.10 Equivalent $Z_{LSL, LT}$, Equivalent $Z_{USL, LT}$, and Equivalent $Z_{MIN, LT}$

The $Z_{MIN, LT}$ index is defined for processes with normally distributed outputs as:

$$Z_{MIN, LT} = \text{Minimum} (Z_{LSL,LT}, Z_{USL,LT}) = \text{Minimum} \left(\frac{\mu - \text{LSL}}{\sigma_{LT}}, \frac{\text{USL} - \mu}{\sigma_{LT}} \right)$$

For non-normal distributions, the equivalent lower $1\sigma_{LT}$ tail is defined as:

$$\text{Equivalent Lower } 3\sigma_{LT} \text{ Tail} = x_{50} - x_{.135} \quad \Rightarrow \quad \text{Equivalent Lower } 1\sigma_{LT} \text{ Tail} = \frac{x_{50} - x_{.135}}{3}$$

Likewise, the equivalent upper $1\sigma_{LT}$ tail is defined as:

$$\text{Equivalent Upper } 3\sigma_{LT} \text{ Tail} = x_{99.865} - x_{50} \quad \Rightarrow \quad \text{Equivalent Upper } 1\sigma_{LT} \text{ Tail} = \frac{x_{99.865} - x_{50}}{3}$$

In a manner similar to that for transforming P_{PK} to equivalent P_{PK}, the formula for $Z_{MIN.LT}$ can be generalized into the formula for equivalent $Z_{MIN.LT}$:

$$\text{Equivalent } Z_{MIN. LT} = \text{Minimum} (\text{Equivalent } Z_{LSL,LT}, \text{Equivalent } Z_{USL,LT})$$

$$= \text{Minimum} \left[\frac{3(x_{50} - \text{LSL})}{x_{50} - x_{.135}}, \frac{3(\text{USL} - x_{50})}{x_{99.865} - x_{50}} \right]$$

$$= \text{Minimum} (3 \, \text{Equivalent } P_{PL}, 3 \, \text{Equivalent } P_{PU})$$

$$= 3 \, \text{Minimum} \, (\text{Equivalent } P_{PL}, \text{Equivalent } P_{PU})$$

$$= 3 \, \text{Equivalent } P_{PK}$$

This formula applies equally well to both normal and non-normal distributions while maintaining the identical rating scale as that for the standard $Z_{MIN, LT}$ measure: less than 3 implies poor performance capability; equal to 3, marginal capability; greater than 3, superior capability. Other than having a rating scale not based on 1.0, this equivalent index has all the same advantages and disadvantages of equivalent P_{PK}.

8.11 Equivalent $Z^*_{LSL, LT}$, Equivalent $Z^*_{USL, LT}$, and Equivalent $Z^*_{MIN, LT}$

In a manner similar to that for transforming P^*_{PK} to equivalent P^*_{PK}, the formula for $Z^*_{MIN, LT}$ can be generalized into the formula for equivalent $Z^*_{MIN, LT}$ as follows:

$$\text{Equivalent } Z^*_{MIN, LT} = \text{Minimum} \, (\text{Equivalent } Z^*_{LSL,LT}, \text{Equivalent } Z^*_{USL,LT})$$

$$= \text{Minimum} \left[\frac{3(x_{50} - \text{LSL}_T)}{x_{50} - x_{.135}}, \frac{3(\text{USL}_T - x_{50})}{x_{99.865} - x_{50}} \right]$$

$$= \text{Minimum} \, (3 \, \text{Equivalent } P^*_{PL}, 3 \, \text{Equivalent } P^*_{PU})$$

$$= 3 \, \text{Minimum} \, (\text{Equivalent } P^*_{PL}, \text{Equivalent } P^*_{PU})$$

$$= 3 \, \text{Equivalent } P^*_{PK}$$

When T is less than x'_{50}, LSL_T and USL_T are defined as:

$$\text{LSL}_T = \text{LSL} \qquad\qquad \text{USL}_T = T + \frac{x_{99.865} - x_{50}}{x_{50} - x_{.135}} (T - \text{LSL})$$

If T is greater than x'_{50}, LSL_T and USL_T are found with these formulas:

$$\text{LSL}_T = T - \frac{x_{50} - x_{.135}}{x_{99.865} - x_{50}} (\text{USL} - T) \qquad\qquad \text{USL}_T = \text{USL}$$

8.12 Equivalent P_{PM}, Equivalent P_{PG}, and Equivalent P_{PMK}

Pearn and Kotz define an equivalent P_{PM} index as follows:

$$\text{Equivalent } P_{PM} = \frac{\text{USL - LSL}}{6\sqrt{[(x_{99.865} - x_{.135})/6]^2 + (x_{50} - M)^2}}$$

This formula equates one sixth of the distance between $x_{99.865}$ and $x_{.135}$ to an "equivalent σ_{LT}" value. However, this formula does not work properly for non-normal distributions.

Consider the three process output distributions presented in Figure 8.34. Although each distribution has a different shape, all three have the same .135, 50 and 99.865 percentiles (10, 15 and 20, respectively). The specification limits of 10 and 20 are also identical for all three.

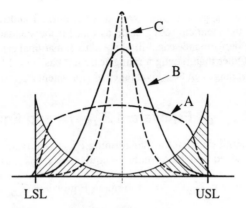

Fig. 8.34 Three output distributions with unequal losses.

Process C, with an unbounded output distribution, has a greater percentage of its output at the target value and, therefore, should have the least loss of the three. On the other hand, process A, with a bounded distribution, has a greater percentage of parts in higher loss areas and, therefore, is expected to have the largest loss. Process B, with a normal distribution, ought to have a loss somewhere in between these two. Yet, with the Pearn-Kotz formula, all three turn out to have *identical* equivalent P_{PM} indexes, each of which should be proportional to the loss of its respective process.

$$\text{Equiv. } P_{PM,A} = \frac{\text{USL - LSL}}{6\sqrt{[(x_{99.865} - x_{.135})/6]^2 + (x_{50} - M)^2}} = \frac{20 - 10}{6\sqrt{[(20 - 10)/6]^2 + (15 - 15)^2}} = 1.00$$

$$\text{Equiv. } P_{PM,B} = \frac{\text{USL - LSL}}{6\sqrt{[(x_{99.865} - x_{.135})/6]^2 + (x_{50} - M)^2}} = \frac{20 - 10}{6\sqrt{[(20 - 10)/6]^2 + (15 - 15)^2}} = 1.00$$

$$\text{Equiv. } P_{PM,C} = \frac{\text{USL - LSL}}{6\sqrt{[(x_{99.865} - x_{.135})/6]^2 + (x_{50} - M)^2}} = \frac{20 - 10}{6\sqrt{[(20 - 10)/6]^2 + (15 - 15)^2}} = 1.00$$

If the process median moves off target, all three suffer an identical decrease in performance capability as measured by this equivalent P_{PM} index. However, process A would have a greater percentage of its output pushed into higher loss areas than either B or C. Because of this larger increase in loss, A's decrease in performance capability should be more rapid than the other two, but with the above formula, it is not. This same argument explains why the formula proposed by Pearn and Kotz for equivalent P_{PMK} is also inappropriate.

$$\text{Equiv. } P_{PMK} = \text{Minimum} \left(\frac{M - \text{LSL}}{3\sqrt{[(x_{50} - x_{.135})/3]^2 + (x_{50} - M)^2}}, \frac{\text{USL} - M}{3\sqrt{[(x_{99.865} - x_{50})/3]^2 + (x_{50} - M)^2}} \right)$$

More work must be done to develop suitable equivalent formulas for these three performance measures. Until then, a better option is to estimate σ_{LT} and μ from the data so $\hat{\tau}_{LT}$ can be calculated.

$$\hat{\tau}_{LT} = 6\sqrt{\hat{\sigma}_{LT}^2 + (\hat{\mu} - M)^2}$$

$\hat{\tau}_{LT}$ should then be used to estimate the standard P_{PM}, P_{PG}, and P_{PMK} indexes.

8.13 Measuring Capability Based on the Percentage Nonconforming

A second method for estimating capability for non-normal distributions employs NOPP to predict the percentage nonconforming rather than the percentiles. With this alternative procedure, these following performance capability measures may be estimated.

1. $ppm_{LSL,\ LT}$, $ppm_{USL,\ LT}$, and $ppm_{TOTAL,\ LT}$

2. Equivalent $P_{PK}^{\%}$

The first measure is essentially the same as the standard $ppm_{TOTAL,\ LT}$ index of performance capability previously defined in Section 6.7 (p. 265), while the second is comparable to the standard P_{PK} method based on an equal percentage of nonconforming parts (Castagliola, Kane, Munechika). Equivalent $P_{PK}^{\%}$ is pronounced "equivalent P_{PK}, based on percent."

8.14 $ppm_{LSL,\ LT}$, $ppm_{USL,\ LT}$, $ppm_{TOTAL,\ LT}$, and $ppm_{MAX,\ LT}$

Suppose NOPP is used to check for normality, and it is discovered the process output is most likely non-normally distributed. A measure of performance capability based on $ppm_{TOTAL,\ LT}$ is made by following these six steps.

1. Plot the cumulative distribution of the sample data on NOPP and fit a curved line through the data (a French curve may help).

2. Draw a vertical line up from the LSL until it intersects the cumulative distribution curve. Then draw a horizontal line from this intersection to the left-hand y axis. From the y-axis scale, read the estimated percentage nonconforming below the LSL, $\hat{p}'_{LSL,\ LT}$. If the 6σ Capability Analysis Worksheet™ is available, the estimated ppm value ($\hat{p}pm_{LSL,\ LT}$) may be read directly from the NOPP.

3. Draw a vertical line up from the USL until it intersects the curve. Then draw a horizontal line from this intersection over to the right-hand y axis. From this y-axis scale, determine the estimated percentage above the USL, $\hat{p}'_{USL,\ LT}$. Again, if the 6σ Capability Worksheet™ is available, $\hat{p}pm_{USL,\ LT}$ may be read directly from the right-hand scale. Note that Flaig offers a radically different approach for estimating these percentages.

4. Sum these two percentages to derive an estimate of the total percentage nonconforming, $\hat{p}'_{TOTAL,\ LT}$. If there is only one specification, the percentage nonconforming for it becomes $\hat{p}'_{TOTAL,\ LT}$.

$$\hat{p}'_{TOTAL,\ LT} = \hat{p}'_{LSL,LT} + \hat{p}'_{USL,LT}$$

5. Multiply this total percentage nonconforming by ten to the sixth power so that $ppm_{TOTAL,\,LT}$ can be determined.

$$\hat{p}\,pm_{TOTAL,LT} = \hat{p}'_{TOTAL.\,LT} \times 10^6$$

If *ppm* values are available from the NOPP, just add them together to derive an estimate of $ppm_{TOTAL.\,LT}$. Find $ppm_{MAX,\,LT}$ by taking the larger of the two.

$$\hat{p}\,pm_{TOTAL,LT} = \hat{p}\,pm_{LSL,LT} + \hat{p}\,pm_{USL,LT}$$

$$\hat{p}\,pm_{MAX.\,LT} = \text{Maximum}\,(\hat{p}\,pm_{LSL,LT}, \hat{p}\,pm_{USL,LT})$$

6. Compare the estimate for $ppm_{TOTAL,\,LT}$ (or $ppm_{MAX,\,LT}$) to the capability goal for this characteristic and make a decision about process capability.

Read through this next example to see how these steps are successfully applied in performing a capability study on a process with a non-normal output distribution.

Example 8.19 Metal Block Length

Recall the case study involving metal block length presented in Example 8.9 on page 463. The cumulative percentage curve for this feature is plotted on the NOPP displayed in Figure 8.35. The resulting S-shaped curve indicates the process output distribution is non-normal and is most likely an unbounded distribution.

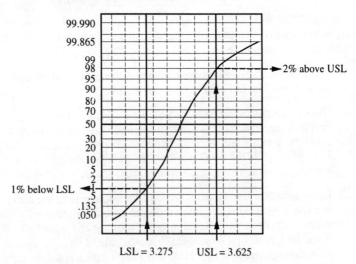

Fig. 8.35 NOPP discloses the percentage of blocks with nonconforming lengths.

Vertical lines from the LSL of 3.275 and the USL of 3.625 are drawn up to the curve and over to the percentage outside the respective specification limit. The total amount of nonconforming blocks is estimated by summing these two percentages.

$$\hat{p}'_{TOTAL.\,LT} = \hat{p}'_{LSL,LT} + \hat{p}'_{USL,LT} = .01 + .02 = .03$$

This combined result is multiplied by ten to the sixth power to estimate $ppm_{TOTAL,LT}$.

$$\hat{p}pm_{TOTAL,LT} = \hat{p}'_{TOTAL,LT} \times 10^6 = .03 \times 10^6 = 30,000$$

Does this process meet the minimum requirement for performance capability? Because the amount of nonconforming parts is independent of distribution shape, the criteria for *ppm* estimated from NOPP is identical to those given in Section 6.7 (p. 269) for *ppm* values estimated from the Z table. With unilateral specifications, the minimum requirement is for $ppm_{TOTAL,LT}$ to be equal to, or less than, 1,350. For bilateral specifications, first calculate $ppm_{MAX,LT}$, which is the larger of $ppm_{LSL,LT}$ and $ppm_{USL,LT}$. Then, evaluate performance capability with the following rating scale.

$ppm_{MAX,LT} < 1,350$ shows that the minimum requirement for long-term performance capability is exceeded, *i.e.*, the process is producing more than 99.73 percent conforming parts;

$ppm_{MAX,LT} = 1,350$ demonstrates the minimum criterion for long-term performance capability is just met;

$ppm_{MAX,LT} > 1,350$ means that the minimum prerequisite for long-term performance capability is *not* met because the process is producing less than 99.73 percent conforming parts.

For this metal block example, $ppm_{MAX,LT}$ is estimated as 20,000.

$$\hat{p}pm_{MAX,LT} = \text{Maximum}(\hat{p}pm_{LSL,LT}, \hat{p}pm_{USL,LT}) = \text{Maximum}(10,000, 20,000) = 20,000$$

Because $ppm_{MAX,LT}$ greatly exceeds 1,350, this process is judged to be severely lacking in long-term performance capability for machining the length of these metal blocks.

Goals

Goals for this *ppm* measure are the same as those for the standard *ppm* measure, with the most common goal being 63 *ppm*, or less. Other often-seen goals are to be less than 3.4 *ppm*, or less than .6 *ppm*.

In the metal block example, the goal for $ppm_{MAX,LT}$ is 32. Due to the estimated $ppm_{MAX,LT}$ being 20,000, this process has a long way to go before meeting its performance capability goal. Because $ppm_{USL,LT}$ is greater than $ppm_{LSL,LT}$, moving the median lower will reduce $ppm_{MAX,LT}$. Regrettably, there is no easy way with this measure of determining exactly how much lower the median should be shifted, other than by trial and error.

Example 8.20 Cam Roller Waviness

In Example 7.4 (p. 383), measurements from a control chart monitoring the waviness of a cam roller were plotted on a piece of NOPP, a copy of which is displayed in Figure 8.36. Because the process output distribution appears skewed right, a manufacturing engineer concludes performance capability should be estimated with the percentage out-of-specification method rather than with the standard approach. As waviness has only an USL of 75, the capability goal is for $ppm_{USL,LT}$ to be no more than 3.4.

From the NOPP, the percentage of parts above the 75 is determined as .008, which represents the total percentage of nonconforming rollers.

$$\hat{p}pm_{USL,LT} = \hat{p}'_{USL,LT} \times 10^6 = .008 \times 10^6 = 8,000$$

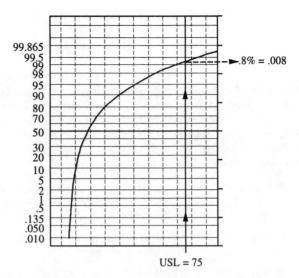

Fig. 8.36 NOPP showing .8 percent above USL of 75.

This result indicates the process does not even meet the minimum stipulation for performance capability ($ppm_{TOTAL, LT} \leq 1,350$), much less any goal better than this minimum. Work should concentrate on lowering the process median and/or reducing process spread.

Example 8.21 Automatic Temperature Controller

A study is undertaken to determine the capability of an automatic controller installed to maintain proper preheat temperature of liner board being pasted to a corrugating medium. The resulting product is then cut, folded and glued to make corrugated shipping boxes. The controller is set at 365 degrees Fahrenheit (185 degrees Centigrade), with a temperature reading taken every 10 minutes. These readings are plotted on an *IX & MR* chart to determine if the process is stable. When stability is demonstrated, the temperature measurements are plotted on NOPP to check for normality. This check reveals the process output is most likely bounded (non-normal). The control limits of the *IX* chart are revised with $x_{.135}$, x_{50}, and $x_{99.865}$, and control is reestablished.

A second analysis with NOPP estimates the amount of readings below the LSL of 360 degrees as .0216 percent, while the amount above the USL of 370 is predicted to be .0047 percent. Estimates of $ppm_{LSL, LT}$, $ppm_{USL, LT}$, $ppm_{TOTAL, LT}$, and $ppm_{MAX, LT}$ may now be made from this information. It is hoped there will be no more than 500 out of a million temperature readings outside either specification, *i.e.*, $ppm_{MAX, LT}$ is less than 500.

$$\hat{p}pm_{LSL,LT} = \hat{p}'_{LSL,LT} \times 10^6 = .000216 \times 10^6 = 216$$

$$\hat{p}pm_{USL,LT} = \hat{p}'_{USL,LT} \times 10^6 = .000047 \times 10^6 = 47$$

$$\hat{p}pm_{TOTAL, LT} = \hat{p}pm_{LSL,LT} + \hat{p}pm_{USL,LT} = 216 + 47 = 263$$

$$\hat{p}pm_{MAX, LT} = \text{Maximum}(\hat{p}pm_{LSL,LT}, \hat{p}pm_{USL,LT}) = \text{Maximum}(216, 47) = 216$$

In view of $ppm_{MAX, LT}$ being substantially less than the goal of 500, this temperature controller appears to be quite capable.

Advantages and Disadvantages

This measure expresses capability in its easiest to understand form, the amount of non-conforming parts. In addition, it works equally well with bilateral and unilateral specifications. Unfortunately, there is no way to estimate potential capability. This disadvantage makes it difficult to determine if shifting the process median will help improve performance capability and, if it does, by how much.

8.15 Equivalent $P^{\%}_{PK}$

A second measure of performance capability based on the percentage nonconforming is made by following these six steps (Bothe). Note how the first four are similar to those for finding $ppm_{TOTAL, LT}$.

1. Plot the cumulative distribution of the sample data on NOPP and fit a curved line through the data.

2. Draw a vertical line up from the LSL until it intersects the cumulative distribution curve, then continue with a horizontal line to the y axis on the left-hand side. Read off the estimated percentage nonconforming below the LSL, $\hat{p}'_{LSL, LT}$.

3. Draw a vertical line up from the USL until it intersects the curve. From the y axis on the right-hand side, determine the estimated percentage above the USL, $\hat{p}'_{USL, LT}$.

4. Select the maximum of these two percentages, $\hat{p}'_{MAX, LT}$. If there is only one specification, use the percentage nonconforming for it.

$$\hat{p}'_{MAX.LT} = \text{Maximum} \, (\, \hat{p}'_{LSL, LT} \, , \hat{p}'_{USL, LT} \,)$$

5. Find the "equivalent $\hat{Z}^{\%}_{MIN, LT}$" value for $\hat{p}'_{MAX, LT}$ from the Z table given in Appendix Table III (Polansky *et al*, Somerville and Montgomery). Equivalent $\hat{Z}^{\%}_{MIN, LT}$ is defined as the Z value a normal distribution would have for this same percentage of nonconforming parts, which is just $Z(\hat{p}'_{MAX.LT})$.

$$\text{Equivalent } \hat{Z}^{\%}_{MIN, LT} = Z(\, \hat{p}'_{MAX, LT} \,)$$

If using the 6σ Capability Analysis Worksheet™, read the Z value for each percentage directly from the Z column scale, then select the smaller of these two as equivalent $\hat{Z}^{\%}_{MIN.LT}$.

$$\text{Equivalent } \hat{Z}^{\%}_{MIN. LT} = \text{Minimum} \, (\, \text{Equivalent } \hat{Z}^{\%}_{LSL. LT} \, , \text{Equivalent } \hat{Z}^{\%}_{USL. LT} \,)$$

6. Estimate equivalent $P^{\%}_{PK}$ by dividing equivalent $\hat{Z}^{\%}_{MIN, LT}$ by 3 (James). If using the 6σ Capability Analysis Worksheet™, read the equivalent $P^{\%}_{PK}$ value directly from the "Z/3" column.

$$\text{Equivalent } \hat{P}^{\%}_{PK} = \frac{\text{Equivalent } \hat{Z}^{\%}_{MIN,LT}}{3} = \frac{Z(\hat{p}'_{MAX,LT})}{3}$$

Compare equivalent $P^{\%}_{PK}$ to the capability goal for this characteristic and make a decision about process capability.

Example 8.22 Metal Block Length

To see how this particular measure is applied, recall Examples 8.9 and 8.18 concerning metal block length. An analysis of the NOPP for block length displayed back in Figure 8.35 (p. 496) indicates the process output distribution is most likely non-normal. Vertical lines from the LSL of 3.275 and the USL of 3.625 are drawn up to the curve, and then over to the y axis where the percentage outside the each specification limit is read. From the curve in Figure 8.35, $\hat{p}'_{LSL,\,LT}$ is .01 and $\hat{p}'_{USL,\,LT}$ is .02. For estimating equivalent $P^{\%}_{PK}$, the equivalent $\hat{Z}^{\%}_{MIN,\,LT}$ value for the *maximum* percentage outside either specification limit is found first.

$$\hat{p}'_{MAX,LT} = \text{Maximum}\,(\hat{p}'_{LSL,LT},\hat{p}'_{USL,LT}) = \text{Maximum}\,(.01,.02) = .02$$

This maximum percentage determines equivalent $\hat{Z}^{\%}_{MIN,LT}$.

$$\text{Equivalent } \hat{Z}^{\%}_{MIN,\,LT} = Z(\hat{p}'_{MAX,\,LT}) = Z(.02) = 2.05$$

Finally, equivalent $P^{\%}_{PK}$ is estimated from this equivalent $\hat{Z}^{\%}_{MIN,\,LT}$ for two percent.

$$\text{Equivalent } \hat{P}^{\%}_{PK} = \frac{\text{Equivalent } \hat{Z}^{\%}_{MIN,LT}}{3} = \frac{2.05}{3} = .683$$

Because the estimated equivalent $P^{\%}_{PK}$ index must be at least 1.00 to meet the minimum capability requirement of having no more than .135 percent outside either specification, this process definitely lacks long-term performance capability.

Concept of Equivalent $P^{\%}_{PK}$

Consider a process with a normally distributed output like the one presented in Figure 8.37. As the process average is closer to the USL, the estimate of P_{PK} is based on the number of long-term standard deviations that will fit between the average of the process output and the USL. This, of course, is just the definition of $Z_{USL,\,LT}$.

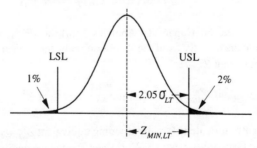

Fig. 8.37 A normal distribution with 2 percent above the USL.

For a process with a normal distribution, having only $2.05\sigma_{LT}$ from its average to the USL, and a greater number to the LSL, causes the standard P_{PK} index to equal .683.

$$P_{PK} = \frac{Z_{MIN,\,LT}}{3} = \frac{Z_{USL,\,LT}}{3} = \frac{USL - \mu}{3\sigma_{LT}} = \frac{2.05\sigma_{LT}}{3\sigma_{LT}} = .683$$

A $Z_{USL,\,LT}$ value of 2.05 implies that 2 percent of the process output lies above the USL. Conversely, if it is known from NOPP that this process is producing 2 percent nonconforming parts above the USL, the corresponding $Z_{USL,\,LT}$ value for this percentage can easily be looked up as 2.05 in the body of the Z table given in Appendix Table III. With the 6σ Capability Analysis Worksheet™, this Z value may be read directly from the scale on the right-hand side of the sheet. P_{PK} is then determined by dividing $Z_{USL,\,LT}$ for 2 percent by 3. Again, this analysis assumes $Z_{USL,\,LT}$ is less than $Z_{LSL,\,LT}$, which makes it equal to $Z_{MIN,\,LT}$.

$$P_{PK} = \frac{Z_{MIN,\,LT}}{3} = \frac{Z(.02)}{3} = \frac{2.05}{3} = .683$$

Now consider a process with a non-normal distribution. The unbounded output distribution for the metal block example is displayed in Figure 8.38.

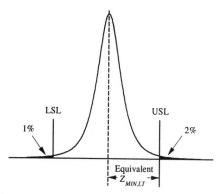

Fig. 8.38 A non-normal distribution with 2 percent above the USL.

Like the normal distribution in Figure 8.37, this process also has 2 percent of its output above the USL, with less than this below the LSL. Kane argues that the *same* level of performance capability should be assigned to this process as given to the one displayed in Figure 8.37, because its performance is identical as far as meeting specifications.

$$\text{Equivalent } P^{\%}_{PK} = \frac{\text{Equivalent } Z^{\%}_{MIN,LT}}{3} = \frac{Z(.02)}{3} = \frac{2.05}{3} = .683$$

This "equivalent capability" concept generates a measure similar in value to that for the normal case, based on the percentage of nonconforming product. This is more in line with the original definition of capability, which is based on the percentage of parts within (and outside) print specifications. In addition, if the P_{PK} goal is set based on a certain percentage of nonconforming product, this method provides an index consistent with this desire. For example, a goal of having no more than .3 ppm outside either specification translates into a standard P_{PK} goal of 1.67. When $ppm_{MAX,\,LT}$ (the maximum of $ppm_{LSL,\,LT}$ and $ppm_{USL,\,LT}$) is .3

for a process with a normally distributed output, P_{PK} will be 1.67. If $ppm_{MAX, LT}$ is .3 for a process with a *non-normally* distributed output, equivalent $P^{\%}_{PK}$ will also be 1.67, signifying both P_{PK} and *ppm* goals are met. As this new measure applies to both normal and non-normal distributions, and is identical to the standard P_{PK}, which only applies to normal distributions, equivalent $P^{\%}_{PK}$ is a more general index.

Notice the equivalent $\hat{P}^{\%}_{PK}$ of .683 for metal block length differs substantially from the equivalent P_{PK} value of .48 estimated for this same process in Example 8.9 (p. 464) using the percentile approach. A P_{PK} of .48 for a normal distribution implies there is 7.5 percent nonconforming product outside at least one specification. With the possibility of having up to this same percentage outside the other specification, the total amount of nonconforming blocks could be as high as 15 percent. This is clearly not true for this situation, where the *total* percentage of nonconforming blocks is only 3 percent.

Being evolved from different concepts, these two methods will typically generate dissimilar capability assessments. However, they both agree that less than 1.00 means more than .135 percent of the process output is outside at least one specification, whereas greater than 1.0 means less than .135 percent is outside either specification (recall Figure 8.16, p. 461).

The formula for equivalent $\hat{P}^{\%}_{PK}$ may also written as:

$$\text{Equivalent } P^{\%}_{PK} = \text{Minimum}\,(\,\text{Equivalent } P^{\%}_{PL}, \text{Equivalent } P^{\%}_{PU}\,)$$

$$= \text{Minimum}\left(\frac{\text{Equivalent } Z^{\%}_{LSL,LT}}{3}, \frac{\text{Equivalent } Z^{\%}_{USL,LT}}{3}\right)$$

$$= \frac{1}{3}\text{Minimum}\,[\,Z(p'_{LSL,LT}), Z(p'_{USL,LT})\,]$$

$$= \frac{1}{3}Z[\,\text{Maximum}\,(\,p'_{LSL,LT}, p'_{USL,LT}\,)]$$

$$= \frac{1}{3}Z(\,p'_{MAX. LT}\,) = \frac{Z(\,p'_{MAX. LT}\,)}{3}$$

Example 8.23 Bearing Support Hole Size

Measurements for the inner diameter of a hole drilled in a bearing support bracket (Figure 8.2) for a missile's inertial guidance system are analyzed on NOPP (Figure 8.3). The curve drawn through the plotted points indicates the process output distribution for this characteristic appears to be skewed right. Using percentiles taken from this graph, equivalent P_P was estimated in Example 8.3 as 1.56 (p. 441), and equivalent P_{PK} as 1.31 in Example 8.8 (p. 458).

Equivalent $P^{\%}_{PK}$ is computed by first finding the percentage of holes below the LSL, as well as that above the USL. As the output distribution for hole size is skewed right, Figure 8.3 reveals there are very few, if any, holes below the LSL. The cumulative curve crosses the USL at about 99.985 percent, implying .015 percent of hole sizes are above the USL.

$$\hat{p}'_{MAX.LT} = \text{Maximum}\,(\,\hat{p}'_{LSL,LT}, \hat{p}'_{USL,LT}\,) = \text{Maximum}\,(\,.00000, .00015\,) = .00015 = 1.5 \times 10^{-4}$$

Equivalent $P^{\%}_{PK}$ can be estimated directly from $\hat{p}'_{MAX. LT}$.

$$\text{Equivalent } \hat{P}^{\%}_{PK} = \frac{Z(\hat{p}'_{MAX, LT})}{3} = \frac{Z(1.5 \times 10^{-4})}{3} = \frac{3.62}{3} = 1.21$$

As $\hat{p}'_{LSL, LT}$ is less than $\hat{p}'_{USL, LT}$, $\hat{p}'_{MAX, LT}$ can be decreased and, thus, equivalent $P^{\%}_{PK}$ increased, by lowering the process median. Unfortunately, there is no simple way, other than simulating several shifts on NOPP, of discerning exactly how far. All that is known, is that x_{50} should be lowered until $\hat{p}'_{LSL, LT}$ equals $\hat{p}'_{USL, LT}$. At this point, equivalent $P^{\%}_{PK}$ will be maximized. However, this may not offer the minimum *total* percentage of nonconforming product! Consider the output distribution exhibited in Figure 8.39, where 6 percent is above the USL and another 3 percent below the LSL. The equivalent $P^{\%}_{PK}$ for this process is .52.

$$\text{Equivalent } \hat{P}^{\%}_{PK} = \frac{Z(.06)}{3} = \frac{1.55}{3} = .52$$

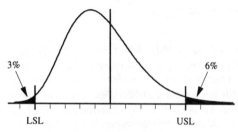

Fig. 8.39 A skewed-right distribution with 9 percent nonconforming parts.

By moving the median lower, the 6 percent is reduced to only 5 percent, and the 3 percent is increased to 5 percent (Figure 8.40). Because the distribution is skewed right, the percentage below the LSL increases at a faster rate than the percentage above the USL decreases.

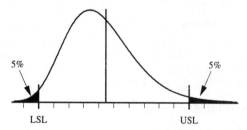

Fig. 8.40 Lowering the median increases the amount of nonconforming parts from 9 to 10 percent.

Equivalent $P^{\%}_{PK}$ for this revised process is .55, indicating a slight improvement in performance capability.

$$\text{Equivalent } \hat{P}^{\%}_{PK} = \frac{Z(.05)}{3} = \frac{1.64}{3} = .55$$

But now there is a total of 10 percent nonconforming product, when originally, there was only 9 percent. Lowering the median increased equivalent $P^{\%}_{PK}$, but unfortunately, also increased $p'_{TOTAL, LT}$.

Maximizing equivalent $P^{\%}_{PK}$ will minimize $p'_{TOTAL,\,LT}$ if the process output distribution is symmetrical, or the characteristic has a unilateral specification. For all other distributions, there is no way of knowing for sure if $p'_{TOTAL,\,LT}$ will decrease when equivalent $P^{\%}_{PK}$ is maximized.

Example 8.24 Cam Roller Waviness

Example 8.5 involved the waviness of an automotive cam roller, with the NOPP for this process displayed in Figure 8.11 (p. 447). Because the process output appears skewed right, a manufacturing engineer concludes the performance capability should be estimated with the percentage out-of-specification method. The capability goal for this characteristic is an equivalent $P^{\%}_{PK}$ of at least 1.50, which means a maximum of 3.4 *ppm* above the USL of 75. From the NOPP, an estimate of rollers with waviness above 75 is .8 percent.

$$\text{Equivalent } \hat{P}^{\%}_{PK} = \frac{Z(\hat{p}'_{USL,LT})}{3} = \frac{Z(8.0 \times 10^{-3})}{3} = \frac{2.41}{3} = .803$$

An equivalent $P^{\%}_{PK}$ of only .803 means this process fails to meet even the minimum requirement of performance capability, so part containment and 100 percent inspection must be initiated. Major sources of common-cause variation must be identified and reduced or eliminated, with additional consideration given to lowering the process median. After improvements are introduced, control must be reestablished and the capability study repeated. This procedure continues until the capability goal of 1.50 is achieved, or exceeded.

Note that the estimate of .583 for equivalent P_{PK} obtained in Example 8.5 is substantially different than the estimate of .803 just derived for equivalent $P^{\%}_{PK}$.

$$\text{Equivalent } \hat{P}_{PK} = \text{Equivalent } \hat{P}_{PU} = \frac{USL - \hat{x}_{50}}{\hat{x}_{99.865} - \hat{x}_{50}} = \frac{75 - 26}{110 - 26} = .583$$

Although both indicate a severe lack of performance capability, the noticeable difference in values arises because the first measure is based on the concept of equivalent spreads, while the second is founded on the notion of equivalent percentages nonconforming. When analyzing the results of a capability study, practitioners should select one or the other to measure performance capability, but not both, as confusion is sure to arise.

Example 8.25 Forging Hardness

After examining the NOPP in Figure 8.6 on page 438, a department supervisor concludes the process output for the hardness of an aluminum forging follows a non-normal distribution. In order to gain information about this process's ability to produce conforming forgings, the supervisor decides to apply the percentage nonconforming method for estimating performance capability. The goal is to have no more than 32 *ppm* outside either hardness specification, which translates into an equivalent $P^{\%}_{PK}$ goal of 1.33.

The first step in assessing capability under this method is determining the percentage of parts with hardness less than the LSL, as well as those with hardness above the USL from the NOPP. Vertical lines are drawn from the LSL and USL to the curve, then to the *y* axis where the percentage nonconforming is estimated for both these specifications.

$$\hat{p}'_{MAX,LT} = \text{Maximum}(\hat{p}'_{LSL,LT}, \hat{p}'_{USL,LT}) = \text{Maximum}(.0003, .0000) = .0003 = 3 \times 10^{-4}$$

Equivalent $P_{PK}^{\%}$ for hardness is based on this $\hat{p}'_{MAX,LT}$ value.

$$\text{Equivalent } \hat{P}_{PK}^{\%} = \frac{Z(\hat{p}'_{MAX,LT})}{3} = \frac{Z(3 \times 10^{-4})}{3} = \frac{3.44}{3} = 1.15$$

Because equivalent $P_{PK}^{\%}$ is greater than 1.00, this process meets the minimum requirement for long-term performance capability, but falls far short of the 1.33 goal. Currently, .03 percent of the parts, or 300 *ppm*, have a hardness below the LSL, whereas achieving the goal would mean there should be no more than 32 *ppm* outside this specification. By shifting the process median higher, this percentage of soft forgings will be reduced, causing the performance capability to improve. Unfortunately, there is no easy way to determine exactly how far the median should be moved. In fact, with this method there is no way of knowing how much better performance capability can be, because, unlike the percentile method, there is no way to estimate potential capability.

If the median is increased to 166, as was done in Figure 8.21 (p. 468), the percentage of soft forgings becomes essentially zero. Notice that the percentage of forgings above the USL is also effectively zero, creating a computational problem for this index of performance capability. As both $\hat{p}'_{LSL,LT}$ and $\hat{p}'_{USL,LT}$ are zero, $\hat{p}'_{MAX,LT}$ is also zero.

$$\hat{p}'_{MAX,LT} = \text{Maximum}(\hat{p}'_{LSL,LT}, \hat{p}'_{USL,LT}) = \text{Maximum}(.0, .0) = .0$$

The equivalent $\hat{Z}_{MIN,LT}^{\%}$ for 0 percent is plus infinity, which divided by 3, makes equivalent $P_{PK}^{\%}$ also equal to plus infinity.

$$\text{Equivalent } \hat{P}_{PK}^{\%} = \frac{Z(\hat{p}'_{MAX,LT})}{3} = \frac{Z(0)}{3} = \frac{+\infty}{3} = +\infty$$

In addition to this nonsensical answer, the process median could be centered at quite a few different locations, all of which would result in the same equivalent $P_{PK}^{\%}$ of infinity. Obviously, some centering locations are preferred over others, such as having more product near the target, but there is no way of distinguishing these with this index. Thus, caution is advised when computing equivalent $P_{PK}^{\%}$ for bounded distributions. Fortunately, these types of distributions are fairly rare in most industries.

Example 8.26 Flange Thickness

A process fabricating composite aircraft components is finally brought into a good state of statistical control for the thickness of a critical flange. When the thickness measurements are plotted on NOPP, a J-shaped curve appears, indicating the process output has a distribution which is skewed left. Due to this being a non-normal distribution, the standard performance capability measures do not apply. A decision is made to assess performance capability with the equivalent $P_{PK}^{\%}$ index. As no more than 32 *ppm* are desired outside either specification, a goal is established to have equivalent $P_{PK}^{\%}$ be at least 1.33.

From the NOPP, .004 percent of the flanges are predicted to have a thickness less than the LSL of 12.640 mm, while there is practically no chance of a flange being fabricated too thick (USL = 12.670 mm).

$$\hat{p}'_{MAX,LT} = \text{Maximum}(\hat{p}'_{LSL,LT}, \hat{p}'_{USL,LT}) = \text{Maximum}(.00004, .00000) = .00004 = 4 \times 10^{-5}$$

The equivalent $P_{PK}^{\%}$ index for thickness is computed from $\hat{p}'_{MAX, LT}$.

$$\text{Equivalent } \hat{P}_{PK}^{\%} = \frac{Z(\hat{p}'_{MAX.LT})}{3} = \frac{Z(4 \times 10^{-5})}{3} = \frac{3.95}{3} = 1.32$$

As this is just slightly below the goal of 1.33, the median should be pushed higher, until $\hat{p}'_{LSL, LT}$ equals $\hat{p}'_{USL, LT}$. Note that this may, or may not, occur when the median is located at the specified target of 12.660 mm. With this measure, there is no incentive to center the process at any particular target value.

Advantages and Disadvantages

The equivalent $P_{PK}^{\%}$ index is directly comparable to the standard P_{PK} measure, based on the percentage of nonconforming parts (Littig and Lam). This is not true for the equivalent P_{PK} version, which is derived from the percentile approach (review the discussion about Figure 8.15 on page 459). The generalized formulas for equivalent $P_{PK}^{\%}$ apply to any type of distribution, and the same formula works for both bilateral and unilateral specifications.

Despite these benefits, there are several distinct disadvantages. NOPP must be available for estimating the percentage of nonconforming parts. As mentioned before, hand fitting a curved line through the cumulative percentage plot points is an inexact science and subject to possible variation between different analysts. Even if a reasonably good fit is obtained, the line usually must be extrapolated quite a distance until it reaches the specification limits. Then the estimated percentage nonconforming must be accurately read off the NOPP and correctly looked up in the body of a Z table. These extra steps make this procedure more complex and time consuming, and therefore, less attractive to users. In addition, there is no way to estimate the potential capability (equivalent $P_{P}^{\%}$ or $P_{R}^{\%}$) with this approach. Thus, it cannot be determined how much of an improvement in performance capability (if any) is available by shifting the process center.

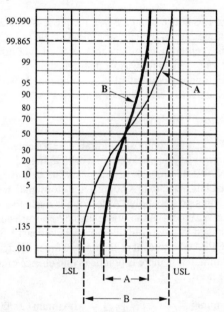

Fig. 8.41 Two distinctly different bounded distributions.

A bounded distribution with its bounds well inside the specification limits has no percentage of nonconforming product outside either specification. As this percentage is zero, equivalent $\hat{Z}^{\%}_{MIN,\ LT}$ is infinite, making equivalent $P^{\%}_{PK}$ infinite as well, which is not very meaningful. In addition, a process output with its bounds just inside the specification limits, such as curve A in Figure 8.41 (shown on previous page), has the same equivalent $P^{\%}_{PK}$ index as one with its bounds well within the tolerance, like the bold curve for process B. B is doing a much better job than A of meeting the specifications, has more parts near the target, has a greater safety margin, and most customers would prefer its output, yet equivalent $P^{\%}_{PK}$ rates it the same as A.

This index does not foster centering the process median at a specified target value. In fact, in order to minimize the percentage of nonconforming product, this measure may encourage users to deliberately move the median away from the target. Figure 8.42 presents a process whose median is centered at the target.

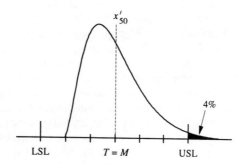

Fig. 8.42 Process output centered on target.

However, with 4 percent of its output above the USL, equivalent $P^{\%}_{PK}$ is only .58.

$$\text{Equivalent } P^{\%}_{PK} = \frac{Z\,(.04)}{3} = \frac{1.75}{3} = .58$$

Let the median be lowered until both $p'_{LSL,\ LT}$ and $p'_{USL,\ LT}$ equal 1 percent, as is shown in Figure 8.43. This maneuver increases equivalent $P^{\%}_{PK}$ to .77.

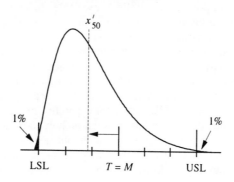

Fig. 8.43 Process output centered below target.

$$\text{Equivalent } P_{PK}^{\%} = \frac{Z(.01)}{3} = \frac{2.32}{3} = .77$$

Unfortunately, it also shifts the median quite a distance from T. Now the majority of parts are not produced at the desired dimension.

Perhaps for all these reasons, this second approach for determining performance capability is not as popular as the percentile method. Although not recognized in any ISO standard, it is mentioned in the proposed ANSI standard for capability studies (see also Krishnamoorthi and Khatwani, and Levinson)

8.16 Summary

This chapter explained two general approaches for estimating capability for non-normal distributions: one based on percentiles, the other on the percentage nonconforming. Under the first approach, modified formulas applying to non-normal distributions were developed for many of the standard capability measures. The formulas of these equivalent measures apply to any type of distribution, including a process output having a normal distribution. These equivalent measures are relatively easy to manually calculate and can be clearly visualized. When NOPP is available, no messy transformations are necessary to "normalize" the data, plus the percentage of nonconforming product is readily available for analysis.

The equivalent performance measures estimated with the percentile method consider both process spread and centering. However, because many non-normal distributions are non-symmetric, locating the process output at the middle of the tolerance does not necessarily provide the highest performance capability. Maximum performance capability occurs when the process median is positioned at x'_{50}, which for symmetrical distributions equals the middle of the tolerance. For asymmetrical distributions, x'_{50} is usually not equal to M.

The second approach considers only the percentage nonconforming for estimating capability for non-normal output distributions. Applied less often, it produces formulas for just the $ppm_{TOTAL, LT}$ and equivalent $P_{PK}^{\%}$ measures. $ppm_{TOTAL, LT}$ is estimated from NOPP, and functions in a manner similar to that for processes with normally distributed outputs, as described in Section 6.7 (p. 265). Equivalent $P_{PK}^{\%}$ is equivalent in the sense of having the same percentage of nonconforming parts as a process with an identical P_{PK} value, meaning this index may also be applied to processes having normally distributed outputs. Unfortunately, the associated calculations are somewhat more complex than with the first method, and no estimate of potential capability is available. In addition, maximizing equivalent $P_{PK}^{\%}$ may not minimize the percentage of nonconforming parts, nor does this index encourage centering the process at a particular target value, if one is specified.

Although both approaches rely heavily on NOPP, the equivalent P_{PK} index for a given process will usually be different than the equivalent $P_{PK}^{\%}$ index. When one is equal to 1.00, so will the other, but they will be different at nearly all other times. The percentile approach is referenced in the ISO standards, whereas the percentage nonconforming method is mentioned in the proposed ANSI standard for capability studies.

Up to this point, discussion has centered on how to measure capability for variable-type data, with either normal, or non-normal output distributions. However, there are many manufacturing and assembly processes that generate attribute-type data. Capability measures for this special family of industrial data are formulated in the next chapter.

8.17 Derivations

Derivation 1: x'_{50}

Maximum performance capability occurs when the process median is centered at a location x'_{50} such that equivalent P_{PL} equals equivalent P_{PU}. If the median moves higher or lower, one of these two indexes would decrease, reducing the overall performance capability as measured by equivalent P_{PK}. To derive x'_{50}, the formula for equivalent P_{PL} is set equal to the formula for equivalent P_{PU}. This derivation assumes shifting the process median does not cause changes to the process's spread or shape.

$$\text{Equivalent } P_{PL} = \text{Equivalent } P_{PU}$$

$$\frac{x'_{50} - \text{LSL}}{x_{50} - x_{.135}} = \frac{\text{USL} - x'_{50}}{x_{99.865} - x_{50}}$$

Let a equal x_{50} minus $x_{.135}$ and b equal $x_{99.865}$ minus x_{50}.

$$\frac{x'_{50} - \text{LSL}}{a} = \frac{\text{USL} - x'_{50}}{b}$$

$$b x'_{50} - b\,\text{LSL} = a\,\text{USL} - a x'_{50}$$

$$(b + a) x'_{50} = a\,\text{USL} + b\,\text{LSL}$$

$$x'_{50} = \frac{a\,\text{USL} + b\,\text{LSL}}{b + a}$$

Substituting the full values for a and b back into the above equation yields the final equation for computing x'_{50} from the percentiles of any distribution.

$$x'_{50} = \frac{(x_{50} - x_{.135})\,\text{USL} + (x_{99.865} - x_{50})\,\text{LSL}}{x_{99.865} - x_{50} + x_{50} - x_{.135}} = \frac{(x_{50} - x_{.135})\,\text{USL} + (x_{99.865} - x_{50})\,\text{LSL}}{x_{99.865} - x_{.135}}$$

Derivation 2: Equivalent k Factor

The formula relating equivalent P_{PK}, equivalent P_P, and equivalent k is given below:

$$\text{Equivalent } P_{PK} = \text{Equivalent } P_P\,(1 - \text{Equivalent } k\,)$$

$$\text{Minimum} \left(\frac{x_{50} - \text{LSL}}{x_{50} - x_{.135}}, \frac{\text{USL} - x_{50}}{x_{99.865} - x_{50}} \right) = \frac{\text{USL - LSL}}{x_{99.865} - x_{.135}}\,(1 - \text{Equivalent } k\,)$$

For the first part of this derivation, assume equivalent P_{PU} is less than equivalent P_{PL}.

$$\frac{USL - x_{50}}{x_{99.865} - x_{50}} = \frac{USL - LSL}{x_{99.865} - x_{.135}} (1 - \text{Equivalent } k)$$

$$\text{Equivalent } k = 1 - \left(\frac{x_{99.865} - x_{.135}}{USL - LSL}\right)\left(\frac{USL - x_{50}}{x_{99.865} - x_{50}}\right)$$

$$= \frac{(USL - LSL)(x_{99.865} - x_{50}) - (x_{99.865} - x_{.135})(USL - x_{50})}{(USL - LSL)(x_{99.865} - x_{50})}$$

The terms in the numerator are multiplied out, then combined as follows:

$$\text{Equivalent } k = \frac{x_{50}(x_{99.865} - x_{.135}) - (x_{50} - x_{.135})USL - (x_{99.865} - x_{50})LSL}{(USL - LSL)(x_{99.865} - x_{50})}$$

$$= \frac{x_{50}(x_{99.865} - x_{.135}) - [(x_{50} - x_{.135})USL + (x_{99.865} - x_{50})LSL]}{(USL - LSL)(x_{99.865} - x_{50})}$$

Recall the formula for x'_{50}.

$$x'_{50} = \frac{(x_{50} - x_{.135})USL + (x_{99.865} - x_{50})LSL}{x_{99.865} - x_{.135}}$$

$$x'_{50}(x_{99.865} - x_{.135}) = (x_{50} - x_{.135})USL + (x_{99.865} - x_{50})LSL$$

This result replaces the quantity in brackets of the prior equation for equivalent k.

$$\text{Equivalent } k = \frac{x_{50}(x_{99.865} - x_{.135}) - x'_{50}(x_{99.865} - x_{.135})}{(USL - LSL)(x_{99.865} - x_{50})}$$

$$= \frac{(x_{50} - x'_{50})(x_{99.865} - x_{.135})}{(USL - LSL)(x_{99.865} - x_{50})}$$

$$= (x_{50} - x'_{50})\left[\frac{x_{99.865} - x_{.135}}{(USL - LSL)(x_{99.865} - x_{50})}\right]$$

The quantity in brackets above is equal to 1 over $(USL - x'_{50})$.

$$\frac{x_{99.865} - x_{.135}}{(USL - LSL)(x_{99.865} - x_{50})} = \frac{1}{USL - x'_{50}}$$

$$USL - x'_{50} = \frac{(USL - LSL)(x_{99.865} - x_{50})}{x_{99.865} - x_{.135}}$$

$$x'_{50} = USL - \frac{(USL - LSL)(x_{99.865} - x_{50})}{x_{99.865} - x_{.135}}$$

A common denominator is found for the right-hand side of this equation.

$$x'_{50} = \frac{\text{USL}(x_{99.865} - x_{.135}) - (\text{USL - LSL})(x_{99.865} - x_{50})}{x_{99.865} - x_{.135}}$$

$$= \frac{\text{USL}\,x_{99.865} - \text{USL}\,x_{.135} - \text{USL}\,x_{99.865} + \text{USL}\,x_{50} + \text{LSL}\,x_{99.865} - \text{LSL}\,x_{50}}{x_{99.865} - x_{.135}}$$

$$= \frac{(x_{50} - x_{.135})\,\text{USL} + (x_{99.865} - x_{50})\,\text{LSL}}{x_{99.865} - x_{.135}}$$

Substituting 1 over (USL - x'_{50}) into the previous formula for equivalent k yields:

$$\text{Equiv. } k = (x_{50} - x'_{50})\left[\frac{x_{99.865} - x_{.135}}{(\text{USL - LSL})(x_{99.865} - x_{50})}\right] = (x_{50} - x'_{50})\left(\frac{1}{\text{USL} - x'_{50}}\right) = \frac{x_{50} - x'_{50}}{\text{USL} - x'_{50}}$$

In a like manner, it can be shown that when equivalent P_{PL} is smaller than equivalent P_{PU}, equivalent k equals:

$$\text{Equivalent } k = \frac{x_{50} - x'_{50}}{\text{LSL} - x'_{50}}$$

As equivalent P_{PK} is the *minimum* of equivalent P_{PU} and equivalent P_{PL}, equivalent k is equal to the *maximum* of:

$$\text{Equivalent } k = \text{Maximum}\left(\frac{x_{50} - x'_{50}}{\text{LSL} - x'_{50}}, \frac{x_{50} - x'_{50}}{\text{USL} - x'_{50}}\right)$$

Derivation 3: USL$_T$ and LSL$_T$ for Non-Normal Distributions

Below is the equation for equivalent P^*_{PK}.

$$\text{Equivalent } P^*_{PK} = \text{Minimum}\,(\text{Equivalent } P^*_{PL}, \text{Equivalent } P^*_{PU})$$

$$= \text{Minimum}\left(\frac{x_{50} - \text{LSL}_T}{x_{50} - x_{.135}}, \frac{\text{USL}_T - x_{50}}{x_{99.865} - x_{50}}\right)$$

The maximum for this index occurs when equivalent P^*_{PU} equals equivalent P^*_{PL}.

$$\text{Equivalent } P^*_{PU} = \text{Equivalent } P^*_{PL}$$

$$\frac{\text{USL}_T - x_{50}}{x_{99.865} - x_{50}} = \frac{x_{50} - \text{LSL}_T}{x_{50} - x_{.135}}$$

If this is to happen when the process median is located at the target value, x_{50} must equal T, making the above equation look like:

$$\frac{\text{USL}_T - T}{x_{99.865} - x_{50}} = \frac{T - \text{LSL}_T}{x_{50} - x_{.135}}$$

To simplify this derivation, let:

$$a = x_{50} - x_{.135} \quad \text{and} \quad b = x_{99.865} - x_{50}$$

Substituting these into the above equation yields:

$$\frac{USL_T - T}{b} = \frac{T - LSL_T}{a}$$

When T is less than x'_{50}, LSL_T equals LSL, and the above can be solved for USL_T.

$$\frac{USL_T - T}{b} = \frac{T - LSL}{a}$$

$$a\,USL_T - aT = bT - b\,LSL$$

$$a\,USL_T = aT + bT - b\,LSL$$

$$a\,USL_T = aT + b(T - LSL)$$

$$USL_T = T + \frac{b}{a}(T - LSL)$$

Substituting the original values for a and b back into the equation yields:

$$USL_T = T + \frac{x_{99.865} - x_{50}}{x_{50} - x_{.135}}(T - LSL)$$

When T is greater than x'_{50}, this same procedure will derive the following formula for computing LSL_T.

$$LSL_T = T - \frac{x_{50} - x_{.135}}{x_{99.865} - x_{50}}(USL - T)$$

Derivation 4: Equivalent P^*_P as a Function of the Effective Tolerance

Equivalent P^*_P is defined as:

$$\text{Equivalent } P^*_P = \text{Minimum} \left(\text{Equivalent } P^T_{PL}, \text{Equivalent } P^T_{PU} \right)$$

$$= \text{Minimum} \left(\frac{T - LSL}{x_{50} - x_{.135}}, \frac{USL - T}{x_{99.865} - x_{50}} \right)$$

Assuming T is less than x'_{50}, then:

$$\text{Equivalent } P^*_P = \frac{T - LSL}{x_{50} - x_{.135}}$$

For this situation, LSL_T and USL_T are defined as:

$$LSL_T = LSL \qquad USL_T = T + \frac{x_{99.865} - x_{50}}{x_{50} - x_{.135}}(T - LSL)$$

This makes the effective tolerance equal to:

$$\text{Effective Tolerance} = USL_T - LSL_T$$

$$= T + \frac{x_{99.865} - x_{50}}{x_{50} - x_{.135}}(T - LSL) - LSL$$

$$= T - LSL + \frac{x_{99.865} - x_{50}}{x_{50} - x_{.135}}(T - LSL)$$

$$= (T - LSL)\left(1 + \frac{x_{99.865} - x_{50}}{x_{50} - x_{.135}}\right)$$

$$= (T - LSL)\left(\frac{x_{50} - x_{.135} + x_{99.865} - x_{50}}{x_{50} - x_{.135}}\right)$$

$$= (T - LSL)\left(\frac{x_{99.865} - x_{.135}}{x_{50} - x_{.135}}\right)$$

Equivalent P^{*}_{P} may also be defined as a function of the effective tolerance:

$$\text{Equivalent } P^{*}_{P} = \frac{\text{Effective Tolerance}}{x_{99.865} - x_{.135}}$$

Substituting in the relationship just derived for the effective tolerance yields:

$$\text{Equivalent } P^{*}_{P} = \frac{\text{Effective Tolerance}}{x_{99.865} - x_{.135}}$$

$$= \frac{(T - LSL)(x_{99.865} - x_{.135}) / (x_{50} - x_{.135})}{x_{99.865} - x_{.135}} = \frac{T - LSL}{x_{50} - x_{.135}}$$

Repeating this procedure, but now assuming T is greater than x'_{50}, will reveal that:

$$\text{Equivalent } P^{T}_{PU} = \frac{\text{Effective Tolerance}}{x_{99.865} - x_{.135}} = \frac{USL - T}{x_{99.865} - x_{50}}$$

8.18 Exercises

8.1 Transistor Gain

A Taiwanese manufacturer of computer monitors is charting the transistor gain of an important circuit. Once in control, a χ^2 goodness-of-fit test discloses the process output for gain is non-normally distributed. Plotting the measurements on NOPP reveals the output has a skewed-right distribution, with $x_{.135}$ estimated as 108, x_{50} as 126, and $x_{99.865}$ as 152. Given a LSL of 100 and an USL of 150, estimate long-term potential capability with the equivalent $6\sigma_{LT}$ spread, equivalent P_R, equivalent P_P, and equivalent P_{ST} measures. Then estimate performance capability with equivalent P_{PK}. Also estimate equivalent k. Considering that the goal is 1.33, what should be done to improve performance capability? Where should the process median being centered to maximize performance capability?

Engineering wants the median centered at M. How does this additional demand alter the above estimates for potential and performance capability?

8.2 Spring Height

Suspension assemblies for computer disk drives require dimensionally precise springs that can hold recording heads at microscopic distances above the recording disks. One of the most critical dimensions, spring height, must be maintained within a tolerance of only $\pm.080$ mm. Once the forming process for these springs is in control, a normality check discovers the output for height has a bounded distribution. From the NOPP, $x_{.135}$ is found to be 1.979 mm, x_{50} is 2.106 mm, and $x_{99.865}$ is 2.058 mm. Estimate the equivalent $6\sigma_{LT}$ spread and equivalent P_P.

If the nominal spring length requirement is 2.000 mm, predict equivalent P_{PK} and equivalent k. Where should the process median be shifted to maximize equivalent P_{PK}? Assume the process spread does not change when the median is moved. What will this maximum value be? Does it meet a goal of 2.00? What would equivalent P_{PK} be if the median is centered at the middle of the tolerance?

8.3 Container Fill

A Mexican food company fills plastic containers with a butter substitute. An important characteristic is fill weight, which has a LSL of 450 grams. After stabilizing this process, a normality check on NOPP shows the output distribution for weight is skewed left. $x_{.135}$ is read off the NOPP as 451 gm, x_{50} as 455 gm, and $x_{99.865}$ as 457 gm. Estimate potential capability with equivalent P'_{PL} and performance capability with equivalent P_{PL}. Is this process capable? Estimate equivalent k'_L. Why do these two measures of capability produce identical results? Management sets a goal of having an average of only 1 underweight container out of every 10,000 containers filled. Where should the process median be located to achieve this goal (assume process spread does not change when median is shifted)? From the NOPP, $x_{.01}$ is estimated as 449. With this new target for the median, recalculate equivalent P'_{PL} and equivalent P_{PL}.

8.4 Contact Closing Speed

A South Korean manufacturer checks circuit breakers for contact motion to verify they move at the proper closing speed. Closing too quickly causes premature wearout of the contacts, while closing too slowly could weld the contacts so the breaker could not be reopened. Their engineering department has developed a window for closing speed to optimize the life of this product:

$$LSL = .0140 \text{ seconds} \qquad USL = .0250 \text{ seconds} \qquad T = .0170 \text{ seconds}$$

Once this process is in control, a normality check on NOPP shows the output distribution for closing speed is bounded. From the NOPP, $x_{.135}$ is estimated as .0150 seconds, x_{50} as .0189 seconds, and $x_{99.865}$ as .0228 seconds. Estimate equivalent P_P, equivalent $Z_{MIN.\ LT}$ and equivalent P_{PK}. Why is the answer for equivalent $Z_{MIN.\ LT}$ similar to the result of 3.80 obtained in Example 6.8 (p. 200), where the output for closing time was assumed to be normally distributed? Where should the process median be positioned to maximize equivalent P_{PK}?
What are the values for LSL_T and USL_T? Find the value for equivalent $\hat{k}^*$. Estimate equivalent P^*_P, equivalent P^*_{ST}, equivalent $Z^*_{MIN.\ LT}$, and equivalent P^*_{PK}. Compare these to the estimate of .78 for C^*_P in Example 6.15 (p. 217) and .28 for C^*_{PK} in Example 6.24 (p. 256). Why are they also similar?
How would you determine the percentage of nonconforming contacts for this process? How would that differ from the method used when the process is assumed to have a normal distribution? Would you expect these percentages to be similar or significantly different?

8.5 Plating Thickness

Costume jewelry, such as earrings, rings, pendants, and chains, is plated with a thin coating of palladium. The thickness of this coating is checked using the eddy-current principle. Once in control, the thickness readings are plotted on NOPP. From this analysis, the process output is judged to have an unbounded distribution, with .0271 percent having a thickness less than the LSL and .0093 percent with a thickness more than the USL. Using the percentage method, estimate $ppm_{TOTAL.\ LT}$, $ppm_{MAX.\ LT}$, and equivalent P^*_{PK} for this process. Does this coating process meet the minimum prerequisite for performance capability? In which direction should the median be moved to increase equivalent P^*_{PK}? By how much should it be shifted?

8.6 Gap Size

During the development phase of a new car program, 16 prototype vehicles are sent down a simulated assembly line to check various fits (door, hood, trunk, window). Gap size measurements taken at one particular location of the windshield are plotted on an *IX & MR* chart. Although there are limited data, the chart indicates the process appeared to operate in a state of statistical control during this trial run. Before estimating any measures of capability, the process development engineers decide to check the normality assumption by plotting the cumulative percentage distribution on NOPP, which has already been done in Figure 7.42 (p. 391). Using the percentage method, estimate equivalent P^*_{PK}, $ppm_{MAX.\ LT}$, and $ppm_{TOTAL.\ LT}$. Does this assembly process satisfy a 1.33 goal for equivalent P^*_{PK}?

8.7 Potato Slices

Potatoes are sliced prior to being fryed into potato chips. One critical characteristic of making the perfect chip is the thickness at the center of the slice. After stabilizing the slicing operation, thickness measurements are plotted on NOPP. Because automatic controls are constantly adjusting this operation, it came as no surprise the output for slice thickness is a bounded distribution. From NOPP, $x_{.135}$ is estimated as 1.079 mm, x_{50} as 1.112 mm, and $x_{99.865}$ as 1.144 mm. Given a thickness specification of $1.10 \pm .05$ mm, estimate equivalent P_P and equivalent P_{PK}. Assuming no change in spread when x_{50} is shifted, where should this process be centered to maximize equivalent P_{PK}?

8.8 Tuft Retention

A bundle of nylon threads, called a "tuft," is inserted into each of the 34 holes molded into the head of a child's toothbrush (Figure 8.44). Every hour, one toothbrush is taken to the lab where one of the 34 tufts is randomly selected. This tuft is pulled until it comes out of its hole. The force required to remove the tuft measures what is referred to as "tuft retention," which is the most important characteristic of the brush. The minimum requirement for tuft retention is 7.8 kilograms, with a target of 8.8 kilograms.

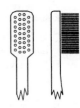

Fig. 8.44

Once the readings for tuft retention display good control, they are plotted on NOPP to check for normality. The output distribution appears to be skewed left, with $x_{.135}$ equaling 7.6 kg, x_{50} is 8.4 kg, and $x_{99.865}$ is 8.9 kg. Estimate the equivalent $6\sigma_{LT}$ spread, equivalent P'_{PL}, and equivalent P_{PK}. The goal for equivalent $6\sigma_{LT}$ spread is to be less than 75 percent of $2(T$ minus LSL), whereas the goal for equivalent P'_{PL} is 1.50, and the goal for equivalent P_{PK} is 1.33. How well does this process meet these capability goals?

A change in the type of nylon thread alters the output distribution such that $x_{.135}$ now equals 8.3 kg, x_{50} becomes 8.9 kg, and $x_{99.865}$ moves to 9.3 kg. What effect does this have on the prior estimates of process capability?

8.9 Powdered Metal Size

In Exercise 7.2 on page 420, the process output for the particle size of a powdered metal was discovered to have a bounded distribution. Estimate equivalent P_{PK} and equivalent P^{*}_{PK}. Why the difference in these two? Assuming the median can be shifted higher by .75 without affecting the output's shape, re-estimate equivalent P_{PK} and equivalent P^{*}_{PK}.

8.10 Cam Roller Waviness

In Exercise 5.8 (p. 162), the P'_{PU} index was estimated as .94 for cam roller waviness. In Example 8.5 (p. 448), equivalent P'_{PU} was estimated as .583. Why this difference? Which estimate is a "better" indicator of actual potential capability?

In Exercise 6.8 (p. 345), P_{PK} was estimated as .83. Then in Example 8.24 (p. 504), equivalent P_{PK} was estimated as .583, while equivalent P^{*}_{PK} was estimated as .803. Why the difference between these three? Which is the "best" predictor of the true performance capability of this process? Why?

8.11 Brake Lining Hardness

Brake linings (shoes) are manufactured by mixing asbestos fibers and resin to the proper consistency, then extruding this mixture into a continuous 6 cm wide strip. The strip is water-jet cut into 25 cm long pieces, which are then baked in an oven for 20 minutes.

Fig. 8.45

When cooled, the finished lining is checked for hardness by pressing a steel tip on the lining with a predetermined force (Figure 8.45). The depth of the resulting indentation is measured, then converted into a hardness measurement. Deeper indentations reflect softer linings, whereas shallow indentations imply harder linings. As the sole concern is that linings are too soft, there is only an USL of 35.0 given for indentation depth, with a target of 15.5.

After control is established for hardness, a NOPP analysis exposes the process output to have a bounded distribution. This makes engineering sense, as indentation depth cannot be less zero, nor greater than the thickness of the lining. From the NOPP, x_{50} and $x_{99.865}$ are estimated as 17.3 and 24.6, respectively. Armed with this information, the capability study team decides to estimate potential capability with the equivalent P'_{PU} index, while employing equivalent P_{PK} to measure performance capability. Compute equivalent $\hat{k}'$, then combine it with equivalent P'_{PU} to estimate equivalent P_{PK}.

Assume the mixing process is altered such that the process median for hardness is lowered to 15.5, while the 99.865 percentile decreases to 22.6. How does this change the above estimates of capability?

One of the team members is curious about the equivalent P^{*}_{PK} index. After the process median is lowered to 15.5, this person estimates from NOPP that the percentage of linings with a hardness reading over 35.0 is essentially zero. What is the estimate of equivalent P^{*}_{PK} based on this percentage? What does this mean? What effect would lowering the process median even further have on this index?

8.12 Thermostat Opening Time

The NOPP analysis conducted in Exercise 7.3 (p. 420) indicated the distribution of thermostat opening times has a skewed-left distribution. Use the estimates of $x_{.135}$, $x_{99.865}$, and $x_{99.865}$ from the NOPP to predict equivalent P_P and equivalent $Z_{MIN. LT}$. The LSL is 75 seconds, while the USL is 105 seconds. What is $\hat{x}'_{50}$?

If a target opening time of 95 seconds is desired, estimate equivalent P^{*}_{P} and equivalent $Z^{*}_{MIN. LT}$. Computed an estimate of equivalent k^{*}, then combine it with equivalent P^{*}_{P} to estimate equivalent P^{*}_{PK}. How much potential capability is lost by specifying a target not equal to x'_{50}?

8.13 Sealing Strength

Plastic bags are filled with a snack food, then sealed, with the integrity of this seal being very important (Figure 8.46). If too weak, the bag will open during shipment, causing the contents to become stale. On the other hand, overly strong seals cause customers difficulty when attempting to open the bag and they may spill the contents. Due to these dual concerns, both a LSL of 9.00 kg and an USL of 12.00 kg are imposed on seal strength. However, in an effort to keep bags fairly easy to open, a target of 9.75 kg is specified.

Fig. 8.46

This seal is made by inserting the bag between two metal bars, with the top one being heated. When these bars are pressed together, the resulting force and heat create the seal. Seal strength is checked by measuring the force required to open a bag. After charting this characteristic and bringing the process into control, it is discovered via NOPP that the process output distribution is skewed right. With $x_{.135}$, x_{50}, and $x_{99.865}$ estimated as 9.13, 10.12, and 11.38, respectively, estimate the following equivalent potential capability measures: $6\sigma_{LT}$ spread, P^*_R, P^*_P, and P^*_{ST}.

Calculate LSL_T and USL_T and then an estimate for x'_{50}. Estimate the following equivalent performance capability measures: $Z^*_{LSL, LT}$, $Z^*_{USL, LT}$, $Z^*_{MIN, LT}$, P^*_{PL}, P^*_{PU}, and P^*_{PK}. Now estimate the equivalent k^* factor and combine it with equivalent P^*_P to estimate the equivalent P^*_{PK} index.

How can performance capability be improved? By how much should the median be shifted? If the median is moved to the target of 9.75 kg with no change in the output's shape, by how much do the equivalent P^*_{PL}, equivalent P^*_{PU}, and equivalent P^*_{PK} measures increase?

8.14 Can Crush Strength

Aluminum cans for soft drinks undergo a crush test to check their strength. Once in control, a plot of the individual measurements on NOPP reveals can strength to have a highly skewed-left distribution. From this graph, $x_{.135}$ is found to be 122 kg while x_{50} is 154 kg. Given a specification of 115 kg for the minimum crush strength, estimate the equivalent P_{PK} index for this feature of the can-forming process.

8.15 Mold Hardness

A quality-improvement team working at a grey-iron foundry discovers the majority of customer complaints are caused by excessive variation in the hardness of the sand molds used to make castings. To better under this process, the team decides to conduct a capability study. A control chart monitoring mold hardness provides evidence the molding process is in a good state of control. When its cumulative percentage distribution is plotted on NOPP, the team learns the output distribution for mold hardness is skewed left. With $x_{.135}$ estimated as 83.2, x_{50} as 89.1, and $x_{99.865}$ as 92.6, estimate both the equivalent P_P and equivalent P_{PK} indexes for this process. The LSL for mold hardness is 84.5 and the USL is 94.5. Where should the median be positioned to maximize equivalent P_{PK}? What will be one of the team's recommendations for improving this molding process?

8.19 References

Bissell, Derek, *Statistical Methods for SPC and TQM*, 1994, Chapman & Hall, London, United Kingdom, p. 260

Bittanti, S., Lovera, M., and Moiraghi, L., "Application of Non-Normal Process Capability Indices to Semi-Conductor Quality Control," *SEEE Transactions of Semiconductor Manufacturing*, Vol. 11, 1998, pp. 296-302

Bothe, Davis R., "A Capability Study for an Entire Product," *46th ASQC Annual Quality Congress Transactions*, Nashville, TN, May 1992, pp. 172-178

Castagliola, Philippe, "Evaluation of Non-Normal Process Capability Indices Using Burr's Distributions," *Quality Engineering*, Vol. 8, No. 4, 1996, pp. 587-593

Chan, Lai K., Cheng, S.W., and Spiring, Fred A., "A Graphical Technique for Process Capability," *42nd ASQC Annual Quality Congress Trans.*, Dallas, TX, May 1988, pp. 268-275

Chan, Lai K., Cheng, S.W., and Spiring, Fred A., "The Robustness of the Process Capability Index, Cp, to Departures from Normality," *Proceedings of the 2nd Pacific Area Statistical Conference*, 1986, Tokyo, Japan

Clements, John A., "Process Capability Calculations for Non-Normal Distributions," *Quality Progress*, Vol. 22, No. 9, September 1989, pp. 95-100

Deleryd, M., "The Effect of Skewness on Estimates of Some Process Capability Indices," *International Journal for Applied Quality Management*, Vol. 2, No. 2, 1999, pp. 153-186

English, J.R., and Taylor, G.D., "Process Capability Analysis - A Robustness Study," *International Journal of Production Research*, Vol. 31, No. 7, July 1993, pp. 1621-1635

English, J.R., and Taylor, G.D., "Process Capability Analysis in Departures from Normality", *Proceedings of the 1st IE Research Conference*, 1992, Chicago, IL, pp. 335-338

Farnum, Nicholas R., "Using Johnson Curves to Describe Non-Normal Process Data," *Quality Engineering*, Vol. 9, No. 2, 1996-97, pp. 329-336

Flaig, John J., "A New Approach to Process Capability Analysis," *Quality Engineering*, Vol. 9, No. 2, 1996-97, pp. 205-211

Gunter, Bert, "The Use and Abuse of Cpk, Part 2" *Quality Progress*, Vol. 22, No. 3, March 1989, pp. 109-109

James, Paul C., "Cpk Equivalencies," *Quality*, Vol. 28, No. 9, September 1989, p. 75

Kane, Victor E., "Process Capability Indices," *Journal of Quality Technology*, Vol. 18, No. 1, January 1986, p. 49

Krishnamoorthi, K.S. and Khatwani, Suraj, "A Capability Index for All Occasions," *54th ASQ Annual Quality Congress Transactions*, May 2000, pp. 77-81

Kocherlakota, S., Kocherlakota, K., and Kirmani, S.N.U.A, "Process Capability Indices Under Non-Normality," *International Journal of Mathematical Statistics*, Vol. 1, 1992

Levinson, William A., "SPC for Real-World Processes: What to Do When the Normality Assumption Doesn't Work," *54th ASQ Annual Quality Congress Transactions*, May 2000, pp. 82-89

Littig, Steven J., and Lam, C. Teresa, "Case Studies in Process Capability Measurement," *47th ASQC Annual Quality Congress Transactions*, Boston, MA, May, 1993, pp. 569-575

McCormack, Jr., D.W., Hurwitz, Arnon M., Spagon, Patrick D., Harris, Ian R., "Capability Indices for Nonnormal Data," *Proceedings of the Section of Quality and Productivity*, American Statistical Association, 1996, pp. 108-113

Mukherjee, S.P. and Singh, N.K., "Sampling Properties of an Estimator of a New Process Capability Index for Weibull Distributed Quality Characteristics," *Quality Engineering*, Vol. 10, No. 2, 1997-1998, pp. 291-294

Munechika, M., "Evaluation of Process Capability for Skew Distributions," *Proceedings of the 30th European Organization of Quality Annual Conference*, 1986

Pearn, W. L., and Kotz, Samuel, "Application of Clement's Method for Calculating Second- and Third-generation Process Capability Indices for Non-Normal Pearsonian Populations," *Quality Engineering*, Vol. 7, No. 1, 1994-1995, p. 142

Polansky, Alan M., "A Smooth Nonparametric Approach to Process Capability," *Quality and Reliability Engineering International*, Vol. 14, 1998, pp. 43-48

Polansky, Alan M., "An Algorithm for Computing a Smooth Nonparametric Capability Estimate," *Journal of Quality Technology*, Vol. 32, 2000, pp. 284-289

Polansky, Alan, Chou, Youn-Min, and Mason, Robert, "Estimating Process Capability Indices for a Truncated Distribution," *Quality Engineering*, Vol. 11, No. 2, 1998-1999, pp. 257-265

Pyzdek, Thomas, "Process Capability Analysis Using Personal Computers," *Quality Engineering*, Vol. 4, No. 3, 1992, pp. 419-440

Somerville, Steven E., and Montgomery, Douglas C., "Process Capability Indices and Non-Normal Distributions," *Quality Engineering*, Vol. 9, No. 2, 1996-97, pp. 305-316

Wu, Hsin-Hung, "Performance Problems of Families of Non-Normal Process Capability Indices," *Quality Engineering*, Vol. 13, No. 3, 2001, pp. 383-388

Yeh, A.B. and Bhattacharya, S., "A Robust Capability Index," *Communications in Statistics - Simulation Computations*, Vol. 27, 1999, pp. 565-589

Chapter 9

Measuring Capability With Attribute Data

"The goal for quality is making all good parts."

Up to this point, capability measures have been presented for variable data exclusively. This type of data is easily quantified, such as 1.5 cm, 245 kg, or 16.3 gm. However, in some instances a quality characteristic can be defined in only a "presence of " or "absence of " sense. Some examples are the presence of surface flaws on sheet metal, solder bridges on circuit boards, burrs on machined edges, porosity in castings, the absence of a component in a finished assembly, or the absence of the date code stamp on a food product. These quality problems are usually determined by sight, where an observer either concludes "yes" or "no" to their presence (or absence). At other times, a count is kept of the number of times they occur. These types of quality assessments generate what is commonly referred to as *attribute* data.

For these situations, control charts monitoring attribute data are a helpful way of identifying problems and establishing priority areas for initiating corrective action. This approach may highlight critical process inputs where control charts for variable-type data should be introduced, or where additional problem-solving attention should be focused. Zaciewski explains how attribute charts may also be applied to monitor administrative functions such as shipping nonconformities, preventative maintenance, and plant performance.

Prior to presenting measures of capability based on attribute data, a brief review of the four major attribute charts is in order. But, before reviewing these charts, it is important to explain the difference between a *nonconformity* (formerly termed defect) and a *nonconforming unit* (formerly called a defective).

Nonconformity - This term describes a fault on, or with, a unit such as a dent, a scratch, a paint blemish, a missing part, or a leak. It is possible to have more than one nonconformity on a single unit, *e.g.*, two missing components on the same circuit board. The number of nonconformities found in subgroups taken from a stable process can often be approximated by the Poisson distribution. This is the output distribution on which control limits for the *c* and *u* charts are based.

Nonconforming unit - This term characterizes a unit that fails to satisfy a product requirement due to the presence of one or more nonconformities. The presence of just one nonconformity makes the entire unit nonconforming. Conversely, a conforming unit has no nonconformities. This definition is applied later to help develop capability measures for attribute data. In these situations, units are considered either "good" (conforming), or "bad" (nonconforming). The number (or percentage) of nonconforming units found in subgroups taken from a stable process usually follows a binomial distribution, which is the theoretical basis for the *np* and *p* charts.

9.1 Review of Attribute Charts

Each of the four attribute control charts mentioned on the previous page is explained in the next few sections, beginning with the *np* chart. A flowchart for selecting the correct attribute chart for any application is presented in Section 9.6 (p. 554).

The *np* Chart

For this chart, each unit in a subgroup is inspected and judged to be either conforming or nonconforming: it either passes or fails the inspection. The number of nonconforming units, called *np*, in a subgroup of *n* units is totaled and plotted. To allow the number of nonconforming items in one subgroup to be directly compared to those in another, *n* must remain constant for all subgroups.

$$\text{Plot point } = np = \text{Number of nonconforming units in the subgroup}$$

When at least 20 subgroups are collected, the centerline, $n\overline{p}$, is calculated by finding the average of the *np* plot points. As before, *k* represents the number of in-control subgroups collected during the capability study.

$$\text{Estimate of the Process Average } = n\hat{p}' = n\overline{p} = \frac{\sum_{i=1}^{k} (np)_i}{k}$$

The important process parameter needed for measuring capability in this situation is np', which represents the true average number of nonconforming units per subgroup of size *n* for this process. np' is labeled $n\hat{p}'$ when it is estimated from $n\overline{p}$, the centerline of the *np* chart. Control limits around the centerline are calculated with these formulas:

$$LCL_{np} = n\overline{p} - 3\sqrt{n\overline{p}\left(1 - \frac{n\overline{p}}{n}\right)} \qquad UCL_{np} = n\overline{p} + 3\sqrt{n\overline{p}\left(1 - \frac{n\overline{p}}{n}\right)}$$

Interpretation is then done for points outside of control limits, runs, trends and cycles. Just as for variable-data charts, the process must first exhibit a reasonably good state of statistical control before any type of capability study can be undertaken. Remember, being in control does *not* imply all parts are conforming, only that the process is producing parts in a consistent manner (at a stable nonconformance rate).

Example 9.1 Cracked Pistons

Hydraulic pistons are checked in a certain area for the presence of cracks (Figure 9.1). When a crack is detected, the piston is immediately scrapped. No time is spent looking for a second or third crack since one crack is all it takes to make the piston nonconforming. As each piston is judged to be either conforming (no cracks) or nonconforming (cracked), the binomial output model is appropriate.

Fig. 9.1

Fifty pistons ($n = 50$) are inspected twice each day with the results for the first 12 subgroups ($k = 12$) given below. Normally a k of 20 is preferred, but for the purposes of providing a clearer explanation, fewer subgroups are used in this example. Remember, np is the *number* of nonconforming (cracked) pistons in a given subgroup of size 50 and becomes the plot point for this chart, as is displayed in Figure 9.2 on the next page.

Subgroup #	1	2	3	4	5	6	7	8	9	10	11	12
np	2	3	0	1	1	7	0	4	1	2	0	2

The centerline for this np chart is calculated as follows:

$$n\hat{p}' = n\overline{p} = \frac{\sum_{i=1}^{k} (np)_i}{k} = \frac{\sum_{i=1}^{12} (np)_i}{12} = \frac{23}{12} = 1.92$$

Control limits are then determined with this centerline, as shown here:

$$UCL_{np} = n\overline{p} + 3\sqrt{n\overline{p}\left(1 - \frac{n\overline{p}}{n}\right)} = 1.92 + 3\sqrt{1.92\left(1 - \frac{1.92}{50}\right)} = 6.00$$

$$LCL_{np} = n\overline{p} - 3\sqrt{n\overline{p}\left(1 - \frac{n\overline{p}}{n}\right)} = 1.92 - 3\sqrt{1.92\left(1 - \frac{1.92}{50}\right)} = -2.16$$

In this example, the *LCL* is set at 0, as a negative *LCL* does not make sense on a chart for the number of nonconforming units. Because this is not the "true" *LCL*, a plot point falling on 0 is *not* considered an out-of-control signal.

For the np chart in Figure 9.2, the sixth plot point is above the *UCL*. This is an undesirable out-of-control condition because it indicates the number of nonconforming units has significantly increased. A search for the assignable source of variation should be initiated at once and, when the cause is located, corrective action implemented to prevent its reoccurrence.

Because out-of-control points do not represent the process under normal operating conditions, they should be removed from the calculations of control limits and centerlines, as well as from any type of capability study concerning this process. This is the same procedure followed for variable-data charts (review discussion on page 20).

What would happen if only 10 pistons are produced today and two cracked pistons are discovered? As the np chart requires a constant subgroup size, in this case, 50, a "2" cannot simply be plotted on the chart and compared to the number of cracked pistons found in other subgroups. A different type of attribute chart is needed to handle situations where the subgroup size can vary.

The *p* Chart

If the subgroup size varies, the *percentage* nonconforming, p, is charted instead of the *number* of nonconforming units. This percentage is calculated by dividing the number of nonconforming units in the subgroup, np, by the subgroup size, n.

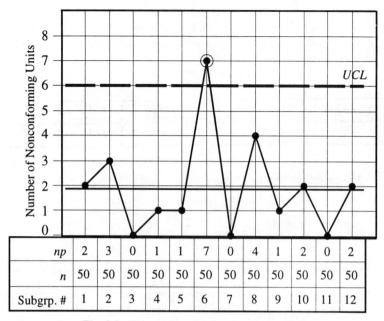

Fig. 9.2 The *np* chart for the piston example.

Plot point = p = Percentage nonconforming in the subgroup

$$= \frac{\text{\# Nonconforming units}}{n} = \frac{np}{n}$$

The centerline of this chart, $\bar{p}$, is an estimate of the population percentage nonconforming, p'. The estimate of this process parameter plays a major role in measuring the capability of this process to meet customer requirements.

$$\text{Estimate of the Process Average} = \hat{p}' = \bar{p} = \frac{\displaystyle\sum_{i=1}^{k} (np)_i}{\displaystyle\sum_{i=1}^{k} n_i}$$

Whenever the subgroup size changes, control limits must be recalculated because they are a function of n, as can be seen in the formulas below. This requirement makes the calculations much more involved and time consuming, so this chart should be implemented only if n varies from subgroup to subgroup, and then only if the calculations can be automated.

$$LCL_p = \bar{p} - 3\sqrt{\frac{\bar{p}(1-\bar{p})}{n}} \qquad UCL_p = \bar{p} + 3\sqrt{\frac{\bar{p}(1-\bar{p})}{n}}$$

Control limits vary when the subgroup size varies, but $\bar{p}$ remains the same for all subgroups because it is an estimate of the *overall* process percentage nonconforming.

Example 9.2 Cracked Pistons

As a result of fluctuations in daily production, the number of pistons checked for cracks varies from subgroup to subgroup, as is displayed in Table 9.1. After the percentage non-conforming in each subgroup is calculated, the results are plotted on a p chart, as presented in Figure 9.3. The centerline of this chart is derived from the total number of cracked pistons for all subgroups, 43, divided by the total number of pistons inspected, which is 717.

$$\bar{p} = \frac{\sum_{i=1}^{12} (np)_i}{\sum_{i=1}^{12} n_i} = \frac{43}{717} = .06$$

Table 9.1 p chart data for cracked piston example.

Subgroup	# Nonconform-ing Units (np)	n	p	UCL	LCL
1	1	14	.07	.25	.00
2	3	66	.05	.15	.00
3	1	37	.03	.18	.00
4	6	91	.07	.13	.00
5	3	78	.04	.14	.00
6	1	10	.10	.29	.00
7	1	50	.02	.16	.00
8	9	103	.09	.13	.00
9	3	66	.05	.15	.00
10	1	33	.03	.19	.00
11	2	21	.10	.22	.00
12	12	151	.08	.12	.00
Totals	43	717			

After $\bar{p}$ is calculated from the data contained in all the in-control subgroups, control limits must be calculated for each subgroup with a different subgroup size. The following calculations are for only the first subgroup, where n is 14.

$$UCL_p = \bar{p} + 3\sqrt{\frac{\bar{p}(1-\bar{p})}{n}} = .06 + 3\sqrt{\frac{.06(1-.06)}{14}} = .25$$

Because the *LCL* is calculated as a negative number, it is set equal to 0.

$$LCL_p = \bar{p} - 3\sqrt{\frac{\bar{p}(1-\bar{p})}{n}} = .06 - 3\sqrt{\frac{.06(1-.06)}{14}} = -.13 \rightarrow .00$$

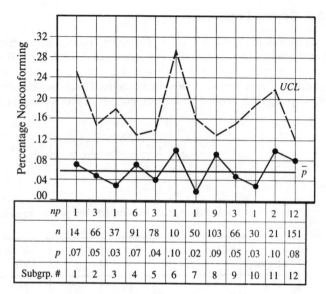

Fig. 9.3 The *p* chart for the piston example.

This calculation must be repeated for each subsequent subgroup with a different subgroup size. Notice how when *n* increases, the control limits are closer to the centerline. For larger subgroup sizes, the subgroup statistic is expected to be closer to the process average. Conversely, whenever *n* decreases, the limits are farther away from the centerline, as is exhibited above in Figure 9.3. To aid chart interpretation when control limits vary, Nelson (1989) recommends using the standardized *p* chart.

When the *p* chart indicates the process is in a good state of control, the average percentage nonconforming for the process, labeled p', is estimated from $\bar{p}$. Notice that "in-control" does not imply the process is producing all conforming parts. The *p* chart in Figure 9.3 is in control, which means the process is *consistently* manufacturing cracked pistons at a rate of about 6 percent. Since a process must generate no more than .27 percentage nonconforming parts to meet the minimum requirement of capability, this process is considered to be in control, but is certainly *not* capable. Part containment and 100 percent inspection of these pistons must be initiated immediately.

If the subgroup size is quite large (over 500), the control limits are very close to the centerline, and almost every plot point appears to be out of control. In these situations, the percentage nonconforming for each subgroup may be plotted as an individual value on an *IX & MR* chart. Control limits computed from the average moving range will better portray true process variation between subgroups. Heimann offers a more detailed analysis of this topic.

The *c* Chart

Often there are many nonconformities associated with a single unit, like paint blemishes on a car or various solder-related faults in a circuit board, *and* there is a desire to track the total number of nonconformities found in each subgroup. As this type of process output typically follows a Poisson distribution, the *c* chart is appropriate. For this chart, each unit in a subgroup is examined, with the total number of nonconformities detected in the subgroup, called *c*, plotted and monitored.

Plot point $= c =$ Number of nonconformities in the subgroup

The average of all in-control plot points, $\bar{c}$, is an estimate of c', the population average number of nonconformities per subgroup of size n. c' plays a pivotal role in determining the capability for this type of process.

$$\text{Estimate of the Process Average} = \hat{c}' = \bar{c} = \frac{\sum\limits_{i=1}^{k} c_i}{k}$$

For Poisson distributions, $\bar{c}$ is an estimate of the average, while $\sqrt{\bar{c}}$ is an estimate of the standard deviation. Thus, the control limits for this chart are derived solely from $\bar{c}$.

$$LCL_c = \bar{c} - 3\sqrt{\bar{c}} \qquad UCL_c = \bar{c} + 3\sqrt{\bar{c}}$$

Because the c chart control limits do not depend on n, they do not vary from subgroup to subgroup. However, n must remain constant for all subgroups.

Example 9.3 Circuit Board Inspection

Printed circuit boards (Figure 9.4) are inspected for three types of assembly-related nonconformities: missing component, wrong-valued component, component mounted incorrectly. Five boards ($n = 5$) are checked every 2 hours. The number of nonconformities found in each subgroup is recorded in Table 9.2, then plotted on the c chart displayed in Figure 9.5. In subgroups 3 and 6, more nonconformities are discovered than boards checked, which is possible because several nonconformities may be detected on each board. In fact, all seven nonconformities in subgroup 3 could have been found on just one of the five boards.

Table 9.2 Number of nonconformities found in each subgroup.

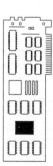

Fig. 9.4 A printed circuit board.

Subgroup Number	# Nonconformities (c)
1	2
2	4
3	7
4	0
5	1
6	8
7	1
8	2
9	1
10	1
11	0
12	3
Total	30

The centerline for this chart is calculated by dividing the total number of nonconformities found in all subgroups, 30, by the number of subgroups, 12. This represents the average number of nonconformities expected in a typical subgroup of five boards. Again, collecting at least 20 subgroups is recommended before calculating $\bar{c}$.

$$\bar{c} = \frac{\sum\limits_{i=1}^{k} c_i}{k} = \frac{30 \text{ Nonconformities}}{12 \text{ Subgroups}} = 2.5 \text{ Nonconformities per subgroup}$$

From this average of 2.5, control limits are computed for the c chart in Figure 9.5.

$$UCL_c = \bar{c} + 3\sqrt{\bar{c}} = 2.5 + 3\sqrt{2.5} = 7.2$$

$$LCL_c = \bar{c} - 3\sqrt{\bar{c}} = 2.5 - 3\sqrt{2.5} = -2.2 \rightarrow 0.0$$

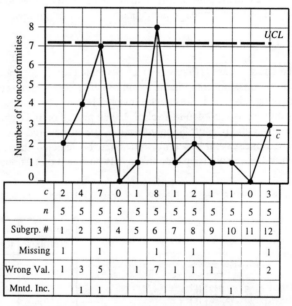

c	2	4	7	0	1	8	1	2	1	1	0	3
n	5	5	5	5	5	5	5	5	5	5	5	5
Subgrp. #	1	2	3	4	5	6	7	8	9	10	11	12
Missing	1		1			1		1				1
Wrong Val.	1	3	5		1	7	1	1	1			2
Mntd. Inc.		1	1								1	

Fig. 9.5 The c chart for the circuit board example.

Subgroup number six on this chart is above the *UCL*, indicating a significantly higher number of problems for that subgroup. An examination of the list of problems at the bottom of the chart reveals this subgroup is out of control due to a large increase in wrong-valued components. Action must be taken to discover the reason for this increase and a fix implemented to prevent a future occurrence.

When the process is finally in control, there is an average of $\bar{c}$ nonconformities per *subgroup*. The average nonconformities per *unit*, referred to as $\bar{u}$, is computed by dividing $\bar{c}$ by n.

$$\bar{u} = \frac{\bar{c}}{n} = \frac{\text{Nonconformities / Subgroup}}{\text{Units / Subgroup}} = \text{Nonconformities / Unit}$$

Because customers buy single units (rather than subgroups), this value becomes very important when estimating capability based on the results of a *c* chart (see Section 9.4, p. 540).

After the process achieves stability, the number of incidences for one particular type of nonconformity may be summed across all subgroups. This result can be compared to the totals for all other types of nonconformities on a Pareto diagram to help identify the largest contributor to the overall level of nonconformities. Based on this type of analysis, if any significant improvements are to be achieved in the manufacture of circuit boards in Example 9.3, the causes of "wrong-valued" components must be addressed.

As *n* is constant for all subgroups, only one set of control limits needs to be calculated and drawn on the *c* chart. If the subgroup size is subject to change, then the nonconformities per unit in each subgroup must be tracked on a *u* chart.

The *u* Chart

This is a variation of the *c* chart in which several units are checked and the average number of nonconformities per unit for the subgroup is monitored on the control chart. The *u* plot point is calculated by dividing the number of nonconformities found in a subgroup, *c*, by the subgroup size, *n*.

$$\text{Plot point} = u = \text{Average nonconformities per unit in the subgroup}$$

$$= \frac{\text{Number of nonconformities}}{\text{Subgroup size}} = \frac{c}{n}$$

The centerline, $\bar{u}$, is an estimate of u', the process parameter for the average nonconformities per unit. This parameter is important in determining the capability for process outputs having a Poisson distribution.

$$\text{Estimate of the Process Average} = \hat{u}' = \bar{u} = \frac{\sum\limits_{i=1}^{k} c_i}{\sum\limits_{i=1}^{k} n_i}$$

Like the *p* chart, the control limits are dependent on the subgroup size and change when *n* changes. Due to all these additional complex calculations, this chart should be selected only when subgroup size cannot be held constant. If chosen, try automating these calculations to ease the computational burden placed on the user.

$$LCL_u = \bar{u} - 3\sqrt{\frac{\bar{u}}{n}} \qquad\qquad UCL_u = \bar{u} + 3\sqrt{\frac{\bar{u}}{n}}$$

These control limits vary inversely as *n* varies: when *n* increases, the control limits are closer to the centerline; when *n* decreases, the limits are farther away from the centerline.

Example 9.4 Circuit Board Inspection

Suppose the number of circuit boards checked per subgroup in the previous *c* chart example varies anywhere from 1 to 10, as is evident in Table 9.3. This makes the *u* chart appropriate for monitoring this assembly process. Note the large number of additional calculations required for this chart compared to the simpler *c* chart.

Table 9.3 Calculations required for the u chart monitoring circuit board assembly.

Subgroup Number	# Nonconformities (c)	n	u	UCL	LCL
1	2	4	.50	1.58	.00
2	3	7	.43	1.32	.00
3	1	3	.33	1.75	.00
4	1	5	.20	1.47	.00
5	5	6	.83	1.38	.00
6	4	10	.40	1.19	.00
7	3	9	.33	1.22	.00
8	3	2	1.50	2.02	.00
9	4	6	.67	1.38	.00
10	2	3	.67	1.75	.00
11	0	1	.00	2.65	.00
12	1	1	1.00	2.65	.00
Totals	29	57			

As before, the centerline is calculated when at least 20 subgroups are collected. For illustrative purposes, k is only 12 in this example.

$$\bar{u} = \frac{\sum\limits_{i=1}^{12} c_i}{\sum\limits_{i=1}^{12} n_i} = \frac{29 \text{ Nonconformities}}{57 \text{ Units}} = .51 \text{ Nonconformities per unit}$$

The upper control limit calculation for just the first subgroup is determined next.

$$UCL_u = \bar{u} + 3\sqrt{\frac{\bar{u}}{n}} = .51 + 3\sqrt{\frac{.51}{4}} = 1.58$$

Being calculated as negative, the *LCL* is set equal to zero.

$$LCL_u = \bar{u} - 3\sqrt{\frac{\bar{u}}{n}} = .51 - 3\sqrt{\frac{.51}{4}} = -.56 \rightarrow .00$$

Control limits must be recalculated for each subgroup with a different n. As was the case with the p chart, these varying control limits also make chart interpretation much more difficult, as Figure 9.6 on the facing page clearly illustrates.

To have statistical validity, there should be an average of at least two nonconforming units (or nonconformities) in each subgroup. As the nonconformance rate is reduced through problem-solving efforts, the subgroup size must be increased to maintain this average (however, Ryan and Schwertman have developed a method for modifying the control limits instead). For instance, if the current subgroup size for a p chart is 100 and $\bar{p}$ is .05 (5 percent), an average of 5 (100 × .05) nonconforming items are expected in each subgroup. This combination satisfies the minimum requirement of at least two. However, if the percentage is reduced to only .01, the subgroup size must be increased to at least 200 in order to maintain an average of two nonconforming items per subgroup ($200 \times .01 = 2$). Thus, as quality levels improve, subgroup sizes must increase. When a process is averaging .135 percent nonconforming parts (the minimum capability requirement), the subgroup size must be at least 1481.

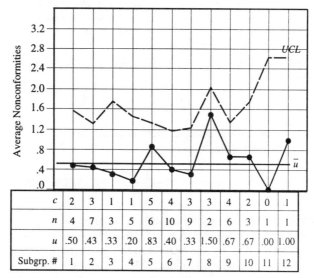

c	2	3	1	1	5	4	3	3	4	2	0	1
n	4	7	3	5	6	10	9	2	6	3	1	1
u	.50	.43	.33	.20	.83	.40	.33	1.50	.67	.67	.00	1.00
Subgrp. #	1	2	3	4	5	6	7	8	9	10	11	12

Fig. 9.6 The *u* chart for the circuit board example.

If the capability goal is a nonconformance rate of only 32 *ppm*, a subgroup size of 62,500 is needed for the proper functioning of the control chart. As this is not practical for most industrial situations, variable-data control charts should replace attribute-data charts wherever possible. With a subgroup size of only five, an $\overline{X}, R$ chart can easily demonstrate achievement of capability goals as high as a C_{PK} of 2.00. Unlike attribute-data charts, the subgroup size of an $\overline{X}, R$ chart does not have to be increased when process variation is reduced, as long as there is no problem with gage discrimination (see Figure 7.26, p. 371). In addition, attribute data is not as sensitive to process changes as variable data for a given subgroup size (Gunter).

Alternative to Attribute Charts

When a process enjoys a very low nonconformance rate (which is desirable), almost all subgroups will have no nonconforming units. A *np* chart monitoring this process will display mostly zero plot points, a very uninteresting chart which is difficult to interpret. Detecting process improvement, such as a decreasing nonconformance rate, is next to impossible because just more zeroes will be seen on the chart.

As an alternative, instead of collecting subgroups at certain intervals, wait until the first nonconforming item is identified. Then plot the number of *conforming* parts made prior to this item as the *IX* plot point on an *IX & MR* chart (for a slightly different approach to this problem, see Wheeler). Continue counting the number of parts produced until the second nonconforming item is discovered. Installing a part counter on the machine will help in this task. Plot the number of conforming pieces manufactured between the first and second nonconforming items as the second plot point on the *IX* chart. The moving range is the absolute value of the difference between these two *IX* plot points.

To illustrate this approach, 2,677 castings are poured in a foundry before the first casting with porosity is found. 2,912 castings are produced from this first nonconforming unit until the next casting with porosity is detected. Thus, the first plot point on the *IX* chart is 2,677 and the second is 2,912. This makes the first moving range 235 (|2,677 - 2,912|).

When at least twenty *IX* values are plotted, calculate control limits and interpret the chart. A point above the *UCL* (or an upward trend) indicates process improvement as significantly more conforming pieces are produced before encountering a nonconforming one. Conversely, a point below the *LCL* (or a downward trend) implies a decrease in process performance as nonconforming units are appearing more frequently. After the process demonstrates stability, the average percentage of nonconforming items, *p'*, for this process is estimated as 1 over $\overline{X}$.

$$\hat{p}' = \frac{1}{\overline{X}}$$

In the porous casting example, there is an average of 2,834 acceptable castings poured before a porous one is made. *p'* is estimated as .0003528, which is .03528 percent.

$$\hat{p}' = \frac{1}{\overline{X}} = \frac{1}{2,834} = .0003528 = .03528\,\%$$

Because the distribution for the number of conforming units between nonconforming ones may not be normally distributed, the *IX* values should be plotted on NOPP. If found to have a non-normal distribution, the NOPP can help to establish control limits for the *IX* chart, as was explained in Section 7.4 on page 401. As an alternative to the NOPP method, Nelson (1993) recommends the following data transformation be applied to first "normalize" the individual measurements.

$$Y_i = X_i^{1/3.6}$$

These *Y* values are then plotted on a regular *IX & MR* chart and control limits are calculated with the standard formulas.

Whichever charting method is applied, once a process that generates attribute data is stabilized, its capability may be assessed.

9.2 Capability Measures for Attribute Data

The capability measures for processes generating attribute data are formulated to allow a direct comparison to the corresponding measures for processes characterized with variable data. The four capability measures for attribute data covered in this book are listed here:

1. Equivalent $P_{PK}^{\%}$

2. Equivalent $P_{P}^{\%}$

3. Equivalent $P_{R}^{\%}$

4. $ppm_{TOTAL\ LT}$

With attribute capability measures, the labels "*P*" and "*LT*" are appropriate because attribute charts combine *all* subgroup data together, both within and between subgroups, to estimate their process parameters. As a result, all capability estimates made from these parameters reflect expected long-term process performance.

Capability measures 1 through 3 are called "equivalent" because they match the comparable measures for variable-data processes with regard to the percentage of nonconforming parts. This concept is analogous to that for the equivalent $P^\%_{PK}$ index introduced in Section 8.15 on page 499 (recall that equivalent $P^\%_{PK}$ is pronounced "equivalent P_{PK}, based on percent"). With attribute capability measures defined in this fashion, the capability of a process generating attribute data can be directly compared to those for any other process. A different approach is offered by Marcucci and Beazley, however, it is not commonly used. Lehrman developed a third alternative, called "process safety," and Hoadley a fourth, referred to as the "quality measurement plan," but these are also seldom seen in practice.

The general procedure for estimating equivalent capability measures for attribute data is outlined below. As with all capability studies, the process must first be in a good state of control before attempting to estimate any process parameters or capability measures.

1. Estimate $\hat{p}'$, the percentage of nonconforming parts generated by the process under study.

2. Locate $\hat{p}'$ in the body of the Z table in Appendix Table III (p. 845), then find the corresponding, or equivalent, Z value for this percentage.

3. Calculate the desired "equivalent" capability measure from the equivalent Z value.

4. Compare this equivalent measure to the capability goal for this characteristic and make a decision about process capability.

Before demonstrating how these measures are estimated from the various attribute-data charts, the concepts behind these equivalent attribute capability measures are explained in the next four sections.

Definition of Equivalent $P^\%_{PK}$

For attribute capability studies, the percentage nonconforming, p', is first estimated from all subgroup data. An "equivalent" $P^\%_{PK}$ index equal to the P_{PK} index for a variable-data process having this *same* percentage of nonconforming parts is then determined with the following logic. Figure 9.7 shows a variable-data process producing .135 percent of its output above the USL. As $Z_{MIN,\ LT}$ is 3.00 for .135 percent, the P_{PK} index is 1.00 for this process.

$$P_{PK} = \frac{Z_{MIN,\ LT}}{3} = \frac{3.00}{3} = 1.00$$

P_{PK} measures how well this process meets the customer requirement in terms of percentage nonconforming parts, in this case, a total of .135 percent.

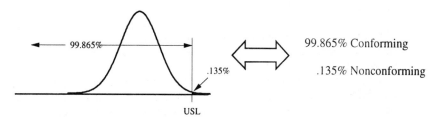

Fig. 9.7 A variable-data process generating .135 percent nonconforming product.

Suppose a second process generating attribute data is also producing a total of .135 percent nonconforming parts, as illustrated in Figure 9.8. This process is meeting the customer requirement to the same extent as the variable-data process displayed in Figure 9.7. Because the customer receives equivalent quality in both cases (99.865 percent conforming parts), the "equivalent" $P^{\%}_{PK}$ index for this attribute process should also be 1.00, identical to the P_{PK} index for the variable-data process (Bothe, 1989; Harry and Lawson, p. 5-13).

$$\text{Equivalent } P^{\%}_{PK} = 1.00$$

The capability measures of these two radically different processes may be compared via the amount of nonconforming parts produced, even if one generates attribute data and the other variable data. As both have identical abilities to satisfy the customer requirement, their measures of capability under this logic are identical as well.

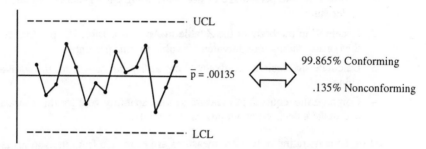

Fig. 9.8 An attribute-data process generating .135 percent nonconforming product.

However, by assuming the entire percentage nonconforming is beyond just one specification limit, the equivalent $P^{\%}_{PK}$ index generates the *lowest* value possible for this percentage. This low estimate represents a very conservative (worst-case scenario) characterization of capability for a process generating attribute data (Harry).

Definition of Equivalent $P^{\%}_{P}$

The equivalent $P^{\%}_{PK}$ index is based on a variable-data process where the entire percentage nonconforming is located outside the same specification limit, which would be true for a unilateral specification. However, for all bilateral specifications, there is usually some percentage nonconforming beyond each specification limit. A P_{PK} of 1.00 means there is a maximum of .135 percent nonconforming parts outside *one* specification limit. But there could also be up to .135 percent outside the *other* specification limit depending on where the process is centered, as is illustrated in Figure 9.9. Thus, the total percentage nonconforming could actually be as high as .27 percent.

As an alternative to equivalent $P^{\%}_{PK}$, Lam and Littig propose a measure, called equivalent $P^{\%}_{P}$, that matches an attribute-data process to a variable-data process that is centered at the middle of the tolerance. In this situation, there is an identical percentage of nonconforming parts outside both the LSL and USL. The $Z_{MIN.\ LT}$ value calculated for this process represents just *half* the total percentage nonconforming, not the entire total, as with equivalent $P^{\%}_{PK}$. Because Z values increase as the percentage decreases, the equivalent $P^{\%}_{P}$ index produces a larger result than does equivalent $P^{\%}_{PK}$ for the same attribute-data process.

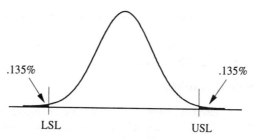

Fig. 9.9 Splitting the percentage nonconforming equally outside the two specification limits.

For example, the percentage of computer monitors failing final test is estimated as 1 percent. The equivalent $P^{\%}_{PK}$ measure is based on the equivalent Z value for 1 percent (2.32), whereas equivalent $P^{\%}_{P}$ is calculated from the Z value for .5 percent (2.58). Since the Z value for .5 percent is greater than the one for 1 percent, equivalent $P^{\%}_{P}$ will be greater than equivalent $P^{\%}_{PK}$. In general:

$$\text{Equivalent } P^{\%}_{P} > \text{Equivalent } P^{\%}_{PK}$$

A measure based on this assumption presents an optimistic view (best case) of attribute capability. That's why the equivalent $P^{\%}_{P}$ designation is appropriate, because in situations with variable data, potential capability indexes represent the highest possible capability for a process (Harry). For normally distributed outputs, this occurs when the process is centered at the middle of the tolerance.

Due to being a more conservative assessment of capability, the equivalent $P^{\%}_{PK}$ index is more widely accepted in practice (Harry and Lawson, p. 7-9). For this reason, most examples in this chapter cover only this measure. But readers should be aware of the assumptions and differences of these two measures and employ the one best suited for their own particular application.

Definition of Equivalent $P^{\%}_{R}$

For attribute capability studies, the equivalent $P^{\%}_{R}$ ratio is determined as usual: taking the inverse of the equivalent $P^{\%}_{P}$ index.

$$\text{Equivalent } P^{\%}_{R} = \frac{1}{\text{Equivalent } P^{\%}_{P}}$$

This measure of attribute-data capability is utilized least of all.

Definition of $ppm_{TOTAL, LT}$

As done for processes having variable data, the *ppm* capability measure is simply an estimate of the percentage nonconforming, times ten to the sixth power.

$$ppm_{TOTAL,LT} = p' \times 10^{6}$$

If the percentage of computer monitors failing final test is estimated as 1 percent, the estimate for $ppm_{TOTAL, LT}$ is 10,000.

$$\hat{p}\,pm_{TOTAL,LT} \;=\; \hat{p}' \times 10^6 \;=\; .01 \times 10^6 \;=\; 10,000$$

This is the easiest to understand of all the attribute-data capability measures and is applied most often in practice (Gitlow *et al*).

Goals for Attribute Capability

As a result of these equivalent measures being analogous to those for variable data, their goals are also similar. Unfortunately, to have the same degree of confidence in the results, the amount of attribute data collected must be much greater than for variable data (for more information on this topic, see Section 11.29, p. 686).

All attribute-data capability measures have $\hat{p}'$ as their foundation. Unfortunately, the exact procedure for deriving this estimate depends on whether the capability study uses a chart based on the binomial or Poisson distribution. Thus, a unique procedure is necessary for each type of chart. These are explained in the next several sections, beginning with the procedure for binomial charts.

9.3 Estimating Process Capability from *p* or *np* Charts

For *p* charts, an estimate of p' comes directly from $\bar{p}$, the centerline of the chart. The $\bar{p}$ statistic is the overall percentage of nonconforming units found in all in-control subgroups.

$$\hat{p}' \;=\; \bar{p}$$

This represents the best estimate of p' that can be made from the chart data, and therefore, is used to calculate the desired attribute capability measure.

Example 9.5 Water Pump Leaks

A *p* chart monitoring the percentage of water pumps failing a leak test is presented in Figure 9.10. It is in a good state of control, with $\bar{p}$ equal to .0017, or .17 percent. This value becomes an estimate of the process percentage nonconforming, p'.

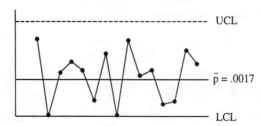

Fig. 9.10 An in-control *p* chart for water pump leaks.

To estimate $ppm_{TOTAL,\ LT}$, multiply .0017 by 10 to the sixth power.

$$\hat{p}\,pm_{LT,TOTAL} \;=\; \hat{p}' \times 10^6 \;=\; \bar{p} \times 10^6 \;=\; .0017 \times 10^6 \;=\; 1,700$$

The equivalent $P^{\%}{}_{PK}$ capability measure is estimated by finding the corresponding Z value for $\bar{p}$ in Appendix Table III. The operator $Z(x)$ means find the Z value for the quantity enclosed within the parentheses. Sometimes the symbol $\Phi^{-1}(x)$ is used in place of $Z(x)$.

$$\text{Equivalent } \hat{Z} = Z(\hat{p}') = Z(\bar{p}) = Z(.0017) = Z(1.7 \times 10^{-3}) = 2.93$$

Because this measure assumes the entire percentage nonconforming lies outside one specification limit, the equivalent Z value for .0017 becomes the estimated $Z_{MIN,\ LT}$ value.

$$\text{Equivalent } \hat{Z}_{MIN.LT} = \text{Equivalent } \hat{Z}$$

Dividing this special $Z_{MIN,\ LT}$ value of 2.93 by 3 produces an equivalent $P^{\%}{}_{PK}$ index that rates the ability of this assembly process to make a pump without leaks.

$$\text{Equivalent } \hat{P}^{\%}_{PK} = \frac{\text{Equivalent } \hat{Z}_{MIN.\ LT}}{3} = \frac{2.93}{3} = .977$$

This estimated equivalent $P^{\%}{}_{PK}$ value of .977 can be directly compared to the P_{PK} indexes of other processes, whether they are generating attribute data or variable data, based on the percentage of nonconforming parts. In practice, the following more direct version of the above formula is used for estimating equivalent $P^{\%}_{PK}$.

$$\text{Equivalent } \hat{P}^{\%}_{PK} = \frac{Z(\bar{p})}{3}$$

If the equivalent $P^{\%}{}_{P}$ index is preferred, first divide the estimate of p' by 2, then find the equivalent Z value.

$$\text{Equivalent } \hat{P}^{\%}_{P} = \frac{Z(\hat{p}' \div 2)}{3} = \frac{Z(\bar{p} \div 2)}{3} = \frac{Z(.0017 \div 2)}{3} = \frac{Z(.00085)}{3} = \frac{3.14}{3} = 1.047$$

The equivalent $P^{\%}{}_{R}$ measure is estimated by taking the inverse of the estimate for the equivalent $P^{\%}_{P}$ index.

$$\text{Equivalent } \hat{P}^{\%}_{R} = \frac{1}{\text{Equivalent } \hat{P}^{\%}_{P}} = \frac{1}{1.047} = .955$$

This attribute process could deliver the same percentage of nonconforming parts as a variable-data process having the potential to take up no more than 95.5 percent of the tolerance.

Example 9.6 Cracked Pistons

For the p chart monitoring the percentage of cracked pistons discussed during the review of attribute charts (Example 9.2, p. 525), $\bar{p}$ was calculated as follows:

$$\bar{p} = \frac{43}{717} = .05997 = 5.997 \times 10^{-2}$$

This $\overline{p}$ value is an estimate of the percentage of cracked pistons, p', from which an estimate can be made for $ppm_{TOTAL, LT}$.

$$\hat{p}\,pm_{TOTAL,LT} \ = \ \hat{p}' \times 10^6 \ = \ .05997 \times 10^6 \ = \ 59,970$$

Equivalent $P_{PK}^{\%}$ may also be determined from this estimate of p'.

$$\text{Equivalent } \hat{P}_{PK}^{\%} \ = \ \frac{Z(\hat{p}')}{3} \ = \ \frac{Z(5.997 \times 10^{-2})}{3} \ = \ \frac{1.56}{3} \ = \ .52$$

Equivalent $P_P^{\%}$ and equivalent $P_R^{\%}$ are based on one-half of $\hat{p}'$.

$$\text{Equivalent } \hat{P}_P^{\%} \ = \ \frac{Z(\hat{p}' \div 2)}{3} \ = \ \frac{Z(5.997 \times 10^{-2} \div 2)}{3} \ = \ \frac{Z(2.9985 \times 10^{-2})}{3} \ = \ \frac{1.88}{3} \ = \ .627$$

$$\text{Equivalent } \hat{P}_R^{\%} \ = \ \frac{1}{\text{Equivalent } \hat{P}_P^{\%}} \ = \ \frac{1}{.627} \ = \ 1.595$$

As all of these measures indicate, this process is far from being capable and 100 percent inspection should be started immediately.

Estimating p' from np Charts

For np charts, p' is estimated from $n\overline{p}$, the estimate of the average number of nonconforming units in a subgroup of size n. To estimate p', divide $n\overline{p}$ by n.

$$\hat{p}' \ = \ \overline{p} \ = \ \frac{n\overline{p}}{n}$$

Estimates of attribute-data capability measures may now be made with the same procedure for p charts, as is demonstrated in this next example.

Example 9.7 Porous Transmission Housings

Porosity in aluminum transmission housings produced at a major defense contractor has long been a chronic quality problem. As an initial attempt to better understand this process, the np chart displayed in Figure 9.11 was started to monitor the number of porous housings in a subgroup of 3000. After eliminating several assignable causes of variation, the chart now provides evidence this process is in a good state of control.

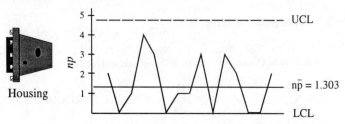

Fig. 9.11 The np chart for porous transmission housings.

The $n\overline{p}$ of 1.303 means an average of 1.303 porous housings are detected in each subgroup of 3000. The capability goals are for $ppm_{TOTAL,\ LT}$ to be less than 90, equivalent $P^{\%}_{PK}$ to be at least 1.25, and equivalent $P^{\%}_{P}$ to be at least 1.32, which implies equivalent $P^{\%}_{R}$ should be less than .76. To determine an estimate of p', which is the process percentage nonconforming, divide $n\overline{p}$ by n.

$$\hat{p}' = \overline{p} = \frac{n\overline{p}}{n} = \frac{1.303}{3000} = .0004343 = 4.343 \times 10^{-4}$$

Once p' is estimated, the identical concept previously described for the p chart applies for estimating the attribute capability measures for this process.

$$\hat{p}pm_{LT,TOTAL} = \hat{p}' \times 10^6 = .0004343 \times 10^6 = 434.3$$

This estimate meets the minimum requirement for process capability, but not the mandated goal of 90 *ppm*.

$$\text{Equivalent } \hat{P}^{\%}_{PK} = \frac{Z(\hat{p}')}{3} = \frac{Z(4.343 \times 10^{-4})}{3} = \frac{3.33}{3} = 1.11$$

Again the minimum requirement for capability has been demonstrated with an equivalent $P^{\%}_{PK}$ greater than 1.00, but the goal of 1.25 is not met.

$$\text{Equivalent } \hat{P}^{\%}_{P} = \frac{Z(\hat{p}' \div 2)}{3} = \frac{Z(2.1715 \times 10^{-4})}{3} = \frac{3.52}{3} = 1.17$$

$$\text{Equivalent } \hat{P}^{\%}_{R} = \frac{1}{\text{Equivalent } \hat{P}^{\%}_{P}} = \frac{1}{1.17} = .85$$

When these last two measures are compared to their respective capability goals, a decision is reached that, although the minimum condition for capability has been achieved, the capability goals for this process are not met. Part containment and 100 percent inspection of the castings must continue until process improvements to reduce the incidence of porosity are successfully implemented.

Example 9.8 Surgical Sponges

An *np* chart is monitoring a process producing surgical sponges. Periodically, subgroups of 750 sponges are visually inspected for proper shape. After several weeks of charting and improving the process, the chart indicates a good state of control, with $n\overline{p}$ equal to 1.20. Steve, the quality engineer assigned to this department, wishes to estimate process capability with both $ppm_{TOTAL,\ LT}$ and equivalent $P^{\%}_{PK}$. The respective capability goals are 1,350 *ppm* and 1.00. First, Steve needs an estimate of p', which can be derived from $n\overline{p}$.

$$\hat{p}' = \overline{p} = \frac{n\overline{p}}{n} = \frac{1.20}{750} = .0016 = 1.6 \times 10^{-3}$$

This estimate allows $ppm_{TOTAL,\ LT}$ to be determined.

$$\hat{p}\,pm_{TOTAL,LT} \;=\; \hat{p}' \times 10^6 \;=\; .0016 \times 10^6 \;=\; 1,600$$

In addition, equivalent $P^{\%}_{PK}$ can also be estimated from $\hat{p}'$.

$$\text{Equivalent } \hat{P}^{\%}_{PK} \;=\; \frac{Z(\hat{p}')}{3} \;=\; \frac{Z(1.6 \times 10^{-3})}{3} \;=\; \frac{2.95}{3} \;=\; .983$$

Both of these measures indicate the specified capability goals have not been achieved for this process. Steve must work on reducing common-cause variation to decrease the number of incorrectly shaped sponges formed by this process.

9.4 Estimating Process Capability from *u* or *c* Charts

For attribute-data charts based on the Poisson distribution, the percentage of *conforming parts, i.e.*, those with "0" nonconformities, is estimated first. This percentage is then subtracted from 100 percent to derive the percentage of nonconforming parts (Bothe, 1989). The formula for the Poisson distribution given below will estimate the percentage of units having a given number of nonconformities (Snedecor and Cochran).

$$P(\text{Number of nonconformities on a single unit } = x) = \frac{(u')^x\, e^{-u'}}{x!}$$

In this formula, x represents the number of nonconformities, u' is the average number of nonconformities per unit, e is a constant equal to 2.7182818 and $x!$ (pronounced x factorial) is defined as $x(x-1)(x-2)(x-3) \cdots (3)(2)(1)$. Typically u' is not known, so it must be estimated from either $\bar{u}$, or $\bar{c}$ divided by n if working with a c chart.

$$\hat{u}' = \bar{u} \text{ or } \hat{u}' = \frac{\bar{c}}{n}$$

Let the average nonconformities per unit, u', be .58. The amount of units having exactly three such nonconformities ($x = 3$) is calculated as .0182, or 1.82 percent.

$$P(x=3) = \frac{(.58)^3\, e^{-.58}}{3!} = \frac{(.1951)(.5599)}{3 \times 2 \times 1} = .0182 = 1.82\%$$

In a like manner, the percentage of units having a certain number of nonconformities can be calculated for x equal 1 through 6.

$P(x=0) = .5599$	$P(x=4) = .0026$
$P(x=1) = .3247$	$P(x=5) = .0003$
$P(x=2) = .0942$	$P(x \geq 6) = .0001$
$P(x=3) = .0182$	

A histogram of this Poisson distribution is displayed in Figure 9.12.

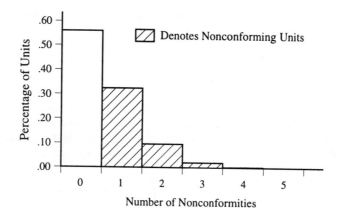

Fig. 9.12 The Poisson distribution for $u' = .58$.

Any unit having one or more nonconformities ($x \geq 1$) is considered a nonconforming unit because it does not meet customer requirements in at least one regard. From Figure 9.12, one could determine the percentage of units with 1 nonconformity, then of units with 2 nonconformities, then of units with 3, and so on. Once these percentages are determined, they must be summed to derive p', the total percentage of nonconforming units.

$$p' = \sum_{i=1}^{+\infty} P(x = i)$$

In this example, p' is calculated as .4401.

$$p' = \sum_{i=1}^{6} P(x = i) = .3247 + .0942 + .0182 + .0026 + .0003 + .0001 = .4401$$

This obviously involves a tremendous amount of work to perform all the associated calculations. Fortunately, there is a much quicker way to find the answer. Merely calculate the percentage of units with *no* nonconformities (the conforming units), then subtract this percentage of good units from 1 to determine the percentage of nonconforming units. To make things even easier, the Poisson formula for 0 nonconformities becomes quite simple when you recall that 0! is defined as 1.

$$P(x = 0) = \frac{(u')^0 e^{-u'}}{0!} = \frac{1 \, e^{-u'}}{1} = e^{-u'}$$

Due to $P(x=0)$ being the percentage of *conforming* parts, the percentage of *nonconforming* parts, p', is estimated as 1 minus $P(x = 0)$.

$$\hat{p}' = 1 - P(x = 0) = 1 - e^{-u'}$$

In the above example, $P(x = 0)$ is .5599. Subtracting this result from 1 provides the percentage of nonconforming parts. This answer of .4401 is identical to that obtained by summing all percentages of parts with at least one nonconformity.

$$\hat{p}' = 1 - P(x = 0) = 1 - .5599 = .4401$$

Once $\hat{p}'$ is calculated from u', $ppm_{TOTAL,\ LT}$, equivalent $P_P^{\%}$, $P_R^{\%}$, and $P_{PK}^{\%}$ are estimated in the same manner as was done for charts based on the binomial distribution.

Example 9.9 Liquid Crystal Displays

A u chart is monitoring the average number of scratches found per unit on the upper surface of a liquid crystal display manufactured for a notebook computer. When the process is finally brought into control, a total of 56 scratches were found on the 933 displays checked during the study. The first step in estimating process capability is to compute $\hat{u}'$ from $\overline{u}$.

$$\hat{u}' = \overline{u} = \frac{56 \text{ Scratches}}{933 \text{ Displays}} = .060 \text{ Scratches per display}$$

Before any predictions of process capability can be made, p' must be estimated from $\hat{u}'$.

$$\hat{p}' = 1 - e^{-\hat{u}'} = 1 - e^{-.060} = 1 - .94176 = .05824 = 5.824 \times 10^{-2}$$

Multiplying $\hat{p}'$ by one million gives an estimate of $ppm_{TOTAL,\ LT}$.

$$\hat{p}\,pm_{TOTAL,LT} = \hat{p}' \times 10^6 = .05824 \times 10^6 = 58,240$$

This measure of capability reveals the process is not capable ($ppm_{TOTAL,\ LT} > 1,350$) and in desperate need of improvements to reduce the high rate of scratches.

Equivalent $P_{PK}^{\%}$ for this process is calculated as follows:

$$\text{Equivalent } \hat{P}_{PK}^{\%} = \frac{Z(\hat{p}')}{3} = \frac{Z(5.824 \times 10^{-2})}{3} = \frac{1.57}{3} = .52$$

With an equivalent $P_{PK}^{\%}$ index considerably less than 1.0, this process is definitely not capable of producing scratch-free displays.

Example 9.10 Fiber-Optic Cable

A c chart for the number of nicks per 1000 meters of fiber-optic cable is in a good state of control with $\overline{c}$ equal to 3.2. The capability of this attribute-data process to produce a single meter of nick-free cable is measured by first finding an estimate of u'.

$$\hat{u}' = \overline{u} = \frac{\overline{c}}{n} = \frac{3.2 \text{ Nicks}}{1000 \text{ Meters}} = .0032 \text{ Nicks / Meter}$$

The percentage of cable with at least one nick per meter is derived from this estimate of u' as follows:

$$\hat{p}' = 1 - P(x = 0) = 1 - e^{-\hat{u}'} = 1 - e^{-.0032} = 1 - .996805 = .003195 = 3.195 \times 10^{-3}$$

From this estimate of p', any of the various attribute-data capability measures may be calculated. For example:

$$\hat{p}\,pm_{TOTAL,LT} = \hat{p}' \times 10^6 = .003195 \times 10^6 = 3,195$$

$$\text{Equivalent } \hat{P}^{\%}_{PK} = \frac{Z(\hat{p}')}{3} = \frac{Z(3.195 \times 10^{-3})}{3} = \frac{2.73}{3} = .91$$

This process does not meet the minimum requirement for capability due to $ppm_{TOTAL,\,LT}$ being more than 1,350, which in turn causes the estimated equivalent $P^{\%}_{PK}$ to be less than 1.0.

Example 9.11 Copying Machine Repairs

A u chart is tracking the average number of repairs made to copying machines (see Figure 9.13) at final test. When a good state of control is demonstrated on the chart, $\overline{u}$ is calculated to be 1.041, meaning this assembly process is averaging slightly over one nonconformity per copying machine.

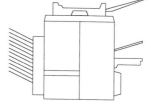

$$\hat{u}' = \overline{u} = 1.041$$

Fig. 9.13

Even though the average is more than one, this does not imply *every* copier has a nonconformity. Some copiers will have 3 nonconformities, some will have 2, some will have 1, and some will have 0. It is the percentage of copiers with none that is of interest for measuring process capability. This percentage is calculated from the Poisson formula as .6469.

$$\hat{p}' = 1 - P(x=0) = 1 - e^{-\hat{u}'} = 1 - e^{-1.041} = 1 - .3531 = .6469$$

From this result, the parts per million is estimated as 646,900.

$$\hat{p}\,pm_{TOTAL,LT} = \hat{p}' \times 10^6 = .6469 \times 10^6 = 646,900$$

Well over half of the copiers, almost 65 percent, are projected to be nonconforming as they have at least one nonconformity. When working with a variable-data process, more than 50 percent nonconforming product is associated with a *negative P_{PK}* value. If this attribute-data capability concept is to provide comparable results, the equivalent $P^{\%}_{PK}$ for this process should also turn out to be negative.

The way Appendix Table III is constructed, a Z value of 0.0 means there is 50 percent from that point in the normal distribution to plus infinity. As the Z values increase, this percentage in the upper tail decreases. In this example, there are 64.69 percent nonconforming parts, which corresponds to a Z value less than zero, as seen in Figure 9.14. However, there are no negative Z values listed in Table III. Due to the symmetry of the normal distribution, simply look up 35.31 percent (100 percent minus 64.69 percent) in the table, locate the corresponding Z value, then make it negative. As with variable data, the resulting negative equivalent $P^{\%}_{PK}$ indicates a serious quality problem with this process.

$$\text{Equivalent } \hat{P}^{\%}_{PK} = \frac{Z(\hat{p}')}{3} = \frac{Z(.6469)}{3} = \frac{-Z(.3531)}{3} = \frac{-.38}{3} = -.13$$

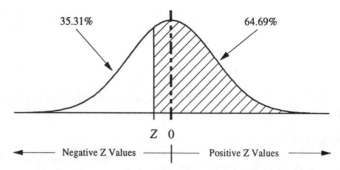

35.31%   64.69%

Z 0

◄——— Negative Z Values ———┼——— Positive Z Values ———►

Fig. 9.14 Negative Z values are needed when $\hat{p}'$ exceeds 50 percent.

Had the percentage nonconforming been exactly 50 percent, the associated Z value is 0.0, making equivalent $P_{PK}^{\%}$ equal to 0.0 (0/3), in perfect agreement with the variable-data case.

Example 9.12 Ski Poles

A u chart for the number of cosmetic paint blemishes found on ski poles is in a good state of control, with $\bar{u}$ equal to .084. A ski pole is considered acceptable if it has *no more than two* paint blemishes. The equivalent $P_{PK}^{\%}$ and $ppm_{TOTAL,\ LT}$ are found for this process by first using the Poisson formula to determine what percentage of the poles are expected to have either 0, 1, or 2 paint blemishes (all of these would be conforming units).

$$P\ (x = 0)\ =\ e^{-.084}\ =\ .919431 \qquad P\ (x = 1)\ =\ \frac{(.084)^1\ e^{-.084}}{1!}\ =\ .077232$$

$$P\ (x = 2)\ =\ \frac{(.084)^2\ e^{-.084}}{2!}\ =\ .003244$$

The sum of these three percentages is the probability a pole is acceptable, that is, has 2 or less paint blemishes. To derive the percentage of unacceptable poles, this sum must be subtracted from 1.

$$\hat{p}'\ =\ 1 - (\ .919431 + .077232 + .003244\ )\ =\ 1 - .999907\ =\ .000093\ =\ 9.3 \times 10^{-5}$$

Both equivalent $P_{PK}^{\%}$ and $ppm_{TOTAL,\ LT}$ are calculated from this estimate of p'.

$$\text{Equivalent } \hat{P}_{PK}^{\%}\ =\ \frac{Z\ (\hat{p}')}{3}\ =\ \frac{Z\ (9.3 \times 10^{-5})}{3}\ =\ \frac{3.74}{3}\ =\ 1.25$$

$$\hat{p}\,pm_{TOTAL,LT}\ =\ \hat{p}' \times 10^6\ =\ 9.3 \times 10^{-5} \times 10^6\ =\ 93$$

These measures indicate the process meets the minimum requirement for performance capability. Since the goal for equivalent $P_{PK}^{\%}$ is 1.20 and the goal for $ppm_{TOTAL,\ LT}$ is 160, this process meets its capability goals as well.

Example 9.13 Aerospace Castings

A complex aerospace casting is checked with a flexible fiberscope hooked up to a color video monitor. An inspector examines the core passageways for indications of core thinning, core breakage, burned-in core sand, porosity, or poor chaplets. After stability is demonstrated, the average nonconformities found per casting is .00667. If the requirement for this casting is to have no more than one nonconformity, does this process meet a customer-imposed goal for equivalent $P^%{}_{PK}$ of 1.33? To find out, the probability of a casting with zero nonconformities, as well as the probability of a casting having one nonconformity, must be estimated.

$$P\,(x=0) \; = \; e^{-.00667} \; = \; .9933522 \qquad P\,(x=1) \; = \; \frac{(.00667)^1\,e^{-.00667}}{1!} \; = \; .0066256$$

The sum of these two is the probability a casting is acceptable, that is, has no more than one nonconformity. To derive the percentage of nonconforming castings, this sum must be subtracted from 1.

$$\hat{p}' \; = \; 1 - (\,.9933522 + .0066256\,) \; = \; 1 - .9999778 \; = \; .0000222 \; = \; 2.22 \times 10^{-5}$$

The required estimate of process capability can now be computed.

$$\text{Equivalent } \hat{P}^%{}_{PK} \; = \; \frac{Z\,(\hat{p}')}{3} \; = \; \frac{Z\,(2.22 \times 10^{-5})}{3} \; = \; \frac{4.08}{3} \; = \; 1.36$$

Comparing this result to the goal of 1.33 implies this process just meets the customer's specified capability requirement.

Deriving u' from p'

In Example 9.2 (p. 525), a p chart monitored the number of cracked pistons found in a machining department. As 43 cracked pistons were found on the 717 inspected, the average percentage of nonconforming pistons was estimated as being just under 6 percent.

$$\hat{p}' \; = \; \frac{43}{717} \; = \; .05997$$

When a crack is discovered, the piston is immediately discarded as being nonconforming. There is little reason to continue checking the piston to find a second or third crack (nonconformity). However, there may be times when an estimate of the average number of cracks per piston is needed. This estimate can be derived from the relationship between the percentage of nonconforming parts, p', and the average nonconformities per unit, u'.

$$p' \; = \; 1 - e^{-u'}$$

To solve this equation for u', isolate u' on one side of the equal sign, then take the natural log of both sides.

$$e^{-u'} \; = \; 1 - p'$$

$$\ln(e^{-u'}) \; = \; \ln(1 - p')$$

$$-u' = \ln(1 - p')$$

$$u' = -\ln(1 - p')$$

For this example, the .05997 percent of cracked pistons turns into an average of .06184 cracks per piston.

$$u' = -\ln(1 - p') = -\ln(1 - .05997) = .06184$$

9.5 Nonconformities per Unit (*dpu*) vs. Nonconformities per Opportunity (*dpo*)

Assume 3 defects (nonconformities) are found in a subgroup of 5 units. The defects per unit, abbreviated *dpu*, are determined as follows:

$$u = dpu = \frac{d}{n} = \frac{3}{5} = .6$$

Here, *d* is the number of defects (nonconformities) and *n* is the number of units inspected. However, if each unit has 200 possible things, or opportunities, that could go wrong, then there is a total of 1000 (200 × 5) opportunities for a defect to occur in this subgroup of 5 units. Therefore, the defects per opportunity, called *dpo*, is calculated as:

$$dpo = \frac{d}{o} = \frac{3}{1000} = .003$$

In this equation, *d* is again the number of defects (nonconformities), while *o* is the number of opportunities for a defect (nonconformity). This quantity can also be expressed as defects per million opportunities by multiplying *dpo* by one million, which is 10^6. Called *dpmo*, this value is very similar to the *ppm* measure for variable data.

$$dpmo = \frac{d}{o} \times 10^6 = \frac{3}{1000} \times 10^6 = 3000$$

dpu looks at capability on the *unit* level while *dpo* measures capability at the *opportunity* level. An important reason for making this distinction is highlighted in the next section.

Measuring Capability at the Unit Level

As seen earlier in this chapter, the capability of a process generating nonconformities is measured by first estimating p' from $\bar{u}$. Because $\bar{u}$ is similar to *dpu*, *dpu* can also be used to estimate p', which in turn is used to estimate $ppm_{TOTAL\ LT}$, or equivalent P^*_{PK}, for this process. For the example above, where 3 defects were found on 5 units, the equivalent capability measures for the **unit** level are calculated by first estimating p'.

$$\hat{p}' = 1 - P(x = 0) = 1 - e^{-dpu} = 1 - e^{-.6} = 1 - .5488 = .4512 = 4.512 \times 10^{-1}$$

With this estimate of p', $ppm_{TOTAL\ LT}$ and equivalent P^*_{PK} can be determined.

$$\hat{p}pm_{TOTAL,LT} = \hat{p}' \times 10^6 = .4512 \times 10^6 = 451,200$$

$$\text{Equivalent } \hat{P}^{\%}_{PK} = \frac{Z(\hat{p}')}{3} = \frac{Z(4.512 \times 10^{-1})}{3} = \frac{.12}{3} = .04$$

Measuring Capability at the Opportunity Level

On the other hand, process capability at the **opportunity** level is determined by first estimating the probability a *single* opportunity is nonconforming. This probability is called p'_O, with the subscript "O" representing opportunity. Because each opportunity can be either conforming or nonconforming, *dpo* is an estimate of this percentage of nonconforming opportunities and should follow a binomial probability distribution, assuming *dpo* is constant for all opportunities and the defects are independent.

$$\hat{p}'_O = dpo = \frac{d}{o} = \frac{3}{1000} = .003 = 3.0 \times 10^{-3}$$

With this percentage, capability at the opportunity level is estimated by looking up the equivalent Z value, then dividing it by 3. Remember that the subscript "O" denotes capability measures for the opportunity level.

$$\text{Equivalent } \hat{P}^{\%}_{PK,O} = \frac{Z(\hat{p}'_O)}{3} = \frac{Z(3.0 \times 10^{-3})}{3} = \frac{2.75}{3} = .92$$

This index of .916 for the opportunity level is much higher than that for the unit level of only .041. For both levels, there is the same number of defects (3), however, the "subgroup" size is considerably different, 5 versus 1000. Measuring capability at the opportunity level gives credit for the number of items done correctly. This means capability results for the opportunity level will always be better than for the unit level because there are typically many opportunities per unit (there must always be at least one). Note that this is not just a way to make process capability "look better," as the goal for capability at the opportunity level will be correspondingly much higher than the goal for the unit level. So why even bother with measuring capability at the opportunity level? One of the main reasons is for allowing a fair comparison between two dissimilar products, as is illustrated in the next example.

Example 9.14 Soldering Circuit Boards

After one week of production, a process soldering two models of printed circuit boards is finally brought into control via a short-run *u* chart (Bothe, 1991). The two models, A and B, are displayed in Figure 9.15.

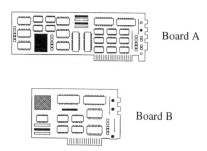

Fig. 9.15 Circuit boards A and B have different levels of complexity.

Once stabilized, 15 solder-related defects are found on a total of 489 boards (units) for model A. Each board has 2617 solder joints (opportunities) that could trigger a problem. Supplied with this information, *dpu*, *dpo*, and *dpmo* can be estimated for this first model.

$$dpu = \frac{d}{n} = \frac{15}{489} = .030675 \qquad\qquad dpo = \frac{d}{o} = \frac{15}{489\,(2617)} = .0000117$$

$$dpmo = dpo \times 10^6 = .0000117 \times 10^6 = 11.7$$

From these results, an estimate of equivalent $P^{\%}_{PK}$ for both the unit (board) and opportunity (solder joint) levels can be made. Beginning with the unit level, equivalent $P^{\%}_{PK}$ is found by first computing the percentage of nonconforming model A boards.

$$\hat{p}' = 1 - P(X = 0) = 1 - e^{-dpu} = 1 - e^{-.030675} = .03021$$

This percentage is used in estimating equivalent $P^{\%}_{PK}$ at the unit level.

$$\text{Equivalent } \hat{P}^{\%}_{PK} = \frac{Z\,(\hat{p}')}{3} = \frac{Z\,(.03021)}{3} = \frac{1.88}{3} = .627$$

In order to determine equivalent $P^{\%}_{PK}$ at the opportunity level, the percentage of nonconforming opportunities must be found from *dpo*.

$$\hat{p}'_o = dpo = .0000117 = 1.17 \times 10^{-5}$$

Based on this percentage, equivalent $P^{\%}_{PK}$ at the opportunity level for model A is 1.410.

$$\text{Equivalent } \hat{P}^{\%}_{PK,O} = \frac{Z\,(\hat{p}'_o)}{3} = \frac{Z\,(1.17 \times 10^{-5})}{3} = \frac{4.23}{3} = 1.410$$

The second model of circuit board (B) produced on this same soldering line has 3 defects on 243 boards, with each board having 997 solder joints. Is the capability of this line to solder model B better or worse than that for model A? To answer this question, begin by estimating *dpu* and *p'* for model B.

$$dpu = \frac{d}{n} = \frac{3}{243} = .0123456$$

$$\hat{p}' = 1 - P(X = 0) = 1 - e^{-.0123456} = 1.227 \times 10^{-2}$$

From these results, equivalent $P^{\%}_{PK}$ at the unit (board) level is assessed as:

$$\text{Equivalent } \hat{P}^{\%}_{PK} = \frac{Z\,(\hat{p}')}{3} = \frac{Z\,(1.227 \times 10^{-2})}{3} = \frac{2.25}{3} = .750$$

Comparing this result with the equivalent $P^{\%}_{PK}$ (unit level) of .627 for model A would lead one to believe the soldering line is more capable for soldering model B. But model A is much more complex, having more than twice the number of solder joints of B (2617 versus 997). This difference is not factored in to any measure of capability at the unit level. However, it plays a prominent role when evaluating capability at the opportunity level.

Equivalent $P^{\%}_{PK}$ at the opportunity level depends on the percentage of nonconforming opportunities, which in turn, depends on the *dpo* for this model. To calculate *dpo* for model B, the total number of opportunities must be determined. With 243 boards, each having 997 opportunities, this total is:

$$\text{Number of opportunities } = 243 \text{ Boards} \times 997 \text{ Opportunities per board } = 242,271$$

From this, p' can be estimated as the *dpo*.

$$\hat{p}'_O = dpo = \frac{d}{o} = \frac{3}{242,271} = .0000123 = 1.23 \times 10^{-5}$$

Then finally, equivalent $P^{\%}_{PK}$ for the opportunity level of model B can be ascertained.

$$\text{Equivalent } \hat{P}^{\%}_{PK,O} = \frac{Z(\hat{p}'_O)}{3} = \frac{Z(1.23 \times 10^{-5})}{3} = \frac{4.22}{3} = 1.407$$

The capability of this soldering line at the opportunity level for model B is practically identical to that of model A. The soldering line is performing at the same quality level for each solder joint. The reason model B's capability looks better at the board level is a result of it having fewer opportunities (solder joints) for a nonconformity to occur. This lower number of opportunities means fewer units will experience a nonconformity.

Most companies specify goals for only *unit*-level capability, and frequently set an identical goal for all products, regardless of complexity. Many OEMs impose the same capability goal on all their suppliers. As just seen, it is impossible for circuit boards with differing numbers of opportunities to have identical capability at the unit level. To be meaningful, capability goals under these circumstances should be established and tracked at the *opportunity* level. It would be reasonable to expect the soldering line to perform the same when soldering a single connection, no matter which board is being run.

When capability goals at the opportunity level are identical for all products, the expected capability at the unit level will be different for products having unequal numbers of opportunities. The next section explains how to establish a unit-level capability goal tailored for each product, based on the specified opportunity-level goal.

Establishing Capability Goals at the Unit Level

Although most companies continue to do charting and capability reporting at the unit level, when different products assembled on one process have varying numbers of opportunities, it makes much more sense to set capability goals at the *opportunity* level. Once a goal is established at the opportunity level in terms of *dpo*, it can be converted into a capability goal for equivalent $P^{\%}_{PK}$ at the unit level. But first, the goal for the probability a single opportunity is conforming must be derived.

$$P(\text{One opportunity is conforming})_{GOAL} = 1 - dpo_{GOAL}$$

Let j be the number of opportunities per unit. For an *entire* unit to be conforming, *i.e.*, to have no nonconformities, *all j* opportunities on the unit must be conforming. Assuming the probability of any particular opportunity to be acceptable is constant and independent of all the others, this relationship is expressed as follows:

$$P(\text{Unit is conf.})_{GOAL} = P(\text{All } j \text{ opportunity on one unit are good})_{GOAL}$$

$$= P(\text{1st opp. is good})_{GOAL} \times P(\text{2nd opp. is good})_{GOAL} \times \cdots \times P(j \text{ th opp. is good})_{GOAL}$$

$$= (1 - dpo_{GOAL}) \times (1 - dpo_{GOAL}) \times \cdots \times (1 - dpo_{GOAL})$$

$$= (1 - dpo_{GOAL})^j$$

Once determined from the above equation, the goal for the percentage of conforming units is subtracted from 1 to derive the goal for p', the percentage of nonconforming units.

$$p'_{GOAL} = P(\text{Unit is nonconforming})_{GOAL} = 1 - P(\text{Unit is conforming})_{GOAL} = 1 - (1 - dpo_{GOAL})^j$$

The goal for equivalent $P_{PK}^{\%}$ at the unit level is derived from this goal for p'.

$$\text{Equivalent } P_{PK,GOAL}^{\%} \text{ for unit level} = \frac{Z(p'_{GOAL})}{3} = \frac{Z[1 - (1 - dpo_{GOAL})^j]}{3}$$

Obviously, products having an unlike number of defect opportunities (j), manufactured on the same process with equal goals for *dpo*, will have different unit-level capability goals. For j equal to 1, which means the unit has only one opportunity that could make it nonconforming, p'_{GOAL} equals dpo_{GOAL}.

$$\text{For } j = 1, \quad p'_{GOAL} = 1 - (1 - dpo_{GOAL})^1 = 1 - (1 - dpo_{GOAL}) = dpo_{GOAL}$$

This situation results in the goal for equivalent $P_{PK}^{\%}$ at the unit level equaling the goal for equivalent $P_{PK,O}^{\%}$ at the opportunity level.

$$\text{Equivalent } P_{PK,O,GOAL}^{\%} = \frac{Z(dpo_{GOAL})}{3} = \frac{Z(p'_{GOAL})}{3} = \text{Equivalent } P_{PK,GOAL}^{\%}$$

Anytime j is greater than 1, the goal for equivalent $P_{PK}^{\%}$ will be less than the goal for equivalent $P_{PK,O}^{\%}$.

Example 9.15 Soldering Circuit Boards

In the last example (Example 9.14) involving circuit board models A and B, suppose the capability goal for the opportunity level is set at a *dpo* level of .00001 for both boards. The unit-level capability goals for each board can be found with the formula developed in the last section. For model A, there are 2,617 opportunities, making the goal for equivalent $P_{PK}^{\%}$ (unit level) equal to .650, as is shown on the top of the facing page.

$$\text{Equivalent } P^{\%}_{PK,GOAL} \text{ for model A } = \frac{Z[1-(1-.00001)^{2617}]}{3}$$

$$= \frac{Z(1-.97417)}{3}$$

$$= \frac{Z(.02583)}{3} = \frac{1.95}{3} = .650$$

For model B, the number of opportunities is only 997, making its unit-level goal equal to:

$$\text{Equivalent } P^{\%}_{PK,GOAL} \text{ for model B } = \frac{Z[1-(1-.00001)^{997}]}{3}$$

$$= \frac{Z(1-.99008)}{3}$$

$$= \frac{Z(.00992)}{3} = \frac{2.33}{3} = .777$$

With the same goal of .00001 for *dpo*, the equivalent $P^{\%}_{PK}$ goal for model B is .777. This unit-level goal is quite a bit higher than the one for model A, due to the considerably lower number of opportunities on model B.

Setting Unit-Level Goals When *dpo* is Not Identical for All Opportunities

A product will usually have several different kinds of detectable nonconformities, which will typically have unequal *dpo* rates. For example, product X has 20 opportunities for nonconformity type A to occur and only 10 for type B. This makes a total of 30 opportunities for a nonconformity to occur, but the *dpo* for type A is .005, while just .003 for type B. How is equivalent $P^{\%}_{PK}$ (unit level) computed for this particular mixture of dissimilar opportunities?

To answer this question, the probability nonconformity A does *not* occur on a unit, g'_A, must be determined. In this formula, j_A represents the number of opportunities for type A.

$$g'_A = (1-dpo_A)^{j_A} = (1-.005)^{20} = .9046$$

Likewise, the probability nonconformity type B does not appear on a unit, g'_B, is:

$$g'_B = (1-dpo_B)^{j_B} = (1-.003)^{10} = .9704$$

In order to be acceptable, a unit must have no nonconformities of either kind present. The chance of this occurring, called g', is the product of g'_A and g'_B.

$$g' = g'_A \times g'_B = .9046(.9704) = .8778$$

The probability a unit is *unacceptable*, p', is 1 minus the probability it is acceptable.

$$p' = 1 - g' = 1 - .8778 = .1222$$

As before, p' is now used to compute equivalent $P_{PK}^{\%}$ for the unit level.

$$\text{Equivalent } P_{PK}^{\%} = \frac{Z(p')}{3} = \frac{Z(1.222 \times 10^{-1})}{3} = \frac{1.16}{3} = .387$$

In general, when m types of nonconformities have different goals for their dpo rates, the goal for equivalent $P_{PK}^{\%}$ at the unit level is found with this formula:

$$\text{Equivalent } P_{PK,GOAL}^{\%} = \frac{Z(p'_{GOAL})}{3}, \text{ where}$$

$$p'_{GOAL} = 1 - \left[(1 - dpo_{1,GOAL})^{j_1} (1 - dpo_{2,GOAL})^{j_2} (1 - dpo_{3,GOAL})^{j_3} \cdots (1 - dpo_{m,GOAL})^{j_m} \right]$$

$dpo_{1,GOAL}$ is the dpo goal for defect type 1 and j_1 is the number of opportunities for that type, while $dpo_{2,GOAL}$ represents the dpo goal for defect type 2 with j_2 being the number of opportunities for that type, and so on, until $dpo_{m,GOAL}$ is the dpo goal for defect category m with j_m opportunities.

Comparison of Equivalent $P_{PK}^{\%}$ at Unit and Opportunity Levels

Equivalent $P_{PK}^{\%}$ for the unit level is related to equivalent $P_{PK}^{\%}$ at the opportunity level and the number of opportunities per unit. The graph in Figure 9.16 reveals how for the same equivalent $P_{PK}^{\%}$ at the opportunity level, the equivalent $P_{PK}^{\%}$ index at the unit level decreases as the number of opportunities per unit increases.

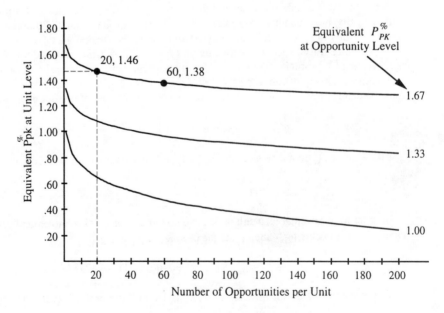

Fig. 9.16 Unit level capability versus number of opportunities.

For example, if equivalent $P^*_{PK, o}$ at the opportunity level is 1.67, and there is only a single opportunity, equivalent P^*_{PK} at the unit level is also 1.67. If there are 20 opportunities for a nonconformity, then equivalent P^*_{PK} at the unit level drops to 1.46. When the number of opportunities increases to 60, equivalent P^*_{PK} decreases to 1.38.

Comparison of *ppm* at Unit and Opportunity Levels

Just like equivalent P^*_{PK}, *ppm* at the unit level is related to *dpmo* at the opportunity level and the number of opportunities per unit. Table 9.4 reveals how, for identical capability at the opportunity level (far left column), *ppm* at the unit level increases as the number of opportunities per unit increases from 1 to 300.

This table demonstrates why it is so very important for complex products to have extremely high capability at the opportunity level. For instance, assume each opportunity has an equivalent $P^*_{PK, o}$ of 1.00. If there is only one opportunity per unit, then there will be 1,350 *ppm* at the unit level. However, if the number of opportunities per unit increases to 50, *ppm* increases to 65,306. When the number of opportunities grows to 300, *ppm* balloons to 333,166, which means over 33 percent of the process output will be nonconforming.

Table 9.4 *ppm* at unit level versus number of opportunities per unit.

Number of Opportunities per Unit

Capability at Oppor. Level	1	50	100	150	200	300
$P_{PK.O} = 1.67$ *dpmo* = .3	.3	15	30	45	60	90
$P_{PK.O} = 1.33$ *dpmo* = 32	32	1,589	3,175	4,759	6,340	9,495
$P_{PK.O} = 1.00$ *dpmo* = 1350	1,350	65,306	126,346	183,401	236,729	333,166

If quality at the opportunity level can be improved to an equivalent $P^*_{PK, o}$ of 1.67, *ppm* is only 90, which is less than 1 nonconforming unit in 10,000, even with 300 opportunities per unit. Of course, another way to increase quality levels is to decrease the number of opportunities. That's why so much effort is expended in designing products with a minimum part count.

9.6 Flowchart for Selecting Proper Attribute-Data Chart

The flowchart given in Figure 9.17 should help in choosing the correct attribute-data chart. First decide if you are dealing with nonconformities or nonconforming units. If it is desired to keep track of all nonconformities found on a unit, then follow the "nonconformities" branch. If the unit is rejected as soon as the first nonconformity is discovered and the search for additional nonconformities on that unit ceases, then pursue the "nonconforming units" branch.

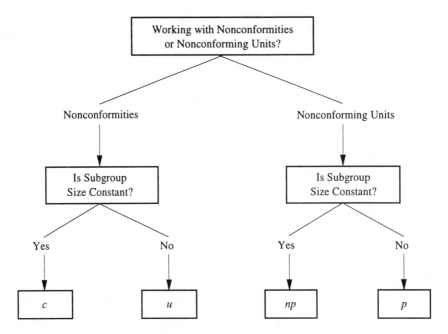

Fig. 9.17 Flowchart for choosing attribute-data charts.

Next, determine if a constant subgroup size can be maintained. If so, choose the "yes" option. Whenever the subgroup size is expected to vary, select the "no" branch. The correct chart for this application has now been properly identified.

9.7 Summary

Capability can be assessed for processes generating attribute data with four different capability measures: equivalent $P^{\%}_{PK}$, equivalent $P^{\%}_{P}$, equivalent $P^{\%}_{R}$, and $ppm_{TOTAL,\ LT}$. All these measures are derived from the percentage of nonconforming parts generated by the process, which is estimated from the centerlines of attribute charts. As with variable data, the chart must show good control before any estimates of process parameters or capability are made.

Equivalent $P^{\%}_{PK}$ is comparable to the P_{PK} measure for variable-data situations, based on the percentage of nonconforming parts. Equivalent $P^{\%}_{P}$ is analogous to the variable-data case where the process is centered at the middle of the tolerance, with half the total percentage nonconforming outside either specification. Used infrequently, the equivalent $P^{\%}_{R}$ ratio is just the inverse of the equivalent $P^{\%}_{P}$ index. The equivalent $P^{\%}_{P}$ and $P^{\%}_{R}$ attribute-data capability measures assume the "best" case, *i.e.*, the percentage nonconforming outside one specification limit is only half the total. On the other hand, the equivalent $P^{\%}_{PK}$ assumes the worse case, that the entire percentage nonconforming is outside one specification limit. In practice, the equivalent $P^{\%}_{PK}$ index is applied most often of these three measures.

Perhaps the most common capability measure for attribute data is $ppm_{TOTAL,\ LT}$, the percentage of nonconforming parts times one million. Expressing process capability as the number of nonconforming parts produced per million, it is the easiest of the attribute capability measures to understand, as well as calculate.

Because of their similarity to the variable-data measures, capability goals for attribute-data measures are established in the same manner.

Some companies look at *dpo* rather than *dpu* to compare the capability of products having different levels of complexity. Goals are established at the opportunity level, then extended to the unit level, based on the number of opportunities.

The method of studying capability presented so far in this book is for an entire process. Quite often there is interest in knowing the capability of just one portion of the overall process, like the machine. Thus, the next chapter takes up the subject of machine capability studies.

9.8 Exercises

9.1 Print Voids

Certain "paste areas," defined as regions covered with solder paste, on circuit board pads assembled with surface mount technology are examined for print voids. If a void is detected, the board is rejected. A subgroup of 40 boards is taken every 4 hours and the number of rejected boards plotted on a *np* chart. When 30 subgroups are collected, the total number of rejected boards is 71. Calculate the centerline and control limits for this chart. Assuming the process is in control, estimate $ppm_{TOTAL, LT}$. Then estimate the equivalent $P^{\%}{}_{P}$ and equivalent $P^{\%}_{PK}$ measures of capability.

9.2 Aerosol Cans

After being filled with hair spray, aerosol cans are checked for improper crimping, bulged dome, leakage, missing date code stamp, dents, and rust. To improve the quality of this filling line, a subgroup of 200 cans is collected every other hour, inspected, and the number of nonconformities identified plotted on a *c* chart. The total number of nonconformities found in 25 subgroups is 83. Calculate the centerline and control limits for this chart.

Given that the process is in a good state of control, measure its performance capability with the equivalent $P^{\%}_{PK}$ index. Does this process meet an equivalent $P^{\%}_{PK}$ goal of 1.00?

9.3 Faucet Assemblies

A plumbing hardware manufacturer in Indonesia checks a faucet and handle assembly for any problems with the seal, top plunger, bottom plunger, spout O-ring, or body O-ring. With $\bar{u}$ equal to .0021, estimate the equivalent $P^{\%}_{R}$ ratio for this assembly process.

9.4 Camera Shutters

Shutter subassemblies for electronic cameras are built up on an automated assembly line in Taiwan. 100 percent of each day's production is checked for several types of nonconformities. The average nonconformities per subassembly is calculated for the day by dividing the total nonconformities detected by the number of units checked. This daily result is then plotted on a *u* chart. When the process is finally stabilized, $\bar{u}$ is calculated as .000509. Does this assembly line meet a capability goal of 100 $ppm_{TOTAL, LT}$? What is the equivalent $P^{\%}_{PK}$ index?

9.5 Car Hoods

A Canadian sheet-metal stamping company produces hoods (bonnets) for a popular sports car. Once an hour, three consecutive hoods are pulled from production and have their surfaces inspected for dents and deformations. The total number found is plotted on a c chart kept near the stamping press. Once process stability is demonstrated, the centerline is calculated as 1.76. If stability is maintained, how many hoods are expected to have no nonconformities? How many will have exactly one nonconformity? How many will have exactly two? How many will have at most two nonconformities? How many will have more than two nonconformities? If a hood is considered nonconforming when it has three or more nonconformities, estimate the equivalent $P^\%_{PK}$ index for this stamping process. What is the estimate for $ppm_{TOTAL, LT}$?

9.6 Windshield Bubbles

A Middle East company produces windshields (windscreens) for automobiles. After being formed to the proper shape, the glass is checked for the presence of air bubbles, or "seeds." The specification allows a maximum of 3 seeds to appear in any $1,000 \text{ cm}^2$ area. Every hour, twenty consecutive windshields are collected. Each is checked for seeds in a $1,000 \text{ cm}^2$ area near the middle of the windshield. The total number of seeds discovered on these twenty windshields is plotted on a c chart. Over six weeks of monitoring, several improvements are made and the process becomes stabilized with $\bar{c}$ equal to 1.82. The goal is to have equivalent $P^\%_{PK}$ for this $1,000 \text{ cm}^2$ area be at least 1.20. Does this process meets its capability goal?

9.7 Pouch Seals

A C-node scanning acoustic microscope examines the seal of a small plastic pouch used for storing disposable contact lenses. If seal integrity has been compromised for any reason, the pouch is rejected. Because of large differences in daily production volumes, the percentage of rejected pouches is plotted on a p chart. When the process is finally stabilized, $\bar{p}$ is calculated to be .000321. Estimate both $ppm_{TOTAL, LT}$ and equivalent $P^\%_{PK}$. Does the sealing process exceed the minimum requirement for long-term performance capability? Does it meet an equivalent $P^\%_{PK}$ goal of 1.33?

9.8 Flywheel Castings

A grey-iron foundry in Malaysia checks flywheel castings for the presence of flash and burnt-in core sand. The total number of such nonconformities in each subgroup of 10 castings is plotted on a c chart. When in control, $\bar{c}$ is found to be 5.24. Estimate the percentage of nonconforming flywheel castings (those with at least one nonconformity), as well as the associated equivalent $P^\%_{PK}$, equivalent $P^\%_{P}$, and $ppm_{TOTAL, LT}$ values.

9.9 Pixel Nonconformities

A Japanese firm has installed a machine vision system to check flat-panel displays for several types of pixel nonconformities relating to the display's background. When stability is achieved, the centerline of the u chart monitoring this process is .0044. If a display with 3 or more pixel nonconformities is considered to be nonconforming, estimate both $ppm_{TOTAL, LT}$ and equivalent $P^\%_{PK}$.

Suppose the customer tightens the quality requirement to allow at most only 1 pixel nonconformity. What effect does this change have on $ppm_{TOTAL, LT}$ and equivalent $P^\%_{PK}$?

9.10 Color Matching

While assembling automobiles, an Australian company uses a portable, hand-held spectro-photometer to quickly gauge the color of plastic interior trim pieces. This measuring device makes a rapid pass / fail color evaluation, with the results of these checks recorded on an np chart ($n = 50$) having a centerline ($n\bar{p}$) of 1.976 when the process is in control. Estimate p' and equivalent $P^{\%}_p$ for the ability of this process to correctly match trim color.

After several substantial improvements, the nonconformance rate has decreased so much that n had to be increased to 100 in order to catch any mismatches in trim color. With this subgroup size, $n\bar{p}$ now equals 2.107. What is the equivalent $P^{\%}_p$ index for this revised process? Why is it better than before, even though $n\bar{p}$ has increased from 1.976 to 2.107?

9.11 Copper Thickness

An automated production system receives stacks of printed circuit boards from a press room, then unstacks and loads the untrimmed sheets onto a conveyor. While loading the laminated sheets, the equipment utilizes the eddy current principle to check the thickness of a copper layer. A "go/no-go" decision is made for each sheet, with the results for each shift summarized on a p chart. When in control, $\bar{p}$ is calculated as .000098 for the first shift, and .000211 for the second. From this information, estimate an equivalent $P^{\%}_{PK}$ value for each shift.

9.12 Pace Makers

Prior to being welded together during final assembly in a titanium casing, the internal circuitry of heart pace makers is X-rayed for flaws: cracked battery, missing components, cracked PC board, broken wire, foreign particles. Only one nonconformity is discovered in 200 subgroups of size 25 each. Estimate $ppm_{TOTAL,\ LT}$ as well as equivalent $P^{\%}_{PK}$ for this process.

9.13 Press Line Downtime

A p chart is kept on the percentage of down time for a press line. After several modifications, the chart stabilizes at a $\bar{p}$ value of .0187. What is the capability of this line to be available for running parts (measure with equivalent $P^{\%}_{PK}$)? How does this compare to its goal of having equivalent $P^{\%}_{PK}$ equal to at least .684, which represents 98 percent up time?

9.14 Wooden Pallet Repair

A company repairs wooden pallets for a major shipping firm. To help improve quality, pallets are checked at final inspection for various problems. For a recent six-month period, the following average number of nonconformities per pallet were discovered: .00401 for protruding fasteners; .00028 for split deck boards; .00009 for 1 / 2 companion stringer. Calculate the equivalent $P^{\%}_{PK}$ index for each type of problem, then estimate equivalent $P^{\%}_{PK}$ for the entire pallet-repairing process.

9.15 Police Radios

Two models of police radios (deluxe and standard) are assembled in a JIT fashion within the same U-shaped work cell. At final test, 2 nonconforming deluxe radios were discovered out of the 312 tested. Of the 546 standard models tested, 3 were unacceptable. Estimate dpu and equivalent $P^{\%}_{PK}$ (at the unit level) for both models.

Each deluxe radio has 173 opportunities for a nonconformity, whereas the standard model has only 84. Compute the *dpo* and *dpmo* values for each model. Use these results to estimate equivalent $P^{\%}_{PK}$ at the opportunity level. Which style of radio does the work cell have a better capability of producing?

9.16 Electronic Clocks

A desktop electronic clock is made up of 6 independent components. If the goal for *dpo* is .00021, and each component is considered an opportunity, what are the goals for equivalent $P^{\%}_{PK,O}$ (component level) and equivalent $P^{\%}_{PK}$ (product level)? What would be the expected $ppm_{TOTAL,LT}$ if this goal is met?
Three components can be eliminated due to a better design. How much would the goal for the equivalent $P^{\%}_{PK}$ index increase? What is the expected $ppm_{TOTAL,LT}$ if this new goal is met?

9.17 Circuit Boards

A particular style of circuit board is designed with 17 opportunities for nonconformity type A to occur, 3 opportunities for type B, and 5 for type C. The dpo_{GOAL} for type A is .00003; for B, .00011; for C, .00006. What is the goal for equivalent $P^{\%}_{PK}$ at the unit level?

9.18 Print Voids

In Exercise 9.1, solder paste areas were checked for print voids. If there are 156 paste areas that could have a void, what are *dpo* and *dpmo* based on the data given in that exercise? From these numbers, estimate equivalent $P^{\%}_{PK,O}$ at the opportunity level, then compare it to equivalent $P^{\%}_{PK}$ at the unit level, which was estimated in Exercise 9.1. If a newly designed circuit board has only 81 paste areas, and the *dpo* is expected to be the same as the first board, what is the predicted equivalent $P^{\%}_{PK}$ at the unit level for this new board?

9.19 Gimbal Assemblies

Gimbal assemblies for navigational systems are assembled from five different subassemblies. The *dpo* rates given in Table 9.5 were computed from the results of extensive testing on each subassembly. A gimbal is made using one each of subassemblies 1, 2, 4, and 5, and two of subassembly 3.

Table 9.5 *dpo* rates for gimbal subassemblies.

Sub-Assembly	Gimbal Usage	*dpo*
1	1	.000026
2	1	.000104
3	2	.000008
4	1	.000075
5	1	.000003

Predict equivalent $P^{\%}_{PK}$ for the entire gimbal assembly when it is built from these components.

9.9 References

Bothe, Davis R., *SPC for Short Production Runs*, 1989, International Quality Institute, Cedarburg, WI, pp. E-27 to E-30

Bothe, Davis R., *SPC for Short Production Runs Handbook*, 1991, International Quality Institute, Cedarburg, WI, pp. 40-41

Gitlow, Howard, Gitlow, Shelly, Oppenheim, Alan, and Oppenheim, Rosa, *Tools and Methods for the Improvement of Quality*, 1989, Irwin, Homewood, IL, p. 428

Gunter, Bert, "Go/No-Go Testing: Going, Going, Gone?," *Quality Progress*, Vol. 23, No. 10, October 1990, pp. 110-114

Harry, Mikel J., *The Nature of Six Sigma Quality*, Motorola University, Schaumberg, IL, pp. 4 and 12

Harry, Mikel J., and Lawson, J. Ronald, *Six Sigma Producibility Analysis and Process Characterization*, 1990, Motorola University, Schaumberg, IL

Heimann, Peter A., "Attributes Control Charts with Large Sample Sizes," *Journal of Quality Technology*, Vol. 28, No. 4, October 1996, pp. 451-459

Hoadley, B., "Quality Measurement Plan," *Encyclopedia of Statistical Sciences, Volume 7*, Kotz and Johnson, eds., John Wiley, New York NY, pp. 393-397

Lam, C.T., and Littig, S.J., "A New Standard in Process Capability Analysis," *Technical Report 92-23*, University of Michigan, Ann Arbor, MI, 1992

Lehrman, Karl-Henry, "Control Charts for Variables and Attributes with Process Safety Analyses," *Quality Engineering*, Vol. 4, No. 2, 1991-92, pp. 243-318

Marcucci, Mark O., and Beazley, Charles C., "Capability Indices: Process Performance Measures," *42nd ASQC Annual Quality Congress Trans.*, Dallas, TX, May 1988, pp. 520-521

Nelson, Lloyd, S., "Standardization of Shewhart Control Charts," *Journal of Quality Technology*, Vol. 21, No. 4, October 1989, pp. 287-289

Nelson, Lloyd, S., "A Control Chart for Parts-Per-Million Nonconforming Items," *Journal of Quality Technology*, Vol. 26, No. 3, July 1993, pp. 239-240

Ryan, Thomas P., and Schwertman, Neil C., "Optimal Control Limits for Attributes Control Charts," *Journal of Quality Technology*, Vol. 29, No. 1, January 1997, pp. 86-98

Snedecor, George W., and Cochran, William G., *Statistical Methods*, 7th edition, 1980, Iowa State University Press, Ames, IA, pp. 130-133

Wheeler, Donald J., "Charts for Rare Events," *Quality Digest*, December, 1996, p. 43

Zaciewski, Robert D., "Attribute Control Charts: Opportunities for Application," *Quality Digest*, Vol. 15, No. 5, May 1995, pp. 36-41

Chapter 10

Conducting Machine Capability Studies

Measure twice, buy once.

Process capability studies compare the total amount of inherent variation to the customer requirements. Inherent process variation originates from the five "M"s: material, machine, method, manpower, measurement. The purpose of a *machine* capability study is to isolate the amount of process variation contributed solely by the machine. This is done by attempting to eliminate, or at least minimize, all other sources of process variation coming from the other four "M"s, as shown in Figure 10.1.

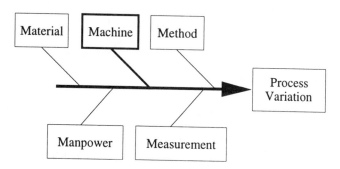

Fig. 10.1 Five sources of inherent process variation.

Even though it is labeled a "machine" capability study, Bayer suggests a similar type of approach can be applied equally well to machine tools (both new and rebuilt), fixtures, pallets, stations, or even test equipment. Some texts refer to machine capability studies as a machine runoff, machine tryout, validation, or short-term capability study. In the remainder of this chapter, the term machine capability study should be considered a generic term for determining the ability of a piece of equipment to meet some requirement. By providing information about expected machine performance, these types of studies are also very helpful when developing tolerances for new products.

If the machine being studied has multiple process streams, such as pallets, spindles, fixtures, stations, filling heads, or cavities, each stream must be studied separately because each may have a different average and/or standard deviation. Lumping data from all process streams together inflates the standard deviation estimate and leads to an incorrect assessment of machine capability. In addition, combining data from several process streams increases the difficulty of identifying which streams are responsible for the majority of variation, and thereby hinders attempts at problem solving.

Usually machine capability studies are conducted on the machine builder's shop floor, before the equipment is purchased. However, this type of study can be performed on your shop floor to determine the capability of an existing machine, or for a recently rebuilt or reconditioned machine. Grandzol and Gershon provide a thorough list of criteria for deciding when machines should be replaced, with machine capability being one of the top considerations. Some companies utilize this technique to assist in scheduling work to appropriate machines, assigning products with tighter tolerances to machines with better capability.

It's very helpful to form a machine capability team, consisting of members from product engineering, manufacturing, reliability engineering, maintenance, quality engineering, production operators, the equipment manufacturer, and the material supplier. One of the team's initial tasks is the development of a machine capability analysis form. This form assures nothing is missed during the study and becomes an essential part of the machine's documentation file. As a minimum, the form should include the machine identification number, characteristic (or variable) studied and the study number, *e.g.*, "1" for the initial study, "2" for the second time the machine is studied. An example of such a form is exhibited later in Figure 10.2. Sometimes special identification numbers are assigned so the form can be stored in a computer file. This simplifies tracking procedures and provides a convenient retrieval system for future comparison studies (Perez-Wilson).

There are at least two recognized procedures available for estimating machine capability: the control chart method and the sequential s test. Both are explained in this chapter, beginning with the more prevalent control chart approach.

10.1 The Control Chart Method

With this method, parts are produced on a machine set up to operate under "ideal" conditions, which means only a minimum contribution of variation from the other four "M"s. The output characteristic of interest is measured and then plotted on a control chart to determine the degree of process stability. Once control is established, machine capability is estimated using the appropriate capability measure(s). This estimate of machine capability is compared to the goal and a decision made to either accept or reject the machine (De Grote). In cases where the focus is on new product/process development, this kind of study provides estimates of the process parameters, which can then be used to establish realistic product tolerances or to track improvements in process development.

Note that several characteristics can be checked during the same study. Simply measure the parts for all characteristics of interest, then repeat the capability calculations for each set of data. For example, one study for a lathe cutting metal shafts could provide conclusions about outer diameter size, surface finish, concentricity, and taper. Lucas and Pilkington present an example of a study conducted on five critical characteristics of a lead-forming operation (see also Haggar, Ouellette and D'Souza).

It is quite possible to accept a machine for some characteristics and reject it for others. If some fail, make appropriate changes to the equipment and repeat the study. All characteristics should be tested again since changes to fix the unacceptable characteristics may adversely affect those judged acceptable in the original study.

In order to measure the total amount of common-cause variation, regular process capability studies are run under normal, every day operating conditions. In contrast, machine capability studies are conducted under specially controlled operating conditions, which are outlined next. These steps are taken to isolate variation associated with only the machine.

Specific Steps for Control Chart Method

1. Begin the study by first identifying the machine and characteristic(s) of interest. Make a sketch of the part showing these characteristics. Write out a complete description of the machine and its intended function. Determine appropriate capability goals for each characteristic and communicate these goals to the machine builder. Verify that the manner of measuring the selected characteristic is correct and that all gages are properly calibrated and have acceptable repeatability and reproducibility (gage R&R). It's best to have one person perform all measurements under consistent environmental conditions such as temperature and humidity. Record these details on a machine capability analysis form similar to the one given in Figures 10.2 and 10.3 on the next two pages.

2. To minimize "Manpower" variation, select a qualified and experienced operator. It's best to bring along an operator who will actually run this machine and has been properly trained. Don't rely on the machine builder to provide an operator as this person could be a highly-trained technician with extensive experience running this type of equipment, or he/she could have very little knowledge concerning this machine. Either scenario could seriously bias the outcome of the study.

3. Verify all stock (or raw material) chosen for the study is within print specification for *every* feature affecting production of the characteristics being studied. Use a single batch of raw material, coil of steel, bag of pellets, pouring run of castings, etc. Check all parts for machinability, cleanliness, stock distribution, and flash. If different types or grades of material are to be processed on the machine when installed, select the expected worst-case material for this study. Even though they are cheaper, resist the temptation to run scrap pieces from your plant as this would not be a fair test of the machine's true capability. Correctly completing this step helps diminish variation arising from differences in "Material."

4. Confirm the machine has proper oil and coolant levels and is correctly set up. Adjust tooling, feeds, speeds, pressures, temperatures, and all other pertinent operating parameters to their proper settings. Make sure tooling, oil, coolant, holding fixtures, and location devices are identical to those expected to be used in regular production. Record all levels and settings on the machine capability form. These recommendations should be closely followed to minimize variation contributed by the "Methods" branch.

5. Allow the machine to warm up by producing several parts. In many situations, machine warm-up, and/or cool down, is one of the greatest contributors to process variation. Also make sure the cycle time of the machine run during the study is the same as that scheduled for regular production. Keep all locating surfaces clean and remove chips from the work piece area. Use the anticipated production method of loading and unloading parts.

6. Make only normal adjustments to the machine while the study is in progress. If the plan for running this machine under regular conditions calls for the operator to "make five parts, then rotate the tool," the machine capability study should be conducted in the same manner. A detailed log, in time order, of all changes and adjustments (planned or unplanned) to *any* machine setting must be kept for each run. This log should also include any tooling maintenance, *e.g.*, shaping

---

| **Machine Capability Study Form** | I.D. #: _____ |

Purpose of Study:

Type of Machine (equipment):

Machine Identification #:

Machine Builder: Date of study: Study #:

Part Name: Part Number:

Characteristic: LSL =   USL =

Target Average: Units of Measurement:

Type of Gage: Gage Number:

Gage R&R:

Method of Calibration:

Person Measuring Parts: Person Running Machine:

Material Identification:

Condition of Material: Method of Loading:

Machine Feed Rate: Speed:

Cycle Time: Tooling:

Coolant: Type of Oil:

Log of Machine Adjustments Made During Study (include explanation):

1.

2.

3.

4.

5.

6.

7.

_____   _____
Signature of person completing this form   Date completed

Fig. 10.2   Front of machine capability study form.

or polishing. Keep notes of any unusual or unexpected occurrences, as well as documenting any signs of tool or die wear.

During the course of the study, verify the list of standard operating procedures that will be followed once the machine is installed at its final destination. These detailed work instructions will reduce the learning curve for new operators, as well as minimize differences between operators.

7. For variable-data characteristics, run at least 50 consecutive parts from each process stream, measure the characteristic of interest, and record all measurements on the backside of a machine capability study form like the one displayed in Figure 10.3. Circle all measurements that are outside print specifications.

Data Collection and Control Chart Calculation Form for $n = 5$ and $k = 10$						
Measurements of Characteristic Under Study					Subgroup Statistics	
					$\overline{X}$	Range
1 ____	2 ____	3 ____	4 ____	5 ____	____	____
6 ____	7 ____	8 ____	9 ____	10 ____	____	____
11 ____	12 ____	13 ____	14 ____	15 ____	____	____
16 ____	17 ____	18 ____	19 ____	20 ____	____	____
21 ____	22 ____	23 ____	24 ____	25 ____	____	____
26 ____	27 ____	28 ____	29 ____	30 ____	____	____
31 ____	32 ____	33 ____	34 ____	35 ____	____	____
36 ____	37 ____	38 ____	39 ____	40 ____	____	____
41 ____	42 ____	43 ____	44 ____	45 ____	____	____
46 ____	47 ____	48 ____	49 ____	50 ____	____	____
				Totals		

$$\overline{\overline{X}} = \frac{\sum_{i=1}^{10} \overline{X}_i}{10} = \frac{}{10} = \qquad UCL_R = 2.115 \overline{R} =$$

$$\overline{R} = \frac{\sum_{i=1}^{10} R_i}{10} = \frac{}{10} = \qquad LCL_R = 0$$

$$UCL_{\overline{X}} = \overline{\overline{X}} + .577 \overline{R} =$$

$$LCL_{\overline{X}} = \overline{\overline{X}} - .577 \overline{R} =$$

$$\hat{\mu}_{MACHINE} = \overline{\overline{X}} = \qquad \hat{\sigma}_{MACHINE} = \frac{\overline{R}}{d_2} = \frac{}{2.326} =$$

Normality Check: Capability Measure:

Capability Goal: Recommendation:

Fig. 10.3 Back of machine capability study form.

For equipment with extremely short cycle times, it may be better to collect a small sample of parts at half-hour (or hour) intervals until at least 50 are collected. Some authors suggest using only 10 parts in a "mini-capability" study (Strongrich *et al*). An estimate of 6σ is derived by doubling the range of the 10 measurements on these parts.

$$6\hat{\sigma} = 2(X_{MAX} - X_{MIN})$$

However, as will be demonstrated in Chapter 11, 10 pieces do not provide very reliable results. This simplified approach should not be considered unless the cost of producing a single piece is very expensive, and then with only extreme caution. Besides, if just 10 pieces are available, it would be better to estimate the machine standard deviation with the S_{TOT} method.

If there is more than one process stream (cavity, mold, spindle, pattern), a separate chart, and capability study, is required for each stream. In case the measurement of a part needs to be rechecked, mark every part, or attach labels, to preserve the chronological order of production and process stream number, if appropriate (see also Chrysler *et al*).

Calculate subgroup averages and ranges on the form, then plot them on an $\overline{X}$ and R (or S) control chart. A subgroup size of 5 consecutive parts is typically selected, meaning there will be 10 subgroup averages and ranges to plot. However, to better detect the presence of any time-related changes such as runs, trends, or cycles that may occur during the study, manufacture 51 pieces and form 17 subgroups of size three.

Wheeler and Chambers recommend first plotting the 50 individual measurements on a run chart to detect any obvious runs, trends, or jumps in the data. If none are detected, moving ranges are calculated and control limits determined for an *IX & MR* chart. If this chart shows good control, the measurements are arranged into subgroups and replotted on an $\overline{X}$, R chart.

8. Evaluate the control chart for stability. Application of the Western Electric rules for runs will help identify out-of-control conditions sooner. If any are detected, discover the cause, initiate corrective action, and repeat the study. Any evidence of instability is a major concern. If this equipment cannot maintain stability over a relatively short period of time under the closely controlled conditions of this study, what will happen when it is installed at the final equipment location and run under normal operating conditions?

9. When control is established, estimate the average of the machine's output from $\overline{\overline{X}}$, and its standard deviation from $\overline{R}/d_2$ (or $\overline{S}/c_4$).

$$\hat{\mu}_{MACHINE} = \overline{\overline{X}} \qquad\qquad \hat{\sigma}_{MACHINE} = \frac{\overline{R}}{d_2}$$

Construct a histogram of the measurements and draw on the print specifications. Calculate the percentage of sample measurements actually outside tolerance. The individual measurements should also be plotted on normal probability paper to check for normality, as was done in Chapter 7. If non-normal, apply the alternative capability formulas given in Chapter 8. In addition, this plot will predict the percentage of nonconforming parts expected in the population.

10. Estimate the desired measure of machine capability, such as C_R, C_P, C_{PK}, C_{PM}, or C_{PMK}. The "C" capability measures are applicable because machine capability studies are conducted under very ideal conditions, run for a relatively short period of time, and the estimate of $\sigma_{MACHINE}$ is based on only within-subgroup variation. To denote a capability measure is just for the machine component of the process, it is prefaced with the word "machine." For example, machine C_P, machine C_{PK}. Sometimes machine C_P is labeled C_M (the "M" for machine replacing the "P" for process), while machine C_{PK} becomes C_{MK}. However, use of this second method of notation is not very widespread.

Compare this estimate of machine capability to the goal and make a decision whether or not to purchase the equipment. Obviously, the goal for the machine must be much higher than for the entire process. For example, a process capability goal of 1.33 may require a machine capability goal of over 1.90. The establishment of machine capability goals is discussed in Section 10.3 (p. 578).

If a machine does not meet the capability goal, improvements must be made (and a second capability study run for verification) before purchasing it. When machine improvements are not feasible, try requesting a tolerance change from the appropriate engineering personnel. Feedback should be given to the process and product development engineers so they can avoid these types of problems in future designs. A third alternative is to adjust the average so scrap and rework are minimized, then initiate 100 percent inspection (review Section 7.3, p. 395).

Work through the next example to see how these steps are applied to make a decision concerning the capability of a new milling machine (see also Gregerson).

Example 10.1 Keyway Width

Following the procedure just described for performing a machine capability study, 50 shafts have a keyway cut on a milling machine. The width of the keyway is measured at one end of the shaft and is the characteristic of interest for this study. These 50 width measurements are divided into 10 subgroups of 5 each and plotted on the $\overline{X}, R$ chart displayed in Figure 10.4. This chart indicates good control and the individual measurements plot as a reasonably straight line on normal probability paper. The average, $\mu_{MACHINE}$, and standard deviation, $\sigma_{MACHINE}$, for width are estimated from the control chart summary statistics.

$$\hat{\mu}_{MACHINE} = \overline{X} = 72.9 \qquad \hat{\sigma}_{MACHINE} = \frac{\overline{R}}{d_2} = \frac{18.5}{2.326} = 7.95$$

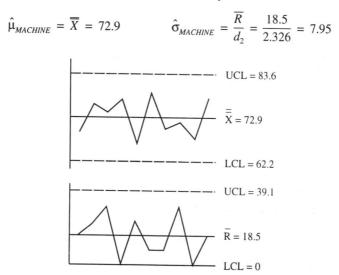

Fig. 10.4 The control chart for the machine capability study.

Given a LSL of 20 and an USL of 100, machine C_P and C_{PK} for this characteristic may now be estimated with the standard formulas. The potential machine capability goal for width is a C_P of at least 1.667. As the output has a normal distribution, this goal means at least ten

$\sigma_{MACHINE}$s $(6 \times 1.667 = 10)$ should be able to fit into the tolerance. From this machine capability study, the potential capability is estimated as 1.677, which is slightly greater than the goal. Based on this result, a decision should be made to purchase the milling machine.

$$\text{Machine } \hat{C}_p = \frac{\text{Tolerance}}{6\hat{\sigma}_{MACHINE}} = \frac{100 - 20}{6(7.95)} = 1.677$$

If interest is in performance capability, the machine C_{PK} index for this feature can be estimated and compared to the required machine C_{PK} goal of 1.667.

$$\text{Machine } \hat{C}_{PK} = \text{Minimum} \left(\frac{\hat{\mu}_{MACHINE} - \text{LSL}}{3\hat{\sigma}_{MACHINE}}, \frac{\text{USL} - \hat{\mu}_{MACHINE}}{3\hat{\sigma}_{MACHINE}} \right)$$

$$= \text{Minimum} \left(\frac{72.9 - 20}{3(7.95)}, \frac{100 - 72.9}{3(7.95)} \right) = 1.136$$

Because the estimate of machine C_{PK} is only 1.136, a decision should be made *not* to purchase this machine. This analysis leads to a completely different decision about buying this milling machine than the one based on machine C_P. Why the conflicting conclusions?

If the machine can be easily adjusted to move the center of the process output, then potential capability measures, such as C_R and C_P, are more important for assessing machine capability. Such is the case for a cut-off operation where the length of cut can easily be changed by a simple machine adjustment. Under these circumstances, concern is limited to the amount of variation around this specified length. Another such situation is a process equipped with real-time gaging, which provides feedback control for automatic tool compensation.

In this example, machine C_P is estimated as 1.677, with a goal of 1.667, so a decision should be made to purchase the machine *only if* adjusting the process average is relatively easy, as would be the case for a milling machine.

However, if moving the process average is quite difficult, or concern involves both centering and spread, then performance capability measures, like C_{PK} and C_{PM}, must be given more consideration in the evaluation of machine capability. For this example, the estimate of machine C_{PK} is only 1.136 compared to a goal of 1.667. If there is considerable apprehension about centering, a decision should be made to hold off buying this machine until this issue is satisfactorily addressed. A formal plan for improving this machining process must be developed and implemented (Huttunen). After all changes are completed, a second machine capability study should be conducted to verify improvements to the milling machine have successfully increased performance capability.

Example 10.2 Cylinder Heads

A machine drills and counterbores a hole in a cylinder head, similar to the one illustrated in Figure 10.5. There are two critical features that must meet their respective capability goals before this machine is purchased. The first is the diameter of the drilled hole, which has a machine C_{PK} goal of 1.67. The second is the depth of the counterbore, with a machine C_{PK} goal of 2.00. The specifications given for hole diameter are $1.90 \pm .03$ cm, while those listed for depth are $1.45 \pm .05$ cm.

The proper preparations for a machine capability study are followed, with 50 holes drilled and counterbored. Each hole is measured for these two characteristics, then each characteristic is plotted on its own $\overline{X}, R$ chart. When control is achieved, the following estimates of machine parameters are made for each feature.

Fig. 10.5 A cylinder head.

$$\hat{\mu}_{HOLESIZE} = 1.904 \qquad \hat{\sigma}_{HOLESIZE} = .0057 \qquad \hat{\mu}_{DEPTH} = 1.446 \qquad \hat{\sigma}_{DEPTH} = .0071$$

From these results, estimates of machine capability for hole size and depth can be calculated and compared to their respective goals.

$$\text{Machine } \hat{C}_{PK,HOLESIZE} = \text{Minimum} \left(\frac{\hat{\mu}_{HOLESIZE} - \text{LSL}}{3\hat{\sigma}_{HOLESIZE}}, \frac{\text{USL} - \hat{\mu}_{HOLESIZE}}{3\hat{\sigma}_{HOLESIZE}} \right)$$

$$= \text{Minimum} \left(\frac{1.904 - 1.87}{3(.0057)}, \frac{1.93 - 1.904}{3(.0057)} \right) = 1.52$$

$$\text{Machine } \hat{C}_{PK,DEPTH} = \text{Minimum} \left(\frac{\hat{\mu}_{DEPTH} - \text{LSL}}{3\hat{\sigma}_{DEPTH}}, \frac{\text{USL} - \hat{\mu}_{DEPTH}}{3\hat{\sigma}_{DEPTH}} \right)$$

$$= \text{Minimum} \left(\frac{1.446 - 1.40}{3(.0071)}, \frac{1.50 - 1.446}{3(.0071)} \right) = 2.16$$

The machine's capability for depth exceeds the goal of 2.00, but its capability for hole size falls somewhat short of the required minimum of 1.67. Therefore, this machine cannot be approved for purchase. Changes must be made and a second machine capability performed to verify their effectiveness. Note that *both* characteristics must be included in the second study, because changes to improve the machine's ability to produce hole size may be detrimental to its performance for counterbore depth.

Control Chart Method for Attribute Data

When the characteristic of interest generates attribute-type data, the capability measures developed in Chapter 9 may be employed. Unfortunately, many more pieces must be run to accurately estimate machine capability when using attribute-type data. The minimum number of required pieces would be for the ideal situation where absolutely *no* nonconforming pieces are produced in the course of the study. If 1 minus α is the desired confidence level (see p. 703) and p'_{GOAL} is the goal for the nonconformance rate, then the required number of pieces, *kn*, to be run *without encountering a problem* is derived from this formula (Hurayt).

$$kn = \frac{\log \alpha}{\log (1 - p'_{GOAL})}$$

For instance, to demonstrate a machine is producing no more than .135 percent nonconforming parts (which is the maximum allowed to meet the minimum criterion for performance capability) with .90 confidence (making α equal to .10), the required minimum number of pieces to be run is 1,705.

$$kn = \frac{\log .10}{\log (1 - .00135)} = 1,705$$

This total represents approximately 17 subgroups of size 100. Not a single nonconforming part may be produced during this entire run if the machine capability goal is to be met. If even one nonconforming part is observed, the machine under study fails the test. With α set at .10, there is only a 10 percent chance of seeing any nonconforming units, and thus rejecting the machine, when the true p' of this machine actually meets the goal.

Note that demonstrating a goal of 100 *ppm* at .90 confidence would necessitate processing a minimum of 23,025 pieces without detecting a problem.

$$kn = \frac{\log .10}{\log (1 - .0001)} = 23,025$$

This translates into 230 subgroups of size 100 each to be collected without a single nonconformance, thus making for quite a lengthy, and expensive, machine capability study.

The formula for the number of required pieces may also be expressed in terms of the equivalent $P^{\%}_{PK}$ goal. Recall that the symbol $\Phi(Z)$ means find the corresponding percentage for the given Z value in Appendix Table III on page 845.

$$\text{Equivalent } P^{\%}_{PK,GOAL} = \frac{Z(p'_{GOAL})}{3}$$

$$3 \text{ Equivalent } P^{\%}_{PK,GOAL} = Z(p'_{GOAL})$$

$$\Phi(3 \text{ Equivalent } P^{\%}_{PK,GOAL}) = p'_{GOAL}$$

This expression for p'_{GOAL} is substituted into the formula for kn.

$$kn = \frac{\log \alpha}{\log (1 - p'_{GOAL})} = \frac{\log \alpha}{\log [1 - \Phi(3 \text{ Equivalent } P^{\%}_{PK,GOAL})]}$$

This version of the kn formula appears in the next example.

Example 10.3 Rivet Installation

The purchase of a new machine for installing rivets in the roof of a line-haul trailer is under consideration (Figure 10.6). To help management make this decision, a machine capability study will be conducted to check how well rivets are installed (no "bucking" is desired). The required goal for this machine characteristic is an equivalent $P^{\%}_{PK}$ of at least 1.333 at .95 confidence ($\alpha = .05$).

Fig. 10.6 The rivet assembly.

To determine the minimum number of rivet installations that must be checked in the study, the following computation is made.

$$kn = \frac{\log \alpha}{\log [\, 1 - \Phi(\, 3 \text{ Equivalent } P^{\%}_{PK, GOAL}\,)\,]} = \frac{\log .05}{\log [\, 1 - \Phi(\, 3 \times 1.333\,)\,]}$$

$$= \frac{\log .05}{\log (\, 1 - .0000318\,)} = 94,204$$

With a subgroup size of 1,000 rivet installations, a total of 94 subgroups must be collected *without encountering a single nonconforming rivet*. After the run is completed, 2 nonconforming rivets are discovered. Thus, this insertion machine does not meet the specified capability goal at .95 confidence. Work must be done on this machine to improve its performance. When completed, the machine capability study must be repeated.

Comments on the Control Chart Method

It is very important the machine be in a good state of control. If not, estimates of μ and σ are unreliable, making all estimates of machine capability based on these parameters questionable. If stability cannot be maintained over such a relatively short period of time, and under ideal operating conditions, what will happen when full production begins? There's little hope shop floor personnel will be able to keep the machine in control over the long term.

The traditional capability measures of C_R, C_P, C_{PK}, and C_{PM} all assume the process output is close to a normal distribution. This assumption must be tested before estimating these measures. Normal probability paper is usually the choice for this task, but the goodness-of-fit tests mentioned in Section 7.5 (p. 402) may be applied as well. If these procedures indicate the machine's output is not normally distributed, machine capability should be estimated with the equivalent capability measures covered in Chapter 8.

If a target value is specified for the machine average, the C^{*}_{R}, C^{*}_{P}, C^{*}_{PK}, and C^{*}_{PM} measures should be applied. When the characteristic under study has a unilateral specification, employ either C'_{RL}, C'_{RU}, C'_{PL}, or C'_{PU} for measuring potential machine capability.

Once an estimate of machine capability is calculated, confidence bounds may be determined as explained in Section 11.28 (p. 682). Some companies require the lower confidence bound to exceed the machine capability goal before agreeing to accept the equipment.

The control chart method may also be applied to general purpose machines. When many different part numbers will be run on a machine, test the one having the "tightest" tolerance, or the worst case as far as material, feeds and speeds, setup, or cycle time is concerned. If a study can be repeated for several materials, a matrix of machine capability versus material type may be developed, as is exhibited below in Table 10.1.

Table 10.1 Machine C_{PK} values for different characteristics and types of materials.

Part Characteristic	Machine C_{PK} by Material	
	Soft Brass	Stainless Steel
Length	1.54	1.91
Taper	1.83	2.07
Width	1.66	1.89

Remember that each process stream must be checked separately. One stream having acceptable capability does not imply the others do as well. Even if all streams have acceptable capability, the results of each stream should be compared to the others. Any differences found may lead to ideas for improving all streams. A measure of the combined capability of all streams can be made with the Average C_{PK} measure explained in Section 12.1 (p. 722).

Machine capability studies for new machines should always be done at the machine builder's facility, *before* purchasing the equipment. It's much easier to resolve problems here because the required engineering expertise and equipment is close at hand. Due to different conditions, *e.g.*, type of foundation or tie-down system, amount of climate control, degree of machine break in, differences in cooling systems, lots of materials, different operators, chip removal systems, amount of break in, shipping or installation damage, Kane warns that results of a second machine capability study performed on the equipment after arrival at its intended destination may deviate substantially from the original study. The effect of many important operating conditions are not usually seen in a short trial run, but are unavoidable in even the best controlled production environment. Sikorsky Aircraft has discovered that if these significant factors are not correctly identified, the machine may pass the runoff test, but perform inadequately once installed at the final equipment location. On the other hand, studies by Down and Marshall have found these varying operating conditions occasionally result in an *improvement* in machine performance.

Make sure all decision makers are aware of these potential differences that may affect the capability results (either positively or negatively) and factor them into any final judgment concerning this piece of equipment. In addition to the quality issue, other considerations involving machine purchase are oil and coolant consumption, tool wear, maintenance costs, percent uptime, cycle times, cost, noise levels, and safety-related items. Always request a preventive maintenance schedule from the equipment supplier to ensure capability can be kept at its highest level over the lifetime of the equipment.

This first method for conducting machine capability studies is very popular, especially in the automotive industry. However, one major disadvantage with the control chart approach is the large number of pieces involved, *i.e.*, at least 50. The next section describes a relatively new procedure that may significantly reduce the required number of runoff pieces in certain situations.

10.2 The Sequential *s* Test

Sometimes it is difficult to produce enough pieces to perform a machine capability study using the control chart method, *e.g.*, long cycle times, expensive materials or destructive testing. If the ability of the machine greatly exceeds, or falls far short of, its requirement, the sequential *s* test will help make a decision about the machine's capability with a much smaller number of pieces than needed with the control chart approach (Nelson, Burr).

To apply the sequential *s* test, follow all the preparations for conducting a machine capability study as specified for the control chart approach. After machine warm up, calculate s_n (the sample standard deviation) from the first eight pieces produced. Then calculate a "test ratio" by dividing s_n by the tolerance of the characteristic under study (n is the sample size).

$$\text{Test Ratio}_n = \frac{s_n}{\text{Tolerance}} = \frac{s_n}{\text{USL - LSL}} \qquad \text{where } s_n = \sqrt{\frac{\sum\limits_{i=1}^{n}(x_i - \bar{x})^2}{n-1}}$$

If this test ratio were multiplied by 6, it would be very similar to the C_R ratio introduced on page 95. When this test ratio is very "small," *i.e.*, less than the lower critical value for a sample of n (called "LCV$_n$"), the machine is considered capable because its variation is only a small percentage of the tolerance. On the other hand, if the test ratio is very "large," exceeds the upper critical value ("UCV$_n$"), then the machine is judged not capable. Lower and upper critical values for n from 8 to 30 are found in Table 10.2 for .90, .95, and .99 confidence.

One of the following three decisions is made based on the comparison of the test ratio to the critical values. If the test ratio is:

1. less than the LCV$_n$, assume machine meets capability goal;
2. greater than the UCV$_n$, assume machine does *not* meet capability goal;
3. between the LCV$_n$ and UCV$_n$, continue testing.

A test ratio falling somewhere in between the two critical values means there is not enough information from the measurements collected so far to make a decision with the desired level of confidence. Two additional pieces must be run off the machine and measured. These last two measurements are combined with the first eight to calculate a new s_n value, making n now equal to 10. A revised test ratio incorporating this updated s_{10} value is calculated and compared to two new critical values to make a decision. If either the new maximum or minimum critical value is exceeded, then a decision about the machine's capability can be made. When the new test ratio is between the critical values, two more pieces must be run, s_{12} calculated (n is 12 now), and the test ratio computed and interpreted as previously described.

This sequential testing procedure continues until either a decision about machine capability is made, or 30 pieces are measured. If no decision can be made with 30 pieces, divide the 30 measurements into 10 subgroups of size 3 each, and switch to the control chart approach for estimating machine capability.

Table 10.2   Critical values for sequential s test.

Confidence Level for Goal of $10\sigma_{MACHINE} \leq$ Tolerance
(Which Is Identical to a Machine C_P Goal ≥ 1.67)

	.90		.95		.99	
n	LCV$_n$	UCV$_n$	LCV$_n$	UCV$_n$	LCV$_n$	UCV$_n$
8	.0636	.1310	.0556	.1418	.0421	.1625
10	.0680	.1277	.0608	.1371	.0482	.1552
12	.0712	.1253	.0645	.1337	.0527	.1499
14	.0736	.1234	.0673	.1312	.0562	.1459
16	.0755	.1220	.0696	.1291	.0590	.1428
18	.0770	.1207	.0714	.1274	.0614	.1402
20	.0783	.1197	.0730	.1260	.0634	.1380
22	.0794	.1188	.0743	.1247	.0651	.1362
24	.0804	.1180	.0754	.1237	.0666	.1346
26	.0812	.1173	.0764	.1227	.0679	.1331
28	.0819	.1167	.0773	.1219	.0691	.1319
30	.0825	.1161	.0781	.1211	.0701	.1308

The critical values in Table 10.2 are based on demonstrating a machine capability goal of having the potential to fit a minimum of $10\sigma_{MACHINE}$ within the tolerance. This equates to a machine C_P goal of 1.67 ($\pm 5\sigma_{MACHINE}$). To test for a different machine capability requirement, multiply the critical values in Table 10.2 by $10/h$, where h is the number of $\sigma_{MACHINE}$s desired to fit within the tolerance. For example, if the goal is to have $12\sigma_{MACHINE}$ equal the tolerance ($h = 12$), multiply the critical values in this table by $10/12$.

Formulas for the critical values are given below. The χ^2 values are found in Appendix Table VI (p. 854) and explained further in Section 11.2 (p. 609).

$$\text{LCV}_n = \frac{1}{10} \sqrt{\frac{\chi^2_{n-1,.90}}{n-1}} \qquad\qquad \text{UCV}_n = \frac{1}{10} \sqrt{\frac{\chi^2_{n-1,.10}}{n-1}}$$

Notice how, for higher levels of confidence, the LCV_n for a given n becomes smaller and the UCV_n becomes greater. As an example, the LCV_8 for .90 confidence ($n = 8$) is .0636, while the LCV_8 for .99 confidence ($n = 8$) is .0421. In order to have a higher degree of confidence in making the correct decision, a more extreme outcome must be observed for the test ratio.

Example 10.4 Cut-off Length

To demonstrate how this test works, assume the goal is to have .90 confidence that machine C_P for the cut-off length of the metal block shown in Figure 10.7 is at least 1.67. This means a minimum of 1.67 six $\sigma_{MACHINE}$ spreads must be able to fit into the tolerance, which is equivalent to saying $10\sigma_{MACHINE}$ (6 × 1.67) must be equal to or less than the tolerance.

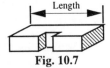

Fig. 10.7

$$\text{Machine } C_p \geq 1.67$$

$$\frac{\text{Tolerance}}{6\sigma_{MACHINE}} \geq 1.67$$

$$\text{Tolerance} \geq 1.67\,(6\sigma_{MACHINE})$$

$$\geq 10\,\sigma_{MACHINE}$$

Doing the best to minimize variation from the remainder of the process (material, operator, etc.) so only machine variation is present, eight pieces are cut and their lengths measured. These first eight consecutive pieces generate a s_8 value of .3972. Specification limits for cut-off length are 125 mm $\pm$ 3 mm, making the tolerance 6 mm. With this information, the first sequential s test ratio is calculated to be .0662.

$$\text{Test Ratio}_8 = \frac{s_8}{\text{Tolerance}} = \frac{.3972}{6} = .0662$$

This first test ratio can be compared to the lower and upper critical values displayed in Table 10.2 for n equal to 8 and .90 confidence. Since the goal is to have $10\sigma_{MACHINE}$ equal the tolerance, no adjustment to the critical values is necessary.

$$\text{LCV}_8 = .0636 \leq .0662 \leq .1310 = \text{UCV}_8$$

The test ratio of .0662 falls between the critical values of .0636 and .1310, so no decision can be made at this point. Two more pieces must be cut and measured. These two new measurements are combined with the original eight and s_{10} calculated (assume it is .4146). An updated test ratio is computed from this as follows:

$$\text{Test Ratio}_{10} = \frac{s_{10}}{\text{Tolerance}} = \frac{.4146}{6} = .0691$$

Compare this ratio to the lower and upper critical values for n equal to 10 (.0680 and .1277).

$$.0680 \le .0691 \le .1277$$

Again this test ratio falls between the critical values, so two additional pieces are cut and measured. When these measurements are combined with the other ten, s_{12} is calculated as .4032, making the third test ratio .0672.

$$\text{Test Ratio}_{12} = \frac{s_{12}}{\text{Tolerance}} = \frac{.4032}{6} = .0672$$

For n equal 12, the lower critical test value is .0712.

$$.0672 < \text{LCV}_{12} = .0712$$

Because this test ratio is less than the lower critical value, the machine has demonstrated 10σ capability with .90 confidence, *i.e.*, machine C_P meets the goal of 1.67. This decision is made after running only 12 pieces, substantially less than the 50 required for the control chart method. Measurements collected for the sequential s machine capability test can be written on a form similar to that displayed in Figure 10.9 on page 578.

Example 10.5 Centerless Grinder

A manufacturing department is considering whether or not to buy a new centerless grinder. The purchase depends on the outcome of a machine capability study for grinding the outer diameter of the wrist pin shown in Figure 10.8. A machine C_P of at least 2.00, with .95 confidence, must be demonstrated for this important feature. This goal means at least one $12\sigma_{MACHINE}$ (6×2.00) spread must be able to fit within the tolerance ($h = 12$).

Fig. 10.8 A wrist pin.

The machine builder gives his assurance, that with a tolerance of 6, there is really nothing to worry about as his machines have always held a much tighter tolerance than that. You inform him that his opinion can be quickly verified with the sequential s test. After following all set-up procedures for conducting a machine capability study, 8 pieces are ground and the O.D. sizes measured. These first eight measurements are listed below.

$$2.9 \quad 3.2 \quad 1.7 \quad 1.9 \quad 2.1 \quad 2.8 \quad 2.5 \quad 3.1$$

To determine whether or not to purchase the machine, the first test ratio must be calculated.

$$\text{Test Ratio}_8 = \frac{s_8}{\text{Tolerance}} = \frac{.5676}{6} = .0946$$

Because the goal is to demonstrate $12\sigma_{MACHINE}$ equals the tolerance at .95 confidence, the critical values in Table 10.2 must be modified by a factor of 10 over 12.

$$\text{LCV}_8 = \frac{10}{12}(.0556) = .0463 \qquad \text{UCV}_8 = \frac{10}{12}(.1418) = .1182$$

As this initial test ratio falls between these two modified critical values, a decision cannot be made at this time. Two more pieces must be run, measured for O.D. size and s_{10} calculated. If these next two measurements are 3.4 and 1.3, the new test ratio is calculated as:

$$\text{Test Ratio}_{10} = \frac{s_{10}}{\text{Tolerance}} = \frac{.7078}{6} = .1180$$

Critical values for n equal 10 must also be multiplied by 10 over 12.

$$\text{LCV}_{10} = \frac{10}{12}(.0608) = .0507 \qquad \text{UCV}_{10} = \frac{10}{12}(.1371) = .1142$$

Due to the test ratio of .1180 for these 10 measurements being greater than the UCV_{10} of .1142, the machine does not appear to meet the required capability goal at .95 confidence. A .95 confidence level implies there is only a 5 percent chance the machine actually meets the goal and a wrong decision has been made in rejecting the machine.

The machine builder must now attempt to improve his machine so the amount of variation can be significantly reduced. After adjustments are completed, the sequential s test should be repeated to verify the effectiveness of these changes and determine if the machine now meets the capability goal at the desired confidence level.

Comments on the Sequential s Test

Prior to using the sequential s test, follow the same set-up procedures for the machine under study as required by the control chart method. To test for a different machine potential capability requirement, multiply the critical values in Table 10.2 by 10 divided by h, where h is the number of machine standard deviations desired to fit within the tolerance. To test for a machine C_P goal of 2.50, at least one $15\sigma_{MACHINE}$ spread must equal the tolerance. With h equal to 15, each critical value must be multiplied by $10/15$.

$$\text{Machine } C_P \geq 2.50$$

$$\frac{\text{Tolerance}}{6\sigma_{MACHINE}} \geq 2.50$$

$$\text{Tolerance} \geq 2.50\,(6\sigma_{MACHINE})$$

$$\geq 15\,\sigma_{MACHINE}$$

To test for a machine C_P goal of 1.33 (one $8\sigma_{MACHINE}$ spread equals the tolerance), multiply each critical value by $10/8$.

This test for machine potential capability applies best where concern is primarily with output variation since the process average is not considered in the calculation of the test ratio. The sequential s test assumes the center of the process output can be easily adjusted to the desired target value. Other assumptions are that the machine output has a normal distribution and is in a good state of control during the testing period. If unstable, this test will most likely indicate the machine lacks potential capability because s_n is inflated due to being calculated from both common *and* assignable causes of machine variation.

If a unilateral specification is given, use the "tolerance" given below for an USL, where T is the targeted process average for this characteristic.

$$\frac{1}{2}\text{Tolerance} = \text{USL} - T \quad \Rightarrow \quad \text{Tolerance} = 2(\text{USL} - T)$$

When only a LSL is specified, calculate the tolerance as $2(T - \text{LSL})$.

$$\frac{1}{2}\text{Tolerance} = T - \text{LSL} \quad \Rightarrow \quad \text{Tolerance} = 2(T - \text{LSL})$$

A similar approach to the sequential s test has been developed by Barnett and Andrews for performing an analysis of gage capability. They discovered that the sequential testing method can dramatically reduce the number of required trials in these types of studies.

It's easy to apply this test procedure if a calculator, preprogrammed to compute the sample standard deviation, is available. Once the first eight measurements are entered, press the s (or σ_{n-1}) key to compute s_n. Divide by the tolerance to derive the test ratio. If a decision cannot be made and two more measurements are taken, just enter them into the calculator, which already has the first eight in memory, and press the s key again. Dividing by the tolerance yields the new test ratio. You may continue entering additional measurements and getting the s_n values all the way up to the 30th measurement, if necessary.

A form for the collection of data and calculation of test ratios required for the sequential s test is displayed in Figure 10.9. As the preparation and setup for this type of machine capability study is similar to that for the control chart method, the front side of this form is identical to the one displayed in Figure 10.2. For a quicker comparison, the upper and lower critical values may be calculated before hand.

A rough estimate of machine C_R is easily derived by multiplying the last test ratio by 6.

$$\text{Machine } \hat{C}_R \cong 6 \times \text{Test Ratio} = \frac{6s}{\text{Tolerance}}$$

Machine C_P is then approximated by taking the inverse of machine C_R.

$$\text{Machine } \hat{C}_P = \frac{1}{\text{Machine } \hat{C}_R} \cong \frac{\text{Tolerance}}{6s}$$

If a sample size of 30 is reached without making a decision to accept or reject the machine, halt the sequential s test. Plot the 30 measurements on an $\overline{X}$ and R chart ($n = 3$, $k = 10$) and employ the control chart method to evaluate machine capability. This type of test outcome will happen often when the actual machine capability is very close to its goal.

| **Data Collection and Test Ratio Calculation Form** | | | | |
| Tolerance =   Confidence Level = C_P Goal = | | | | |
Measurement of Characteristic	s_n	Test Ratio	LCV_n	UCV_n
1_____ 2_____				
3_____ 4_____				
5_____ 6_____				
7_____ 8_____	_____	_____	_____	_____
9_____ 10_____	_____	_____	_____	_____
11_____ 12_____	_____	_____	_____	_____
13_____ 14_____	_____	_____	_____	_____
15_____ 16_____	_____	_____	_____	_____
17_____ 18_____	_____	_____	_____	_____
19_____ 20_____	_____	_____	_____	_____
21_____ 22_____	_____	_____	_____	_____
23_____ 24_____	_____	_____	_____	_____
25_____ 26_____	_____	_____	_____	_____
27_____ 28_____	_____	_____	_____	_____
29_____ 30_____	_____	_____	_____	_____
Test Ratio = s_n / Tolerance	Decision:			

Fig. 10.9 Back of sequential *s* test form.

10.3 Determining Machine Capability Goals

Capability goals are typically specified for an entire process. As the machine is only one of many elements making up a process, its capability goal must be considerably higher than the one for the whole process. But how much higher? Suppose that from prior experience with similar machining operations, the machine by itself is discovered to contribute no more than 40 percent of the total process variation. This means:

$$\sigma_{MACHINE}^2 \leq .40\,\sigma_{TOTAL}^2$$

By performing a series of algebraic manipulations, the above relationship may be rewritten to express the machine capability goal as a percentage of the process capability goal. This is accomplished by first taking the square root of each side of the previous inequality.

$$\sqrt{\sigma^2_{MACHINE,GOAL}} \leq \sqrt{.40\, \sigma^2_{TOTAL,GOAL}}$$

$$\sigma_{MACHINE,GOAL} \leq \sqrt{.40}\, \sigma_{TOTAL,GOAL}$$

Next, both sides of this relationship are multiplied by six, then divided by the tolerance.

$$\frac{6\, \sigma_{MACHINE,GOAL}}{\text{Tolerance}} \leq \sqrt{.40}\; \frac{6\, \sigma_{TOTAL,GOAL}}{\text{Tolerance}}$$

The left-hand side is now the goal for machine C_R, while the second part of the right-hand side is the C_R goal for the entire process.

$$\text{Machine } C_{R,GOAL} \leq \sqrt{.40}\; C_{R,GOAL}$$

If the C_R goal for this entire process is .75, the C_R goal for just the machine portion is determined as .474.

$$\text{Machine } C_{R,GOAL} \leq \sqrt{.40}\; C_{R,GOAL}$$

$$\leq \sqrt{.40}\,(.75)$$

$$\leq .474$$

In order to meet this machine potential capability goal, machine variation, as measured by one $6\sigma_{MACHINE}$ spread, must be no greater than 47.4 percent of the tolerance. The control chart method provides an estimate of the actual machine C_R, which is then compared to this goal so a decision can be reached concerning machine capability. For the sequential s test approach, divide 6 by machine C_R to determine h.

$$h = \frac{6}{\text{Machine } C_{R,GOAL}}$$

For this example, h is 12.66, meaning the machine must have the potential to fit at least one $12.66\sigma_{MACHINE}$ spread within the tolerance.

$$h = \frac{6}{\text{Machine } C_{R,GOAL}} = \frac{6}{.474} = 12.66$$

Critical values in Table 10.2 corresponding to the chosen confidence level must be multiplied by .79 ($10/h = 10/12.66$) to convert them for testing to this goal.

In general, if "V" represents the percentage that machine variation should be of the total process variation, then:

$$\text{Machine } C_{R,GOAL} \leq \sqrt{V}\; C_{R,GOAL}$$

To save time and calculations with the control chart approach, the estimate for $\sigma_{MACHINE}$ extracted from the range (or S) chart can be directly compared to the maximum allowed, referred to as $\sigma_{MACHINE, GOAL}$. This goal for $\sigma_{MACHINE}$ is determined as is shown here:

$$\text{Machine } C_{R,GOAL} \leq \sqrt{V}\, C_{R,GOAL}$$

$$\frac{6\sigma_{MACHINE,GOAL}}{\text{Tolerance}} \leq \sqrt{V}\, C_{R,GOAL}$$

$$\sigma_{MACHINE,GOAL} \leq \frac{\sqrt{V}}{6}(\text{Tolerance})\, C_{R,GOAL}$$

$$\leq \frac{\sqrt{V}}{6}(\text{USL - LSL})\, C_{R,GOAL}$$

When the range chart is in control, divide $\overline{R}$ by the appropriate d_2 factor, then compare this result to $\sigma_{MACHINE, GOAL}$. If the estimate of $\sigma_{MACHINE}$ is less than $\sigma_{MACHINE, GOAL}$, the machine meets its capability goal. If greater, the machine requires additional improvement work.

For an even faster check, a goal can be specified for the maximum $\overline{R}$ (called $\overline{R}_{MACHINE, GOAL}$) that may be seen on the range chart and still meet the machine C_R goal.

$$\sigma_{MACHINE,GOAL} \leq \frac{\sqrt{V}}{6}(\text{USL} - \text{LSL})\, C_{R,GOAL}$$

$$\frac{\overline{R}_{MACHINE,GOAL}}{d_2} \leq \frac{\sqrt{V}}{6}(\text{USL} - \text{LSL})\, C_{R,GOAL}$$

$$\overline{R}_{MACHINE,GOAL} \leq \frac{d_2\sqrt{V}}{6}(\text{USL} - \text{LSL})\, C_{R,GOAL}$$

In a similar manner, machine capability goals can be derived from other process capability measures. Again, let "V" be the desired maximum percentage machine variation should be of the total process variation. For machine C_P:

$$\text{Machine } C_{P,GOAL} \geq \frac{C_{P,GOAL}}{\sqrt{V}} \qquad\qquad h = 6\,\text{Machine } C_{P,GOAL}$$

$$\sigma_{MACHINE,GOAL} \leq \frac{\sqrt{V}\,\text{Tolerance}}{6\,C_{P,GOAL}} \qquad\qquad \overline{R}_{MACHINE,GOAL} \leq \frac{d_2\sqrt{V}\,(\text{USL} - \text{LSL})}{6\,C_{P,GOAL}}$$

For machine C_{PK}:

$$\text{Machine } C_{PK,GOAL} \geq \frac{C_{PK,GOAL}}{\sqrt{V}}$$

For machine C_{PM}:

$$\text{Machine } C_{PM,GOAL} \geq \frac{C_{PM,GOAL}}{\sqrt{V}}$$

For machine C_{PMK}:

$$\text{Machine } C_{PMK,GOAL} \geq \frac{C_{PMK,GOAL}}{\sqrt{V}}$$

If process centering is important to machine performance, then estimating C_{PK}, C_{PM} (Spiring), or C_{PMK} should be the main focus of the machine capability study. Because the sequential s test does not apply when performance capability is being evaluated, no formula for h is given for these last three measures.

When there is no prior experience available to determine V, Bissell recommends setting the goal for machine C_P as 1.33 times the C_P goal for the entire process. Koons mentions setting a machine C_P goal of 1.50 for new equipment.

$$\text{Machine } C_{P,GOAL} = 1.33\, C_{P,GOAL}$$

Example 10.6 Bend Angle

The process capability goal for the bend angle of computer housings (Figure 10.10) formed on a press brake is a C_P of at least 1.67. From studies on similar machines, the variation of the press brake has contributed no more than 65 percent (.65) of the total process variation for bend angle. The machine C_P goal for this characteristic is determined as:

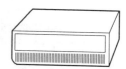

Fig. 10.10

$$\text{Machine } C_{P,GOAL} \geq \frac{C_{P,GOAL}}{\sqrt{V}} = \frac{1.67}{\sqrt{.65}} = 2.071$$

If the sequential s test is to be applied, the h factor for adjusting the lower and upper critical values is found to be 12.426.

$$h = 6\,\text{Machine } C_{P,GOAL} = 6\,(\,2.071\,) = 12.426$$

All critical values from Table 10.2 for the chosen confidence level must be multiplied by .8046 (10 over 12.426) to correctly test for this particular machine capability goal.

10.4 Eliminating Gage Variation from Estimates of Machine Capability

One major source of variation not removed so far from the estimates of machine capability is that due to the measurement system (review Figure 10.1 on page 561). Because pieces produced during the course of a capability study must be measured in some manner before they can be analyzed, the resulting estimates of machine capability reflect variation coming from not only the machine, but the measurement device (usually some type of gage) as well. Because interest is in measuring only machine variation, this additional source of variation must be eliminated from any estimate of machine capability. However, in those cases where the measurement system is an integral part of the machine, its variation should *not* be removed from the estimate of machine capability.

A "pure" estimate of just machine capability is acquired by subtracting gage variation from the total variation measured in the machine capability study, called σ^2_{STUDY}. McNeese and

Klein (1991) contend there are several components to σ^2_{STUDY}, one of which is the true machine variation, $\sigma^2_{MACHINE}$, and another that is measurement system variation, σ^2_{GAGE}. Their analysis assumes all components of variation are independent.

$$\sigma^2_{STUDY} = \sigma^2_{MACHINE} + \sigma^2_{GAGE} + \sigma^2_{OTHERS}$$

Because of the special steps taken when the machine capability study is conducted, σ^2_{OTHERS} is held to 0. This reduces the above equation to the following (Grant and Leavenworth):

$$\sigma^2_{STUDY} = \sigma^2_{MACHINE} + \sigma^2_{GAGE}$$

Rearranging terms to solve for $\sigma_{MACHINE}$ yields this relationship:

$$\sigma^2_{MACHINE} = \sigma^2_{STUDY} - \sigma^2_{GAGE}$$

$$\sigma_{MACHINE} = \sqrt{\sigma^2_{STUDY} - \sigma^2_{GAGE}}$$

This last equation allows a better estimate of the true amount of machine variation by not penalizing the machine for gage variation. Note that gage variation is usually not removed from estimates of *process* capability, because the measurement system is one of the five "M"s, and is quite often a significant part of process variation. However, the inclusion of gage variation puts a lower bound on the process standard deviation since process variation can never be less than gage variation.

Estimates of gage variation typically come from gage R&R (repeatability and reproducibility) studies. For two excellent references on calculating gage variation, see Barrentine's book as well as *Measurement System Analysis*, by the Automotive Industry Action Group.

Example 10.7 Keyway Width

The control chart case study at the beginning of this chapter assessed the capability of a milling machine to produce the width of a keyway (LSL = 20, USL = 100). The range chart in Figure 10.4 has an $\overline{R}$ of 18.5, with n equal to 5. Dividing $\overline{R}$ by d_2 provided an estimate of 7.954 (18.5/2.326) for σ_{STUDY}, making the estimate of machine C_p equal to 1.677.

However, after performing a gage R&R study, gage variation, as measured by $6\sigma_{GAGE}$, is estimated as 20 percent of the tolerance. What is the estimate of machine C_p *without* gaging variation? To find out, an estimate of σ_{GAGE} is needed first. Because $6\hat{\sigma}_{GAGE}$ equals .20 of the tolerance, $\hat{\sigma}_{GAGE}$ is found as follows:

$$6\,\hat{\sigma}_{GAGE} = .20\,(\text{Tolerance})$$

$$\hat{\sigma}_{GAGE} = \frac{.20\,(\text{USL} - \text{LSL})}{6}$$

$$= \frac{.20\,(100 - 20)}{6} = 2.667$$

Substituting this value for $\hat{\sigma}_{GAGE}$ into the formula for $\hat{\sigma}_{MACHINE}$ derived in the last section yields an estimate of 7.4935 for $\sigma_{MACHINE}$.

$$\hat{\sigma}_{MACHINE} = \sqrt{\hat{\sigma}^2_{STUDY} - \hat{\sigma}^2_{GAGE}} = \sqrt{(7.954)^2 - (2.667)^2} = 7.4935$$

With this refined estimate of machine variation, the revised machine C_P becomes 1.779.

$$\text{Machine } \hat{C}_P = \frac{\text{Tolerance}}{6\,\hat{\sigma}_{MACHINE}} = \frac{100 - 20}{6\,(7.4935)} = 1.779$$

The original estimate of 1.677, which contained gage variation, was considered just barely acceptable. This second analysis indicates the true machine capability of 1.779 actually exceeds the required goal of 1.667 by a substantial margin.

The relationship between these three components of variation can be portrayed graphically, as is depicted in Figure 10.11 (Persijn and VanNuland).

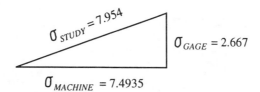

Fig. 10.11   The three components of variation in a machine capability study.

The variation associated with a machine capability study is the combined result of both machine and gage variation. If gage variation can be reduced, the study variation more closely represents just machine variation. In fact, if σ_{STUDY} ever equals 0, then:

$$\sigma_{MACHINE} = \sqrt{\sigma^2_{STUDY} - 0} = \sigma_{STUDY}$$

However, if gage variation is quite large, it can dominate the total variation observed in a machine capability study (Bisgaard *et al*), as is the case in the next example.

Example 10.8 Paint Thickness

A paint booth sprays a coat of enamel paint on various types of metal office furniture, an example of which is displayed in Figure 10.12. One important outcome of this operation is the thickness of the paint once it dries, with a C_{PK} goal of 1.80. The results of a "machine" capability study performed on the paint booth determine $\hat{\mu}_{MACHINE}$ to be 170.13, while σ_{STUDY} is predicted as 2.57. With this information, and knowing the specification limits for paint thickness are 170 ± 10, an estimate of machine C_{PK} (including gage variation) may be calculated as follows, assuming paint thickness has a normal distribution.

Fig. 10.12

$$\text{Original Machine } \hat{C}_{PK} = \text{Minimum}\left(\frac{\hat{\mu}_{MACHINE} - \text{LSL}}{3\,\hat{\sigma}_{STUDY}} , \frac{\text{USL} - \hat{\mu}_{MACHINE}}{3\,\hat{\sigma}_{STUDY}} \right)$$

$$= \text{Minimum}\left(\frac{170.13 - 160}{3(2.57)} , \frac{180 - 170.13}{3(2.57)} \right)$$

$$= \text{Minimum}(1.314 , 1.280) = 1.280$$

This estimate of machine C_{PK} would cause rejection of the painting operation because it is significantly less than the goal of 1.80. But the estimate of variation involved in this calculation is a combination of both machine and gage variation. A gage R&R study done prior to conducting the capability study estimates σ_{GAGE} as 1.99. Given this information, the contribution of variation from the gage can be factored out to derive an estimate of only $\sigma_{MACHINE}$.

$$\hat{\sigma}_{MACHINE} = \sqrt{\hat{\sigma}^2_{STUDY} - \hat{\sigma}^2_{GAGE}} = \sqrt{(2.57)^2 - (1.99)^2} = 1.626$$

$\hat{\sigma}_{MACHINE}$ is now used to re-estimate machine C_{PK}.

$$\text{Revised Machine } \hat{C}_{PK} = \text{Minimum}\left(\frac{\hat{\mu}_{MACHINE} - \text{LSL}}{3\,\hat{\sigma}_{MACHINE}} , \frac{\text{USL} - \hat{\mu}_{MACHINE}}{3\,\hat{\sigma}_{MACHINE}} \right)$$

$$= \text{Minimum}\left(\frac{170.13 - 160}{3(1.626)} , \frac{180 - 170.13}{3(1.626)} \right)$$

$$= \text{Minimum}(2.077 , 2.023) = 2.023$$

This revised estimate for equipment capability of 2.023 more accurately portrays what the painting operation alone is capable of doing. In fact, this second analysis indicates the paint booth exceeds the 1.80 goal by a rather wide margin. Had the decision been based on the original estimate of machine capability, the equipment builder would have been unfairly accused of providing a paint booth with unsatisfactory capability for paint thickness. Valuable resources would have been expended improving equipment that really didn't need the improvement. In addition, no efforts would have been directed to fixing the measurement system. With the revised estimate of machine capability, a correct decision is made and resources are properly focused on the major source of variation.

It's apparent that gage variation is rather large in this example. An estimate of what percentage gage variation is of the tolerance is found as shown here:

$$\frac{6\hat{\sigma}_{GAGE}}{\text{Tolerance}} = \frac{6(1.99)}{180 - 160} = .597 = 59.7\%$$

As many companies want this measure of gage variation to be no more than 20 percent of the tolerance, much work needs to be done on improving this particular measurement system.

The highest machine C_{PK} value occurs when the average paint thickness is centered directly on the target of 170. In this situation, machine C_{PK} equals machine C_P. A machine C_P index equal to the goal of 1.80 is the same as a machine C_R ratio of .556, the inverse of 1.80. Thus, in order to meet the specified machine capability goal, machine C_R must be less than .556. If the decision about paint booth capability is based on $\hat{\sigma}_{STUDY}$, this equipment would *never* be accepted, even if it had absolutely *no* variation whatsoever, because the percentage of the tolerance taken up by gage variation alone (.597) is greater than the maximum percentage of .556 required to satisfy the goal for the entire painting system.

Table 10.3 displays the maximum machine C_{PK} possible (meaning $\sigma_{MACHINE}$ is 0) for various amounts of gage variation. In this table, gage R&R is expressed as the ratio of $5.15\sigma_{GAGE}$ to the tolerance since this ratio is more commonly used in gage R&R studies.

Table 10.3 Maximum machine C_{PK} for various levels of gage variation.

Gage R&R % of Tol.	Maximum Machine C_{PK}
5	17.17
10	8.58
15	5.72
20	4.29
25	3.43
30	2.86
35	2.45
40	2.15
45	1.91
50	1.72
55	1.56
60	1.43

Because many goals for machine C_{PK} are typically around 2.00, or even higher, there isn't much chance for a machine to demonstrate acceptable capability if the measurement system has a gage R&R greater than 35 percent. For better decision making, it's always important to remove the effects of gage variation from the results of machine capability studies.

If a goal for machine variation has been established, and the amount of gage variation is known, a specific goal can be derived for the amount of variation observed in the machine capability study.

$$\sigma_{STUDY,GOAL} = \sqrt{\sigma^2_{MACHINE,GOAL} + \sigma^2_{GAGE}}$$

The estimate of σ_{STUDY} from the machine capability study can be compared directly to this goal, as it has already been adjusted for the contribution of gage variation.

Instead of comparing gage variation to the tolerance, McNeese and Klein (1992) recommend gage variation be responsible for no more than ten percent of the total study variance.

$$\frac{\sigma^2_{GAGE}}{\sigma^2_{STUDY}} \leq .10$$

$$\frac{\sigma^2_{STUDY}}{\sigma^2_{GAGE}} \geq 10$$

$$\frac{\sigma_{STUDY}}{\sigma_{GAGE}} \geq 3.16$$

$$\frac{\sigma_{STUDY}}{3.16\,\sigma_{GAGE}} \geq 1$$

Along the lines proposed by Persijn and Van Nuland, a measurement index, *MI*, may be defined as follows:

$$MI = \frac{\sigma_{STUDY}}{3.16\,\sigma_{GAGE}}$$

The rating scale for *MI* is similar to that for the C_P index: when *MI* is equal to 1.00, gage variability is barely acceptable; values of *MI* greater than 1.00 imply good gage capability; values less than 1.00 mean poor gage capability. For the paint booth example, *MI* is only .41, revealing an extreme lack of gage capability.

$$MI = \frac{\sigma_{STUDY}}{3.16\,\sigma_{GAGE}} = \frac{2.57}{3.16(1.99)} = .41$$

Example 10.9 Paint Thickness

In Example 10.8, assume the C_P goal for the paint booth is also 1.80. This translates into the following goal for the equipment standard deviation:

$$\text{Equipment } C_{P,GOAL} = 1.80$$

$$\frac{\text{Tolerance}}{6\sigma_{EQUIPMENT,GOAL}} = 1.80$$

$$\frac{\text{Tolerance}}{6(1.80)} = \sigma_{EQUIPMENT,GOAL}$$

The specification for paint thickness is 170 ± 10, making the tolerance 20.

$$\frac{20}{6(1.80)} = \sigma_{EQUIPMENT,GOAL}$$

$$1.852 = \sigma_{EQUIPMENT,GOAL}$$

Given that σ_{GAGE} is 1.99, the goal for the overall study standard deviation can now be computed.

$$\sigma_{STUDY,GOAL} = \sqrt{\sigma^2_{EQUIPMENT,GOAL} + \sigma^2_{GAGE}} = \sqrt{1.852^2 + 1.99^2} = 2.72$$

Thus, if the estimate of σ_{STUDY} is less than 2.72, the equipment meets the capability goal. If it is greater, the equipment fails the runoff test and should not be approved for purchase. In this example, $\hat{\sigma}_{STUDY}$ was 2.57. As this result is less than the goal of 2.72, the painting equipment has demonstrated acceptable capability.

Eliminating Gage Variation from Sequential *s* Test

A similar procedure can be applied to remove the influence of gage variation on decisions about machine capability when using the sequential *s* test. The $s_{n,\,STUDY}$ values must first be converted into an estimate of σ_{STUDY} by dividing $s_{n,\,STUDY}$ by the c_4 factor for *n*. c_4 factors are found in Appendix Table II.

$$\hat{\sigma}_{STUDY} = \frac{s_{n,STUDY}}{c_4}$$

This estimate of σ_{STUDY} is substituted into the following equation to predict the standard deviation for just machine variation.

$$\hat{\sigma}_{MACHINE} = \sqrt{\hat{\sigma}^2_{STUDY} - \hat{\sigma}^2_{GAGE}} = \sqrt{\left(\frac{s_{n,STUDY}}{c_4}\right)^2 - \hat{\sigma}^2_{GAGE}}$$

Finally, this estimate of $\sigma_{MACHINE}$ must be multiplied by c_4 to convert back to $s_{n,\,MACHINE}$, which is the value needed to calculate each test ratio.

$$\text{Test Ratio}_n = \frac{s_{n,MACHINE}}{\text{Tolerance}} \qquad \text{where } s_{n,MACHINE} = c_4 \hat{\sigma}_{MACHINE}$$

By comparing this test ratio to the appropriate critical values, a decision unaffected by gage variation can be made whether or not to purchase this machine.

The only drawback to removing gage variation is the extra amount of calculations (and time) needed. Usually the sequential *s* test is performed in real time: as pieces are produced on the machine, they are measured, the test ratio computed, and a decision made to continue testing or to stop. Increasing the amount of required calculations slows down this procedure, causing it to lose some of its advantage over the control chart approach. To alleviate this problem somewhat, the above equation for the test ratio may be rewritten to help reduce the amount of required calculations.

$$s_{n,MACHINE} = c_4 \hat{\sigma}_{MACHINE} = c_4 \sqrt{\hat{\sigma}^2_{STUDY} - \hat{\sigma}^2_{GAGE}} = c_4 \sqrt{\left(\frac{s_{n,STUDY}}{c_4}\right)^2 - \hat{\sigma}^2_{GAGE}}$$

$$= \sqrt{s^2_{n,STUDY} - (c_4 \hat{\sigma}_{GAGE})^2}$$

This last result can be substituted into the test ratio equation.

$$\text{Test Ratio}_n = \frac{s_{n,MACHINE}}{\text{Tolerance}} = \frac{\sqrt{s^2_{n,STUDY} - (c_4 \hat{\sigma}_{GAGE})^2}}{\text{Tolerance}}$$

Because gage variation is known before the sequential *s* test is undertaken, the $c_4 \hat{\sigma}_{GAGE}$ values for *n* of 8, 10, 12, and so on can be determined prior to running the test. Once the test begins, the values of $s_{n,\,STUDY}$ are calculated and substituted into the above equation to quickly derive the test ratios.

Example 10.10 Cut-off Length

In the first example of the sequential s test (Example 10.4), the goal was to have .90 confidence the machine C_P for cut-off length is at least 1.67. Specification limits for cut-off length were 125 mm $\pm$ 3 mm, making the tolerance 6 mm. The first eight pieces generated a s_8 value of .3972, which resulted in a test ratio of .0662.

$$\text{Test Ratio}_8 = \frac{s_8}{\text{Tolerance}} = \frac{.3972}{6} = .0662$$

However, s_8 contains gage variation in addition to machine variation. Before comparing the test ratio to the critical values, gage variation should be removed. A previous gage R&R study reveals σ_{GAGE} is .10.

$$\text{Test Ratio}_8 = \frac{\sqrt{s_{8,STUDY}^2 - (c_{48}\,\hat{\sigma}_{GAGE})^2}}{\text{Tolerance}} = \frac{\sqrt{(.3972)^2 - [.9650\,(.10)]^2}}{6} = .0642$$

This revised test ratio of .0642 is now compared to the lower and upper critical values from Table 10.2 for n equal to 8 and .90 confidence.

$$\text{LCV}_8 = .0636 \le .0642 \le .1310 = \text{UCV}_8$$

Because the test ratio falls between the critical values of .0636 and .1310, no decision can be made at this point. Two more pieces are cut and measured. These two new measurements are combined with the original eight and s_{10} is calculated as .4146. A second revised test ratio is calculated and compared to the critical values for n of 10.

$$\text{Test Ratio}_{10} = \frac{\sqrt{s_{10,STUDY}^2 - (c_{410}\,\hat{\sigma}_{GAGE})^2}}{\text{Tolerance}} = \frac{\sqrt{(.4146)^2 - [.9727\,(.10)]^2}}{6} = .0672$$

$$.0672 < \text{LCV}_{10} = .0680$$

As this test ratio of .0672 is less than the lower critical value of .0680, the machine has demonstrated with .90 confidence it meets the machine C_P goal of 1.67. In the original test, this machine was not accepted until after 12 pieces were run. By making s_n smaller through the elimination of gage variation, the sequential s test is able to accept "good" machines sooner, thereby reducing the total number of pieces, and cost, needed to make a decision. However, this same "benefit" of a smaller s_n will also delay rejecting a machine that does not meet the capability goal.

10.5 Summary

Machine capability studies are a convenient and effective tool for evaluating how much variation is generated by just a single piece of equipment. This information helps management make better decisions about the following: buying new machines; identifying machines in need of repair; determining whether old machines have been properly overhauled; assigning jobs to machines; and establishing realistic tolerances. Some machine builders are already becoming proactive in this regard by providing machine capability documentation, even when customers don't specifically request it. Litsikas has reported that the Universal Instruments Corporation automatically provides a "Machine Performance Certificate" that includes various machine performance statistics, such as *ppm* and C_{PK}, as well as definitions for the terms contained within this document.

Two major procedures for appraising machine capability are the control chart approach and the sequential *s* test. The first allows estimates of both potential and performance capability, but usually requires producing and measuring a large number of test pieces. The second frequently needs fewer pieces run to make a decision concerning capability, and allows decisions to be made at some specified confidence level, but can evaluate only potential machine capability. Specific methods for evaluating the performance of CNC machining centers are covered in the ANSI/ASME B5.54 standard.

Machine capability goals are derived directly from process capability goals. In order to achieve the process capability goal, the machine must have a much higher capability goal.

To make a meaningful assessment of machine capability, the effects of gage variation should be removed from the estimate of machine variation. However, if the measurement system is part of the equipment under study, measurement variation should be included in the estimate of machine capability.

Results of a machine capability study, performed under one set of controlled conditions on a machine builder's shop floor, may not accurately predict machine performance when placed into service at its final destination. Differences in environment and foundations, or even shipping damage, may significantly impact machine performance.

All measures of capability, both process and machine, presented so far are "point estimators" of the true capability. Because of sampling variation, these estimators will vary from study to study, even when the process is in control. The next chapter explains how this expected variation in sample results can be quantified in the form of confidence bounds.

10.6 Exercises

10.1 Burst Strength

A German company is considering purchasing a machine that molds plastic bottles for sparkling mineral water. By conducting a machine capability study, a manufacturing team charged with making this decision hopes to verify this machine is capable for burst strength, which is one of the most important properties of the bottle. 15 subgroups of size 5 are run and plotted on an $\overline{X}$, R chart, which indicates good control for burst strength. $\overline{\overline{X}}$ is 143.5 psi and $\overline{R}$ is 11.2 psi. Given a LSL of 125 psi and a target of 150 psi, help the team estimate machine C'_{PL} and C_{PK} (a goodness-of-fit test validates the assumption of normality). Which index would be more important for making a decision about purchasing this machine?

A gage R&R study reveals gage variation, $\hat{\sigma}_{GAGE}$, to be 1.4 psi for the device measuring burst strength. By how much does this alter the estimates of machine C'_{PL} and C_{PK}?

Based on past experience with similar molding operations, machine capability must be no more than 82 percent of the total process variation. If the goal for the overall process is 2.00 for C'_{PL} and 1.75 for C_{PK}, what should the goals be for machine C'_{PL} and C_{PK}?

After several machine modifications, the process average for burst strength is increased to 151 psi, with no change in $\hat{\sigma}_{MACHINE}$. How does this change the estimates of machine C'_{PL} and C_{PK}? What effect does this have on the decision to purchase this new bottle-molding machine?

10.2 Fill Volume

A dispensing machine has four filling stations which fill aluminum cans with a popular fruit drink. A critical characteristic is the volume of liquid delivered into a can. As each filling station could be performing at a different level, a study is designed to collect 100 consecutive cans from each station for the purpose of measuring equipment capability. After the data from each station is plotted on a separate chart, all stations are judged to be in a good state of control. Although they have different averages and standard deviations, as listed in Table 10.4, the output for each station appears to be normally distributed when plotted on NOPP.

Table 10.4 Capability study results for each station.

Stream	$\overline{\overline{X}}$	$\hat{\sigma}_{ST}$	$\hat{C}_P$	$\hat{C}_{PK}$
1	341.24	.952		
2	340.19	.813		
3	339.19	1.035		
4	342.07	.724		

Given that the specification for fill volume is 340 ml ± 4 ml, analyze potential and performance capability for each station, then make a decision about accepting this dispensing machine. The goals for machine C_P and machine C_{PK} are 1.90 and 1.67, respectively.
Assume $\hat{\sigma}_{GAGE}$ is .173 ml. How much of a change is made to the estimates of machine C_P and machine C_{PK} if the effect of gage variation is removed?

10.3 Connecting Rods

A boring machine is built to produce connecting rods (Figure 10.13). Table 10.5 lists seven critical features that are selected to be part of a machine capability study. 50 rods are machined, measurements made and plotted on control charts ($n = 5$) to check for stability. Because adjusting the process average for each characteristic is relatively easy, interest is in only potential machine capability.

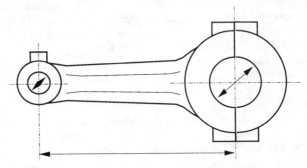

Fig. 10.13 A connecting rod.

Table 10.5 Results of capability study for each feature.

#	Characteristic	Spec. Limits	Control Status	Normality Check	$\overline{R}$	C_P Goal
1	Crank Diameter, Top	±.0005	In	Normal	.000198	1.67
2	Crank Dia., Bottom	±.0005	Out	?	.000151	2.00
3	Pin Diameter, Top	±.0007	In	Normal	.000395	1.50
4	Pin Diameter, Bottom	±.0007	In	Normal	.000364	1.67
5	Dist. Betw. Centerlines	±.0010	Out	?	.000205	2.50
6	Bend	±.0016	In	Normal	.000457	1.90
7	Twist	±.0023	In	Skewed Right	$x_{.135} = -.0011$ $x_{99.865} = .0014$	1.75

What decision can be reached about buying this piece of equipment based on the results of this machine capability study? What follow-up actions must be taken?

10.4 Circuit Board Positioning

An automated production system receives stacks of printed circuit boards from a press room, then unstacks and loads the untrimmed boards onto a conveyor. Processing problems occur if boards are positioned incorrectly on the conveyor. How many boards must be positioned correctly without a single problem to demonstrate that a goal of 1.15 for equipment equivalent $P^{\%}_{PK}$ has been achieved with .75 confidence ($\alpha = .25$)?

10.5 Baking Plates

Chinaware plates are fired in one of six molds as they pass through an oven. Afterwards, they are examined for warping, interior particles, spotty color, and poor finish. After an extensive refurbishing of the oven and molds, a study is conducted to verify oven capability. 6,938 loads are fired and inspected, with the results recorded below in Table 10.6.

Table 10.6 Inspection results for each mold.

Mold #	Nonconf. Found
1	2
2	1
3	11
4	3
5	2
6	9

Make a point estimate of equipment $ppm_{TOTAL, LT}$ for each mold. The goal is 500 ppm. Estimate equipment $ppm_{TOTAL, LT}$ for the output of the entire oven, which is the combination of all six molds. How does this compare to a goal of 500 ppm?

10.6 Punch Press

The goal for a punch press is to have an overall process C_{PM} of at least 1.50 for forming the width of a particular slot. Previous studies indicate that presses involved with this type of operation contribute 72.8 percent of the total process variation. What should be the C_{PM} goal for just the machine?
Suppose a machine capability study finds the machine average equals 289, while the study standard deviation is 1.0. With a LSL of 286, an USL of 294, and a target average of 290, does this punch press meet its capability goal for machine C_{PM}?
You discover the gage used in this study has a standard deviation of .50. Does this change your conclusion about machine capability?
After improvements, the estimated machine average for width becomes 290. Does the press now meet its capability goal? What additional course(s) of action should be initiated?

10.7 Tread Length

In order to manufacture tire tread, synthetic rubber is extruded in a continuous strip onto a moving conveyor belt. After being cooled with a water spray, cement is applied to the bottom of this continuous tread. Rotating blades then cut the tread to the required length. The specific length requirement depends on which type of tire tread is being produced.
To evaluate the capability of just the cutting operation, a quality-improvement team decides to apply the sequential s test. After following all the rules for properly conducting a machine capability study, the first eight lengths are measured and listed here:

$$191.38, \; 191.58, \; 191.40, \; 191.37, \; 191.49, \; 191.53, \; 191.42, \; 191.51$$

The tightest tolerance of all the different tire treads expected to be run on this process is $\pm.3$ cm. Given the goal for machine C_p is 1.75, calculate the first test ratio and make a decision about the potential capability of this cutting machine at the 99 percent confidence level.
If the next two tread length measurements are 191.31 and 191.49, what is your decision?
If the next two tread length measurements are 191.56 and 191.41, what is your decision?
If the next two tread length measurements are 191.35 and 191.54, what is your decision?
From a gage R&R study, $\hat{\sigma}_{GAGE}$ turns out to be .021 cm. How does this additional information affect the outcome of the sequential s test?

10.7 References

Automotive Industry Action Group, *Measurement System Analysis Reference Manual*, 1990, Southfield, MI

Barnett, Andy, and Andrews, Richard, "Measurement Error Study Procedure Featuring Variable Sample Sizes," *Quality Engineering*, Vol. 9, No. 2, 1996-97, pp. 259-267

Barrentine, Larry B., *Concepts for R & R Studies*, 1991, ASQC Quality Press, Milw., WI

Bayer, Harmon S., "How to Determine the Quality Capability of Machine and Tools," *Tooling and Production*, May and October, 1954

Bisgaard,Soren, Graves, Spencer, and Valverde, Rene, "Quality Quandaries," *Quality Engineering*, Vol. 11, No. 2, 1998-1999, pp. 331-335

Bissell, Derek, *Statistical Methods for SPC and TQM*, 1994, Chapman & Hall, London, UK, p. 238

Burr, Irving W., *Statistical Quality Control Methods*, 1976, Marcel Dekker, Inc., New York, NY, p. 371

Chrysler Corp., Ford Motor Co., and General Motors Corp., *Quality System Requirements: Tooling & Equipment Supplement*, Automotive Industry Action Group, Southfield, MI, 1996

De Grote, Irwin A., "Machine Quality Capability Studies," *16th ASQC Midwest Quality Control Conference Transactions*, 1961, St. Louis, MO, pp. 155-169

Down, Michael, and Marshall, Sheri, "What Can We Really Learn from Machine Run-off?," *50th ASQC Annual Quality Congress Transactions*, Chicago, IL, May 1996, pp. 619-624

Gregerson, David A., "Buying Capital Equipment the Right Way," *Quality in Manufacturing*, Vol. 8, No. 1, Jan./Feb., 1997, pp. 34-35

Grandzol, John, and Gershon, Mark, "Multiple Criteria Decision Making," *Quality Progress*, Vol. 27, No. 1, January 1994, pp. 69-73

Grant, Eugene L., and Leavenworth, Richard S., *Statistical Quality Control*, 6th ed., 1988, McGraw-Hill, New York, NY, p. 377

Haggar, Bruce, "Process Capability as a Statistic Used in Process Validation," *Quality Controller*, Vol. 26, No. 2, p. 3

Hurayt, Gerald, "Sample Size for PPM," *Quality*, Vol. 25, No. 5, March 1986, p. 75

Huttunen, Edward K., "Quality Planning for a Precision Machining Process," *46th ASQC Annual Quality Congress Transactions*, Nashville, TN, May 1992, pp. 405-411

Kane, Victor E., "Process Capability Indices," *Journal of Quality Technology*, Vol. 18, No. 1, January 1986, p. 43

Koons, George F., *Indices of Capability: Classical and Six Sigma Tools*, Addison-Wesley Publishing Co., Reading, MA, p. 10

Litsikas, Mary, "Universal Instruments Reaches Out to Customers, Suppliers," *Quality*, Vol. 34, No. 12, December 1995, pp. 48-50

Lucas, Thomas J., and Pilkington, Richard, "Selecting Equipment Using SPC," *Quality*, Vol. 34, No. 2, February 1995, p. 8

McNeese, William H., and Klein, Robert A., *Statistical Methods for the Process Industries*, 1991 ASQC Quality Press, Milwaukee, WI, p. 433

McNeese, William H., and Klein, Robert A., "Measurement Systems, Sampling, and Process Capability," *Quality Engineering*, Vol. 4, No. 1, 1992, pp. 21-39

Nelson, Lloyd, "Sequential Range Capability Test," *Journal of Quality Technology*, Vol. 17, No. 1, January 1985, pp. 57-58

Ouellette, Steven M. and D'Souza, Althea, "Optimum Equipment Start-Ups: Conquering the Commissioning Blues," *54th ASQ Annual Quality Congress Transactions*, May 2000, pp. 54-64

Perez-Wilson, Mario, *Machine/Process Capability Study*, 1989, Advanced Systems Consultants, Scottsdale, AZ, p. 114

Persijn, Marcel, and Van Nuland, Yves, "Relation Between Measurement System Capability and Process Capability," *Quality Engineering*, Vol. 9, No. 1, 1996-97, pp. 95-98

Sikorsky Aircraft, "Characterizing Machines - Shortcut to Quality," *Quality in Manufacturing*, Vol. 5, No. 5, July-August 1994, pp. 40-41

Spiring, Fred A., "A Unifying Approach to Process Capability Indices," *Journal of Quality Technology*, Vol. 29, No. 1, January 1997, pp. 49-58

Strongrich, Andrew L., Herbert, Gerald E., and Jacoby, Thomas J., "Simple Statistical Methods," *Statistical Process Control, SP-547*, 1983, Society of Automotive Engineers, Inc., Warrendale, PA, pp. 31-38

Wheeler, Donald J., and Chambers, David S., *Understanding Statistical Process Control*, 2nd ed., 1992, SPC Press, Inc., Knoxville, TN, pp. 138-139

Chapter 11

Calculating Confidence Bounds for Capability Measures

Uncertainty exists in all measurements.

As seen throughout this book, there are many methods available for estimating process capability. However, no matter which measure is used, it is only an *estimate* of the true process capability. Because of random sampling variation, measurements included in an initial capability study to estimate μ and σ will be slightly different than those collected in a second study on this same process. This is true even when the process is stable. If the estimates of μ and σ change, the estimate of process capability also changes. As an example, recall the formula for estimating the P_R ratio.

$$\hat{P}_R = \frac{6\,\hat{\sigma}_{LT}}{\text{Tolerance}}$$

$\hat{P}_R$ is called a *point estimator* of P_R. In general, a point estimator is a sample statistic used to estimate a process parameter. $\overline{X}$ is a point estimator of μ, $\overline{R}/d_2$ is a point estimator of σ_{ST}, $\overline{p}$ is a point estimator of p'. Because point estimators are calculated from sample measurements taken from a much larger number of possible measurements, it is risky to say a point estimator is *exactly* equal to the true value of the process parameter it is trying to predict.

Assume the true P_R ratio for a certain process is .50. P_R is estimated from an initial process capability study as .53. A week later, without any alterations to this process, $\hat{P}_R$ from a second study is calculated as .48. Which is the "correct" estimate of P_R?

$$\hat{P}_{R,1} = .53 \qquad \hat{P}_{R,2} = .48$$

The answer is neither is "correct." They are both just estimates of the true P_R measure. In fact, they are only two of many such estimates that could result from a capability study conducted on this process. So what "number" should be reported to top management and customers to accurately describe the capability of this process?

Unfortunately, a certain amount of uncertainty in sample estimates is unavoidable and any approach to properly measuring process capability must comprehend this fact. Fortunately, there are methods for statistically quantifying this inherent uncertainty. These procedures establish "confidence bounds" for process parameters, as well as capability measures, and are explained in this chapter. In fact, several companies, *e.g.*, Boeing Commercial Airplane Group, are now actively encouraging their suppliers to provide confidence bounds as part of all capability study reports. As this topic involves more advanced material, the reader may wish to skip this chapter during a first reading of this book and return to it at a later time.

11.1 Confidence Bounds for the Process Average, μ

If many estimates of the process average are made while a process is in control, they may all be slightly different, but will form a certain distribution. Recall from the initial review in Chapter 2 (pp. 10-12) that subgroup averages, $\overline{X}$, vary from subgroup to subgroup, but over time, follow a normal distribution due to the central limit theorem (Figure 11.1).

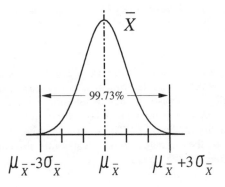

Fig. 11.1 Sampling distribution for subgroup averages.

This is called the *sampling distribution* of the subgroup averages. The average of this distribution of $\overline{X}$s is labeled $\mu_{\overline{X}}$, which is equal to μ, the process average.

$$\mu_{\overline{X}} = \mu$$

Because the sampling distribution of $\overline{X}$ is normally distributed (review Section 2.2), 99.73 percent of the $\overline{X}$s fall between $\mu_{\overline{X}} - 3\sigma_{\overline{X}}$ and $\mu_{\overline{X}} + 3\sigma_{\overline{X}}$. The probability the next $\overline{X}$ collected from this process falls between these two values is summarized with the following equation:

$$P\,(\mu_{\overline{X}} - 3\sigma_{\overline{X}} \leq \overline{X} \leq \mu_{\overline{X}} + 3\sigma_{\overline{X}}) = .9973$$

Since $\mu_{\overline{X}}$ is equal to μ, it may be replaced by μ in the above equation.

$$P\,(\mu - 3\sigma_{\overline{X}} \leq \overline{X} \leq \mu + 3\sigma_{\overline{X}}) = .9973$$

Thus, there is only a very small chance (.27 percent) that an $\overline{X}$ from any subgroup is more than a distance of $3\sigma_{\overline{X}}$ from the true process average, μ.

The relationship inside the parentheses of the last equation may be rearranged as follows:

$$P(\mu - 3\sigma_{\overline{X}} \leq \overline{X} \leq \mu + 3\sigma_{\overline{X}}) = .9973$$

$$P(-3\sigma_{\overline{X}} \leq \overline{X} - \mu \leq +3\sigma_{\overline{X}}) =$$

$$P(-\overline{X} - 3\sigma_{\overline{X}} \leq -\mu \leq -\overline{X} + 3\sigma_{\overline{X}}) =$$

Multiplying through this last inequality by -1 reverses the direction of the inequality signs.

$$P(\overline{X} + 3\sigma_{\overline{X}} \geq \mu \geq \overline{X} - 3\sigma_{\overline{X}}) = .9973$$

The above inequality can be rewritten as follows:

$$P(\overline{X} - 3\sigma_{\overline{X}} \leq \mu \leq \overline{X} + 3\sigma_{\overline{X}}) = .9973$$

Recall that the standard deviation of the subgroup averages is equal to the process standard deviation divided by the square root of the subgroup size (p. 11).

$$\sigma_{\overline{X}} = \frac{\sigma}{\sqrt{n}}$$

Substituting the above for $\sigma_{\overline{X}}$ in the prior relationship produces this result:

$$P\left(\overline{X} - 3\frac{\sigma}{\sqrt{n}} \leq \mu \leq \overline{X} + 3\frac{\sigma}{\sqrt{n}}\right) = .9973$$

As $\overline{X}$ is an estimate of the process average based on a sample of n measurements, the symbol $\hat{\mu}$ can replace it in the above inequality to make it more general.

$$P\left(\hat{\mu} - 3\frac{\sigma}{\sqrt{n}} \leq \mu \leq \hat{\mu} + 3\frac{\sigma}{\sqrt{n}}\right) = .9973$$

When working with capability studies there are kn measurements available to estimate μ. Remember that k is the number of subgroups in the study and n is the subgroup size.

$$P\left(\hat{\mu} - 3\frac{\sigma}{\sqrt{kn}} \leq \mu \leq \hat{\mu} + 3\frac{\sigma}{\sqrt{kn}}\right) = .9973$$

In statistics, the span between $\hat{\mu} - 3\sigma/\sqrt{kn}$ and $\hat{\mu} + 3\sigma/\sqrt{kn}$ is known as a .9973 *confidence interval* for μ. It is called this because the probability this interval includes the actual process average is 99.73 percent. In practice, it is usually expressed as:

$$\text{A .9973 confidence interval for } \mu = \hat{\mu} \pm 3\frac{\sigma}{\sqrt{kn}}$$

Only a .135 percent chance exists that the lower end of this interval is greater than μ. Therefore, $\hat{\mu} - 3\sigma/\sqrt{kn}$ is referred to as a lower .99865 confidence bound for μ because the probability is .99865 this lower bound is less than the true process average.

A lower .99865 confidence bound for $\mu = \hat{\mu} - 3\dfrac{\sigma}{\sqrt{kn}}$

Instead of writing out this long phrase, the lower confidence bound is commonly given the following notation, where the tilde (~) *below* the process parameter indicates it is a *lower* confidence bound (Hahn and Meeker, p. 46), while the numerical subscript denotes the level of confidence.

$$\underset{\sim}{\mu}_{.99865} = \hat{\mu} - 3\dfrac{\sigma}{\sqrt{kn}}$$

Likewise, there is only a .135 percent chance the upper end of the .9973 confidence interval is less than μ, so $\hat{\mu} + 3\sigma/\sqrt{kn}$ is referred to as an upper .99865 confidence bound for μ. Note that for an upper confidence bound, the tilde is above $\mu_{99.865}$.

$$\tilde{\mu}_{.99865} = \text{An upper .99865 confidence bound for } \mu = \hat{\mu} + 3\dfrac{\sigma}{\sqrt{kn}}$$

To gain a better understanding of this confidence concept, work through the next example.

Example 11.1 Valve O.D. Size

During the course of a capability study conducted on a grinding operation producing an automotive exhaust valve (Figure 11.2), μ is estimated from $\overline{\overline{X}}$ to be 10.0041 (p. 40).

Fig. 11.2 An automotive exhaust valve.

This estimate is based on measurements collected in 20 subgroups of size 5 each. The process standard deviation is known to be .01610 cm from previous studies on this process. A confidence interval with a .9973 probability of containing μ is computed as:

$$\text{A .9973 confidence interval for } \mu = \hat{\mu} \pm 3\dfrac{\sigma}{\sqrt{kn}}$$

$$= 10.0041 \text{ cm} \pm 3\dfrac{.01610 \text{ cm}}{\sqrt{20(5)}}$$

$$= 10.0041 \text{ cm} \pm .00483 \text{ cm}$$

$$= 9.99927 \text{ cm to } 10.00893 \text{ cm}$$

The probability is .9973 this interval from 9.99927 cm to 10.00893 cm covers the true process average. Thus, this distance is referred to as a .9973 confidence interval for μ, and is shown in Figure 11.3. There is only a .0027 chance that this interval does not cover μ. In general, this chance is called the α (alpha) risk.

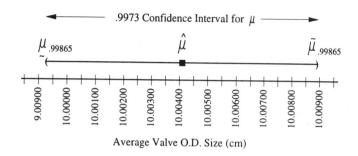

Fig. 11.3 A .9973 confidence interval for μ.

For this example, the α risk equals .0027, which is the risk of believing the interval includes μ, when in fact, it doesn't. As the probability is .99865 that 9.99927 cm is less than μ, this lower limit of the confidence interval is called a lower .99865 confidence bound for μ.

$$\underset{\sim}{\mu}_{.99865} = 9.99927 \text{ cm}$$

Half of the α risk (.00135) is that this lower confidence bound is greater than μ.

$$\alpha/2 = .0027/2 = .00135$$

Likewise, 10.00893 cm is an upper .99865 confidence bound for μ because the probability is .99865 this value is greater than μ.

$$\tilde{\mu}_{.99865} = 10.00893 \text{ cm}$$

The other half of the α risk is that this upper confidence bound of 10.00893 is lower than the actual process average, μ. If this level of risk is not acceptable, it may be easily changed, as is explained in the next section.

Adjusting the Risk Level, α

Instead of requiring a .9973 confidence interval ($\alpha = .0027$) for μ, assume a .99 confidence interval ($\alpha = .01$) is desired. The .9973 confidence interval is equal to a distance of ± 3 standard deviations around the estimated process average. The "3" represents the Z value for .00135, which is half of the risk ($\alpha/2 = .00135$). By changing this Z value, confidence intervals with different risk levels are obtained.

If a .99 confidence interval is needed, the α risk is .01. To construct an interval (two tails) with this particular risk, replace the Z value of 3 with one corresponding to .005, which is half the α risk. Looking for .005 in the body of Appendix Table III, the new Z value is found to be about 2.58, making the formula for a .99 confidence interval as follows:

$$\text{A .99 confidence interval for } \mu = \hat{\mu} \pm 2.58 \frac{\sigma}{\sqrt{kn}}$$

For the valve O.D. size example, this interval becomes:

$$\hat{\mu} \pm 2.58 \frac{\sigma}{\sqrt{kn}} \;=\; 10.0041 \text{ cm} \pm 2.58 \frac{.01610 \text{ cm}}{\sqrt{20(5)}}$$

$$= \; 10.0041 \text{ cm} \pm .00415 \text{ cm}$$

$$= \; 9.99995 \text{ cm to } 10.00825 \text{ cm}$$

Note how in Figure 11.4 this .99 confidence interval is *shorter* than the .9973 confidence interval of 9.99927 cm to 10.00893 cm originally calculated for this characteristic.

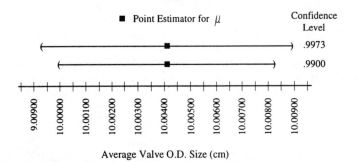

Fig. 11.4 Comparing .9973 and .99 confidence intervals for μ.

In general, the formula for a 1 minus α confidence interval is written as follows, where $Z_{\alpha/2}$ is the Z value for half of α, *i.e.*, $Z_{\alpha/2} = Z(\alpha/2)$.

$$\text{A } 1 - \alpha \text{ confidence interval for } \mu \;=\; \hat{\mu} \pm Z_{\alpha/2} \frac{\sigma}{\sqrt{kn}}$$

Use these next formulas if only a one-sided 1 minus α confidence interval, called a confidence bound, is needed.

$$\underset{\sim}{\mu}_{1-\alpha} \;=\; \text{Lower } 1 - \alpha \text{ confidence bound for } \mu \;=\; \hat{\mu} - Z_{\alpha} \frac{\sigma}{\sqrt{kn}}$$

$$\tilde{\mu}_{1-\alpha} \;=\; \text{Upper } 1 - \alpha \text{ confidence bound for } \mu \;=\; \hat{\mu} + Z_{\alpha} \frac{\sigma}{\sqrt{kn}}$$

When constructing a 1 minus α confidence bound, all the α risk goes into just one tail, so $Z_{\alpha/2}$ is replaced with Z_{α}.

Example 11.2 Valve O.D. Size

A .9973 confidence interval for μ was found previously for the exhaust valve example where μ is estimated from $\overline{X}$ as 10.0041 cm.

$$\hat{\mu} \;=\; \overline{X} \;=\; 10.0041 \text{ cm}$$

What if a .95 confidence interval ($\alpha = .05$) for μ is desired instead? From Appendix Table III, $Z_{.025}$ is found to be 1.96.

$$Z_{\alpha/2} = Z_{.05/2} = Z_{.025} = 1.96$$

From past studies, the process standard deviation is known to be .01610 cm. Given that 20 in-control subgroups of size 5 each are collected during the study, a .95 confidence interval for μ is computed as is done here:

$$\hat{\mu} \pm Z_{.025}\frac{\sigma}{\sqrt{kn}} = 10.0041 \text{ cm} \pm 1.96\frac{.01610 \text{ cm}}{\sqrt{20(5)}} = 10.00094 \text{ cm to } 10.00726 \text{ cm}$$

There is .95 confidence the interval from 10.00094 cm to 10.00726 cm covers the true process average. As illustrated in Figure 11.5, this is a smaller distance than the interval for .9973 confidence (9.99927 to 10.00893) as well as the interval for .99 confidence (9.99995 to 10.00825). For comparison, a .90 confidence interval is also shown.

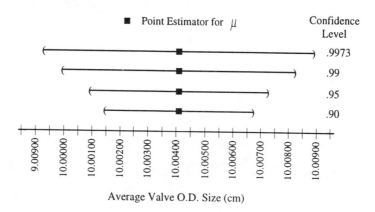

Fig. 11.5 A comparison of several confidence intervals for μ.

Obviously, the wider an interval, the greater the confidence it captures the process average. Higher confidence corresponds to a lower risk level because the two are related as follows:

$$\text{Confidence} = 1 - \text{Risk} = 1 - \alpha$$

As the risk decreases, the $Z_{\alpha/2}$ value becomes larger (see Table 11.1 on the next page), thus generating a wider confidence interval. In fact, to be 100 percent confident, the interval must stretch from minus infinity to plus infinity. Although there is 100 percent probability the process average is *somewhere* in this interval, this statement does not provide much helpful information. Thus, some certainty must be sacrificed for usefulness. For many capability studies, a .95 confidence interval offers a reasonable trade-off between these two, providing a fairly good idea of where the process average is located, with only a relatively small risk of error.

When additional capability studies are performed on the valve grinding operation, assuming it remains in good control, slightly different estimates of μ will be obtained. Even if kn and

Table 11.1 Z_α and $Z_{\alpha/2}$ values for several α risk levels.

$1 - \alpha$	α	Z_α	$\alpha/2$	$Z_{\alpha/2}$
.80	.20	.842	.10	1.282
.90	.10	1.282	.05	1.645
.95	.05	1.645	.025	1.960
.98	.02	2.054	.01	2.326
.99	.01	2.326	.005	2.576
1.00	.00	$+\infty$	.00	$+\infty$

σ remain unchanged for all these extra studies, the confidence intervals will vary because they are centered on $\hat{\mu}$, which is usually slightly different for each study. However, over time, approximately $(1 - \alpha) \times 100$ percent of these intervals should include the true process average. For example, Figure 11.6 displays twenty .90 confidence intervals calculated for the valve O.D. size example, based on the twenty estimated process averages obtained from twenty different capability studies, each with kn equal to 100.

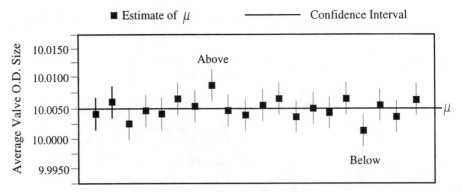

Fig. 11.6 A comparison of twenty .90 confidence intervals for μ.

A .90 confidence interval implies that of many such intervals computed for this process parameter, approximately 90 percent will include the true process average (which is actually 10.0050 cm), while 10 percent ($\alpha = .10$) will miss it. Half of the "misses" ($\alpha/2 = .05$) will err by being too high, while the other half will err by being too low.

In the example displayed in Figure 11.6, 1 of the 20 intervals, or 5 percent, has its lower confidence bound above 10.0050 cm. If by bad luck this just happens to be the one interval calculated from the sample data collected during the capability study, a misleading conclusion will be drawn that the process average is located higher than 10.0050 cm. If the goal is to keep the process centered at 10.0050 cm, action will be taken to lower the process average based on the results of this study, which imply the average is significantly higher than 10.0050 cm. When a centering adjustment is made, the process will actually be shifted to a lower average than desired. This is the *wrong* decision as the process was really centered at 10.0050 cm, but unfortunately this particular study just happened to have an unusually high sample average.

The risk of making this type of false conclusion was set at .05 ($\alpha/2 = .10/2$) for this example. There is also a .05 risk of obtaining a confidence interval where the upper bound is *lower* than the true process average, and then falsely deciding to raise μ. If this combined 10 percent risk is too great, α must be reduced and a corresponding confidence interval calculated and analyzed. The decision about what α risk to use must be made prior to the collecting of data. It's not fair to first see the results of the capability study, then select a convenient risk level.

Remember, for any single estimate of μ, there is 50 percent confidence the actual process average is greater than this point estimate and 50 percent confidence it is less.

Example 11.3 Particle Contamination

A capability study is conducted on particle contamination of a cleaning bath for printed circuit boards coming off a wave soldering operation. There are 25 in-control plot points on the *IX* chart ($n = 1$) available for analysis. From these 25 measurements, μ is estimated from $\overline{X}$ as 127.30 *ppm*. The process standard deviation is known to be 12.90 *ppm* from past studies.

$$\hat{\mu} = \overline{X} = 127.30\ ppm \qquad \sigma = 12.90\ ppm$$

The team investigating this cleaning bath realizes 127.30 *ppm* is only an estimate of the true particle contamination. A second set of 25 contamination readings will most likely generate another estimate of μ that is different from the first. To quantify how much variation in these estimates can be expected, the team decides to construct a confidence interval for μ. However, as concern is only about how bad (high) particle contamination could actually be, the team members are only interested in an *upper* confidence bound. After selecting an α risk of .05, they calculate an upper .95 confidence bound for the true average particle contamination based on the above study average of 127.30 *ppm*. Notice that the entire alpha risk of .05 is in just this upper tail ($Z_{.05}$ is from Table 11.1 on the previous page).

$$\tilde{\mu}_{.95} = \hat{\mu} + Z_{.05}\frac{\sigma}{\sqrt{kn}} = 127.30\ ppm + 1.645\frac{12.90\ ppm}{\sqrt{25(1)}} = 131.54\ ppm$$

The team has .95 confidence the actual average particle contamination of this cleaning bath is less than 131.54 *ppm*. If the USL is 135 *ppm*, the team is quite confident the process meets this specification, as there is only a .05 risk that the process average is greater than this upper confidence bound of 131.54, which is less than the USL of 135.

Quite often when confidence intervals are applied to measures of process capability, interest is also in only one side of the interval. For instance, when process capability is estimated with measures such as P_{PK}, where larger results indicate better capability, concern centers on how low the true P_{PK} might be for this process. Thus, only a lower confidence bound is usually calculated, as is displayed in Figure 11.7.

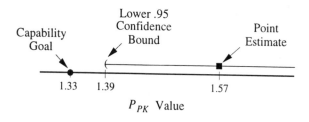

Fig. 11.7 Comparing a lower .95 confidence bound for P_{PK} to the goal for P_{PK}.

Hopefully, this lower bound for P_{PK} is at, or even above, the goal for P_{PK}, as this would provide 1 minus α confidence the goal is met. So, instead of calculating a confidence *interval* for a capability measure, almost all capability studies use either a lower or an upper confidence *bound*, but seldom both. Because the lower .95 confidence bound is greater than the goal of 1.33 shown in Figure 11.7, there is at least .95 confidence this process exceeds its mandated capability objective.

Student's *t* Distribution

The confidence interval formula developed in the last section is based on the normal distribution. This distribution is appropriate *if* the process standard deviation, σ, is known from previous studies. Unfortunately, this is *not* true in most capability studies. σ is usually unknown and must also be estimated from the same sample measurements collected during the study that are used to estimate μ. When σ is estimated with S_{TOT}, the appropriate sampling distribution for constructing a confidence interval is the Student's *t* distribution displayed in Figure 11.8 (McNeese and Klein). Similar in shape to the normal distribution, it is somewhat "flatter" near the middle, with slightly more area in the tail regions.

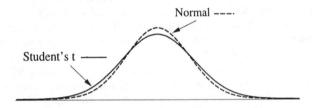

Fig. 11.8   The Student's *t* distribution compared to the normal distribution.

As the sample size increases, the sample standard deviation, S_{TOT}, becomes closer to the true process standard deviation, σ, and the shape of the Student's *t* distribution becomes more like that of the normal distribution. Thus, the spread of the Student's *t* distribution depends on the sample size, or more correctly, the number of *degrees of freedom*, as is revealed in Figure 11.9 on the facing page. There is a unique Student's *t* distribution for each sample size: the smaller the sample size, the greater the spread of this distribution. Selecting the correct Student's *t* distribution for constructing a confidence interval depends on the degrees of freedom, which is the topic of the next section.

Number of Degrees of Freedom, ν

One important aspect of the Student's *t* distribution is the concept of "degrees of freedom." The number of degrees of freedom for this distribution is the number of *independent* measurements involved in computing the estimate of σ. For instance, in studies concentrating on long-term process capability, S_{TOT} is used in the estimation of σ_{LT}. The calculation of S_{TOT} involves taking the deviation of each individual measurement from the sample average.

$$S_{TOT} = \sqrt{\frac{\sum_{i=1}^{kn} (X_i - \hat{\mu})^2}{kn - 1}}$$

There are *kn* measurements in the study, and therefore, *kn* deviations of individual measurements from the sample average. The sum of these *kn* deviations always equals 0. This allows the first *kn* minus 1 deviations to be any value, but fixes the last one because the sum of all *kn* deviations must be zero. As the first *kn* minus 1 measurements have the "freedom" to be any value, the degrees of freedom for S_{TOT} is *kn* minus 1. For example, the measurements of a study are 37, 33, 30, and 36. The average of this sample of four is 34.

$$\overline{X} = \frac{\sum_{i=1}^{kn} X_i}{kn} = \frac{(37+33+30+36)}{4} = \frac{136}{4} = 34$$

The deviation of each measurement from this average is listed in Table 11.2.

Table 11.2 Deviations from the sample average.

i	X_i	$X_i - \overline{X}$
1	37	+3
2	33	-1
3	30	-4
4	36	+2
Sum	136	0

Because the sum of deviations from the sample average must always be zero, the fourth deviation is known after the first three are determined. In this example, the sum of the first three deviations is -2 (+3 + -1 + -4 = -2). Since the sum of all four is zero, the fourth deviation has no choice but to be 2. As this last deviation has no freedom to be any other value, one degree of freedom is "lost."

The number of degrees of freedom is extremely important because it influences the shape of the Student's *t* distribution. Quite often, the Greek letter ν (pronounced "new") is employed to represent the degrees of freedom. The degrees of freedom associated with the calculation of S_{TOT} is equal to one less than the total number of measurements.

$$\nu = kn - 1$$

In Figure 11.9, *t* distributions are displayed for ν equal to 20, 50, and 100. Notice how as ν increases, the Student's *t* distribution becomes more and more like a normal distribution.

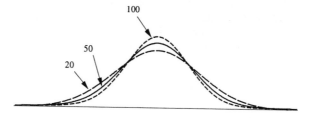

Fig. 11.9 The Student's *t* distribution for various degrees of freedom.

$t_{v,\alpha}$ is defined as the point on the x axis of the Student's t distribution for v degrees of freedom to the right of which lies an area equal to α underneath this curve.

$$P(X > t_{v,\alpha}) = \alpha$$

For example, $t_{50,.05}$ represents the x value to the right of which lies 5 percent of a Student's t distribution for 50 degrees of freedom, as is shown below in Figure 11.10.

$$P(X > t_{50,.05}) = P(X > 1.676) = .05$$

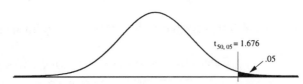

Fig. 11.10   The upper 5 percent tail of a Student's t distribution for $v = 50$.

Values for $t_{v,\alpha}$ are provided in Appendix Table VII for v from 1 through 500 (by 1) and α values of .10, .05, and .01. For example, $t_{50,.05}$ is found to be 1.676, as is seen in Figure 11.11.

v	.10	.05	.01
26	1.315	1.706	2.479
⋮	⋮	⋮	⋮
48	1.299	1.677	2.407
49	1.299	1.677	2.405
50	1.299	1.676	2.403

Fig. 11.11   A section of Appendix Table VII.

Notice how the $t_{50,.05}$ value of 1.676 is quite close to the $Z_{.05}$ value of 1.645. In fact, as the number of degrees of freedom increases, the shape of the t distribution becomes more and more like that of a normal distribution, and $t_{v,\alpha}$ approaches Z_{α}. For 250 degrees of freedom, $t_{250,.05}$ equals 1.651. For 500 degrees of freedom, $t_{500,.05}$ is 1.648. For an infinite number of degrees of freedom, $t_{\infty,.05}$ is 1.645, equaling $Z_{.05}$. This occurs because S_{TOT} becomes closer to the true process standard deviation as the sample size increases.

When σ_{LT} is estimated from the S_{TOT} value calculated from sample measurements, the Z values in the confidence interval formula for μ must be replaced with t values.

$$1 - \alpha \text{ Confidence interval for } \mu = \hat{\mu} \pm t_{v,\alpha/2} \frac{S_{TOT}}{\sqrt{kn}}$$

The same applies to the confidence bound formulas.

$$\underset{\sim}{\mu}_{1-\alpha} = \hat{\mu} - t_{v,\alpha}\frac{S_{TOT}}{\sqrt{kn}} \qquad \tilde{\mu}_{1-\alpha} = \hat{\mu} + t_{v,\alpha}\frac{S_{TOT}}{\sqrt{kn}}$$

Remember, when constructing a single confidence bound, all of α occurs in just one tail.

Example 11.4 Valve O.D. Size

A .95 confidence interval for μ of 10.00094 cm to 10.00726 cm was found previously for the exhaust valve example where μ was estimated as 10.0041 cm from 20 subgroups of size 5 each ($kn = 100$). However, this assumed σ_{LT} was known from previous studies to be .01610 cm. Actually, it is not, so instead, S_{TOT} is calculated from these 100 sample measurements and found to be .01606 cm.

$$S_{TOT} = \sqrt{\frac{\sum\limits_{i=1}^{100}(X_i - \hat{\mu})^2}{100 - 1}} = .01606 \text{ cm}$$

This outcome for S_{TOT} is then used with the appropriate Student's t distribution in the confidence interval formula for μ. Selecting the correct t distribution requires determining the number of degrees of freedom for S_{TOT}.

$$v = kn - 1 = 20(5) - 1 = 99$$

From Appendix Table VII, the t value for 99 degrees of freedom and $\alpha/2$ equal to .025 is found to be 1.984, which is just slightly greater than the $Z_{.025}$ value of 1.960.

$$\hat{\mu} \pm t_{99,.05/2}\frac{S_{TOT}}{\sqrt{kn}} = 10.0041 \text{ cm} \pm 1.984\frac{.01606 \text{ cm}}{\sqrt{20(5)}} = 10.00091 \text{ cm to } 10.00729 \text{ cm}$$

When the process standard deviation is unknown, the probability is .95 that the interval from 10.00091 cm to 10.00729 cm covers the true process average. Notice that this is a slightly wider interval than the one calculated assuming the standard deviation is known (Figure 11.12).

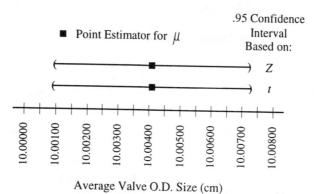

Fig. 11.12 A comparison of confidence intervals for μ.

Due to the increased uncertainty of not knowing the exact spread of the process output distribution, the *t* distribution is wider than the normal distribution. For any *x* value located above the average, there is more area under the curve to the right of this value for the *t* distribution than for the normal. In Figure 11.13, 10 percent of the *t* distribution (with 18 degrees of freedom) is to the right of an *x* value of 1.33. For a normal distribution, this area is .0918.

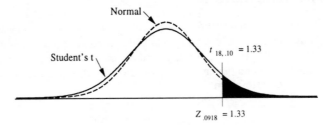

Fig. 11.13 The *t* distribution for $\nu = 18$ versus a normal distribution.

Therefore, to have an identical area to the right, the *Z* value for a given α risk must be *less* than the *t* value for the same α. For example, Figure 11.14 shows how the *Z* value for .10 is 1.28, somewhat less than the *t* value of 1.33 for .10. This relationship between these two distributions causes the confidence interval for μ derived with the *t* value to always be larger than the one derived from the *Z* value. This disparity is a direct result of the increased uncertainty in the expected width of the distribution due to estimating the standard deviation.

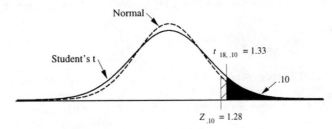

Fig. 11.14 The upper .10 tails of a normal and *t* distribution for $\nu = 18$.

Example 11.5 Particle Contamination

The capability study for particle contamination of a cleaning solution for circuit boards has 25 in-control plot points on the *IX* chart ($n = 1$). From these 25 readings, μ is estimated from $\overline{X}$ as 127.30 *ppm*. As the process standard deviation is unknown, S_{TOT} is also calculated from these 25 readings and turns out to be 12.80 *ppm*.

$$\hat{\mu} = \overline{X} = 127.30 \, ppm \qquad\qquad S_{TOT} = 12.80 \, ppm$$

Because concern is focused on only how high particle contamination could actually be, the team members are interested in just an *upper* confidence bound for μ. As a first step in constructing this bound, the number of degrees of freedom for S_{TOT} are determined.

$$\nu = kn - 1 = 25 \, (1) - 1 = 24$$

After deciding on an α risk of .05, they are able to look up the appropriate *t* value for an .05 risk in Appendix Table VII as 1.711.

$$t_{24,.05} = 1.711$$

Armed with all this information, the team calculates an upper .95 confidence bound for the average particle contamination using the formula based on the Student's *t* distribution.

$$\tilde{\mu}_{1-\alpha} = \hat{\mu} + t_{24,.05}\frac{S_{TOT}}{\sqrt{kn}}$$

$$= 127.30\ ppm + 1.711\frac{12.80\ ppm}{\sqrt{25\,(1)}}$$

$$= 131.68\ ppm$$

The team members have .95 confidence the average particle contamination of this cleaning bath is less than 131.68 *ppm*. Due to the added uncertainty associated with having to estimate σ_{LT}, this upper bound is slightly greater than the one of 131.54 *ppm* calculated with $Z_{.05}$ when the standard deviation was known to be 12.90 *ppm* (review page 603).

Just how much uncertainty is involved with using S_{TOT} as an estimate of the process standard deviation? One way to find out is by constructing a confidence bound for σ_{LT}.

11.2 Upper Confidence Bound for σ_{LT}

Quite often, interest is in constructing a confidence bound for the long-term standard deviation of the process, σ_{LT}. The ratio of ν times S^2_{TOT} divided by σ^2_{LT} follows a χ^2 (pronounced chi squared) distribution with ν degrees of freedom (Duncan, 1986).

$$\frac{\nu\,S^2_{TOT}}{\sigma_{LT}^{\,2}} \sim \chi^2_\nu$$

Just as for the Student's *t* distribution, ν represents the number of degrees of freedom for S_{TOT}, which is *kn* minus 1.

$$\nu = kn - 1$$

Like the Student's *t* distribution, there is a distinct χ^2 distribution for each value of ν. Three of these (for ν = 20, 30, and 50) are displayed in Figure 11.15. The average of a given χ^2 distribution is its degrees of freedom, ν.

Because ν, S_{TOT}, and σ_{LT} are all positive, the χ^2 distribution is skewed to the right, as any ratio of these three cannot be less than zero. However, the shape of the χ^2 distribution begins to look somewhat like a normal distribution as ν increases. The *x* coordinates for various areas underneath the χ^2 distribution (.99, .95, .90, .10, .05, .01) for a given ν (1 through 500) are found in Appendix Table VI. The values in the body of the table are the *x* coordinates for areas under the χ^2 curve to the *right* of that value.

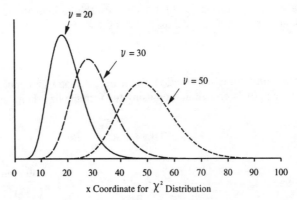

Fig. 11.15 χ^2 distributions for various degrees of freedom (v).

As presented in Figure 11.16, the x coordinate having an area of .01 to its right is about 50.89 for a χ^2 distribution with 30 degrees of freedom (see Figure 11.17 on the next page). In general, this x coordinate is written as $\chi^2_{v,\alpha}$. For this particular example, it becomes $\chi^2_{30,.01}$.

$$P(X > \chi^2_{v,\alpha}) = \alpha$$

$$P(X > \chi^2_{30,.01}) = .01$$

$$P(X > 50.89218) = .01$$

		α	
v	.10	.05	.01
26	35.56317	38.88514	45.64168
⋮	⋮	⋮	⋮
29	39.08747	42.55697	49.58789
30	40.25602	43.77297	**50.89218**
31	41.42174	44.98535	52.19140

Fig. 11.16 A portion of Appendix Table VI.

When constructing an upper confidence bound for σ_{LT}, it will soon be demonstrated that interest is in the *lower* tail of the χ^2 distribution. Because this distribution is not symmetrical, the x coordinate for this lower tail is *not* identical to the negative of the x coordinate for the upper tail. This is different than both the normal and Student's t distributions, which are symmetrical. If v equals 30, the x coordinate having an area of .01 to its *left* (or equivalently, .99 to its right), is about 14.95, as is seen in Figure 11.18 on the facing page.

The value of this x coordinate is written as $\chi^2_{v,1-\alpha}$ and is tabulated on the left-hand side of each page in Appendix Table VI for 1 minus α of .99, .95, and .90. In this example where v is 30 and α is .01:

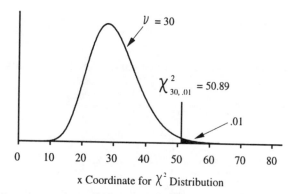

Fig. 11.17 The upper .01 tail of the χ^2 distribution for $\nu = 30$.

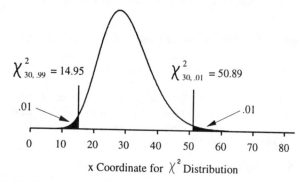

Fig. 11.18 The lower and upper .01 tails of the χ^2 distribution for $\nu = 30$.

$$P(X < \chi^2_{\nu, 1-\alpha}) = \alpha$$

$$P(X < \chi^2_{30, .99}) = .01$$

$$P(X < 14.95) = .01 \quad \Rightarrow \quad \chi^2_{30, 1-.01} = \chi^2_{30, .99} = 14.95$$

A confidence interval for σ_{LT} is derived from this sampling distribution for S_{TOT} in much the same manner as was previously done for μ with the Student's t distribution. The following relationship is used to derive a 1 minus α confidence interval for σ_{LT}.

$$P\left(\chi^2_{\nu, 1-\alpha/2} \le \frac{\nu S^2_{TOT}}{\sigma^2_{LT}} \le \chi^2_{\nu, \alpha/2} \right) = \alpha$$

However, in most capability studies, interest is in obtaining just an *upper* 1 minus α confidence bound for σ_{LT} as this is the condition of most concern. The associated probability statement is given below. The *lower* tail is used because σ_{LT} starts out in the denominator.

$$P\left(\chi_{v,\,1-\alpha}^2 \le \frac{v\,S_{TOT}^2}{\sigma_{LT}^2} \right) = \alpha$$

Rearranging terms to isolate σ_{LT} on one side produces the following relationship:

$$P\left(\sigma_{LT}^2 \le \frac{v\,S_{TOT}^2}{\chi_{v,\,1-\alpha}^2} \right) = \alpha$$

$$P\left(\sigma_{LT} \le S_{TOT}\sqrt{\frac{v}{\chi_{v,\,1-\alpha}^2}} \right) =$$

This makes the upper 1 minus α confidence bound formula appear as is given here:

Upper $1 - \alpha$ confidence bound for $\sigma_{LT} = S_{TOT}\sqrt{\dfrac{v}{\chi_{v,\,1-\alpha}^2}}$

To simplify notation, the upper 1 minus α confidence bound for σ_{LT} is written as $\tilde{\sigma}_{LT,1-\alpha}$.

$$\tilde{\sigma}_{LT,\,1-\alpha} = \text{Upper } 1 - \alpha \text{ confidence bound for } \sigma_{LT}$$

Be aware that this confidence interval formula for σ_{LT} assumes the process output (individuals) has a normal distribution, which can be checked with the methods presented in Chapter 7. To become more familiar with applying this formula, read through the valve O.D. size and particle contamination examples presented next.

Example 11.6 Valve O.D. Size

S_{TOT} is calculated as .01606 cm from a capability study involving 100 measurements of valve O.D. size. An upper .90 confidence bound ($\alpha = .10$) for σ_{LT} is found by first determining the appropriate number of degrees of freedom for S_{TOT}.

$$v = kn - 1 = 100 - 1 = 99$$

v is needed to locate the correct χ^2 value in Appendix Table VI for an α risk of .10.

$$\chi_{v,\,1-\alpha}^2 = \chi_{99,\,1-.10}^2 = \chi_{99,\,.90}^2 = 81.4492$$

Now the upper .90 confidence bound for σ_{LT} may be computed.

$$\tilde{\sigma}_{LT,\,1-.10} = \tilde{\sigma}_{LT,\,.90} = S_{TOT}\sqrt{\frac{v}{\chi_{v,\,1-.10}^2}} = .01606\,\text{cm}\sqrt{\frac{99}{\chi_{99,\,.90}^2}}$$

$$= .01606\,\text{cm}\sqrt{\frac{99}{81.4492}} = .017706\,\text{cm}$$

There is .90 confidence the actual process standard deviation is less than .017706 cm, and only a .10 risk this bound is less than the true σ_{LT}. Of course, this is true only as long as the process remains in control and the process output has a normal distribution.

Example 11.7 Particle Contamination

The capability study for particle contamination of a cleaning bath has 25 in-control plot points on the *IX* chart. From these 25 readings, S_{TOT} is calculated to be 12.80 *ppm*. A team investigating the performance of this process is concerned about how large the true process variation might be based on the results of this one study. After choosing on an α risk of .05, they begin to determine an upper .95 confidence bound for the long-term process standard deviation by first finding the number of degrees of freedom for S_{TOT}.

$$\nu = kn - 1 = 25(1) - 1 = 24$$

With this information, they are able to look up in Appendix Table VI the corresponding χ^2 value for an .05 risk as 13.8484, and then construct the desired confidence bound.

$$\tilde{\sigma}_{LT,.95} = S_{TOT}\sqrt{\frac{\nu}{\chi^2_{\nu,1-.05}}} = 12.80\ ppm\sqrt{\frac{24}{\chi^2_{24,.95}}} = 12.80\ ppm\sqrt{\frac{24}{13.8484}} = 16.85\ ppm$$

The team has .95 confidence the long-term standard deviation for particle contamination of this cleaning bath is less than 16.85 *ppm*.

11.3 Upper Confidence Bound for σ_{ST}

Confidence bound formulas for the short-term standard deviation depend if $\overline{R}$ or $\overline{S}$ is used in estimating σ_{ST}. When $\overline{R}$ is used:

$$\hat{\sigma}_{ST} = \hat{\sigma}_{\overline{R}} = \frac{\overline{R}}{d_2} \quad \Rightarrow \quad P\left(\chi^2_{\nu,1-\alpha} \le \frac{\nu\hat{\sigma}^2_{\overline{R}}}{\sigma^2_{ST}}\right) \cong \alpha$$

This is only an approximate probability statement because $\hat{\sigma}_{\overline{R}}$ is used in place of S_{TOT} (Patnaik). As before, the terms of this inequality are rearranged to isolate σ_{ST} on one side.

$$P\left(\sigma^2_{ST} \le \frac{\nu\hat{\sigma}^2_{\overline{R}}}{\chi^2_{\nu,1-\alpha}}\right) \cong \alpha$$

$$P\left(\sigma_{ST} \le \hat{\sigma}_{\overline{R}}\sqrt{\frac{\nu}{\chi^2_{\nu,1-\alpha}}}\right) \cong \alpha$$

Thus, the formula for an approximate upper 1 minus α confidence bound for σ_{ST} becomes:

$$\tilde{\sigma}_{ST,1-\alpha} \cong \hat{\sigma}_{\overline{R}}\sqrt{\frac{\nu}{\chi^2_{\nu,1-\alpha}}} = \frac{\overline{R}}{d_2}\sqrt{\frac{\nu}{\chi^2_{\nu,1-\alpha}}}$$

If the capability study is based on an $\overline{X}$, S chart, $\sigma_{\overline{R}}$ is replaced with $\sigma_{\overline{S}}$, and the corresponding formula becomes:

$$\tilde{\sigma}_{ST,1-\alpha} \cong \hat{\sigma}_{\bar{S}} \sqrt{\frac{v}{\chi^2_{v,1-\alpha}}} = \frac{\bar{S}}{c_4} \sqrt{\frac{v}{\chi^2_{v,1-\alpha}}}$$

These two formulas are very similar to the one for determining an upper confidence bound for σ_{LT}. The only difference is in the calculation of v, the degrees of freedom. When σ_{ST} is estimated from $\hat{\sigma}_{\bar{R}}$, Duncan (1955, 1958) determined that v is approximately $.9k(n-1)$ for n between 2 and 6, which are the most commonly used subgroup sizes for $\bar{X}$, R charts.

$$v \text{ for } \hat{\sigma}_{\bar{R}} \cong .9k(n-1)$$

Degrees of freedom are lost in the calculation of the subgroup ranges where only the highest and lowest measurements are used to calculate each subgroup range. Thus, only about $.9(n-1)$ degrees of freedom are available from each of the k subgroups for a total of $.9k(n-1)$.

When the estimate of σ_{ST} comes from $\hat{\sigma}_{\bar{S}}$, D. Bissell shows that v is approximately equal to $f_n k(n-1)$, where f_n is a discount factor based on n.

$$v \text{ for } \hat{\sigma}_{\bar{S}} \cong f_n k(n-1)$$

f_n factors are listed in Table 11.3 for subgroup sizes 2 through 10, which are the ones most often chosen for $\bar{X}$, S charts.

Table 11.3 Table of discount factors for $n = 2$ to 10.

n	f_n
2	.88
3	.92
4	.94
5	.95
6	.96
7	.96
8	.97
9	.97
10	.98

Note how both of these methods for estimating σ_{ST} offer fewer degrees of freedom than the one for estimating σ_{LT}. Assume 30 in-control subgroups of size 5 each are collected in a capability study. The process standard deviation may be estimated in one of three ways: S_{TOT}/c_4; $\bar{S}/c_4$; or $\bar{R}/d_2$. There are 149 degrees of freedom associated with the S_{TOT}/c_4 method.

$$v = kn - 1 = 30(5) - 1 = 149$$

Estimating σ_{ST} with $\bar{S}/c_4$ generates only 114 degrees of freedom, a loss of 35 compared to the first method.

$$v \cong f_n k(n-1) = (.95)30(5-1) = 114$$

The least degrees of freedom (only 108) occur when σ_{ST} is estimated from $\overline{R}/d_2$.

$$\nu \cong .9k(n-1) = .9(30)(5-1) = 108$$

For the same sample size, a lower ν means smaller χ^2 values, which result in *larger* upper confidence bounds for σ_{ST} than for σ_{LT}, as will be discovered in the next example.

Example 11.8 Valve O.D. Size

In Example 11.6, an upper .90 confidence bound for the σ_{LT} associated with valve size was computed as .017706 cm, based on the 99 degrees of freedom associated with calculating S_{TOT}. This upper bound is 10.2 percent above the point estimate of .01606 cm for σ_{LT}.

$$\frac{\tilde{\sigma}_{LT,.90}}{\hat{\sigma}_{LT}} = \frac{.017706}{.01606} = 1.102$$

σ_{ST} for valve size is estimated with $\overline{S}$ to be .01552 cm from an $\overline{X}$, S chart involving 20 subgroups of size 5 each.

$$\hat{\sigma}_{ST} = \frac{\overline{S}}{c_4} = .01552 \text{ cm}$$

An upper .90 confidence bound for σ_{ST} is computed by first determining the number of degrees of freedom for this method ($n = 5$), where f_s is from Table 11.3.

$$\nu \cong f_n k(n-1) = .95(20)(5-1) = 76$$

With 76 degrees of freedom, an upper .90 confidence bound for σ_{ST} is found as:

$$\tilde{\sigma}_{ST,1-.10} = \tilde{\sigma}_{ST,.90} \cong \hat{\sigma}_{ST}\sqrt{\frac{\nu}{\chi^2_{\nu,1-.10}}} = .01552 \text{ cm}\sqrt{\frac{76}{\chi^2_{76,.90}}}$$

$$= .01552 \text{ cm}\sqrt{\frac{76}{60.68985}} = .01737 \text{ cm}$$

There is .90 confidence the actual short-term process standard deviation is less than .01738 cm, and only a .10 risk it is greater than this value. This upper bound is 11.9 percent greater than the estimate for σ_{ST}.

$$\frac{\tilde{\sigma}_{ST,.90}}{\hat{\sigma}_{ST}} = \frac{.01737}{.01552} = 1.119$$

If the estimate of σ_{ST} is derived from $\overline{R}$ instead of $\overline{S}$, the number of degrees of freedom decreases to 72.

$$\nu \cong .9k(n-1) = .9(20)(5-1) = .9(80) = 72$$

Assuming the same estimate of .01552 cm for σ_{ST}, an upper .90 confidence bound ($\alpha = .10$) for σ_{ST} becomes .01743 cm.

$$\tilde{\sigma}_{ST,1-.10} = \tilde{\sigma}_{ST,.90} \cong \hat{\sigma}_{ST}\sqrt{\frac{v}{\chi^2_{v,1-.10}}} = .01552\,\text{cm}\sqrt{\frac{72}{\chi^2_{72,.90}}}$$

$$= .01552\,\text{cm}\sqrt{\frac{72}{57.1129}} = .01743\,\text{cm}$$

There is .90 confidence the actual short-term process standard deviation is less than .01743 cm, with a .10 risk it is greater than this value. Note that this upper bound is 12.3 percent greater than the point estimate for σ_{ST}.

$$\frac{\tilde{\sigma}_{ST,.90}}{\hat{\sigma}_{ST}} = \frac{.01743}{.01552} = 1.123$$

A comparison of these last three outcomes is summarized in Table 11.4. With a greater number of degrees of freedom, the upper confidence bound is closer to the point estimate of the standard deviation. The closer the bound, the more useful the information. This is why it is always best to estimate the process standard deviation with a method that maximizes v.

Table 11.4   Comparison of confidence bound sizes for σ.

Method of Estimation	Degrees of Freedom, v	% Above Estimate of σ
S_{TOT}	99	10.2
$\overline{S}$	76	11.9
$\overline{R}$	72	12.3

These formulas for constructing confidence bounds on σ play a crucial role in determining the appropriate confidence bounds for measures of potential process capability, such as the 6σ spread.

11.4 Upper Confidence Bound for the 6σ Spread

As the $6\sigma_{LT}$ spread is just 6 times the long-term process standard deviation, the upper confidence bound for it is just 6 times the upper confidence bound for σ_{LT}. Therefore, an upper 1 minus α confidence bound for the $6\sigma_{LT}$ spread is determined by the following formula, where v equals kn minus 1.

$$6\tilde{\sigma}_{LT,1-\alpha}\,\text{Spread} = 6\tilde{\sigma}_{LT,1-\alpha} = 6S_{TOT}\sqrt{\frac{v}{\chi^2_{v,1-\alpha}}}$$

If interest is in obtaining an upper confidence bound for the $6\sigma_{ST}$ spread, use $f_n k(n-1)$ degrees of freedom if $\overline{S}$ is used to estimate σ_{ST}, and $.9k(n-1)$ degrees of freedom if $\overline{R}$ is used.

$$6\tilde{\sigma}_{ST,1-\alpha}\,\text{Spread} \cong 6\hat{\sigma}_{\overline{S}}\sqrt{\frac{v}{\chi^2_{f_n k(n-1),1-\alpha}}}$$

$$6\tilde{\sigma}_{ST, 1-\alpha} \text{ Spread } \cong 6\hat{\sigma}_{\bar{R}}\sqrt{\frac{v}{\chi^2_{.9k(n-1), 1-\alpha}}}$$

These calculations are similar to those for the upper confidence bound for σ, as becomes evident in this next example.

Example 11.9 Valve O.D. Size

S_{TOT} is calculated to be .01606 cm from a capability study involving 100 measurements of valve O.D. size ($k = 20$, $n = 5$). From this, the $6\sigma_{LT}$ spread is estimated to be .09660 cm.

$$6\hat{\sigma}_{LT} \text{ Spread } = 6\hat{\sigma}_{LT} = 6\frac{S_{TOT}}{c_4} = 6\frac{.01606 \text{ cm}}{.9975} = .09660 \text{ cm}$$

An upper .90 confidence bound ($\alpha = .10$) for the true $6\sigma_{LT}$ spread is found by first determining the number of degrees of freedom for S_{TOT}.

$$v = kn - 1 = 20(5) - 1 = 99$$

With this value for v, the χ^2 value for α equal to .10 is found in Appendix Table VI, and the desired confidence bound can be calculated.

$$6\tilde{\sigma}_{LT, 1-.10} \text{ Spread } = 6\tilde{\sigma}_{LT, .90} \text{ Spread } = 6S_{TOT}\sqrt{\frac{v}{\chi^2_{v, 1-.10}}}$$

$$= 6(.01606 \text{ cm})\sqrt{\frac{99}{\chi^2_{99, .90}}}$$

$$= .09636 \text{ cm}\sqrt{\frac{99}{81.4492}} = .10624 \text{ cm}$$

There is .90 confidence the actual $6\sigma_{LT}$ spread for this process is less than .10624 cm. Conversely, there is only a .10 risk this upper bound of .10624 cm is less than the true $6\sigma_{LT}$ spread for this process. Note that even with this large sample size of 100 measurements, the upper confidence bound is still 10 percent greater than the point estimate for this measure of potential capability.

$$\frac{6\tilde{\sigma}_{LT, .90} \text{ Spread}}{6\hat{\sigma}_{LT} \text{ Spread}} = \frac{.10624 \text{ cm}}{.09660 \text{ cm}} = 1.10$$

With a tolerance for O.D. size of .12000 cm, there is .90 confidence this process meets the minimum requirement for long-term potential capability, that is, the $6\sigma_{LT}$ spread is less than the tolerance.

$$6\tilde{\sigma}_{LT, .90} \text{ Spread } \leq \text{ Tolerance}$$

$$.10624 \text{ cm } \leq .1200 \text{ cm}$$

However, if the goal for this process is to have .90 confidence the $6\sigma_{LT}$ spread is less than 85 percent of the tolerance, there is a problem. .85 times .12000 cm is .10200 cm, and the upper bound for the $6\sigma_{LT}$ spread is greater than this.

$$6\tilde{\sigma}_{LT,.90} \text{ Spread } \geq .85 \text{ Tolerance}$$

$$.10624 \text{ cm } \geq .10200 \text{ cm}$$

Thus, a statement *cannot* be made that there is .90 confidence the process meets this potential capability goal. This is true even though the point estimate of .09660 cm is less than the goal. A second capability study on this process has more than a .10 chance of producing an estimate of the $6\sigma_{LT}$ spread which is greater than .10200 cm.

Note that this upper .90 confidence bound for the $6\sigma_{LT}$ spread is just 6 times the upper .90 confidence bound for σ_{LT}, which was computed back in Example 11.6 on page 612.

$$6\tilde{\sigma}_{LT,.90} \text{ Spread } = 6(\tilde{\sigma}_{LT,.90}) = 6(.017706 \text{ cm}) = .10624 \text{ cm}$$

In Example 5.4 (p. 93), the $6\sigma_{ST}$ spread was estimated from $\hat{\sigma}_{\bar{R}}$ as .09312 cm. If an upper .90 confidence bound for this short-term potential capability measure is desired, the following procedure applies, beginning with determining the degrees of freedom:

$$v \cong .9k(n-1) = .9(20)(5-1) = 72$$

From the control chart monitoring O.D. size, $\bar{R}$ is found to be .03610 cm, while the d_2 factor for n equal to 5 is 2.326.

$$6\tilde{\sigma}_{ST,1-.10} \text{ Spread } \cong 6\frac{\bar{R}}{d_2}\sqrt{\frac{v}{\chi^2_{v,1-.10}}} = 6\frac{.03610 \text{ cm}}{2.326}\sqrt{\frac{72}{\chi^2_{72,.90}}}$$

$$= .09312 \text{ cm}\sqrt{\frac{72}{57.1129}} = .10455 \text{ cm}$$

When $\bar{S}$ is used instead of $\bar{R}$ to estimate σ_{ST}, this next set of formulas applies for calculating degrees of freedom and an upper confidence bound. The degrees of freedom are determined from k, n, and the f_s factor of .95 from Table 11.3.

$$v \cong f_n k(n-1) = .95(20)(5-1) = 76$$

From the control chart, $\bar{S}$ is found to be .01459, while c_4 is .9400 for n equal to 5.

$$6\tilde{\sigma}_{ST,1-.10} \text{ Spread } \cong 6\frac{\bar{S}}{c_4}\sqrt{\frac{v}{\chi^2_{v,1-.10}}} = 6\frac{.01459 \text{ cm}}{.9400}\sqrt{\frac{80}{\chi^2_{80,.90}}}$$

$$= .09312 \text{ cm}\sqrt{\frac{80}{64.2778}} = .10389 \text{ cm}$$

Again, these results are just 6 times the respective upper .90 confidence bounds for σ_{ST} computed back in Example 11.8 on page 615.

With these confidence bound formulas for the 6σ spread, confidence bound formulas for both P_R and C_R can be easily derived.

11.5 Upper Confidence Bound for P_R

The P_R measure of long-term potential process capability is equal to the $6\sigma_{LT}$ spread divided by the tolerance.

$$\hat{P}_R = \frac{6\hat{\sigma}_{LT}}{\text{Tolerance}}$$

Likewise, an upper 1 minus α confidence bound for P_R is the upper 1 minus α confidence bound for the $6\sigma_{LT}$ spread, divided by the tolerance.

$$\tilde{P}_{R, 1-\alpha} = \frac{6\tilde{\sigma}_{LT, 1-\alpha} \text{ Spread}}{\text{Tolerance}} = \frac{6S_{TOT}}{\text{Tolerance}} \sqrt{\frac{v}{\chi^2_{v, 1-\alpha}}}$$

Because c_4 is so close to 1, especially when kn is large, σ_{LT} is approximately equal to S_{TOT}, and the following relationship holds true:

$$\hat{P}_R = \frac{6\hat{\sigma}_{LT}}{\text{Tolerance}} = \frac{6S_{TOT}}{c_4 \text{Tolerance}} \cong \frac{6S_{TOT}}{\text{Tolerance}}$$

Substituting this result into the upper confidence bound formula for P_R yields the following:

$$\tilde{P}_{R, 1-\alpha} \cong \hat{P}_R \sqrt{\frac{v}{\chi^2_{v, 1-\alpha}}}$$

This formula relates the confidence bound directly to the estimate of P_R. Once an estimate of P_R is made from measurements collected in a capability study, an upper 1 minus α confidence bound may be determined, as is demonstrated with the valve O.D. size example.

Example 11.10 Valve O.D. Size

S_{TOT} is calculated to be .01606 cm from a capability study involving 100 measurements of valve O.D. size. From this information, σ_{LT} may be estimated (c_4 for 100 is .9975).

$$\hat{\sigma}_{LT} = \frac{S_{TOT}}{c_4} = \frac{.01606 \text{ cm}}{.9975} = .01610 \text{ cm}$$

Given a tolerance of .1200 cm, a point estimate of the P_R measure of potential capability is computed as .805.

$$\hat{P}_R = \frac{6\hat{\sigma}_{LT}}{\text{Tolerance}} = \frac{6(.01610 \text{ cm})}{.1200 \text{ cm}} = .805$$

The number of degrees of freedom for this estimate is based on S_{TOT}.

$$v = kn - 1 = 100 - 1 = 99$$

With v equal to 99 and an α risk equal to .10, the appropriate χ^2 value is found as 81.4492 in Appendix Table VI. Now an approximate upper .90 confidence bound for the true value of P_R can be calculated.

$$\tilde{P}_{R, 1 - .10} = \tilde{P}_{R, .90} \cong \hat{P}_R \sqrt{\frac{v}{\chi^2_{v, 1 - .10}}}$$

$$= .805 \sqrt{\frac{99}{\chi^2_{99, .90}}}$$

$$= .805 \sqrt{\frac{99}{81.4492}} = .888$$

There is .90 confidence the actual P_R value for this process is less than .888. As the minimum requirement for long-term potential capability is having P_R less than 1.0, there is .90 confidence the process meets this benchmark.

An exact upper 1 minus α confidence bound for P_R can be determined by dividing the upper 1 minus α confidence bound for the $6\sigma_{LT}$ spread calculated in Example 11.9 by the tolerance for valve O.D. size.

$$\tilde{P}_{R, 1 - \alpha} = \frac{6\tilde{\sigma}_{LT, 1 - \alpha} \text{ Spread}}{\text{Tolerance}} = \frac{.10624 \text{ cm}}{.1200 \text{ cm}} = .885$$

Because it is based on S_{TOT} instead of $\hat{\sigma}_{LT}$, this second bound of .885 varies slightly from the approximate bound of .888. When .885 is divided by the c_4 factor of .9975 (for $kn = 100$), .888 is obtained (.885 / .9975 = .888).

Concept of an Upper Confidence Bound

Review the example given on the first page of this chapter where the true P_R ratio is .50 for a given process. An initial capability study, with k equal to 20 and n of 3, estimated P_R as .53. An upper .95 confidence bound for the actual P_R ratio is calculated by first figuring v, the number of degrees of freedom.

$$v = kn - 1 = 20(3) - 1 = 59$$

With 59 degrees of freedom, an upper .95 confidence bound for P_R is determined to be .63.

$$\tilde{P}_{R, .95} \cong \hat{P}_R \sqrt{\frac{v}{\chi^2_{v, .95}}} = .53 \sqrt{\frac{59}{\chi^2_{59, .95}}} = .53 \sqrt{\frac{59}{42.33927}} = .63$$

There is .95 confidence the actual P_R ratio is less than .63, with only a .05 risk it is greater than .63. Note that the true P_R ratio of .50 for this process is less than this upper confidence bound of .63.

Based on the 60 sample measurements of a second capability study on this process, P_R was estimated as .48. Performing the same set of calculations as was done for the first study, the upper .95 confidence bound for P_R is found to be .57. Again, this confidence bound is greater than the true P_R value of .50. Figure 11.19 displays upper .95 confidence bounds calculated from an additional 18 capability studies performed on this process, all having the same v.

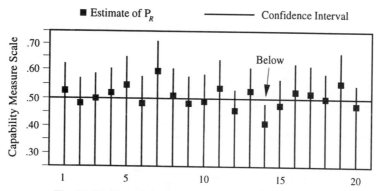

Fig. 11.19 Twenty upper .95 confidence bounds for P_R.

Over a long period of time, 5 percent (1 out of 20) of the upper .95 confidence bounds calculated for this process will be *less* than the true P_R value and, thus, provide an *erroneous* conclusion concerning the potential capability of this process. An example of such a "miss" is the upper confidence bound of .48 computed from the data collected during the course of the 14th capability study. This outcome would have led one to falsely conclude the true P_R ratio is less than .48, when it actually is .50.

If this risk of 5 percent is too large for comfort, a lower α must be selected. However, this change will widen the confidence interval, thereby diminishing its usefulness.

Maximum Observed Estimate of P_R to Demonstrate Goal

Sometimes it is desired to know *before* undertaking an actual capability study what is the largest P_R estimate that may occur, yet still provide sufficient evidence the process meets its capability goal with some stated degree of confidence. If the upper confidence bound just equals the goal for P_R, there is 1 minus α confidence the process capability equals or exceeds the stated goal. The confidence bound formula for P_R can be solved for this special condition to find the largest $\hat{P}_R$ that may be observed that still demonstrates attainment of the desired capability goal at the specified confidence.

$$\tilde{P}_{R,1-\alpha} \cong \hat{P}_R \sqrt{\frac{\nu}{\chi^2_{\nu,1-\alpha}}}$$

$\hat{P}_R$ is at the maximum allowed for this level of confidence when $\tilde{P}_{R,1-\alpha}$ equals the goal for P_R. This special estimate of P_R is called the "maximum observed $\hat{P}_{R,1-\alpha}$." Substituting $P_{R,GOAL}$ for $\tilde{P}_{R,1-\alpha}$, and the maximum observed $\hat{P}_{R,1-\alpha}$ for $\hat{P}_R$, in the above equation generates the following result:

$$P_{R,GOAL} \cong \text{Maximum Observed } \hat{P}_{R,1-\alpha} \sqrt{\frac{\nu}{\chi^2_{\nu,1-\alpha}}}$$

$$P_{R,GOAL} \sqrt{\frac{\chi^2_{\nu,1-\alpha}}{\nu}} \cong \text{Maximum Observed } \hat{P}_{R,1-\alpha}$$

Switching sides in this equation produces the desired result.

$$\text{Maximum Observed } \hat{P}_{R,1-\alpha} \cong P_{R,GOAL} \sqrt{\frac{\chi^2_{v,1-\alpha}}{v}}$$

This is the largest estimate of P_R that may be observed in a capability study and yet provide 1 minus α confidence the process meets the P_R goal. In order to demonstrate a high level of confidence the goal is met or exceeded, this maximum observed point estimate must be considerably *less than the goal* for P_R, as is discovered in the next example.

Example 11.11 Candle Height

Molten wax is poured into a mold to make votive candles, an example of which is given in Figure 11.20. A critical characteristic of the finished candle is its height. Given that the $P_{R,\,GOAL}$ for height is .75, Peter, the quality engineer of this department, wishes to determine the largest $\hat{P}_R$ that may be seen and still retain .99 confidence the process meets this capability goal.

Fig. 11.20 A votive candle.

This task is accomplished by first finding the expected degrees of freedom associated with using S_{TOT} to estimate the long-term process standard deviation. If the plan calls for 25 subgroups of size 5 to be included in the capability study for candle height, v can be forecast as 124.

$$v = kn - 1 = 25(5) - 1 = 124$$

With 124 degrees of freedom, the largest estimate of P_R this engineer can observe in the capability study that still demonstrates the P_R goal of .75 is met with .99 confidence turns out to be .64, as is shown here.

$$\text{Maximum Observed } \hat{P}_{R,.99} \cong P_{R,GOAL} \sqrt{\frac{\chi^2_{124,.99}}{124}} = .75 \sqrt{\frac{90.32722}{124}} = .64$$

If v is at least 124 and the estimated P_R from the study is less than or equal to .64, Peter has at least .99 confidence the process meets the specified capability goal. If the point estimate for the P_R ratio turns out to be greater than .64, there is less than .99 confidence the process has the desired level of potential capability.

Example 11.12 Valve O.D. Size

The potential capability goal for the exhaust valve O.D. size example is to have .90 confidence P_R is no more than .85. Before undertaking a study for potential capability on this process, a manufacturing engineer would like an idea of how large an estimate of P_R may be observed from the results of the study, and yet furnish .90 confidence the process meets the required potential capability goal. Assuming 20 subgroups of size 3 each will be collected during the course of the capability study, this maximum point estimate of P_R is calculated by first finding the anticipated number of degrees of freedom.

$$v = kn - 1 = 20(3) - 1 = 59$$

With 59 degrees of freedom, and a $P_{R,\,GOAL}$ of .85, the maximum observed estimate of P_R turns out to be .747.

$$\text{Maximum Observed } \hat{P}_{R,.90} \cong P_{R,GOAL} \sqrt{\frac{\chi^2_{59,.90}}{59}} = .85 \sqrt{\frac{45.57695}{59}} = .747$$

After the study is completed, σ_{LT} is estimated (with $v = 59$) as .01646 cm. With a tolerance of .1200 cm, this makes the point estimate of P_R equal to:

$$\hat{P}_R = \frac{6\hat{\sigma}_{LT}}{\text{Tolerance}} = \frac{6(.01646 \text{ cm})}{.1200 \text{ cm}} = .823$$

Because the point estimate of .823 for P_R is not less than the "bogie" of .747, this sample result indicates the process does not meet the capability goal of .85 with the desired level of confidence. This is true even though the point estimate of .823 is less than the goal of .85.

Maximum Observed $\hat{P}_{R,1-\alpha}$ versus v

The graph in Figure 11.21 illustrates how the maximum observed P_R estimate at .95 confidence increases as the number of degrees of freedom, v, grows. As v approaches plus infinity, the maximum observed estimate of P_R approaches the P_R goal of .75.

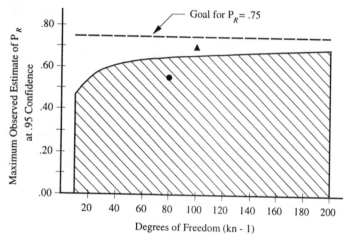

Fig. 11.21 Maximum observed $\hat{P}_{R,.95}$ versus degrees of freedom.

Estimates of P_R in the shaded area indicate the goal for P_R has been met with .95 confidence. Suppose P_R is estimated as .55 with v equal to 80, a condition represented by the dot in the shaded area of Figure 11.21. There is .95 confidence the true P_R ratio for this process is less than, or equal to, the goal of .75.

If P_R is estimated to be .70 based on v of 100 (see triangle on graph), there is less than .95 confidence the $P_{R,GOAL}$ of .75 has been demonstrated. Again, this is true even though the point estimate of .70 is less than the goal of .75. Improvements aimed at reducing σ_{LT} must be made to enhance potential capability.

The graph in Figure 11.22 compares the maximum observed P_R estimate at .95 confidence for v equal to 10 through 200 to those for .90 and .99 confidence. As the confidence level increases, the curve for the maximum observed P_R estimate moves farther below the P_R goal. In order to have a higher degree of confidence the desired goal is met, a smaller estimate of

P_R must be observed in the capability study. Or, for the same point estimate, a greater number of pieces, which generates a larger v value, must be included in the study in order to have a higher degree of confidence the process meets the goal.

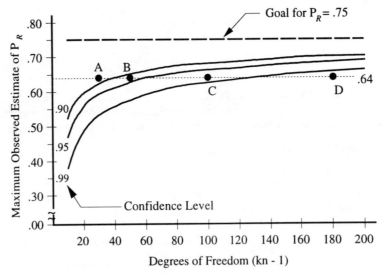

Fig. 11.22 Maximum observed $\hat{P}_{R,1-\alpha}$ for .90, .95, and .99 confidence.

For example, point A in Figure 11.22 represents a P_R estimate of .64 based on v equal to 30. As this point lies above the .90 confidence line, there is less than .90 confidence the process meets the P_R goal of .75.

What if v had instead equaled 50 in this study, with the estimate of P_R still being .64 (see point B)? Now there would .90 confidence the process meets the goal. However, there is less than .95 confidence, as this point still lies above the .95 line. If v had equaled 100, an estimate of .64 would provide enough evidence to believe the goal is met with .95 confidence (point C), but not .99 confidence. And finally, had v equaled 180 (point D), this would have shown the goal was met with .99 confidence, as this point falls below all three lines.

11.6 Upper Confidence Bound for C_R

As the C_R ratio is very much like the P_R ratio, the formula for the upper confidence bound is also very similar (Li *et al*).

$$\hat{C}_R = \frac{6\,\hat{\sigma}_{ST}}{\text{Tolerance}} \qquad \tilde{C}_{R,1-\alpha} \cong \hat{C}_R \sqrt{\frac{v}{\chi^2_{v,1-\alpha}}}$$

The only difference is in the calculation of v, the degrees of freedom. Recall that v depends on the number of independent measurements used to calculate σ. The C_R ratio is based on σ_{ST}, which is estimated in one of two ways, depending on the type of control chart monitoring the process under study.

$$\hat{\sigma}_{ST} = \hat{\sigma}_{\overline{R}} = \frac{\overline{R}}{d_2} \text{ for } \overline{X}, R \text{ charts} \qquad \hat{\sigma}_{ST} = \hat{\sigma}_{\overline{S}} = \frac{\overline{S}}{c_4} \text{ for } \overline{X}, S \text{ charts}$$

When $\hat{\sigma}_{\overline{S}}$ is used as an estimate of σ_{ST}, ν is approximately equal to $f_n k(n-1)$, where the f_n factors are given in Table 11.3. If $\hat{\sigma}_{\overline{R}}$ is used to estimate σ_{ST}, ν is about $.9k(n-1)$. Note how both of these estimates of σ_{ST} offer fewer degrees of freedom than estimates of σ_{LT}. As already seen with the upper confidence bound for the $6\sigma_{ST}$ spread, this difference in degrees of freedom means *higher* upper confidence bounds for C_R than for P_R, as is seen in the next example.

Example 11.13 Voltage Regulators

An $\overline{X}, R$ chart monitoring a process assembling voltage regulators for marine applications has an $\overline{R}$ of 53 millivolts. $\overline{R}$ is calculated from 21 (k) in-control subgroups of size 2 (n). From this information, an estimate of σ_{ST} may be generated.

$$\hat{\sigma}_{ST} = \frac{\overline{R}}{d_2} = \frac{53 \text{ mv}}{1.128} = 47 \text{ mv}$$

With a tolerance of 400 mv, a point estimate of C_R is calculated as follows:

$$\hat{C}_R = \frac{6\,\hat{\sigma}_{ST}}{\text{Tolerance}} = \frac{6\,(47 \text{ mv})}{400 \text{ mv}} = .705$$

The best single-number estimate of C_R is .705. However, this number will vary from study to study due to random sampling variation. To quantify how much variation is expected in this estimate, an upper confidence bound may be calculated for C_R. This procedure begins with determining the approximate number of degrees of freedom.

$$\nu \cong .9k(n-1) = .9(21)(2-1) = 18.9 \cong 19$$

If α is chosen as .05, an upper .95 confidence bound for C_R can now be calculated for this number of degrees of freedom.

$$\tilde{C}_{R,.95} \cong \hat{C}_R \sqrt{\frac{19}{\chi^2_{19,.95}}} = .705 \sqrt{\frac{19}{10.117}} = .966$$

Based on the results of this study, there is .95 confidence the true C_R measure for this process is no greater than .966. Thus, this process most likely meets the minimum criterion for potential process capability, i.e., C_R is less than 1.0. Note that due to the small sample size, this upper bound for C_R is slightly more than 37 percent greater than the point estimate $(.705 \times 1.3704 = .966)$.

Suppose P_R is estimated instead of C_R, with the point estimate for P_R identical to that for C_R, namely .705. The number of degrees of freedom associated with the estimate of P_R is based on the calculation of S_{TOT}.

$$\nu = kn - 1 = 21(2) - 1 = 41$$

With α also equal to .05, an upper .95 confidence bound for P_R is then computed as .864.

$$\tilde{P}_{R,.95} \cong \hat{P}_R \sqrt{\frac{41}{\chi^2_{41,.95}}} = .705\sqrt{\frac{41}{27.32554}} = .864$$

Due to the higher number of degrees of freedom, this upper confidence bound is closer to the point estimate for P_R (.864 versus .705, a 22.5 percent difference) than the upper confidence bound for C_R is to the point estimate of C_R (.966 versus .705, a 37 percent difference). For a given number of measurements in a capability study, P_R provides a more precise estimate of the true potential capability than does C_R.

Maximum Observed Estimate

In a manner similar to the P_R ratio, the maximum observed C_R estimate is computed with this formula:

$$\text{Maximum Observed } \hat{C}_{R,1-\alpha} \cong C_{R,GOAL}\sqrt{\frac{\chi^2_{v,1-\alpha}}{v}}$$

v is determined with $f_n k(n-1)$ if $\sigma_{\bar{s}}$ is used in estimating C_R, while v is about $.9k(n-1)$ when $\sigma_{\bar{R}}$ is used. For the previous voltage example, if the goal for C_R is to be less than 1.00 with .95 confidence, the maximum observed estimate of C_R becomes .730. This analysis assumes v will equal 19 when the capability study is completed.

$$\text{Maximum Observed } \hat{C}_{R,.95} \cong 1.00\sqrt{\frac{\chi^2_{19,.95}}{19}} = 1.00\sqrt{\frac{10.117}{19}} = .730$$

Because the estimate for C_R of .705 observed in the study is lower than the maximum of .730 allowed, there is .95 confidence the capability goal for this process is met. If the estimate had exceeded .730, there would have been less than .95 confidence the process capability meets the desired goal of 1.00.

Example 11.14 Floppy Disks

A field problem with floppy disks is traced to variation in the diameter of the inner hole (Figure 11.23). From a capability study conducted on the hole cutting process ($k = 21$, $n = 4$), C_R is estimated as 1.078, while P_R is estimated as 1.099 (pp. 112-113).

To construct an upper confidence bound for C_R, the degrees of freedom must be determined first. Because $\overline{R}$ was used to estimate σ_{ST}, v is computed to be 57.

Fig. 11.23

$$v \cong .9k(n-1) = .9(21)(4-1) = 56.7 \cong 57$$

With the C_R ratio estimated as 1.078, an upper .95 confidence bound for C_R is calculated as 1.277, which is about 18.4 percent higher than the point estimate.

$$\tilde{C}_{R,.95} \cong \hat{C}_R\sqrt{\frac{57}{\chi^2_{57,.95}}} = 1.078\sqrt{\frac{57}{40.64588}} = 1.277$$

Based on the results of this capability study, there is .95 confidence the true C_R for this process is less than 1.277. Unfortunately, this is greater than 1.00, which is the minimum requirement for short-term potential capability, meaning there is little likelihood the process is capable. Of course, this conclusion should be fairly obvious. Due to the point estimate for C_R of 1.078 being greater than 1.00, there is little confidence the true C_R ratio for this hole-cutting process is actually less than 1.00.

An upper confidence bound may also be constructed for long-term potential capability based on the estimate of P_R. As S_{TOT} is used in estimating σ_{LT}, ν is determined as 83.

$$\nu = kn - 1 = 21(4) - 1 = 83$$

With P_R estimated as 1.099, an upper .95 confidence bound for this parameter is 1.261, which is 14.8 percent higher than the point estimate.

$$\tilde{P}_{R,.95} \cong \hat{P}_R \sqrt{\frac{83}{\chi^2_{83,.95}}} = 1.099 \sqrt{\frac{83}{63.00382}} = 1.261$$

From the results of this capability study, there is .95 confidence the true P_R for this process is less than 1.261. There is little prospect this process meets the minimum requirement for long-term potential process capability.

The goal for this operation is to have .95 confidence the C_R ratio is less than .60. The maximum estimate for C_R that may be observed in a capability study done on this process, that still demonstrates with .95 confidence the goal is met, is computed as is shown here, assuming ν will be 57:

$$\text{Maximum Observed } \hat{C}_{R,1-\alpha} \cong C_{R,GOAL} \sqrt{\frac{\chi^2_{\nu,1-\alpha}}{\nu}}$$

$$= .60 \sqrt{\frac{\chi^2_{57,.95}}{57}}$$

$$= .60 \sqrt{\frac{40.64588}{57}} = .507$$

The goal for P_R is to be less than .75 with .95 confidence. Predicting ν will be 83 in a capability study, the maximum observed estimate of P_R is .653.

$$\text{Maximum Observed } \hat{P}_{R,1-\alpha} \cong P_{R,GOAL} \sqrt{\frac{\chi^2_{\nu,1-\alpha}}{\nu}}$$

$$= .75 \sqrt{\frac{\chi^2_{83,.95}}{83}}$$

$$= .75 \sqrt{\frac{63.00382}{83}} = .653$$

Due to the higher ν associated with its calculation, the maximum observed estimate for P_R of .653 is considerably larger than the .507 maximum computed for the estimate of C_R.

After the study is conducted, the estimate of C_R turns out to be 1.078, while the estimate of P_R comes in at 1.099. Based on the results of this capability study, the hole-cutting operation doesn't even come close to demonstrating attainment of the respective capability goals at the desired .95 confidence level.

Minimum Required Sample Size

Sometimes it is desired to have the upper confidence bound for P_R be no more than a certain distance above the estimate for P_R. The next formula provides the minimum sample size which must be collected during the course of a capability study to satisfy this condition. The original derivation of this formula is given in a paper by Franklin (1994).

$$kn \cong 1 + .5 \left(\frac{Z_\alpha}{e_{PR}} \right)^2 \qquad \text{where } e_{PR} = 1 - \frac{\hat{P}_R}{\tilde{P}_{R, 1-\alpha}}$$

For example, a customer asks to have the upper .95 confidence bound be no more than 10 percent larger than $\hat{P}_R$.

$$\tilde{P}_{R, 1-\alpha} \le 1.10 \, \hat{P}_R \quad \Rightarrow \quad \frac{1}{1.10} \le \frac{\hat{P}_R}{\tilde{P}_{R, 1-\alpha}}$$

This ratio is inserted into the formula for e_{PR}.

$$e_{PR} = 1 - \frac{\hat{P}_R}{\tilde{P}_{R, 1-\alpha}} = 1 - \frac{1}{1.10} = .0909$$

With e_{PR} equal to .0909 and α equal to .05, the minimum required sample size can now be computed as follows:

$$kn \cong 1 + .5 \left(\frac{Z_{.05}}{e_{PR}} \right)^2 = 1 + .5 \left(\frac{1.645}{.0909} \right)^2 = 164.75 \cong 165$$

Thus, at least 165 pieces (always round to the next higher integer) must be measured in the capability study to satisfy this request.

After the capability study is completed, 165 pieces are collected from in-control subgroups, and an estimate of .500 is made for P_R. An upper .95 confidence bound is found for this point estimate by first determining v.

$$v = kn - 1 = 165 - 1 = 164$$

With v equal to 164 and α equal to .05, the upper .95 confidence bound is .550.

$$\tilde{P}_{R, .95} \cong \hat{P}_R \sqrt{\frac{v}{\chi^2_{v, .95}}} = .500 \sqrt{\frac{164}{135.38997}} = .550$$

As originally desired, this upper bound of .550 is exactly 10 percent above the .500 estimate for P_R (.500 × 1.10 = .550).

11.7 Upper Confidence Bounds for C^*_R and P^*_R

When the target for the process average is not located at the middle of the tolerance, the C_R and P_R ratios are not appropriate for measuring potential capability. Instead, they should be replaced with C^*_R and P^*_R, as was revealed in Section 5.5 on page 113

$$C^*_R = \text{Maximum}\left(\frac{3\sigma_{ST}}{T - LSL}, \frac{3\sigma_{ST}}{USL - T}\right) = \frac{6\sigma_{ST}}{\text{Effective Tolerance}}$$

$$P^*_R = \text{Maximum}\left(\frac{3\sigma_{LT}}{T - LSL}, \frac{3\sigma_{LT}}{USL - T}\right) = \frac{6\sigma_{LT}}{\text{Effective Tolerance}}$$

These modified measures are point estimators of the true potential capability for these special situations. An upper 1 minus α confidence bound for C^*_R is derived in a manner similar to that done for C_R.

$$\text{Upper } 1 - \alpha \text{ confidence bound for } C^*_R = \tilde{C}^*_{R, 1-\alpha} \cong \hat{C}^*_R \sqrt{\frac{v}{\chi^2_{v, 1-\alpha}}}$$

where $v \cong f_n k(n-1)$ for $\hat{\sigma}_{\bar{S}}$ or $v \cong .9k(n-1)$ for $\hat{\sigma}_{\bar{R}}$

Likewise, the upper 1 minus α confidence bound formula for P^*_R is similar to that for P_R.

$$\text{Upper } 1 - \alpha \text{ confidence bound for } P^*_R = \tilde{P}^*_{R, 1-\alpha} \cong \hat{P}^*_R \sqrt{\frac{v}{\chi^2_{v, 1-\alpha}}} \quad \text{where } v = kn - 1$$

These are applied and interpreted in the same manner as the confidence bounds for C_R and P_R, as this next example illustrates.

Example 11.15 Gold Plating

The LSL for the thickness of gold being plated on a silicon wafer is 50 angstroms and the USL is 100 angstroms. For cost reasons, a decision is made to target the average thickness at 65 angstroms, as is portrayed in Figure 11.24.

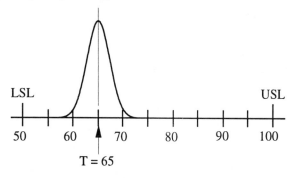

Fig. 11.24 The distribution of gold plating thickness when centered on target.

Measurements from 22 subgroups of size 3 each are combined to derive an estimate of 2.5 angstroms for σ_{LT}. With this information, a point estimate of .50 is made for P_R^*.

$$\hat{P}_R^* = \text{Maximum}\left(\frac{3\hat{\sigma}_{LT}}{T-\text{LSL}}, \frac{3\hat{\sigma}_{LT}}{\text{USL}-T}\right)$$

$$= \text{Maximum}\left(\frac{3(2.5)}{65-50}, \frac{3(2.5)}{100-65}\right)$$

$$= \text{Maximum}(.50, .21) = .50$$

An upper .99 confidence bound for P_R^* is determined by first finding the number of degrees of freedom involved with estimating σ_{LT}.

$$\nu = kn - 1 = 22(3) - 1 = 65$$

With 65 degrees of freedom, the χ^2 value for an α risk equal to .01 is selected from Appendix Table VI and inserted into the confidence bound formula for P_R^*.

$$\tilde{P}_{R,1-.01}^* \cong \hat{P}_R^* \sqrt{\frac{\nu}{\chi_{\nu,1-.01}^2}} = .50\sqrt{\frac{65}{\chi_{65,.99}^2}} = .50\sqrt{\frac{65}{41.4434}} = .626$$

There is .99 confidence the long-term potential capability for this plating process meets the minimum requirement of P_R^* being less than 1.0. If the goal for P_R^* is .75, there is .99 confidence this goal is met as well.

11.8 Upper Confidence Bounds for C'_{RL}, C'_{RU}, P'_{RL}, and P'_{RU}

The C_R and P_R ratios have modified formulas to handle unilateral specifications, which also require a target average be specified for the process center. For a characteristic like tensile strength, where only a LSL is given, the following formulas estimate C'_{RL} and P'_{RL}:

$$\hat{C}'_{RL} = \text{Minimum}\left(\frac{3\hat{\sigma}_{ST}}{T-\text{LSL}}, \frac{3\hat{\sigma}_{ST}}{\hat{\mu}-\text{LSL}}\right) \qquad \hat{P}'_{RL} = \text{Minimum}\left(\frac{3\hat{\sigma}_{LT}}{T-\text{LSL}}, \frac{3\hat{\sigma}_{LT}}{\hat{\mu}-\text{LSL}}\right)$$

If $\hat{\mu}$ is less than or equal to T, upper 1 minus α confidence bounds for C'_{RL} and P'_{RL} are found in a manner similar to that done for C_R and P_R.

$$\tilde{C}'_{RL,1-\alpha} \cong \hat{C}'_{RL}\sqrt{\frac{\nu}{\chi_{\nu,1-\alpha}^2}} \quad \text{where } \nu \cong f_n\,k(n-1) \text{ for } \hat{\sigma}_{\tilde{S}} \text{ or } \nu \cong .9k(n-1) \text{ for } \hat{\sigma}_{\bar{R}}$$

$$\tilde{P}'_{RL,1-\alpha} \cong \hat{P}'_{RL}\sqrt{\frac{\nu}{\chi_{\nu,1-\alpha}^2}} \qquad \text{where } \nu = kn-1$$

When $\hat{\mu}$ is greater than T, no confidence bound can be obtained for C'_{RL}. However, an upper confidence bound for P'_{RL} can be computed using a method similar to that presented in Section 11.17 for P_{PL} (p. 653).

For those situations where the characteristic being studied has only an USL (like surface finish), apply these next formulas to estimate C'_{RU} and P'_{RU}.

$$\hat{C}'_{RU} = \text{Minimum}\left(\frac{3\hat{\sigma}_{ST}}{\text{USL} - T}, \frac{3\hat{\sigma}_{ST}}{\text{USL} - \hat{\mu}}\right) \qquad \hat{P}'_{RU} = \text{Minimum}\left(\frac{3\hat{\sigma}_{LT}}{\text{USL} - T}, \frac{3\hat{\sigma}_{LT}}{\text{USL} - \hat{\mu}}\right)$$

Whenever $\hat{\mu}$ is greater than, or equal to, T, the upper 1 minus α confidence bounds for C'_{RU} and P'_{RU} are similar to those for C_R and P_R.

$$\tilde{C}'_{RU, 1-\alpha} \cong \hat{C}'_{RU}\sqrt{\frac{\nu}{\chi^2_{\nu, 1-\alpha}}} \text{ where } \nu \cong f_n k(n-1) \text{ for } \hat{\sigma}_{\bar{S}} \text{ or } \nu \cong .9k(n-1) \text{ for } \hat{\sigma}_{\bar{R}}$$

$$\tilde{P}'_{RU, 1-\alpha} \cong \hat{P}'_{RU}\sqrt{\frac{\nu}{\chi^2_{\nu, 1-\alpha}}} \qquad \text{where } \nu = kn - 1$$

If $\hat{\mu}$ is less than T, no confidence bound can be obtained for C'_{RU}. However, an upper confidence bound for P'_{RU} can be found with a method like that explained in Section 11.17.

Example 11.16 Particle Contamination

An *IX & MR* chart monitors particle contamination in *ppm* of a cleaning bath for printed circuit boards after passing through a wave soldering operation. The following calculations provide estimates of μ and σ_{ST} from 25 in-control measurements collected on the chart.

$$\hat{\mu} = \bar{X} = \frac{\sum\limits_{i=1}^{25} X_i}{25} = 127.32 \, ppm \qquad \overline{MR} = \frac{\sum\limits_{i=1}^{24} MR_i}{24} = 13.625 \, ppm \qquad \hat{\sigma}_{ST} = \frac{\overline{MR}}{d_2} = 12.079 \, ppm$$

If the USL for contamination is 165 *ppm*, and T is 120 *ppm* or less, the C'_{RU} ratio is estimated as follows, assuming a normal distribution for this process characteristic:

$$\hat{C}'_{RU} = \text{Minimum}\left(\frac{3\hat{\sigma}_{ST}}{\text{USL} - T}, \frac{3\hat{\sigma}_{ST}}{\text{USL} - \hat{\mu}}\right)$$

$$= \text{Minimum}\left(\frac{3(12.079)}{165 - 120}, \frac{3(12.079)}{165 - 127.32}\right)$$

$$= \text{Minimum}(.805, .962) = .805$$

To determine if a goal of .85 is met with .90 confidence, an upper .90 confidence bound for C'_{RU} is needed. Because the estimated process average of 127.32 is greater than the target average of 120 *ppm*, the formula given near the beginning of this section is chosen to construct the upper confidence bound. As before, the first step is always to determine the number of degrees of freedom. There are 24 moving ranges (25 minus 1), with the subgroup size for each moving range being 2.

$$\nu \cong .9k(n-1) = .9(24)(2-1) \cong 21$$

Once ν is known, an upper confidence bound may be calculated. Due to the small number of degrees of freedom, the resulting confidence bound of 1.014 is almost 26 percent greater than the point estimate of .805.

$$\tilde{C}'_{RU,1-.10} = \tilde{C}'_{RU,.90} \cong \hat{C}'_{RU} \sqrt{\frac{\nu}{\chi^2_{\nu,1-.10}}} = .805 \sqrt{\frac{21}{\chi^2_{21,.90}}} = .805 \sqrt{\frac{21}{13.2396}} = 1.014$$

Although $\hat{C}'_{RU}$ is less than .85, there is less than .90 confidence the process meets this goal. In fact, with an upper bound of 1.014, there is less than .90 confidence this process satisfies even the minimum requirement for potential capability of C'_{RU} being no greater than 1.00.

11.9 Lower Confidence Bounds for C_P and P_P

Because C_P and P_P are just the inverse of C_R and P_R, their confidence bound formulas are approximately the inverse of those for C_R and P_R (Kotz and Johnson, p. 45). However, recall that numerically *higher* measures are better for C_P and P_P. So in order to meet their capability goals, they must equal, or be *greater* than, some minimum number, *e.g.* $C_P \geq 1.33$. To have a certain level of confidence the goal is met for one of these indexes, its *lower* confidence bound must be at, or above, the goal (Nagata).

$$\text{Lower } 1-\alpha \text{ confidence bound for } C_P = C_{P,1-\alpha} \cong \hat{C}_P \sqrt{\frac{\chi^2_{\nu,1-\alpha}}{\nu}}$$

$$\text{where } \nu \cong f_n k(n-1) \text{ for } \hat{\sigma}_{\bar{S}} \quad \text{or } \nu \cong .9k(n-1) \text{ for } \hat{\sigma}_{\bar{R}}$$

If interest is in long-term potential capability, Montgomery (p. 374) recommends applying this formula, where ν equals kn minus 1.

$$\text{Lower } 1-\alpha \text{ confidence bound for } P_P = P_{P,1-\alpha} \cong \hat{P}_P \sqrt{\frac{\chi^2_{\nu,1-\alpha}}{\nu}}$$

The confidence bound formulas for both C_P and P_P assume the process output is approximately a normal distribution (Lehrman).

Minimum Observed Estimate

Below are formulas for determining the *smallest* estimate of these measures that still provide 1 minus α confidence the process has the desired level of potential capability (Chou *et al*).

$$\text{Minimum Observed } \hat{C}_{P,1-\alpha} \cong C_{P,GOAL} \sqrt{\frac{\nu}{\chi^2_{\nu,1-\alpha}}}$$

$$\text{Minimum Observed } \hat{P}_{P,1-\alpha} \cong P_{P,GOAL} \sqrt{\frac{\nu}{\chi^2_{\nu,1-\alpha}}}$$

These values function in the same manner as the maximum observed C_R and P_R ratios, only that concern is now with the *minimum* observed estimate.

Minimum Observed P_p versus ν

The graph in Figure 11.25 shows how the minimum observed P_p estimate at .95 confidence decreases as the number of degrees of freedom increases. In fact, as ν approaches plus infinity, the minimum observed estimate of P_p approaches the $P_{P,\ GOAL}$.

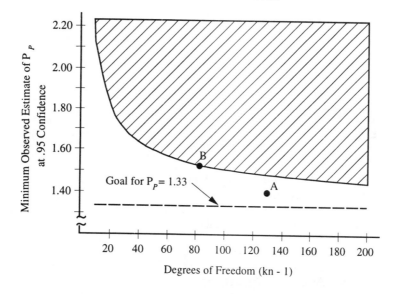

Fig. 11.25 Minimum observed P_p for .95 confidence versus ν.

Estimates of P_p falling in the shaded area indicate the P_p goal of 1.33 is demonstrated with at least .95 confidence. Estimates falling elsewhere imply the goal is not met with this level of confidence. This is true even if the estimate exceeds the goal, but is not within the shaded area. Point A on this graph represents a point estimate for P_p of 1.40, which is greater than the goal of 1.33. But with only 130 degrees of freedom, not enough evidence is provided to believe the true P_p index is at least 1.33 with .95 confidence.

Example 11.17 Floppy Disks

A field problem is traced to variation in the diameter of an inner hole of a floppy disk. From a capability study ($k = 21$ and $n = 4$) performed on the hole cutting process, C_p is estimated from $\overline{R}$ as .928. A lower .95 confidence bound for C_p is calculated by first figuring ν.

$$\nu \cong .9k(n-1) = .9(21)(4-1) = 56.7 \cong 57$$

Now a lower bound can be computed.

$$C_{P,.95} \cong \hat{C}_P \sqrt{\frac{\chi^2_{57,.95}}{57}} = .928 \sqrt{\frac{40.64588}{57}} = .783$$

There is .95 confidence the C_P index for this process is at least .783. Because it must be greater than 1.0 to meet the minimum requirement of short-term potential capability, it is extremely unlikely this process is capable.

In Example 11.14 (p. 626), an upper .95 confidence bound for the C_R ratio of this process was calculated as 1.277, the inverse of which is equal to this lower bound for C_P.

$$C_{P,.95} = \frac{1}{\tilde{C}_{R,.95}} = \frac{1}{1.277} = .783$$

The P_P index for this process is estimated as .909. A lower confidence bound can be constructed for this measure of long-term potential capability by first determining v.

$$v = kn - 1 = 21(4) - 1 = 83$$

With α equal to .05, the appropriate χ^2 value is obtained from Appendix Table VI, and a lower bound computed for P_P.

$$P_{P,.95} \cong \hat{P}_P \sqrt{\frac{\chi^2_{83,.95}}{83}} = .909 \sqrt{\frac{63.004}{83}} = .792$$

Based on the point estimate for P_P of .909, there is .95 confidence the true P_P measure for this process is greater than .792. As this lower confidence bound (as well as the point estimate) is much less than 1.00, there is very little confidence this process meets the minimum requirement for long-term potential capability.

Prior to this capability study being undertaken, the capability goals are established as a C_P of at least 1.667 with .95 confidence and a P_P of at least 1.333 with .95 confidence. The minimum estimates that may be observed at the conclusion of the study, and still furnish .95 confidence these goals are met, are deduced by first predicting the total number of measurements to be collected during the study. Assuming k will equal 21 and n will be 4, and $\sigma_{\bar{x}}$ is to be estimated, the expected number of degrees of freedom for the minimum observed C_P value is forecasted as 57.

$$v \cong .9k(n-1) = .9(21)(4-1) = 56.7 \cong 57$$

For α equal to .05, the minimum observed C_P estimate is computed to be 1.974.

$$\text{Min. Obs. } \hat{C}_{P,.95} \cong C_{P,GOAL} \sqrt{\frac{v}{\chi^2_{v,1-\alpha}}} = 1.667 \sqrt{\frac{57}{\chi^2_{57,.95}}} = 1.667 \sqrt{\frac{57}{40.646}} = 1.974$$

After the study is completed, the point estimate of C_P is .928. Due to the minimum observed C_P for .95 confidence being only 1.974, a conclusion is reached that the process does not meet its specified capability goal with the desired level of confidence.

Anticipating 21 subgroups of size 4 each, the expected number of degrees of freedom for the minimum observed P_P value is 83.

$$v = kn - 1 = 21(4) - 1 = 83$$

$$\text{Min. Obs. } \hat{P}_{P,1-\alpha} \cong P_{P,GOAL} \sqrt{\frac{v}{\chi^2_{v,1-\alpha}}} = 1.333 \sqrt{\frac{83}{\chi^2_{83,.95}}} = 1.333 \sqrt{\frac{83}{63.004}} = 1.530$$

When the data collection is completed, P_P is estimated to be .909, which is much less than the minimum observed P_P value of 1.530. Thus, this process also does not meet the long-term potential capability goal at the desired confidence level. Note that this minimum observed estimate could also have been obtained by reading it off of the graph in Figure 11.25 (see point B). As with the confidence bounds, these results are just the inverse of the maximum observed C_R and P_R ratios previously calculated in Example 11.14.

Minimum Required Sample Size

The approximate sample size required to have the lower confidence a chosen distance below the estimate of P_P has been determined by Franklin (1994) (a more precise, but more complicated, formula was proposed by Franklin in 1995).

$$kn \cong 1 + \frac{1}{2}\left(\frac{Z_\alpha}{e_{PP}}\right)^2$$

The factor e_{PP} reflects the desired difference between the point estimate of P_P and the lower confidence bound for P_P, expressed as a percentage of the point estimate.

$$e_{PP} = \frac{\hat{P}_P - \underset{\sim}{P}_{P,1-\alpha}}{\hat{P}_P} = 1 - \frac{\underset{\sim}{P}_{P,1-\alpha}}{\hat{P}_P}$$

In practice, the second form of the above equation is seen more often, as it expresses the difference between the lower confidence bound and $\hat{P}_P$ as a percentage. For example, it is desired to have the lower .95 confidence limit for P_P in the floppy disk case study be no more than 10 percent below $\hat{P}_P$. This request means there is .95 confidence the estimated P_P value for this process is within 10 percent of the true P_P value. To satisfy this requirement, the lower confidence bound should be 90 percent of $\hat{P}_P$, making their ratio equal to .90.

$$\underset{\sim}{P}_{P,1-\alpha} = .90\,\hat{P}_P \quad \Rightarrow \quad \frac{\underset{\sim}{P}_{P,1-\alpha}}{\hat{P}_P} = .90$$

Subtracting this ratio from 1 yields an e_{PP} factor of .10.

$$e_{PP} = 1 - \frac{\underset{\sim}{P}_{P,1-\alpha}}{\hat{P}_P} = 1 - .90 = .10$$

The Z_α value for α equal to .05 is found in Table 11.1 back on page 602.

$$Z_\alpha = Z_{.05} = 1.645$$

Substituting these values into the sample size formula reveals that kn must be at least 137 (always round up to the next highest integer) if these conditions are to be met.

$$kn \cong 1 + \frac{1}{2}\left(\frac{1.645}{.10}\right)^2 = 136.3 \cong 137$$

Because the subgroup size was selected as 4, a minimum of 35 in-control subgroups must be collected before estimating P_p and calculating a lower .95 confidence bound.

$$kn \cong 137$$

$$\frac{kn}{n} \cong \frac{137}{4}$$

$$k \cong 34.25 \cong 35$$

The estimate for P_p with these 140 measurements turns out to be 1.20. The number of degrees of freedom is computed as 139.

$$v = kn - 1 = 35(4) - 1 = 139$$

The lower .95 confidence bound for P_p is 1.08, which is the desired 90 percent of $\hat{P}_p$.

$$P_{P,.95} \cong \hat{P}_P \sqrt{\frac{\chi^2_{139,.95}}{139}} = 1.20 \sqrt{\frac{112.758}{139}} = 1.08$$

A similar approach works for estimating the minimum sample size required for establishing a lower 1 minus α confidence bound at some prescribed distance from the estimate of C_p. First the total number of degrees of freedom required is calculated with this formula:

$$v \cong \frac{1}{2} \left(\frac{Z_\alpha}{e_{CP}} \right)^2 \quad \text{where } e_{CP} = 1 - \frac{C_{P,1-\alpha}}{\hat{C}_P}$$

With this value for v, the number of in-control subgroups (k) needed to be collected can be determined once the subgroup size is decided for the control chart.

$$v \cong .9k(n-1) \quad \Rightarrow \quad k \cong \frac{v}{.9(n-1)}$$

For example, if it is desired to have the lower .99 confidence limit for C_p in the floppy disk example be 80 percent of the value for $\hat{C}_P$ (20 percent below it), the ratio between them becomes .80, making e_{CP} equal to .20.

$$e_{CP} = 1 - \frac{C_{P,1-\alpha}}{\hat{C}_P} = 1 - .80 = .20$$

The value for $Z_{.01}$ is found to be 2.326 in Table 11.1 (p. 602), which means at least 68 degrees of freedom are needed.

$$v \cong \frac{1}{2} \left(\frac{2.326}{.20} \right)^2 = 67.63 \cong 68$$

Assuming a subgroup size of four will be used in the capability study, k must be at least 25 in order to generate this number of degrees of freedom.

$$k \cong \frac{v}{.9(n-1)} = \frac{67.63}{.9(4-1)} \cong 25$$

Suppose the estimate for C_P turns out to be 1.60. The lower .99 confidence bound for C_P is computed as 1.28, which is the desired 20 percent below $\hat{C}_P$.

$$v \cong .9k(n-1) = .9(25)(4-1) = 67.5 \cong 68$$

$$C_{P,.99} \cong \hat{C}_P \sqrt{\frac{\chi^2_{68,.99}}{68}} = 1.60 \sqrt{\frac{43.838}{68}} = 1.28$$

A very large sample size must be collected if a high level of confidence is desired that the lower confidence bound is fairly close to the estimate of P_P. Table 11.5 on the next page furnishes the minimum sample sizes for lower .90, .95, and .99 confidence bounds for P_P when the bound should be no more than 15, 10, and 5 percent away from the estimate of P_P.

A scan of this table quickly reveals that fairly large sample sizes, over 1,000 in some cases, are necessary to create meaningful lower confidence bounds for P_P. For example, to have .99 confidence that the estimate of P_P is within 5 percent of the true value of P_P, at least 1,078 measurements must be collected in the capability study. Unfortunately, most capability studies usually involve somewhere between only 100 to 125 observations. The e_{PP} factor can be calculated for this typical sample size of around 112 measurements.

$$112 \cong 1 + \frac{1}{2}\left(\frac{Z_\alpha}{e_{PP}}\right)^2$$

$$112 - 1 \cong \frac{1}{2}\left(\frac{Z_\alpha}{e_{PP}}\right)^2$$

$$2(111) \cong \left(\frac{Z_\alpha}{e_{PP}}\right)^2$$

$$\sqrt{222} \cong \frac{Z_\alpha}{e_{PP}}$$

$$e_{PP} \cong \frac{Z_\alpha}{\sqrt{222}}$$

For a confidence level of .99, the e_{PP} factor is about 15.6 percent.

$$e_{PP} \cong \frac{Z_{.01}}{\sqrt{222}} \cong \frac{2.326}{14.900} = .156$$

This calculation reveals the true P_P value could be as much as 15.6 percent less than the point estimate generated for P_P, a fairly substantial difference.

Table 11.5 *kn* requirements for various combinations of confidence and e_{pp} levels.

	e_{pp} Factor		
Confidence Level	.15	.10	.05
.90	38	83	329
.95	62	136	542
.99	121	270	1078

Confidence Level versus Estimate of Potential Capability

There are three ways to estimate σ: $\hat{\sigma}_{\bar{R}}$, $\hat{\sigma}_{\bar{s}}$, and $\hat{\sigma}_S$. The method chosen determines the appropriate capability measure, as well as the *size* of the confidence bound. Assume a capability study involves 20 subgroups of size 5 each, making *kn* equal to 100. Table 11.6 below presents the distance, expressed as a percentage of the estimate, a lower confidence bound would be below the estimate of C_P (based on $\hat{\sigma}_{\bar{R}}$), C_P (based on $\hat{\sigma}_{\bar{s}}$), and P_P (based on $\hat{\sigma}_S$). Even though all three point estimates of potential capability are derived from the same 100 measurements, the distances between them and their respective confidence bounds are distinctly dissimilar. This is due to the different number of degrees of freedom associated with the various methods of estimating σ.

Table 11.6 Distance confidence bound is below capability index for $k = 20$, $n = 5$.

Capability Index	$C_{P.\bar{R}}$	$C_{P.\bar{s}}$	P_P
Degrees of Freedom	72	76	99
e_{pp} Factor for .90 Confidence	.109	.106	.093
e_{pp} Factor for .95 Confidence	.138	.135	.118
e_{pp} Factor for .99 Confidence	.192	.187	.164

For the same amount of data, a better idea of the true process capability is derived with the P_P estimate because it utilizes more of the available information. This is reflected in the higher degrees of freedom and the corresponding "tighter" confidence bound associated with this measure. Notice how the lower .99 confidence bound is 19.2 percent less than the estimate for C_P (based on $\bar{R}$), while for P_P it's only 16.4 percent less.

Maximum Observed $\bar{R}$ to Meet C_p Goal

C_p can be estimated from $\hat{\sigma}_{\bar{R}}$, which is in turn estimated from $\bar{R}$.

$$\hat{C}_p = \frac{\text{Tolerance}}{6\hat{\sigma}_{ST}} \quad \text{where} \quad \hat{\sigma}_{ST} = \hat{\sigma}_{\bar{R}} = \frac{\bar{R}}{d_2}$$

After the minimum observed C_p value is determined, one may derive the *largest* $\bar{R}$ value that may be seen on the range chart during the course of a capability study that still demonstrates attainment of the capability goal with 1 minus α confidence. This $\bar{R}$ value, called maximum observed $\bar{R}_{1-\alpha}$, is found from the minimum observed C_p value.

$$\text{Minimum Observed } \hat{C}_{P,1-\alpha} \cong C_{P,GOAL} \sqrt{\frac{v}{\chi^2_{v,1-\alpha}}}$$

The *maximum* estimate of $\sigma_{\bar{R}}$ that still allows attainment of the C_p goal at the specified 1 minus α confidence level occurs for this *minimum* observed estimate of C_p.

$$\text{Minimum Observed } \hat{C}_{P,1-\alpha} = \frac{\text{Tolerance}}{6\,\text{Maximum Observed } \hat{\sigma}_{\bar{R},1-\alpha}}$$

Substituting these results into the equation for the minimum observed C_p value yields:

$$\frac{\text{Tolerance}}{6\,\text{Maximum Observed } \hat{\sigma}_{\bar{R},1-\alpha}} \cong C_{P,GOAL} \sqrt{\frac{v}{\chi^2_{v,1-\alpha}}}$$

$$\frac{1}{\text{Maximum Observed } \hat{\sigma}_{\bar{R},1-\alpha}} \cong \frac{6\,C_{P,GOAL}}{\text{Tolerance}} \sqrt{\frac{v}{\chi^2_{v,1-\alpha}}}$$

Taking the inverse of both sides produces a formula for finding the maximum observed estimate for $\sigma_{\bar{R}}$.

$$\text{Maximum Observed } \hat{\sigma}_{\bar{R},1-\alpha} \cong \frac{\text{Tolerance}}{6\,C_{P,GOAL}} \sqrt{\frac{\chi^2_{v,1-\alpha}}{v}}$$

Using the relationship between $\hat{\sigma}_{\bar{R}}$ and $\bar{R}$, the following formula is derived for the maximum observed $\bar{R}_{1-\alpha}$. This is the largest $\bar{R}$ value which still demonstrates attainment of the potential capability goal with 1 minus α confidence.

$$\text{Maximum Observed } \hat{\sigma}_{\bar{R},1-\alpha} \cong \frac{\text{Tolerance}}{6\,C_{P,GOAL}} \sqrt{\frac{\chi^2_{v,1-\alpha}}{v}}$$

$$\frac{\text{Maximum Observed } \bar{R}_{1-\alpha}}{d_2} \cong \frac{\text{Tolerance}}{6\,C_{P,GOAL}} \sqrt{\frac{\chi^2_{v,1-\alpha}}{v}}$$

$$\text{Maximum Observed } \overline{R}_{1-\alpha} \cong \frac{d_2 \text{ Tolerance}}{6 \, C_{P, GOAL}} \sqrt{\frac{\chi^2_{v, 1-\alpha}}{v}}$$

Once the subgroup size for the capability study is chosen, which then determines d_2, the maximum observed $\overline{R}_{1-\alpha}$ may be computed for any desired α risk. When the study is conducted and the centerline of the range chart calculated, $\overline{R}$ can be quickly compared to this maximum observed $\overline{R}_{1-\alpha}$ value, with an immediate decision made concerning potential capability. If $\overline{R}$ is less than, or equal to, $\overline{R}_{1-\alpha}$, the process meets the capability goal with the selected level of confidence. When $\overline{R}$ is greater than this value, the process does not meet the goal with the chosen level of confidence.

Example 11.18 Fuse Amperage

The $C_{P, GOAL}$ is set at 2.00 for the amperage of a certain type of industrial fuse. k is planned to be 20, with n being 5, which means $d_2 = 2.326$. From this, v can be predicted.

$$v \cong .9k(n-1) = .9(20)(5-1) = 72$$

With the tolerance given as 10 amps, the maximum observed $\overline{R}$ at .95 confidence is determined to be 1.670 amps.

$$\text{Max. Obs. } \overline{R}_{.95} \cong \frac{2.326 \,(10 \text{ amps})}{6 \,(2.00)} \sqrt{\frac{\chi^2_{72, .95}}{72}} = \frac{23.26 \text{ amps}}{12} \sqrt{\frac{53.46226}{72}} = 1.670 \text{ amps}$$

There is .95 confidence this process meets the $C_{P, GOAL}$ of 2.00 if the $\overline{R}$ calculated from measurements collected during the capability study (after removing data from out-of-control subgroups) is less than 1.670 amps (and v is at least 72). If $\overline{R}$ is greater than 1.670 amps, it cannot be stated that this process meets the goal with the desired level of confidence.

Maximum Observed $\overline{S}_{1-\alpha}$ to Meet C_P Goal

A similar concept to deriving $\overline{R}_{1-\alpha}$ is applied to finding the maximum observed $\overline{S}$ when plans call for using an S chart in the capability study.

$$\text{Maximum Observed } \overline{S}_{1-\alpha} \cong \frac{c_4 \text{ Tolerance}}{6 \, C_{P, GOAL}} \sqrt{\frac{\chi^2_{v, 1-\alpha}}{v}} \qquad \text{where } v \cong f_n k(n-1)$$

If the $\overline{S}$ calculated for the control chart in the capability study is less than this maximum, the process meets its potential capability goal at the specified confidence level.

Example 11.19 Shaft Diameter

A C_P goal of 1.50 is required by the customer for the outer diameter of a process turning steel shafts. Jim, the supervisor of this department, wishes to find the maximum observed $\overline{S}$ possible that demonstrates attainment of the goal with .99 confidence. The plan is to collect 25 subgroups of size 4 each (which means c_4 equals .9213) on an $\overline{X}$, S chart. The expected number of degrees of freedom is determined first, with f_4 taken from Table 11.3 (p. 614).

$$v \cong f_n k(n-1) = .94(25)(4-1) = 70.5 \cong 71$$

With a tolerance for outer diameter size of .050 centimeters, the maximum observed $\overline{S}$ value is found to be .004131 cm.

$$\text{Maximum Observed } \overline{S}_{.99} \cong \frac{c_4 \text{Tolerance}}{6\, C_{P.\,GOAL}} \sqrt{\frac{\chi^2_{v,\,1-\alpha}}{v}}$$

$$= \frac{.9213\,(.050\,\text{cm})}{6\,(1.50)} \sqrt{\frac{\chi^2_{71,\,.99}}{71}}$$

$$= \frac{.046065\,\text{cm}}{9.0} \sqrt{\frac{46.24559}{71}} = .004131\,\text{cm}$$

Shaft diameter measurements are collected, plotted, and control limits and centerlines calculated. Because $\overline{S}$ turns out to be only .003987 cm, which is less than the maximum observed $\overline{S}$ of .004131 cm, Jim has .99 confidence the process meets the C_P goal of 1.50.

11.10 Lower Confidence Bounds for C^*_P and P^*_P

When the target for the process average is not located at the middle of the tolerance, the C_P and P_P indexes are inappropriate for measuring potential capability. In these cases, C^*_P and P^*_P will provide a much better measure of potential capability.

$$C^*_P = \text{Minimum}\left(\frac{T-LSL}{3\sigma_{ST}}, \frac{USL-T}{3\sigma_{ST}}\right) \qquad P^*_P = \text{Minimum}\left(\frac{T-LSL}{3\sigma_{LT}}, \frac{USL-T}{3\sigma_{LT}}\right)$$

These modified measures are point estimators of the true potential capability for these special situations. A lower 1 minus α confidence bound for C^*_P is derived in a manner similar to that done for the C_P index.

$$\text{Lower } 1-\alpha \text{ confidence bound for } C^*_P = C^*_{P,\,1-\alpha} \cong \hat{C}^*_P \sqrt{\frac{\chi^2_{v,\,1-\alpha}}{v}}$$

$$\text{where } v \cong f_n k(n-1) \text{ for } \hat{\sigma}_{\overline{S}} \quad \text{or } v \cong .9k(n-1) \text{ for } \hat{\sigma}_{\overline{R}}$$

Likewise, the lower 1 minus α confidence bound for P^*_P is similar to that for P_P.

$$P^*_{P,\,1-\alpha} \cong \hat{P}^*_P \sqrt{\frac{\chi^2_{v,\,1-\alpha}}{v}} \qquad \text{where } v = kn-1$$

Minimum Observed Estimate

The minimum observed estimates for these measures are derived in a manner similar to those for C_P and P_P.

$$\text{Minimum Observed } \hat{C}^*_{P,\,1-\alpha} \cong C^*_{P.\,GOAL} \sqrt{\frac{v}{\chi^2_{v,\,1-\alpha}}}$$

$$\text{Minimum Observed } \hat{P}^*_{P,1-\alpha} \cong P^*_{P,GOAL} \sqrt{\frac{\nu}{\chi^2_{\nu,1-\alpha}}}$$

The maximum observed $\overline{R}$ and $\overline{S}$ are found with the same formula as for C_P, but with the effective tolerance replacing the actual tolerance.

$$\text{Maximum Observed } \overline{R}_{1-\alpha} \cong \frac{d_2 \text{ (Effective Tolerance)}}{6 \, C^*_{P,GOAL}} \sqrt{\frac{\chi^2_{\nu,1-\alpha}}{\nu}}$$

$$\text{Maximum Observed } \overline{S}_{1-\alpha} \cong \frac{c_4 \text{ (Effective Tolerance)}}{6 \, C^*_{P,GOAL}} \sqrt{\frac{\chi^2_{\nu,1-\alpha}}{\nu}}$$

These formulas are applied and interpreted in an identical manner to those for C_P and P_P, as is demonstrated in the following example.

Example 11.20 Gold Plating

The LSL for the thickness of gold being plated on a silicon wafer is 50 angstroms and the USL is 100 angstroms. For cost reasons, a decision is made to target the average thickness at 65 angstroms. Measurements from 22 subgroups of size 3 each predict σ_{LT} to be 2.5 angstroms. From this information, P^*_P is estimated as 2.00.

$$\hat{P}^*_P = \text{Minimum} \left(\frac{T - \text{LSL}}{3\hat{\sigma}_{LT}}, \frac{\text{USL} - T}{3\hat{\sigma}_{LT}} \right)$$

$$= \text{Minimum} \left(\frac{65 - 50}{3(2.5)}, \frac{100 - 65}{3(2.5)} \right)$$

$$= \text{Minimum} \, (2.00 , 4.67) = 2.00$$

A lower .99 confidence bound for P^*_P is determined by first finding the number of degrees of freedom for the estimate of σ_{LT}.

$$\nu = kn - 1 = 22(3) - 1 = 65$$

With 65 degrees of freedom, the χ^2 value for an α risk of .01 is found in Appendix Table VI to be 41.4434.

$$P^*_{P,1-.01} \cong \hat{P}^*_P \sqrt{\frac{\chi^2_{\nu,1-.01}}{\nu}} = 2.00 \sqrt{\frac{\chi^2_{65,.99}}{65}} = 2.00 \sqrt{\frac{41.4434}{65}} = 1.60$$

There is .99 confidence the long-term potential capability for this plating process meets the minimum capability requirement of P^*_P being greater than 1.00. If the goal for P^*_P is 1.33, there is .99 confidence this higher requirement is met as well.

Assuming ν will be 65, the minimum observed estimate for P^*_P which still demonstrates attainment of a 1.33 goal at .99 confidence is discovered to be 1.67.

$$\text{Min. Observed } \hat{P}^*_{P,.99} \cong P^*_{P,GOAL} \sqrt{\frac{v}{\chi^2_{v,.99}}} = 1.33 \sqrt{\frac{65}{\chi^2_{65,.99}}} = 1.33 \sqrt{\frac{65}{41.4434}} = 1.67$$

Minimum Required Sample Size

The formula for figuring the minimum sample size required to have the lower confidence bound a chosen distance below the estimate of P^*_P is similar to the one for P_P.

$$kn \cong 1 + .5\left(\frac{Z_\alpha}{e_{P^*_P}}\right)^2$$

The e factor reflects the expected difference between the lower confidence bound for P^*_P and the point estimate of P^*_P, expressed as a percentage of the point estimate.

$$e_{P^*_P} = 1 - \frac{\underset{\sim}{P^*_{P,1-\alpha}}}{\hat{P}^*_P}$$

11.11 Lower Confidence Bounds for C'_{PL}, C'_{PU}, P'_{PL}, and P'_{PU}

The C_P and P_P indexes have different formulas to handle unilateral specifications that require a target average be specified for the process center. For dimensions like tensile strength where only a LSL is given, the following formulas estimate C'_{PL} and P'_{PL}:

$$\hat{C}'_{PL} = \text{Maximum}\left(\frac{T-LSL}{3\hat{\sigma}_{ST}}, \frac{\hat{\mu}-LSL}{3\hat{\sigma}_{ST}}\right) \qquad \hat{P}'_{PL} = \text{Maximum}\left(\frac{T-LSL}{3\hat{\sigma}_{LT}}, \frac{\hat{\mu}-LSL}{3\hat{\sigma}_{LT}}\right)$$

If $\hat{\mu}$ is less than, or equal to, T, lower 1 minus α confidence bounds for C'_{PL} and P'_{PL} are found in a manner similar to that done for C_P and P_P.

$$\underset{\sim}{C'_{PL,1-\alpha}} \cong \hat{C}'_{PL} \sqrt{\frac{\chi^2_{v,1-\alpha}}{v}} \quad \text{where } v \cong f_n k(n-1) \text{ for } \hat{\sigma}_{\bar{S}} \quad \text{or } v \cong .9k(n-1) \text{ for } \hat{\sigma}_{\bar{R}}$$

$$\underset{\sim}{P'_{PL,1-\alpha}} \cong \hat{P}'_{PL} \sqrt{\frac{\chi^2_{v,1-\alpha}}{v}} \qquad \text{where } v = kn - 1$$

When $\hat{\mu}$ is greater than T, no confidence bound can be obtained for C'_{PL}. However, a lower confidence bound for P'_{PL} can be computed using a method to be presented in Section 11.17 (p. 653) for calculating a lower confidence bound for P_{PL}.

For situations where a characteristic has only an USL, like concentricity, apply these formulas to compute point estimators of C'_{PU} and P'_{PU}.

$$\hat{C}'_{PU} = \text{Maximum}\left(\frac{USL-T}{3\hat{\sigma}_{ST}}, \frac{USL-\hat{\mu}}{3\hat{\sigma}_{ST}}\right) \qquad \hat{P}'_{PU} = \text{Maximum}\left(\frac{USL-T}{3\hat{\sigma}_{LT}}, \frac{USL-\hat{\mu}}{3\hat{\sigma}_{LT}}\right)$$

Whenever $\hat{\mu}$ is greater than or equal to T, the lower 1 minus α confidence bounds for C'_{PU} and P'_{PU} are similar to those for C_P and P_P.

$$C'_{PU,1-\alpha} \cong \hat{C}_{PU} \sqrt{\frac{\chi^2_{\nu,1-\alpha}}{\nu}} \quad \text{where } \nu \cong f_n k(n-1) \text{ for } \hat{\sigma}_{\bar{S}} \quad \text{or } \nu \cong .9k(n-1) \text{ for } \hat{\sigma}_{\bar{R}}$$

$$P'_{PU,1-\alpha} \cong \hat{P}_{PU} \sqrt{\frac{\chi^2_{\nu,1-\alpha}}{\nu}} \quad \text{where } \nu = kn-1$$

If $\hat{\mu}$ is less than T, no confidence bound can be obtained for C'_{PU}. However, a lower confidence bound for P'_{PU} may be constructed using a method covered in Section 11.17.

Example 11.21 Particle Contamination

An *IX & MR* chart monitors particle contamination in *ppm* of a cleaning bath for printed circuit boards after they pass through a wave soldering operation. The following calculations provide estimates of μ and σ_{ST} from the 25 in-control measurements collected on this chart.

$$\hat{\mu} = \overline{X} = 127.32 \, ppm \qquad \overline{MR} = 13.625 \, ppm \qquad \hat{\sigma}_{ST} = \frac{\overline{MR}}{d_2} = 12.079 \, ppm$$

Given that the USL for contamination is 165 *ppm*, while T is 120 *ppm*, the C'_{PU} index is estimated as follows, assuming a normal distribution for this process characteristic:

$$\hat{C}'_{PU} = \text{Maximum}\left(\frac{\text{USL} - T}{3\hat{\sigma}_{ST}}, \frac{\text{USL} - \hat{\mu}}{3\hat{\sigma}_{ST}}\right)$$

$$= \text{Maximum}\left(\frac{165 - 120}{3(12.079)}, \frac{165 - 127.32}{3(12.079)}\right)$$

$$= \text{Maximum}(1.24, 1.04) = 1.24$$

To determine if a goal of 1.20 is met with .90 confidence, a lower .90 confidence bound for C'_{PU} is needed. Because the estimated process average (127.32) is greater than the target average of 120 *ppm*, the formula given at the beginning of this section is selected to construct a lower confidence bound. As before, the first step in this construction is determining the number of degrees of freedom. There are 24 moving ranges (25 minus 1), with the subgroup size for each moving range being 2.

$$\nu \cong .9k(n-1) = .9(24)(2-1) \cong 21$$

Once the number of degrees of freedom is known, the desired confidence bound may be calculated.

$$C'_{PU,.90} \cong \hat{C}_{PU} \sqrt{\frac{\chi^2_{\nu,.90}}{\nu}} = 1.24 \sqrt{\frac{\chi^2_{21,.90}}{21}} = 1.24 \sqrt{\frac{13.2396}{21}} = .98$$

Although the point estimate of C'_{PU} is greater than the goal of 1.20, there is less than .90 confidence the process meets this goal. In fact, with a lower confidence bound of only .98, there is less than .90 confidence this process meets even the minimum prerequisite for potential capability, which requires C'_{PU} to be greater than 1.00.

11.12 Lower Confidence Bounds for C_{ST} and P_{ST}

C_{ST} and P_{ST} are very similar to C_p and P_p.

$$C_{ST} = 6C_p \qquad P_{ST} = 6P_p$$

Because of this similarity, their confidence bound formulas are practically identical.

$$\underline{C}_{ST,1-\alpha} \cong \hat{C}_{ST} \sqrt{\frac{\chi^2_{v,1-\alpha}}{v}} \quad \text{where } v \cong f_n k(n-1) \text{ for } \hat{\sigma}_{\bar{S}} \text{ or } v \cong .9k(n-1) \text{ for } \hat{\sigma}_{\bar{R}}$$

$$\underline{P}_{ST,1-\alpha} \cong \hat{P}_{ST} \sqrt{\frac{\chi^2_{v,1-\alpha}}{v}} \quad \text{where } v = kn-1$$

These formulas assume the process output is approximately a normal distribution.

Minimum Observed Estimate

Below are formulas for determining the *smallest* estimate of these measures that still provide 1 minus α confidence the process has the desired level of potential capability.

$$\text{Minimum Observed } \hat{C}_{ST,1-\alpha} \cong C_{ST,GOAL} \sqrt{\frac{v}{\chi^2_{v,1-\alpha}}}$$

$$\text{Minimum Observed } \hat{P}_{ST,1-\alpha} \cong P_{ST,GOAL} \sqrt{\frac{v}{\chi^2_{v,1-\alpha}}}$$

These function in the same manner as the minimum observed C_p and P_p indexes.

Minimum Required Sample Size

The approximate sample size required to have the lower confidence bound a chosen distance below the estimate of P_{ST} is determined with this formula:

$$kn \cong 1 + .5\left(\frac{Z_\alpha}{e_{PST}}\right)^2$$

The factor e_{PST} reflects the desired difference between the lower confidence bound for P_{ST} and its point estimate, expressed as a percentage of the point estimate.

$$e_{PST} = 1 - \frac{\underline{P}_{ST,1-\alpha}}{\hat{P}_{ST}}$$

11.13 Lower Confidence Bounds for C^*_{ST} and P^*_{ST}

When the target for the process average is not located at the middle of the tolerance, the C_{ST} and P_{ST} indexes are inappropriate for measuring potential capability. Instead, the C^*_{ST} and P^*_{ST} indexes should be evaluated.

$$C^*_{ST} = 2\,\text{Minimum}\left(\frac{T-\text{LSL}}{\sigma_{ST}}, \frac{\text{USL}-T}{\sigma_{ST}}\right) \qquad P^*_P = 2\,\text{Minimum}\left(\frac{T-\text{LSL}}{\sigma_{LT}}, \frac{\text{USL}-T}{\sigma_{LT}}\right)$$

Lower 1 minus α confidence bounds for these measures are derived in a manner similar to that for C_{ST} and P_{ST}.

$$C^*_{\underset{\sim}{ST},1-\alpha} \cong \hat{C}^*_{ST}\sqrt{\frac{\chi^2_{v,1-\alpha}}{v}} \quad \text{where } v \cong f_n k(n-1) \text{ for } \hat{\sigma}_{\bar{s}} \text{ or } v \cong .9k(n-1) \text{ for } \hat{\sigma}_{\bar{R}}$$

$$P^*_{\underset{\sim}{ST},1-\alpha} \cong \hat{P}^*_{ST}\sqrt{\frac{\chi^2_{v,1-\alpha}}{v}} \quad \text{where } v = kn-1$$

Minimum Observed Estimate

The minimum observed estimate formulas for these measures are similar to those for C_{ST} and P_{ST}.

$$\text{Minimum Observed } \hat{C}^*_{ST,1-\alpha} \cong C^*_{ST,GOAL}\sqrt{\frac{v}{\chi^2_{v,1-\alpha}}}$$

$$\text{Minimum Observed } \hat{P}^*_{ST,1-\alpha} \cong P^*_{ST,GOAL}\sqrt{\frac{v}{\chi^2_{v,1-\alpha}}}$$

Minimum Required Sample Size

The formula for figuring the minimum sample size required to have the lower confidence bound a specified distance below the estimate of P^*_{ST} is similar to the one for P_{ST}.

$$kn \cong 1 + .5\left(\frac{Z_\alpha}{e_{P^*ST}}\right)^2$$

This e factor reflects the desired difference between the lower confidence bound and the point estimate for P^*_P, expressed as a percentage of the point estimate.

$$e_{P^*ST} = 1 - \frac{P^*_{\underset{\sim}{ST},1-\alpha}}{\hat{P}^*_{ST}}$$

11.14 Lower Confidence Bounds for C'_{STL}, C'_{STU}, P'_{STL}, and P'_{STU}

The C_{ST} and P_{ST} indexes have different formulas to handle unilateral specifications that require a target average (T) be specified for the process center. For features like weld strength, where only a LSL is given, the following formulas estimate C'_{STL} and P'_{STL}:

$$\hat{C}'_{STL} = 2\,\text{Maximum}\left(\frac{T-\text{LSL}}{\hat{\sigma}_{ST}}, \frac{\hat{\mu}-\text{LSL}}{\hat{\sigma}_{ST}}\right) \qquad \hat{P}'_{STL} = 2\,\text{Maximum}\left(\frac{T-\text{LSL}}{\hat{\sigma}_{LT}}, \frac{\hat{\mu}-\text{LSL}}{\hat{\sigma}_{LT}}\right)$$

If $\hat{\mu}$ is less than, or equal to, T, lower 1 minus α confidence bounds for these measures are found in a manner similar to that done for C_{ST} and P_{ST}.

$$\underset{\sim}{C}'_{STL,1-\alpha} \cong \hat{C}_{STL}\sqrt{\frac{\chi^2_{v,1-\alpha}}{v}} \quad \text{where } v \cong f_n k(n-1) \text{ for } \hat{\sigma}_{\bar{s}} \text{ or } v \cong .9k(n-1) \text{ for } \hat{\sigma}_{\bar{R}}$$

$$\underset{\sim}{P}'_{STL,1-\alpha} \cong \hat{P}_{STL}\sqrt{\frac{\chi^2_{v,1-\alpha}}{v}} \quad \text{where } v = kn-1$$

When $\hat{\mu}$ is greater than T, no confidence bound can be obtained for C'_{STL}. However, a lower confidence bound for P'_{STL} may be computed with a method to be explained in the very next section (11.15).

For situations where a characteristic has only an USL, like squareness, apply these formulas to compute point estimates of C'_{STU} and P'_{STU}.

$$\hat{C}'_{STU} = 2\,\text{Maximum}\left(\frac{\text{USL}-T}{\hat{\sigma}_{ST}}, \frac{\text{USL}-\hat{\mu}}{\hat{\sigma}_{ST}}\right) \qquad \hat{P}'_{STU} = 2\,\text{Maximum}\left(\frac{\text{USL}-T}{\hat{\sigma}_{LT}}, \frac{\text{USL}-\hat{\mu}}{\hat{\sigma}_{LT}}\right)$$

Whenever $\hat{\mu}$ is greater than or equal to T, the lower 1 minus α confidence bounds are similar to those for C_{ST} and P_{ST}.

$$\underset{\sim}{C}'_{STU,1-\alpha} \cong \hat{C}_{STU}\sqrt{\frac{\chi^2_{v,1-\alpha}}{v}} \quad \text{where } v \cong f_n k(n-1) \text{ for } \hat{\sigma}_{\bar{s}} \text{ or } v \cong .9k(n-1) \text{ for } \hat{\sigma}_{\bar{R}}$$

$$\underset{\sim}{P}'_{STU,1-\alpha} \cong \hat{P}_{STU}\sqrt{\frac{\chi^2_{v,1-\alpha}}{v}} \quad \text{where } v = kn-1$$

If $\hat{\mu}$ is less than T, no confidence bound can be obtained for C'_{STU}. However, a lower confidence bound for P'_{STU} can be constructed using a method presented for $Z_{USL,LT}$ in the next section.

11.15 Lower Confidence Bounds for $Z_{MIN,LT}$, $Z_{LSL,LT}$, and $Z_{USL,LT}$

The sampling distribution for $Z_{MIN,LT}$ is quite complex due to both μ and σ_{LT} being involved in its calculation.

$$Z_{MIN,LT} = \text{Minimum}\left(\frac{\mu-\text{LSL}}{\sigma_{LT}}, \frac{\text{USL}-\mu}{\sigma_{LT}}\right)$$

A very good approximation for a lower 1 minus α confidence bound for $Z_{MIN,LT}$, labeled $\underset{\sim}{Z}_{MIN,LT,1-\alpha}$, was developed by A. F. Bissell.

$$Z_{MIN.\,LT,\,1-\alpha} = \hat{Z}_{MIN.\,LT} - Z_\alpha \sqrt{\frac{1}{kn} + \frac{\hat{Z}^2_{MIN.\,LT}}{2(kn-1)}}$$

Typically, only a *lower* confidence bound is computed for the $Z_{MIN.\,LT}$ measure because the goal is to be equal to, or *greater* than, some number, *e.g.*, $Z_{MIN.\,LT} \geq 4$. To have a certain level of confidence the $Z_{MIN.\,LT}$ goal is surpassed, its lower confidence bound must be above the specified goal.

Example 11.22 Valve O.D. Size

The $Z_{MIN.\,LT}$ measure for the O.D. size of an exhaust valve run on a grinding machine was estimated in Example 6.2 as 3.472 (p. 174). With k equal to 20 and n equal to 5, a lower .90 confidence bound ($\alpha = .10$) for $Z_{MIN.\,LT}$ is found as shown here:

$$Z_{MIN.\,LT,.90} = \hat{Z}_{MIN.\,LT} - Z_\alpha \sqrt{\frac{1}{kn} + \frac{\hat{Z}^2_{MIN.\,LT}}{2(kn-1)}} = 3.472 - Z_{.10}\sqrt{\frac{1}{100} + \frac{3.472^2}{2(100-1)}} = 3.131$$

Based on a point estimate for $Z_{MIN.\,LT}$ of 3.472 and the number of pieces in this capability study, there is .90 confidence the true $Z_{MIN.\,LT}$ measure for this process is at least 3.131. Thus, there is a reasonably high degree of confidence this process meets the minimum requirement for performance capability, *i.e.*, $Z_{MIN.\,LT}$ is greater than 3.0.

Minimum Observed Estimate

The lower confidence bound formula for $Z_{MIN.\,LT}$ can be rearranged to derive the smallest $Z_{MIN.\,LT}$ value that may be observed in a capability study, yet offer 1 minus α confidence the goal for $Z_{MIN.\,LT}$ is met. As $1/kn$ is very close to zero, this term is dropped to simplify the formula, whose full derivation is provided in Section 11.32 (see Derivation 1, p. 708).

$$\text{Minimum Observed } \hat{Z}_{MIN.\,LT,\,1-\alpha} \cong \frac{Z_{MIN.\,LT.\,GOAL}}{1 - Z_\alpha\sqrt{1 \div 2(kn-1)}}$$

For the valve O.D. size example (assuming k will be 20 and n will be 5), the minimum estimate of $Z_{MIN.\,LT}$ that may be observed in a capability study, and still indicate with .90 confidence the process has achieved a $Z_{MIN.\,LT}$ goal of 3.00, is calculated as 3.301.

$$\text{Minimum Observed } \hat{Z}_{MIN.LT,.90} \cong \frac{3.00}{1 - Z_{.10}\sqrt{1 \div 2(100-1)}} = \frac{3.00}{1 - 1.282\sqrt{1 \div 198}} = 3.301$$

If the estimate of $Z_{MIN.\,LT}$ turns out to be 3.472, there would be .90 confidence the true $Z_{MIN.\,LT}$ is greater than or equal to the goal of 3.00, *i.e.*, the process meets, or even exceeds, the capability requirement.

Example 11.23 Floppy Disks

A field problem is traced to variation in the diameter of the inner hole of a floppy disk. From a capability study involving 21 subgroups of size 4 each ($kn = 84$), $Z_{MIN.\,LT}$ is estimated as 1.528. A lower .95 confidence bound ($\alpha = .05$) for $Z_{MIN.\,LT}$ is calculated as 1.263.

$$Z_{MIN.\,LT,.95} = \hat{Z}_{MIN.\,LT} - Z_{.05}\sqrt{\frac{1}{kn} + \frac{\hat{Z}^2_{MIN.\,LT}}{2(kn-1)}} = 1.528 - 1.645\sqrt{\frac{1}{84} + \frac{1.528^2}{2(84-1)}} = 1.263$$

With a point estimate for $Z_{MIN.\,LT}$ of 1.528, there is .95 confidence the true $Z_{MIN.\,LT}$ for this process is somewhere above 1.263. Because this lower confidence bound must at least 3.0 to meet the minimum requirement for long-term performance capability at .95 confidence, this process is judged not capable. In fact, as the point estimate of 1.526 is also less than 3.0, there is very little likelihood the hole-cutting operation is capable of holding the given tolerance.

Suppose that before the capability study begins, it is desired to establish a "bogie" for the smallest $\hat{Z}_{MIN.\,LT}$ that may be seen, and yet provide .95 confidence ($\alpha = .05$) a $Z_{MIN.\,LT}$ goal of 4.50 is met. This bogie is calculated as follows, assuming k will be 21 and n will be 4:

$$\text{Min. Observed } \hat{Z}_{MIN.LT,.95} \cong \frac{Z_{MIN.LT,\,GOAL}}{1 - Z_{.05}\sqrt{1 \div 2(kn-1)}} = \frac{4.50}{1 - 1.645\sqrt{1 \div 2(84-1)}} = 5.159$$

The estimate of $Z_{MIN.\,LT}$ turned out to be only 1.528, which is so far from the minimum required of 5.159 that there is very little hope this process meets a $Z_{MIN.\,LT}$ goal of 4.50 at the desired .95 confidence level.

Unfortunately, no formula for computing a lower confidence bound for $Z_{MIN.\,ST}$ has been developed so far. This is due to the very complex sampling distribution of $\hat{\sigma}_{\bar{R}}$ (or $\hat{\sigma}_{\bar{S}}$) when it is used in conjunction with $\hat{\mu}$. Hopefully, a good approximation will be derived some day.

Minimum Required Sample Size

A formula for approximating the sample size required to have the lower confidence a chosen distance below the estimate of $Z_{MIN.\,LT}$ can be deduced from work presented in a paper by Franklin (1994).

$$kn \cong \left(\frac{1}{\hat{Z}^2_{MIN.\,LT}} + \frac{1}{2}\right)\left(\frac{Z_\alpha}{e_{ZMIN}}\right)^2$$

The factor e_{ZMIN} expresses the percentage difference between the lower confidence bound and the estimate of $Z_{MIN.\,LT}$.

$$e_{ZMIN} = 1 - \frac{Z_{MIN.\,LT,\,1-\alpha}}{\hat{Z}_{MIN.\,LT}}$$

The above formula for kn requires the estimate of $Z_{MIN.\,LT}$ to be known *before* conducting the capability study. As this is unlikely, the formula isn't of much practical use. However, as the estimate of $Z_{MIN.\,LT}$ increases, the quantity one over $\hat{Z}^2_{MIN.\,LT}$ decreases, causing the required sample size to also decrease. So, as a conservative estimate of the minimum sample size, let $\hat{Z}_{MIN.\,LT}$ be fairly small, say equal to 3.0. This greatly simplifies the kn formula.

$$kn \cong \left(\frac{1}{\hat{Z}^2_{MIN.\,LT}} + \frac{1}{2}\right)\left(\frac{Z_\alpha}{e_{ZMIN}}\right)^2 = \left(\frac{1}{3^2} + \frac{1}{2}\right)\left(\frac{Z_\alpha}{e_{ZMIN}}\right)^2 = \left(\frac{1}{9} + \frac{1}{2}\right)\left(\frac{Z_\alpha}{e_{ZMIN}}\right)^2 = .611\left(\frac{Z_\alpha}{e_{ZMIN}}\right)^2$$

For example, the lower .95 confidence limit for $Z_{MIN, LT}$ in the floppy disk case study is desired to be 10 percent below $\hat{Z}_{MIN, LT}$, which means there is .95 confidence the true $Z_{MIN, LT}$ value for this process is within 10 percent of the estimated $Z_{MIN, LT}$ value. This requires the lower confidence bound to be 90 percent of $\hat{Z}_{MIN, LT}$, making their ratio equal to .90.

$$\underset{\sim}{Z}_{MIN, LT, 1-\alpha} = .90\,\hat{Z}_{MIN, LT} \quad \Rightarrow \quad \frac{\underset{\sim}{Z}_{MIN, LT, 1-\alpha}}{\hat{Z}_{MIN, LT}} = .90$$

Subtracting this ratio from 1 yields an e_{ZMIN} factor of .10.

$$e_{ZMIN} = 1 - \frac{\underset{\sim}{Z}_{MIN, LT, 1-\alpha}}{\hat{Z}_{MIN, LT}} = 1 - .90 = .10$$

Z_α for α equal to .05 is found in Table 11.1 (p. 602) as 1.645. Substituting these values into the minimum sample size formula reveals that kn must be at least 166 to achieve these objectives (always round up to the next highest integer).

$$kn \cong .611\left(\frac{Z_\alpha}{e_{ZMIN}}\right)^2 = .611\left(\frac{1.645}{.10}\right)^2 = 165.34 \cong 166$$

Because the subgroup size (n) was selected as 4, at least 42 in-control subgroups must be collected before estimating $Z_{MIN, LT}$ and calculating a lower .95 confidence bound.

$$kn \cong 166$$

$$\frac{kn}{n} \cong \frac{166}{4}$$

$$k \cong 41.5 = 42$$

The estimate for $Z_{MIN, LT}$ with these 168 (42 × 4) measurements turns out to be 3.450. A lower .95 confidence bound for $Z_{MIN, LT}$ would be computed as 3.115.

$$\underset{\sim}{Z}_{MIN, LT, .95} = \hat{Z}_{MIN, LT} - Z_{.05}\sqrt{\frac{1}{kn} + \frac{\hat{Z}^2_{MIN, LT}}{2(kn - 1)}}$$

$$= 3.450 - 1.645\sqrt{\frac{1}{168} + \frac{3.450^2}{2(168 - 1)}} = 3.115$$

Due to the conservative nature of the minimum sample size formula, this lower bound is just 9.7 percent below the point estimate, rather than the specified 10 percent.

$$3.115 = (1 - .097)\,3.450 = (.903)\,3.450$$

Lower Confidence Bounds for $Z_{LSL, LT}$ and $Z_{USL, LT}$

Lower confidence bound formulas for these measures are similar to the one just developed for $Z_{MIN, LT}$.

$$\underset{\sim}{Z}_{LSL,\,LT,\,1-\alpha} = \hat{Z}_{LSL,\,LT} - Z_\alpha \sqrt{\frac{1}{kn} + \frac{\hat{Z}^2_{LSL,\,LT}}{2(kn-1)}} \qquad \underset{\sim}{Z}_{USL,\,LT,\,1-\alpha} = \hat{Z}_{USL,\,LT} - Z_\alpha \sqrt{\frac{1}{kn} + \frac{\hat{Z}^2_{USL,\,LT}}{2(kn-1)}}$$

In fact, the lower confidence bound for $Z_{MIN,\,LT}$ is equal to the smaller of the lower confidence bounds for these two measures.

$$\underset{\sim}{Z}_{MIN,\,LT,\,1-\alpha} = \text{Minimum}\left(\underset{\sim}{Z}_{LSL,\,LT,\,1-\alpha}, \underset{\sim}{Z}_{USL,\,LT,\,1-\alpha}\right)$$

Their formulas for the minimum observed estimate are also similar to the one for $Z_{MIN,\,LT}$.

$$\text{Minimum Observed } \hat{Z}_{LSL,\,LT,\,1-\alpha} \cong \frac{Z_{LSL,\,LT,\,GOAL}}{1 - Z_\alpha \sqrt{1 \div 2(kn-1)}}$$

$$\text{Minimum Observed } \hat{Z}_{USL,\,LT,\,1-\alpha} \cong \frac{Z_{USL,\,LT,\,GOAL}}{1 - Z_\alpha \sqrt{1 \div 2(kn-1)}}$$

The minimum sample size formula is identical to that for $Z_{MIN,\,LT}$, with either $Z_{LSL,\,LT}$ and e_{ZLSL} (or $Z_{USL,\,LT}$ and e_{ZUSL}) replacing $Z_{MIN,\,LT}$ and e_{ZMIN}.

$$e_{ZLSL} = 1 - \frac{\underset{\sim}{Z}_{LSL,\,LT,\,1-\alpha}}{\hat{Z}_{LSL,\,LT}} \qquad\qquad e_{ZUSL} = 1 - \frac{\underset{\sim}{Z}_{USL,\,LT,\,1-\alpha}}{\hat{Z}_{USL,\,LT}}$$

11.16 Lower Confidence Bounds for $Z^*_{MIN,\,LT}$, $Z^*_{LSL,\,LT}$, and $Z^*_{USL,\,LT}$

When the target for μ is not the middle of the tolerance, $Z^*_{MIN,\,LT}$ should replace $Z_{MIN,\,LT}$ as the measure of long-term performance capability. $Z^*_{MIN,\,LT}$ was defined on page 197 as:

$$Z^*_{MIN,\,LT} = \text{Minimum}\,(Z^*_{LSL,\,LT}, Z^*_{USL,\,LT}) = \text{Minimum}\left(\frac{\mu - LSL_T}{\sigma_{LT}}, \frac{USL_T - \mu}{\sigma_{LT}}\right)$$

If T is less than M, use these formulas to compute LSL_T and USL_T:

$$LSL_T = LSL \qquad\qquad USL_T = 2T - LSL$$

For situations where T is greater than M:

$$LSL_T = 2T - USL \qquad\qquad USL_T = USL$$

Lower 1 minus α confidence bound formulas for these measures are similar to those for the corresponding Z measure.

$$\underset{\sim}{Z}^*_{MIN,\,LT,\,1-\alpha} = \hat{Z}^*_{MIN,\,LT} - Z_\alpha \sqrt{\frac{1}{kn} + \frac{\hat{Z}^{*2}_{MIN,\,LT}}{2(kn-1)}}$$

$$Z^*_{\underline{LSL.\,LT},\,1-\alpha} = \hat{Z}^*_{LSL.\,LT} - Z_\alpha \sqrt{\frac{1}{kn} + \frac{\hat{Z}^{*2}_{LSL.\,LT}}{2\,(kn-1)}}$$

$$Z^*_{\underline{USL.\,LT},\,1-\alpha} = \hat{Z}^*_{USL.\,LT} - Z_\alpha \sqrt{\frac{1}{kn} + \frac{\hat{Z}^{*2}_{USL.\,LT}}{2\,(kn-1)}}$$

Minimum Observed Estimate

The lower confidence bound formula for $Z^*_{MIN.\,LT}$ can be rearranged to derive the smallest $\hat{Z}^*_{MIN.\,LT}$ value that still offers 1 minus α confidence the goal for $Z^*_{MIN.\,LT}$ is met. As was done for $Z_{MIN.\,LT}$, the $1/kn$ term is dropped to simplify the formula's derivation.

$$\text{Minimum Observed } \hat{Z}^*_{MIN.\,LT,\,1-\alpha} \cong \frac{Z^*_{MIN.\,LT.\,GOAL}}{1 - Z_\alpha \sqrt{1 \div 2(kn-1)}}$$

Likewise, the minimum observed estimates for $Z^*_{LSL.\,LT}$ and $Z^*_{USL.\,LT}$ are:

$$\text{Minimum Observed } \hat{Z}^*_{LSL.\,LT,\,1-\alpha} \cong \frac{Z^*_{LSL.\,LT.\,GOAL}}{1 - Z_\alpha \sqrt{1 \div 2(kn-1)}}$$

$$\text{Minimum Observed } \hat{Z}^*_{USL.\,LT,\,1-\alpha} \cong \frac{Z^*_{USL.\,LT.\,GOAL}}{1 - Z_\alpha \sqrt{1 \div 2(kn-1)}}$$

Example 11.24 Gold Plating Thickness

The LSL for the thickness of gold plated on a silicon wafer is 50 angstroms, while the USL is 100 angstroms. Because the target average for thickness is 65, which is lower than the middle of the tolerance, LSL_T equals the LSL of 50, while USL_T is determined as:

$$USL_T = 2T - LSL = 2(65) - 50 = 80$$

If the estimates of σ_{LT} and μ are 2.5 angstroms and 70 angstroms respectively, $Z^*_{MIN.\,LT}$ is estimated as 4.0.

$$\hat{Z}^*_{MIN.LT} = \text{Minimum}\left(\frac{\hat{\mu} - LSL_T}{\hat{\sigma}_{LT}}, \frac{USL_T - \hat{\mu}}{\hat{\sigma}_{LT}}\right) = \text{Minimum}\left(\frac{70-50}{2.5}, \frac{80-70}{2.5}\right) = 4.0$$

From this estimate, a lower .99 confidence bound can now be derived for the true $Z^*_{MIN.\,LT}$ value ($kn = 66$).

$$Z^*_{\underline{MIN.LT},.99} = \hat{Z}^*_{MIN.LT} - Z_\alpha \sqrt{\frac{1}{kn} + \frac{\hat{Z}^{*\,2}_{MIN.LT}}{2(kn-1)}} = 4.0 - Z_{.01}\sqrt{\frac{1}{66} + \frac{(4.0)^2}{2(66-1)}} = 3.135$$

Because this lower confidence bound is greater than 3.0, there is a fairly high degree of confidence this process meets the minimum performance capability requirement.

Minimum Required Sample Size

Formulas for approximating the minimum sample size for having the lower confidence bound a certain distance below the point estimate are virtually identical to the formula developed for $Z_{MIN, LT}$ in the previous section.

$$kn \cong .611 \left(\frac{Z_\alpha}{e_{z^*MIN}} \right)^2 \qquad kn \cong .611 \left(\frac{Z_\alpha}{e_{z^*LSL}} \right)^2 \qquad kn \cong .611 \left(\frac{Z_\alpha}{e_{z^*USL}} \right)^2$$

e factors are defined as follows:

$$e_{z^*MIN} = 1 - \frac{Z^*_{MIN. LT, 1-\alpha}}{\hat{Z}^*_{MIN. LT}} \qquad e_{z^*LSL} = 1 - \frac{Z^*_{LSL. LT, 1-\alpha}}{\hat{Z}^*_{LSL. LT}} \qquad e_{z^*USL} = 1 - \frac{Z^*_{USL. LT, 1-\alpha}}{\hat{Z}^*_{USL. LT}}$$

11.17 Lower Confidence Bounds for P_{PL} and P_{PU}

On account of these indexes being directly related to their respective Z values, their confidence bound formulas are very similar to those for $Z_{LSL. LT}$ and $Z_{USL. LT}$ (Franklin *et al*, Levinson).

$$\hat{P}_{PL} = \frac{\hat{Z}_{LSL.LT}}{3} \qquad \underset{\sim}{P}_{PL, 1-\alpha} = \hat{P}_{PL} - Z_\alpha \sqrt{\frac{1}{9kn} + \frac{\hat{P}^2_{PL}}{2(kn-1)}}$$

$$\hat{P}_{PU} = \frac{\hat{Z}_{USL.LT}}{3} \qquad \underset{\sim}{P}_{PU, 1-\alpha} = \hat{P}_{PU} - Z_\alpha \sqrt{\frac{1}{9kn} + \frac{\hat{P}^2_{PU}}{2(kn-1)}}$$

Notice the addition of a "9" under the radical sign in the confidence bound equation. This comes from the "3" in the relationship between the two measures ($3^2 = 9$).

Example 11.25 Floppy Disks

A field problem is traced to variation in the diameter of the inner hole of a floppy disk. From a capability study involving 21 subgroups of size 4 each ($kn = 84$), P_{PL} was estimated as .509, while P_{PU} was estimated as 1.310. Lower .95 confidence bounds ($\alpha = .05$) for these two measures are calculated as follows:

$$\underset{\sim}{P}_{PL,.95} = \hat{P}_{PL} - Z_{.05} \sqrt{\frac{1}{9kn} + \frac{\hat{P}^2_{PL}}{2(kn-1)}} = .509 - 1.645 \sqrt{\frac{1}{9(84)} + \frac{(.509)^2}{2(84-1)}} = .421$$

$$\underset{\sim}{P}_{PU,.95} = \hat{P}_{PU} - Z_{.05} \sqrt{\frac{1}{9kn} + \frac{\hat{P}^2_{PU}}{2(kn-1)}} = 1.310 - 1.645 \sqrt{\frac{1}{9(84)} + \frac{(1.310)^2}{2(84-1)}} = 1.132$$

Notice that the lower confidence bound for P_{PL} is 17.7 percent less than the P_{PL} point estimate, while the lower confidence bound for P_{PU} is only 13.7 percent less than the P_{PU} point estimate. If the point estimate had been 2.00, the lower confidence bound of 1.738 would have been only 13.1 percent less. Had the point estimate been 3.00, the lower bound of 2.612 would have been just 12.9 percent less. In general, the larger the point estimate, the closer the lower confidence bound is to the point estimate.

Minimum Observed Estimate

Formulas for the minimum observed estimate of these two measures are also similar to those derived for $Z_{LSL, LT}$ and $Z_{USL, LT}$. To simplify these formulas, the quantity $1/9kn$ is assumed to be zero.

$$\text{Minimum Observed } \hat{P}_{PL, 1-\alpha} \cong \frac{P_{PL, GOAL}}{1 - Z_\alpha \sqrt{1 \div 2(kn - 1)}}$$

$$\text{Minimum Observed } \hat{P}_{PU, 1-\alpha} \cong \frac{P_{PU, GOAL}}{1 - Z_\alpha \sqrt{1 \div 2(kn - 1)}}$$

Minimum Required Sample Size

Minimum sample size formulas are identical to those for $Z_{LSL, LT}$ and $Z_{USL, LT}$.

$$kn \cong .611 \left(\frac{Z_\alpha}{e_{PPL}} \right)^2 \qquad kn \cong .611 \left(\frac{Z_\alpha}{e_{PPU}} \right)^2$$

Where the e factors are defined as:

$$e_{PPL} = 1 - \frac{P_{PL, 1-\alpha}}{\hat{P}_{PL}} \qquad e_{PPU} = 1 - \frac{P_{PU, 1-\alpha}}{\hat{P}_{PU}}$$

11.18 Lower Confidence Bound for P_{PK}

As explained in Chapter 6, P_{PK} is a function of $Z_{MIN, LT}$.

$$\hat{P}_{PK} = \frac{\hat{Z}_{MIN.LT}}{3}$$

Thus, the formula for its confidence bound is closely related to the one developed by A.F. Bissell for $Z_{MIN, LT}$.

$$P_{PK, 1-\alpha} = \hat{P}_{PK} - Z_\alpha \sqrt{\frac{1}{9kn} + \frac{\hat{P}_{PK}^2}{2(kn - 1)}}$$

Alternative formulas have been derived by Barnes and Pierce, Chou *et al*, Greenwich and Chen, Kahle *et al*, Sarkar and Pal, as well as Zhang *et al*. However, studies done by Kushler and Hurley recommend the method proposed by Bissell for general use. An alternative approach for constructing a lower confidence bound using the bootstrap methodology, which does not assume normality, is presented by Price and Price, and Franklin and Wasserman.

A lower confidence bound for P_{PK} can also be found as the smaller of the lower bounds for P_{PL} and P_{PU}.

$$P_{PK, 1-\alpha} = \text{Minimum} \left(P_{PL, 1-\alpha}, P_{PU, 1-\alpha} \right)$$

Ordinarily, only a lower confidence bound for P_{PK} is computed in practice. However, there are times when a costly improvement is being considered for a process. In this situation, there may be interest in knowing the absolute *best* performance capability that could be expected from the current process without the improvement. This upper limit is quantified by computing an upper 1 minus α confidence bound, where a plus sign replaces the negative sign in the formula for a lower bound.

$$\tilde{P}_{PK, 1-\alpha} = \hat{P}_{PK} + Z_\alpha \sqrt{\frac{1}{9kn} + \frac{\hat{P}_{PK}^2}{2(kn-1)}}$$

Example 11.26 Valve O.D. Size

For the exhaust valve O.D. size example, P_{PK} is estimated as 1.157.

$$\hat{P}_{PK} = \frac{\hat{Z}_{MIN.LT}}{3} = \frac{3.472}{3} = 1.157$$

With kn equal to 100, a lower .90 confidence bound for P_{PK} is calculated as follows:

$$\underset{\sim}{P}_{PK,.90} = 1.157 - Z_{.10} \sqrt{\frac{1}{9kn} + \frac{1.157^2}{2(kn-1)}} = 1.157 - 1.282 \sqrt{\frac{1}{9(100)} + \frac{1.157^2}{2(100-1)}} = 1.043$$

Based on the estimate for P_{PK} of 1.157, there is .90 confidence the true P_{PK} measure of capability for this process is greater than 1.043. If the goal for P_{PK} is to be at least 1.00, there is .90 confidence the process meets this goal. Note this lower confidence bound for P_{PK} is equal to the .90 lower confidence bound for $Z_{MIN. LT}$ found in Example 11.22 divided by 3.

$$\underset{\sim}{P}_{PK,.90} = \frac{\underset{\sim}{Z}_{MIN.LT,.90}}{3} = \frac{3.131}{3} = 1.043$$

Minimum Observed Estimate

P_{PK} is estimated for a process as 1.44, with a goal of 1.33. Does this sample result imply the goal is met? It would seem so, but a second study on this process will most likely generate a different estimate for P_{PK} that could be larger, or smaller, than 1.44. This second estimate may even be less than 1.33, indicating the goal is *not* met.

What minimum P_{PK} estimate must be observed from the sample data of just one study to provide a reasonably high degree of confidence that almost every additional study on this process will also conclude the capability goal has been demonstrated? Obviously, this minimum P_{PK} estimate must be somewhat larger than the goal. Exactly how much larger depends on the goal, the level of confidence, and the sample size, as revealed by this formula:

$$\text{Minimum Observed } \hat{P}_{PK, 1-\alpha} \cong \frac{P_{PK. GOAL}}{1 - Z_\alpha \sqrt{1 \div 2(kn-1)}}$$

Table 11.7 lists the minimum observed P_{PK} estimate required to demonstrate a desired P_{PK} goal is met with a certain confidence for various goals and number of measurements included in the study. The top number in each box is for .99 confidence, the middle number for .95, and the bottom number is for .90. For example, if kn equals 80 and the goal for P_{PK} is 1.33, the P_{PK} index estimated in the study must be at least 1.48 to have .90 confidence the specified capability goal of 1.33 is met (see also Pearn and Chen).

Table 11.7 Minimum P_{PK} to demonstrate various capability goals.

Goal for P_{PK}

kn	1.00	1.33	1.50	1.67	2.00
20	1.60	2.13	2.41	2.67	3.21
	1.36	1.82	2.05	2.27	2.73
	1.26	1.68	1.89	2.10	2.52
40	1.36	1.81	2.03	2.26	2.71
	1.23	1.64	1.84	2.05	2.46
	1.17	1.56	1.75	1.95	2.34
60	1.27	1.70	1.91	2.12	2.54
	1.18	1.57	1.77	1.96	2.36
	1.13	1.51	1.70	1.89	2.27
80	1.23	1.64	1.84	2.04	2.45
	1.15	1.53	1.73	1.92	2.30
	1.11	**1.48**	1.67	1.86	2.23
100	1.20	1.60	1.80	2.00	2.39
	1.13	1.51	1.70	1.89	2.26
	1.10	1.47	1.65	1.83	2.20
125	1.17	1.56	1.76	1.95	2.35
	1.12	1.49	1.67	1.86	2.23
	1.09	1.45	1.63	1.81	2.18
150	1.16	1.54	1.73	1.93	2.31
	1.11	1.47	1.66	1.84	2.21
	1.08	1.44	1.62	1.80	2.16

With a small number of measurements, demonstrating a capability goal with a high degree of confidence requires seeing a very high point estimate of P_{PK}. For instance, a point estimate of 1.60 must be obtained in a capability study involving 20 measurements to provide sufficient evidence a goal of 1.00 is met with .99 confidence. These conditions require the point estimate to be at least 60 percent greater than the goal. Note that for a capability goal of 2.00 (twice as high as a 1.00 goal), the point estimate must be at least 3.21, which is also 60 percent greater than the goal.

Even with 150 measurements, the point estimate for P_{PK} must be 16 percent greater than the goal in order to have .99 confidence the process meets the desired goal. A quick examination of this table shows that due caution must always be observed when interpreting any capability measure, especially if the capability study incorporates a small sample size.

Example 11.27 Floppy Disks

Warranty problems are traced to variation in the diameter of the inner hole of a floppy disk. From a capability study having a sample size of 84, P_{PK} is estimated as .509. A lower .95 confidence ($\alpha = .05$) bound for P_{PK} is calculated as follows:

$$P_{PK,.95} = \hat{P}_{PK} - Z_{.05}\sqrt{\frac{1}{9kn} + \frac{\hat{P}_{PK}^2}{2(kn-1)}} = .509 - 1.645\sqrt{\frac{1}{9(84)} + \frac{.509^2}{2(84-1)}} = .421$$

Based on the estimate for P_{PK} of .509, there is .95 confidence the true process P_{PK} index is greater than .421. Conversely, there is only a .05 risk that the true P_{PK} is less than .421. Thus, it is highly unlikely this process meets the minimum capability requirement ($P_{PK} > 1.00$) for holding hole size within the current tolerance.

Note that this lower bound is equal to the smaller of the two lower bounds calculated for P_{PL} and P_{PU} in Example 11.25 on page 653.

$$P_{PK,1-\alpha} = \text{Minimum}\left(P_{PL,1-\alpha}, P_{PU,1-\alpha}\right) = \text{Minimum}(.421, 1.132) = .421$$

A team contemplating conducting a future capability study for this process plans to collect 21 subgroups of size 4 each. From this, they wish to calculate the smallest P_{PK} estimate that may be observed and still allow the them to have .95 confidence the P_{PK} goal of 1.50 is met.

$$\text{Minimum Observed } \hat{P}_{PK,.95} \cong \frac{P_{PK.\,GOAL}}{1 - Z_{.95}\sqrt{1 \div 2(kn-1)}} = \frac{1.50}{1 - 1.645\sqrt{1 \div 2(84-1)}} = 1.72$$

Once the capability study is completed, if the estimate of P_{PK} turns out to be only .509, there is little chance this process meets the long-term performance capability goal of having a P_{PK} index of at least 1.50 with .95 confidence.

Minimum Required Sample Size

The approximate sample size required to have the lower confidence a chosen distance below the estimate of P_{PK} has been developed in a paper by Franklin (1994).

$$kn \cong \left(\frac{1}{9\hat{P}_{PK}^2} + \frac{1}{2}\right)\left(\frac{Z_\alpha}{e_{PPK}}\right)^2$$

The factor e_{PPK} expresses the percentage difference between the lower confidence bound and the estimate of P_{PK}.

$$e_{PPK} = 1 - \frac{P_{PK,1-\alpha}}{\hat{P}_{PK}}$$

This formula for kn requires the estimate of P_{PK} to be known before conducting the capability study. As this is unlikely, the formula as given isn't much help. However, as the estimate of P_{PK} increases, the quantity $1/9\,\hat{P}_{PK}^2$ decreases, causing the required sample size to also decrease. So, as a conservative estimate of the minimum sample size, let $\hat{P}_{PK}$ be fairly small, say equal to 1.0. This greatly simplifies the above formula for kn.

$$kn \cong \left(\frac{1}{9\hat{P}_{PK}^2}+\frac{1}{2}\right)\left(\frac{Z_\alpha}{e_{PPK}}\right)^2 = \left(\frac{1}{9(1)^2}+\frac{1}{2}\right)\left(\frac{Z_\alpha}{e_{PPK}}\right)^2 = \left(\frac{1}{9}+\frac{1}{2}\right)\left(\frac{Z_\alpha}{e_{PPK}}\right)^2 = .611\left(\frac{Z_\alpha}{e_{PPK}}\right)^2$$

For example, the lower .95 confidence limit for P_{PK} in the above floppy disk case study is desired to be no more than 10 percent below $\hat{P}_{PK}$, which means there is .95 confidence the true P_{PK} index for this process is within 10 percent of the estimated P_{PK} index. This requires the lower confidence bound to be 90 percent of $\hat{P}_{PK}$, making their ratio equal to .90.

$$\underset{\sim}{P}_{PK,1-\alpha} = .90\,\hat{P}_{PK} \quad \Rightarrow \quad \frac{\underset{\sim}{P}_{PK,1-\alpha}}{\hat{P}_{PK}} = .90$$

Subtracting this ratio from 1 yields an e_{PPK} factor of .10.

$$e_{PPK} = 1 - \frac{\underset{\sim}{P}_{PK,1-\alpha}}{\hat{P}_{PK}} = 1 - .90 = .10$$

Z_α for α equal to .05 is found in Table 11.1 to be 1.645. Substituting these values into the minimum sample size formula reveals that kn must be at least 166 to achieve these objectives.

$$kn \cong .611\left(\frac{Z_\alpha}{e_{PPK}}\right)^2 = .611\left(\frac{1.645}{.10}\right)^2 \cong 166$$

Because the subgroup size was selected as 4, at least 42 in-control subgroups must be collected before estimating P_{PK} and calculating a lower .95 confidence bound.

$$kn \cong 166$$

$$\frac{kn}{n} \cong \frac{166}{4}$$

$$k \cong 41.5 \cong 42$$

Quite a few more subgroups are needed than the 35 that were required to estimate P_P in Example 11.17 (p. 633). This is due to the increased uncertainty caused by having to estimate not only the process standard deviation, but also the process average. In order to provide the same level of confidence, a larger sample size is needed to compensate for this larger amount of uncertainty.

Assume the estimate for P_{PK} with these 168 (42 × 4) measurements turns out to be 1.150. The lower .95 confidence bound for P_{PK} is 1.038.

$$\underset{\sim}{P}_{PK,.95} = \hat{P}_{PK} - Z_{.05}\sqrt{\frac{1}{9kn}+\frac{\hat{P}_{PK}^2}{2(kn-1)}}$$

$$= 1.150 - 1.645\sqrt{\frac{1}{9(168)}+\frac{1.150^2}{2(168-1)}} = 1.038$$

Due to the conservative nature of the minimum sample size formula, this lower bound is just 9.7 percent below the point estimate, rather than the specified 10 percent.

$$\text{Actual } e_{PPK} = 1 - \frac{P_{PK,.95}}{\hat{P}_{PK}} = 1 - \frac{1.038}{1.150} = .097 = 9.7\%$$

Notice that a very large sample size must be collected if a high level of confidence is desired to have the lower confidence bound fairly close to the point estimate of P_{PK}. Table 11.8 displays the minimum sample sizes for .90, .95, and .99 confidence when the lower bound for the P_{PK} index should be no more than 15, 10, and 5 percent away from the estimate of P_{PK}. For example, to have the lower .95 confidence bound no more than 10 percent below the estimate of P_{PK}, a sample size of at least 166 must be included in the capability study. These results assume the point estimate of P_{PK} is greater than 1.00.

Table 11.8 Required sample size for various confidence and e_{PPK} levels.

Confidence Level	e_{PPK}		
	.15	.10	.05
.90	45	101	402
.95	74	166	662
.99	147	331	1323

Due to the added uncertainty in estimating μ as well as σ_{LT}, these sample sizes are around 20 percent larger than those presented in Table 11.5 (p. 638) for estimating P_P. The closer a confidence bound is to the point estimate, the more meaningful the result. However, in order to have a lower confidence bound close to the estimate, exceptionally large sample sizes are needed. Most capability studies involve 20 to 25 subgroups with a subgroup size of 4 or 5, with kn then ranging from 80 to 125. For this number of measurements, confidence bounds will be at least 10 percent below the point estimate for the majority of capability studies. Thus, utmost care is advised when evaluating the results of any such capability study.

11.19 Lower Confidence Bounds for P^*_{PL}, P^*_{PU}, and P^*_{PK}

When the target for the process average is not equal to M, P^*_{PL} and P^*_{PU} must replace the P_{PL} and P_{PU} indexes.

$$P^*_{PL} = \frac{\mu - LSL_T}{3\sigma_{LT}} = \frac{Z^*_{LSL,LT}}{3} \qquad P^*_{PU} = \frac{USL_T - \mu}{3\sigma_{LT}} = \frac{Z^*_{USL,LT}}{3}$$

In these situations, P^*_{PK} must replace P_{PK}.

$$P^*_{PK} = \text{Minimum } (P^*_{PL}, P^*_{PU})$$

Lower confidence bound formulas for these measures of performance capability are:

$$P^*_{\underline{PL},1-\alpha} = \hat{P}^*_{PL} - Z_\alpha \sqrt{\frac{1}{9kn} + \frac{\hat{P}^{*\,2}_{PL}}{2(kn-1)}} \qquad P^*_{\underline{PU},1-\alpha} = \hat{P}^*_{PU} - Z_\alpha \sqrt{\frac{1}{9kn} + \frac{\hat{P}^{*\,2}_{PU}}{2(kn-1)}}$$

Minimum Observed Estimate

The minimum observed estimates for these measures are calculated in a manner similar to that for the P_{PL}, P_{PU}, and P_{PK} indexes.

$$\text{Minimum Observed } \hat{P}^*_{PL,1-\alpha} \cong \frac{P^*_{PL.\,GOAL}}{1 - Z_\alpha \sqrt{1 \div 2(kn-1)}}$$

$$\text{Minimum Observed } \hat{P}^*_{PU,1-\alpha} \cong \frac{P^*_{PU.\,GOAL}}{1 - Z_\alpha \sqrt{1 \div 2(kn-1)}}$$

$$\text{Minimum Observed } \hat{P}^*_{PK,1-\alpha} \cong \frac{P^*_{PK.\,GOAL}}{1 - Z_\alpha \sqrt{1 \div 2(kn-1)}}$$

Minimum Required Sample Size

Minimum sample size formulas are similar to those for P_{PL}, P_{PU}, and P_{PK}.

$$kn \cong .611 \left(\frac{Z_\alpha}{e_{P^*PL}} \right)^2 \qquad kn \cong .611 \left(\frac{Z_\alpha}{e_{P^*PU}} \right)^2 \qquad kn \cong .611 \left(\frac{Z_\alpha}{e_{P^*PK}} \right)^2$$

Where the *e* factors are defined as:

$$e_{P^*PL} = 1 - \frac{P^*_{\underline{PL},1-\alpha}}{\hat{P}^*_{PL}} \qquad e_{P^*PU} = 1 - \frac{P^*_{\underline{PU},1-\alpha}}{\hat{P}^*_{PU}} \qquad e_{P^*PK} = 1 - \frac{P^*_{\underline{PK},1-\alpha}}{\hat{P}^*_{PK}}$$

Example 11.28 Gold Plating Thickness

The LSL for the thickness of gold plated on a silicon wafer is 50 angstroms, while the USL is 100 angstroms. Because the target average for thickness is 65, which is lower than the middle of the tolerance, LSL_T equals the LSL of 50, while USL_T is determined as .80.

$$USL_T = 2T - LSL = 2(65) - 50 = 80$$

If $\hat{\sigma}_{LT}$ is 2.5 angstroms and $\overline{\overline{X}}$ is 70 angstroms, P^*_{PL} and P^*_{PU} are estimated as:

$$\hat{P}^*_{PL} = \frac{\hat{\mu} - LSL_T}{3\hat{\sigma}_{LT}} = \frac{70 - 50}{3(2.5)} = 2.667 \qquad \hat{P}^*_{PU} = \frac{USL_T - \hat{\mu}}{3\hat{\sigma}_{LT}} = \frac{80 - 70}{3(2.5)} = 1.333$$

From these estimates, and knowing *kn* equals 66, lower .99 confidence bounds can now be derived for the true values of the P^*_{PL} and P^*_{PU} indexes for this plating process.

$$\underset{\sim}{P^*_{PL,.99}} = \hat{P}^*_{PL} - Z_{.01}\sqrt{\frac{1}{9kn} + \frac{\hat{P}^{*\,2}_{PL}}{2(kn-1)}} = 2.667 - 2.326\sqrt{\frac{1}{9(66)} + \frac{2.667^2}{2(66-1)}} = 2.115$$

$$\underset{\sim}{P^*_{PU,.99}} = \hat{P}^*_{PU} - Z_{.01}\sqrt{\frac{1}{9kn} + \frac{\hat{P}^{*\,2}_{PU}}{2(kn-1)}} = 1.333 - 2.326\sqrt{\frac{1}{9(66)} + \frac{1.333^2}{2(66-1)}} = 1.045$$

As a result of both lower confidence bounds being greater than the minimum requirement for performance capability of 1.00, there is a high degree of confidence this plating process is capable. Note that the smaller of these two lower confidence bounds is equal to the lower .99 confidence bound for $Z^*_{MIN.\,LT}$ divided by 3 (see Example 11.24).

$$\text{Minimum}\left(\underset{\sim}{P^*_{PL,.99}}, \underset{\sim}{P^*_{PU,.99}}\right) = \frac{\underset{\sim}{Z^*_{MIN.\,LT,.99}}}{3}$$

$$\text{Minimum}\,(2.115, 1.045) = \frac{3.135}{3}$$

$$1.045 = 1.045$$

In the next capability study on this process, it is desired to have the lower .99 confidence bound no more than 10 percent below the point estimate for the P^*_{PU} index. This required distance makes the e factor for P^*_{PU} equal to .10. With this factor, the minimum number of measurements that must be collected during the course of the proposed capability study may be determined.

$$kn \cong .611\left(\frac{Z_{.01}}{e_{P^*PU}}\right)^2 \cong .611\left(\frac{2.326}{.10}\right)^2 \cong 331$$

11.20 Confidence Bounds for *k* Factors

Confidence bounds may also be established for the various k factors, beginning with k_U.

$$k_U = \frac{\mu - M}{\text{Tolerance}/2} \qquad\qquad \hat{k}_U = \frac{\hat{\mu} - M}{\text{Tolerance}/2}$$

The first formula can be rewritten to express μ as a function of k_U, and the second rewritten to show $\hat{\mu}$ as a function of $\hat{k}_U$.

$$\mu = M + \frac{\text{Tolerance}}{2}k_U \qquad\qquad \hat{\mu} = M + \frac{\text{Tolerance}}{2}\hat{k}_U$$

With these two equations, the confidence interval formula for μ, which was derived on pages 606 and 607, can be modified to provide confidence bounds for k_U.

$$\hat{\mu} - t_{v,\alpha/2}\frac{S_{TOT}}{\sqrt{kn}} \le \mu \le \hat{\mu} + t_{v,\alpha/2}\frac{S_{TOT}}{\sqrt{kn}}$$

Substituting for μ and $\hat{\mu}$ yields the following:

$$M + \frac{\text{Tolerance}}{2}\hat{k}_U - t_{v,\alpha/2}\frac{S_{TOT}}{\sqrt{kn}} \le M + \frac{\text{Tolerance}}{2}k_U \le M + \frac{\text{Tolerance}}{2}\hat{k}_U + t_{v,\alpha/2}\frac{S_{TOT}}{\sqrt{kn}}$$

Subtracting M throughout this inequality, then multiplying through by 2 over the tolerance, produces a 1 minus α confidence interval formula for k_U.

$$\hat{k}_U - t_{v,\alpha/2}\frac{2S_{TOT}}{\text{Tolerance}\sqrt{kn}} \le k_U \le \hat{k}_U + t_{v,\alpha/2}\frac{2S_{TOT}}{\text{Tolerance}\sqrt{kn}}$$

If just a lower 1 minus α confidence bound for k_U is desired, the formula becomes:

$$\underset{\sim}{k}_{U,1-\alpha} = \hat{k}_U - t_{v,\alpha}\frac{2S_{TOT}}{\text{Tolerance}\sqrt{kn}}$$

An upper 1 minus α confidence bound for k_U can be derived in a similar fashion:

$$\tilde{k}_{U,1-\alpha} = \hat{k}_U + t_{v,\alpha}\frac{2S_{TOT}}{\text{Tolerance}\sqrt{kn}}$$

Example 11.29 Floppy Disks

In Example 6.19 (p. 235), k_U for the hole size of the floppy disks was estimated as minus .44. In addition, the process average was estimated as 40.9, S_{TOT} was calculated as .913, and kn was found to be 84. Constructing a .90 confidence interval for k_U requires first figuring out the number of degrees of freedom for S_{TOT}.

$$v = kn - 1 = 84 - 1 = 83$$

From this value of v, the lower bound of the .90 confidence interval is calculated as .374.

$$\underset{\sim}{k}_{U,1-.10/2} = \underset{\sim}{k}_{U,.95} = \hat{k}_U - t_{83,.10/2}\frac{2S_{TOT}}{\text{Tolerance}\sqrt{84}} = -.44 - 1.663\frac{2(.913)}{5.0\sqrt{84}} = -.506$$

The upper bound for this same interval is found to be .506.

$$\tilde{k}_{U,1-.10/2} = \tilde{k}_{U,.95} = \hat{k}_U + t_{83,.10/2}\frac{2S_{TOT}}{\text{Tolerance}\sqrt{84}} = -.44 + 1.663\frac{2(.913)}{5.0\sqrt{84}} = -.374$$

There is .90 confidence the interval from -.506 to -.374 captures the true k_U factor for this hole-making process.

Other *k* Factors

Confidence intervals for other k factors are derived in a similar manner to those for k_U. For example, k_L is defined as:

$$k_L = \frac{M - \mu}{\text{Tolerance} / 2}$$

The two bounds of a 1 minus α confidence interval become:

$$\underset{\sim}{k}_{L,1-\alpha/2} = \hat{k}_L - t_{v,\alpha/2} \frac{2S_{TOT}}{\text{Tolerance} \sqrt{kn}} \qquad \tilde{k}_{L,1-\alpha/2} = \hat{k}_L + t_{v,\alpha/2} \frac{2S_{TOT}}{\text{Tolerance} \sqrt{kn}}$$

The k factor is estimated with the following formula:

$$\hat{k} = \frac{|M - \hat{\mu}|}{\text{Tolerance} / 2}$$

An upper bound of a 1 minus α confidence interval for k is defined as:

$$\tilde{k}_{1-\alpha/2} = \hat{k} + t_{v,\alpha/2} \frac{2S_{TOT}}{\text{Tolerance} \sqrt{kn}}$$

The lower bound of this confidence interval for k depends on the relationship between $\hat{k}$ and t_k, where t_k is defined as:

$$t_k = t_{v,\alpha/2} \frac{2S_{TOT}}{\text{Tolerance} \sqrt{kn}}$$

When $\hat{k}$ is greater than t_k, the lower confidence bound is:

$$\underset{\sim}{k}_{1-\alpha/2} = \hat{k} - t_{v,\alpha/2} \frac{2S_{TOT}}{\text{Tolerance} \sqrt{kn}} \quad \text{for } \hat{k} > t_k$$

Because the absolute value operator in the formula for k guarantees a nonnegative result, if $\hat{k}$ is less than, or equal to, t_k, then:

$$\underset{\sim}{k}_{1-\alpha/2} = 0 \quad \text{for } \hat{k} \le t_k$$

Lower and upper bounds of confidence intervals for the factors k^*, k^*_L, k^*_U, k'_L, and k'_U are similar to those above, but with the effective tolerance replacing the actual tolerance. For example, the bounds of a 1 minus α confidence interval for k^* are:

$$\underset{\sim}{k}^*_{1-\alpha/2} = \hat{k}^* - t_{v,\alpha/2} \frac{2S_{TOT}}{(\text{Effective Tolerance}) \sqrt{kn}} \qquad \tilde{k}^*_{1-\alpha/2} = \hat{k}^* + t_{v,\alpha/2} \frac{2S_{TOT}}{(\text{Effective Tolerance}) \sqrt{kn}}$$

The k_T factor does not need a confidence interval because no estimates of process parameters are included in its calculation and, thus, there is no uncertainty in its calculated value.

$$k_T = \frac{|T - M|}{\text{Tolerance} / 2}$$

Example 11.30 Valve O.D. Size

In Example 6.18 (p. 234), the k factor for exhaust valve O.D. size was found to be .0683. As this estimate is quite close to zero, some team members believe the process is centered right at the middle of the tolerance, and this small non-zero value of k is due to just sampling variation. This belief can be checked by constructing a confidence interval for k. If this interval extends down to zero, then it is possible the process is located at M, and the observed k factor of .0683 could be explained by sampling variation. On the other hand, an interval not including zero provides strong evidence the process average is not centered at M.

With 20 subgroups of n equal to 5, the number of degrees of freedom is 99.

$$\nu = kn - 1 = 20(5) - 1 = 99$$

Given S_{TOT} is .01606 cm and the tolerance is .12 cm, a .90 confidence interval for k is computed as follows:

$$.90 \text{ confidence interval for } k = \hat{k} \pm t_{\nu, .10/2} \frac{2S_{TOT}}{\text{Tolerance } \sqrt{kn}}$$

$$= .0683 \pm t_{99, .05} \frac{2(.01606)}{.12 \sqrt{100}}$$

$$= .0683 \pm 1.660 \, (.026767)$$

$$= .0683 \pm .0444$$

$$= .0239 \text{ to } .1127$$

Thus, there is .90 confidence the interval from .0239 to .1127 contains the true k factor for this process. Only a 5 percent chance exists that the lower bound of .0239 is greater than the true k value, meaning the process is most likely *not* centered at M, and efforts to locate it there should be initiated.

Example 11.31 Contact Closing Speed

In Example 6.26 (p. 262), k^* was estimated as .633. S_{TOT} is calculated as .00136 seconds, while LSL_T equals .0140 seconds and USL_T equals .0200 seconds. With the total number of measurements collected during the capability study equal to 75, the number of degrees of freedom for S_{TOT} is 74.

$$\nu = kn - 1 = 75 - 1 = 74$$

An upper .99 confidence bound ($\alpha = .01$) for k^* is computed as .882. Note that when a single bound is calculated, all of α goes into the one tail associated with that bound.

$$\tilde{k}^*_{.99} = \hat{k}^* + t_{\nu, \alpha} \frac{2S_{TOT}}{\text{Effective Tol. } \sqrt{kn}} = .633 + t_{74, .01} \frac{2(.00136)}{(.0200 - .0170)\sqrt{75}} = .882$$

There is .99 confidence the true k^* factor for this process is less than .882.

Minimum Required Sample Size

The following derivation provides a formula determining the sample size required to have the difference between the upper 1 minus α confidence bound for k and the estimate of k be no more than a specified amount.

$$\tilde{k}_{1-\alpha} = \hat{k} + t_{v,\alpha}\frac{2S_{TOT}}{\text{Tolerance}\sqrt{kn}}$$

$$\tilde{k}_{1-\alpha} - \hat{k} = t_{v,\alpha}\frac{2S_{TOT}}{\text{Tolerance}\sqrt{kn}}$$

$$\sqrt{kn} = t_{v,\alpha}\frac{2S_{TOT}}{\text{Tolerance}(\tilde{k}_{1-\alpha} - \hat{k})}$$

$$kn = \left(\frac{2t_{v,\alpha}S_{TOT}}{e_k\text{Tolerance}}\right)^2$$

As $t_{v,\alpha}$ is very close to Z_α, the approximate sample size formula may be rewritten as:

$$kn \cong \left(\frac{2Z_\alpha S_{TOT}}{e_k\text{Tolerance}}\right)^2$$

With the above formula, the e_k factor is defined as shown here:

$$e_k = \tilde{k}_{1-\alpha} - \hat{k}$$

Once the difference between the upper bound and the estimate for k is decided upon, the minimum sample size needed can be computed. However, this formula for kn requires the quantity S_{TOT} to be known in advance of the study, which in some cases, is not possible.

Suppose a capability study team wishes to have a difference of no more than .08 between the upper .99 confidence bound for k and its point estimate.

$$e_k = \tilde{k}_{1-\alpha} - \hat{k} = .08$$

The specification for the characteristic under study is 93.50 ± 1.35, making the tolerance equal to 2.70. From similar processes in their factory, the team believes S_{TOT} will be about .3805. With α equal to .01, the minimum sample required under these conditions is 68.

$$kn \cong \left(\frac{2Z_{.01}S_{TOT}}{e_k\text{Tolerance}}\right)^2 = \left(\frac{2(2.326).3805}{.08(2.70)}\right)^2 \cong 68$$

Selecting a subgroup size of 3, the team must collect at least 13 subgroups during the course of their capability study ($13 \times 3 = 69$). When the study is completed, 69 in-control measurements are in fact collected. From this data, S_{TOT} turns out to be .3829, while the process average is estimated as 93.77. This makes the estimate of the k factor equal to .20.

$$\hat{k} = \frac{|M - \hat{\mu}|}{\text{Tolerance}/2} = \frac{|93.50 - 93.77|}{2.70/2} = .20$$

Based on these results, an upper .99 confidence interval for k is found as follows, where ν is equal to 69 minus 1, or 68.

$$\tilde{k}_{1-\alpha} = \hat{k} + t_{68,.01}\frac{2S_{TOT}}{\text{Tolerance}\sqrt{kn}} = .20 + 2.382\frac{2(.3829)}{2.70\sqrt{69}} = .281$$

Because S_{TOT} was somewhat greater than anticipated, .3829 versus .3805, this upper bound is slightly more than the prescribed .08 above $\hat{k}$ (.281 minus .20 equals .081).

11.21   Upper Confidence Bounds for $ppm_{LSL,\,LT}$ and $ppm_{USL,\,LT}$

To determine an *upper* 1 minus α confidence bound for $ppm_{LSL,\,LT}$, calculate the *ppm* value for the *lower* 1 minus α confidence bound for $Z_{LSL,\,LT}$ (Lewis). As there is a certain level of confidence there are at least these many standard deviations from the process mean to the lower specification limit, the corresponding *ppm* for this lower 1 minus α confidence bound on $Z_{LSL,\,LT}$ is the highest expected $ppm_{LSL,\,LT}$. Recall that the operator Φ means to look up in Appendix Table III the percentage associated with the Z value in parentheses.

$$\underset{\sim}{Z}_{LSL,LT,1-\alpha} = \hat{Z}_{LSL,LT} - Z_\alpha\sqrt{\frac{1}{kn} + \frac{\hat{Z}_{LSL,LT}^2}{2(kn-1)}} \qquad \tilde{p}pm_{LSL,LT,1-\alpha} = \Phi\left(\underset{\sim}{Z}_{LSL,LT,1-\alpha}\right) \times 10^6$$

Likewise, the upper 1 minus α confidence bound for $ppm_{USL,\,LT}$ is based on the *ppm* value for the lower 1 minus α confidence bound for $Z_{USL,\,LT}$.

$$\underset{\sim}{Z}_{USL,LT,1-\alpha} = \hat{Z}_{USL,LT} - Z_\alpha\sqrt{\frac{1}{kn} + \frac{\hat{Z}_{USL,LT}^2}{2(kn-1)}} \qquad \tilde{p}pm_{USL,LT,1-\alpha} = \Phi\left(\underset{\sim}{Z}_{USL,LT,1-\alpha}\right) \times 10^6$$

Both *ppm* confidence bound formulas assume the process output has a normal distribution.

Example 11.32   Valve O.D. Size

An estimate of $Z_{USL,\,LT}$ for the O.D. size of a valve run on a grinding machine was computed as 3.472 in Example 6.2 on page 174. The estimated $ppm_{USL,\,LT}$ associated with this estimated Z value is calculated as is shown here:

$$\hat{p}pm_{USL,LT} = \Phi(\hat{Z}_{USL,LT}) \times 10^6 = \Phi(3.472) \times 10^6 = (.000258) \times 10^6 = 258$$

A lower .90 confidence bound for $Z_{USL,\,LT}$ is calculated to be 3.131.

$$\underset{\sim}{Z}_{USL,LT,.90} = \hat{Z}_{USL,LT} - Z_{.10}\sqrt{\frac{1}{kn} + \frac{\hat{Z}_{USL,LT}^2}{2(kn-1)}} = 3.472 - 1.282\sqrt{\frac{1}{100} + \frac{3.472^2}{2(100-1)}} = 3.131$$

The *ppm* value associated with this *lower* .90 confidence bound for $Z_{USL,\,LT}$ becomes an *upper* .90 confidence bound for $ppm_{USL,\,LT}$.

$$\tilde{p}pm_{USL,LT,.90} = \Phi\left(\underset{-}{Z_{USL,LT,.90}}\right) \times 10^6 = \Phi(3.131) \times 10^6 = (8.74 \times 10^{-4}) \times 10^6 = 874$$

Based on this estimate for $ppm_{USL\ LT}$ of 258 from the results of the capability study, there is .90 confidence the actual $ppm_{USL\ LT}$ for this process is less than 874, with only a .10 risk it is greater. This analysis assumes the process output has a normal distribution.

Maximum Observed Estimate

The largest *ppm* estimate that may be observed outside *either* specification in a capability study, and still provide 1 minus α confidence the goal for the P_{PK} index has been met, is determined directly from the minimum observed $Z_{MIN\ LT}$ value for the same confidence level.

$$\text{Maximum Observed } \hat{p}pm_{MAX.\ LT.\ 1-\alpha} \cong \Phi(\text{Minimum Observed } \hat{Z}_{MIN.\ LT,1-\alpha}) \times 10^6$$

$$\text{where Minimum Observed } \hat{Z}_{MIN.\ LT.\ 1-\alpha} \cong \frac{Z_{MIN.\ LT.\ GOAL}}{1 - Z_\alpha \sqrt{1 \div 2(kn-1)}}$$

If both the actual $ppm_{LSL\ LT}$ and the actual $ppm_{USL\ LT}$ are less than this maximum observed value, the goal for $Z_{MIN\ LT}$ has been met with 1 minus α confidence. Because P_{PK} is just $Z_{MIN\ LT}$ divided by three, the goal for P_{PK} would also be satisfied.

The maximum observed estimate for $ppm_{LSL\ LT}$ and $ppm_{USL\ LT}$ may also be found from their associated Z values. For $ppm_{LSL\ LT}$:

$$\text{Maximum Observed } \hat{p}pm_{LSL,LT,1-\alpha} \cong \Phi(\text{Minimum Observed } \hat{Z}_{LSL.\ LT,1-\alpha}) \times 10^6$$

$$\text{where Minimum Observed } \hat{Z}_{LSL.\ LT,1-\alpha} \cong \frac{Z_{LSL.\ LT.\ GOAL}}{1 - Z_\alpha \sqrt{1 \div 2(kn-1)}}$$

For $ppm_{USL.\ LT}$:

$$\text{Maximum Observed } \hat{p}pm_{USL,LT,1-\alpha} \cong \Phi(\text{Minimum Observed } \hat{Z}_{USL.\ LT,1-\alpha}) \times 10^6$$

$$\text{where Minimum Observed } \hat{Z}_{USL.\ LT,1-\alpha} \cong \frac{Z_{USL.\ LT.\ GOAL}}{1 - Z_\alpha \sqrt{1 \div 2(kn-1)}}$$

Note that the goals for $Z_{LSL\ LT}$ and $ppm_{LSL\ LT}$ must be consistent, *e.g.*, if the goal for $Z_{LSL\ LT}$ is 4, the goal for $ppm_{LSL\ LT}$ must be 32.

Example 11.33 Valve O.D. Size

In Example 11.22 (p. 648), the smallest $Z_{MIN\ LT}$ estimate that could be observed in a capability study for valve O.D. size, and still indicate with .90 confidence the process has achieved a $Z_{MIN\ LT}$ goal of 3.00, was found to be 3.301.

$$\text{Min. Observed } \hat{Z}_{MIN.\ LT.\ .90} \cong \frac{Z_{MIN.\ LT.\ GOAL}}{1 - Z_{.10}\sqrt{1 \div 2(kn-1)}} = \frac{3.00}{1 - 1.282\sqrt{1 \div 2(100-1)}} = 3.301$$

A $Z_{MIN,\,LT}$ goal of 3.00 translates into a $ppm_{MAX,\,LT}$ goal of 1,350.

$$ppm_{MAX,\,LT,\,GOAL} = \Phi(Z_{MIN,LT,GOAL}) \times 10^6 = \Phi(3.00) \times 10^6 = .001350 \times 10^6 = 1,350$$

The maximum point estimate of $ppm_{MAX,\,LT}$ that may be observed in the capability study for either specification, and yet furnish at least .90 confidence the $ppm_{MAX,\,LT}$ goal of 1,350 is met or exceeded, is found to be 484.

$$\text{Maximum Observed } \hat{p}pm_{MAX,\,LT,\,.90} \cong \Phi(\text{Minimum Observed } \hat{Z}_{MIN,LT,.90}) \times 10^6$$

$$= \Phi(3.301) \times 10^6$$

$$= .000484 \times 10^6 = 484$$

The estimate of *ppm* for each specification from the ensuing capability study can be compared to this bogie of 484 to determine if the process meets its capability goal for $ppm_{MAX,\,LT}$ with .90 confidence.

Example 11.34 Floppy Disks

A field problem is traced to variation in the diameter of the inner hole of a floppy disk. From a capability study where *kn* equaled 84, μ is estimated as 40.9 while σ_{LT} is predicted to be .916. Given a specification of 42.0 ± 2.5 for hole size, $Z_{LSL,\,LT}$ and $Z_{USL,\,LT}$ may be estimated.

$$\hat{Z}_{LSL,LT} = \frac{\hat{\mu} - USL}{\hat{\sigma}_{LT}} = \frac{40.9 - 39.5}{.916} = 1.528$$

$$\hat{Z}_{USL,LT} = \frac{USL - \hat{\mu}}{\hat{\sigma}_{LT}} = \frac{44.5 - 40.9}{.916} = 3.930$$

Lower .95 confidence bounds can also be computed for these capability measures.

$$Z_{LSL,\,LT,\,.95} = \hat{Z}_{LSL,\,LT} - Z_{.05} \sqrt{\frac{1}{kn} + \frac{\hat{Z}^2_{LSL,\,LT}}{2(kn-1)}} = 1.528 - 1.645 \sqrt{\frac{1}{84} + \frac{1.528^2}{2(84-1)}} = 1.263$$

$$Z_{USL,\,LT,\,.95} = \hat{Z}_{USL,\,LT} - Z_{.05} \sqrt{\frac{1}{kn} + \frac{\hat{Z}^2_{USL,\,LT}}{2(kn-1)}} = 3.930 - 1.645 \sqrt{\frac{1}{84} + \frac{3.930^2}{2(84-1)}} = 3.397$$

From the point estimates of $Z_{LSL,\,LT}$ and $Z_{USL,\,LT}$, $ppm_{LSL,\,LT}$ is estimated as 63,010, while $ppm_{USL,\,LT}$ is estimated as 43.

$$\hat{p}pm_{LSL,LT} = \Phi(\hat{Z}_{LSL,LT}) \times 10^6 = \Phi(1.528) \times 10^6 = 63,010$$

$$\hat{p}pm_{USL,LT} = \Phi(\hat{Z}_{USL,LT}) \times 10^6 = \Phi(3.930) \times 10^6 = 43$$

From the lower .95 confidence bounds for $Z_{LSL,\,LT}$ and $Z_{USL,\,LT}$, upper .95 confidence bounds for $ppm_{LSL,\,LT}$ and $ppm_{USL,\,LT}$ can be determined, beginning with the one for $ppm_{LSL,\,LT}$.

$$\tilde{p}pm_{LSL,LT,.95} = \Phi\left(\underset{\sim}{Z}_{LSL,LT,.95}\right) \times 10^6 = \Phi(1.263) \times 10^6 = .1033 \times 10^6 = 103,300$$

Based on an estimate for $ppm_{LSL,\ LT}$ of 63,010 from the capability study on hole size, there is .95 confidence the actual $ppm_{LSL,\ LT}$ for this process is less than 103,300. Recall that 103,300 ppm translates into about 10.3 percent undersized holes.

An upper .95 confidence bound for $ppm_{USL,\ LT}$ is computed next.

$$\tilde{p}pm_{USL,LT,.95} = \Phi\left(\underset{\sim}{Z}_{USL,LT,.95}\right) \times 10^6$$

$$= \Phi(3.397) \times 10^6$$

$$= (3.42 \times 10^{-4}) \times 10^6 = 342$$

Based on an estimate for $ppm_{USL,\ LT}$ of 43 from the capability study, there is .95 confidence the true $ppm_{USL,\ LT}$ for this hole-making process is less than 342, with a .05 risk it is greater.

Minimum Required Sample Size

The following formula determines the minimum sample size required to have the upper confidence bound for the true ppm value (for either specification) a certain distance above the estimate for ppm. However, a reasonable idea of the estimate for ppm must be known in advance, which may not be possible in all cases. For $ppm_{LSL,\ LT}$:

$$kn \cong \left[\frac{1}{Z(\hat{p}pm_{LSL,LT} \times 10^{-6})^2} + .5\right]\left(\frac{Z_\alpha}{e_{ZLSL}}\right)^2$$

Here, the $e_{Z\,LSL}$ factor is defined as:

$$e_{ZLSL} = 1 - \frac{Z_{LSL,LT,1-\alpha}}{\underset{\sim}{\hat{Z}}_{LSL,LT}} = 1 - \frac{Z(\tilde{p}pm_{LSL,LT,1-\alpha} \times 10^{-6})}{Z(\hat{p}pm_{LSL,LT} \times 10^{-6})}$$

Formulas for $ppm_{USL,\ LT}$ are identical to those above, but with "USL" replacing "LSL." Assume the estimate of $ppm_{LSL,\ LT}$ is expected to be 10,000. If it is desired to have the upper .95 confidence bound no higher than 20,000, the $e_{Z\,LSL}$ factor becomes:

$$e_{ZLSL} = 1 - \frac{Z(\tilde{p}pm_{LSL,LT,.95} \times 10^{-6})}{Z(\hat{p}pm_{LSL,LT} \times 10^{-6})}$$

$$= 1 - \frac{Z(20,000 \times 10^{-6})}{Z(10,000 \times 10^{-6})}$$

$$= 1 - \frac{Z(2.0 \times 10^{-2})}{Z(1.0 \times 10^{-2})} = 1 - \frac{2.054}{2.326} = .1169$$

This result is then inserted into the formula for kn.

$$kn \cong \left[\frac{1}{Z(\hat{p} pm_{LSL,LT} \times 10^{-6})^2} + .5 \right] \left(\frac{Z_{.05}}{e_{Z\,LSL}} \right)^2$$

$$\cong \left[\frac{1}{Z(10,000 \times 10^{-6})^2} + .5 \right] \left(\frac{1.645}{.1169} \right)^2$$

$$\cong \left[\frac{1}{(2.326)^2} + .5 \right] (14.07186)^2 \cong 135.61 \cong 136$$

If a subgroup size of 5 is chosen for the control chart, at least 28 (28 × 5 = 140) in-control subgroups must be collected during the course of the capability study to accumulate a minimum of 136 individual measurements.

11.22 Upper Confidence Bound for *ppm*$_{TOTAL, LT}$

For characteristics with bilateral specifications, total *ppm* is estimated by adding the *ppm* below the LSL to the *ppm* above the USL.

$$\hat{p} pm_{TOTAL,LT} = \hat{p} pm_{LSL,LT} + \hat{p} pm_{USL,LT} = [\Phi(\hat{Z}_{LSL,LT}) + \Phi(\hat{Z}_{USL,LT})] \times 10^6$$

A method for determining an upper 1 minus α confidence bound for *ppm*$_{TOTAL,\ LT}$ is proposed in a paper by Wharton and Czachor (1994). A version of their formula is presented here:

$$\tilde{p} pm_{TOTAL,LT,1-\alpha} = [\Phi(h_{LSL,1-\alpha}) + \Phi(h_{USL,1-\alpha})] \times 10^6$$

The quantities h_{LSL} and h_{USL} are defined as follows, where γ is the Greek letter gamma:

$$h_{LSL,1-\alpha} = \hat{Z}_{LSL,LT} - \gamma_{kn,i-\alpha} \sqrt{\frac{2 + \hat{Z}^2_{LSL,LT}}{kn}} \qquad h_{USL,1-\alpha} = \hat{Z}_{USL,LT} - \gamma_{kn,1-\alpha} \sqrt{\frac{2 + \hat{Z}^2_{USL,LT}}{kn}}$$

Table 11.9 on the facing page lists values of $\gamma_{kn,1-\alpha}$ for α equal to .10, .05, and .01, which were derived from work done by Cheng and Iles (see also Wharton and Czachor, 1996).

Example 11.35 Valve O.D. Size

For the automotive exhaust valve example, $Z_{LSL,\ LT}$ and $Z_{USL,\ LT}$ were estimated as 3.981 and 3.472, respectively (p. 174). This made the estimate of *ppm*$_{LSL,\ LT}$ equal to 34, and the estimate of *ppm*$_{USL,\ LT}$ equal to 258. The estimate for *ppm*$_{TOTAL,\ LT}$ is the sum of these two.

$$\hat{p} pm_{TOTAL,LT} = \hat{p} pm_{LSL,LT} + \hat{p} pm_{USL,LT} = 34 + 258 = 292$$

An upper .95 confidence bound for *ppm*$_{TOTAL,\ LT}$ is constructed by first calculating the quantities h_{LSL} and h_{USL}. For this example, kn equals 100.

$$h_{LSL,.95} = \hat{Z}_{LSL,LT} - \gamma_{100,.95} \sqrt{\frac{2 + \hat{Z}^2_{LSL,LT}}{kn}} = 3.981 - 1.733 \sqrt{\frac{2 + 3.981^2}{100}} = 3.249$$

Table 11.9 Values of $\gamma_{kn,1-\alpha}$ for various sample size ranges.

kn	$\gamma_{kn,.90}$	$\gamma_{kn,.95}$	$\gamma_{kn,.99}$
20-29	1.529	1.742	2.158
30-39	1.526	1.739	2.154
40-49	1.524	1.737	2.153
50-59	1.523	1.736	2.151
60-79	1.522	1.735	2.150
80-99	1.521	1.734	2.150
100-199	1.520	1.733	2.149
200-399	1.520	1.733	2.148
400-599	1.519	1.732	2.148

$$h_{USL,.95} = \hat{Z}_{USL,LT} - \gamma_{100,.95} \sqrt{\frac{2 + \hat{Z}_{USL,LT}^2}{kn}} = 3.472 - 1.733 \sqrt{\frac{2 + 3.472^2}{100}} = 2.822$$

These values are then substituted into the confidence bound formula for $ppm_{TOTAL,LT}$.

$$\tilde{p}pm_{TOTAL,LT,.95} = [\Phi(h_{LSL,.95}) + \Phi(h_{USL,.95})] \times 10^6$$

$$= [\Phi(3.249) + \Phi(2.822)] \times 10^6$$

$$= (.00058 + .00240) \times 10^6 = 2980$$

Based on the estimate for $ppm_{TOTAL,LT}$ of 292, there is .95 confidence the true $ppm_{TOTAL,LT}$ for this process is less than 2980. Note that this confidence bound is over 10 times the point estimate of $ppm_{TOTAL,LT}$.

If the feature under study has a unilateral specification, $ppm_{TOTAL,LT}$ is equal to either $ppm_{LSL,LT}$ or $ppm_{USL,LT}$. Thus, the upper 1 minus α confidence bound for $ppm_{TOTAL,LT}$ is identical to that developed for these measures in Section 11.21. For example, when only a LSL is given, select this confidence bound formula:

$$\tilde{p}pm_{TOTAL,LT,1-\alpha} = \tilde{p}pm_{LSL,LT,1-\alpha} = \Phi\left(\underset{\sim}{Z}_{LSL,LT,1-\alpha}\right) \times 10^6$$

In situations where only an USL is provided, the upper 1 minus α confidence bound formula for $ppm_{TOTAL,LT}$ is:

$$\tilde{p}pm_{TOTAL,LT,1-\alpha} = \tilde{p}pm_{USL,LT,1-\alpha} = \Phi\left(\underset{\sim}{Z}_{USL,LT,1-\alpha}\right) \times 10^6$$

For information on constructing a lower confidence bound for the percentage conforming, see Wang and Lam.

11.23 Lower Confidence Bound for P_{PM}

The P_{PM} index incorporates the Taguchi loss function as part of its measure for the performance capability of a process .

$$P_{PM} = \frac{\text{Tolerance}}{6\tau_{LT}} \qquad \text{where} \quad \tau_{LT} = \sqrt{\sigma_{LT}^2 + (\mu - M)^2}$$

A conservative formula for this measure's lower 1 minus α confidence bound is derived in an article by Boyles. An alternative approach is offered by Cheng (1992, 1994-1995). However, studies by Kushler and Hurley indicate a preference for Boyles' formula.

$$\underset{\sim}{P}_{PM, 1-\alpha} \cong \hat{P}_{PM} \sqrt{\frac{\chi_{v, 1-\alpha}^2}{v}} \qquad \text{where} \quad v = \frac{kn(1 + b^2)^2}{1 + 2b^2} \quad \text{and} \quad b = \frac{\hat{\mu} - M}{\hat{\sigma}_{LT}}$$

Recall that M is the middle of the tolerance, kn is the total number of measurements collected during the course of the capability study, and $\hat{\sigma}_{LT}$ is an estimate of the long-term process standard deviation. The sampling distribution for estimates of C_{PM} is very complex because it involves estimates of both the process average and short-term standard deviation (either $\sigma_{\bar{R}}$ or $\sigma_{\bar{S}}$). Due to this complexity, no formula as of yet has been derived for figuring a lower confidence for C_{PM} (the interested reader can see the work done by Zimmer and Hubele).

Minimum Observed Estimate

The smallest estimate of P_{PM} that may be seen in a capability study, and still demonstrate a given capability goal is met with a certain amount of confidence, is calculated with the following formula:

$$\text{Min. Obs. } \hat{P}_{PM, 1-\alpha} \cong P_{PM, GOAL} \sqrt{\frac{v}{\chi_{v, 1-\alpha}^2}} \qquad \text{where} \quad v = \frac{kn(1 + b^2)^2}{1 + 2b^2} \quad \text{and} \quad b = \frac{\hat{\mu} - M}{\hat{\sigma}_{LT}}$$

Example 11.36 Valve O.D. Size

In Example 6.29 (p. 277), P_{PM} was estimated as 1.204 for the process machining valve O.D. size. The middle of the tolerance is 10.0000, while $\hat{\mu}$ is 10.0041, and $\hat{\sigma}_{LT}$ equals .01610. A lower .99 confidence bound for P_{PM} is calculated by first computing b.

$$b = \frac{\hat{\mu} - M}{\hat{\sigma}_{LT}} = \frac{10.0041 - 10.0000}{.01610} = .25466$$

Given that kn is 100 for this capability study, this value of b determines the approximate number of degrees of freedom.

$$v = \frac{kn(1 + b^2)^2}{1 + 2b^2} = \frac{100(1 + .25466^2)^2}{1 + 2(.25466)^2} \cong 100$$

With 100 degrees of freedom, the appropriate χ^2 value may be selected from Appendix Table VI on page 857 and, with it, a lower confidence bound computed for P_{PM}.

$$P_{\underline{PM},.99} = 1.204 \sqrt{\frac{\chi^2_{100,.99}}{100}} = 1.204 \sqrt{\frac{70.06472}{100}} = 1.008$$

Based on the estimate for P_{PM} of 1.204, there is .99 confidence the true P_{PM} for this process is greater than 1.008.

Anticipating that the degrees of freedom for a future capability study will also be 100, the minimum observed estimate of P_{PM} demonstrating attainment of the P_{PM} goal of 1.20 at .99 confidence is calculated as:

$$\text{Min. Obs. } \hat{P}_{PM,.99} \cong P_{PM,GOAL} \sqrt{\frac{v}{\chi^2_{v,1-.01}}} = 1.20 \sqrt{\frac{100}{\chi^2_{100,.99}}} = 1.20 \sqrt{\frac{100}{70.06472}} = 1.434$$

If the actual P_{PM} estimate turns out to be 1.204, there would be little confidence the P_{PM} goal of 1.20 for this process was met, even though the point estimate is greater than the goal.

Example 11.37 Floppy Disks

A field problem is traced to variation in the diameter of the inner hole of a floppy disk. In Example 6.32 (p. 301), P_{PM} was estimated as .582 for this hole-making process. Given estimates for μ and σ_{LT} of 40.9 and .916 respectively, the b value may be computed.

$$b = \frac{\hat{\mu} - M}{\hat{\sigma}_{LT}} = \frac{40.9 - 42.0}{.916} = -1.2009$$

This b value of -1.2009 determines the number of degrees of freedom ($kn = 84$).

$$v = \frac{kn(1+b^2)^2}{1+2b^2} = \frac{84[1 + (-1.2009)^2]^2}{1 + 2(-1.2009)^2} \cong 129$$

After the degrees of freedom are known, a lower confidence bound may be calculated.

$$P_{\underline{PM},.95} \cong \hat{P}_{PM} \sqrt{\frac{\chi^2_{129,.95}}{v}} = .582 \sqrt{\frac{103.7646}{129}} = .522$$

Based on the results of the capability study that estimated P_{PM} as .582, there is .95 confidence the true P_{PM} for this process is at least .522.

Minimum Required Sample Size

A conservative minimum required sample size for having the lower confidence bound for P_{PM} at a chosen percentage below the estimate of P_{PM} is determined as is demonstrated below. The full derivation of this formula is provided in Section 11.32 (see Derivation 2, p. 709), while Franklin (1995) offers a second approach.

$$kn \cong .5 \left(\frac{Z_\alpha}{e_{PPM}} \right)^2$$

The factor e_{PPM} is based on the desired ratio between the confidence bound and the point estimate for P_{PM}.

$$e_{PPM} = 1 - \frac{\underset{\sim}{P}_{PM,1-\alpha}}{\hat{P}_{PM}}$$

A decision is made to have the lower .95 confidence bound for P_{PM} be no more than 10 percent below the estimate for P_{PM} in the floppy disk example.

$$\underset{\sim}{P}_{PM,1-\alpha} = .90\,\hat{P}_{PM} \quad \Rightarrow \quad \frac{\underset{\sim}{P}_{PM,1-\alpha}}{\hat{P}_{PM}} = .90$$

e_{PPM} is then found as follows:

$$e_{PPM} = 1 - \frac{\underset{\sim}{P}_{PM,1-\alpha}}{\hat{P}_{PM}} = 1 - .90 = .10$$

The minimum number of floppy disks that must be included in the capability study is 136.

$$kn \cong .5\left(\frac{Z_{.05}}{e_{PPM}}\right)^2 = .5\left(\frac{1.645}{.10}\right)^2 \cong 136$$

As the subgroup size is 4, measurements from at least 34 in-control subgroups must be collected before estimating P_{PM}.

$$kn \cong 136$$

$$\frac{kn}{n} \cong \frac{136}{4}$$

$$k \cong 34$$

Once the capability study is completed with kn equal to 136 (34 × 4), the estimate of P_{PM} is 1.40, while b turns out to be .3. This value of b makes the approximate number of degrees of freedom equal to 137.

$$v = \frac{kn(1+b^2)^2}{1+2b^2} = \frac{136[1+(.3)^2]^2}{1+2(.3)^2} \cong 137$$

After the degrees of freedom are computed, a lower confidence bound can be calculated for P_{PM}.

$$\underset{\sim}{P}_{PM,.95} \cong \hat{P}_{PM}\sqrt{\frac{\chi^2_{137,.95}}{v}} = 1.40\sqrt{\frac{110.95628}{137}} = 1.40(.90) = 1.26$$

The lower confidence bound of 1.26 is the desired 90 percent of the P_{PM} estimate, meaning there is only a .05 risk the true P_{PM} for this process is less than 1.26.

Notice that as the absolute value of b increases, v increases as well. For b equal to 1, v increases to 182 when kn equals 136.

$$v = \frac{kn(1+b^2)^2}{1+2b^2} = \frac{136(1+1^2)^2}{1+2(1)^2} = 182$$

When b equals 2, v jumps to 378.

$$v = \frac{136(1+2^2)^2}{1+2(2)^2} = 378$$

A larger v value creates a "tighter" lower confidence bound for P_{PM}. With v equal to 182 and α equal to .05:

$$P_{PM,.95} \cong \hat{P}_{PM} \sqrt{\frac{\chi^2_{v,.95}}{v}} = \hat{P}_{PM} \sqrt{\frac{151.79559}{182}} = \hat{P}_{PM}(.9133)$$

This lower bound is 8.67 percent (1 - .9133) below the estimate of P_{PM}. However, if v is increased to 378, the lower bound will be just 6.01 percent (1 - .9399) below the estimate.

$$P_{PM,.95} \cong \hat{P}_{PM} \sqrt{\frac{\chi^2_{378,.95}}{378}} = \hat{P}_{PM} \sqrt{\frac{333.93925}{378}} = \hat{P}_{PM}(.9399)$$

For a given $\hat{\sigma}_{LT}$, the absolute value of b is larger when $\hat{\mu}$ is farther from M, the middle of the tolerance.

$$b = \frac{\hat{\mu} - M}{\hat{\sigma}_{LT}}$$

As τ_{LT} is also a function of this difference between μ and M, larger b values are associated with larger estimates of τ_{LT}.

$$\hat{\tau}_{LT} = \sqrt{\hat{\sigma}^2_{LT} + (\hat{\mu} - M)^2}$$

Of course, a larger estimate of τ_{LT} causes the estimate of P_{PM} to be smaller, which is undesirable. But with a tighter confidence bound for P_{PM}, there is less uncertainty in the results of the capability study.

Comparison of Lower Confidence Bounds

Assume for a given process, the average is centered at M, and P_P, P_{PK}, and P_{PM} are all equal to 1.00. With kn equal to 125, lower .95 confidence bounds for these capability measures are computed as follows:

$$P_{P,.95} \cong \hat{P}_P \sqrt{\frac{\chi^2_{124,.95}}{124}} = 1.00 \sqrt{\frac{99.28261}{124}} = .8948$$

$$P_{PK,.95} = \hat{P}_{PK} - Z_{.05} \sqrt{\frac{1}{9kn} + \frac{\hat{P}^2_{PK}}{2(kn-1)}} = 1.00 - 1.645 \sqrt{\frac{1}{9(125)} + \frac{1.00^2}{2(125-1)}} = .8846$$

$$b = \frac{\hat{\mu} - M}{\hat{\sigma}_{LT}} = 0 \qquad\qquad v = \frac{125(1 + 0^2)^2}{1 + 2(0)^2} = 125$$

$$\underset{\sim}{P}_{PM,.95} \cong \hat{P}_{PM} \sqrt{\frac{\chi^2_{125,.95}}{125}} = 1.00 \sqrt{\frac{100.17811}{125}} = .8952$$

Due to the added uncertainty of estimating both the average and standard deviation, it comes as no shock that the lower bound for P_{PK} of .8846 is less than the one for P_P of .8948. However, it is surprising to discover that the lower bound of .8952 for P_{PM} is not only greater than the lower bound for P_{PK}, but even greater than the one for P_P. This is true despite P_{PM} being a function of two estimated process parameters, whereas P_P is a function of only one. Because "tighter" confidence bounds (for a given sample size) are desirable, the P_{PM} index has a definite statistical advantage over both P_P and P_{PK}.

11.24 Lower Confidence Bound for P^*_{PM}

The P^*_{PM} measure of long-term performance capability is calculated as is shown below. Recall that T is the target average for the process average.

$$P^*_{PM} = \text{Minimum}\left(\frac{T - \text{LSL}}{3\tau^*_{LT}}, \frac{\text{USL} - T}{3\tau^*_{LT}}\right) \qquad \text{where} \quad \tau^*_{LT} = \sqrt{\sigma^2_{LT} + (\mu - T)^2}$$

The lower 1 minus α confidence bound formula for this index is directly related to the one previously given for P_{PM}.

$$\underset{\sim}{P}^*_{PM,1-\alpha} \cong \hat{P}^*_{PM} \sqrt{\frac{\chi^2_{v,1-\alpha}}{v}} \qquad \text{where} \quad v = \frac{kn(1 + b^2)^2}{1 + 2b^2} \quad \text{and} \quad b = \frac{\hat{\mu} - T}{\hat{\sigma}_{LT}}$$

Minimum Observed Estimate

The smallest estimate of P^*_{PM} that can be observed in a capability study, and still provide 1 minus α confidence the capability goal for the P^*_{PM} index is met, is calculated with the following formula.

$$\text{Minimum Observed } \hat{P}^*_{PM,1-\alpha} \cong P^*_{PM,GOAL} \sqrt{\frac{v}{\chi^2_{v,1-\alpha}}}$$

Minimum Required Sample Size

A conservative minimum required sample size for having the lower confidence bound for P^*_{PM} at a chosen percentage below $\hat{P}^*_{PM}$ is given by the formula below. This is similar to the one developed for P_{PM} in the previous section.

$$kn \cong .5\left(\frac{Z_\alpha}{e_{P^*PM}}\right)^2$$

The e factor is based on the desired ratio between the confidence bound and $\hat{P}^*_{PM}$.

$$e_{P^*_{PM}} = 1 - \frac{P^*_{PM,1-\alpha}}{\hat{P}^*_{PM}}$$

Example 11.38 Gold Plating Thickness

The LSL for the thickness of gold plating on a silicon wafer is 50 angstroms, while the USL is 100 angstroms. It is decided to target the average thickness at 65 angstroms, as is displayed in Figure 11.26.

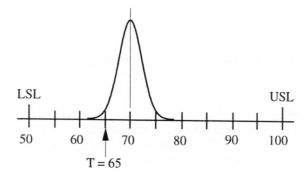

LSL   USL

50   60   70   80   90   100

T = 65

Fig. 11.26 Thickness of gold plating on silicon wafers.

For $\hat{\sigma}_{LT}$ of 2.5 and $\hat{\mu}$ of 70, τ^*_{LT} is estimated as 5.5902.

$$\hat{\tau}^*_{LT} = \sqrt{\hat{\sigma}^2_{LT} + (\hat{\mu} - T)^2} = \sqrt{(2.5)^2 + (70 - 65)^2} = 5.5902$$

With this estimate of τ^*_{LT}, P^*_{PM} is estimated as .894.

$$\hat{P}^*_{PM} = \text{Minimum}\left(\frac{T - LSL}{3\hat{\tau}^*_{LT}}, \frac{USL - T}{3\hat{\tau}^*_{LT}}\right)$$

$$= \text{Minimum}\left(\frac{65 - 50}{3(5.5902)}, \frac{100 - 65}{3(5.5902)}\right)$$

$$= \text{Minimum}(.894, 2.087) = .894$$

A lower .99 confidence bound for P^*_{PM} is found by first computing b.

$$b = \frac{\hat{\mu} - T}{\hat{\sigma}_{LT}} = \frac{70 - 65}{2.5} = 2$$

This b value is then used to determine the number of degrees of freedom. Recall that kn equals 66 for this example.

$$v = \frac{kn(1 + b^2)^2}{1 + 2b^2} = \frac{66(1 + 2^2)^2}{1 + 2(2)^2} = \frac{66(5)^2}{1 + 2(4)} \cong 184$$

With 184 degrees of freedom, a .99 confidence is calculated as .786.

$$P^*_{PM,.99} \cong \hat{P}^*_{PM} \sqrt{\frac{\chi^2_{184,.99}}{v}} = .894 \sqrt{\frac{142.33235}{184}} = .786$$

Based on the estimate of .894 derived from the capability study for P^*_{PM}, there is .99 confidence the true P^*_{PM} for this process is at least .786.

11.25 Upper Confidence Bound for P_{PG}

P_{PG} is equal to the inverse of P_{PM} squared, making it directly proportional to the average loss.

$$P_{PG} = \frac{1}{P^2_{PM}}$$

As smaller P_{PG} values indicate higher performance capability, interest is in obtaining an *upper* confidence bound for this measure. A conservative formula for this upper bound has been derived by Marcucci and Beazley.

$$\tilde{P}_{PG,1-\alpha} \cong \hat{P}_{PG} \frac{v}{\chi^2_{v,1-\alpha}} \quad \text{where } v = \frac{kn(1 + b^2)^2}{1 + 2b^2} \quad \text{and } b = \frac{\hat{\mu} - M}{\hat{\sigma}_{LT}}$$

Recall that M is the middle of the tolerance, kn is the total number of measurements collected during the course of the capability study, and $\hat{\sigma}_{LT}$ is an estimate of the long-term process standard deviation.

Maximum Observed Estimate

The largest estimate of P_{PG} that may be seen in a capability study, and still demonstrate a given capability goal is met with a certain level of confidence, is computed with the following formula:

$$\text{Max. Observed } \hat{P}_{PG,1-\alpha} \cong P_{PG,GOAL} \frac{\chi^2_{v,1-\alpha}}{v} \quad \text{where } v = \frac{kn(1 + b^2)^2}{1 + 2b^2} \text{ and } b = \frac{\hat{\mu} - M}{\hat{\sigma}_{LT}}$$

Example 11.39 Valve O.D. Size

In Example 6.29 (p. 277), P_{PM} was estimated as 1.204 for the process machining the valve outer diameter size. From this result, P_{PG} is estimated as .690.

$$\hat{P}_{PG} = \frac{1}{\hat{P}^2_{PM}} = \frac{1}{1.204^2} = .690$$

An upper .99 confidence bound for P_{PG} is calculated by first computing b. The middle of the tolerance, M, is 10.0000, while $\hat{\mu}$ is 10.0041 and $\hat{\sigma}_{LT}$ equals .01610.

$$b = \frac{\hat{\mu} - M}{\hat{\sigma}_{LT}} = \frac{10.0041 - 10.0000}{.01610} = .25466$$

This value of b determines the approximate number of degrees of freedom. The total number of measurements, kn, is 100 for this capability study.

$$v = \frac{kn(1 + b^2)^2}{1 + 2b^2} = \frac{100(1 + .25466^2)^2}{1 + 2(.25466)^2} \cong 100$$

With 100 degrees of freedom, the appropriate χ^2 value may be selected from Appendix Table VI and an upper .99 confidence bound for P_{PG} computed.

$$\tilde{P}_{PG,.99} \cong \hat{P}_{PG} \frac{v}{\chi^2_{v,.99}} = .690 \frac{100}{70.06472} = .985$$

Given a point estimate for P_{PG} of .690, there is .99 confidence the true P_{PG} for this process is less than .985. Note that this result is the squared inverse of the lower .99 confidence bound for P_{PM} computed in Example 11.36.

$$\tilde{P}_{PG,.99} = \frac{1}{P^2_{PM,.99}} = \frac{1}{1.008^2} = .984$$

Minimum Required Sample Size

A conservative minimum required sample size for having the upper confidence bound for P_{PG} at a chosen percentage above the estimate of P_{PG} is calculated via this formula:

$$kn \cong .5 \left(\frac{Z_\alpha}{e_{PPG}} \right)^2$$

The factor e_{PPG} is based on the desired ratio between the upper confidence bound and the estimate for P_{PG}.

$$e_{PPG} = 1 - \sqrt{\frac{\hat{P}_{PG}}{\tilde{P}_{PG,1-\alpha}}}$$

A capability study team decides to have the upper .95 confidence bound for P_{PG} be no more than 15 percent above the estimate for P_{PG}.

$$\tilde{P}_{PG,1-\alpha} = 1.15 \hat{P}_{PG} \quad \Rightarrow \quad \frac{1}{1.15} = \frac{\hat{P}_{PG}}{\tilde{P}_{PG,1-\alpha}}$$

e_{PPG} is then computed as follows:

$$e_{PPG} = 1 - \sqrt{\frac{\hat{P}_{PG}}{\tilde{P}_{PG,1-\alpha}}} = 1 - \sqrt{\frac{1}{1.15}} = .0675$$

To achieve the team's objective, at least 297 pieces must be included in the study.

$$kn \cong .5\left(\frac{Z_\alpha}{e_{PPG}}\right)^2 = .5\left(\frac{1.645}{.0675}\right)^2 \cong 297$$

If the planned subgroup size is 5, measurements from at least 60 in-control subgroups must be collected before estimating P_{PG}.

$$kn \cong 297$$

$$\frac{kn}{n} \cong \frac{297}{5}$$

$$k \cong 60$$

Once the capability study is completed with 60 subgroups of size 5 each, the estimate of P_{PG} is .70, while b turns out to be .3. This value of b makes the number of degrees of freedom approximately equal to 302.

$$v = \frac{kn(1+b^2)^2}{1+2b^2} = \frac{300[1+(.3)^2]^2}{1+2(.3)^2} \cong 302$$

After v is computed, an upper .95 confidence bound can be calculated for P_{PG}.

$$\tilde{P}_{PG,.95} \cong \hat{P}_{PG}\frac{v}{\chi^2_{302,.95}} = .70\frac{302}{262.744} = .805$$

This upper confidence bound of .805 is 1.15 percent above the P_{PG} estimate of .70.

$$\tilde{P}_{PG,.95} = 1.15\hat{P}_{PG} = 1.15(.70) = .805$$

11.26 Upper Confidence Bound for P^*_{PG}

The P^*_{PG} measure of long-term performance capability is calculated as given below. Recall that T is the target average for the process average.

$$P^*_{PG} = \text{Minimum}\left[\left(\frac{3\tau^*_{LT}}{T-LSL}\right)^2, \left(\frac{3\tau^*_{LT}}{USL-T}\right)^2\right] \quad \text{where } \tau^*_{LT} = \sqrt{\sigma^2_{LT}+(\mu-T)^2}$$

The formula for P^*_{PG} may also be written as:

$$P^*_{PG} = \left(\frac{6\tau^*_{LT}}{\text{Effective Tolerance}}\right)^2 = \frac{36\tau^{*2}_{LT}}{(\text{Effective Tolerance})^2} \text{ where Effective Tolerance} = USL_T - LSL_T$$

The upper confidence bound formula for this index is directly related to the one for P_{PG}.

$$\tilde{P}^*_{PG,1-\alpha} \cong \hat{P}^*_{PG}\frac{v}{\chi^2_{v,1-\alpha}} \quad \text{where } v = \frac{kn(1+b^2)^2}{1+2b^2} \quad \text{and } b = \frac{\hat{\mu}-T}{\hat{\sigma}_{LT}}$$

Maximum Observed Estimate

The largest estimate of $P^*{}_{PG}$ that can be observed in a capability study, and yet offer 1 minus α confidence the capability goal for $P^*{}_{PG}$ is met, is calculated with this formula:

$$\text{Maximum Observed } \hat{P}^*_{PG,1-\alpha} \cong P^*_{PG,GOAL} \frac{\chi^2_{v,1-\alpha}}{v}$$

Minimum Required Sample Size

A conservative minimum required sample size for having the upper confidence bound for $P^*{}_{PG}$ at a chosen percentage above $\hat{P}^*{}_{PG}$ is calculated by the formula given here:

$$kn \cong .5 \left(\frac{Z_\alpha}{e_{P^*PG}} \right)^2$$

The e factor is based on the desired ratio between the upper confidence bound and $\hat{P}^*_{PG}$.

$$e_{P^*PG} = 1 - \sqrt{\frac{\hat{P}^*_{PG}}{\hat{P}^*_{PG,1-\alpha}}}$$

Example 11.40 Gold Plating Thickness

The LSL for the thickness of gold plating on wafers is 50 angstroms, the USL is 100 angstroms, while T equals 65 angstroms. For $\hat{\sigma}_{LT}$ of 2.5 and $\hat{\mu}$ of 70, $\hat{\tau}^*_{LT}$ is found to be 31.25.

$$\hat{\tau}^{*\,2}_{LT} = \hat{\sigma}^2_{LT} + (\hat{\mu} - T)^2 = (2.5)^2 + (70 - 65)^2 = 31.25$$

As the target average for plating thickness is 65 angstroms, LSL_T equals 50, while USL_T equals 80 (review page 660). This makes the effective tolerance equal to 30.

$$\text{Effective Tolerance} = USL_T - LSL_T = 80 - 50 = 30$$

Now P^*_{PG} may be estimated.

$$\hat{P}^*_{PG} = \frac{36\hat{\tau}^{*\,2}_{LT}}{(\text{Effective Tolerance})^2} = \frac{36(31.25)}{(30)^2} = 1.25$$

An upper .99 confidence bound for P^*_{PG} is computed by first calculating b.

$$b = \frac{\hat{\mu} - T}{\hat{\sigma}_{LT}} = \frac{70 - 65}{2.5} = 2$$

Given that kn equals 66, this b value helps determine v.

$$v = \frac{kn(1 + b^2)^2}{1 + 2b^2} = \frac{66(1 + 2^2)^2}{1 + 2(2)^2} \cong 184$$

With 184 degrees of freedom, an upper .99 confidence bound is calculated as 1.616.

$$\tilde{P}^*_{PG,.99} \cong \hat{P}^*_{PG} \frac{\nu}{\chi^2_{184,.99}} = 1.25 \frac{184}{142.33235} = 1.616$$

Based on the estimate made from the capability study for P^*_{PG} of 1.25, there is .99 confidence the true P^*_{PG} ratio for this process is no more than 1.616. Notice that this upper confidence bound for P^*_{PG} is just the inverse of the square of the lower confidence bound for P^*_{PM}. This lower bound was calculated as .786 in Example 11.38 (p. 677).

$$\tilde{P}^*_{PG,.99} \cong \frac{1}{P^*_{PM,.99}{}^2} = \frac{1}{(.786)^2} = 1.618$$

11.27 Confidence Bounds for P_{PMK} and P^*_{PMK}

Both the P_{PMK} and P^*_{PMK} measures of performance capability have very complicated sampling distributions. As of yet, no simple approximation for their confidence intervals exists. The interested reader can review the work done to date on this subject by Kotz and Johnson, as well as Pearns *et al.*

11.28 Confidence Bounds for Machine Capability Measures

To make statements about machine capability at some particular confidence level under the control chart approach, apply the confidence bound formulas previously developed for C_R, P_R, C_P, P_P, and P_{PK}. These are summarized below, beginning with the formulas for machine C_R. Recall that ν is the degrees of freedom, while 1 minus α is the chosen confidence level.

$$\text{Machine } \tilde{C}_{R,1-\alpha} \cong \text{Machine } \hat{C}_R \sqrt{\frac{\nu}{\chi^2_{\nu,1-\alpha}}}$$

$$\text{Maximum Observed Machine } \hat{C}_{R,1-\alpha} \cong \text{Machine } C_{R,GOAL} \sqrt{\frac{\chi^2_{\nu,1-\alpha}}{\nu}}$$

$$\text{Machine } \tilde{P}_{R,1-\alpha} \cong \text{Machine } \hat{P}_R \sqrt{\frac{\nu}{\chi^2_{\nu,1-\alpha}}}$$

$$\text{Maximum Observed Machine } \hat{P}_{R,1-\alpha} \cong \text{Machine } P_{R,GOAL} \sqrt{\frac{\chi^2_{\nu,1-\alpha}}{\nu}}$$

The corresponding confidence bound formulas for machine C_P and P_P are given next.

$$\text{Machine } C_{P,1-\alpha} \cong \text{Machine } \hat{C}_P \sqrt{\frac{\chi^2_{\nu,1-\alpha}}{\nu}}$$

$$\text{Minimum Observed Machine } \hat{C}_{P,1-\alpha} \cong \text{Machine } C_{P,GOAL} \sqrt{\frac{\nu}{\chi^2_{\nu,1-\alpha}}}$$

$$\text{Machine } \underset{\sim}{P}_{P,1-\alpha} \cong \text{Machine } \hat{P}_P \sqrt{\frac{\chi^2_{\nu,1-\alpha}}{\nu}}$$

$$\text{Minimum Observed Machine } \hat{P}_{P,1-\alpha} \cong \text{Machine } P_{P,GOAL} \sqrt{\frac{\nu}{\chi^2_{\nu,1-\alpha}}}$$

For all the above formulas, ν is $f_n k(n-1)$ if the capability measure is estimated from $\overline{S}$, $.9k(n-1)$ if estimated from $\overline{R}$, and $k(n-1)$ if estimated from S_{TOT}.

No confidence bound can be computed for machine C_{PK}. However, one may be constructed for machine P_{PK}. Remember, this index requires S_{TOT} to be calculated from the measurements collected during the course of the machine capability study.

$$\text{Machine } \underset{\sim}{P}_{PK,1-\alpha} \cong \text{Machine } \hat{P}_{PK} - Z_\alpha \sqrt{\frac{1}{9kn} + \frac{(\text{Machine } \hat{P}_{PK})^2}{2(kn-1)}}$$

$$\text{Minimum Observed Machine } \hat{P}_{PK,1-\alpha} \cong \frac{\text{Machine } P_{PK,GOAL}}{1 - Z_\alpha \sqrt{1 \div 2(kn-1)}}$$

The formula for the lower confidence bound for machine P_{PM} is listed next:

$$\text{Machine } \underset{\sim}{P}_{PM,1-\alpha} \cong \text{Machine } \hat{P}_{PM} \sqrt{\frac{\chi^2_{\nu,1-\alpha}}{\nu}}$$

$$\text{Minimum Observed Machine } \hat{P}_{PM,1-\alpha} \cong \text{Machine } P_{PM,GOAL} \sqrt{\frac{\nu}{\chi^2_{\nu,1-\alpha}}}$$

The number of degrees of freedom are found with these formulas:

$$\nu = \frac{kn(1+b^2)^2}{1+2b^2} \quad \text{where } b = \frac{\hat{\mu} - M}{\hat{\sigma}_{LT}}$$

If T is not equal to M, replace the above measures with their modified versions, *i.e.*, C_R^*, P_R^*, C_P^*, P_P^*, P_{PK}^*, and P_{PM}^*.

Example 11.41 Width of Keyway

In Example 10.1 (p. 567), the machine C_P for the width of a keyway at one end of a shaft was estimated as 1.676. Adria, the manufacturing engineer in charge of this capability study, wants a lower .90 confidence bound for machine C_P. In this study, n equaled 5, and k was 10. Because a range chart was used to determine process stability, the number of degrees of freedom is approximately 36.

$$\nu \cong .9\,k\,(n-1) = .9\,(10)(5-1) = 36$$

This information helps her calculate a lower .90 confidence bound for machine C_P.

$$\text{Machine } \underline{C}_{P,1-.10} \cong \text{Machine } \hat{C}_P \sqrt{\frac{\chi^2_{\nu,1-.10}}{\nu}}$$

$$\text{Machine } C_{P,.90} \cong 1.676 \sqrt{\frac{\chi^2_{36,.90}}{36}} = 1.676 \sqrt{\frac{25.64329}{36}} = 1.415$$

This lower bound of 1.415 is a fair distance below the point estimate of 1.676 because of the small number of pieces run (only 50) in the machine capability study.

The machine is required to meet a capability goal of having C_P at least 1.67 with .90 confidence. The *smallest* C_P value that Adria may observed in the machine capability study (assuming $n = 5$ and $k = 10$), and still have .90 confidence this goal is met, is found with these formulas:

$$\nu \cong .9\,k\,(n-1) = .9\,(10)(5-1) = 36$$

$$\text{Min. Obs. Machine } \hat{C}_{P,.90} \cong \text{Machine } C_{P,GOAL} \sqrt{\frac{\nu}{\chi^2_{36,.90}}} = 1.67 \sqrt{\frac{36}{25.64329}} = 1.979$$

Because the estimate of machine C_P turned out to be only 1.676, the machine does not meet the capability goal for keyway width at the desired level of confidence. Note that even though the point estimate exceeds the numerical goal of 1.67, this is not enough evidence to accept the machine with .90 confidence.

In the past, some authors have proposed a "mini" machine capability study consisting of only 10 pieces (Strongrich *et al*). This is not a good recommendation for two important reasons. First, it would be extremely difficult to determine if a process is in control (stable) with only 10 measurements. Second, the confidence bound for any measure of capability would be quite large. Suppose these 10 measurements are used to estimate σ_{LT}, which is in turn is used to estimate machine P_P. There are only 9 degrees of freedom associated with this estimate of potential capability.

$$\nu = kn - 1 = 10 - 1 = 9$$

With 9 degrees of freedom, the lower .95 confidence bound for machine P_P would be 39 percent below the estimate of this index (only 61 percent of it).

$$\text{Machine } \underline{P}_{P,.95} = \text{Machine } \hat{P}_P \sqrt{\frac{\chi^2_{9,.95}}{9}} = \text{Machine } \hat{P}_P \sqrt{\frac{3.32511}{9}} = \text{Machine } \hat{P}_P(.61)$$

If the estimate for machine P_P turned out to be 1.33, the true machine P_P could be as low as .81 (1.33 × .61). A sample size of only 10 provides so little information about what the actual machine capability could be, that it is hardly worth doing. In order to have .95 confidence the estimate for machine P_P is within 15 percent of the true machine P_P, the ratio of the lower confidence bound to the point estimate must be .85, making e_{pp} equal to .15.

$$e_{PP} = 1 - \frac{\text{Machine } P_{P,1-\alpha}}{\text{Machine } \hat{P}_P} = 1 - .85 = .15$$

Z_α is 1.645 for α equal to .05. Substituting these values into the sample size formula reveals that kn must be at least 62.

$$kn = 1 + .5\left(\frac{1.645}{.15}\right)^2 = 61.13 \cong 62$$

Because the subgroup size for most machine capability studies is 5, at least 13 in-control subgroups ($13 \times 5 = 65$) must be collected before estimating machine P_P, and then calculating its lower .95 confidence bound. This is close to the 10 subgroups of five often recommended for machine capability studies.

Confidence Bounds for the Sequential *s* Test

Table 10.2 (p. 573) is set up for testing machine capability when the P_P goal is 1.67, or conversely, if the P_R goal is .60. Both goals imply that 10 $\sigma_{MACHINE}$s could fit within the tolerance. If a different goal is specified, apply one of these formulas to determine h, the number of $\sigma_{MACHINE}$s required to equal the tolerance.

$$h = \frac{6}{\text{Machine } P_{R,GOAL}} \qquad h = 6 \text{ Machine } P_{P,GOAL}$$

To demonstrate attainment of this machine capability goal at some chosen level of confidence, select the appropriate columns from Table 10.2 for the desired confidence level. Each critical value in these columns is then multiplied by $10/h$ to convert them for testing to the specified capability goal.

Example 11.42 Exhaust Manifold Hole Size

The machine P_P goal for a process drilling holes in exhaust manifolds is 1.90. From this goal, h is determined as 11.4.

$$h = 6 \text{ Machine } P_{P,GOAL} = 6(1.90) = 11.4$$

This value for h is used to convert the lower and upper critical values of Table 10.2 to the chosen capability goal. If a .95 confidence level is needed, multiply critical values from the fourth and fifth columns by $10/11.4$. The first set of critical values given in this table (.0556 and .1418) is converted as is done here:

$$\text{LCV}_8\left(\frac{10}{h}\right) = .0556\left(\frac{10}{11.4}\right) = .0488 \qquad \text{UCV}_8\left(\frac{10}{h}\right) = .1418\left(\frac{10}{11.4}\right) = .1244$$

If the first test ratio is less than .0488, there is .95 confidence the piece of equipment being tested meets the machine P_P goal of 1.90. If this first ratio is greater than .1244, there is .95 confidence the piece of equipment does not meet this goal. When the test ratio falls between these two critical values, there is not enough information to make a decision either way at this level of confidence. More data must be collected, a second test ratio calculated, and the result compared to two appropriately adjusted critical values from Table 10.2. This procedure continues until a decision is reached about machine capability.

For higher machine P_P goals, smaller sample standard deviations are expected by the sequential s test. This is reflected in LCV_8 decreasing from .0556 for a machine P_P goal of 1.67, to only .0488 for a goal of 1.90. In order to have the same confidence level for accepting the machine as meeting a higher goal, a much *smaller* test ratio must be observed. Conversely, if a lower machine P_P goal than 1.67 is specified, the lower critical values increase for each test ratio since a larger machine standard deviation is expected.

Also notice from Table 10.2, that in order for a machine to meet a given capability goal at a higher confidence level, a smaller test ratio must be observed. To demonstrate a machine P_P goal of 1.67 at .90 confidence, the test ratio for the first eight pieces run must be less than .0636. In order to provide sufficient evidence this same goal is met at .99 confidence, the first test ratio must be less than .0421.

For a given level of confidence, lower critical values increase as the number of pieces run in the study increases, while upper critical values decrease.

11.29 Confidence Bounds for Attribute Capability Measures

Due to ever-present random sampling variation, measures of capability for attribute data will vary slightly from study to study. In a manner similar to that done for variable data, appropriate confidence bounds can be established for attribute capability measures. The exact procedure depends on whether the capability study incorporated a control chart based on the Poisson or binomial distribution.

Charts Based on the Poisson Distribution

When a capability study is founded on either a c or u chart, an upper confidence bound is first determined for the total number of nonconformities observed in the study (c_{TOTAL}). This worst case for c_{TOTAL} is then used to calculate either an equivalent P^a_{PK}, P^a_P, or P^a_R value, which then becomes the appropriate confidence bound for that measure of capability. If desired, an upper confidence bound for $ppm_{TOTAL, LT}$ may also be calculated.

Assume a c chart is in a good state of control with $\bar{c}$ equal to 1.30, k equal to 20, while n is 30. This makes the estimate of u' equal to .04333 nonconformities per unit.

$$\hat{u}' = \frac{\bar{c}}{n} = \frac{1.30}{30} = .04333$$

With this, an estimate of 4.241 percent is made for p', and 42,410 for ppm_{TOTAL}.

$$\hat{p}' = 1 - e^{-\hat{u}'} = 1 - e^{-.04333} = .04241$$

$$\hat{p}pm_{TOTAL, LT} = \hat{p}' \times 10^6 = .04241 \times 10^6 = 42,410$$

Finally, an estimate of equivalent P^a_{PK} may be computed.

$$\text{Equivalent } \hat{P}^a_{PK} = \frac{Z(\hat{p}')}{3} = \frac{Z(.04241)}{3} = \frac{1.72}{3} = .573$$

The capability team now decides to calculate a lower .95 confidence bound for equivalent $P^{\%}_{PK}$. This procedure begins by finding the total number of nonconformities (c_{TOTAL}) observed during the study. This total is just the sum of all nonconformities detected in every one of the k in-control subgroups.

$$c_{TOTAL} = \sum_{i=1}^{k} c_i$$

This result is the same total used to calculate $\bar{c}$ when the c chart was constructed for the capability study.

$$\bar{c} = \frac{\sum_{i=1}^{k} c_i}{k} = \frac{c_{TOTAL}}{k}$$

Because this centerline must already be calculated before estimating any measure of capability, an easier method for obtaining c_{TOTAL} is simply multiplying $\bar{c}$ by k.

$$c_{TOTAL} = \bar{c}\, k$$

In this example, $\bar{c}$ equals 1.30 and k is 20, so the total number of nonconformities found in the 20 subgroups must have been 26.

$$c_{TOTAL} = \bar{c}\, k = 1.30\,(20) = 26$$

The formula for computing an upper 1 minus α confidence bound for c_{TOTAL} is from an article by Nelson.

$$\tilde{c}_{TOTAL,1-\alpha} = \frac{\chi^2_{v,\alpha}}{2} \qquad \text{where } v = 2\,(c_{TOTAL} + 1)$$

According to this formula for v, the applicable number of degrees of freedom is 54 for this example.

$$v = 2\,(c_{TOTAL} + 1) = 2\,(26 + 1) = 54$$

An upper .95 confidence bound for c_{TOTAL} can be constructed with this value of v.

$$\tilde{c}_{TOTAL,.95} = \frac{\chi^2_{v,\alpha}}{2} = \frac{\chi^2_{54,.05}}{2} = \frac{72.15322}{2} = 36.0766$$

As there is .95 confidence that no more than 36.0766 nonconformities would have been seen in these 20 subgroups, there is .95 confidence the true average number of nonconformities per subgroup (c') is less than 1.8038.

$$\frac{\tilde{c}_{TOTAL,.95}}{k} = \frac{36.0766}{20} = 1.8038$$

This last result becomes an upper .95 confidence bound for c' and is labeled $\tilde{c}'_{.95}$, or in general, $\tilde{c}'_{1-\alpha}$.

An upper $1 - \alpha$ confidence bound for $c' = \tilde{c}'_{1-\alpha} = \dfrac{\tilde{c}_{TOTAL, 1-\alpha}}{k}$

The relationship between these bounds is displayed in Figure 11.27.

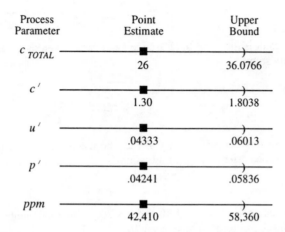

Fig. 11.27   Progression of upper .95 confidence bounds from c_{TOTAL} to *ppm*.

The average number of nonconformities per subgroup divided by n yields an estimate of the average nonconformities per unit. In this example, $\bar{c}$ is 1.30, while n is 30.

$$\hat{u}' = \frac{\hat{c}'}{n} = \frac{\bar{c}}{n} = \frac{1.30}{30} = .04333$$

An upper .95 confidence bound for the true u' is derived by dividing the upper .95 confidence bound for c' by n.

$$\tilde{u}'_{.95} = \frac{\tilde{c}'_{.95}}{n} = \frac{1.8038}{30} = .06013$$

A point estimate for the average nonconformities per unit is .04333, with an upper .95 confidence bound of .06013. This upper .95 confidence bound for u' is now inserted into the Poisson formula to find the probability of a unit having zero nonconformities, which would make it a conforming unit. As explained in Chapter 9, this result is then subtracted from 1 to derive an upper .95 confidence bound for p'.

$$\tilde{p}'_{.95} = 1 - e^{-\tilde{u}'_{.95}} = 1 - e^{-.06013} = .05836$$

Based on the original point estimate of 4.241 percent nonconforming parts, there is .95 confidence the true p' for this process is less than 5.836 percent. This upper bound of .05836 for p' may now be used to determine an upper .95 confidence bound for $ppm_{TOTAL, LT}$, as is displayed above in Figure 11.27.

$$\tilde{p}pm_{TOTAL, LT, .95} = \tilde{p}'_{.95} \times 10^6 = .05836 \times 10^6 = 58,360$$

With its point estimate being 42,410, there is .95 confidence the true $ppm_{TOTAL,\,LT}$ for this process is less than 58,360. This upper bound for p' of .05836 is also needed to compute the *lower* .95 confidence bound for equivalent $P^{\%}_{PK}$ (a larger p' value means a smaller $P^{\%}_{PK}$ index).

$$\text{Equivalent } P^{\%}_{\underline{PK},.95} = \frac{Z(\tilde{p}'_{.95})}{3} = \frac{Z(.05836)}{3} = \frac{1.57}{3} = .523$$

With equivalent $P^{\%}_{PK}$ estimated as .573 for this process, there is .95 confidence the true equivalent $P^{\%}_{PK}$ is at least .523. The relationship between these two confidence bounds is presented below in Figure 11.28.

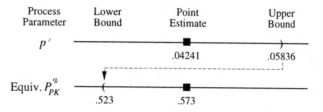

Process Parameter	Lower Bound	Point Estimate	Upper Bound

p' .04241   .05836

Equiv. $P^{\%}_{PK}$ .523   .573

Fig. 11.28 The lower confidence bound for equivalent $P^{\%}_{PK}$ corresponds to the upper confidence bound for p'.

A lower .95 confidence bound for equivalent $P^{\%}_{P}$ and an *upper* .95 confidence bound for equivalent $P^{\%}_{R}$ may be derived from $\tilde{p}'_{.95}$ in a similar manner.

$$\text{Equivalent } P^{\%}_{\underline{P},.95} = \frac{Z(\tilde{p}'_{.95} \div 2)}{3} = \frac{Z(.05836 \div 2)}{3} = \frac{1.89}{3} = .630$$

$$\text{Equivalent } \tilde{P}^{\%}_{R,.95} = \frac{1}{\text{Equivalent } P^{\%}_{P,.95}} = \frac{1}{.630} = 1.587$$

As equivalent $P^{\%}_{P}$ and $P^{\%}_{R}$ are rarely seen in practice, these last two confidence bound formulas are infrequently employed.

To simplify the calculation of the confidence bounds for $ppm_{TOTAL,\,LT}$ and equivalent $P^{\%}_{PK}$, use these formulas, where $v = 2(c_{TOTAL} + 1)$:

$$\tilde{p}\,pm_{TOTAL,LT,1-\alpha} = \left[1 - e\exp\left(-\frac{\chi^2_{v,\alpha}}{2m}\right)\right] \times 10^6 \qquad \text{Equivalent } P^{\%}_{\underline{PK},.95} = \frac{1}{3}Z\left[1 - e\exp\left(-\frac{\chi^2_{v,\alpha}}{2m}\right)\right]$$

Example 11.43 Display Scratches

A u chart monitoring the number of scratches found per unit on the upper surface of a liquid crystal display used for notebook computers is finally brought into a good state of control with $\bar{u}$ equal to .060. The percentage of nonconforming displays is estimated by first subtracting the number of displays with no scratches from 1.

$$\hat{p}' = 1 - e^{-\bar{u}} = 1 - e^{-.060} = 1 - .941764 = .058236$$

From this estimate of p', point estimates of $ppm_{TOTAL, LT}$ and equivalent P^*_{PK} may be calculated for this process.

$$\hat{p}pm_{TOTAL, LT} = \hat{p}' \times 10^6 = .058236 \times 10^6 = 58,236$$

$$\text{Equivalent } \hat{P}^\%_{PK} = \frac{Z(\hat{p}')}{3} = \frac{Z(.058236)}{3} = \frac{1.57}{3} = .523$$

56 nonconformities are found ($c_{TOTAL} = 56$) on the 933 units checked during the study, thus making v equal 114.

$$v = 2(c_{TOTAL} + 1) = 2(57) = 114$$

v is needed to derive an upper .90 confidence bound for c_{TOTAL}.

$$\tilde{c}_{TOTAL, .90} = \frac{\chi^2_{114, .10}}{2} = \frac{133.72858}{2} = 66.8643$$

Knowing $\tilde{c}_{TOTAL, .90}$, along with the total number of displays examined during the capability study (933), allows an upper confidence bound to be determined for the average scratches per display.

$$\tilde{u}'_{.90} = \frac{\tilde{c}_{TOTAL, .90}}{\sum_{i=1}^{k} n_i} = \frac{66.8643}{933} = .071666$$

An upper bound for the percentage of scratched displays is calculated by inserting this upper bound for u' in the Poisson formula for zero nonconformities.

$$\tilde{p}'_{.90} = 1 - e^{-\tilde{u}'_{.90}} = 1 - e^{-.071666} = 1 - .93084 = .06916$$

With a point estimate of about 5.8 percent, there is .90 confidence the true p' is no more than 6.9 percent. Multiplying this upper .90 confidence bound for p' by 10^6 produces an upper .90 confidence bound for $ppm_{TOTAL, LT}$.

$$\tilde{p}pm_{TOTAL, LT, .90} = \tilde{p}'_{.90} \times 10^6 = .06916 \times 10^6 = 69,160$$

Based on the point estimate of 58,236, there is .90 confidence the true $ppm_{TOTAL, LT}$ measure for this process is less than 69,160.

In this last step, the upper .90 confidence bound for p' is also used to construct a *lower* .90 confidence bound for equivalent $P^\%_{PK}$.

$$\text{Equivalent } P^\%_{PK, .90} = \frac{Z(\tilde{p}'_{.90})}{3} = \frac{Z(.06916)}{3} = \frac{1.48}{3} = .493$$

From the point estimate of .523, there is .90 confidence the actual equivalent P^*_{PK} index for this process at least .493. Note this result could have been computed directly with this formula:

$$\text{Equivalent } P^{\%}_{PK,.95} = \frac{1}{3}Z\left[1 - e\exp\left(-\frac{\chi^2_{114,.10}}{2m}\right)\right]$$

$$= \frac{1}{3}Z\left[1 - e\exp\left(-\frac{133.72858}{2(933)}\right)\right]$$

$$= \frac{1}{3}Z(1 - e^{-.071666})$$

$$= \frac{Z(.06916)}{3} = \frac{1.48}{3} = .493$$

Example 11.44 Glass Bubbles

A c chart monitoring the number of bubbles, or "seeds," found in a square meter of glass molded for windshields (windscreens) records no problems in any of its 25 subgroups. As n is 50 windshields, m (the total number of windshields checked) is 1,250 (25 × 50).

$$c_{TOTAL} = \sum_{i=1}^{25} c_i = 0$$

With an absence of nonconformities, the estimate of u' is zero, which makes the estimate of p' also zero.

$$\hat{u}' = \frac{\sum_{i=1}^{25} c_i}{\sum_{i=1}^{25} n_i} = \frac{0}{m} = \frac{0}{1,250} = 0 \qquad \hat{p}' = 1 - e^{-\hat{u}'} = 1 - e^0 = 1 - 1 = 0$$

The point estimate of equivalent $P^{\%}_{PK}$ for this result of 0 cannot be determined.

$$\text{Equivalent } \hat{P}^{\%}_{PK} = \frac{Z(\hat{p}')}{3} = \frac{Z(0)}{3} = \frac{\infty}{3} = \infty$$

However, the lack of bubbles in the 25 subgroups collected for constructing the c chart does not necessarily mean there are no bubbles present in the *entire* process output. Even with zero bubbles, .99 confidence bounds can be calculated for both $ppm_{TOTAL\,LT}$ and equivalent $P^{\%}_{PK}$. Begin by finding the number of degrees of freedom from c_{TOTAL}.

$$\nu = 2(c_{TOTAL} + 1) = 2(0 + 1) = 2$$

For this example, m equals kn, which is equal to 25 times 50, or 1,250. An upper .99 confidence bound for $ppm_{TOTAL\,LT}$ is computed with the following formula:

$$\tilde{p}pm_{TOTAL,LT,.99} = \left[1 - e\exp\left(-\frac{\chi^2_{2,.01}}{2m}\right)\right] \times 10^6$$

$$= \left[1 - e\exp\left(-\frac{9.21034}{2(1,250)}\right)\right] \times 10^6$$

$$= (1 - e^{-.003684}) \times 10^6 = 3,677$$

There is .99 confidence the true $ppm_{TOTAL, LT}$ for this process is less than 3,677. In order to satisfy the minimum requirement of long-term performance capability with the desired level of confidence, the upper confidence bound for $ppm_{TOTAL, LT}$ must be less than 1,350. Even with zero bubbles found in the study, this process cannot be considered capable at the specified level of confidence.

A lower .99 confidence bound for equivalent $P^{\%}_{PK}$ is calculated with this formula:

$$\text{Equivalent } \underset{\sim}{P}^{\%}_{PK,.99} = \frac{1}{3}Z\left[1 - e \exp\left(-\frac{\chi^2_{2,.01}}{2m}\right)\right]$$

$$= \frac{1}{3}Z\left[1 - e \exp\left(-\frac{9.21034}{2(1,250)}\right)\right]$$

$$= \frac{Z(.003677)}{3} = \frac{2.68}{3} = .893$$

There is .99 confidence the true equivalent $P^{\%}_{PK}$ index for this process is at least .893, while there is a .01 risk it is less than .893. So even with no nonconformities observed during the study, the process must be considered not capable of meeting the minimum requirement for long-term performance capability at this confidence level. In order to provide sufficient evidence this process does meet the minimum capability requirement at the chosen confidence level, *i.e.*, equivalent $P^{\%}_{PK,1-\alpha}$ is greater than 1.00, more windshields without bubbles must be seen. This formula for determining m is derived on page 704.

$$m = \frac{\log \alpha}{\log(1 - p'_{GOAL})}$$

Meeting this minimum capability requirement requires a ppm value of 1,350, which means the p' goal must be .00135. To have .99 confidence this goal is achieved, 3,409 windshields must be sampled *without observing a single bubble*.

$$m = \frac{\log .01}{\log(1 - .00135)} = 3408.934 \cong 3,409$$

If this happens, the above confidence bounds would be recalculated by first finding v.

$$v = 2(c_{TOTAL} + 1) = 2(0 + 1) = 2$$

The revised upper .99 confidence bound for equivalent $P^{\%}_{PK}$ is found with this formula, with m now equaling 3,409:

$$\text{Equivalent } \underset{\sim}{P}^{\%}_{PK,.99} = \frac{1}{3}Z\left[1 - e \exp\left(-\frac{\chi^2_{2,.01}}{2m}\right)\right]$$

$$= \frac{1}{3}Z\left[1 - e \exp\left(-\frac{9.21034}{2(3,409)}\right)\right]$$

$$= \frac{Z(.0013499)}{3} = \frac{3.00}{3} = 1.00$$

As this lower bound is 1.00, there would be .99 confidence this process meets the minimum prerequisite for long-term performance capability, given that zero bubbles were found on the 3,409 windshields.

Minimum Required Sample Size

The following formula provides a conservative estimate of the number of pieces that must be run during the course of an attribute capability study to provide a certain level of confidence the upper bound for u' is no more than a preselected distance above the point estimate.

$$m \cong .25 \left(\frac{Z_\alpha}{d_u} \right)^2 \quad \text{where } d_u = \sqrt{\tilde{u}'_{1-\alpha}} - \sqrt{\hat{u}' + .00015}$$

Unfortunately, before m can be computed, the outcome for $\hat{u}'$ must be predicted. If a reasonable guess cannot be made, set $\hat{u}'$ much lower than expected. This conservative approach makes m larger than necessary, but guarantees the difference between $\hat{u}'$ and its upper bound will be no more than the desired amount at the specified confidence level. For example, a capability study team forecasts $\hat{u}'$ to be at least .008. They also want the upper .95 confidence bound to be no more than 25 percent above this estimate, which would be .010 (1.25 × .008). These stipulations make d_u equal to .009723.

$$d_u = \sqrt{\tilde{u}'_{.95}} - \sqrt{\hat{u}' + .00015} = \sqrt{.010} - \sqrt{.008 + .00015} = .009723$$

As seen below, the approximate number of pieces needed to be run in the capability study is 7,156.

$$m \cong .25 \left(\frac{Z_{.05}}{d_u} \right)^2 \cong .25 \left(\frac{1.645}{.009723} \right)^2 \cong 7,156$$

Assume 58 nonconformities are found on these 7,156 pieces. The estimate for u' becomes .0081, while v is 118.

$$v = 2(c_{TOTAL} + 1) = 2(58 + 1) = 118$$

The upper .95 confidence bound is found as .0101, which is 24.7 percent greater than the point estimate of .0081 for u'.

$$\tilde{u}'_{.95} = \frac{\chi^2_{118,.05}}{2m} = \frac{144.35367}{2(7,156)} = .0101$$

If the confidence bound for equivalent P^*_{PK} should be no more than a certain distance below the point estimate of P^*_{PK}, m is found by first making an educated guess at the expected estimate for equivalent P^*_{PK}. Select the desired lower bound for this point estimate, then apply these formulas to determine d_u.

$$\tilde{u}'_{1-\alpha} = -\ln \left[1 - \Phi \left(3 \, \text{Equivalent} \, P^{\%}_{PK} \right) \right] \qquad \hat{u}' = -\ln [1 - \Phi (3 \, \text{Equivalent} \, \hat{P}^{\%}_{PK})]$$

$$d_u = \sqrt{\tilde{u}'_{1-\alpha}} - \sqrt{\hat{u}' + .00015}$$

When Z_α and d_u are substituted into the formula for m, the minimum required sample size can be computed.

$$m \cong .25\left(\frac{Z_\alpha}{d_u}\right)^2$$

For example, the estimate of equivalent $P^\%_{PK}$ is predicted to be around 1.00. This makes the expected $\hat{u}'$ equal to .0013509.

$$\hat{u}' = -\ln[1 - \Phi(3\,\text{Equivalent}\,\hat{P}^\%_{PK})]$$

$$= -\ln[1 - \Phi(3 \times 1.00)]$$

$$= -\ln[1 - \Phi(3.00)]$$

$$= -\ln(1 - .00135) = .0013509$$

The lower .95 confidence bound should be no more than 10 percent below the point estimate for equivalent $P^\%_{PK}$, thus making it a minimum of .90. This requirement causes the upper .95 confidence bound for u' to equal .003473.

$$\bar{u}'_{.95} = -\ln\left[1 - \Phi\left(3\,\text{Equivalent}\,\underset{\sim}{P}^\%_{PK}\right)\right]$$

$$= -\ln[1 - \Phi(3 \times .90)]$$

$$= -\ln[1 - \Phi(2.70)]$$

$$= -\ln(1 - .003467) = .003473$$

From this next relationship, d_u is found as .020189.

$$d_u = \sqrt{\bar{u}'_{.95}} - \sqrt{\hat{u}' + .00015} = \sqrt{.003473} - \sqrt{.001351 + .00015} = .020189$$

The minimum required sample size is then determined with this value of d_u.

$$m \cong .25\left(\frac{Z_{.05}}{d_u}\right)^2 \cong .25\left(\frac{1.645}{.020189}\right)^2 \cong 1,660$$

If a subgroup size of 50 will be used, at least 34 subgroups must be collected to accumulate a minimum of 1,660 units, as 34 times 50 is 1,700.

Charts Based on the Binomial Distribution

For process capability studies involving np and p charts, an estimate of the population percentage defective, p', must first be made in order to estimate $ppm_{TOTAL\ LT}$ or equivalent $P^\%_{PK}$. Likewise, an upper confidence bound for p' must be calculated before confidence bounds for either of these measures can be obtained. However, in order to construct meaningful bounds, m must be at least 400. With k usually at 20 or more, and subgroup sizes typically around 25, this should not be a serious problem ($m = kn = 20(25) = 500 > 400$).

The particular formula for determining an upper confidence bound for p' depends on the magnitude of $\hat{p}'$.

For $\hat{p}' \geq .03$

If $\hat{p}'$ is greater than 3 percent, Hahn and Meeker (p. 106) recommend the following formula for calculating an approximate upper 1 minus α confidence bound for p'. Recall that Z_α equals $Z(\alpha)$, which is the Z value for the chosen α level.

$$\tilde{p}'_{1-\alpha} = \hat{p}' + Z_\alpha \sqrt{\frac{\hat{p}'(1-\hat{p}')}{m}}$$

m represents the total number of items inspected during the capability study.

$$m = \sum_{i=1}^{k} n_i \quad \text{and} \quad \hat{p}' = \overline{p} \text{ for a } p \text{ chart}$$

$$m = kn \quad \text{and} \quad \hat{p}' = \frac{n\overline{p}}{n} \text{ for a } np \text{ chart}$$

This upper confidence bound for p' is used to calculate a lower 1 minus α confidence bound for equivalent $P_{PK}^{\%}$.

$$\text{Equivalent } \underset{\sim}{P}_{PK,1-\alpha}^{\%} = \frac{Z(\tilde{p}'_{1-\alpha})}{3} = \frac{1}{3} Z\left(\hat{p}' + Z_\alpha \sqrt{\frac{\hat{p}'(1-\hat{p}')}{m}} \right)$$

Although equivalent $P_{PK}^{\%}$ is applied most often in practice, confidence bounds for these other measures may also be computed.

$$\text{Equivalent } \underset{\sim}{P}_{P,1-\alpha}^{\%} = \frac{Z(\tilde{p}'_{1-\alpha} \div 2)}{3} \qquad \text{Equivalent } \tilde{P}_{R,1-\alpha}^{\%} = \frac{1}{\text{Equivalent } \underset{\sim}{P}_{P,1-\alpha}^{\%}}$$

The upper confidence bound for p' also allows the calculation of an upper 1 minus α confidence bound for $ppm_{TOTAL,\,LT}$.

$$\tilde{p}pm_{TOTAL,LT,1-\alpha} = \tilde{p}'_{1-\alpha} \times 10^6$$

Example 11.45 Radiator Leaks

Data from a p chart monitoring the percentage of radiators having a leak at final test is finally brought into control with $\overline{p}$ equal to .12123.

$$\hat{p}' = \overline{p} = .12123$$

First, a point estimate of equivalent $P_{PK}^{\%}$ is determined.

$$\text{Equivalent } \hat{P}_{PK}^{\%} = \frac{Z(\hat{p}')}{3} = \frac{Z(.12123)}{3} = \frac{1.17}{3} = .39$$

Second, an upper .95 confidence bound for the percentage of leaking radiators is found. During the course of the capability study, 1,592 radiators were checked for leaks ($m = 1,592$).

$$\tilde{p}'_{.95} = \hat{p}' + Z_{.05} \sqrt{\frac{\hat{p}'(1 - \hat{p}')}{m}} = .12123 + 1.645 \sqrt{\frac{.12123(1 - .12123)}{1,592}} = .13469$$

Finally, a lower .95 confidence bound for equivalent $P^{\%}_{PK}$ is computed.

$$\text{Equivalent } P^{\%}_{\underline{P}K,.95} = \frac{Z(\tilde{p}'_{.95})}{3} = \frac{Z(.13469)}{3} = \frac{1.11}{3} = .37$$

Because this lower .95 confidence bound for equivalent $P^{\%}_{PK}$ is only .37, there is little prospect of this process satisfying even the minimum requirement for long-term performance capability. Note that due to the very high m, this lower confidence bound is very close to the point estimate of .39.

An upper .95 confidence bound for $ppm_{TOTAL, LT}$ may also be calculated from the upper .95 confidence bound for p'. With $\hat{p}'$ equal to .12123, the estimate of $ppm_{TOTAL, LT}$ is 121,230.

$$\tilde{p}pm_{TOTAL, LT, .95} = \tilde{p}'_{.95} \times 10^6 = .13469 \times 10^6 = 134,690$$

With an upper bound of 134,690 *ppm*, there is little chance this process meets the minimum requirement of performance capability at .99 confidence. This should be no surprise, since this method of determining the confidence bound is for situations where the estimated percentage nonconforming is more than 3 percent.

Example 11.46 Cracked Welds

Flat sheet metal blanks are butt welded into motor compartment rails with the help of a laser. Nancy, a quality technician, selected an *np* chart to monitor the number of rails found with a cracked weld ($n = 75$). After establishing control of this process, she calculates $n\bar{p}$ as 2.636. From this centerline, an estimate of p' can be made.

$$p' = \frac{n\bar{p}}{n} = \frac{2.636}{75} = .03515$$

$\hat{p}'$ is then used to estimate equivalent $P^{\%}_{PK}$.

$$\text{Equivalent } \hat{P}^{\%}_{PK} = \frac{Z(\hat{p}')}{3} = \frac{Z(.03515)}{3} = \frac{1.81}{3} = .603$$

Knowing $\hat{p}'$ also allows Nancy to construct a lower .90 confidence bound for equivalent $P^{\%}_{PK}$. The number of rails checked in this example, m, is equal to 1,650 ($kn = 22 \times 75 = 1,650$).

$$\text{Equivalent } P^{\%}_{\underline{P}K,.90} = \frac{1}{3} Z \left(\hat{p}' + Z_{.10} \sqrt{\frac{\hat{p}'(1 - \hat{p}')}{m}} \right)$$

$$= \frac{1}{3} Z \left(.03515 + 1.282 \sqrt{\frac{.03515(1 - .03515)}{1,650}} \right)$$

$$= \frac{Z(.04096)}{3} = \frac{1.74}{3} = .580$$

With a point estimate of .603, there is .90 confidence the actual equivalent P^*_{PK} index for this laser welding process is at least .580.

Maximum Observed Estimate

The formula for the largest estimate of p' that may be observed in a capability study, and still provide 1 minus α confidence the goal is met, is fairly complicated. Its full derivation is given in Section 11.32 (see Derivation 3, p. 711). To simplify the formula somewhat, the following definitions are used:

$$a = \frac{Z_\alpha^2}{m} \qquad c = 2p'_{GOAL}$$

The formula then becomes:

$$\text{Maximum Observed } \hat{p}'_{1-\alpha} = \frac{a + c - \sqrt{a^2 + 2ac - ac^2}}{2(1+a)}$$

Assume the goal for p' is .10, and 1,000 units will be checked. The largest estimate of p' that may be seen in this study which still furnishes .95 confidence the goal is met, is determined as follows:

$$a = \frac{Z_{.05}^2}{m} = \frac{1.645^2}{1,000} = .002706 \qquad c = 2p'_{GOAL} = 2(.10) = .20$$

Inserting these values into the maximum observed estimate formula yields:

$$\text{Max. Observed } \hat{p}'_{.95} = \frac{.002706 + .20 - \sqrt{.002706^2 + 2(.002706)(.20) - .002706(.20)^2}}{2(1 + .002706)}$$

$$= \frac{.1713767}{2.005412} = .0855$$

If the maximum observed estimate is calculated by hand, note that this formula is very sensitive to round off errors. The above answer can be verified by substituting .0855 into the upper confidence bound formula for p'.

$$\tilde{p}'_{.95} = \hat{p}' + Z_{.05}\sqrt{\frac{\hat{p}'(1-\hat{p}')}{m}} = .0855 + 1.645\sqrt{\frac{.0855(1-.0855)}{1,000}} = .1000$$

This upper .95 confidence bound is right at the goal of .10 for p'.

Minimum Required Sample Size

The confidence bound formula for p' can be rearranged to derive a formula for computing the minimum required sample size required to have the upper confidence bound be a predetermined distance above $\hat{p}'$.

$$\tilde{p}'_{1-\alpha} = \hat{p}' + Z_\alpha \sqrt{\frac{\hat{p}'(1-\hat{p}')}{m}}$$

$$\tilde{p}'_{1-\alpha} - \hat{p}' = Z_\alpha \sqrt{\frac{\hat{p}'(1-\hat{p}')}{m}}$$

$$(\tilde{p}'_{1-\alpha} - \hat{p}')^2 = Z_\alpha^2 \frac{\hat{p}'(1-\hat{p}')}{m}$$

$$m = Z_\alpha^2 \frac{\hat{p}'(1-\hat{p}')}{(\tilde{p}'_{1-\alpha} - \hat{p}')^2}$$

$$m = \hat{p}'(1-\hat{p}')\left(\frac{Z_\alpha}{\tilde{p}'_{1-\alpha} - \hat{p}'}\right)^2$$

Let d_p be defined as the desired difference between the upper 1 minus α confidence bound for p' and the point estimate of p':

$$d_p = \tilde{p}'_{1-\alpha} - \hat{p}'$$

d_p is then substituted into the minimum required sample size formula.

$$m = \hat{p}'(1-\hat{p}')\left(\frac{Z_\alpha}{d_p}\right)^2$$

Before m can be computed, a guess for $\hat{p}'$ must be made. This is rather unfortunate, as most capability studies are undertaken to determine this estimate. If no reasonable speculation can be ventured, assume the most conservative case by setting $\hat{p}'$ higher than expected. This precaution will make m larger than necessary, but guarantees the difference between $\hat{p}'$ and its upper bound will be no more than the desired amount at the specified confidence level.

Suppose $\hat{p}'$ is predicted to come out at .04. The upper .90 confidence bound should be no more than 30 percent above this estimate, which would be a maximum of .052 (.40 × 1.30). This makes d_p equal to .012 (.052 minus .40).

$$m = \hat{p}'(1-\hat{p}')\left(\frac{Z_{.10}}{d_p}\right)^2 = .04(1-.04)\left(\frac{1.282}{.012}\right)^2 = 438.27 \cong 439$$

Thus, at least 439 parts (always round to the next largest integer) must be run to achieve these objectives. If $\hat{p}'$ ends up being less than .04, the upper .90 confidence bound will be less than .012 above $\hat{p}'$. For example, if $\hat{p}'$ turns out to be .036, its upper .90 confidence bound of .0474 is .0114 (.0474 minus .036) above this point estimate.

$$\tilde{p}'_{.90} = \hat{p}' + Z_{.10}\sqrt{\frac{\hat{p}'(1-\hat{p}')}{m}} = .036 + 1.282\sqrt{\frac{.036(1-.036)}{439}} = .0474$$

On the other hand, when $\hat{p}'$ turns out to be greater than .04, the upper bound will be more than .012 above $\hat{p}'$. Let the point estimate come out to be .048. At .0611, the upper bound would be .0131 greater than $\hat{p}'$.

$$\tilde{p}'_{.90} = \hat{p}' + Z_{.10} \sqrt{\frac{\hat{p}'(1-\hat{p}')}{m}} = .048 + 1.282 \sqrt{\frac{.048(1-.048)}{439}} = .0611$$

If the confidence bound for equivalent $P^{\mathcal{q}}_{PK}$ should be no more than a certain distance below the point estimate of $P^{\mathcal{q}}_{PK}$, m is found in the following manner. First, make an educated guess at the expected estimate for equivalent $P^{\mathcal{q}}_{PK}$, then use it to find $\hat{p}'$ with this formula:

$$\hat{p}' = \Phi(3 \text{ Equivalent } \hat{P}^{\mathcal{q}}_{PK})$$

Decide on the desired difference between the lower .90 bound and $\hat{p}'$, and express it as a e_{PPK} value (review Example 11.27 on page 657 if necessary).

$$e_{PPK} = 1 - \frac{\text{Equivalent } P^{\mathcal{q}}_{PK, 1-\alpha}}{\text{Equivalent } \hat{P}^{\mathcal{q}}_{PK}}$$

Use this formula to determine the desired upper 1 minus α confidence bound for p'.

$$\tilde{p}'_{1-\alpha} = \Phi[3(1 - e_{PPK}) \text{ Equivalent } \hat{P}^{\mathcal{q}}_{PK}]$$

d_p can now be determined.

$$d_p = \tilde{p}'_{1-\alpha} - \hat{p}'$$

When $\hat{p}'$ and d_p are substituted into the formula for m, the minimum required sample size can be found. For example, the estimate of equivalent $P^{\mathcal{q}}_{PK}$ is predicted to be around 1.00, making the expected $\hat{p}'$ equal .00135.

$$\hat{p}' = \Phi(3 \text{ Equivalent } \hat{P}^{\mathcal{q}}_{PK}) = \Phi(3 \times 1.00) = \Phi(3.00) = .00135$$

The lower .95 confidence bound for equivalent $P^{\mathcal{q}}_{PK}$ should be at least 90 percent of the point estimate of equivalent $P^{\mathcal{q}}_{PK}$ (no more than 10 percent below it).

$$e_{PPK} = 1 - \frac{\text{Equivalent } P^{\mathcal{q}}_{PK, .95}}{\text{Equivalent } \hat{P}^{\mathcal{q}}_{PK}} = 1 - \frac{.90 \text{ Equivalent } \hat{P}^{\mathcal{q}}_{PK}}{\text{Equivalent } \hat{P}^{\mathcal{q}}_{PK}} = 1 - .90 = .10$$

This makes the upper .95 confidence bound for p' equal to .003467.

$$\tilde{p}'_{.95} = \Phi[3(1 - e_{PPK}) \text{ Equivalent } \hat{P}^{\mathcal{q}}_{PK}] = \Phi[3(1-.10)1.00] = \Phi(2.70) = .003467$$

From this relation, d_p can be found as .002117.

$$d_p = \tilde{p}'_{.95} - \hat{p}' = .003467 - .00135 = .002117$$

With the expected $\hat{p}'$ and d_p, the minimum required sample size can be determined.

$$m = \hat{p}'(1-\hat{p}')\left(\frac{Z_{.05}}{d_p}\right)^2 = .00135(1-.00135)\left(\frac{1.645}{.002117}\right)^2 \cong 815$$

If a subgroup size of 40 is chosen, at least 21 subgroups must be collected to accumulate a minimum of 815 units ($21 \times 40 = 840$).

For $0 < \hat{p}' < .03$

The confidence bound formula for p' presented in the previous section becomes a poor approximation when $\hat{p}'$ is less than 3 percent. For this range of $\hat{p}'$ (and $kn > 400$), the method of calculating a confidence bound for the Poisson charts is a much better approximation. First obtain the total number of nonconforming units in the study (np_{TOTAL}) by summing the number of nonconforming units in each of the k in-control subgroups.

$$np_{TOTAL} = \sum_{i=1}^{k} np_i$$

When the capability study involves a p chart, calculate np_{TOTAL} by multiplying the chart's centerline by the total number of units checked.

$$\bar{p} = \frac{\sum_{i=1}^{k} np_i}{\sum_{i=1}^{k} n_i} \quad \Rightarrow \quad np_{TOTAL} = \sum_{i=1}^{k} np_i = \bar{p} \times \sum_{i=1}^{k} n_i$$

If an np chart is involved:

$$n\bar{p} = \frac{\sum_{i=1}^{k} np_i}{k} \quad \Rightarrow \quad np_{TOTAL} = \sum_{i=1}^{k} np_i = n\bar{p} \times k$$

Once the total number of nonconforming units is obtained, apply this next formula to calculate an upper 1 minus α confidence bound for np_{TOTAL}.

$$n\tilde{p}_{TOTAL, 1-\alpha} = \frac{\chi^2_{v,\alpha}}{2} \quad \text{where } v = 2(np_{TOTAL} + 1)$$

This upper confidence bound for np_{TOTAL} is used in turn to determine an upper 1 minus α confidence bound for p'.

$$\hat{p}'_{1-\alpha} = \frac{n\tilde{p}_{TOTAL, 1-\alpha}}{m} = \frac{\chi^2_{v,\alpha}}{2m}$$

In this formula, m is the total number of items checked in the course of the capability study.

$$m = \sum_{i=1}^{k} n_i \text{ for a } p \text{ chart}, \quad m = kn \text{ for an } np \text{ chart}$$

This confidence bound for p' may be used to derive 1 minus α confidence bounds for the equivalent capability measures in the same manner as is done when $\hat{p}'$ is greater than .10.

$$\text{Equivalent } P^{\%}_{PK, 1-\alpha} = \frac{Z(\hat{p}'_{1-\alpha})}{3} = \frac{1}{3} Z\left(\frac{\chi^2_{v,\alpha}}{2m}\right)$$

$$\text{Equivalent } \underset{\sim}{P}\,{}^{\%}_{P,1-\alpha} = \frac{Z\,(\tilde{p}'_{1-\alpha} \div 2)}{3} = \frac{1}{3}Z\left(\frac{\chi^2_{v,\alpha}}{4m}\right)$$

$$\text{Equivalent } \tilde{P}\,{}^{\%}_{R,1-\alpha} = \frac{1}{\text{Equivalent } \underset{\sim}{P}\,{}^{\%}_{P,1-\alpha}} = \frac{3}{Z\,(\chi^2_{v,\alpha}/4m)}$$

Most practitioners rely on only the equivalent $P^{\%}_{PK}$ index for attribute capability studies.

Example 11.47 Sponge Shape

Example 9.8 (p. 539) described a process producing surgical sponges, which were inspected for proper shape. Those with the wrong shape are discarded. Once control is established, a point estimate for equivalent $P^{\%}{}_{PK}$ is desired, as well as a lower .90 confidence bound. From the np chart monitoring this process, $n\bar{p}$ equals 1.20, n is 750, and k is 20.

$$\hat{p}' = \frac{n\bar{p}}{n} = \frac{1.20}{750} = .0016$$

Now a point estimate of equivalent $P^{\%}_{PK}$ may be made.

$$\text{Equivalent } \hat{P}^{\%}_{PK} = \frac{Z\,(\hat{p}')}{3} = \frac{Z\,(.0016)}{3} = \frac{2.95}{3} = .983$$

A lower .90 confidence bound ($\alpha = .10$) for this measure is determined by following these steps. First, the total number of defectives is found.

$$np_{TOTAL} = n\bar{p} \times k = 1.20 \times 20 = 24$$

From this, the number of degrees of freedom for this study are determined.

$$v = 2\,(np_{TOTAL} + 1) = 2\,(24 + 1) = 50$$

The total number of sponges checked, m, during the course of the capability study is equal to kn, which is 20 times 750, or 15,000. With this information, an upper .90 confidence bound for p' is computed as .002106.

$$\tilde{p}'_{.90} = \frac{\chi^2_{v,.10}}{2m} = \frac{\chi^2_{50,.10}}{2\,(15,000)} = \frac{63.16712}{30,000} = .002106$$

This upper confidence bound for p' is used to determine a lower .90 confidence bound for equivalent $P^{\%}_{PK}$.

$$\text{Equivalent } \underset{\sim}{P}^{\%}_{PK,.90} = \frac{Z\,(\tilde{p}'_{.90})}{3} = \frac{Z\,(.002106)}{3} = \frac{2.86}{3} = .953$$

If a lower confidence bound for equivalent $P^{\%}_P$ is preferred, it is calculated as follows:

$$\text{Equivalent } P_{\underset{\sim}{P},.90}^{\%} = \frac{Z(\tilde{p}'_{.90} \div 2)}{3} = \frac{Z(.002106 \div 2)}{3} = \frac{3.07}{3} = 1.023$$

As usual, the upper confidence bound for equivalent $P_R^{\%}$ is the inverse of the lower confidence bound for equivalent $P_P^{\%}$.

$$\text{Equivalent } \tilde{P}_{R,.90}^{\%} = \frac{1}{\text{Equivalent } P_{\underset{\sim}{P},.90}^{\%}} = \frac{1}{1.023} = .978$$

In addition, the upper confidence bound for p' helps compute an upper .90 confidence bound for $ppm_{TOTAL, LT}$.

$$\tilde{p}pm_{TOTAL,LT,.90} = \tilde{p}'_{.90} \times 10^6 = .002106 \times 10^6 = 2106$$

For $\hat{p}' = 0$

In the special case where no nonconforming units are detected during the course of a capability study (meaning $\hat{p}' = 0$), the formula below computes an upper bound for p' at any 1 minus α confidence level. The derivation for this formula is given on page 712.

$$\tilde{p}'_{1-\alpha} = 1 - \alpha^{1/m} \qquad \text{where } m = \sum_{i=1}^{k} n_i \text{ for a } p \text{ chart, while } m = kn \text{ for an } np \text{ chart}$$

This upper confidence bound for p' is then entered into the appropriate confidence bound formula for the desired capability measure.

$$\text{Equivalent } P_{\underset{\sim}{PK},1-\alpha}^{\%} = \frac{Z(\tilde{p}'_{1-\alpha})}{3} = \frac{Z(1-\alpha^{1/m})}{3}$$

$$\tilde{p}pm_{TOTAL,LT,1-\alpha} = \tilde{p}'_{1-\alpha} \times 10^6 = (1 - \alpha^{1/m}) \times 10^6$$

Example 11.48 PC Boards

After several iterations of problem solving to eliminate quality problems at final test for printed circuit boards, 2,642 boards are tested with no rejects.

$$\hat{p}' = \overline{p} = \frac{\sum_{i=1}^{k} np_i}{\sum_{i=1}^{k} n_i} = \frac{0}{2,642} = 0$$

An upper .99 confidence bound ($\alpha = .01$) for this percentage of rejected boards is:

$$\tilde{p}'_{.99} = 1 - \alpha^{1/m} = 1 - .01^{1/2642} = 1 - .998258 = .001742$$

$\tilde{p}'_{.99}$ is then entered into the appropriate confidence bound formula for the desired equivalent capability measure.

$$\text{Equivalent } P^{\%}_{\underset{\sim}{PK}, .99} = \frac{Z(\tilde{p}'_{.99})}{3} = \frac{Z(.001742)}{3} = \frac{2.92}{3} = .973$$

$$\text{Equivalent } P^{\%}_{\underset{\sim}{P}, .99} = \frac{Z(\tilde{p}'_{.99} \div 2)}{3} = \frac{Z(.001742 \div 2)}{3} = \frac{Z(.000871)}{3} = \frac{3.13}{3} = 1.043$$

$$\text{Equivalent } \tilde{P}^{\%}_{R, .99} = \frac{1}{\text{Equivalent } P^{\%}_{\underset{\sim}{P}, .99}} = \frac{1}{1.043} = .958$$

An upper .99 confidence bound can also be constructed for $ppm_{TOTAL, LT}$.

$$\tilde{p}pm_{TOTAL, LT, .99} = \tilde{p}'_{.99} \times 10^6 = .001742 \times 10^6 = 1,742$$

When these confidence bounds are compared to their respective capability goals, a decision about process capability can be made.

Example 11.49 Transducers

Over a one-week period, 2,503 transducers are tested without any rejects, making the estimate of p' equal to zero.

$$\hat{p}' = \frac{0}{2,503} = 0$$

A lower .90 confidence bound ($\alpha = .10$) for the equivalent $P^{\%}_{PK}$ index of this process is calculated by first finding the upper .90 confidence bound for p'.

$$\tilde{p}'_{.90} = 1 - \alpha^{1/m} = 1 - .10^{1/2503} = 1 - .9990804 = 9.195 \times 10^{-4}$$

Inserting this upper bound for p' into this next formula provides a lower .90 confidence bound for equivalent $P^{\%}_{PK}$.

$$\text{Equivalent } P^{\%}_{\underset{\sim}{PK}, .90} = \frac{Z(\tilde{p}'_{.90})}{3} = \frac{Z(9.195 \times 10^{-4})}{3} = \frac{3.12}{3} = 1.04$$

As this lower bound is greater than 1.00, there is .90 confidence this process meets the minimum requirement of performance capability.

If desired, an upper .90 confidence bound for $ppm_{TOTAL, LT}$ can be calculated as well:

$$\tilde{p}pm_{TOTAL, LT, .90} = \tilde{p}'_{.90} \times 10^6 = .0009195 \times 10^6 = 919.5$$

This result also indicates there is .90 confidence this process meets the minimum requirement for long-term performance capability ($ppm_{TOTAL, LT} < 1,350$).

Minimum Required Sample Size

Quite often a capability goal is specified and the capability study must provide sufficient evidence a process generating attribute data meets, or exceeds, this goal. For example, a customer wishes to have .90 confidence the equipment installing rivets into an airplane wing (Figure 11.29) has an equivalent $P^{\%}_{PK}$ of at least 1.33.

As the installation of each rivet is judged either successful or unsuccessful, this is an attribute-data situation involving binomial data. Of course, the smallest number of rivet installations required to demonstrate attainment of this goal would be for the case where no failures are observed. So how many rivets must be successfully installed by this piece of equipment *without a single failure* so that the lower confidence bound for equivalent $P^{\%}_{PK}$ is equal to 1.33? To determine this number, reexamine the confidence bound formula for equivalent $P^{\%}_{PK}$.

Fig. 11.29

$$\text{Equivalent } P^{\%}_{PK,1-\alpha} = \frac{Z(\tilde{p}'_{1-\alpha})}{3}$$

When this lower confidence bound for equivalent $P^{\%}_{PK}$ is equal to (or greater than) the goal for equivalent $P^{\%}_{PK}$, the process has demonstrated attainment of the desired capability goal with 1 minus α confidence.

$$\text{Equivalent } P^{\%}_{PK,1-\alpha} = \frac{Z(\tilde{p}'_{1-\alpha})}{3} = \text{Equivalent } P^{\%}_{PK,GOAL}$$

If the desired capability goal for equivalent $P^{\%}_{PK}$ is given, this last equation may be solved for $\tilde{p}'_{1-\alpha}$. This would be the largest upper 1 minus α confidence bound for p' that could be seen in the capability study, which still shows the capability goal is met with the specified level of confidence.

$$\frac{Z(\tilde{p}'_{1-\alpha})}{3} = \text{Equivalent } P^{\%}_{PK,GOAL}$$

$$Z(\tilde{p}'_{1-\alpha}) = 3 \text{ Equivalent } P^{\%}_{PK,GOAL}$$

$$\tilde{p}'_{1-\alpha} = \Phi(3 \text{ Equivalent } P^{\%}_{PK,GOAL})$$

The Φ operator means to find the area under the normal curve to the right of the value within the parentheses. For example, $\Phi(3.0)$ equals .00135. This operator is the opposite of the operator Z(.00135).

In order to witness this upper 1 minus α confidence bound for p' with a minimum sample size, m items must be tested with *no* failures, where m is determined from this equation (see Derivation 4 on page 712).

$$\tilde{p}'_{1-\alpha} = 1 - \alpha^{1/m}$$

$$1 - \tilde{p}'_{1-\alpha} = \alpha^{1/m}$$

$$\log(1 - \tilde{p}'_{1-\alpha}) = \frac{1}{m} \log \alpha$$

$$m = \frac{\log \alpha}{\log(1 - \tilde{p}'_{1-\alpha})}$$

Previously, this relationship for $\tilde{p}'_{1-\alpha}$ was derived.

$$\tilde{p}'_{1-\alpha} = \Phi(3 \text{ Equivalent } P^{\%}_{PK,GOAL})$$

Substituting this result for $\tilde{p}'_{1-\alpha}$ into the prior equation produces a formula for m based solely on α and the equivalent $P^{\%}_{PK}$ goal.

$$m = \frac{\log \alpha}{\log[1 - \Phi(3 \text{ Equivalent } P^{\%}_{PK,GOAL})]}$$

For this rivet-installation example, α is chosen as .10 and the equivalent $P^{\%}_{PK,GOAL}$ as 1.33. Placing these values into the above equation produces m, the minimum number rivets that must be installed without any problems.

$$m = \frac{\log .10}{\log[1 - \Phi(3 \times 1.33)]} = \frac{\log .10}{\log[1 - \Phi(4)]} = \frac{\log .10}{\log(1 - .00003179)} = \frac{-1.00}{-.0000138} = 72,407$$

At 9,179 rivets per wing, 8 wings ($72,407/9,179 = 7.89$) must be completed by the machine without improperly installing even one, single rivet in order to demonstrate attainment of the required capability goal at the chosen confidence level. If the test is begun, and a failure is detected after only 11,239 rivet installations, the equipment would fail to provide sufficient evidence it meets this capability goal with the desired confidence.

Example 11.50 Web Breakage

A personal care company is evaluating a new type of polystyrene material for labeling shampoo bottles. A machine cuts the labels from a large roll of this material and applies them to plastic bottles passing through the machine at a high rate of speed. If a break occurs in the roll of polystyrene material (called web breakage), considerable downtime is incurred to get the machine back up and running. Although the new material is less expensive than the current material, Dan, the manufacturing manager, will not approve the switch unless the new material can demonstrate a breakage rate of no more than once per 100,000 labels with at least .90 confidence. These test criteria mean the new material will be accepted only if the upper .90 confidence bound for the rate of web breaks, p', does not exceed .00001.

$$\tilde{p}'_{.90} = \frac{1}{100,000} = .00001$$

A quality engineer running the capability study to evaluate this new material must determine the number of labels to be run without a single break to confirm attainment of this objective.

$$m = \frac{\log \alpha}{\log(1 - \tilde{p}'_{1-\alpha})} = \frac{\log .10}{\log(1 - .00001)} = 230,257$$

As approximately 80,000 bottles are run each day, the new material must run for at least three straight days without a web break to demonstrate achievement of the specified goal with .90 confidence.

$$\text{Days required without web break} = \frac{230,257 \text{ labels}}{80,000 \text{ labels per day}} = 2.88 \text{ days}$$

If even one break occurs during this test period, the new material cannot be said to have met manufacturing's stipulated capability goal at the desired level of confidence.

11.30 Statistical Tolerance Limits

When the process average and standard deviation are known, the middle 99.73 percent of the process output (assuming normality) lies between these two limits:

$$\text{Boundaries for middle 99.73 percent} = \mu \pm 3\sigma_{LT}$$

Unfortunately, μ and σ_{LT} are seldom known at the conclusion of a capability study and must be estimated with $\hat{\mu}$ and $\hat{\sigma}_{LT}$, respectively. Due to random sampling variation, these estimates will vary from study to study. In light of this uncertainty, one would conservatively predict the middle 99.73 percent to lie between boundaries somewhat *larger* than those calculated when both process parameters are known. A factor K, which is greater than 3, is used to determine these expanded boundaries, which are commonly called statistical tolerance limits (*STL*).

$$\text{Estimated boundaries for middle 99.73 percent} = \hat{\mu} \pm K\hat{\sigma}_{LT}$$

The value of K depends on both the confidence level desired for predicting these limits and the sample size involved with estimating μ and σ_{LT}. Wald and Wolfowitz described a method for computing approximate K factors for any percentage of the process output at any 1 minus α confidence level (Owen has tabulated several of these), while Keenan and Kim employed their procedure to derive K values specifically for the middle 99.73 percent at .95 confidence. A few of those values, which depend on kn, are displayed in Table 11.10. Notice how they approach 3.00 as the sample size increases.

Table 11.10 *K* factors for 99.73 percent at .95 confidence.

kn	*K*
25	4.02
50	3.64
100	3.42
120	3.38
150	3.35
300	3.23

Both Montgomery and Natrella show how statistical tolerance limits are obtained for processes with unilateral specifications as well as for those with non-normal output distributions. However, the latter technique requires unrealistically large sample sizes.

Example 11.51 Valve O.D. Size

For the automotive exhaust valve example, P_{PK} was estimated as 1.157 (p. 655). With a sample size of 100, a lower .95 confidence bound is found to be 1.011. This result means the valve grinding process just barely meets the minimum performance capability criteria (99.73 percent conforming parts) at .95 confidence.

$$P_{PK,.95} = 1.157 - Z_{.05} \sqrt{\frac{1}{9kn} + \frac{1.157^2}{2(kn-1)}}$$

$$= 1.157 - 1.645 \sqrt{\frac{1}{9(100)} + \frac{1.157^2}{2(100-1)}} = 1.011$$

The statistical tolerance limits at .95 confidence (labeled $STL_{.95}$) for the middle 99.73 percent of the valve sizes are:

$$STL_{.95} \text{ for middle } 99.73\,\% = \hat{\mu} \pm K\,\hat{\sigma}_{LT}$$

$$= 10.0041\,\text{cm} \pm 3.42\,(.01610)\,\text{cm}$$

$$= 9.9490\,\text{cm to } 10.0592\,\text{cm}$$

As the LSL is 9.9400 cm, and the USL is 10.0600 cm, the above interval just fits between these specifications, implying this grinding operation is barely capable for valve O.D. size. Capability team members can state they are .95 confident this process can produce the middle 99.73 percent of its output within tolerance, a conclusion agreeing with that based on the .95 lower confidence bound for P_{PK}.

There are some disadvantages to this method. For example, how is capability improvement for this process tracked over time? Smaller intervals are better, but location of the interval within the tolerance is also important. In addition, because these limits are not dimensionless, and must be interpreted in relationship to only this one tolerance, comparison between other characteristics with different tolerances is impossible.

11.31 Summary

Very infrequently does a capability team possess the resources needed to gather unlimited amounts of data. Therefore, to make valid decisions, the uncertainty inherent in point estimates of process capability must be quantified in some fashion. Assume the P_{PK} index for a particular process is estimated as 1.33. If, based on this study, the true P_{PK} value is known to be somewhere between 1.26 and 1.41, we have a fairly precise idea of how this process will actually perform. On the other hand, if the true P_{PK} index is known only to be somewhere between 1.02 and 1.68, we have very little knowledge about the real process capability. Thus, by itself, the point estimate of 1.33 provides scant information about a process's true character.

This chapter has explained how confidence bounds are computed for every measure of potential capability covered in this book so far, as well as for almost all measures of performance capability. In addition, confidence intervals may be established for the process average and standard deviation.

As an alternative to confidence bounds, Wheeler recommends plotting the desired capability measure for a process on an *IX & MR* chart. Once in control, if the centerline of the *IX* chart is at, or above, the goal for the measure, the process has demonstrated achievement of that goal. However, no confidence level can be stated.

Formulas were provided to calculate a minimum (or maximum for some measures) observed estimate, which demonstrates a specified capability goal has been met at some confidence level. As an aid for planning, procedures were presented for determining the minimum required sample size of a capability study to control the difference between the point estimate and the appropriate confidence bound.

Because confidence bounds (and intervals) provide decision makers with valuable information about process capability, there is a growing movement among customers to prefer suppliers who can demonstrate achievement of capability goals with a certain level of confidence. Although not widespread now, this trend will most likely accelerate in the future as customers become more statistically sophisticated.

A graph for visually comparing the confidence bounds for several different characteristics to their respective goals is presented in Example 12.10 on page 748.

11.32 Derivations

Derivation 1: Minimum Observed $\hat{Z}_{MIN, LT, 1-\alpha}$

The lower confidence bound formula for $Z_{MIN, LT}$ can be rearranged to derive the smallest $Z_{MIN, LT}$ value that still provides 1 minus α confidence the goal for $Z_{MIN, LT}$ is met.

$$Z_{\underset{\sim}{MIN,\ LT,1-\alpha}} \cong \hat{Z}_{MIN,\ LT} - Z_\alpha \sqrt{\frac{1}{kn} + \frac{\hat{Z}^2_{MIN.LT}}{2(kn-1)}}$$

As $1/kn$ is very close to zero, this term is dropped to simplify the formula's derivation.

$$Z_{\underset{\sim}{MIN.LT,1-\alpha}} \cong \hat{Z}_{MIN.LT} - Z_\alpha \sqrt{\frac{\hat{Z}^2_{MIN.LT}}{2(kn-1)}}$$

$$= \hat{Z}_{MIN.LT} - Z_\alpha \frac{\hat{Z}_{MIN.LT}}{\sqrt{2(kn-1)}}$$

$$= \hat{Z}_{MIN.LT} \left(1 - Z_\alpha \frac{1}{\sqrt{2(kn-1)}} \right)$$

$$= \hat{Z}_{MIN.LT} \left(1 - Z_\alpha \sqrt{\frac{1}{2(kn-1)}} \right)$$

The minimum observed estimate of $Z_{MIN, LT}$ at a given level of confidence (minimum observed $\hat{Z}_{MIN.LT,1-\alpha}$) occurs when the lower confidence bound is equal to the goal for $Z_{MIN, LT}$ ($Z_{MIN.LT.\ GOAL}$). Thus, $Z_{MIN.LT,1-\alpha}$ in the above formula is replaced with $Z_{MIN.LT.\ GOAL}$, and $\hat{Z}_{MIN.LT}$ is replaced with the minimum observed $\hat{Z}_{MIN.LT,1-\alpha}$.

$$Z_{MIN.LT.\ GOAL} \cong \text{Minimum Observed } \hat{Z}_{MIN.LT,1-\alpha} \left(1 - Z_\alpha \sqrt{\frac{1}{2(kn-1)}} \right)$$

$$\frac{Z_{MIN.LT.\ GOAL}}{1 - Z_\alpha \sqrt{1 \div 2(kn-1)}} \cong \text{Minimum Observed } \hat{Z}_{MIN.LT,1-\alpha}$$

Or, switching sides, the final formula becomes:

$$\text{Minimum Observed } \hat{Z}_{MIN.LT, 1-\alpha} \cong \frac{Z_{MIN.LT. \, GOAL}}{1 - Z_\alpha \sqrt{1 \div 2(kn - 1)}}$$

Derivation 2: Minimum Sample Size for P_{PM}

An approximate lower confidence bound for P_{PM} is given as follows:

$$P_{\underset{\sim}{PM}, 1-\alpha} \cong \hat{P}_{PM} \sqrt{\frac{\chi^2_{v, 1-\alpha}}{v}} \qquad \text{where } v = \frac{kn(1 + b^2)^2}{1 + 2b^2} \quad \text{and } b = \frac{\hat{\mu} - M}{\hat{\sigma}_{LT}}$$

To solve this relationship for the sample size, the normal approximation to the χ^2 distribution developed by Fisher is applied. This is a reasonably good approximation when v is greater than 30.

$$P_{\underset{\sim}{PM}, 1-\alpha} \cong \hat{P}_{PM} \left(\frac{-Z_\alpha}{\sqrt{2v}} + \sqrt{1 - \frac{1}{2(v - 1)}} \right)$$

To make the derivation easier, the following quantity is assumed to be zero for large v:

$$\frac{1}{2(v - 1)} \cong 0$$

This simplifies the previous equation as is shown here:

$$P_{\underset{\sim}{PM}, 1-\alpha} \cong \hat{P}_{PM} \left(\frac{-Z_\alpha}{\sqrt{2v}} + \sqrt{1 - 0} \right) = \hat{P}_{PM} \left(\frac{-Z_\alpha}{\sqrt{2v}} + 1 \right) = \hat{P}_{PM} \left(1 - \frac{Z_\alpha}{\sqrt{2v}} \right)$$

Now this equation can be solved for v, the number of degrees of freedom.

$$\frac{P_{\underset{\sim}{PM}, 1-\alpha}}{\hat{P}_{PM}} \cong 1 - \frac{Z_\alpha}{\sqrt{2v}} \quad \Rightarrow \quad \frac{Z_\alpha}{\sqrt{2v}} \cong 1 - \frac{P_{\underset{\sim}{PM}, 1-\alpha}}{\hat{P}_{PM}}$$

The quantity on the right-hand side is defined as e_{PPM}.

$$\frac{Z_\alpha}{\sqrt{2v}} \cong e_{PPM} \qquad \text{where } e_{PPM} = 1 - \frac{P_{\underset{\sim}{PM}, 1-\alpha}}{\hat{P}_{PM}}$$

Finally, the equation for v is derived.

$$\frac{Z_\alpha}{e_{PPM}} \cong \sqrt{2v}$$

$$\left(\frac{Z_\alpha}{e_{PPM}} \right)^2 \cong 2v$$

$$\frac{1}{2} \left(\frac{Z_\alpha}{e_{PPM}} \right)^2 \cong v$$

The formula for ν is a function of kn.

$$\frac{kn(1+b^2)^2}{1+2b^2} = \nu$$

$$\frac{kn(1+b^2)^2}{1+2b^2} \cong \frac{1}{2}\left(\frac{Z_\alpha}{e_{PPM}}\right)^2$$

$$kn \cong \frac{1+2b^2}{2(1+b^2)^2}\left(\frac{Z_\alpha}{e_{PPM}}\right)^2$$

As can be seen from the formula given below, b is the number of long-term standard deviations between the estimate of the process average and the middle of the tolerance.

$$b = \frac{\hat{\mu} - M}{\hat{\sigma}_{LT}}$$

For b equal to 0, kn equals:

$$kn \cong \frac{1+2b^2}{2(1+b^2)^2}\left(\frac{Z_\alpha}{e_{PPM}}\right)^2 \cong \frac{1+2(0)^2}{2(1+0^2)^2}\left(\frac{Z_\alpha}{e_{PPM}}\right)^2 \cong .500\left(\frac{Z_\alpha}{e_{PPM}}\right)^2$$

For b equal to 1, kn equals:

$$kn \cong \frac{1+2(1)^2}{2(1+1^2)^2}\left(\frac{Z_\alpha}{e_{PPM}}\right)^2 \cong .375\left(\frac{Z_\alpha}{e_{PPM}}\right)^2$$

For b equal to 2, kn equals:

$$kn \cong \frac{1+2(2)^2}{2(1+2^2)^2}\left(\frac{Z_\alpha}{e_{PPM}}\right)^2 \cong .180\left(\frac{Z_\alpha}{e_{PPM}}\right)^2$$

For b equal to 3, kn equals:

$$kn \cong \frac{1+2(3)^2}{2(1+3^2)^2}\left(\frac{Z_\alpha}{e_{PPM}}\right)^2 \cong .095\left(\frac{Z_\alpha}{e_{PPM}}\right)^2$$

A graph of b versus kn is displayed in Figure 11.30. From here, it can be seen that the largest required minimum sample size occurs when b is equal to 0. Thus, to be conservative when figuring the minimum sample size for P_{PM}, this formula should always be used:

$$kn \cong .500\left(\frac{Z_\alpha}{e_{PPM}}\right)^2$$

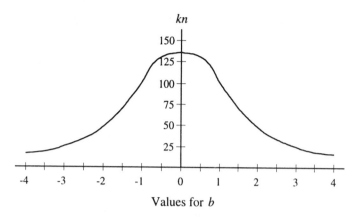

Fig. 11.30 Graph of b versus kn.

Derivation 3: Maximum Observed $\hat{p}'_{1-\alpha}$

This maximum occurs when p'_{GOAL} just equals the upper confidence bound for p'.

$$p'_{GOAL} = \hat{p}'_{1-\alpha} + Z_\alpha \sqrt{\frac{\hat{p}'_{1-\alpha}(1-\hat{p}'_{1-\alpha})}{m}}$$

$$(p'_{GOAL} - \hat{p}'_{1-\alpha})^2 = Z_\alpha^2 \left[\frac{\hat{p}'_{1-\alpha}(1-\hat{p}'_{1-\alpha})}{m} \right]$$

$$p'^2_{GOAL} - 2p'_{GOAL}\hat{p}'_{1-\alpha} + \hat{p}'^2_{1-\alpha} = \frac{Z_\alpha^2}{m}(\hat{p}'_{1-\alpha} - \hat{p}'^2_{1-\alpha})$$

$$p'^2_{GOAL} - 2p'_{GOAL}\hat{p}'_{1-\alpha} = \frac{Z_\alpha^2}{m}\hat{p}'_{1-\alpha} - \frac{Z_\alpha^2}{m}\hat{p}'^2_{1-\alpha} - \hat{p}'^2_{1-\alpha}$$

$$p'^2_{GOAL} - 2p'_{GOAL}\hat{p}'_{1-\alpha} = \frac{Z_\alpha^2}{m}\hat{p}'_{1-\alpha} - \left(1 + \frac{Z_\alpha^2}{m}\right)\hat{p}'^2_{1-\alpha}$$

Moving all terms to the left-hand side of the equation yields the following:

$$p'^2_{GOAL} - 2p'_{GOAL}\hat{p}'_{1-\alpha} - \frac{Z_\alpha^2}{m}\hat{p}'_{1-\alpha} + \left(1 + \frac{Z_\alpha^2}{m}\right)\hat{p}'^2_{1-\alpha} = 0$$

$$p'^2_{GOAL} - \left(2p'_{GOAL} + \frac{Z_\alpha^2}{m}\right)\hat{p}'_{1-\alpha} + \left(1 + \frac{Z_\alpha^2}{m}\right)\hat{p}'^2_{1-\alpha} = 0$$

This equation now has the form of a quadratic equation:

$$a_0 + a_1 x + a_2 x^2 = 0 \qquad \text{where } x = \hat{p}'_{1-\alpha}$$

$$a_0 = p'^2_{GOAL}$$

$$a_1 = -\left(2p'_{GOAL} + \frac{Z_\alpha^2}{m}\right)$$

$$a_2 = 1 + \frac{Z_\alpha^2}{m}$$

It can be solved for $\hat{p}'_{1-\alpha}$ as follows:

$$\hat{p}'_{1-\alpha} = \frac{-a_1 - \sqrt{a_1^2 - 4a_2 a_0}}{2a_2}$$

Substituting for the values of a_0, a_1, and a_2 yields:

$$\hat{p}'_{1-\alpha} = \frac{\left(2p'_{GOAL} + \frac{Z_\alpha^2}{m}\right) - \sqrt{\left(2p'_{GOAL} + \frac{Z_\alpha^2}{m}\right)^2 - 4p'^2_{GOAL}\left(1 + \frac{Z_\alpha^2}{m}\right)}}{2\left(1 + \frac{Z_\alpha^2}{m}\right)}$$

Let a and c equal the following:

$$a = \frac{Z_\alpha^2}{m} \qquad c = 2p'_{GOAL}$$

The formula then becomes:

$$\text{Maximum Observed } \hat{p}'_{1-\alpha} = \frac{(c+a) - \sqrt{(c+a)^2 - c^2(1+a)}}{2(1+a)}$$

$$= \frac{a + c - \sqrt{c^2 + 2ac - a^2 - c^2 - ac^2}}{2(1+a)}$$

$$= \frac{a + c - \sqrt{a^2 + 2ac - ac^2}}{2(1+a)}$$

Derivation 4: Confidence Bound for p' When $\hat{p}'$ Equals 0

Consider the case of determining an upper 1 minus α confidence bound for p' when $\hat{p}'$ equals 0. The associated α risk is the probability the upper confidence bound "misses" (is less than) p'. This could happen if, by chance, zero nonconforming units are produced during the course of the capability, even though p' is quite high. Obviously, the higher p' is, the less the likelihood of finding zero nonconforming units. The greatest risk of observing this outcome, assuming p' is *not* less than the upper confidence bound, occurs when p' is just equal to $\bar{p}'_{1-\alpha}$.

This maximum risk is what is defined as α, the probability of seeing zero nonconforming units and believing p' is less than the upper confidence bound, when in fact, it is not. Letting x represent the number of nonconforming units, and kn the sample size, this probability statement can be made:

$$P(x = 0 \text{ when } kn = m, \text{ given } p' \geq \tilde{p}'_{1-\alpha}) \leq \alpha$$

The full α risk materializes when p' is just equal to $\tilde{p}'_{1-\alpha}$.

$$P(x = 0 \text{ when } kn = m, \text{ given } p' = \tilde{p}'_{1-\alpha}) = \alpha$$

The probability of a process producing a *conforming* unit is 1 minus p', which is also the probability of seeing zero nonconforming units in a sample of 1.

$$P(x = 0 \text{ when } kn = 1, \text{ given } p' = \tilde{p}'_{1-\alpha}) = 1 - p'$$

As the assumption is that p' just equals the upper confidence bound, this last probability statement can be rewritten as:

$$P(x = 0 \text{ when } kn = 1, \text{ given } p' = \tilde{p}'_{1-\alpha}) = 1 - p' = 1 - \tilde{p}'_{1-\alpha}$$

Assuming production units are independent, the probability of having two conforming units produced in a row is the probability the first is conforming times the probability the second is conforming. This result is the square of the above probability, 1 minus $\tilde{p}'_{1-\alpha}$. Therefore, the probability of finding zero nonconforming units in a sample of two is:

$$P(x = 0 \text{ when } kn = 2, \text{ given } p' = \tilde{p}'_{1-\alpha}) = (1 - \tilde{p}'_{1-\alpha}) \times (1 - \tilde{p}'_{1-\alpha}) = (1 - \tilde{p}'_{1-\alpha})^2$$

In general, the probability of producing m consecutive conforming units is the probability $1 - \tilde{p}'_{1-\alpha}$ taken to the mth power. Of course, if all m units are conforming, this is also the probability of having zero nonconforming units in a sample of m.

$$P(x = 0 \text{ when } kn = m, \text{ given } p' = \tilde{p}'_{1-\alpha}) = (1 - \tilde{p}'_{1-\alpha})^m$$

This last probability statement is identical to the second one given in this derivation section, where the above probability was shown to equal α. Thus:

$$(1 - \tilde{p}'_{1-\alpha})^m = \alpha$$

$$1 - \tilde{p}'_{1-\alpha} = \alpha^{1/m}$$

$$\tilde{p}'_{1-\alpha} = 1 - \alpha^{1/m}$$

This then becomes the formula for determining the upper 1 minus α confidence bound for p' when zero nonconforming units are observed in the capability study.

11.33 Exercises ━━━━━━━━━━

11.1 Connecting Force

The force required to join the male and female parts of an electrical connector should be between 1.30 and 1.50 kilograms. After charting on an $\overline{X}$, R chart ($n = 3$, $k = 26$), the process was finally stabilized. In Exercise 5.1 (p. 161), the $6\sigma_{ST}$ spread, C_R, C_P, and C_{ST} were estimated. Compute the appropriate .95 confidence bounds for these short-term potential capability measures. Is there .95 confidence the C_P goal of 1.33 is met?

Find the minimum estimate of C_P that can be observed and still demonstrate the goal of 1.33 has been met with .95 confidence. Assume n will be 3, while k will be 25.

What is the minimum required sample size to have the lower .95 confidence bound for C_P no more than 20 percent away from the point estimate ($e_{CP} = .20$)?

11.2 Bolt Diameter

The diameter of bolts used in the aerospace industry was determined to be a key characteristic. Because of automatic gaging, data recording, and control charting, an $\overline{X}$, S chart ($n = 4$, $k = 29$) was selected to monitor this process. Back in Exercise 5.2 (p. 161), the $6\sigma_{ST}$ spread and C_P were estimated. Compute appropriate .90 confidence bounds for these two capability measures. Calculate .99 confidence bounds for these two and compare results. Why the difference?

11.3 Relay Voltage

In Exercise 5.3 (p. 161), C'_{PU} and C'_{STU} were estimated for a process responsible for the voltage to lift a relay armature. Recalling that an IX & MR chart was employed to help stabilize this process, develop lower .95 confidence bounds for these measures. 31 moving ranges were used to calculate $\overline{MR}$. Remember, n equals 2 for an MR chart.

11.4 Piston Ring Diameter

In Exercise 3.4 (p. 48), the process average and long-term standard deviation for the inside diameter of a piston ring were estimated as 9.02 cm and .02170 cm, respectively. Construct a .99 confidence interval for μ, and an upper .99 confidence bound for σ_{LT} ($n = 6$, $k = 22$).

Estimates of the $6\sigma_{LT}$ spread and P_P were made in Exercise 5.4 (p. 161). Please find appropriate .99 confidence bounds for these two capability measures. Is there .99 confidence the $6\sigma_{LT}$ spread for this process is less than 75 percent of the tolerance (tolerance = .1500 cm)? Does the P_P index for this process meet its goal of 1.33 with .99 confidence?

The following performance capability measures were estimated in Exercise 6.4 (p. 344): $Z_{LSL,\ LT}$, $Z_{USL,\ LT}$, $Z_{MIN,\ LT}$, P_{PL}, P_{PU}, P_{PK}, P_{PM}, P_{PG}, P_{PMK}, $ppm_{LSL,\ LT}$, $ppm_{USL,LT}$, and $ppm_{TOTAL,\ LT}$. Construct appropriate .99 confidence bounds for these measures. Does this process meet the P_{PK} goal of 1.33 with .99 confidence?

What is the minimum estimate of P_{PK} that can be observed in a capability study, yet still offer .99 confidence this process satisfies the goal of 1.33? Assume n will be 6, while k will be 25. In addition to using the formula for directly calculating the minimum estimate, use Table 11.7 (p. 656) to check this answer.

What is the minimum required sample size to have the lower .99 confidence bound for P_{PK} no more than 10 percent away from the point estimate ($e_{PPK} = .10$)?

k, k_L, and k_U were also estimated in Exercise 6.4 (p. 344). Calculate .99 confidence intervals for these factors.

11.5 Shaft Length

A target of minus .1 mm for average shaft length was specified in Exercise 5.5 (p. 162). C^*_P and C^*_{ST} were then estimated using $\hat{\sigma}_{\bar{R}}$. Construct lower .90 confidence bounds for these two indexes ($n = 5$, $k = 25$). Does this process meet a C^*_P goal of 1.67 with .90 confidence?

11.6 Tab Locator Length

The process output for tab locator length could be assumed to be normally distributed, as its cumulative percentage distribution plotted as a reasonably straight line on NOPP in Exercise 7.5 (p. 422). The process average and long-term standard deviation were estimated in this exercise from the 23 measurements for tab length. Construct a .90 confidence interval for μ, and an upper .95 confidence bound for σ_{LT}.

Estimates were also made for P_P, P_{PK}, and $ppm_{TOTAL,\ LT}$. Construct appropriate .95 confidence bounds for these measures. Why are there such large distances between the point estimates and their confidence bounds?

11.7 Connector Length

Electrical connectors are cut to length on a shearing operation. A target of 67.100 was specified in Exercise 5.7 (p. 162) for connector length, and P^*_P was estimated. Assuming n was 4 and k was 25, develop a lower .90 confidence bound for P^*_P.

In Exercise 6.7 (p. 345), P^*_{PK}, P^*_{PM}, and P^*_{PMK} were estimated. Establish lower .90 confidence bounds for these three measures of performance capability.

What is the minimum estimate of P^*_{PM} that can be observed in a future capability study where n is planned to be 4 and k will be 25, and still show this process satisfies a P^*_{PM} goal of 1.40 with .90 confidence?

What is the minimum required sample size to have the lower .99 confidence bound for P^*_{PMK} be no more than 15 percent away from the point estimate?

k^* was also estimated in Exercise 6.7. Calculate a .90 confidence interval for this factor.

11.8 Clevis Width

Cockpit levers controlling the landing gear and wing flaps on commercial aircraft are machined in a small job shop. During one critical operation, a clevis is cut into the connecting end of the lever, with the minimum clevis width being a key characteristic. After control is established via a control chart ($n = 3$, $k = 37$), estimates were made in Exercise 6.12 (p. 346) for P'_{PL}, P_{PL}, and $ppm_{LSL,\ LT}$. Come up with .95 confidence bounds for theses measures. Does this process meet the P_{PK} goal of 1.33 with .95 confidence?

k'_L was also estimated for this process. Calculate a .95 confidence interval for this factor.

11.9 Press Brake

A capability study team decides to employ the sequential *s* test to verify the capability of a new press brake. If the C_P goal for this process is 1.33, and the press brake by itself should contribute no more than 53.6 percent of the total process variation, what should be the first four sets of critical values (LCV$_n$ and UCV$_n$) for checking the test ratios? The team wishes to conduct this test at the .99 confidence level.

The team changes its mind, and decides to go with the traditional control chart approach to verify machine capability. If it is still desired that the press brake meets its capability goal with .99 confidence, how many pieces must be run to have the lower confidence bound for C_P be no more than 15 percent below the point estimate? Given a cycle time of 45 seconds for the press brake, how long will it take to run the necessary pieces?

11.10 Print Voids

Certain "paste areas" (regions covered with solder paste) on circuit board pads assembled with surface mount technology are examined for print voids. If a void is detected, the board is rejected. A subgroup of 40 boards is taken every 4 hours and the number of rejected boards plotted on an *np* chart. When 30 subgroups are collected, the total number of rejected boards is 71. In Exercise 9.1 (p. 555), estimates were made for $ppm_{TOTAL\ LT}$, equivalent P^*_P, and equivalent P^*_{PK}. Construct appropriate .90 confidence bounds for these measures of capability.

11.11 Aerosol Cans

After being filled with hair spray, aerosol cans are checked for improper crimping, bulged dome, leakage, missing date code stamp, dents, and rust. To improve the quality of this filling line, a subgroup of 200 cans is collected every other hour, inspected, and the number of nonconformities found plotted on a *c* chart. The total number of nonconformities found in 25 subgroups is 83. Equivalent P^*_{PK} was estimate in Exercise 9.2 (p. 555). Does this process meet an equivalent P^*_{PK} goal of 1.00 with .95 confidence?

11.12 Camera Shutters

Shutter subassemblies for electronic cameras are built up on an automated assembly line. The average nonconformities per subassembly is calculated for the day by dividing the total nonconformities detected by the number of units checked. The result is plotted on a *u* chart. When the process is finally stabilized, $\bar{u}$ is .000509, while *m* is 13,752. In Exercise 9.4 (p. 555), both $ppm_{TOTAL\ LT}$ and equivalent P^*_{PK} were estimated. Find appropriate .99 confidence bounds for these two measures. Does this process meet a goal of 100 $ppm_{TOTAL\ LT}$ with .99 confidence? Does this process meet an equivalent P^*_{PK} goal of 1.238 with .99 confidence?

11.13 Pouch Seals

A C-node scanning acoustic microscope examines the seal of a small plastic pouch used for storing disposable contact lenses. If seal integrity has been compromised for any reason, the pouch is rejected. Because of large differences in daily production volumes, the percentage of rejected pouches is plotted on a *p* chart. When the process is finally stabilized, $\bar{p}$ is calculated to be .000321 (*m* = 12,461). $ppm_{TOTAL\ LT}$ and equivalent P^*_{PK} were estimated in Exercise 9.7 (p. 556). Construct .90 confidence bounds for both measures. Does the sealing process meet the minimum requirement for long-term performance capability with .90 confidence?

11.14 Color Matching

While assembling automobiles, a company uses a portable, hand-held spectrophotometer to rapidly measure the color of plastic interior trim pieces. This measuring device makes a quick pass/fail color evaluation, with the results recorded on an np chart ($n = 50$, $k = 25$) that has a centerline ($n\bar{p}$) of 1.976 when the process is in control. In Exercise 9.10 (p. 557), p' was estimated for this process. Establish an upper .80 confidence bound for this point estimate. Use this to compute a lower .80 confidence bound for equivalent $P^{\%}_p$. Is there at least .80 confidence $P^{\%}_p$ is greater than 1.00?

11.15 Copper Thickness

An automated production system receives stacks of printed circuit boards from a press room, then unstacks and loads the untrimmed sheets onto a conveyor. While loading the laminated sheets, the equipment utilizes the eddy current principle to check the thickness of a copper layer. A "go/no-go" decision is made for each sheet with the results for each shift summarized on a separate p chart. When in control, $\bar{p}$ is calculated as .000098 for the first shift ($m = 20,408$), and .000211 for the second ($m = 19,048$). In Exercise 9.11 (p. 557), a point estimate of equivalent $P^{\%}_{PK}$ was made for each shift. From this information, estimate appropriate lower .95 confidence bounds.

11.16 Pace Makers

Prior to being welded together during final assembly in a titanium casing, the internal circuitry of heart pace makers is X-rayed for flaws such as: cracked battery, missing components, cracked PC board, broken wire, foreign particles. Assume zero nonconformities are discovered in 200 subgroups of size 25 each. $ppm_{TOTAL, LT}$ and equivalent $P^{\%}_{PK}$ for this process were estimated on page 557. Construct .90 confidence bounds for each of these two measures.

11.17 Graph Paper

Sheets of normal probability graph paper are bound to a piece of cardboard backing. There should be a minimum of 50 sheets per pad. To check the binding process, the number of sheets on each of 11,111 pads are counted. Although a few had more than 50 sheets, none had less than 50. With zero nonconforming pads, p' for this process is estimated as 0. From this information, compute an upper .90 confidence bound for p', then use this value to construct a lower .90 confidence bound for equivalent $P^{\%}_{PK}$.

Does this process meet a equivalent $P^{\%}_{PK}$ goal of 1.20 with .90 confidence? If not, how many additional pads must be checked without finding one with less than 50 sheets to meet this capability goal?

11.34 References

Barnes, Ron and Pierce, R.C., "Inference and Capability Indexes," *Proceedings of the Section on Quality and Productivity*, American Statistical Association, 1997, pp.101-102

Bissell, A.F., "How Reliable is Your Capability Index," 1990, *Applied Statistics*, 39, pp. 331-340

Bissell, Derek, *Statistical Methods for SPC and TQM*, 1994, Chapman & Hall, London, England, p. 306

Boeing Aircraft Company, *Advanced Quality Systems for Boeing Suppliers: D1-9000*, The Boeing Co., Seattle, WA, p. 2-27

Boyles, Russel A., "The Taguchi Capability Index," *Journal of Quality Technology*, Vol. 23, No. 1, January 1991, pp. 22-23

Chan, Lai K., Cheng, Smiley W., and Spiring, Fred A., "A Graphical Technique for Process Capability," *42nd Annual Quality Congress Transactions*, Dallas, TX, May 1988, pp. 268-275

Cheng, R.C.H., and Iles, T.C., "Confidence Bands for Cumulative Distribution Functions of Continuous Random Variables," *Technometrics*, Vol. 25, 1983, pp. 77-86

Cheng, Smiley W., "Is the Process Capable?," *Quality Engineering*, 4, 4, 1992, pp. 563-576

Cheng, Smiley W., "Practical Implementation of the Process Capability Indices," *Quality Engineering*, Vol. 7, No. 2, 1994-1995, pp. 239-259

Chou, Youn-Min, Owen, D.B., and Borrego, Salvador A., "Lower Confidence Limits on Process Capability Indices," *Journal of Quality Technology*, 22, 3, July 1990, p. 224

Duncan, Acheson J., "The Use of Ranges in Comparing Variabilities," *Industrial Quality Control*, Vol. 11, No. 5, February, 1955

Duncan, Acheson J., "Design and Operation of a Double-Limit Variables Sampling Plan," *Journal of the American Statistical Association*, Vol. 53, 1958, p. 548

Duncan, Acheson J., *Quality Control and Industrial Statistics*, 5th edition, 1986, Richard D. Irwin, Inc., Homewood, IL, p. 141

Fisher, R.A., "On the Interpretation of χ^2 from Contingency Tables and Calculation of P," 1922, *Journal of the Royal Statistical Society, Series A*, 85, pp. 87-94

Franklin, LeRoy A., "Improved Sample Size Determination Formula for Lower Confidence Limits for Cp and Cpm," *Proceedings of the Section on Quality and Productivity*, American Statistical Association, 1995, pp. 55-58

Franklin, LeRoy A., "Determining Sample Sizes for Lower Confidence Limits for Cp and Cpk," *Proceedings of the Section on Quality and Productivity*, American Statistical Association, 1994, pp. 23-27

Franklin, LeRoy A., Cooley, Belva J., and Elrod, Gary, "Comparing the Importance of

Variation and Mean of a Distribution," *Quality Progress*, 32, 10, Oct. 1999, pp. 90-94

Franklin, LeRoy A., and Wasserman, Gary S., "Bootstrap Lower Confidence Limit Estimates for Capability Indices," *Proceedings of the Section on Quality and Productivity*, American Statistical Association, 1992, pp. 169-179

Greenwich, Michael, and Chen, Hwa-Nien, "Confidence Intervals for *Cpk*," *40th Annual Minnesota Quality Conference Proceedings*, March 1993, pp. 2A-1 to 2A-8

Hahn, Gerald J., and Meeker, William Q., *Statistical Intervals: A Guide for Practitioners*, 1991, John Wiley & Sons, New York, NY

Kahle, W., Collani, E.V., Franz, J., and Jensen, U., *Advances in Stochastic Models for Reliability, Quality and Safety*, 1998, Birkhauser, Boston, MA

Keenan, Thomas M., and Kim, John S., "Normal Tolerance Limits: An Alternative to Cpk," *40th Annual Minnesota Quality Conference Transactions*, Minneapolis, MN, March 1993, pp. 2B-1 to 2B-8

Kotz, Samuel, and Johnson, Norman L., *Process Capability Indices*, 1993, Chapman & Hall, New York, NY, pp. 117-121

Kushler, Robert H., and Hurley, Paul, "Confidence Bounds for Capability Indices," *Journal of Quality Technology*, Vol. 24, No. 4, October 1992, pp. 192-194

Lehrman, K. H., "Process Capability Analyses and Index Interval Estimates: Quantitative Characteristics (Variables)," *Quality Engineering*, Vol. 4, No. 1, 1991-1992, pp. 93-130

Levinson, William A., "Exact Confidence Limits for Process Capabilities," *Quality Engineering*, Vol. 9, No. 3, 1997, pp. 521-528

Lewis, Sydney S., "Process Capability Estimates from Small Samples," *Quality Engineering*, Vol. 3, No. 3, 1991, p. 387

Li, H., Owen, D.B., and Borrego, A.S.A., "Lower Confidence Limits on Process Capability Indices Based on the Range," *Communications in Statistics - Simulation Comp.*, Vol. 19, 1990, pp. 1-24

Marcucci, Mark O., and Beazley, Charles C., "Capability Indices: Process Performance Measures," *42nd ASQC Annual Quality Congress Trans.*, Dallas, TX, May 1988, p. 518

McNeese, William H., and Klein, Robert A., *Statistical Methods for the Process Industries*, 1991, ASQC Quality Press, Milwaukee, WI, p. 444

Montgomery, Douglas C., *Introduction to Statistical Quality Control*, 2nd ed., 1991, John Wiley & Sons, New York, NY, pp. 403-405

Nagata, Y., Interval Estimation for the Process Capability Indices," *Journal of Japanese Society for Quality Control*, Vol. 21, 1991, pp. 109-114

Natrella, Mary Gibbons, *Experimental Statistics: National Bureau of Standards Handbook 91*, 1963, U.S. Government Printing Office, Washington, DC, pp. 2-13 to 2-15

Nelson, Lloyd, "Upper Confidence Limits on Average Numbers of Occurrences," *Journal of Quality Technology*, Vol. 21, No. 1, January 1989, pp. 71-72

Owen, D. B., *Handbook of Statistical Tables*, 1962, Addison-Wesley, Reading MA

Patnaik, P.B., "The Use of Mean Range as an Estimator of Variance in Statistical Tests," *Biometrika*, Vol. 37, 1950, pp. 78-87

Pearn, W.L. and Chen, K.S., "A Practical Implementation of the Process Capability Index C_{PK}," *Quality Engineering*, Vol. 9, No. 4, 1997, pp. 721-737

Pearn, W.L., Kotz, Samuel, and Johnson, Norman L., "Distributional and Inferential Properties of Process Capability Indices," *Journal of Quality Technology*, Vol. 24, No. 4, October 1992, pp. 216-231

Price, Barbara, and Price, Kelly, "A Methodology to Estimate the Sampling Variability of the Capability Index, Cpk," *Quality Engineering*, Vol. 5, No. 4, 1993, pp. 527-544

Sarkar, Ashok and Pal, Surajit, "Estimation of Process Capability Index for Concentricity," *Quality Engineering*, Vol. 9, No. 4, 1997, pp. 665-671

Strongrich, Andrew L., Herbert, Gerald E., and Jacoby, Thomas J., "Simple Statistical Methods," *Statistical Process Control, SP-547*, Society of Automotive Engineers, 1983, Warrendale, PA, pp. 31-38

Wald, A. and Wolfowitz, J., "Tolerance Limits for a Normal Distribution," *Annals of Mathematical Statistics*, Vol. 17, 1946, pp. 208-215

Wang, C. Ming and Lam, C. Teresa, "Confidence Limits for Proportion of Conformance," *Journal of Quality Technology*, Vol. 28, No. 4, October 1996, pp. 439-445

Wharton, Robert and Czachor, James, "An Upper Bound for the Proportion of Process Output Not Meeting Specifications," *Proceedings of the Section on Quality and Productivity*, American Statistical Association, 1994, pp. 264-266

Wharton, Robert and Czachor, James, "A Comparative Study of Upper Bound Calculations for the Proportion of Output Not Meeting Specifications," *Proceedings of the Section on Quality and Productivity*, American Statistical Association, 1996, pp. 135-137

Wheeler, Donald J., *Advanced Topics in Statistical Process Control*, 1995, SPC Press, Inc., Knoxville, TN, p. 193

Zhang, N.F., Stenback, G.A., and Wardrop, D.M., "Interval Estimation of Process Capability Index Cpk," *Communications in Statistics - Theoretical Methods*, Vol. 19, 1990, pp. 4455-4470

Zimmer, L.S. and Hubele, N.F., "Quantiles of the Sampling Distribution of C_{PM}," *Quality Engineering*, Vol. 10, No. 2, 1997-1998, pp. 309-329

Chapter 12

Combining Capability Measures

Customers want everything perfect.

There are certain occasions when it may be desirable to measure the capability of a combination of several process outputs. This chapter develops three unique measures of capability to address these special situations. The first measure is referred to as "Average" capability, the second as "Product" capability, and the third as "Normalized" capability.

12.1 Average Capability for Multiple Process Streams

Suppose the process selected for a capability study is a four-cavity die producing molded parts. Each cavity (in general, process stream) demonstrates good control for a given characteristic, but the output parameters vary from cavity to cavity, as is illustrated in Figure 12.1. After parts are produced, they fall into one common container, which is then shipped to a customer. How can capability measures for these individual cavities be combined into one overall, "average" measure of what a customer purchasing the mixed output of all four can expect to receive?

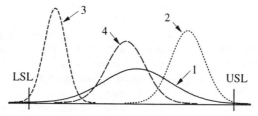

Fig. 12.1 The output distributions for each of the four cavities.

The average capability measure presented in this section is applied where all process streams are producing the *same* characteristic (with the same tolerance) and the capability of the combined output is desired. Separate process streams are created when a process's output splits into several presumably identical paths during an operation, *e.g.*, different molds, spindles, stations, fixtures, or pallets. Of course, the first prerequisite is that all process streams are in a good state of statistical control. They may have different averages and/or standard deviations, but every stream must have a stable, predictable output.

In this example of a plastic injection molding machine, there are four cavities, each producing supposedly identical parts. The important quality characteristic is the I.D. (inner diameter) size of a molded hole. Once control is established for all four cavities, what should the customer receiving the output of this process, which is a mixture of the output from all four cavities, expect in terms of parts meeting the specification requirements for hole size?

Unfortunately, the C_{PK} values for each stream cannot simply be averaged, as C_{PK} values are not linear in their relationship with the percentage of nonconforming product. For example, imagine an operation having two process streams, the first has a C_{PK} of 2.00, and the second has a C_{PK} of 0. The average of these two C_{PK} indexes is 1.00, indicating the combined output of the two streams meets the minimum requirement of performance capability.

$$\text{"Average" } C_{PK} = \frac{2.00 + .00}{2} = 1.00$$

Meeting this minimum requirement implies there are at least 99.73 percent conforming parts produced by this process, assuming a bilateral specification. In this particular situation, the first process stream is producing essentially 100 percent conforming parts, whereas the second is making just 50 percent conforming parts. If both streams produce an equal number of parts, there will be an overall average of 75 percent conforming parts, which is far short of 99.73 percent. Assigning an "average" C_{PK} of 1.00 to this process would be quite misleading.

Windham has developed the concept of an "acceptable quality distribution" (AQD), which he hopes will provide a monetary incentive for encouraging part suppliers to center a process's output at the target average and then continuously work to reduce process variation. However, one of the steps for determining the quality distribution involves averaging C_{PK} values for several part characteristics in order to derive an overall, weighted average C_{PK} for that part. As just seen above, this step could produce very deceptive results.

Some quality practitioners estimate an overall average and standard deviation by combining the measurements from all process streams. The fallacy of this approach was already discussed in Section 3.3 on page 42 (Zhang).

Average C_{PK} and P_{PK}

To correctly measure the "Average C_{PK}" of these four cavities, first calculate the percentage (or *ppm*) of parts out of specification for each cavity. Suppose this generates the results displayed in Table 12.1. The percentages of holes below the LSL for each cavity sum to a total of 1.6 percent. This total is then divided by four (in general, the number of process streams) to come up with an average of .4 percent below the LSL for the *entire* process output. This weighted average, where each cavity contributes 1/4 the total percentage outside the LSL, is the percentage of nonconforming parts a customer receives and, therefore, is an indication of the process's ability to mold hole sizes above the LSL.

$$\hat{p}'_{LSL,ST} = \frac{.1\% + .0\% + 1.5\% + .0\%}{4} = .4\% = .004 = 4 \times 10^{-3}$$

Table 12.1 Averaging the percentage nonconforming from the four cavities.

Cavity #	% Below LSL	% Above USL
1	.1	.4
2	.0	.8
3	1.5	.0
4	.0	.0
% "out" for all 4 Cavities	1.6	1.2
Avg. % "out" for Process	.4	.3

An estimate of the $Z_{LSL,ST}$ value for the average percentage below the LSL ($p'_{LSL,ST}$) is determined with the Z table given in Appendix Table III. Find 4×10^{-3} in the body of this table, then read off the corresponding Z value of 2.65 in the left-hand column, as was done in Chapter 9 for attribute capability studies. The short-term version of the Z value is appropriate in this example because the percentage of nonconforming holes for each cavity is estimated from $\hat{\sigma}_{ST}$.

$$\hat{Z}_{LSL,ST} = Z(4 \times 10^{-3}) = 2.65$$

This same procedure is repeated to estimate $Z_{USL,ST}$ for the average percentage of holes above the USL ($p'_{USL,ST}$). In this case, the Z value for .3 percent (or 3×10^{-3}) is 2.75.

$$\hat{Z}_{USL,ST} = Z(3 \times 10^{-3}) = 2.75$$

As before, $Z_{MIN,ST}$ is the smaller of these two estimated Z values.

$$\hat{Z}_{MIN,ST} = \text{Minimum}\,(\hat{Z}_{LSL,ST}, \hat{Z}_{USL,ST}) = \text{Minimum}\,(2.65, 2.75) = 2.65$$

Average C_{PK} is defined as $Z_{MIN,ST}$ divided by 3 (Bothe, 1993). If the out-of-specification percentage for each process stream is derived from σ_{LT}, the resulting combined measure is referred to as Average P_{PK}.

$$\text{Average } \hat{C}_{PK} = \frac{\hat{Z}_{MIN,ST}}{3} = \frac{2.65}{3} = .883$$

Thus, an estimate of the Average C_{PK} for hole I.D. size produced by this molding process is .883. As this is quite a bit less than 1, the Average C_{PK} reveals the four cavities producing this combined output are *not* capable of consistently providing the customer at least 99.73 percent parts with conforming hole sizes. In fact, with .4 percent of hole sizes below the LSL and .3 percent above the USL, only 99.3 percent are within tolerance. 100 percent inspection must be implemented to assure this output meets customer expectations. From examining Table 12.1, improvement efforts should first concentrate on moving the average of cavity 3 up to the middle of the tolerance. Then the average of cavity 2 needs to be lowered so it is also centered at M. If the results of a new combined capability study indicate the process is still not capable, work must be done to reduce the variation of cavity 1.

Notice that this combined measure of capability does not require the output from all process streams to be centered at the same value, nor have identical process spreads (which is good since in most cases, they don't). Statistical tests for determining if the process parameters for separate streams are significantly different are provided by Larsson, as well as Snedecor and Cochran.

Formulas for determining Average C_{PK} in general are given next, where j is the number of process streams and $p'_{LSL,ST,i}$ ($p'_{USL,ST,i}$) is the percentage of product below (above) the LSL (USL) for stream i.

$$p'_{LSL,ST} = \frac{\sum\limits_{i=1}^{j} p'_{LSL,ST,i}}{j} \qquad p'_{USL,ST} = \frac{\sum\limits_{i=1}^{j} p'_{USL,ST,i}}{j}$$

$$p'_{MAX,ST} = \text{Maximum}\,(p'_{LSL,ST}, p'_{USL,ST})$$

Corresponding Z values for the LSL and USL are determined from these percentages.

$$Z_{LSL,ST} = Z(p'_{LSL,ST}) \qquad Z_{USL,ST} = Z(p'_{USL,ST})$$

As before, the smaller of these two becomes the $Z_{MIN,ST}$ value.

$$Z_{MIN,ST} = \text{Minimum}(Z_{LSL,ST}, Z_{USL,ST}) = Z(p'_{MAX,ST})$$

Average C_{PK} is then defined as:

$$\text{Average } C_{PK} = \frac{Z_{MIN,ST}}{3} = \frac{Z(p'_{MAX,ST})}{3}$$

For estimating Average P_{PK}, the formulas become:

$$p'_{LSL,LT} = \frac{\sum_{i=1}^{j} p'_{LSL,LT,i}}{j} \qquad p'_{USL,LT} = \frac{\sum_{i=1}^{j} p'_{USL,LT,i}}{j}$$

$$p'_{MAX,LT} = \text{Maximum}(p'_{LSL,LT}, p'_{USL,LT})$$

Corresponding Z values are found as before.

$$Z_{LSL,LT} = Z(p'_{LSL,LT}) \qquad Z_{USL,LT} = Z(p'_{USL,LT})$$

$$Z_{MIN,LT} = \text{Minimum}(Z_{LSL,LT}, Z_{USL,LT}) = Z(p'_{MAX,LT})$$

Average P_{PK} is then calculated as follows:

$$\text{Average } P_{PK} = \frac{Z_{MIN,LT}}{3} = \frac{Z(p'_{MAX,LT})}{3}$$

Example 12.1 Five-Spindle Boring Machine

Results of the output for hole size from each of the five spindles on a multispindle boring machine are listed in Table 12.2. The stability of all five streams is demonstrated on a Group control chart (Boyd; Bothe, 1985). As these *ppm* values were estimated with $\hat{\sigma}_{LT}$, calculating an estimate of the Average P_{PK} for this process is appropriate. The goal established for Average P_{PK} is 1.33.

The percentage of parts outside each specification limit are given in *ppm*. These *ppm* values must first be converted to percentages before looking them up in Appendix Table III. After this, an estimate of $Z_{LSL,LT}$ is derived for the average percentage below the LSL.

$$\hat{p}'_{LSL,LT} = \hat{p}pm_{LSL,LT} \times 10^{-6} = 59.7 \times 10^{-6} = 5.97 \times 10^{-5}$$

$$\hat{Z}_{LSL,LT} = Z(\hat{p}'_{LSL,LT}) = Z(5.97 \times 10^{-5}) = 3.85$$

Next, an estimate of $Z_{USL,LT}$ is found for the average percentage above the USL.

Table 12.2 *ppm* results for a five-spindle boring machine.

Spindle #	# Pieces	ppm	
		LSL	USL
1	2,000	.4	124.0
2	2,000	.1	308.2
3	2,000	4.6	132.8
4	2,000	293.1	.7
5	2,000	.3	257.3
Total	10,000	298.5	823.0
Avg. *ppm*		59.7	164.6

$$\hat{p}'_{USL,LT} = \hat{p}\,pm_{USL,LT} \times 10^{-6} = 164.6 \times 10^{-6} = 1.646 \times 10^{-4}$$

$$\hat{Z}_{USL,LT} = Z(\hat{p}'_{USL,LT}) = Z(1.646 \times 10^{-4}) = 3.59$$

With these two estimates, $Z_{MIN,\ LT}$ and Average P_{PK} can be predicted.

$$\hat{Z}_{MIN.LT} = \text{Minimum}\ (3.85, 3.59) = 3.59 \quad \Rightarrow \quad \text{Average } \hat{P}_{PK} = \frac{\hat{Z}_{MIN.LT}}{3} = \frac{3.59}{3} = 1.20$$

Because the estimate of Average P_{PK} is greater than 1.00, the combined output of this machining operation exceeds the minimum prerequisite for long-term performance capability of having at least 99.73 percent conforming hole sizes. With $ppm_{LSL,\ LT}$ equal to 59.7 and $ppm_{USL,\ LT}$ equal to 164.6, there is a total of 99.97757 percent conforming parts. However, it does not meet the goal of 1.33, which for a bilateral specification is a minimum of 99.9937 percent conforming parts. 100 percent inspection must be continued until this process meets the customer's quality requirement. From a look at Table 12.2, initial improvement efforts should focus on centering the output on spindles 2, 4, and 5 closer to the middle of the tolerance. Then work must begin to reduce the variation of all spindles.

Note that there is no stipulation of this index requiring the process output for each process stream to have a normal distribution. If a stream has a non-normal distribution, use NOPP to estimate the percentage of parts outside each specification.

As a short-cut, Average P_{PK} can also be estimated directly from $\hat{p}'_{MAX,\ LT}$.

$$\text{Average } P_{PK} = \frac{Z(p'_{MAX,\ LT})}{3} = \frac{Z(1.646 \times 10^{-4})}{3} = \frac{3.59}{3} = 1.20$$

This alternative formula is less work because the Z value for only one of the two average percentages (the larger) needs to be found in Appendix Table III.

Example 12.2 Uneven Number of Pieces from Each Stream

A three-spindle boring machine has demonstrated fairly good control utilizing a method developed by Mortell and Runger. However, due to maintenance problems, the three spindles do not produce an identical number of parts during the course of a combined capability study. This difference in contribution to the total output must be taken into consideration when calculating the Average P_{PK} index (or Average C_{PK} if σ_{ST} is involved) by weighting the *ppm* from each spindle by the number of parts produced on that spindle.

Table 12.3 lists the number of pieces produced for each of the three spindles. These amounts become the weights applied to each spindle to determine the weighted average of all three, as demonstrated by the following calculations.

Table 12.3 *ppm* results for the three spindles.

Spindle #	# Pieces	ppm	
		LSL	USL
1	10,000	52	.1
2	6,000	211	1
3	12,000	2	1714
Total	28,000		

The weighted average percentage below the LSL and its associated Z value are determined first, as is shown here:

$$\text{Weighted Avg. } ppm \text{ for LSL} = \frac{10,000(52) + 6,000(211) + 12,000(2)}{10,000 + 6,000 + 12,000} = \frac{1,810,000}{28,000} = 64.64$$

$$\hat{p}'_{LSL,LT} = \hat{p}pm_{LSL,LT} \times 10^{-6} = 64.64 \times 10^{-6} = 6.464 \times 10^{-5}$$

$$\hat{Z}_{LSL,LT} = Z(\hat{p}'_{LSL,LT}) = Z(6.464 \times 10^{-5}) = 3.83$$

This procedure is then repeated for the USL.

$$\text{Weighted Avg. } ppm \text{ for USL} = \frac{10,000(.1) + 6,000(1) + 12,000(1714)}{10,000 + 6,000 + 12,000} = \frac{20,575,000}{28,000} = 734.8$$

$$\hat{p}'_{USL,LT} = \hat{p}pm_{USL,LT} \times 10^{-6} = 734.8 \times 10^{-6} = 7.348 \times 10^{-4}$$

$$\hat{Z}_{USL,LT} = Z(\hat{p}'_{USL,LT}) = Z(7.348 \times 10^{-4}) = 3.18$$

As before, Average P_{PK} is found by dividing $Z_{MIN.LT}$ by 3.

$$\text{Average } \hat{P}_{PK} = \frac{\hat{Z}_{MIN.LT}}{3} = \frac{1}{3}\text{Minimum}(3.83, 3.18) = \frac{3.18}{3} = 1.06$$

With 65 ppm below the LSL and 735 above the USL, $ppm_{TOTAL, LT}$ is estimated as 800, which translates into .0800 percent. Therefore, this process has a 99.9200 percent conformance rate (100% minus .0800%). Thus, the combined capability of these three process streams exceeds the minimum requirement for long-term performance capability, which is a 99.7300 percent conformance rate for a bilateral specification. However, the goal for the Average P_{PK} of this process is 1.25 (a conformance rate of 99.9823 percent). Efforts to improve the capability of spindle 3 should begin immediately.

Example 12.3 Pizza Weight

Dough for frozen pizzas is arranged in six rows on a conveyor. The conveyor then moves under several dispensing stations which deposit various vegetables, meats, sauces, and cheeses on the dough. At the end of this line, the completed pizzas are checked for minimum weight, with a LSL of .8 kilograms (28 ounces). The capability goal for this line is to have an Average C_{PK} index of at least 1.33. After achieving stability for each, conformance rates for the six rows are estimated and summarized in Table 12.4.

Table 12.4 Conformance rates for each row of dough.

Row	$ppm_{LSL, ST}$
1	16
2	3
3	22
4	12
5	5
6	14
Total	72
Average	12

The combined capability of this line to produce pizzas satisfying the minimum weight requirement is computed by first determining the average percentage below the LSL.

$$\hat{p}'_{LSL,ST} = \hat{p}pm_{LSL,ST} \times 10^{-6} = 12 \times 10^{-6} = 1.2 \times 10^{-5}$$

As weight has a unilateral specification, the Z value for the LSL is also $Z_{MIN, ST}$.

$$\hat{Z}_{MIN.ST} = \hat{Z}_{LSL,ST} = Z(\hat{p}'_{LSL,ST}) = Z(1.2 \times 10^{-5}) = 4.23$$

Average C_{PK} is just $Z_{MIN, ST}$ divided by 3.

$$\text{Average } \hat{C}_{PK} = \frac{\hat{Z}_{MIN.ST}}{3} = \frac{4.23}{3} = 1.41$$

Because the estimated Average C_{PK} index surpasses the goal of 1.33, this line demonstrates very good capability for producing pizzas above the minimum weight requirement.

Due to this being a unilateral specification, the conformance rate for fulfilling the minimum performance capability requirement is 99.865 percent, while it is 99.9968 for achieving the stated capability goal of 1.33. Based on the capability study results presented above, the estimated percentage of pizzas meeting the specified weight is 99.9988 (100 minus .0012).

Example 12.4 Flash on Die-cast Parts

A die-casting operation has eight cavities producing the same part number. Occasionally flash is detected on a critical sealing surface and causes that part to be rejected. Each cavity is charted on a Group p chart (Bothe, 1991) and is in a good state of control, although the various cavities have different $\overline{p}$ values, which are listed in Table 12.5.

Table 12.5   Percentage of nonconforming parts by cavity.

Cavity #	$\overline{p}$
1	.00012
2	.00073
3	.00046
4	.00019
5	.00011
6	.00039
7	.00024
8	.00027
Total	.00251
Average	.00031375

Average P_{PK} is estimated from the overall average percentage of parts with flash, as is done below. Note that the choice of Average P_{PK} is appropriate because attribute charts generate estimates of long-term process variation.

$$\hat{p}'_{LT} = .00031375 = 3.1375 \times 10^{-4}$$

$$\text{Average } \hat{P}_{PK} = \frac{Z(\hat{p}'_{LT})}{3} = \frac{Z(3.1375 \times 10^{-4})}{3} = \frac{3.42}{3} = 1.14$$

An Average P_{PK} index greater than 1.0 indicates the combined output of all eight cavities exceeds the minimum requirement for performance capability as defined for processes generating attribute data.

12.2 Product Capability for Multiple Product Characteristics

Most texts on capability treat products as if they have only one critical characteristic. However, most products have many features that are important to the customer. For example, a fairly complex part is produced on a flexible machining center. Three different characteristics of this part are done while on this one machine (see Figure 12.2). What is the capability of this machining process to manufacture a "good" product, that is, one where all three characteristics are within their respective print specifications?

On occasion, a single measure is needed to summarize the capability of several *different* characteristics on the same part. For the part shown in Figure 12.2, all three characteristics must be within their respective print specifications if the product is to be acceptable, whereas *only one* has to be out of its specification to make the product unacceptable to the customer.

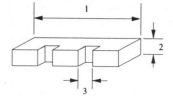

Fig. 12.2 Three features are produced during one machining operation.

Begin this second type of combined capability study by calculating the capability of the first characteristic (length). Assume the process output is normally distributed and this dimension has a bilateral specification. Determine $Z_{LSL, LT}$ and the probability this characteristic is below the LSL by using the Z table (review p. 190). Then do the same for the USL.

For this example, $Z_{LSL, LT}$ equals 3, so the probability of a part having this characteristic below the LSL is .00135, as displayed in Table 12.6. $Z_{USL, LT}$ turns out to be 4, so the probability this characteristic is above the USL is .000032.

Table 12.6 Probability characteristic number one is within specification.

Char. No.	$Z_{LSL, LT}$	Prob. Char. Below LSL	$Z_{USL, LT}$	Prob. Char. Above USL	Prob. Out of Spec.	Probability Within
1	3	.001350	4	.000032	.001382	**.998618**

The second to last column of Table 12.6 ("Prob. Out of Spec.") is the sum of these two probabilities (.001350 + .000032 = .001382), which is the probability this first feature is outside either the LSL or USL. The last column ("Prob. Within") is the probability part length is within specification, which is 1 minus the probability it is "out" (1 - .001382 = .998618).

The second characteristic (thickness) has a unilateral specification (only an USL) and also has a non-normal distribution (one that is bounded). Calculate the probability this second characteristic exceeds its USL by using normal probability paper, as was explained in Section 8.14 on page 495 (let it turn out to be .000032). The probability of a part having its thickness within specification can be figured the same way as done for the first characteristic (1 minus probability it's "out"), as is illustrated below in Table 12.7.

Table 12.7 Probability characteristic number two is within specification.

Char. No.	Prob. Char. Is Below LSL	Prob. Char. Is Above USL	Prob. Char. Is Out of Spec.	Probability Within
1	.001350	.000032	.001382	.998618
2	-	.000032	.000032	**.999968**

The third characteristic (slot width) is measured with a go/no-go gage, thus generating attribute-type data. The in-control p chart monitoring slot width has $\bar{p}$ equal to .008598, which is an estimate of the probability this characteristic is nonconforming (see Table 12.8). Therefore, the probability slot width is within specification is .991402 (1 minus .008598).

Table 12.8 Probability characteristic number three is within specification.

Char. No.	Prob. Char. Is Below LSL	Prob. Char. Is Above USL	Prob. Char. Is Out (p'_{OUT})	Probability Within
1	.001350	.000032	.001382	.998618
2	-	.000032	.000032	.999968
3	-	-	.008598	.991402
			Product =	**.990000**

The probability of the *entire* product being acceptable, *i.e.*, having all three characteristics within specifications, is just the *product* of the three individual probabilities (Bothe, 1992). In this case it is .990000, which is the probability the customer receives an acceptable product, *i.e.*, one with all 3 characteristics within their respective specifications. Note that multiplying these probabilities together is valid only if the probability of each characteristic being within specification is independent of the others. If they are dependent, see Section 13.8 (p. 816).

Product C_{PK} and P_{PK}

To express this as a "Product" P_{PK} value (assuming the percentage nonconforming for each characteristic is derived from σ_{LT}), first subtract the probability all characteristics are "good" from 1 to estimate the percentage of parts ($\hat{p}'_{P,LT}$) having at least one characteristic out of specification (a nonconforming part). The subscript "P" in $\hat{p}'_{P,LT}$ is for "product," while "LT" is for "long term."

$$\hat{p}'_{P,LT} = P(\text{At least one characteristic is out of specification})$$

$$= 1 - P(\text{All characteristics are within specification})$$

$$= 1 - .990000 = .010000 = 1.0 \times 10^{-2}$$

From Appendix Table III, the Z value for this percentage is found to be 2.32. Because Product P_{PK} assumes the worst case, where the *entire* percentage of nonconforming parts is outside the same specification limit (similar to the equivalent P^*_{PK} index for attribute data defined on page 533), this Z value becomes $Z_{MIN.\,LT}$ (Harry and Lawson, p. 7-10). Product P_{PK} is then defined as this $Z_{MIN.\,LT}$ value divided by 3.

$$\text{Product } \hat{P}_{PK} = \frac{Z_{MIN.\,LT}}{3} = \frac{Z(\hat{p}'_{P,LT})}{3} = \frac{Z(1.0 \times 10^{-2})}{3} = \frac{2.32}{3} = .773$$

With this definition for Product P_{PK}, the minimum capability requirement (Product P_{PK} = 1.00) is to have a 99.865 percent conformance rate. Therefore, because .773 is much less than 1.00, the combined output of this process lacks long-term performance capability.

The term "Product" capability was chosen for three reasons: 1) to signify this measure is not calculated with the traditional capability formulas; 2) to indicate it assesses the ability of a process to make an entire *product* which satisfies all customer requirements; 3) to acknowledge that this measure is based on the *product* of the probabilities of conformance for the individual product features (see also Byun *et al*, Mackertich and Stephens, Veevers, Venkatasamy and Agrawal).

If the percentage out of specification for each characteristic is estimated from short-term process variation (σ_{ST}), then this analysis produces a Product C_{PK} value. The general formula is given below, where q is the number of characteristics and $p'_{OUT,ST,i}$ is the percentage of parts having characteristic i outside its specification. The operator Π (pronounced "pie") means to multiply the q results together. The resulting product is the probability a part has all characteristics within specification. Subtracting it from 1 yields the probability at least one feature on the product is nonconforming.

$$\text{Product } C_{PK} = \frac{Z(p'_{P,ST})}{3} \quad \text{where } p'_{P,ST} = 1 - \prod_{i=1}^{q} (1 - p'_{OUT,ST,i})$$

For example, if q equals 3:

$$p'_{P,ST} = 1 - \prod_{i=1}^{3} (1 - p'_{OUT,ST,i}) = 1 - (1 - p'_{OUT,ST,1})(1 - p'_{OUT,ST,2})(1 - p'_{OUT,ST,3})$$

When a measure of long-term capability is desired:

$$\text{Product } P_{PK} = \frac{Z(p'_{P,LT})}{3} \quad \text{where } p'_{P,LT} = 1 - \prod_{i=1}^{q} (1 - p'_{OUT,LT,i})$$

An SPC software program, called Short Run Master™ (International Quality Institute), is available that charts up to 32 separate process streams (or characteristics), automatically checks the normality assumption, calculates all the various capability measures for each stream separately, then estimates Average C_{PK} and Product C_{PK}.

Example 12.5 Toothbrushes

Four critical characteristics of a toothbrush are produced in one operation. Evidence of control for the three variable-data dimensions has been provided by a multivariate profile chart developed by Fuchs and Benjamini. The stability of the attribute-data characteristic is demonstrated by a c chart. All of these features are listed in Table 12.9, along with a summary of the capability study for each characteristic. To estimate the combined capability with the Product P_{PK} index, first determine the percentage of toothbrushes with at least one feature out of specification. This is done by subtracting from 1 the percentage of toothbrushes with all features within specification.

$$\hat{p}'_{P,LT} = 1 - P \text{ (All features are within specification)}$$

$$= 1 - \prod_{i=1}^{4} (1 - \hat{p}'_{OUT,LT,i})$$

$$= 1 - [.99969(.99800)(.99998)(.99999)]$$

$$= 1 - .99766 = .00234 = 2.34 \times 10^{-3}$$

Table 12.9 Capability results of four toothbrush characteristics.

Char. No.	$Z_{LSL,LT}$	Prob. Char. Is Below LSL	$Z_{USL,LT}$	Prob. Char. Is Above USL	Prob. Out (p'_{OUT})	Probability Within
1	-	-	-	-	.00031	.99969
2	3.09	.00100	3.09	.00100	.00200	.99800
3	-	-	4.11	.00002	.00002	.99998
4	4.26	.00001	-	-	.00001	.99999
				P(All are within spec.) =		.99766

The Product P_{PK} index is estimated by first finding the Z value corresponding to this nonconformance percentage, then dividing it by 3.

$$\text{Product } \hat{P}_{PK} = \frac{Z(\hat{p}'_{P,LT})}{3}$$

$$= \frac{Z(2.34 \times 10^{-3})}{3} = \frac{2.83}{3} = .94$$

With an estimated Product P_{PK} of only .94, this operation is not capable of consistently producing toothbrushes having all four characteristics within their respective print specifications. To fulfill the minimum performance capability requirement, there must be at least 99.865 percent conforming product (brushes having all four features within specification) but, for this process, the conformance rate is only 99.766 percent (100% minus .234%).

Notice how the probability all characteristics are within specification is dominated by feature 2. Overall product capability cannot be better than that for the worst feature. Therefore, all improvement efforts must concentrate on this one characteristic if there is to be any hope of making a noticeable increase in overall capability.

Product C_{PK} for a Machining Line

Imagine a part that has seven critical dimensions, each produced on a different machine. The Product C_{PK} index can assess the capability of this entire machining line to manufacture an acceptable product. Simply determine the conformance rate for each operation, then multiply these seven percentages together to obtain the percentage of parts having all seven features within their respective specification limits. Again, this analysis assumes the various conformance rates are independent (see also Kotz and Lovelace).

Subtracting this result from 1 yields the percentage of parts with at least one of the seven characteristics out of specification. Dividing the Z value for this percentage by 3 produces the Product C_{PK} index for this machining line.

This same procedure also applies to an assembly line where many of the assembly operations generate attribute data (see Exercise 12.9, p. 755). Since only the percentage conforming for each feature is needed, the Product C_{PK} index is versatile enough to allow operations generating attribute data to be combined with those producing variable data.

Product C_P, P_P, C_R, and P_R

The Product C_{PK} and P_{PK} indexes consider the "worst" case, where the *total* percentage of nonconforming product determines the Z value, which is very similar to equivalent P^*_{PK} for attribute data. A somewhat more optimistic approach considers only *half* this total percentage for finding Z_{MIN}. A Product C_P (or P_P) index is based on only half the percentage of parts having at least one nonconforming characteristic.

$$\text{One-half the percentage of nonconforming parts} = \frac{p'_P}{2}$$

This division by 2 is done to compute the percentage out of specification for only *one* tail, similar to the logic behind the derivation of equivalent P^*_P for attribute capability studies (p. 534). As this index measures the optimistic (or "potential") capability of a process, the terms "C_P" and "P_P" are fitting. Product C_P is applicable if p'_P is determined from short-term process variation, whereas Product P_P is appropriate when p'_P is based on long-term variation.

$$\text{Product } C_P = \frac{Z(p'_{P,ST} \div 2)}{3} \qquad \text{Product } P_P = \frac{Z(p'_{P,LT} \div 2)}{3}$$

Recall from the toothbrush case study for Product P_{PK} that $\hat{p}'_{P,LT}$ equaled .00234. To estimate the Product P_P index for this operation, first divide this percentage in half, then find the corresponding Z value for the result.

$$\text{Product } \hat{P}_P = \frac{Z(\hat{p}'_{P,LT} \div 2)}{3} = \frac{Z(.00234 \div 2)}{3} = \frac{Z(.00117)}{3} = \frac{3.04}{3} = 1.01$$

This answer compares to the Product P_{PK} index of only .94 computed back in Example 12.5 (previous page). As always, performance capability is less than potential capability. It should also come as no surprise that Product P_R is simply the inverse of Product P_P.

$$\text{Product } \hat{P}_R = \frac{1}{\text{Product } \hat{P}_P} = \frac{1}{1.01} = .99$$

Example 12.6 Alternator Assembly

Having collected the information in Table 12.10 about a particular model of alternator produced on a new assembly line, Chris, the quality manager overseeing this project, wishes to estimate Product P_{PK} and Product P_P for this assembly. The measurements were charted on a multivariable control chart (Wright, Liu) and seen to display good control.

As before, the percentage of alternators with at least one feature out of specification ($p'_{P,LT}$) is estimated first, after verifying all features are independent of each other.

$$\hat{p}'_{P,LT} = P \text{ (At least one feature is bad)}$$

$$= 1 - P \text{ (All features are good)}$$

$$= 1 - .933946 = .066054 = 6.6054 \times 10^{-2}$$

Table 12.10   Four features of an alternator.

Feature # and Type	Process Parameters	LSL	USL	Probability Within
1 (Var.)	$\hat{\mu} = .1 \quad \hat{\sigma}_{LT} = 1.3$	-5	+5	.999874
2 (Att.)	$\hat{u}' = .063$	-	-	.938943
3 (Var.)	$\hat{\mu} = 17.9 \quad \hat{\sigma}_{LT} = 2.2$	-	24	.997197
4 (Att.)	$\hat{p}' = .0024$	-	-	.997600
			$P\,(\text{All are good}) =$	.933946

From this last percentage, Chris estimates Product P_{PK}, P_P, and P_R.

$$\text{Product } \hat{P}_{PK} = \frac{Z\,(\hat{p}'_{P,LT})}{3} = \frac{Z\,(6.6054 \times 10^{-2})}{3} = \frac{1.51}{3} = .503$$

$$\text{Product } \hat{P}_P = \frac{Z\,(\hat{p}'_{P,LT} \div 2)}{3} = \frac{Z\,(3.3027 \times 10^{-2})}{3} = \frac{1.84}{3} = .613$$

$$\text{Product } \hat{P}_R = \frac{1}{\text{Product } \hat{P}_P} = \frac{1}{.613} = 1.631$$

The terms "P_{PK}," "P_P," and "P_R" are appropriate in this situation because estimates of the percentage within specification for each characteristic are based on the combination of *all* measurements from their respective charts. This type of approach provides estimates of long-term process variation.

Product Capability for a Single Characteristic

Product C_{PK} and P_{PK} are different than the standard C_{PK} and P_{PK} values, even when a product has only one characteristic. Consider the process output distribution exhibited in Figure 12.3.

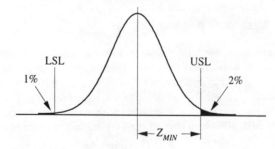

Fig. 12.3   A process output with nonconforming parts.

The standard C_{PK} index is estimated from $Z_{MIN.\ ST}$. Due to the percentage of parts above the USL being larger than that below the LSL, $Z_{USL.\ ST}$ is smaller than $Z_{LSL.\ ST}$ and, thus, becomes the $Z_{MIN.\ ST}$ value. If 2 percent of the process output is above the USL, $Z_{USL.\ ST}$ is 2.05, making the standard C_{PK} index equal to .68.

$$\hat{C}_{PK} = \frac{1}{3} \text{Minimum} \, (\hat{Z}_{LSL,ST}, \hat{Z}_{USL,ST})$$

$$= \frac{\hat{Z}_{USL,ST}}{3} = \frac{2.05}{3} = .68$$

To determine Product C_{PK}, $p'_{P,ST}$ must be estimated first. This is the total percentage of nonconforming product, which includes parts below the LSL (1 percent) as well as parts above the USL (2 percent).

$$\hat{p}'_{P,ST} = \hat{p}_{LSL,ST} + \hat{p}_{USL,ST} = .01 + .02 = .03$$

Product C_{PK} is equal to the corresponding Z value for this percentage, divided by 3.

$$\text{Product } \hat{C}_{PK} = \frac{Z(\hat{p}'_{P,ST})}{3} = \frac{Z(.03)}{3} = \frac{1.88}{3} = .63$$

The Product C_P index is derived from the Z value for *one-half* of this percentage.

$$\text{Product } \hat{C}_P = \frac{Z(\hat{p}'_{P,ST} \div 2)}{3} = \frac{Z(.03 \div 2)}{3} = \frac{2.17}{3} = .72$$

The standard C_{PK} index is calculated from $Z_{MIN.\,ST}$, which is determined by only the 2 percent of parts having this characteristic above the USL. The 1 percent of parts below the LSL is ignored in its .68 rating of process capability.

As a result of Product C_{PK} assuming the worst case, that the *entire* 3 percent of nonconforming parts is outside one specification, it is based on a lower Z value than the standard C_{PK} index and, thus, generates a lower measure of capability. This establishes a lower bound on the estimate of capability for a process with a total of 3 percent nonconforming parts. On the other hand, Product C_P assumes the *best* possible situation; that half of the total percentage nonconforming is outside each specification limit. Thus, it furnishes an upper bound on the estimate of process capability for a given total percentage nonconforming.

When the process output is centered at the middle of the tolerance, C_{PK} and Product C_P are equal (assuming a normal distribution) because one-half the percentage nonconforming is outside each specification. If μ is not centered at M, C_{PK} is less than Product C_P.

$$C_{PK} \leq \text{Product } C_P$$

By assuming the entire percentage nonconforming is outside of one specification, the Product C_{PK} index will typically be less than the standard C_{PK} index. However, if μ is located quite close to a specification such that the entire percentage nonconforming is in fact outside just this one specification, then Product C_{PK} will equal C_{PK}. In general:

$$\text{Product } C_{PK} \leq C_{PK} \leq \text{Product } C_P$$

If the characteristic under study has a unilateral specification, C_{PK} and Product C_{PK} are equal (assuming a normal distribution) because the total percentage nonconforming is outside the single specification. In this situation, Product C_P is always somewhat higher than C_{PK}.

Example 12.9 Comparing Two Suppliers

In the case of only one characteristic, product capability measures do a better job of describing what customers can expect to receive. This is highlighted in Figure 12.4, which presents the (stable) output from two suppliers (A and B) producing the same product. Assuming the decision as to which supplier should be awarded a new contract comes down to which has the best process capability, which one should be selected?

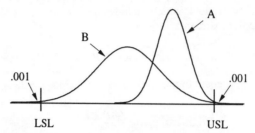

Fig. 12.4 Supplier A's output versus supplier B's.

The standard C_{PK} measure for both A and B is based on the .1 percent of nonconforming parts above the USL. This percentage equates to a $Z_{USL,ST}$ value of 3.09.

$$\hat{C}_{PK} \text{ for A } = \frac{1}{3}\text{Minimum} \left(Z_{LSL,ST}, Z_{USL,ST} \right) = \frac{1}{3}Z_{USL,ST} = \frac{3.09}{3} = 1.03$$

$$\hat{C}_{PK} \text{ for B } = \frac{1}{3}\text{Minimum} \left(Z_{LSL,ST}, Z_{USL,ST} \right) = \frac{1}{3}Z_{USL,ST} = \frac{3.09}{3} = 1.03$$

A comparison of the two suppliers based solely on these C_{PK} values would conclude their capabilities are identical (both are producing .1 percent nonconforming parts above the USL), yet visually there appears to be a significant difference between the two. In fact, supplier B is producing twice the amount of nonconforming product as A.

$$\hat{p}'_{P,ST} \text{ for A } = 1 - P \text{ (Product is conforming)} = 1 - .999 = .001$$

$$\hat{p}'_{P,ST} \text{ for B } = 1 - P \text{ (Product is conforming)} = 1 - .998 = .002$$

These are the percentages which determine the Product C_{PK} indexes for A and B.

$$\text{Product } \hat{C}_{PK} \text{ for A } = \frac{Z(.001)}{3} = \frac{3.09}{3} = 1.03$$

$$\text{Product } \hat{C}_{PK} \text{ for B } = \frac{Z(.002)}{3} = \frac{2.88}{3} = .96$$

The Product C_{PK} index for supplier A is based on .1 percent nonconforming parts, whereas this index for supplier B is based on .2 percent. This measure distinguishes between these two distributions by assigning a lower capability measure to supplier B due to the additional

.1 percent of nonconforming parts below the LSL. Product C_{PK} is similar to $ppm_{TOTAL, ST}$ in that both give a better indication of the overall process capability (in terms of percentage nonconforming parts) for each supplier.

12.3 Normalized C_P, C_{PK}, P_P, and P_{PK} for Product Comparisons

The Normalized C_{PK} (P_{PK}) index is related to the Product C_{PK} (P_{PK}) index. It is defined as the C_{PK} (or P_{PK}) value for a single characteristic of a product under the premise each characteristic has the *same* nonconformance rate. This assumption allows a meaningful capability comparison between processes producing products with different numbers of characteristics (Harry and Lawson, p. 2-3).

Let product **A** with 5 characteristics have a Product P_{PK} of 1.278. Product **B**, with 12 characteristics, is produced on a different process and also has a Product P_{PK} of 1.278. This rating implies 63 *ppm* will have at least one characteristic out of specification, meaning there is a 99.9937 percent chance a part will have all characteristics within their respective specifications. Because both have identical Product P_{PK} indexes, it initially appears the processes producing each product have the same capability. However, the process making product **B** must be much more capable for each individual characteristic because it has to produce many more characteristics than product **A**, yet come up with an identical Product P_{PK}.

Product **A** has only 5 probabilities of being "good" multiplied together (one for each characteristic) in the determination of the $\hat{p}'_{P, LT}$ used to calculate its Product P_{PK} index. Let $\hat{g}'_i$ be the estimated probability that the single characteristic "i" is good. Then for product **A**, the following relationship exists ($\hat{p}'_{P, LT, A}$ denotes the $\hat{p}'_{P, LT}$ value for product **A**).

$$P \text{ (All characteristics are good)} = 1 - \hat{p}'_{P,LT,A} = .999937 = \hat{g}'_1 \times \hat{g}'_2 \times \hat{g}'_3 \times \hat{g}'_4 \times \hat{g}'_5$$

Assume each characteristic of this first product has the *same* probability of being good, labeled $\hat{g}'_{N, LT, A}$. The "N" in the subscript represents a "normalized" (or equalized) probability, while the subscript "A" is for product **A**. Since $\hat{g}'_{N, LT, A}$ equals $\hat{g}'_1$, which equals $\hat{g}'_2$, which equals $\hat{g}'_3$, which equals $\hat{g}'_4$, which equals $\hat{g}'_5$, the equation above can be solved for $\hat{g}'_{N, LT, A}$.

$$\hat{g}'_1 \times \hat{g}'_2 \times \hat{g}'_3 \times \hat{g}'_4 \times \hat{g}'_5 = .999937$$

$$\hat{g}'_{N,LT,A} \times \hat{g}'_{N,LT,A} \times \hat{g}'_{N,LT,A} \times \hat{g}'_{N,LT,A} \times \hat{g}'_{N,LT,A} = .999937$$

$$(\hat{g}'_{N,LT,A})^5 = .999937$$

$$\hat{g}'_{N,LT,A} = (.999937)^{1/5}$$

$$= .9999874$$

In general, $\hat{g}'_{N,LT}$ is calculated as follows, where q is the number of characteristics:

$$\hat{g}'_{N,LT} = (1 - \hat{p}'_{P,LT})^{1/q}$$

For this example, $\hat{g}'_{N, LT, A}$ represents the probability a "typical" characteristic of product **A** is within specification. Conversely, $\hat{p}'_{N, LT, A}$ is defined as the probability this typical (normalized) characteristic is *not* within specification.

$$\hat{p}'_{N,LT,A} = 1 - \hat{g}'_{N,LT,A}$$

For product **A**, the normalized nonconformance rate, $\hat{p}'_{N,LT,A}$, is estimated as .0000126.

$$\hat{p}'_{N,LT,A} = 1 - \hat{g}'_{N,LT,A} = 1 - .9999874 = .0000126 = 1.26 \times 10^{-5}$$

Dividing the corresponding Z value for $\hat{p}'_{N,LT,A}$ by 3 gives an estimate of the "Normalized" P_{PK} index for product **A**.

$$\text{Normalized } \hat{P}_{PK} \text{ for A} = \frac{Z(\hat{p}'_{N,LT,A})}{3} = \frac{Z(1.26 \times 10^{-5})}{3} = \frac{4.215}{3} = 1.405$$

This Normalized P_{PK} index of 1.405 represents what the standard P_{PK} index must be for a single, typical characteristic, assuming all characteristics have the same nonconformance rate. Combining the capability of five such product characteristics, each having a Normalized P_{PK} index of 1.405, results in a Product P_{PK} index of 1.278.

Now consider product **B**, which has a total of 12 characteristics. With a Product P_{PK} index of 1.278, the probability a randomly-selected part has all 12 characteristics within their respective specifications is .999937.

$$P(\text{ All characteristics are good}) = 1 - \hat{p}'_{P,LT,B} = .999937 = \hat{g}'_1 \times \hat{g}'_2 \times \cdots \times \hat{g}'_{12}$$

Making the assumption each characteristic of product **B** has the same conformance rate, then $\hat{g}'_{N,LT,B} = \hat{g}'_1 = \hat{g}'_2 = \ldots = \hat{g}'_{12}$. The above equation can now be solved for $\hat{g}'_{N,LT,B}$.

$$\hat{g}'_1 \times \hat{g}'_2 \times \cdots \times \hat{g}'_{12} = .999937$$

$$(\hat{g}'_{N,LT,B})^{12} = .999937$$

$$\hat{g}'_{N,LT,B} = (.999937)^{1/12}$$

$$= .99999475$$

$\hat{p}'_{N,LT,B}$ is computed by subtracting this result from 1.

$$\hat{p}'_{N,LT,B} = 1 - \hat{g}'_{N,LT,B} = 1 - .99999475 = .00000525 = 5.25 \times 10^{-6}$$

Dividing the Z value associated with $\hat{p}'_{N,LT,B}$ by 3 provides an estimate of the Normalized P_{PK} index for **B**.

$$\text{Normalized } \hat{P}_{PK} \text{ for B} = \frac{Z(\hat{p}'_{N,LT,B})}{3} = \frac{Z(5.25 \times 10^{-6})}{3} = \frac{4.41}{3} = 1.470$$

Now a fairer comparison between the process producing **A** and the one producing **B** can be achieved by analyzing their Normalized P_{PK} indexes, as these measures are "normalized" (standardized) to compensate for the unequal number of product characteristics.

$$\text{Normalized } \hat{P}_{PK} \text{ for A} = 1.405 \qquad \text{Normalized } \hat{P}_{PK} \text{ for B} = 1.470$$

As expected, the process manufacturing product **B** is much more capable for a normalized characteristic than the one for **A**. In fact, the typical characteristic of **B** has only 42 percent (5.25 ppm ÷ 12.6 ppm) of the *ppm* as the typical characteristic of product **A**.

Normalized C_P and P_P

The Normalized C_P (P_P) measure is estimated in a manner similar to that for Normalized C_{PK} (P_{PK}), with only one exception: $\hat{p}'_N$ must be divided by 2 before finding the corresponding Z value. This is how the Normalized P_P index is estimated for product **B** of the last section, where $\hat{p}'_{N.LT.B}$ was 5.25×10^{-6}.

$$\text{Normalized } \hat{P}_P \text{ for B} = \frac{Z\,(\hat{p}'_{N,LT,B} \div 2)}{3} = \frac{Z\,(5.25 \times 10^{-6} \div 2)}{3} = \frac{Z\,(2.625 \times 10^{-6})}{3} = \frac{4.55}{3} = 1.52$$

As usual, the Normalized P_R ratio is found by inverting the Normalized P_P index. For product **B** this is .66, meaning the middle 99.73 percent of the output distribution for each variable-data characteristic can be no wider than 66 percent of the characteristic's tolerance.

$$\text{Normalized } \hat{P}_R \text{ for B} = \frac{1}{\text{Normalized } \hat{P}_P \text{ for B}} = \frac{1}{1.52} = .66$$

To have a Normalized P_R ratio of .66, each attribute-data characteristic of **B** must have a nonconformance rate no larger than one-half of $\hat{p}'_{N.LT.B}$, which would be just 2.625 *ppm*.

When only short-term process variation is involved in the determination of $p'_{N.ST}$, Normalized C_P and Normalized C_R are figured with these formulas:

$$\text{Normalized } C_P = \frac{Z\,(p'_{N,ST} \div 2)}{3} \qquad\qquad \text{Normalized } C_R = \frac{1}{\text{Normalized } C_P}$$

The Normalized C_R, C_P, P_R, and P_P measures are seldom used in practice.

Example 12.7 Video Camera

A portable video camera has 14 independent critical characteristics and an estimated Product P_{PK} of 1.00. Thus, the probability ($\hat{p}'_{P.LT}$) a camera has at least 1 of these 14 features being nonconforming is .00135. The Normalized P_{PK} index for each characteristic is determined by first estimating $g'_{N.LT}$. Remember, $p'_{P.LT}$ is the probability the *product* has at least one nonconforming feature, $p'_{N.LT}$ is the *normalized* probability a single characteristic is nonconforming (assuming all characteristics have equal nonconformance rates), while $g'_{N.LT}$ is the normalized probability a *single characteristic* is conforming.

$$\hat{g}'_{N,LT} = (1 - \hat{p}'_{P,LT})^{1/q} = (1 - .00135)^{1/14} = .9999035$$

$p'_{N.LT}$ is estimated from this result.

$$\hat{p}'_{N,LT} = 1 - \hat{g}'_{N,LT} = 1 - .9999035 = .0000965 = 9.65 \times 10^{-5}$$

Normalized P_{PK} is based on the Z value for $p'_{N.LT}$.

$$\text{Normalized } \hat{P}_{PK} = \frac{Z(\hat{p}'_{N,LT})}{3} = \frac{Z(9.65 \times 10^{-5})}{3} = \frac{3.73}{3} = 1.24$$

This standardized capability index of 1.24 may now be compared to other electronic assemblies with an identical, or a different, number of characteristics. Normalized P_{PK} can also help establish initial capability goals for each characteristic of this assembly operation such that the combination of all 14 features will generate the desired overall capability goal.

Generalized Formulas for Normalized Capability

$p'_{P,LT}$ represents the percentage of product with at least one nonconforming characteristic, while $g'_{P,LT}$ stands for the percentage of product having all characteristics within their respective specification limits.

$$g'_{P,LT} = 1 - p'_{P,LT}$$

With this definition, the Normalized capability formulas of the last few sections can be somewhat simplified. As before, q is the number of product characteristics included in the product capability study.

$$\text{Normalized } P_{PK} = \frac{Z[1 - (g'_{P,LT})^{1/q}]}{3} = \frac{Z[1 - (1 - p'_{P,LT})^{1/q}]}{3}$$

$$\text{Normalized } P_P = \frac{Z\{[1 - (g'_{P,LT})^{1/q}] \div 2\}}{3} = \frac{Z\{[1 - (1 - p'_{P,LT})^{1/q}] \div 2\}}{3}$$

$$\text{Normalized } P_R = \frac{1}{\text{Normalized } P_P}$$

Recall that $p'_{P,LT}$ is related to the Product P_{PK} index. The Φ operator means to find the area under the normal curve to the right of the value within the parentheses. For instance, $\Phi(3)$ is equal to .00135.

$$\text{Product } P_{PK} = \frac{Z(p'_{P,LT})}{3}$$

$$3 \text{ Product } P_{PK} = Z(p'_{P,LT})$$

$$\Phi(3 \text{ Product } P_{PK}) = p'_{P,LT}$$

$p'_{P,LT}$ is also related to the Product P_P index.

$$\text{Product } P_P = \frac{Z(p'_{P,LT} \div 2)}{3}$$

$$3 \text{ Product } P_P = Z(p'_{P,LT} \div 2)$$

$$\Phi(3 \text{ Product } P_P) = p'_{P,LT} \div 2$$

$$2\,\Phi(3 \text{ Product } P_P) = p'_{P,LT}$$

With these last equations, Normalized P_{PK} can now be expressed as a function of Product P_{PK}, while Normalized P_P can be written as a function of Product P_P.

$$\text{Normalized } P_{PK} = \frac{Z\{1-[1-\Phi(3\text{ Product }P_{PK})]^{1/q}\}}{3}$$

$$\text{Normalized } P_P = \frac{Z(\{1-[1-2\Phi(3\text{ Product }P_P)]^{1/q}\}\div2)}{3}$$

The above relationships also apply to Normalized C_{PK} and Normalized C_P by replacing Product P_{PK} with Product C_{PK}, and Product P_P with Product C_P.

Example 12.8 Steering Wheel

A particular model of a plastic injection-molded steering wheel has 11 different (and independent) critical characteristics. With a Product C_{PK} index estimated as 1.28, the Normalized C_{PK} index for this model is derived as follows:

$$\text{Normalized } \hat{C}_{PK} = \frac{Z\{1-[1-\Phi(3\text{ Product }\hat{C}_{PK})]^{1/q}\}}{3}$$

$$= \frac{Z\{1-[1-\Phi(3\times1.28)]^{1/11}\}}{3}$$

$$= \frac{Z\{1-[1-\Phi(3.84)]^{1/11}\}}{3}$$

$$= \frac{Z[1-(1-.0000616)^{1/11}]}{3}$$

$$= \frac{Z[1-(.9999384)^{1/11}]}{3}$$

$$= \frac{Z[1-.9999944]}{3} = \frac{Z[.0000056]}{3} = \frac{4.39}{3} = 1.46$$

This value may be compared to the Normalized C_{PK} for any other steering wheel with a similar, or different, number of characteristics. For example, a second model with 8 characteristics has a Product C_{PK} rating of 1.19.

$$\text{Normalized } \hat{C}_{PK} = \frac{Z\{1-[1-\Phi(3\text{ Product }\hat{C}_{PK})]^{1/q}\}}{3}$$

$$= \frac{Z\{1-[1-\Phi(3\times1.19)]^{1/8}\}}{3}$$

$$= \frac{Z\{1-[1-\Phi(3.57)]^{1/8}\}}{3}$$

$$= \frac{Z\,[\,1-(\,1-.0001786\,)^{1/8}\,]}{3}$$

$$= \frac{Z\,[\,1-(.9998214\,)^{1/8}\,]}{3}$$

$$= \frac{Z\,[.0000223\,]}{3} = \frac{4.08}{3} = 1.36$$

The process producing this second steering wheel is not as capable for its normalized characteristic as the process molding the first wheel, which has a Normalized C_{PK} of 1.46.

Relationship of Normalized C_{PK} to Product C_{PK}

Normalized C_{PK} values for a given Product C_{PK} must become greater as the number of product characteristics increases, as is illustrated by the graph in Figure 12.5 on the facing page. For a product with 20 characteristics to achieve a Product C_{PK} of 1.33, the Normalized C_{PK} for each characteristic must be 1.554. However, if a product has 60 characteristics, Normalized C_{PK} must increase to 1.629 in order to maintain a Product C_{PK} of 1.33.

12.4 Graphically Displaying Capability Comparisons

There are many times when the results of several capability studies must be reported to either top management or to a customer. This could be a comparison of capabilities for: the same characteristic run on several different processes; dissimilar characteristics run on different processes; various part numbers run on the same process (as is the case in short-run situations); competing suppliers; the outputs of multiple process streams; multiple characteristics of the same part number, run on one process or on several.

For example, the four critical characteristics listed in Table 12.11 for a telephone answering machine may be combined to derive a single Product C_{PK} value. This provides a good overall summary of this product's ability to satisfy customers. However, there are times when more detail is desired about how each characteristic is performing relative to the others.

Table 12.11 Four characteristics of a telephone answering machine.

Char. #	k_U	$1 - |k_U|$	C_P	C_{PK}
1	.1	.9	1.67	1.50
2	.1	.9	1.33	1.20
3	.4	.6	2.00	1.20
4	-.5	.5	3.00	1.50

The capabilities of these characteristics may be visually compared by plotting their results on either the A/P graph or on the radar graph.

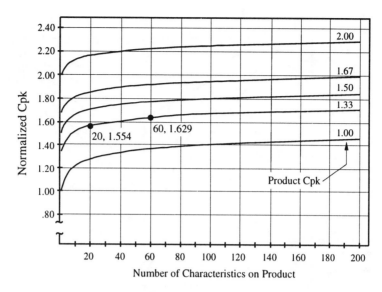

Fig. 12.5 Normalized C_{PK} versus Product C_{PK}.

A/P Graph

Individual variable-data characteristics can be compared on the A/P graph introduced in Section 6.5 (p. 245). Even though the units of measurement for each characteristic may be different, their k_U and C_P values are dimensionless. Thus, for comparison purposes, the four characteristics of the telephone answering machine may be plotted together on the same A/P graph (see Figure 12.6). Recall that the slanted lines represent differing capability levels, and that these characteristics do not have to be independent.

The process averages for characteristics 1 and 2 have the same relative deviation above the middle of their tolerances ($k_U = .1$), but characteristic 1 has a higher C_{PK} as a result of its higher potential capability ($C_{P,1} = 1.67$, $C_{P,2} = 1.33$). Although the process average of characteristic 3 is centered quite a distance above M ($k_U = .4$), it has the same performance capability as 2 ($C_{PK} = 1.20$) because its potential capability is much higher ($C_{P,3} = 2.00$, $C_{P,2} = 1.33$). Characteristic 4 has the most severe centering problem, as indicated by a k_U index of minus .5, meaning its average is located halfway between M and its LSL. The reason its performance capability is so good ($C_{PK} = 1.50$) is due to its extremely high potential capability, with a C_P index of 3.00 (see also Deleryd and Vannman, and Pearn and Chen).

The A/P graph works well if all characteristics have the same capability goal. If they don't, it is extremely difficult to determine which characteristic is in the most immediate need of process improvement. The radar graph solves this problem.

Radar Graph

As demonstrated on the next page in Table 12.12, simply presenting a list of all characteristics studied, their actual capabilities, and then their capability goals can be rather confusing, especially when their goals vary significantly.

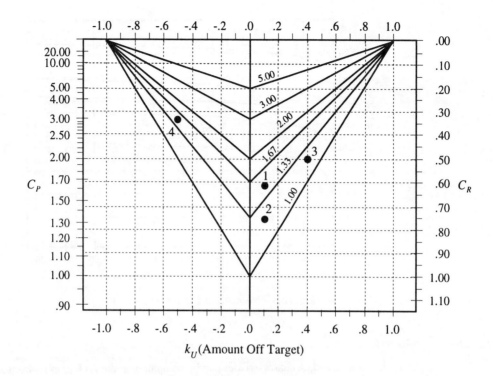

Fig. 12.6 The A/P graph comparing four different characteristics.

Table 12.12 C_P comparison for five characteristics.

Charac- teristic	C_P Actual	C_P Goal
A	1.68	1.33
B	1.79	1.33
C	1.50	1.50
D	1.02	1.67
E	2.93	2.00

How well does each characteristic meet its goal? Which characteristic is doing best in relationship to its goal? Which are in need of help? Should improvement efforts focus on shifting the process average, reducing variation, or both? Finding answers to these important questions among the figures listed in Table 12.12 is quite a daunting task.

One easy method for enhancing the analysis of this information is to visually display these results on a circular "radar" graph (Nagashima). A blank copy of this graph set up for five characteristics (A through E) is presented in Figure 12.7, while a completely blank radar chart is provided at the end of this chapter. Carangelo calls this a "web" chart, while Madigan refers to it as a "measures matrix," or M^2, chart.

The ratio of actual capability to its goal is plotted on this graph for each characteristic. Plot points for comparing C_P indexes are calculated via this formula:

$$\text{Plot Point} = \frac{C_P \ \text{Actual}}{C_P \ \text{Goal}}$$

For example, the plot point for characteristic A in Table 12.12 is calculated as 1.26.

$$\text{Plot Point for A} = \frac{1.68}{1.33} = 1.26$$

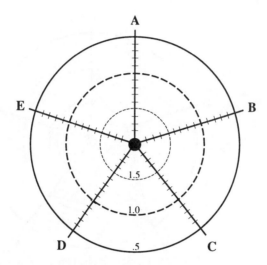

Fig. 12.7 The blank radar graph for comparing 5 characteristics.

This ratio of 1.26 means characteristic A surpasses its C_P goal by 26 percent. A ratio of 1.00 would indicate the feature just met its goal, while a ratio less than 1.00 reveals a characteristic not achieving its C_P goal.

The ratio for each characteristic is then plotted on a separate line radiating from the center of the radar graph to its outermost circle. There are as many lines as characteristics, with equal angles between all lines. In this example, five lines are drawn on the radar graph for the five characteristics (A through E), each being 72 degrees apart (360/5 = 72).

All lines have identical scales, beginning at .5 on the outermost circle and increasing to 2.0 at the graph's center. The concentric circles denote ratio levels of .5 (only 50 percent goal attainment), 1.0 (100 percent goal attainment) and 1.5 (150 percent of goal achieved). The center dot represents over 200 percent of goal attainment. Another outer circle may be added to accommodate ratios less than .50.

After plot points are computed for all five characteristics and listed in column four of Table 12.13, they are plotted on the radar graph. As is shown in Figure 12.8, a shaded polygon is created by drawing a line connecting these five separate plot points.

Table 12.13 C_P comparison for several characteristics.

Charac- teristic	C_P Actual	C_P Goal	Plot Point
A	1.68	1.33	1.26
B	1.79	1.33	1.35
C	1.50	1.50	1.00
D	1.02	1.67	.61
E	2.93	2.00	1.46

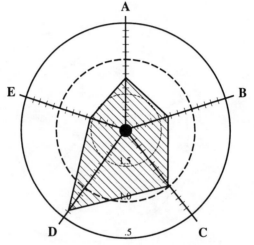

Fig. 12.8 C_P ratios plotted on the radar graph.

The closer a point is to the center of this graph, the better its goal attainment. A point greater than 1.0 indicates the capability of that characteristic exceeds its goal. For example, the point for characteristic A is plotted at 1.26, which being well inside the 1.0 circle, discloses that this feature's C_P index greatly exceeds its goal. The B and E points are also inside the 1.0 circle, indicating the capabilities for these features surpass their respective capability goals as well. Note that their capability goals are substantially different, 1.33 versus 2.00.

The plot point for C falls directly on the 1.0 circle (C's ratio is exactly 1.00), revealing that the potential capability for this feature just matches its requirement.

Points outside the 1.0 circle identify characteristics *not* meeting their goal for C_P. As its plot point is only .61, the process variation for characteristic D must be substantially reduced if it is ever to meet its potential capability goal.

Thus, a quick glance at this radar "screen" immediately reveals which characteristics are in trouble (those outside the 1.0 circle), which are marginal (those on, or near, the 1.0 circle), and which have outstanding capability (those well within the 1.0 circle).

A similar graph can be made for comparing the C_{PK} values of different characteristics. Table 12.14 summarizes the capability study results for these same five features, along with their calculated plot points.

Table 12.14 C_{PK} comparison for several characteristics.

Charac- teristic	C_{PK} Actual	C_{PK} Goal	Plot Point
A	1.64	1.33	1.23
B	1.33	1.33	1.00
C	1.09	1.50	.73
D	1.00	1.67	.60
E	1.28	2.00	.64

The plot point formula is the same as before, but with C_{PK} replacing C_P.

$$\text{Plot Point for A} = \frac{C_{PK}\ \text{Actual}}{C_{PK}\ \text{Goal}} = \frac{1.64}{1.33} = 1.23$$

Ratios for the other characteristics are given in column four of Table 12.14 and are plotted on the radar graph displayed in Figure 12.9. They are then connected with a bold line. Like the analysis of C_P results, points within the 1.0 circle identify characteristics meeting their C_{PK} goal, while points outside the circle flag features failing to achieve their performance capability goal.

Because ratios for both capability indexes are presented on the same radar graph, a comparison can easily be made between potential capability, as measured by C_P, and performance capability, as measured by C_{PK}.

In this example, characteristic A has very good potential and performance capability, and little improvement is needed. Because B has good potential, but only marginal performance, centering the process average closer to the middle of the tolerance is the best way to increase its C_{PK} value. Although C can also benefit from better centering, the best its C_{PK} can be is equal to its C_P value, which is just at the goal. To improve both potential and performance capability beyond their goals, the process variation for C must be reduced.

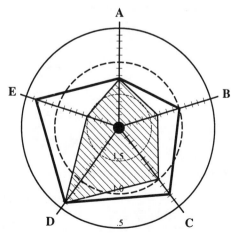

Fig. 12.9 The radar graph with C_{PK} results added.

As its plot points for C_P and C_{PK} are practically identical, characteristic D is properly centered. Unfortunately, the standard deviation for D is too large and must be significantly decreased if this feature is ever to achieve its capability goal. Finally, E has terrific potential capability, but is failing to realize it due to a drastic centering problem (the plot point for C_{PK} is only .64). All process improvement efforts must concentrate on shifting E's process average to the middle of the tolerance.

There is no limit to the number of features that can be portrayed on a radar graph, nor do characteristics have to be independent. This technique also works with the C_{PM} and C_{PMK} measures of process capability, as well as with all long-term capability indexes, such as P_P, P_{PK}, P_{PM}, and P_{PMK}. As was seen in this last example, different types of capability measures (even with unequal goals) may be plotted and compared on the same radar graph.

The rating scale for some capability indexes is such that "smaller is better." When comparing measures like C_R, C_{PG}, or ppm, the plot point formula must be modified. For example, when analyzing C_R results, use this plot point formula, which is just the inverse of the original:

$$\text{Plot Point} = \frac{C_R \text{ Goal}}{C_R \text{ Actual}}$$

As the actual C_R value improves (becomes smaller), this ratio becomes larger. This causes the plot point on the radar graph to move closer to the center, which is now consistent with the behavior for ratios computed with the C_P and C_{PK} indexes.

Example 12.9 Telephone Answering Machine

At the beginning of this section, the results of several capability studies were listed for various features of a telephone answering machine. The C_P indexes of those four characteristics, along with their respective capability goals and calculated plot points, are displayed in Table 12.15 and plotted on the radar graph exhibited in Figure 12.10 (facing page).

Note that characteristic 2, with the lowest C_P value, is actually performing the best with regards to meeting its goal. On the other hand, characteristic 4, with the highest C_P index, is doing the worst as far as meeting its goal. Characteristics 1 and 3 are at, or close to, their respective goals.

Table 12.15 C_P results of four answering machine characteristics.

Char. Number	C_P Actual	C_P Goal	Plot Point
1	1.67	1.70	.98
2	1.33	1.00	1.33
3	2.00	2.00	1.00
4	3.00	3.50	.86

Example 12.10 Plotting Confidence Intervals on a Radar Graph

The radar graph can also simultaneously compare a capability measure and its associated confidence bound to a capability goal for several features. Assume a product has three characteristics. The estimate of C_P for each one, along with a lower .95 confidence bound (review Section 11.9, p. 632), is displayed in Table 12.16.

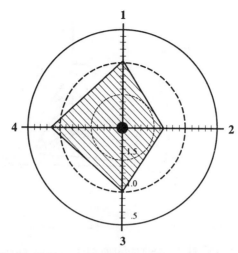

Fig. 12.10 Four C_P ratios plotted on one radar graph.

Table 12.16 C_P estimates and confidence bounds for three characteristics.

Char. Number	C_P Actual	$C_{P_{.95}}$	C_P Goal	Estimated Ratio	Confid. Ratio
1	2.06	1.63	1.50	1.37	1.09
2	1.72	1.17	1.33	1.29	.88
3	1.55	1.32	1.67	.93	.79

The C_P plot point ratios are computed as before and are listed in the "Estimated Ratio" column. Ratios for the confidence bounds are figured with this next formula and are tabulated in the "Confid. Ratio" column.

$$\text{Plot Point} = \frac{C_{P,1-\alpha}}{C_P \text{ Goal}}$$

These two sets of ratios are plotted on the radar graph exhibited in Figure 12.11, which is located on the next page. Because both the C_P and confidence bound plot points for characteristic 1 are within the 1.0 circle, there is .95 confidence this characteristic meets its potential capability goal. Although the C_P plot point for characteristic 2 is within the 1.0 circle, the confidence bound point is not. This means that there is less than .95 confidence this feature satisfies its capability goal. Unfortunately, both plot points for characteristic 3 are outside the 1.0 circle. As the point estimate of C_P is less than its goal, there is very little confidence the true C_P for this feature meets its goal.

When comparing confidence bounds in this manner, the sample sizes used to estimate each C_P index are allowed to vary. Likewise, the C_P goals may also be different for each feature. However, even though the confidence levels could vary between characteristics, the comparison between features is much easier to comprehend if they remain equal.

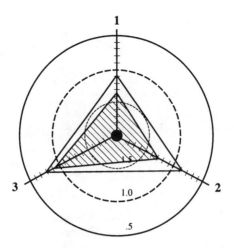

Fig. 12.11 C_p and confidence bound ratios plotted on a radar graph.

12.5 Summary

The Average C_{PK} index allows an assessment of the short-term performance capability of the combined output of several process streams. The identical characteristic must be checked for each stream, and this characteristic must have the same tolerance for all streams. Average P_{PK} provides an estimate of the long-term performance capability for a mixture of several process streams. As only the percentage nonconforming for each stream is used, these measures may be calculated for processes generating normally distributed variable data, non-normally distributed variable data, and even attribute data. Nonconformance rates of the various process streams do not have to be independent. Another approach to this problem involving ANOVA is presented by Traver for an example in the food processing industry.

Product C_{PK} measures the short-term performance capability of a process to produce a part having many different characteristics, each with its own, unique tolerance. Quite often this capability measure is applied to CNC machining operations or flexible machining systems. There is no limit to the number of characteristics that may be included in one product capability study. The features may be a mixture of normally distributed variable data, non-normally distributed variable data, and attribute data. However, their nonconformance rates must be independent of each other.

Based on just half the percentage of nonconforming product, Product C_p and Product P_p furnish optimistic estimates of process capability for multiple characteristics. They are seen very infrequently in practice.

The Normalized C_{PK} and P_{PK} indexes portray the capability of each characteristic, assuming every characteristic has the same capability. This concept allows goals to be established for each characteristic, as well as allowing capability comparisons between products having an unequal number of features. Product C_{PK} and Normalized C_{PK} are somewhat analogous to the *dpu* and *dpo* measures introduced in Section 9.5 on page 546.

The results of several capability studies may be visually compared on either an A/P graph or on a radar graph. The radar graph is the best technique for comparing many capability measures when their goals are different.

12.6 Blank Copy of the Radar Graph

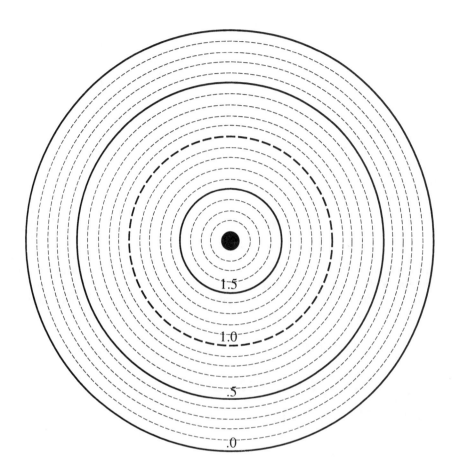

12.7 Exercises

12.1 Cylinder Head Gaskets

Gaskets for cylinder heads are stamped out from a sheet of material with a six-cavity die. Each cycle of the press produces six separate gaskets. A critical characteristic is the width of an inner edge of the gasket. The output from each cavity is charted and, after several weeks, all six process streams are stabilized. After verifying the normality assumption and estimating the capability of each cavity, the C_{PL} and C_{PU} values summarized in Table 12.17 are obtained.

Table 12.17 Capability study results for each cavity.

Cavity	C_{PL}	C_{PU}	C_{PK}
1	1.59	1.71	1.59
2	1.37	2.06	1.37
3	1.64	1.47	1.47
4	1.28	1.27	1.27
5	1.76	1.92	1.76
6	1.54	1.88	1.54

Estimate the Average C_{PK} index for this stamping press. What should be done to improve the overall performance of this process?

The C_{PK} goal for each cavity is 1.40. For each cavity, calculate the ratio of its estimated C_{PK} index to this goal, then plot these six ratios on a radar graph.

12.2 Shampoo Bottles

A Puerto Rican consumer-products company fills bottles of shampoo with an automatic dispenser having 12 filling heads. Control charts monitoring the weight of shampoo dispensed from each head are begun. When stability is demonstrated for all process streams, a capability study is conducted for each head. As the company is only concerned with underweight fills, only $ppm_{LSL,\ LT}$ values are estimated, with the results displayed in Table 12.18.

Table 12.18 Capability study results for each filling head.

Head	$ppm_{LSL,\ LT}$
1	92
2	1301
3	762
4	89
5	57
6	108
7	1142
8	2714
9	153
10	78
11	669
12	504

Estimate the Average P_{PK} index for this filling process, and develop a plan for improving its quality. The goal for the Average P_{PK} rating is 1.33.

12.3 Wrapper Ink

A special ink for printing candy bar wrappers is blended in a batch operation. Each batch is checked for viscosity, weight per gallon, and coefficient of friction (COF). The COF of the ink is very important as the candy bar wrapper should not be so "slippery" that it allows the bar to slide off the conveyor line during manufacturing. However, wrappers must be slippery enough to prevent the ink from smudging when bars rub together during shipment.

After stabilizing the process for all three features, a capability study is performed on each one, with the results presented in Table 12.19. Estimate the Product C_{PK} index for this ink-blending operation. If the goal for Product C_{PK} is 1.67, what should be done?

Table 12.19 Results of capability studies on ink features.

Feature	ppm_{ST}	
	LSL	USL
Viscosity	114	5
Weight/gallon	1058	46
COF	12	2397

12.4 Office Furniture

Office furniture is assembled from five components. The $ppm_{TOTAL, ST}$ for each sub-component is given in Table 12.20. Find the Product C_{PK} and Product C_P indexes for this process.

Table 12.20 Capability study results for each component.

Component	$ppm_{TOTAL, ST}$
Desk Top	2692
Left Side	1378
Right Side	1290
Front	1782
Back	1007

12.5 Office Furniture

For the previous office furniture example, estimate the Normalized C_{PK} index. How does this compare to a credenza (with seven components) having a Product C_{PK} index of 1.22?

12.6 Fill Volume

A dispensing machine has four filling stations which fill aluminum cans with a popular fruit drink. A critical characteristic is the volume of liquid delivered into a can. As each filling station could be performing at a different level, a study is designed to collect 100 consecutive cans from each station for the purpose of measuring equipment capability. After the data from each station is plotted on a separate chart, all stations are judged to be in a good state of control. Although they have different averages and standard deviations, the output for each station appears to be normally distributed when plotted on NOPP.

In Exercise 10.2 (p. 590), the specifications for fill volume were given as 340 ml $\pm$ 4 ml. This information was used to estimate the performance capability for each station, as listed below in Table 12.21.

Table 12.21 Summary of capability study results on four filling stations.

Station	$\overline{\overline{X}}$	$\hat{\sigma}_{ST}$	$\hat{C}_{PK}$
1	341.24	.952	.966
2	340.19	.813	1.562
3	339.19	1.035	1.027
4	342.07	.724	.889

Combine the results of each station into one overall measure of process capability. If the goal is an Average C_{PK} rating of 1.67, what decision would be reached about the performance capability of this multistation dispensing machine?

12.7 Connecting Rods

A boring machine is built to produce the connecting rod displayed in Figure 10.13 on page 590. Seven features deemed critical were selected to be part of a machine capability study in Exercise 10.3, and are listed in Table 12.22. 50 rods are machined; measurements are made and plotted on control charts ($n = 5$) to check for stability (see Flury *et al*). After control is established for all seven characteristics, ppm_{ST} estimates are made for both LSL and USL. Assuming all characteristics are independent of each other, estimate the Product C_{PK} index for this boring machine. How does this estimate compare to its goal of 1.25?

Table 12.22 Capability study results for connecting rods.

#	Characteristic	Spec. Limits	$ppm_{LSL,\,ST}$	$ppm_{USL,\,ST}$
1	Crank Diameter, Top	$\pm$.0005	4	0
2	Crank Diameter, Bottom	$\pm$.0005	0	0
3	Pin Diameter, Top	$\pm$.0007	51	7
4	Pin Diameter, Bottom	$\pm$.0007	0	82
5	Dist. Betw. Centerlines	$\pm$.0010	0	0
6	Bend	$\pm$.0016	1	0
7	Twist	$\pm$.0023	0	204

Estimate the Normalized C_{PK} index for this machine.

12.8 Baking Plates

Chinaware plates are fired in one of six molds in an oven. Afterwards, they are examined for warping, interior particles, spotty color and poor finish. After an extensive refurbishing of the oven and molds, a study is conducted to verify oven capability. 6,938 loads are fired and inspected, with the results tabulated in Table 12.23.

Table 12.23 Capability study results for six molds.

Mold #	Nonconf. Found	Estimated $ppm_{TOTAL, LT}$
1	2	288
2	1	144
3	11	1585
4	3	432
5	2	288
6	9	1297

In Exercise 10.5 (p. 591), a point estimate of the $ppm_{TOTAL, LT}$ for each mold was derived. Combine these estimates to estimate the Average P_{PK} index for this process. If the goal is 1.09, what should be done?

12.9 Oil Pump Housing

An aluminum casting for an oil pump housing is manufactured in one department by passing over three separate machines. Each machine produces one critical characteristic, with the capability results (in *ppm*) summarized in Table 12.24. Assuming these features are independent, what is the Product P_{PK} index of this machining department's ability to produce this part number? Does this rating meet the minimum requirement for long-term performance capability? Estimate the Normalized P_{PK} index for each machine in this line.

Table 12.24 Capability results for three machines.

Machine	$ppm_{LSL, LT}$	$ppm_{USL, LT}$
1	6	17
2	-	36
3	2	12

12.10 Gimbal Assemblies

A gimbal assembly for navigational systems has 22 components. Assuming each component will have the same level of nonconformance, what must $ppm_{TOTAL, ST}$ be for each component if the goal for the entire gimbal assembly is a Product C_{PK} index of 1.50?

12.11 Radar Graph for Cylinder Heads Gaskets

Calculate the ratio of actual C_{PK} to C_{PK} goal for the output of each of the 6 cavities in Exercise 12.1. Plot these results on a radar graph and interpret. The C_{PK} goal for each cavity is the same, 1.50.

12.12 Radar Graph for Oil Pump Housings

In Exercise 12.9, oil pump housings were produced by passing over three different machines, with each machine responsible for one characteristic. The results of capability studies conducted on each machine are listed in Table 12.25, along with the goal for each.

Table 12.25 Actual capability versus goal for three machines.

Machine	$ppm_{TOTAL, LT}$ Actual	Goal
1	23	40
2	36	30
3	14	10

Calculate the ratio of $ppm_{TOTAL, LT}$ goal to $ppm_{TOTAL, LT}$ actual, then plot the results on a radar graph and interpret.

12.13 Steel Coupons

An automotive supplier produces 1.25 cm cold-rolled steel by first rolling hot-rolled steel, then annealing it. Rolling reduces the steel's gage and increases its hardness. Annealing makes the steel pliable enough so it can be properly formed into various parts by customers. A coupon-cutting machine is used to take a sample from each run so various tests can be performed on the steel's mechanical properties. The results from five of these tests (which are independent) are given in Table 12.26.

Table 12.26 Test results for five steel properties.

Property	$\hat{p}'_{LSL,ST}$	$\hat{p}'_{USL,ST}$
1	.00024	-
2	.00009	.00008
3	-	.00019
4	.00076	-
5	-	.00031

Estimate the Product C_{PK} index of this process to manufacture steel meeting all five of these requirements.

12.8 References

Bothe, Davis R., "The Group Control Chart," *Quality*, Vol. 24, No. 3, March 1985, pp. 53-54

Bothe, Davis R., *SPC for Short Runs & JIT*, 1991, International Quality Institute, Inc., Sacramento, CA, 916-933-2318, pp. D-47 to D-53

Bothe, Davis R., "A Capability Study for an Entire Product," *46th ASQC Annual Quality Congress Transactions*, 1992, pp. 172-178

Bothe, Davis R., *Measuring Process Capability Training Manual*, 1993, International Quality Institute, Inc., Sacramento, CA, pp. K-2 to K-5

Boyd, D.F., "The Group Chart," *Industrial Quality Control*, November 1950

Byun, Jai-Hyun, Elsayed, E.A., Chen, Argon C.K., and Bruins, Rieks, "A Producibility Measure for Quality Characteristics With Design Specifications," *Quality Engineering*, Vol. 10, No. 2, 1997-1998, pp. 351-358

Carangelo, Richard M., "Clearly Illustrate Multiple Measures with the Web Chart," *Quality Progress*, Vol. 28, No. 1, January 1995, p. 136

Deleryd, M. and Vannman, K., "Process Capability Plots - A Quality Improvement Tool," *Quality and Reliability Engineering International*, Vol. 15, 1999, pp. 213-227

Flury, Bernard D., Nel, Daan G., and Pienaar, Inet, "Simultaneous Detection of Shift in Means and Variances," *Journal of the American Statistical Association*, Vol. 90, No. 432, December 1995, pp. 1474-1481

Fuchs, Camil, and Benjamini, Yoav, "Multivariate Profile Chart for Statistical Control," *Technometrics*, Vol. 36, No. 2, May 1994, pp. 182-195

Harry, Mikel J., and Lawson, J. Ronald, *Six Sigma Producibility Analysis and Process Characterization*, 1990, Motorola University Press, Schaumberg, IL

International Quality Institute, Inc., Short Run Master™ Software Program, 1995, Sacramento, CA, 916-933-2318

Kotz, S. and Lovelace, C.R., *Process Capability Indices in Theory and Practice*, 1998, Arnold, London, U.K., pp. 122-124

Larsson, Brandon, "Multiple Fixture SPC," *Quality*, Vol. 28, No. 11, Nov. 1989, p. 63

Liu, Regina Y., "Control Charts for Multivariate Processes," *Journal of the American Statistical Association*, Vol. 90, No. 432, December 1995, pp. 1380-1387

Mackertich, Neal A. and Stephens, Vic, "Virtual Process Capability," *ASQ 51st Annnual Quality Congress Transactions*, 1997, pp. 769-773

Madigan, James M., "Measures Matrix Chart: A Holistic Approach to Understanding Operations," *Quality Management Journal*, Vol. 1, Issue 1, October 1993, pp. 77-86

Mortell, Robert R., and Runger, George C., "Statistical Control of Multiple Stream Processes," *Journal of Quality Technology*, Vol. 27, No. 1, January 1995, pp. 1-12

Nagashima, Soichiro, *100 Management Charts*, 1973, Asian Productivity Organization, Tokyo, Japan, pp. 56-57

Pearn, W.L. and Chen, K.S., "Multiprocess Performance Analysis: A Case Study," *Quality Engineering*, Vol. 10, No. 1, 1997-1998, pp. 1-8

Snedecor, George W., and Cochran, William G., *Statistical Methods*, 7th ed., 1980, Iowa State University Press, Ames, IA, pp. 252-254

Traver, Mae-G., "Multistation Process Capability - Filling Equipment," *40th ASQC Annual Quality Congress Transactions*, 1984, pp. 281-289

Veevers A., "Viability and Capability Indices for Multiresponse Processes," *Journal of Applied Statistics*, Vol. 25, 1998, pp. 545-558

Venkatasamy, R. and Agrawal, V.P., "A Digraph Approach to Quality Evaluation of an Automotive Vehicle," *Quality Engineering*, Vol. 9, No. 3, 1997, pp. 405-417

Windham, Jeff, "Implementing Deming's Fourth Point," *Quality Progress*, Vol. 28, No. 12, December 1995, pp. 43-48

Wright, Don, "Multivariable Charting," *Quality*, Vol. 32, No. 3, March 1993, p. 49

Zhang, N.F., "Combining Process Capability Indices," *Proceedings of the Section on Quality and Productivity*, American Statistical Association, 1997, pp. 84-89

Chapter 13

Assessing Process Capability for Special Situations

Necessity helps one devise useful
approaches to uncommon problems.

Conventional capability measures don't readily apply to processes with:

1. repeating trends, like those due to tool wear;
2. setup variation between runs;
3. correlation between consecutive measurements (autocorrelation);
4. limited data due to short production runs;
5. products having within-piece variation;
6. features with geometric tolerances;
7. characteristics without specification limits;
8. correlated characteristics.

This chapter describes an appropriate charting method for each of these special situations, then explains how to derive measures of both potential and performance capability.

13.1 Measuring Capability in the Presence of Repeating Trends

A major benefit of control charts is their ability to help identify sources of assignable variation so steps may be taken to rid the process of them. However, there are times when an assignable cause is known, but cannot be easily eliminated because it stems from an integral part of the process. Consider a machine turning the outer diameter (O.D.) of a piston. As the tool wears, the O.D. size of successively manufactured pistons increases. When the O.D. size approaches the USL, the tool is changed so the O.D. sizes are small again. As the new tool wears, the cycle is repeated. Monitoring this feature on an $\overline{X}$, R chart results in a series of upward trends on the $\overline{X}$ portion of the control chart, as is shown in Figure 13.1.

To eliminate these out-of-control conditions, the tool could be sharpened, rotated, or changed more often. But all these actions add cost to the product. Since the goal is quality improvement at *minimum* cost, it may be worthwhile to let the tool wear for some period of time before taking action. Unfortunately, this decision also guarantees the appearance of trends on the control chart.

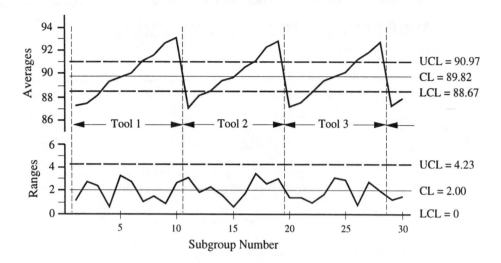

Fig. 13.1 Tool wear causes upward trends in O.D. size.

Comparable trends arise when charting die wear, human fatigue, moisture content, ambient temperature, or the viscosity of a chemical process. Krishnamoorthi provides several additional examples from a foundry setting. Note that these trends can also be downward. For example, as the tool cutting an inner diameter wears, a gradual drop occurs in I.D. size. When the tool is replaced, I.D. size immediately jumps higher. The acid concentration of a pickling bath decreases as more parts are processed through it. When concentrate is added to the bath, the acid concentration returns to a higher level, only to begin declining once again. Tatum, as well as Spurrier and Thombs, describe statistical procedures for detecting the presence of cycles and other periodic changes.

As tool wear is commonplace in industrial settings, that example was chosen to illustrate the following procedures for controlling, and assessing the capability of, this type of process. But these procedures apply to all situations where similar systematic trends exist.

Establishing Stability

If the short-term potential capability is quite high for a process exhibiting tool wear (the tolerance is at least two times greater than 6σ), the resulting trend can be allowed to continue for a brief period of time. The exact length of this interval is determined by the actual potential capability, the distance between the specification limits, and the performance capability goal. To allow a certain amount of tool wear, these factors can be utilized to modify the standard limits of a control chart. On page 18, control limit formulas for an $\overline{X}$ chart were given as:

$$LCL_{\overline{X}} = \overline{\overline{X}} - A_2\overline{R} \qquad\qquad UCL_{\overline{X}} = \overline{\overline{X}} + A_2\overline{R}$$

These formulas were used to calculate control limits for the chart displayed in Figure 13.1, where the subgroup size is five. However, because piece-to-piece (within subgroup) variation is relatively small (assuming minimal tool wear within a subgroup), the subgroup ranges, and hence, $\overline{R}$, are also small. A small $\overline{R}$ results in narrow control limits on the $\overline{X}$ chart. Because of the systematic drifts in subgroup averages, both trends and points outside the control limits materialize on the $\overline{X}$ chart.

To provide sufficient room for tool wear, the regular $\overline{X}$ chart control limits are replaced with "modified" limits. Conservative formulas developed by I.W. Burr for these special limits are given below and included in this procedure, while more "liberal" (wider) formulas are offered by Manuele, Montgomery (pp. 317-319), and Quesenberry (1988). Aerne *et al*, as well as Davis and Woodhall, explore problems associated with interpreting control charts when the process average is undergoing sustained linear shifts.

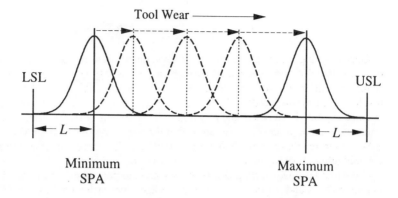

Fig. 13.2 To meet the capability goal, subgroup averages must be no closer than a distance of L to either specification limit.

Figure 13.2 reveals the largest subgroup average that can ever be observed, and still have this process meet its performance capability goal, must be a distance of at least L below the USL. This largest average at which it is still "safe" to operate the process is called the maximum safe process average (SPA). Likewise, the smallest subgroup average that may be observed, and yet have the process achieve the desired capability level, must be no nearer than a distance L to the LSL. This second critical average is referred to as the minimum safe process average, or minimum SPA.

$$\text{Maximum Safe Process Average} = \text{Max SPA} = \text{USL} - L$$

$$\text{Minimum Safe Process Average} = \text{Min SPA} = \text{LSL} + L$$

For example, if a C_{PK} goal of 1.33 is specified for this process, then subgroup averages must never be any closer than $4\sigma_{ST}$ to either specification. This is written as:

$$\text{Max SPA} = \text{USL} - L = \text{USL} - 4\hat{\sigma}_{ST}$$

$$\text{Min SPA} = \text{LSL} + L = \text{LSL} + 4\hat{\sigma}_{ST}$$

Because three times the C_{PK} goal equals this minimum number of standard deviations between the average and either specification limit, the above formulas can be generalized for any C_{PK} goal as follows:

$$\text{Max SPA} = \text{USL} - 3\,C_{PK,GOAL}\,\hat{\sigma}_{ST}$$

$$\text{Min SPA} = \text{LSL} + 3\,C_{PK,GOAL}\,\hat{\sigma}_{ST}$$

Due to the short-term standard deviation being estimated from $\overline{R}$ over d_2, it is extremely important the range chart exhibits a reasonable state of control. The centerline and control limits for this chart are identical to those for the conventional range chart.

$$\overline{R} = \frac{\sum\limits_{i=1}^{k} R_i}{k} \qquad\qquad \hat{\sigma}_{ST} = \frac{\overline{R}}{d_2}$$

$$LCL_R = D_3\overline{R} \qquad\qquad UCL_R = D_4\overline{R}$$

Once this range chart displays good control, the maximum and minimum SPA values may be estimated. To attain the stated performance capability goal, this process must be set up so the first subgroup average is at, or slightly above, the minimum SPA (assuming tool wear results in an upward trend). The process is permitted to continue running until a subgroup average is plotted just below the maximum SPA. At this point, the tool must be replaced or adjusted, and the process aim set up so that the next subgroup average is near the minimum SPA. Observing these steps will maximize tool life, minimize down time for tool maintenance, and still allow this process to satisfy its performance capability goal.

If the trend is downward, begin running when the first subgroup average is just below the maximum SPA and stop when subgroup averages approach the minimum SPA. Then set up to the maximum SPA and repeat this procedure (see also Jang *et al*).

For a more statistically sophisticated approach, Long and De Coste (as well as Mandel) explain how to fit a least squares regression line to the tool wear trend, while Grant and Leavenworth describe a technique for constructing slanting control limits around this line of best-fit. Gibra (1967) presents a least-cost procedure for determining when to replace tooling. For additional articles on economically managing a process with tool wear, see Arcelus and Banerjee (1985, 1987).

Example 13.1 Piston O.D. Size

For the chart displayed in Figure 13.1 where n equals 5, the short-term standard deviation is estimated as .86.

$$\hat{\sigma}_{ST} = \frac{\overline{R}}{d_2} = \frac{2.00}{2.326} = .86$$

Given a C_{PK} goal of 1.50, a LSL of 83.0, and an USL of 97.0, the maximum and minimum SPA limits are calculated as follows (note that the tolerance of 14 is more than 2.7 times the 6σ spread):

$$\text{Max SPA} = \text{USL} - 3\,C_{PK,GOAL}\,\hat{\sigma}_{ST} = 97.0 - 3\,(1.50)\,(.86) = 93.13$$

$$\text{Min SPA} = \text{LSL} + 3\,C_{PK,GOAL}\,\hat{\sigma}_{ST} = 83.0 + 3\,(1.50)\,(.86) = 86.87$$

The minimum and maximum SPA replace the standard $\overline{X}$ chart control limits for monitoring this process, as is illustrated in Figure 13.3. This turning operation is permitted to produce pistons as long as the subgroup averages remain between these modified limits. No centerline is drawn on this chart as runs cannot be meaningfully interpreted in the presence of systematic trends. Notice how the range chart is identical to the one displayed back in Figure 13.1.

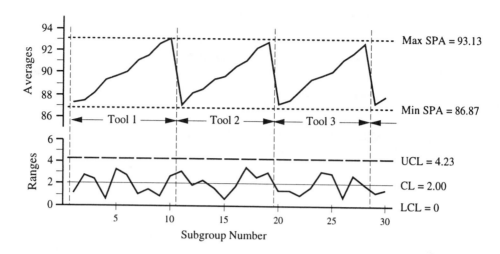

Fig. 13.3 Trends in tool wear are monitored with modified limits.

A few words of caution: as the estimate of σ_{ST} plays a major role in determining these modified limits, the range chart must always be in good state of control. If it is ever out of control, the modified limits are probably misleading and should not form the basis for any type of decision making. In addition, because the minimum and maximum SPA values are not true control limits, Wheeler and Chambers (p. 340) point out that a state of control cannot be achieved with these limits. These modified limits also encourage the production of parts throughout most of the tolerance, which is fine for companies following the goalpost philosophy of quality (see Figure 6.80, p. 278). However, this approach is unsuitable for those companies embracing Taguchi's philosophy, where great emphasis is placed on minimizing product variation around a given target value. Pyzdek (pp. 203-207) explains how the economic loss can be incorporated in determining the minimum and maximum SPA limits.

Measuring Potential Capability

As short-term potential capability is a function of only the short-term standard deviation, with no influence from the process average, estimating potential capability for processes exhibiting tool wear patterns is no different than for the general case. Once an estimate is made for σ_{ST} (typically from $\overline{R}$), various short-term potential capability measures can be estimated, with several examples given below. The subscript "*MOD*" (for modified) has been added to each to distinguish them from the standard measures. All reflect how well a process could meet print specifications assuming its average could be centered, and kept, at M. These formulas also assume negligible tool wear within subgroups. If there is, see Spiring (1989).

$$\hat{\sigma}_{ST} = \frac{\overline{R}}{d_2} \quad \text{or,} \quad \hat{\sigma}_{ST} = \frac{\overline{S}}{c_4} \qquad\qquad \hat{C}_{P.MOD} = \hat{C}_P = \frac{\text{Tolerance}}{6\hat{\sigma}_{ST}}$$

$$\hat{C}_{R.MOD} = \hat{C}_R = \frac{6\hat{\sigma}_{ST}}{\text{Tolerance}} \qquad\qquad \hat{C}_{ST.MOD} = \hat{C}_{ST} = \frac{\text{Tolerance}}{\hat{\sigma}_{ST}}$$

Estimating the long-term standard deviation is not recommended for this type of process. Harding *et al* have shown how significant changes in the process average greatly inflate the estimate of σ_{LT}, rendering it totally useless as a measure of process variability (review Section 3.3, p. 42, and Figure 3.5, p. 34). Therefore, long-term potential capability measures like P_P and P_R should never be estimated when systematic drifts in the process average are present.

Example 13.2 Piston O.D. Size

In Example 13.1, the short-term standard deviation for piston O.D. size was estimated as .86. With a LSL of 83.0 and an USL of 97.0, the potential capability of this process may be estimated with either of these two indexes:

$$\hat{C}_{P.MOD} = \frac{\text{Tolerance}}{6\hat{\sigma}_{ST}} = \frac{97.0 - 83.0}{6(.86)} = 2.71$$

$$\hat{C}_{ST.MOD} = \frac{\text{Tolerance}}{\hat{\sigma}_{ST}} = \frac{97.0 - 83.0}{.86} = 16.3$$

The minimum SPA of 86.87 for this process is set $4.5\sigma_{ST}$ above the LSL, while the maximum SPA of 93.13 is $4.5\sigma_{ST}$ below the USL. The distance between these two modified limits is 6.26 (93.13 - 86.87), which equals $7.3\sigma_{ST}$ (6.26/.86). The sum of these three distances is $16.3\sigma_{ST}$ (4.5 + 4.5 + 7.3), which equals the width of the tolerance (note that $C_{ST.MOD}$ is 16.3).

These potential capability measures reveal what a process can do if the process average is centered at M. However, in the case of tool wear, this assumption is invalid and attention should be focused more on actual performance capability.

Measuring Performance Capability

From work done by Long and De Coste, an estimate of performance capability, labeled $C_{PK.MOD}$, is derived as follows when the trend is moving upward:

$$\hat{C}_{PK.MOD} = \text{Minimum}\left(\hat{C}_{PL.MOD}, \hat{C}_{PU.MOD}\right) = \text{Minimum}\left(\frac{\hat{\mu}_I - \text{LSL}}{3\hat{\sigma}_{ST}}, \frac{\text{USL} - \hat{\mu}_F}{3\hat{\sigma}_{ST}}\right)$$

$$\text{where } \hat{\mu}_I = \frac{\sum\limits_{i=1}^{j} \overline{X}_{I,i}}{j} \quad \text{and} \quad \hat{\mu}_F = \frac{\sum\limits_{i=1}^{j} \overline{X}_{F,i}}{j}$$

$\overline{X}_I$ is the average of the initial subgroup after a tool adjustment, $\overline{X}_F$ is the average of the final subgroup before a tool adjustment, while j is the number of tool adjustments. This makes $\hat{\mu}_I$ the overall average of the beginning subgroup averages for each tool wear trend, whereas $\hat{\mu}_F$ is the average of the ending subgroup averages. Ideally, if all tool setups are done carefully, $\hat{\mu}_I$ will be very close to the minimum SPA (just slightly above), while $\hat{\mu}_F$ is near the maximum SPA (slightly below). An *IX & MR* chart for the $\overline{X}_I$ values (each $\overline{X}_I$ is treated as an *IX*) should be done to verify the setup procedure is in control for setting the initial process aim. Likewise, the stability of when the process is halted is checked with an *IX & MR* chart for the $\overline{X}_F$ values.

If the systematic trend is moving downward, initial setups are now close to the maximum SPA, and process operation continues until the subgroup averages near the minimum SPA. For these situations, the roles of $\hat{\mu}_I$ and $\hat{\mu}_F$ are reversed, and performance capability is assessed with this formula:

$$\hat{C}_{PK.\ MOD} = \text{Minimum}\left(\frac{\hat{\mu}_F - \text{LSL}}{3\hat{\sigma}_{ST}}, \frac{\text{USL} - \hat{\mu}_I}{3\hat{\sigma}_{ST}}\right)$$

As defined above, $C_{PK.\ MOD}$ is quite conservative, because the process average spends the majority of its time between $\hat{\mu}_I$ and $\hat{\mu}_F$. The actual $ppm_{TOTAL\ ST}$ for this process with tool wear will be substantially less than the $ppm_{TOTAL\ ST}$ predicted for a stable process with an identical C_{PK} value. The exact difference depends on the gap between $\hat{\mu}_I$ and $\hat{\mu}_F$. The larger this gap, the greater the difference in *ppm* values. As the gap narrows, $ppm_{TOTAL\ ST}$ for the tool wear situation increases. When $\hat{\mu}_I$ equals $\hat{\mu}_F$, which implies *no* tool wear, $ppm_{TOTAL\ ST}$ for this process will equal the $ppm_{TOTAL\ ST}$ of a stable process with an identical C_{PK} value.

Performance capability can at best be equal to potential capability ($C_{PK.\ MOD} = C_{P.\ MOD}$). This happens only when the process average is centered, and held, at M, which means all drifts in the process average must be eliminated.

As noted above with the potential capability measures, estimating the long-term standard deviation is inappropriate, and long-term capability indexes (such as P_{PK} or P_{PM}) should never be applied to this type of process (see also Sarkar and Pal (1998), and Zhang and Fang).

Example 13.3 Piston O.D. Size

The following estimates of $\hat{\mu}_I$ and $\hat{\mu}_F$ are based on the results of 12 tools ($j = 12$) run on the piston-turning operation:

$$\hat{\mu}_I = \frac{\sum\limits_{i=1}^{j} \overline{X}_{I,i}}{j} = \frac{1044.48}{12} = 87.04 \qquad \hat{\mu}_F = \frac{\sum\limits_{i=1}^{j} \overline{X}_{F,i}}{j} = \frac{1100.28}{12} = 91.69$$

σ_{ST} for piston O.D. size was estimated as .86 in Example 13.1 (p. 762). With a LSL of 83.0 and an USL of 97.0, the performance capability of this machining operation is computed as 1.57. This is a "worst case" scenario as the actual $ppm_{TOTAL\ ST}$ for piston O.D. size will be substantially less than that for a stable process with a C_{PK} index of 1.57.

$$\hat{C}_{PK.\ MOD} = \text{Minimum}\left(\frac{\hat{\mu}_I - \text{LSL}}{3\hat{\sigma}_{ST}}, \frac{\text{USL} - \hat{\mu}_F}{3\hat{\sigma}_{ST}}\right)$$

$$= \text{Minimum}\left(\frac{87.04 - 83.0}{3(.86)}, \frac{97.0 - 91.69}{3(.86)}\right)$$

$$= \text{Minimum}(1.57, 2.06) = 1.57$$

As the goal for $C_{PK.\ MOD}$ is 1.50, the process satisfies this customer requirement. However, $\hat{\mu}_F$ is only 91.69, but could be as high as 93.13 (the maximum SPA) and still meet the capability goal of 1.50. In order to maximize tool life, and thereby reduce operating costs, consideration might be given to allowing each tool to run slightly longer before replacement. As the accuracy of the initial setup and the determination of when to replace tooling improves, $\hat{\mu}_I$ will approach the minimum SPA, while $\hat{\mu}_F$ becomes nearer to the maximum SPA. As this happens, $C_{PK.\ MOD}$ decreases to the C_{PK} goal originally used to compute the two SPA values.

Conversely, one way to improve performance capability is to set up at a higher initial average and change tools more often. This strategy increases $\hat{\mu}_I$ and decreases $\hat{\mu}_F$, resulting in a growth in $C_{PK.\ MOD}$. However, this enhanced capability is paid for by shorter tool life and decreased productivity. Note that replacing the tool more often does not influence the process's potential capability.

Other performance capability measures may also be reported for these types of processes. For instance, $C_{PM, MOD}$ is defined as:

$$\hat{C}_{PM, MOD} = \frac{\text{Tolerance}}{6\hat{\tau}_{ST. MOD}}$$

where $\hat{\tau}_{ST. MOD} = \text{Maximum}\left[\sqrt{\hat{\sigma}_{ST}^2 + (\hat{\mu}_I - M)^2}, \sqrt{\hat{\sigma}_{ST}^2 + (\hat{\mu}_F - M)^2} \right]$

For the piston outer diameter example, $\tau_{ST. MOD}$ is estimated as 3.08.

$$\hat{\tau}_{ST. MOD} = \text{Maximum}\left[\sqrt{\hat{\sigma}_{ST}^2 + (\hat{\mu}_I - M)^2}, \sqrt{\hat{\sigma}_{ST}^2 + (\hat{\mu}_F - M)^2} \right]$$

$$= \text{Maximum}\left[\sqrt{.86^2 + (87.04 - 90)^2}, \sqrt{.86^2 + (91.69 - 90)^2} \right]$$

$$= \text{Maximum}\,(3.08, 1.90) = 3.08$$

From this result, $C_{PM. MOD}$ is determined to be .76.

$$\hat{C}_{PM. MOD} = \frac{\text{Tolerance}}{6\hat{\tau}_{ST. MOD}} = \frac{97.0 - 83.0}{6(3.08)} = .76$$

Unilateral Specifications

The concept described in this section for assessing capability applies to unilateral specifications as well. Consider an operation machining the surface finish of a clutch plate where only an USL is given for this feature. This process is set up so surface finish is initially as low as possible, then as the tool wears, the subgroup averages will drift upward until reaching the maximum SPA. At this point, the tool must be adjusted and the procedure repeated as before. However, as there is no minimum SPA, the process is set up to produce the lowest surface finish possible.

Measures of potential capability are not usually calculated for these cases unless a target average is provided. If one is, use $C^T_{PU. MOD}$ to assess potential capability when only an USL is given (review Section 5.10, p. 144).

$$C^T_{PU. MOD} = \frac{\text{USL} - T}{3\sigma_{ST}}$$

If just a LSL is specified, switch to this formula:

$$C^T_{PL. MOD} = \frac{T - \text{LSL}}{3\sigma_{ST}}$$

For the majority of cases, a target is not furnished and only performance capability may be evaluated. $C_{PK. MOD}$ is the most popular index for this task and is defined as given below, when an USL is the sole specification:

$$\hat{C}_{PK.\,MOD} = \hat{C}_{PU.\,MOD} = \frac{USL - \hat{\mu}_F}{3\hat{\sigma}_{ST}}$$

In those cases where just a LSL is available, run the process until the subgroup averages approach the minimum SPA, then reset as high as possible. A conservative estimate of performance capability is established with this formula:

$$\hat{C}_{PK.\,MOD} = \hat{C}_{PL.\,MOD} = \frac{\hat{\mu}_F - LSL}{3\hat{\sigma}_{ST}}$$

Additional Approaches

Other methods for measuring capability in the presence of tool wear have been proposed. In Europe, the following formulas are often employed, where primes have been added to distinguish these measures from those previously presented.

$$\hat{C}'_{P.\,MOD} = \frac{\text{Tolerance}}{6\hat{\sigma}_{ST} + \hat{\mu}_F - \hat{\mu}_I}$$

$$\hat{C}'_{PK.\,MOD} = \frac{1}{3\hat{\sigma}_{ST} + .5(\hat{\mu}_F - \hat{\mu}_I)}\,\text{Minimum}\,(\hat{\mu} - LSL, USL - \hat{\mu})$$

There are some conceptual problems with both these indexes. $C'_{P.\,MOD}$ is intended to represent a measure of instantaneous, or potential, capability, which should be independent of the process average. Yet, it is defined above as a function of how the process is centered initially and how long it is left to run before adjusting the tool. Suppose the LSL is -10, the USL is +10, and the estimate for σ_{ST} is 1. From page 763, $C_{P.\,MOD}$ is estimated as:

$$\hat{C}_{P.\,MOD} = \frac{\text{Tolerance}}{6\hat{\sigma}_{ST}} = \frac{10 - -10}{6(1)} = 3.33$$

To achieve a minimum C_{PK} index of 1.33, the minimum SPA is set at $4\sigma_{ST}$ above the LSL, making it equal to minus 6.

$$\text{Min SPA} = LSL + 3\,C_{PK.GOAL}\,\hat{\sigma}_{ST} = -10 + 3(1.33)(1) = -6$$

Likewise, the maximum SPA is established at $4\sigma_{ST}$ below the USL, or at +6.

$$\text{Max SPA} = USL - 3\,C_{PK.GOAL}\,\hat{\sigma}_{ST} = 10 - 3(1.33)(1) = 6$$

However, assuming the process is run very carefully such that $\hat{\mu}_I$ equals -6, while $\hat{\mu}_F$ is +6, $C'_{P.\,MOD}$ would turn out to be only 1.11 with the proposed formula, seriously underestimating true potential capability. In fact, it is even less than performance capability ($C_{PK} = 1.33$).

$$\hat{C}'_{P.\,MOD} = \frac{\text{Tolerance}}{6\hat{\sigma}_{ST} + \hat{\mu}_F - \hat{\mu}_I} = \frac{10 - -10}{6(1) + 6 - -6} = 1.11$$

There are also problems with the performance capability measure $C'_{PK. MOD}$. The minimum and maximum SPA values are commonly set at equal distances inside their respective specification limits. If the process is run consistently between these modified limits, the overall average of all subgroups run, $\hat{\mu}$, will be approximately the middle of the tolerance, M. Thus:

$$\hat{\mu} - LSL \cong M - LSL = 1/2 \, \text{Tolerance} \quad \text{and} \quad USL - \hat{\mu} \cong USL - M = 1/2 \, \text{Tolerance}$$

$$\text{Minimum} \, (\hat{\mu} - LSL, USL - \hat{\mu}) \cong \text{Minimum} \, (1/2 \, \text{Tolerance}, 1/2 \, \text{Tolerance})$$

$$= 1/2 \, \text{Tolerance}$$

Substituting this into the formula given for $C'_{PK. MOD}$ yields:

$$\hat{C}'_{PK. MOD} = \frac{1}{3\hat{\sigma}_{ST} + .5 \, (\hat{\mu}_F - \hat{\mu}_I)} \text{Minimum} \, (\hat{\mu} - LSL, USL - \hat{\mu})$$

$$\cong \frac{1/2 \, \text{Tolerance}}{3\hat{\sigma}_{ST} + .5 \, (\hat{\mu}_F - \hat{\mu}_I)}$$

$$= \frac{\text{Tolerance}}{6\hat{\sigma}_{ST} + \hat{\mu}_F - \hat{\mu}_I} = \hat{C}'_{P. MOD}$$

Because this performance capability index equals the potential capability index for the majority of cases, estimating only one of the two is necessary, making the other redundant.

Spiring (1991) has proposed a method for estimating the "dynamic process capability." His procedure calculates a C_{PM} measure for each subgroup, called $C_{PM. SUB}$, which he defines as shown below, where n, R, and $\overline{X}$ are the subgroup size, range, and average, respectively. Spiring recommends a subgroup size of at least 5 to get reasonable estimates, but no more than 25 so as not to include tool wear effects within the subgroup. The term $n/(n-1)$ is added to improve the estimating ability of the regular C_{PM} index when very small samples are taken (Subbaiah and Taam).

$$\hat{C}_{PM. SUB} = \frac{\text{Tolerance}}{6\sqrt{(R/d_2)^2 + n \, (\overline{X} - M)^2/(n-1)}}$$

Right after tool replacement, the subgroup averages are at their greatest distance from M. Because the difference between $\overline{X}$ and M is quite large at this point, $C_{PM. SUB}$ will be relatively small for these first subgroups. As the tool wears, the subgroup averages approach M, causing the corresponding $C_{PM. SUB}$ values to grow, as is illustrated in Figure 13.4 on the facing page. This growth continues until the subgroup averages get to M, where $C_{PM. SUB}$ reaches its peak since the difference between M and $\hat{\mu}$ is zero. Then, as the subgroup averages start to move above M, the subgroup $C_{PM. SUB}$ indexes begin decreasing. This decline continues until the subgroup averages reach the maximum SPA, where the $C_{PM. SUB}$ values recede to their minimum because the difference between $\overline{X}$ and M is again quite large. Thus, dynamic process capability is at its lowest during the beginning and ending of a tool's life and at its highest near the middle of the tool's life.

The $C_{PM. SUB}$ index calculations for the first ten subgroups ($n = 5$) of the piston example (where M is 90.0) are tabulated in Table 13.1. The $C_{PM. SUB}$ values for all 30 subgroups are plotted in Figure 13.4.

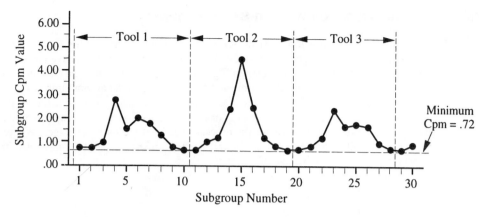

Fig. 13.4 $C_{PM.\,SUB}$ values increase as the subgroup averages approach M, then decrease as they move above M.

Putting aside the issue of how unreliable these capability estimates would be due to the small subgroup size (review Section 11.23, p. 672), how would capability be reported to a customer? The maximum attained during the tool's life, which occurs when subgroup averages are near M? This is an unfair representation of quality since this maximum is achieved for only a brief period and not all customers will receive this high level of quality. As customers are shipped parts produced during the entire life of the tool, the true overall process capability is quite a bit less than this peak. To be safe and make sure no customer is disappointed, the *smallest* capability value measured during the tool's life must be reported. From Figure 13.4, this would be a C_{PM} value of about .72. This approach is now similar to the original concept presented in this section for calculating and reporting $C_{PM.\,MOD}$, which was estimated on page 766 as .76 for piston O.D. size.

Table 13.1 $C_{PM.\,SUB}$ calculations for the piston example.

Sub-group	R	$\overline{X}$	$C_{PM.\,SUB}$
1	1.2	87.2	.74
2	2.8	87.4	.74
3	2.4	88.1	.99
4	.6	89.3	2.83
5	3.3	89.6	1.57
6	2.8	90.0	1.94
7	1.1	91.1	1.77
8	1.5	91.5	1.30
9	.9	92.6	.80
10	2.7	93.0	.66

13.2   Processes Dependent Upon Setup Accuracy

There are times when a job must be set up fairly often. Once set up, the job runs very consistently, however, there are considerable differences in centering the average from setup to setup. The control chart presented in Figure 13.5 tracks a dimensional characteristic of a stamping operation. Almost every time a setup is performed, an abrupt jump in the process average appears on the $\overline{X}$ chart. Quite often, attempting to make small adjustments to better center the process average is impractical or not cost effective, as the setup operation lacks the necessary precision to recenter the process exactly at the target average. Whereas the potential capability of each run is similar, its performance capability may vary greatly, depending on the setup average of that run.

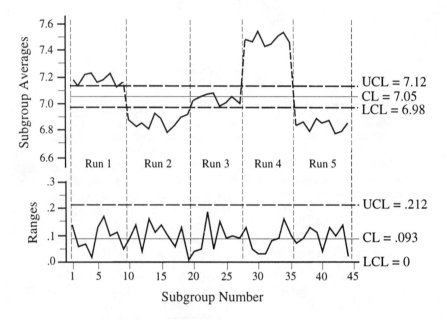

Fig. 13.5   Most runs have significantly different averages.

A similar condition frequently occurs in processing industries where distinct differences are observed from one batch to the next. These unique circumstances call for different approaches to charting, as well as for assessing capability.

Establishing Stability

Measurements from all runs are occasionally combined to derive an overall grand average for this process. However, because the range within any one subgroup is relatively small, the overall average range is also small, causing the $\overline{X}$ chart control limits to be "tight" to the centerline (reexamine Figure 13.5). This situation results in many out-of-control signals to appear, which cannot be addressed because the setup average cannot be easily changed. And when should an operator try another setup? There is no way of knowing if the current setup is as good as can be expected, or if a second attempt might center the process closer to the target average. Even given that an acceptable setup was done, the chart's control limits offer no help for controlling this process throughout the duration of any specific run. These major problems render conventional control charts of little practical value in establishing stability for these types of processes.

Some practitioners combine measurements from several runs to compute an overall standard deviation, which they then use to calculate control limits on the $\overline{X}$ chart. However, this standard deviation includes variation from within each run as well as between runs and, thus, is not the correct amount of variation for either, resulting in incorrect control limits (ones that are too wide).

An acceptable alternative is to employ two control charts: an *IX & MR* chart to check the initial setup, then a Target $\overline{X}$, *R* chart for monitoring the remainder of the run. The first chart is for checking setup (run-to-run) variation, while the second is for tracking within-run variation.

Monitoring Setup (Run-to-Run) Variation

The average of the first subgroup for each run, called the "setup average," is plotted on an *IX* chart, as is shown in Figure 13.6. Even though this is a subgroup average, it is treated as an individual measurement on this special version of the *IX & MR* chart. When a second setup is made, the average of the first subgroup collected from that run is plotted on this chart as the second *IX* plot point. The absolute value of the difference between these two *IX* plot points creates the first moving range, which is a measure of setup-to-setup variation.

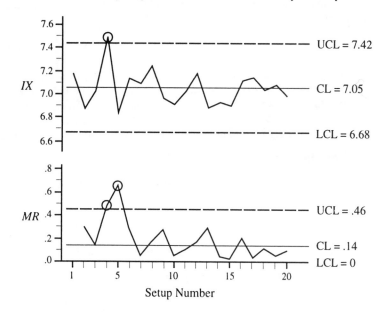

Fig. 13.6 Plotting setup averages on an *IX & MR* chart.

After a sufficient number of points are plotted, centerlines and control limits are calculated, beginning with the moving range chart (here, *m* is the number of setups).

$$\overline{MR} = \frac{\sum_{i=2}^{m} MR_i}{m-1} \qquad UCL_{MR} = 3.267\,\overline{MR} \qquad LCL_{MR} = 0$$

The following estimate of setup variation ($\hat{\sigma}_{SETUP}$) provides information about how precisely the average output of this process can be centered.

$$\hat{\sigma}_{SETUP} = \frac{\overline{MR}}{1.128}$$

The centerline of the *IX* chart is the overall average of all setup subgroup averages and is labeled $\overline{X}_S$, with the subscript "*S*" representing "setup."

$$\text{Overall Setup Average} = \overline{X}_S = \frac{\sum\limits_{i=1}^{m} (\text{Setup Average})_i}{m}$$

If the setup operation has good accuracy, $\overline{X}_S$ should be very close to the desired target average for this process. However, if $\overline{X}_S$ is much higher than this target, the setup personnel have a bias to favor the high side of the tolerance and need to lower their aim. When $\overline{X}_S$ is much lower than the target, there is a preference for the lower half of the tolerance, and setup personnel should raise their aim.

Once control limits for the *IX* chart are established with the following formulas (assuming normality), this chart will let setup personnel know if the process is centered with the typical accuracy for this particular operation.

$$LCL_{IX} = \overline{X}_S - 2.66\,\overline{MR} \qquad\qquad UCL_{IX} = \overline{X}_S + 2.66\,\overline{MR}$$

If a setup average falls below (above) the LCL_{IX} (UCL_{IX}), the current setup has not been done with its normal accuracy. A second setup should be attempted, with the goal of raising (lowering) the process aim. When a setup average falls within these control limits, the setup has been accomplished with its usual precision and accuracy, and trying to recenter the process average will generally not produce a better result. The run may now begin.

Example 13.4 Stamping Size

For the data on the chart in Figure 13.6, the average moving range is calculated as .14, making UCL_{MR} equal to .46.

$$UCL_{MR} = 3.267\,\overline{MR} = 3.267(.14) = .46$$

With an overall setup average of 7.05, control limits for the *IX* chart are computed as:

$$UCL_{IX} = \overline{X}_S + 2.66\,\overline{MR} = 7.05 + 2.66(.14) = 7.42$$

$$LCL_{IX} = \overline{X}_S - 2.66\,\overline{MR} = 7.05 - 2.66(.14) = 6.68$$

Analysis of this chart reveals the setup average for run number 4 was too high. This run should not have begun until another setup was attempted and found to be in control. Note that this conclusion could not have been drawn from the chart in Figure 13.5, because setup averages for almost all runs are outside of the (incorrect) control limits.

Additional insight concerning this setup operation is gained by noting the narrowing of the plot point spread near the end of the chart caused by less differences between setups. This pattern indicates setup personnel are becoming more proficient at their job.

Monitoring Within-Run Variation

Once a setup is in control and the run begins, subgroup ranges acquired during the remainder of this run are plotted on a regular range chart. When k ranges are collected, the centerline and control limits are computed from the following equations, with the centerline, $\overline{R}$, representing within-run (piece-to-piece) variation. When this range chart is in control, operators know the process is producing parts as consistently as possible for this setup.

$$\overline{R} = \frac{\sum\limits_{i=1}^{k} R_i}{k}$$

$$LCL_R = D_3 \overline{R} \qquad\qquad UCL_R = D_4 \overline{R}$$

Subgroup averages gathered during the run are handled in a different manner than with a standard $\overline{X}$ chart. After a subgroup average is calculated, the setup average for this particular run is subtracted from it, and the resulting deviation plotted on a Target $\overline{X}$ chart (Bothe, 1991, pp. 9-11).

$$\text{Target } \overline{X} \text{ Plot Point} = \overline{X} - \text{Setup Average}$$

This calculation is performed to code out differences between setups so subgroup averages from all runs can be plotted on the same chart, with one common set of control limits. Plot points falling within the control limits of this special $\overline{X}$ chart provide evidence that parts produced during a run are still centered at the initial setup average for this run.

Due to the above coding method, the average of the Target $\overline{X}$ plot points for this run, as well as all others, is expected to be zero. Assuming $\overline{R}$ values for all runs are similar, the centerline and control limits for the Target $\overline{X}$ chart become:

$$\text{Centerline} = \frac{\sum\limits_{i=1}^{k} (\text{Target } \overline{X} \text{ Plot Point})_i}{k} \cong 0$$

$$UCL_{\overline{x}} \cong 0 + A_2 \overline{R} = A_2 \overline{R}$$

$$LCL_{\overline{x}} \cong 0 - A_2 \overline{R} = -A_2 \overline{R}$$

When the second setup occurs, plot the new setup average on the *IX & MR* chart to verify that this setup has been done correctly. After the *IX* chart displays control, switch to the Target $\overline{X}$, R chart for all additional subgroups collected during the course of this second run. Plot each subgroup range on the range portion, and the deviation between each subgroup average and the *second* setup average on the Target $\overline{X}$ portion, as is illustrated in Figure 13.7. If these plot points are in control, the second run is still centered at its initial setup average.

Because the initial setup is so important, a larger subgroup size is often chosen for the setup average ($5 \leq n \leq 10$) than for subgroups collected during the run ($1 \leq n \leq 5$). Since this setup average is plotted on the *IX* chart, control limits for the Target $\overline{X}$ chart are not affected by this variance in subgroup size.

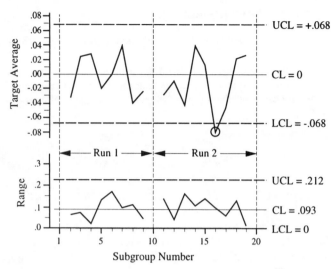

Fig. 13.7 The first two runs are plotted on one Target $\overline{X}$, R chart.

Example 13.5 Stamping Size

The Target range chart of Figure 13.7 is similar to the one displayed in Figure 13.5, with the exception of missing the first subgroup range for each run. With the centerline of the Target $\overline{X}$ chart close to zero, its control limits are approximated as follows ($n = 4$):

$$UCL_{\overline{x}} \cong +A_2\overline{R} = +.729(.093) = +.068$$

$$LCL_{\overline{x}} \cong -A_2\overline{R} = -.729(.093) = -.068$$

Detailed analysis of this second chart identifies the average of subgroup 16 as being below the *LCL*. Some type of assignable-cause variation occurring during the course of the second run is responsible. Note that identifying this point as being out of control is impossible on the original chart given in Figure 13.5.

Measuring Process Capability

After these charts display a reasonable state of control, five different capability measures may be estimated. The first three communicate information about the ability to correctly set up a process, while the last two supply knowledge about the capability of a process to consistently produce parts once properly set up (see also Morris and Watson, Pearn and Chang).

Setup Capability

The first measure is for the potential capability of the setup operation, called $C_{P.\,SETUP}$. This index is estimated as is shown below, where $\overline{MR}$ is the centerline of the moving range chart kept for monitoring the difference between setup averages.

$$\hat{C}_{P.\,SETUP} = \frac{\text{Tolerance}}{6\hat{\sigma}_{SETUP}} \quad \text{where } \hat{\sigma}_{SETUP} = \frac{\overline{MR}}{1.128}$$

If the characteristic being studied has a unilateral specification, switch to the following formulas to assess potential capability of the setup operation. For a LSL only:

$$\hat{C}'_{PL.\,SETUP} = \text{Maximum}\left(\frac{T - LSL}{3\hat{\sigma}_{SETUP}}, \frac{\hat{\mu}_{SETUP} - LSL}{3\hat{\sigma}_{SETUP}} \right)$$

$$\text{where } \hat{\mu}_{SETUP} = \overline{X}_S = \frac{\sum\limits_{i=1}^{m} (\text{Setup Average})_i}{m}$$

When the USL is the sole specification, select this formula:

$$\hat{C}'_{PU.\,SETUP} = \text{Maximum}\left(\frac{USL - T}{3\hat{\sigma}_{SETUP}}, \frac{USL - \hat{\mu}_{SETUP}}{3\hat{\sigma}_{SETUP}} \right)$$

$C_{P.\,SETUP}$ compares variation between setup averages to the tolerance in order to quantify how consistently a job can be set up. As operators become more proficient in setting up this piece of equipment, this index will increase. However, reducing setup variation will not usually improve the potential capability for producing parts *during* a given run.

A second capability measure discloses setup accuracy. The k factor from Chapter 6 (p. 226) given below can quantify this aspect of process capability. In the following formulas, M is the middle of the tolerance, while m is the number of setup averages used to calculate the centerline of the IX chart.

$$\hat{k}_{SETUP} = \frac{|M - \hat{\mu}_{SETUP}|}{(USL - LSL)/2} \qquad \text{where } \hat{\mu}_{SETUP} = \overline{X}_S = \frac{\sum\limits_{i=1}^{m} (\text{Setup Average})_i}{m}$$

The closer k_{SETUP} is to zero (which requires μ_{SETUP} to be near M), the greater the accuracy of the setup personnel. Improvements in this measure indicate a better ability to center the process at M during the setup procedure. k_{SETUP} may also be combined with $C_{P.\,SETUP}$ to generate a single performance capability measure for the setup operation.

$$\hat{C}_{PK.\,SETUP} = \hat{C}_{P.\,SETUP}(1 - \hat{k}_{SETUP}) = \text{Minimum}\left(\frac{\hat{\mu}_{SETUP} - LSL}{3\hat{\sigma}_{SETUP}}, \frac{USL - \hat{\mu}_{SETUP}}{3\hat{\sigma}_{SETUP}} \right)$$

The highest $C_{PK.\,SETUP}$ can ever be is equal to $C_{P.\,SETUP}$, which occurs when k_{SETUP} is zero, meaning the setup averages are centered right at M. At this point, improvements in the setup operation are achieved only through reduction of σ_{SETUP}.

These three measures, $C_{P.\,SETUP}$, k_{SETUP}, and $C_{PK.\,SETUP}$, completely characterize the setup capability of this process. This trio may also be used to benchmark the setup capability of other processes. In cases where T is not equal to M, these measures are replaced with C^*_P, k^*, and C^*_{PK}, respectively.

Example 13.6 Stamping Size

For the chart displayed in Figure 13.6 (p. 771), the setup standard deviation is estimated from the average moving range as .124.

$$\hat{\sigma}_{SETUP} = \frac{\overline{MR}}{1.128} = \frac{.14}{1.128} = .124$$

With a specification of $7.00 \pm .80$, $C_{P.\,SETUP}$ is computed as 2.15.

$$\hat{C}_{P.\,SETUP} = \frac{\text{Tolerance}}{6\hat{\sigma}_{SETUP}} = \frac{7.80 - 6.20}{6(.124)} = 2.15$$

k_{SETUP} is estimated from the $\overline{X}_S$ value of 7.05, M of 7.00, and one-half the tolerance.

$$\hat{k}_{SETUP} = \frac{|M - \overline{X}_S|}{(USL - LSL)/2} = \frac{|7.00 - 7.05|}{(7.80 - 6.20)/2} = .0625$$

Setup performance capability is found by combining the estimates for $C_{P.\,SETUP}$ and k_{SETUP}.

$$\hat{C}_{PK.\,SETUP} = \hat{C}_{P.\,SETUP}(1 - \hat{k}_{SETUP}) = 2.15(1 - .0625) = 2.02$$

Within-Run Capability

The fourth capability measure, $C_{P.\,RUN}$, assesses the potential capability of this process to produce pieces throughout a run. This index is found as follows, where $\overline{R}$ is from the range portion of the Target $\overline{X}$, R chart.

$$\hat{C}_{P.\,RUN} = \frac{\text{Tolerance}}{6\hat{\sigma}_{RUN}} \quad \text{where } \hat{\sigma}_{RUN} = \frac{\overline{R}}{d_2}$$

Recall from Chapter 5 that potential capability is independent of the process average, in this case, the setup average for each run. This index is not a function of centering (which shifts from run to run) and portrays the absolute best this process can do, assuming the process average is centered, and held, at M. $C_{P.\,RUN}$ reveals information about the precision, or inherent capability, of this process to produce parts for a given setup. Improving this measure of capability requires the reduction of common-cause variation (the 5Ms) occurring during a run, after its setup is successfully completed.

If the characteristic being studied has a unilateral specification, switch to the following formulas to assess within-run potential capability. For a LSL only:

$$\hat{C}'_{PL.\,RUN} = \text{Maximum}\left(\frac{T - LSL}{3\hat{\sigma}_{RUN}}, \frac{\hat{\mu}_L - LSL}{3\hat{\sigma}_{RUN}} \right)$$

Apply this formula when just an USL is provided, where μ_L and μ_H are defined in the next few paragraphs:

$$\hat{C}'_{PU.\,RUN} = \text{Maximum}\left(\frac{USL - T}{3\hat{\sigma}_{RUN}}, \frac{USL - \hat{\mu}_H}{3\hat{\sigma}_{RUN}} \right)$$

The last of the five capability measures, $C_{PK.\,CUSTOMER}$, represents what minimum quality level customers can expect from this process. When the setup operation is in control, the lowest allowed setup average (μ_L) is $3\sigma_{SETUP}$ below the overall setup average. If a setup average were any less than this, the point would be out of control on the IX chart (below the LCL), and the setup operation redone until the setup average fell within the control limits.

$$\text{Lowest Allowed Setup Average} = \hat{\mu}_L = \hat{\mu}_{SETUP} - 3\hat{\sigma}_{SETUP}$$

$$= \overline{X}_S - \frac{3\overline{MR}}{1.128}$$

$$= \overline{X}_S - 2.66\,\overline{MR} = LCL_{IX}$$

In a similar manner, the highest allowed setup average, labeled μ_H, can be no more than $3\sigma_{SETUP}$ above the overall setup average.

$$\text{Highest Allowed Setup Average} = \hat{\mu}_H = \hat{\mu}_{SETUP} + 3\hat{\sigma}_{SETUP}$$

$$= \overline{X}_S + \frac{3\overline{MR}}{1.128}$$

$$= \overline{X}_S + 2.66\,\overline{MR} = UCL_{IX}$$

Minimum performance capability occurs when the process is set up and run at one of these extreme averages, as the process average is then closest to one of the specification limits. This case represents the worst possible performance capability a customer would ever see since the process average for the vast majority of setups is between μ_L and μ_H.

$$\hat{C}_{PK.\,CUSTOMER} = \text{Minimum}\left(\frac{\hat{\mu}_L - LSL}{3\hat{\sigma}_{RUN}}, \frac{USL - \hat{\mu}_H}{3\hat{\sigma}_{RUN}} \right)$$

$$= \text{Minimum}\left(\frac{(\hat{\mu}_{SETUP} - 3\hat{\sigma}_{SETUP}) - LSL}{3\hat{\sigma}_{RUN}}, \frac{USL - (\hat{\mu}_{SETUP} + 3\hat{\sigma}_{SETUP})}{3\hat{\sigma}_{RUN}} \right)$$

$$= \text{Minimum}\left(\frac{LCL_{IX} - LSL}{3\hat{\sigma}_{RUN}}, \frac{USL - UCL_{IX}}{3\hat{\sigma}_{RUN}} \right)$$

This procedure doesn't give a precise measure of capability, but establishes a conservative lower bound. Customers will usually get far better quality than indicated by this index, with the exact level depending on where the job was set up when the parts received by that customer were run. But no matter where it was set up, no customer should receive a quality level less than that indicated by $C_{PK.\,CUSTOMER}$. If this "worst-case" capability meets or exceeds the customer's goal, parts run with any other in-control setup will also satisfy the goal.

As variation between setups disappears, σ_{SETUP} approaches zero. This causes $C_{PK.\,CUSTOMER}$ to increase, and eventually become equal to the following:

$$\hat{C}_{PK.\,CUSTOMER} = \text{Minimum}\left(\frac{(\hat{\mu}_{SETUP} - 0) - LSL}{3\hat{\sigma}_{RUN}}, \frac{USL - (\hat{\mu}_{SETUP} + 0)}{3\hat{\sigma}_{RUN}} \right)$$

$$= \text{Minimum}\left(\frac{\hat{\mu}_{SETUP} - LSL}{3\hat{\sigma}_{RUN}}, \frac{USL - \hat{\mu}_{SETUP}}{3\hat{\sigma}_{RUN}} \right)$$

Improving setup accuracy moves μ_{SETUP} closer to M and increases $C_{PK.\,CUSTOMER}$, implying better overall quality. When the setup average is centered at M, performance capability equals the potential capability for a run, namely, $C_{P.\,RUN}$.

$$
\begin{aligned}
\hat{C}_{PK.\,CUSTOMER} &= \text{Minimum}\left(\frac{\hat{\mu}_{SETUP} - \text{LSL}}{3\hat{\sigma}_{RUN}}, \frac{\text{USL} - \hat{\mu}_{SETUP}}{3\hat{\sigma}_{RUN}} \right) \\[2mm]
&= \text{Minimum}\left(\frac{M - \text{LSL}}{3\hat{\sigma}_{RUN}}, \frac{\text{USL} - M}{3\hat{\sigma}_{RUN}} \right) \\[2mm]
&= \text{Minimum}\left(\frac{1/2\,\text{Tolerance}}{3\hat{\sigma}_{RUN}}, \frac{1/2\,\text{Tolerance}}{3\hat{\sigma}_{RUN}} \right) \\[2mm]
&= \text{Minimum}\left(\frac{\text{Tolerance}}{6\hat{\sigma}_{RUN}}, \frac{\text{Tolerance}}{6\hat{\sigma}_{RUN}} \right) \\[2mm]
&= \frac{\text{Tolerance}}{6\hat{\sigma}_{RUN}} = \hat{C}_{P.\,RUN}
\end{aligned}
$$

Example 13.7 Stamping Size

For the chart in Figure 13.7 (where $n = 4$), σ_{RUN} is estimated from $\overline{R}$ as .0452.

$$
\hat{\sigma}_{RUN} = \frac{\overline{R}}{d_2} = \frac{.093}{2.059} = .0452
$$

With a LSL of 6.20, and an USL of 7.80, $C_{P.\,RUN}$ is calculated as 5.90.

$$
\hat{C}_{P.\,RUN} = \frac{\text{Tolerance}}{6\hat{\sigma}_{RUN}} = \frac{7.80 - 6.20}{6(.0452)} = 5.90
$$

A lower bound for the performance capability witnessed by customers is determined by first finding μ_L and μ_H from the *IX* chart control limits in Figure 13.6 (p. 771).

$$
\hat{\mu}_L = \text{LCL}_{IX} = 6.68 \qquad\qquad \hat{\mu}_H = \text{UCL}_{IX} = 7.42
$$

With this information, $C_{PK.\,CUSTOMER}$ is determined to be 2.80, meaning that all customers should receive parts with a quality level at least equivalent to a C_{PK} rating of 2.80.

$$
\begin{aligned}
\hat{C}_{PK.\,CUSTOMER} &= \text{Minimum}\left(\frac{\hat{\mu}_L - \text{LSL}}{3\hat{\sigma}_{RUN}}, \frac{\text{USL} - \hat{\mu}_H}{3\hat{\sigma}_{RUN}} \right) \\[2mm]
&= \text{Minimum}\left(\frac{6.68 - 6.20}{3(.0452)}, \frac{7.80 - 7.42}{3(.0452)} \right) \\[2mm]
&= \text{Minimum}\,(\,3.54\,,\,2.80\,) = 2.80
\end{aligned}
$$

If the setup average is shifted lower so it equals the target of 7.00, LCL_{IX} decreases by .05 to 6.63, while UCL_{IX} falls by .05 to 7.37. This improved accuracy in setup makes $C_{PK.\ CUSTOMER}$ jump to 3.17.

$$\hat{C}_{PK.\ CUSTOMER} = \text{Minimum} \left(\frac{LCL_{IX} - \text{LSL}}{3\hat{\sigma}_{RUN}}, \frac{\text{USL} - UCL_{IX}}{3\hat{\sigma}_{RUN}} \right)$$

$$= \text{Minimum} \left(\frac{6.63 - 6.20}{3(.0452)}, \frac{7.80 - 7.37}{3(.0452)} \right)$$

$$= \text{Minimum} (3.17, 3.17) = 3.17$$

In addition, if the average moving range for setups is reduced to just 80 percent of its former value, UCL_{IX} shrinks to 7.30, while LCL_{IX} increases to 6.70.

$$UCL_{IX} = \overline{X}_S + 2.66(.80)\overline{MR} = 7.00 + 2.66(.80)(.14) = 7.30$$

$$LCL_{IX} = \overline{X}_S - 2.66(.80)\overline{MR} = 7.00 - 2.66(.80)(.14) = 6.70$$

This improvement in setup precision pushes $C_{PK.\ CUSTOMER}$ to 3.69.

$$\hat{C}_{PK.\ CUSTOMER} = \text{Minimum} \left(\frac{LCL_{IX} - \text{LSL}}{3\hat{\sigma}_{RUN}}, \frac{\text{USL} - UCL_{IX}}{3\hat{\sigma}_{RUN}} \right)$$

$$= \text{Minimum} \left(\frac{6.70 - 6.20}{3(.0452)}, \frac{7.80 - 7.30}{3(.0452)} \right)$$

$$= \text{Minimum} (3.69, 3.69) = 3.69$$

If the within-run standard deviation is also decreased by 20 percent, making $\hat{\sigma}_{RUN}$ equal to .0362 (.0452 times .80), $C_{PK.\ CUSTOMER}$ jumps to 4.60.

$$\hat{C}_{PK.\ CUSTOMER} = \text{Minimum} \left(\frac{\hat{\mu}_L - \text{LSL}}{3\hat{\sigma}_{RUN}}, \frac{\text{USL} - \hat{\mu}_H}{3\hat{\sigma}_{RUN}} \right)$$

$$= \text{Minimum} \left(\frac{6.70 - 6.20}{3(.0362)}, \frac{7.80 - 7.30}{3(.0362)} \right)$$

$$= \text{Minimum} (4.60, 4.60) = 4.60$$

Usage Variation

A situation similar to the setup problem happens when customers utilize products in diverse applications that affect their capability. Taylor describes such an example where a company produces a brand of water pump that should be able to deliver a volume of 200 liters per minute, plus or minus 10 liters. A capability study performed at the test stand for pump output verifies the process is in control, with high performance capability (the output distribution with the solid line in Figure 13.8).

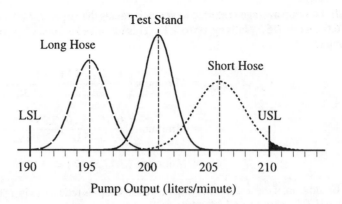

Fig. 13.8 Pump performance at test stand and at customer.

When installed by a customer, the pump is connected to a standard diameter hose, however, hose length can vary depending upon a customer's application. Variation in hose length affects average pump output. As displayed in Figure 13.8, applications with long hoses experience a slightly lower average output than customers using short hoses. Taylor labels this difference in performance "usage variation."

How should capability be reported? As measured at the test stand? This is good information for people manufacturing the pump, but does not reflect the performance level customers will actually experience when using this product in the field. Capability could be reported as a function of hose length, although this may be confusing to customers. If a single capability value is desired, the worst case capability should be reported. This conservative estimate is calculated as follows for the general case where it is not known if the average output of pumps with long hoses is less than that for pumps with short hoses, or whether their standard deviations are similar or not.

$$\hat{C}_{PK. CUSTOMER} = \text{Minimum}\left(\frac{\hat{\mu}_{SHORT} - \text{LSL}}{3\hat{\sigma}_{SHORT}}, \frac{\text{USL} - \hat{\mu}_{SHORT}}{3\hat{\sigma}_{SHORT}}, \frac{\hat{\mu}_{LONG} - \text{LSL}}{3\hat{\sigma}_{LONG}}, \frac{\text{USL} - \hat{\mu}_{LONG}}{3\hat{\sigma}_{LONG}}\right)$$

In order to estimate this index, the relationship between hose length and pump output must be known.

A second capability index, called $C_{P.\ USAGE}$, provides a guideline for pump designers:

$$\hat{C}_{P.\ USAGE} = \frac{\text{Tolerance}}{6\,\text{Maximum}\,(\hat{\sigma}_{SHORT}, \hat{\sigma}_{LONG})}$$

This measure provides a conservative estimate of process precision, which, when factored with the most extreme average output, allows designers to analyze a worst-case scenario.

13.3 Processes with Autocorrelated Measurements

Conventional Shewhart control charts assume measurements collected from a process are independent, *i.e.*, there is no relationship between consecutive measurements. They also assume the process average is stable, which means the best prediction for any future measurement is simply the average of all past measurements. Knowledge that the most recent observation was above the average is not helpful in forecasting the next observation.

For discrete parts manufacturing, the above two assumptions are usually valid. However, dependent, or correlated, measurements are fairly common in continuous operations such as:

- weaving textiles,
- extruding vinyl sheets,
- mixing dry chemicals,
- blending gasoline,
- extruding plastic tubing,
- coating paper,
- applying insulation to wire,
- processing food,
- brewing beer.

For example, if the current sample of beer has a high alcohol content, a sample taken one minute from now will probably also have a high alcohol content. Unlike a process with independent measurements, knowing that the most recent observation is above the average *does* help predict future observations for a process with correlated measurements.

Even data generated in discrete parts manufacturing may display correlation if there is a very small time interval between measurements, like that for automatic gauging systems where every part is measured. Slow movements in ambient plant temperature may cause part size to gradually drift up or down throughout the day. This difficult-to-eliminate cause becomes an integral component of the common-cause variation for this process.

As a result of consecutive measurements being very similar, the average subgroup range is small, resulting in narrow control limits for the $\overline{X}$ chart. In this situation, any inherent drifts in the process average will dramatically increase the false alarm rate of Shewhart charts to the point of rendering them ineffective, and possibly even misleading according to Harris and Ross. In addition, the standard deviation estimated from this $\overline{R}$ value is virtually worthless for estimating process capability (Anderson; Cryer and Ryan; Harding *et al*).

Consider an example from the petrochemical industry where gasoline is blended to a desired octane level. As changes are made to the blending process, the octane level varies also, as witnessed on the *IX & MR* chart displayed in Figure 13.9 (data for this chart are listed in Table 13.2). If the octane measurement taken at time t minus 1 is above the centerline, there is a strong likelihood the next measurement at time t will also be above the centerline. Likewise, when the measurement at t minus 1 is below the centerline, it is very probable the octane measurement for time period t is also below.

Because there is little change from one octane measurement to the next, the moving ranges are relatively small, but display good control on the moving range chart. However, these small moving ranges cause $\overline{MR}$ to also be small, resulting in very "tight" control limits for the *IX* chart. As the process average for octane gradually drifts up and down due to subtle changes in the blending process, a number of octane readings are pushed outside of these narrow control limits, a problem somewhat similar to that for the tool wear and setup situations previously covered. If the cause of this drift is part of the process (as it often is in continuous processes), or cannot be readily adjusted, the operator has an out-of-control signal he cannot correct, leading to either the chart being ignored or the process being over adjusted. The cyclical drift of the process average also negates the usefulness of any Western Electric run rules, thus contributing even more interpretation difficulties.

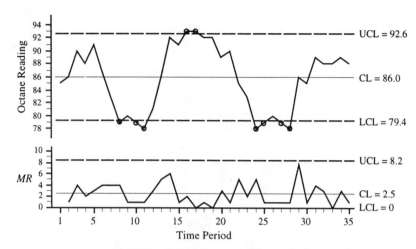

Fig. 13.9 *IX & MR* chart for octane level.

This charting problem is a direct result of high autocorrelation, which is a measure of the degree to which individual measurements from a single process are related to each other. Autocorrelation is occasionally referred to as serial correlation, as it occurs in a time series of measurements, like those produced in the processing industries. Lack of independence in consecutive observations is typically detected in one of two ways: plotting the data on a scatter diagram and looking for a pattern, or calculating the sample autocorrelation and testing to determine if it is significant. Both Bissell (pp. 313-317) and Bingham and Nelson describe a third method, called von Neumann's ratio test (von Neumann), but this technique is seldom applied in practice.

With the first approach, the measurement at time t minus 1 (labeled X_{t-1}) is considered the x coordinate, while the measurement at time t (labeled X_t) becomes the y coordinate for each point plotted on the scatter diagram. X_{t-1} is often referred to in the statistical literature as the measurement for the "lag one" time period. If analysis of the scatter diagram reveals a pattern, then a reasonable prediction of X_t can be made from X_{t-1}, implying that consecutive measurements are *not* independent. If so, standard Shewhart control charts should not be chosen to monitor this process.

The lag two time period, X_{t-2} (or even X_{t-3}), could also be checked as a predictor of X_t. However, the farther apart the measurements, the less likely they are to be correlated, so usually the scatter-diagram check for autocorrelation involves comparing measurements only one time interval apart (Wheeler, 1995, p. 276). In fact, to make autocorrelation negligible, Keller (1993*b*) recommends increasing the sampling interval until consecutive measurements appear independent. Unfortunately, the time between measurements needed under this approach often becomes so large that critical process changes are missed.

The data in Table 13.2 are the first 25 readings from the gasoline-blending process mentioned at the beginning of this section. The first column lists the measurement number in chronological order from time period 1 through 25, while the second column contains the actual octane measurement at time t, labeled as X_t. In the third column, the measurements of column two have been shifted down one row to become X_{t-1}, the lag one time period. Thus, for time period two, X_2 equals 86, while the octane measurement for the lag one time period, X_{2-1}, or X_1, is 85.

Table 13.2 25 octane readings displaying autocorrelation.

Time Period, t	$y = X_t$	$x = X_{t-1}$	$X_t - \overline{X}$	$X_{t-1} - \overline{X}$	$(X_t - \overline{X}) \times (X_{t-1} - \overline{X})$	$(X_t - \overline{X})^2$
1	85	-	-1	-	-	1
2	86	85	0	-1	0	0
3	90	86	4	0	0	16
4	88	90	2	4	8	4
5	91	88	5	2	10	25
6	87	91	1	5	5	1
7	83	87	-3	1	-3	9
8	79	83	-7	-3	21	49
9	80	79	-6	-7	42	36
10	79	80	-7	-6	42	49
11	78	79	-8	-7	56	64
12	81	78	-5	-8	40	25
13	86	81	0	-5	0	0
14	92	86	6	0	0	36
15	91	92	5	6	30	25
16	93	91	7	5	35	49
17	93	93	7	7	49	49
18	92	93	6	7	42	36
19	92	92	6	6	36	36
20	89	92	3	6	18	9
21	90	89	4	3	12	16
22	85	90	-1	4	-4	1
23	83	85	-3	-1	3	9
24	78	83	-8	-3	24	64
25	79	78	-7	-8	56	49
Total	2150				522	658
Average	86					

Each octane measurement, along with its corresponding lag one time period measurement, is plotted as a point on the scatter diagram exhibited in Figure 13.10. Note that the scales on both axes are identical for an autocorrelation scatter diagram. Although data were collected over 25 time periods, there are only 24 plot points because the first measurement (at $t = 1$) has no corresponding lag one time period.

Because the plot points are grouped in a fairly tight ellipsoid around the 45-degree line, the existence of positive autocorrelation is quite likely. Positive autocorrelation means there is a high probability that the current octane reading will be higher than average when the immediately previous reading is higher than average. Negative autocorrelation is the reverse: there is a high probability X_t will be low given that X_{t-1} was high.

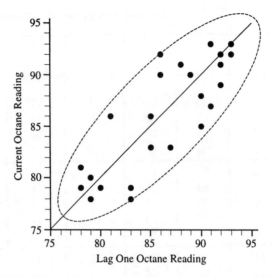

Fig. 13.10   Scatter diagram revealing presence of autocorrelation.

Positive autocorrelation may be due to the "inertia" of a process. The temperature of a solder bath cannot be instantaneously shifted higher or lower. When heated, the bath temperature will slowly rise; when cooling off, the temperature will gradually decrease. If temperature measurements are collected often, they will be quite similar in value, *i.e*, the current temperature is high when the previous reading was high, the current temperature will be low if the previous temperature was low. This type of autocorrelation is far more prevalent in manufacturing than negative autocorrelation.

Had the 45-degree line been the minor axis of the ellipsoid, negative autocorrelation is likely present in the observations. This condition is often caused by "tampering" with the process, *i.e.*, when a high reading is observed, no matter if the resulting plot point is in or out of control, an adjustment is automatically made to lower the process average. When a low reading is seen, the process average is immediately adjusted higher. This situation frequently occurs with many automatic controllers.

The presence of either positive or negative autocorrelation implies the process average is not stable, but moving. In many situations, this movement is inherent to the process, especially in the chemical and processing industries. If so, these drifts in the process average cannot be easily or quickly eliminated and must definitely be taken into consideration when conducting a capability study on this process.

Autocorrelation may be more precisely quantified by computing an estimate of ρ (pronounced "roe"), the coefficient of autocorrelation. Its formula is given below, where j represents the lag period, while m is the number of time periods (Alwan). Thus, ρ_1 measures the amount of correlation between the current reading and the lag one time period reading, whereas ρ_2 would indicate the level of correlation between the current measurement and that for the lag two time period.

$$\hat{\rho}_j = \frac{\sum_{t=j+1}^{m} (X_t - \overline{X})(X_{t-j} - \overline{X})}{\sum_{t=1}^{m} (X_t - \overline{X})^2}$$

The estimate of ρ will always be between -1 and +1. A negative ρ value implies negative autocorrelation, whereas a positive value indicates the presence of positive autocorrelation. ρ values close to zero mean little autocorrelation is present. Autocorrelation begins creating problems for the calculation of meaningful control limits when ρ is greater than .6 (Wheeler, 1995, p. 288). For the octane example presented in Table 13.2, autocorrelation of the current measurement with the lag one measurement is estimated as .793 ($j = 1$, $m = 25$).

$$\hat{\rho}_1 = \frac{\sum_{t=2}^{25} (X_t - \overline{X})(X_{t-1} - \overline{X})}{\sum_{t=1}^{25} (X_t - \overline{X})^2} = \frac{522}{658} = .793$$

With a random process, the sampling distribution of the estimates for ρ is approximately normal, with an average of zero and a standard deviation of 1 over the square root of the number of time periods.

$$\mu_{\hat{\rho}} = 0 \qquad \sigma_{\hat{\rho}} = \frac{1}{\sqrt{m}}$$

Autocorrelation is considered to be significant at the 1 minus α confidence level if the estimate of ρ is more than $Z_{\alpha/2}$ standard deviations from the expected average (a two-sided test, review pages 599-600). With the above average and standard deviation for the sampling distribution of ρ, autocorrelation is significant if the absolute value of the estimate for ρ is greater than $Z_{\alpha/2}$ divided by the square root of m.

$$|\hat{\rho}| > \mu_{\hat{\rho}} + Z_{\alpha/2}\sigma_{\hat{\rho}}$$

$$> 0 + Z_{\alpha/2}\frac{1}{\sqrt{m}}$$

$$> \frac{Z_{\alpha/2}}{\sqrt{m}}$$

For an α level of .05, autocorrelation is significant if the absolute value of the estimate for ρ is greater than 1.96 (review Table 11.1, p. 602) divided by the square root of m. In the octane example, ρ_1 is estimated as .793, while the number of time periods is 25.

$$|\hat{\rho}_1| \overset{?}{>} \frac{Z_{.05/2}}{\sqrt{m}}$$

$$|.793| \overset{?}{>} \frac{1.96}{\sqrt{25}}$$

$$.793 > .392$$

As .793 is substantially greater than .392, autocorrelation with the lag one time period is certainly significant at the α equal .05 level. Thus, traditional control charts should *not* be chosen to monitor the octane level of this blended gasoline because the octane readings cannot be considered independent.

Establishing Stability

When a systematic time-related drift is present in a process, Alwan and Roberts recommend modeling the time series with one of the autoregressive, integrated, moving average (ARIMA) models developed by Box and Jenkins. Unfortunately, fitting an appropriate ARIMA model usually requires complex computer software programs and a level of statistical expertise that is beyond the scope of this text. But once a reasonable model is established, future process measurements may be reliably predicted from the current and previous observations due to the autocorrelated nature of the data. When such a model fits the data well, differences between forecasted and actual measurements, called forecast errors or "residuals," should be relatively small. In addition, a correctly specified model removes autocorrelation from the forecast errors, leaving only random residuals, which may then be correctly monitored on an *IX & MR* chart.

For example, one of the simpler ARIMA models is the exponentially weighted moving average (EWMA), described by Hunter (1986). Montgomery and Mastrangelo have found this model useful for forecasting when there is positive autocorrelation and the process average changes slowly. The EWMA model predicts the measurement for time period t, $\hat{X}_t$, by taking a weighted average of the previous measurement, X_{t-1}, and the forecast for X_{t-1}, labeled $\hat{X}_{t-1}$. The symbol λ (pronounced lam-da) is the weighting factor, which is a positive number less than 1.

$$\hat{X}_t = \lambda X_{t-1} + (1 - \lambda) \hat{X}_{t-1}$$

λ values close to 1 give more weight to the most recent observation and less to previous ones, whereas values close to 0 weigh older observations more than current ones. The precise λ for a particular process is frequently selected as the one which minimizes the sum of the squared residuals (the forecast errors).

Even though the octane level of the blending process is changing quite rapidly, a EWMA model with λ equal to .8 has been applied to the octane data listed in Table 13.2. The expected octane reading for time period 2 is predicted as follows:

$$\hat{X}_2 = .8 X_{2-1} + (1 - .8) \hat{X}_{2-1} = .8 X_1 + .2 \hat{X}_1$$

The octane reading for period 1 is 85. However, as the above forecast for period two is the first to be computed, there is no forecast for period 1. Thus, to begin the model, the overall octane average of 86 becomes $\hat{X}_1$. If this number is not available, substitute the target for the process average. In cases where T is not identified, use the reading for time period 1.

$$\hat{X}_2 = .8 X_1 + .2 \hat{X}_1 = .8(85) + .2(86) = 85.20$$

The actual measurement for period 2 is 86, a difference of .80 from the prediction of 85.20. This .80 value is the residual (forecast error) for period 2 and will be plotted on the *IX* chart for residuals.

The projected reading for period 3 is computed as follows:

$$\hat{X}_3 = .8 X_2 + .2 \hat{X}_2 = .8(86) + .2(85.20) = 85.84$$

With a real octane reading of 90, the residual for period 3 is 4.16 (90 minus 85.84). This residual is also plotted on the *IX* chart, while the absolute value of the difference between residuals for periods 3 and 2 becomes the moving range plot point for the *MR* chart.

$$MR \text{ Plot Point}_t = |\text{Residual}_t - \text{Residual}_{t-1}|$$

$$MR \text{ Plot Point}_3 = |\text{Residual}_3 - \text{Residual}_2|$$

$$= |4.16 - .80| = 3.36$$

For period 4, the anticipated octane level is 89.17.

$$\hat{X}_4 = .8X_3 + .2\hat{X}_3 = .8(90) + .2(85.84) = 89.17$$

As the actual reading for period 4 is 88, the residual is -1.17 (88 minus 89.17), while the moving range is 5.33 (|-1.17 - 4.16|). Results for the remaining time periods are listed in Table 13.3, along with their residuals. Note that in order to develop a reliable prediction model, octane readings should be taken at equal time intervals.

Table 13.3 Application of the EWMA model to octane readings.

Time Period, t	Actual X_t	Forecast $\hat{X}_t$	Residual $X_t - \hat{X}_t$	Moving Range
1	85	86.00	-1.00	-
2	86	85.20	.80	1.80
3	90	85.84	4.16	3.36
4	88	89.17	-1.77	5.33
5	91	88.23	2.77	3.94
6	87	90.45	-3.45	6.22
7	83	87.69	-4.69	1.24
8	79	83.94	-4.94	.25
9	80	79.99	.01	4.95
10	79	80.00	-1.00	1.01
11	78	79.20	-1.20	.20
12	81	78.24	2.76	3.96
13	86	80.45	5.55	2.79
14	92	84.89	7.11	1.56
15	91	90.58	.42	6.69
16	93	90.92	2.08	1.66
17	93	92.58	.42	1.66
18	92	92.92	-.92	1.34
19	92	92.18	-.18	.74
20	89	92.04	-3.04	2.86
21	90	89.61	.39	3.43
22	85	89.92	-4.92	5.31
23	83	85.98	-2.98	1.94
24	78	83.60	-5.60	2.62
25	79	79.12	-.12	5.48
Total			-9.34	70.34
Average			-.37	2.93

Determining stability for a process with autocorrelation necessitates keeping two separate charts. The first is a run chart of the forecasted values, called a "common-cause" chart by Alwan. Forecasted values for the octane example are plotted on the run chart exhibited in Figure 13.11. Visual analysis of this graph hopefully provides a better understanding of the common causes responsible for the systematic drifts in the process average. Eliminating the underlying cause of these drifts will help stabilize the process average and, thereby, substantially increase performance capability.

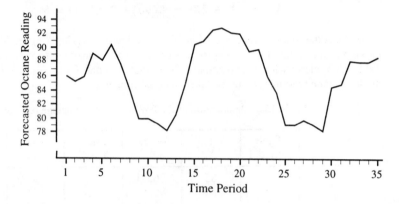

Fig. 13.11 Run chart for octane measurements forecasted by the EWMA model.

Occasionally, action limits are added to signal when an operator should recenter the process average. These limits are not statistically determined, but are based on cost considerations, in particular, when the cost of running off target product exceeds the cost of adjusting the process average back to the target (see Arcelus and Banerjee, 1985, 1987; Gibra, 1967). A point outside one of these action limits does not imply assignable-cause variation is present, only that a predetermined process adjustment should be made. Thus, a hunt for an assignable source of variation should not be initiated.

The second required chart is an *IX & MR* chart for the residuals, referred to as the "special-cause" chart. Analysis of this chart helps identify when assignable-cause variation is present, as the expected variation in the process average due to autocorrelation has already been removed by the EWMA model, leaving only unexpected variation. For a stable process, and a correctly specified model, this unexplained variation should be random, with the residuals being independent. A check of their autocorrelation should be made, just as was done for the original readings. If there is significant autocorrelation among the residuals of this model, a different ARIMA model (perhaps another EWMA model, but with a different value for λ) must be selected and this procedure repeated.

For the octane example, control limits for the special-cause *IX & MR* chart are calculated from the summary statistics of Table 13.3. Beginning with the *MR* chart:

$$\overline{MR} = \frac{\sum\limits_{i=2}^{k} MR_i}{k-1} = \frac{70.34}{24} = 2.93$$

$$UCL_{MR} = 3.267(2.93) = 9.57 \qquad\qquad LCL_{MR} = 0(2.93) = 0$$

For the *IX* chart:

$$\overline{X} = \frac{\sum\limits_{i=1}^{k} X_i}{k} = \frac{-9.34}{25} = -.37$$

$$UCL_{IX} = \overline{X} + 2.66\,\overline{MR} = -.37 + 2.66\,(2.93) = 7.42$$

$$LCL_{IX} = \overline{X} - 2.66\,\overline{MR} = -.37 - 2.66\,(2.93) = -8.16$$

When no significant autocorrelation exists between the residuals, and this special-cause chart does not display good control, assignable-cause variation unrelated to the systematic drifts in the process average is most likely present, and attention must focus on possible sources of this extraneous process disturbance. The *IX* chart for the octane residuals depicted in Figure 13.12 suggests an assignable cause of variation acted on this process at time period 29, which increased the process average significantly above the level forecasted. This abnormal deviation for reading 29 went undetected on the *IX* chart of actual octane readings introduced in Figure 13.9 (p. 782) because assignable-cause variation was confounded with the common-cause variation responsible for drifts in the process average.

When the *IX & MR* chart for residuals indicates a good state of control, process improvement efforts must concentrate on understanding which common causes are responsible for the autocorrelation. Because these causes make the process average drift away from the target, they should be found and eliminated so the process can produce more on-target product. Note that the residuals plotted on the *IX* chart in Figure 13.12 appear cyclic instead of random, suggesting a better forecasting model should be sought (one that better reacts to the rapid transitions in average octane). For additional insight on analyzing this control chart for residuals, see Gitlow and Oppenheim. The above approach may also be applied to processes generating autocorrelated attribute data.

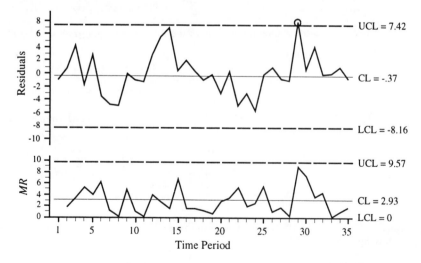

Fig. 13.12 The sequence of residuals is plotted on an *IX & MR* chart.

Other control charting ideas for continuous processes have been proposed by Box, Box and Kramer, Dodson, Ermer, Gilbert *et al*, Hunter (1989), Keller (1993*b*, 1993*c*), Liao *et al*, Lucas and Saccucci, Mamzic and Tucker (1988*a*, 1988*b*), Montgomery (pp. 341-351), Montgomery and Mastrangelo, Negiz and Cinar, Ng and Case, Shaw, Vander Wiel, and finally Wardell *et al*. The following authors provide examples of SPC and quality improvement in continuous processing industries that may or may not have autocorrelation: Abraham and Whitney, ASQC (1987, 1996*a*, 1996*b*), Harding *et al*, Hill and Bishop, Hunter (1986), Juran *et al* (pp. 9-31 to 9-34), Keller (1993*a*), Martin, McNeese and Klein, Montgomery *et al*, Palm, Pugh, Pyzdek (pp. 136-164), and Schneider and Pruett.

Measuring Process Capability

Evaluating performance capability is difficult under these circumstances due to the systematic drifts affecting the process average. Fortunately, potential capability is easily assessed since it is not influenced by the location of the process average.

Potential Capability

The short-term variation is estimated from the moving range chart of residuals, which is relatively free from the effects of drifts in μ. The subscript "*AUTO*" is for autocorrelated and has been added to denote this particular version of σ_{ST}.

$$\hat{\sigma}_{ST,AUTO} = \frac{\overline{MR}}{1.128}$$

This standard deviation is used to estimate potential capability, which is the best this process can do (instantaneous capability), assuming the autocorrelation could be made to completely disappear and the process average centered at M.

$$\hat{C}_{P,AUTO} = \frac{\text{Tolerance}}{6\hat{\sigma}_{ST,AUTO}}$$

When a target other than the middle of the tolerance is specified, replace $C_{P,AUTO}$ with the $C^*_{P,AUTO}$ index.

$$\hat{C}^*_{P,AUTO} = \text{Minimum}\left(\frac{T - LSL}{3\hat{\sigma}_{ST,AUTO}}, \frac{USL - T}{3\hat{\sigma}_{ST,AUTO}} \right)$$

For features with unilateral specifications where T is given, switch to either the $C^T_{PL,AUTO}$ or $C^T_{PU,AUTO}$ index.

$$\hat{C}^T_{PL,AUTO} = \frac{T - LSL}{3\hat{\sigma}_{ST,AUTO}} \qquad\qquad \hat{C}^T_{PU,AUTO} = \frac{USL - T}{3\hat{\sigma}_{ST,AUTO}}$$

Example 13.8 Octane Level

For the *IX & MR* chart of residuals in Figure 13.12, $\sigma_{ST,AUTO}$ is estimated from the average moving range of 2.93.

$$\hat{\sigma}_{ST,AUTO} = \frac{\overline{MR}}{d_2} = \frac{2.93}{1.128} = 2.60$$

With an USL of 98 and a LSL of 74 for octane, $C_{P,AUTO}$ is estimated as:

$$\hat{C}_{P,AUTO} = \frac{\text{Tolerance}}{6\hat{\sigma}_{ST,AUTO}} = \frac{98 - 74}{6(2.60)} = 1.54$$

This measure reveals a fair amount of potential capability, meaning the process average for octane may drift somewhat before being in danger of producing nonconforming gasoline. Note that the overall, "long-term" standard deviation estimated from all 25 original octane readings is more than twice as large as the estimate of 2.60 for $\sigma_{ST,AUTO}$.

$$S_{TOT} = \sqrt{\frac{\sum\limits_{i=1}^{25}(X_i - \overline{X})^2}{25 - 1}} = 5.236 \qquad \hat{\sigma}_{LT} = \frac{S_{TOT}}{c_4} = \frac{5.236}{.9896} = 5.29$$

Anderson describes how, by combining measurement-to-measurement variation with the time-to-time drifts in the process average, σ_{LT} becomes artificially inflated. Because of this problem, Harding *et al* claim this statistic is rendered meaningless as a measure of true process variation and cannot be used for assessing any type of process capability, or even for calculating control limits. Thus, long-term measures estimated from σ_{LT} (like P_P) are not reliable, and certainly not comparable to correctly calculated indexes of long-term capability for other processes (review Section 3.3, p. 42). For the interested reader, Gardiner and Mitra have proposed, and evaluated, other modified estimators of the long-term standard deviation when autocorrelation exists among the observations.

Performance Capability

While no exact estimate for performance capability can be made due to the continual shifting of the process average, a lower bound may be established. This is accomplished by considering the two worst cases for the wandering process average, *i.e.*, the lowest and highest forecasted averages from the common-cause chart, labeled $\hat{\mu}_L$ and $\hat{\mu}_H$, respectively. These two extreme averages are ascertained from the run chart of forecasted values. Combining these averages with an estimate of the short-term standard deviation generates a lower bound for performance capability, called $C_{PK,AUTO}$ (see also Shore, Zhang).

$$\hat{C}_{PK,AUTO} = \text{Minimum}\left(\frac{\hat{\mu}_L - LSL}{3\hat{\sigma}_{ST,AUTO}}, \frac{USL - \hat{\mu}_H}{3\hat{\sigma}_{ST,AUTO}}\right)$$

The above formula computes how many three "sigma" spreads fit between the lowest predicted average and the LSL, as well as between the highest expected average and the USL. The smaller of these two values is the capability that can be promised, and delivered, to all customers. Each customer would receive product manufactured with at least this quality level, and most will receive much better quality because the process average spends the majority of its time between these two extremes. But to ensure no customers are disappointed by getting a lesser quality level than advertised, this worst case must be reported.

In order to improve performance capability, the average moving range for the residuals must be decreased, which will reduce $\sigma_{ST,AUTO}$. A second strategy is to minimize (or eliminate) the cause of the systematic drifts in μ. If this is accomplished and the process centered at M, performance capability will equal potential capability. This substantially higher $C_{PK,AUTO}$ may be publicized to all customers, while operators monitoring this process may return to using a standard *IX & MR* chart for the actual measurements, which is much easier to maintain and interpret.

Herman discusses the importance of quantifying and evaluating how large a role measurement variability plays in affecting capability assessments made for continuous processes (see also Section 10.4 on page 581).

Example 13.9 Octane Level

From the run chart of forecasted octane levels exhibited in Figure 13.11 (p. 788), $\hat{\mu}_L$ and $\hat{\mu}_H$ are found to be 78.24 and 92.92, respectively. With the short-term standard deviation estimated as 2.60 in Example 13.8 (p. 790), $C_{PK,AUTO}$ is estimated as:

$$\hat{C}_{PK,AUTO} = \text{Minimum}\left(\frac{\hat{\mu}_L - \text{LSL}}{3\hat{\sigma}_{ST,AUTO}}, \frac{\text{USL} - \hat{\mu}_H}{3\hat{\sigma}_{ST,AUTO}}\right)$$

$$= \text{Minimum}\left(\frac{78.24 - 74}{3(2.60)}, \frac{98 - 92.92}{3(2.60)}\right)$$

$$= \text{Minimum}(.54, .65) = .54$$

As $C_{PK,AUTO}$ is quite a bit less than 1.0, work must begin to either shrink $\sigma_{ST,AUTO}$ or diminish the amount of drifting in the process average.

13.4 Short Run Processes with Many Different Part Numbers

In most job shops or companies utilizing just-in-time manufacturing, many different part numbers are produced on the same piece of equipment. Because each part number may have a different average and standard deviation, the application of conventional SPC charts requires a separate chart for each part number, meaning hundreds of charts for every operation. And if this isn't bad enough, because of the limited amount of data associated with short production runs, a run is often over before control limits can be calculated. Since proper interpretation can't be done without limits, these charts aren't of much use for real-time process control.

In addition, by dividing process data among several charts, detecting any time-related changes occurring over two (or more) consecutively run part numbers becomes extremely difficult. As the goal of SPC is *process* control, not *part number* control, traditional charts serve very little purpose in these applications and are soon discarded.

Instead of forcing this multitude of separate control charts on shop floor personnel, several charting alternatives have been proposed over the years (Farnum; Hillier; Proschan and Savage; Quesenberry, 1991; Yang and Hillier). One of these, the Short Run $\overline{X}$, R chart (Bothe, 1989) is very practical and has become one of the most popular.

Establishing Stability

The Target chart data transformation mentioned in Section 13.2 (p. 773) works well when the $\overline{R}$ values for different setups or part numbers are similar, but is inappropriate if the $\overline{R}$ values also vary. To handle this possibility, a control chart is needed whose control limits for both the $\overline{X}$ chart and the R chart are independent of $\overline{\overline{X}}$ and $\overline{R}$. As will be seen shortly, the Short Run $\overline{X}$, R chart is designed to meet these criteria.

As might be expected, the formulas for this new chart are unlike those for conventional charts. To understand why, consider the standard range chart, which is in-control when the plotted range of a given subgroup falls between its upper and lower control limits.

$$LCL_R < R < UCL_R$$

$$D_3\overline{R} < R < D_4\overline{R}$$

When $\overline{R}$ changes from part number to part number, so do the control limits because they are a function of $\overline{R}$. However, dividing this inequality through by $\overline{R}$ yields control limits independent of $\overline{R}$.

$$\frac{D_3\overline{R}}{\overline{R}} < \frac{R}{\overline{R}} < \frac{D_4\overline{R}}{\overline{R}}$$

$$D_3 < \frac{R}{\overline{R}} < D_4$$

If the actual range of a subgroup is divided by the expected $\overline{R}$ value (referred to as "target $\overline{R}$") for that part number, no matter which part number is being run, the resulting ratio can be plotted on a special range chart, called the Short Run R chart, with D_3 as the lower control limit and D_4 as the upper. Note that both limits are independent of $\overline{R}$, so even if $\overline{R}$ varies significantly from one part number to the next, these new control limits are not affected as long as the subgroup size remains constant.

The data transformation for plotting points on this Short Run R chart consists of dividing the actual subgroup range by the expected, or target $\overline{R}$ for that part number. This formula is written as follows:

$$\text{Short run } R \text{ plot point} = \frac{R}{\text{Target } \overline{R}}$$

This ratio standardizes (rescales) the range from any part number so it fits on the same Short Run R chart. If the actual range of a subgroup equals the target $\overline{R}$ for that part number, this ratio equals 1. This value becomes the centerline of the Short Run R chart, as is displayed in Figure 13.13. For this chart ($n = 3$), the UCL equals 2.575, which is the D_4 factor for a subgroup size of 3 (see Appendix Table I on page 842).

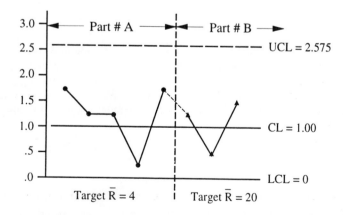

Fig. 13.13 Control limits for a Short Run R chart depend only on n.

For example, the actual range of the first subgroup for part number A is 7. Dividing this by a target $\overline{R}$ of 4 yields the ratio 1.75 (7/4), which is plotted on the Short Run R chart in Figure 13.13. This method of coding continues for as long as A is being produced. When the process is eventually switched over to manufacture part number B, the range of B's first subgroup turns out to be 25, which is substantially different from the ranges for A. However, when this actual range is divided by the target $\overline{R}$ of 20 for part number B, the resulting ratio of 1.25 (25/20) can be plotted on the same Short Run R chart containing A's data. These two part numbers, with significantly different $\overline{R}$ values, may have their scaled range values plotted on the same Short Run R chart because its control limits are not a function of $\overline{R}$. But what about the $\overline{X}$ chart? Its limits are a function of not only $\overline{R}$, but $\overline{\overline{X}}$ as well. Both of these must be eliminated from the $\overline{X}$ control limit formulas before part numbers with significantly different $\overline{\overline{X}}$ and/or $\overline{R}$ values may be plotted on the same chart.

The conventional $\overline{X}$ chart indicates an in-control condition when a subgroup average falls between its control limits.

$$LCL_{\overline{x}} < \overline{X} < UCL_{\overline{x}}$$

$$\overline{\overline{X}} - A_2\overline{R} < \overline{X} < \overline{\overline{X}} + A_2\overline{R}$$

Subtracting $\overline{\overline{X}}$ throughout this inequality yields the following:

$$\left(\overline{\overline{X}} - A_2\overline{R}\right) - \overline{\overline{X}} < \overline{X} - \overline{\overline{X}} < \left(\overline{\overline{X}} + A_2\overline{R}\right) - \overline{\overline{X}}$$

$$-A_2\overline{R} < \overline{X} - \overline{\overline{X}} < +A_2\overline{R}$$

These control limit formulas are now free of $\overline{\overline{X}}$, but remain a function of $\overline{R}$, and thus could still vary from part number to part number. This lingering problem is remedied by dividing all portions of the last inequality by $\overline{R}$.

$$\frac{-A_2\overline{R}}{\overline{R}} < \frac{\overline{X} - \overline{\overline{X}}}{\overline{R}} < \frac{+A_2\overline{R}}{\overline{R}}$$

$$-A_2 < \frac{\overline{X} - \overline{\overline{X}}}{\overline{R}} < +A_2$$

Because these limits of $\pm A_2$ are now also independent of $\overline{R}$, part numbers with different averages *and* different ranges can be plotted on this same special $\overline{X}$ chart, called the Short Run $\overline{X}$ chart. This data transformation (which is similar to finding Z values) subtracts the desired process average, called target $\overline{\overline{X}}$, from the actual subgroup average, then scales this difference by dividing by the expected average range, target $\overline{R}$, for this part number. This is expressed as:

$$\text{Short run } \overline{X} \text{ plot point } = \frac{\overline{X} - \text{Target } \overline{\overline{X}}}{\text{Target } \overline{R}}$$

This coded $\overline{X}$ plot point goes on the Short Run $\overline{X}$ chart, which has an *UCL* of $+A_2$ and a *LCL* of $-A_2$. If a subgroup average equals target $\overline{\overline{X}}$, the coded plot point equals zero. This zero value becomes the centerline of the Short Run $\overline{X}$ chart, an example of which is displayed in Figure 13.14 for n equal to 3. With this subgroup size, A_2 is 1.023.

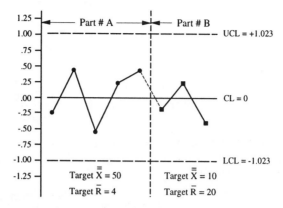

Fig. 13.14 Part numbers with different averages may be plotted on the same Short Run $\overline{X}$ chart.

If the first subgroup average is 49 for part number A, the corresponding plot point for this chart is -.25 (recall that target $\overline{R}$ for A equals 4).

$$\text{Short run } \overline{X} \text{ plot point } = \frac{\overline{X} - \text{Target } \overline{\overline{X}}}{\text{Target } \overline{R}} = \frac{49 - 50}{4} = -.25$$

Control limits for both Short Run $\overline{X}$ and Short Run R charts are independent of $\overline{\overline{X}}$ and $\overline{R}$. Part numbers with significantly different $\overline{\overline{X}}$ or $\overline{R}$ values can now be plotted on the same Short Run $\overline{X}$, R chart via these special data transformations. As long as a constant subgroup size is maintained, an operator needs only one chart to monitor all part numbers produced on his/her process (for other short-run charting ideas, see Wise and Fair).

Another important advantage of the Short Run $\overline{X}$, R chart is that these constant control limits can be used when beginning the first control chart, with the very first plot point. Waiting to collect 20 subgroups is no longer necessary, which is especially ideal for short runs. Control limits never need to be calculated, or recalculated, because they are constant for a given subgroup size. However, if changes are detected through out-of-control conditions in either the grand average or the variation of any part number, target values may need recalculation to accurately reflect the current state of the process when running this part number.

If dealing with attribute data, the traditional c, u, np, and p charts can be converted to handle short-run processes in a manner similar to that done for creating the Short Run $\overline{X}$ chart. For instance, the data transformation for the Short Run c chart is derived as follows (Bothe, 1991, pp. 34-41):

$$LCL < c < UCL$$

$$\overline{c} - 3\sqrt{\overline{c}} < c < \overline{c} + 3\sqrt{\overline{c}}$$

$$(\overline{c} - 3\sqrt{\overline{c}}) - \overline{c} < c - \overline{c} < (\overline{c} + 3\sqrt{\overline{c}}) - \overline{c}$$

$$-3\sqrt{\overline{c}} < c - \overline{c} < +3\sqrt{\overline{c}}$$

$$\frac{-3\sqrt{\overline{c}}}{\sqrt{\overline{c}}} < \frac{c - \overline{c}}{\sqrt{\overline{c}}} < \frac{+3\sqrt{\overline{c}}}{\sqrt{\overline{c}}}$$

$$-3 < \frac{c - \overline{c}}{\sqrt{\overline{c}}} < +3$$

This makes the plot point:

$$\text{Short run } c \text{ plot point} = \frac{c - \text{Target } \overline{c}}{\sqrt{\text{Target } \overline{c}}}$$

Control limits for this standardized chart are ±3, with a centerline of 0. Plot point formulas for the other short-run attribute charts are given below, and all have the same control limits of ±3, with centerlines of 0.

$$\text{Short run } u \text{ plot point} = \frac{u - \text{Target } \overline{u}}{\sqrt{(\text{Target } \overline{u})/n}}$$

$$\text{Short run } np \text{ plot point} = \frac{np - \text{Target } n\overline{p}}{\sqrt{\text{Target } n\overline{p}\,[\,1 - (\text{Target } n\overline{p})/n\,]}}$$

$$\text{Short run } p \text{ plot point} = \frac{p - \text{Target } \overline{p}}{\sqrt{[\,\text{Target } \overline{p}\,(\,1 - \text{Target } \overline{p}\,)\,]/n}}$$

Target values for a given part number are normally determined from:

- Prior traditional control charts for that part number (Gibra, 1971).

- Historical data, usually final inspection records, for this part number (Eibl *et al*).

- Prior experience on similar, or surrogate, part numbers (Bothe, 1991, pp. 49-50).

Some practitioners try to base target values on the specification limits and/or capability requirements for a part number, but this is not advisable for general use, as control limits should be based solely on actual process performance. For additional case studies and applications of SPC to short-run production, see Bothe (1991), J.T. Burr, Cook, Koons and Luner, Thompson, as well as Wheeler (1991).

Measuring Process Capability

Once control is established, process capability must be estimated separately for each part number because the average, standard deviation, and specification limits for one part number could vary considerably from every other part number. Temptation often arises to conduct a capability analysis on the combined data of several part numbers, which is not a recommended practice. If the part numbers are truly different, which is most likely why they have been designated as unique part numbers, then separate capability studies must be performed. Section 3.3 (p. 42) demonstrated the folly of estimating process parameters, and then capability, from the combination of dissimilar output distributions. If, by chance, the process outputs for these part numbers are in fact identical, there is still no sense in combining the data, as a capability study on any one representative part number will suffice for the rest.

By their nature, parts run in small quantities are not on the process for long and, therefore, do not experience all possible shifts and drifts in the process average due to different operators, diverse shipments of raw materials, tool wear, temperature changes, etc. Thus, any measure of variation is unlikely to reflect the full extent of long-term process variation. The outcome of any capability study based on short runs must therefore be treated as optimistic, or the best possible. Mindful of this, some practitioners prefer to report capability measures based on the long-term standard deviation. If σ_{ST} is preferred, a higher capability goal is quite often specified than that for a similar part number produced in a long, continuous run.

Confidence bounds (review Chapter 11) are strongly recommended for any measure computed from limited data because estimates derived from small samples may differ substantially from study to study (see also Kalyanasundaram and Balamurali). Verifying the normality assumption is also much more difficult with small samples. A procedure relying on Monte Carlo simulation for dealing with this particular problem is offered by Constable and Hobbs. Another approach for assessing capability from small samples is presented by Little and Harrelson, who illustrate their method, which incorporates the Student's t distribution, with an example drawn from the aerospace industry.

Example 13.10 Screw Machine

A job shop produces small quantities of various part numbers on a screw machine. After control is demonstrated with a Short Run $\overline{X}$, R chart ($n = 3$), a capability study is undertaken on the length characteristic of part number DW-1301. From the 8 in-control subgroups collected for this part number, the average and long-term standard deviation for length are estimated as 2.171 cm and .034 cm, respectively. With a specification of $2.200 \pm .150$ cm for length, and assuming normality, the long-term performance capability for part number DW-1301 is estimated with P_{PK} as 1.19. It is important to realize this result is meaningful for only part number DW-1301. Other part numbers produced on this same screw machine will likely have different averages, standard deviations, specification limits and, thus, unique levels of performance capability. Even though all part numbers may be plotted on the same chart, each must be evaluated with its own capability study.

$$
\hat{P}_{PK, DW\text{-}1301} = \text{Minimum}\left(\frac{\hat{\mu}_{DW\text{-}1301} - LSL_{DW\text{-}1301}}{3\hat{\sigma}_{LT, DW\text{-}1301}}, \frac{USL_{DW\text{-}1301} - \hat{\mu}_{DW\text{-}1301}}{3\hat{\sigma}_{LT, DW\text{-}1301}} \right)
$$

$$
= \text{Minimum}\left(\frac{2.171 - 2.050}{3(.034)}, \frac{2.350 - 2.171}{3(.034)} \right)
$$

$$
= \text{Minimum}\,(1.19, 1.75) = 1.19
$$

With just eight subgroups of size 3 each, the lower .95 confidence bound for the P_{PK} index (p. 654) of part number DW-1301 is only .90, meaning there is very little confidence this process meets the minimum capability requirement of 1.00 for this product. Improvement efforts must focus on both better centering and reducing process variation when this part number is run in the future.

$$
\underset{\sim}{P}_{PK, DW\text{-}1301} = \hat{P}_{PK, DW\text{-}1301} - Z_{.05} \sqrt{ \frac{1}{9kn} + \frac{\hat{P}_{PK, DW\text{-}1301}}{2(kn - 1)} }
$$

$$
= 1.19 - 1.645 \sqrt{ \frac{1}{9(8)(3)} + \frac{1.19}{2[\,8(3) - 1\,]} } = .90
$$

Machine Capability Studies for Short Runs

At times, a machine capability study must be conducted for a piece of equipment that is scheduled to be employed primarily for short runs of many diverse part numbers (see also Deleryd and Vannman). As only a limited amount of parts can usually be produced during the runoff test, one must study the worst case, *e.g.*, tightest tolerance, longest or shortest part, hardest or softest material, smoothest or roughest surface finish, highest or lowest moisture content. To be even more conservative, base all capability estimates on σ_{LT}, which also makes more efficient use of the limited data available. If warranted, machine capability may be expressed as a function of certain process conditions, *e.g.*, capability equals a function of part length, type of fixture, material, tooling, feed rate, speeds, etc. (Traver and Davis). A matrix can be constructed depicting capability versus these conditions, an example of which is presented in Table 13.4 (see also Tables 6.8 and 6.9 on page 223).

Table 13.4   Machine P_{PK} as a function of material type and part diameter.

Part Diameter	P_{PK} by Material Type	
	Aluminum	Brass
24.00	1.35	1.48
24.25	1.41	1.78
24.50	1.49	1.86
24.75	1.56	1.93

When a very few pieces of a part number are produced, capability estimates based on the resulting small number of measurements may be unreliable. Assuming capability is comparable to that for a surrogate process, or for a similar part number run over this process, may offer a better approach for decision making. This is especially true in new product development work, as there is typically an older version of the product currently in production utilizing a related manufacturing technology.

13.5   Products Having Within-Piece Variation

There are three important families of variation to monitor in some processes: within-piece, piece-to-piece, and time-to-time. For example:

- a metal block is checked for thickness at each end;
- two readings (top and bottom) from each vat of paint are collected to check viscosity;
- the surface finish of every clutch plate is checked at three places;
- three samples are taken from each batch of a certain chemical to monitor purity;
- the plating thickness of a metal shaft is measured in four spots;
- extruded vinyl is measured for thickness at five points across the sheet;
- the weight of solder paste silk screened onto printed circuit boards is measured at six locations per board.

The traditional $\overline{X}$, R chart assumes only *one* measurement per piece is made and recorded. In all the above examples, more than one measurement per piece is collected, usually because there is variation occurring within the piece that could affect product performance and, therefore, customer satisfaction. This section explains how these "extra" measurements can be monitored on a special control chart so stability may be established, which then permits a proper assessment of process capability.

Establishing Stability

Conventional Shewhart charts track information about just two families of variation: piece-to-piece on the range chart, and time-to-time on the average chart. No space is available for charting within-piece variation as these charts expect each piece to generate only one measurement. The 3-D $\overline{X}$, R chart (Bothe, 1990; Janis) allows all three of these important sources of process variation to be simultaneously displayed on a single chart. When there is just one piece per subgroup, Wheeler and Chambers (pp. 221-226) explain how to construct a 3-D *IX & MR* chart.

A process molds ceramic spacers for use in assembling groups of printed circuit boards. A critical quality characteristic is the thickness of this spacer. The inspection routing requires each spacer to be measured for thickness in four locations: the first at random, the others at 90-degree increments from this first point, as is illustrated in Figure 13.15.

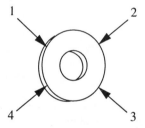

Fig. 13.15 The thickness of a ceramic spacer is measured in four locations.

Three consecutive spacers are taken from the process to form a subgroup of size three ($n = 3$). The thickness of each spacer, as a deviation from the nominal, is measured four times ($j = 4$) at 90-degree increments. Thickness readings from the first subgroup collected are recorded in Table 13.5.

Table 13.5 Thickness readings for the first subgroup of three spacers.

Piece #	Thickness Measurement			
	1st	2nd	3rd	4th
1	3	3	0	2
2	4	4	2	3
3	1	3	0	5

Within-piece variation, labeled R_W, is found for each spacer by calculating the range of the four thickness readings for that piece. For the first spacer, the largest thickness measurement is 3, while the smallest is 0. Thickness variation within this first piece is therefore 3 (3 minus 0), with this value recorded in the second column of Table 13.6. R_W is 2 for the second piece (4 minus 2), and 5 for the third (5 minus 0).

The average thickness of the first spacer, labeled $\overline{X}_P$, is computed by summing its four within-piece measurements, then dividing this total by 4 (in general, by j). The result of 2.0 (8/4) is the best one-number representation of the overall thickness for this first piece. A similar calculation is repeated for the other two spacers in this first subgroup, with their averages listed in column four of Table 13.6.

Table 13.6 3-D analysis of the first subgroup.

Piece #	R_W	ΣX	$\overline{X}_P$
1	3	8	2.0
2	2	13	3.3
3	5	9	2.2
		$\Sigma \overline{X}_P =$	7.5
		$\overline{X} =$	2.5
		$R_P =$	1.3

Average spacer thickness for this subgroup is computed by summing all three $\overline{X}_P$ values, then dividing this total by 3 (the subgroup size). This value of 2.5 (7.5/3) depicts average spacer thickness, or $\overline{X}$, for the entire subgroup.

A second range measuring piece-to-piece variation, labeled R_P, is calculated by taking the difference between the highest and lowest $\overline{X}_P$ values. In this first subgroup, R_P equals 3.3 minus 2.0, or 1.3, and this value is recorded at the bottom of Table 13.6.

Through the above analysis, all three families of variation are quantified for this first subgroup by one of these three subgroup statistics: R_W (of which there are three), R_P, or $\overline{X}$. With the traditional $\overline{X}$, R chart, there is only room for *two* of these five values: R_P is plotted on the range chart, while $\overline{X}$ is monitored on the average chart. To monitor within-piece variation as well, a second range chart for tracking R_W must be added, as was done at the bottom of the chart displayed in Figure 13.16. Because three dimensions of the process are portrayed on this expanded chart, it is referred to as a 3-D $\overline{X}$, R chart.

However, there are three R_W values (3, 2, and 5) calculated for this first subgroup. Of these, only the maximum needs to be plotted, because if it is below the *UCL*, then so are the remaining two. If the maximum R_W is above the *UCL*, the within-piece range has changed significantly and investigative action must begin, no matter if the other two are in or out of control. Thus, only three points ($\overline{X}$, R_P, and maximum R_W) are plotted on the 3-D $\overline{X}$, R chart.

To determine if the process is in control or not, plot points must be compared to their respective control limits and centerlines. Formulas for these values are given below, beginning with the range-within chart.

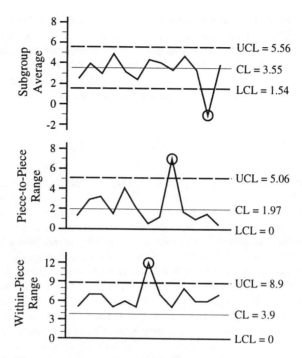

Fig. 13.16 Three families of variation are monitored on a single 3-D $\overline{X}$, R chart.

Three R_W values (one for each piece) are computed for every subgroup. When k subgroups are accumulated, there is a total of kn range-within values, *all* of which are included in figuring their overall average.

$$\overline{R}_W = \frac{\sum\limits_{i=1}^{kn} R_{W,i}}{kn}$$

The subgroup size for the range-within chart, which determines what D_3 and D_4 factors are needed for calculating the control limits, is j, the number of measurements per piece. In this example, j equals 4.

$$LCL = D_3 \overline{R}_W \qquad\qquad UCL = D_4 \overline{R}_W$$

For the piece-to-piece range chart, there is only one R_P per subgroup. Sum these and divide by k to get their overall average.

$$\overline{R}_P = \frac{\sum\limits_{i=1}^{k} R_{P,i}}{k}$$

Control limits for this middle chart are computed as follows, where the subgroup size equals n, the number of pieces in the subgroup.

$$LCL = D_3 \overline{R}_P \qquad\qquad UCL = D_4 \overline{R}_P$$

The centerline and control limits for the $\overline{X}$ portion of the 3-D chart are calculated last. Sum the one $\overline{X}$ from each subgroup, then divide this total by k to get the grand average. The subgroup size for this 3-D $\overline{X}$ chart equals n, the number of pieces in each subgroup, which in this example is 3. This value of n also determines the proper A_2 factor for computing control limits. Be careful to use the average range of piece-to-piece variation, $\overline{R}_P$, rather than within-piece variation, to calculate the $\overline{X}$ control limits.

$$\overline{\overline{X}} = \frac{\sum\limits_{i=1}^{k} \overline{X}_i}{k} \qquad LCL_{\overline{x}} = \overline{\overline{X}} - A_2\overline{R}_P \qquad UCL_{\overline{x}} = \overline{\overline{X}} + A_2\overline{R}_P$$

To interpret this chart, begin at its bottom. If a point is above the *UCL* on the within-piece range chart (see the seventh subgroup in Figure 13.16), something has caused the variation within a single piece to significantly increase. Look for factors that act on the process *during* the cycle time of one part. Note that there will usually be a run above the centerline of this bottom chart, as only the *maximum* of the three R_W values for each subgroup is plotted.

When a point exceeds the *UCL* of the piece-to-piece range chart, a substantial change has happened in the short time interval occurring between pieces. Consider those causes that may affect this process *between* the cycles of two consecutively made pieces.

Should a subgroup average ever breach one of the 3-D $\overline{X}$ chart's control limits, some extraordinary change affecting all pieces in that subgroup occurred in the interval since the last subgroup was taken. Investigative action should focus on assignable causes that are time related (different operator, different lot of material, temperature change).

One note of caution: the 3-D chart should be chosen only when there is more than one measurement of the same characteristic per piece and within-piece variation is expected to be dramatically different from piece-to-piece variation. In the ceramic spacer example shown in Figure 13.16, the average within-piece range is 3.9, almost twice the average piece-to-piece range of just 1.97.

Example 13.11 Zinc Plating of Brake Hose Fittings

Brake hose fittings (Figure 13.17) are sent through a zinc plating tank to provide a corrosion-resistant finish. The thickness of the zinc coating is checked every half hour by sampling three pieces ($n = 3$), and measuring each in two random locations ($j = 2$). Measurements collected for subgroups one, two, three, and twenty five are presented in Table 13.7, along with the appropriate subgroup statistics (maximum R_W values are highlighted in bold).

Fig. 13.17 Each fitting is checked in two spots.

Centerlines and control limits for constructing a 3-D $\overline{X}$, R chart to monitor zinc plating thickness are computed next, starting with the within-piece range chart. With k equal to 25 and n equal to 3, there is a total of 75 R_W values.

$$\overline{R}_W = \frac{\sum\limits_{i=1}^{kn} R_{W,i}}{kn} = \frac{268}{25(3)} = 3.57$$

With j equal to 2, D_3 and D_4 are found in Appendix Table I.

Table 13.7 Subgroup data and statistics for zinc plating thickness.

Piece	Reading No.		Statistics for Subgroup 1			
No.	1	2	R_W	$\overline{X}_P$	$\overline{X}$	R_P
1	13	16	3	14.5		
2	17	11	6	14.0	14.33	.5
3	15	14	1	14.5		

Piece	Reading No.		Statistics for Subgroup 2			
No.	1	2	R_W	$\overline{X}_P$	$\overline{X}$	R_P
1	11	18	7	14.5		
2	15	11	4	13.0	14.17	2.0
3	17	13	4	15.0		

Piece	Reading No.		Statistics for Subgroup 3			
No.	1	2	R_W	$\overline{X}_P$	$\overline{X}$	R_P
1	19	16	3	17.5		
2	14	18	4	16.0	16.50	1.5
3	16	16	0	16.0		

Piece	Reading No.		Statistics for Subgroup 25			
No.	1	2	R_W	$\overline{X}_P$	$\overline{X}$	R_P
1	15	17	2	16.0		
2	13	14	1	13.5	14.67	2.5
3	17	12	5	14.5		

$$UCL = D_4\overline{R}_W = 3.267(3.57) = 11.66$$

$$LCL = D_3\overline{R}_W = 0(3.57) = 0$$

This chart tracks changes influencing the uniformity of plating thickness within individual brake hose fittings. There has been a significant deterioration in this uniformity if a plot point falls above the *UCL* of 11.66.

Calculations for the piece-to-piece range chart are done next, where *n* is 3.

$$\overline{R}_P = \frac{\sum\limits_{i=1}^{k} R_{P,i}}{k} = \frac{39.5}{25} = 1.58$$

$$LCL = D_3\overline{R}_P = 0(1.58) = 0 \qquad\qquad UCL = D_4\overline{R}_P = 2.575(1.58) = 4.07$$

A plot point exceeding the *UCL* of 4.07 for this chart indicates the consistency of plating thickness from one piece to the next has been substantially altered.

The average chart's computations are performed last, where the subgroup size is also 3.

$$\overline{\overline{X}} = \frac{\sum\limits_{i=1}^{k} \overline{X}_i}{k} = \frac{372.17}{25} = 14.89$$

$$UCL_{\overline{X}} = \overline{\overline{X}} + A_2\overline{R}_P = 14.89 + 1.023(1.58) = 16.51$$

$$LCL_{\overline{X}} = \overline{\overline{X}} - A_2\overline{R}_P = 14.89 - 1.023(1.58) = 13.27$$

A subgroup average falling below this *LCL* implies the overall plating thickness has decreased considerably for *all* fittings in that subgroup. Conversely, a point above the *UCL* of this special $\overline{X}$ chart depicts a process whose average plating thickness has significantly increased since the last subgroup.

Measuring Capability

Once control is established, the short-term piece-to-piece ($\sigma_{P,ST}$) and within-piece ($\sigma_{W,ST}$) standard deviations are combined to form the total process variation, $\sigma_{TOTAL,ST}$, from which estimates of process capability are generated (this analysis assumes $\sigma_{W,ST}$ and $\sigma_{P,ST}$ are independent). Recalling that *j* represents the number of measurements per piece, while *n* is the subgroup size:

$$\hat{\sigma}_{TOTAL,ST} = \sqrt{\hat{\sigma}_{W,ST}^2 + \hat{\sigma}_{P,ST}^2} \qquad \text{where } \hat{\sigma}_{W,ST} = \frac{\overline{R}_W}{d_{2_j}} \text{ and } \hat{\sigma}_{P,ST} = \frac{\overline{R}_P}{d_{2_n}}$$

$\sigma_{TOTAL,ST}$ may also be estimated directly as follows:

$$\hat{\sigma}_{TOTAL,ST} = \sqrt{\left(\frac{\overline{R}_W}{d_{2_j}}\right)^2 + \left(\frac{\overline{R}_P}{d_{2_n}}\right)^2}$$

Short-term potential and performance capability are now assessed with any of the various measures covered in this book. For C_P and C_{PK}:

$$\hat{C}_P = \frac{\text{Tolerance}}{6\hat{\sigma}_{TOTAL,ST}}$$

$$\hat{C}_{PK} = \text{Minimum}\left(\frac{\hat{\mu} - LSL}{3\hat{\sigma}_{TOTAL,ST}}, \frac{USL - \hat{\mu}}{3\hat{\sigma}_{TOTAL,ST}}\right) \qquad \text{where } \hat{\mu} = \overline{\overline{X}} = \frac{\sum\limits_{i=1}^{k} \overline{X}_i}{k}$$

Whenever within-piece variation is zero, then $\sigma_{TOTAL,ST}$ equals $\sigma_{P,ST}$ and the capability study becomes just like a standard short-term capability study, as both are now based solely on piece-to-piece variation.

$$\hat{\sigma}_{TOTAL,ST} = \sqrt{\hat{\sigma}^2_{W,ST} + \hat{\sigma}^2_{P,ST}} = \sqrt{0^2 + \hat{\sigma}^2_{P,ST}} = \sqrt{\hat{\sigma}^2_{P,ST}} = \hat{\sigma}_{P,ST}$$

For an alternative approach to measuring capability where within-piece variation exists, see the Span-Plan method developed by Seder and Cowan, which involves a detailed dissection of multiple sources of process variation. Runger explores the robustness of standard deviation estimates when within-batch variation is present.

Example 13.12 Zinc Plating of Brake Hose Fittings

Centerlines for a 3-D chart monitoring the thickness of zinc plating on brake hose fittings were calculated in Example 13.11.

$$\overline{R}_W = 3.57 \qquad \overline{R}_P = 1.58 \qquad \overline{\overline{X}} = 14.89$$

From these, estimates are derived for $\sigma_{W,ST}$ and $\sigma_{P,ST}$ as follows ($j = 2, n = 3$):

$$\hat{\sigma}_{W,ST} = \frac{\overline{R}_W}{d_{2_j}} = \frac{3.57}{1.128} = 3.16 \qquad \hat{\sigma}_{P,ST} = \frac{\overline{R}_P}{d_{2_n}} = \frac{1.58}{1.693} = .93$$

These two are combined to estimate overall short-term process variation.

$$\hat{\sigma}_{TOTAL,ST} = \sqrt{\hat{\sigma}^2_{W,ST} + \hat{\sigma}^2_{P,ST}} = \sqrt{3.16^2 + .93^2} = 3.29$$

Given a plating thickness specification of 15 ± 7 microns, short-term potential capability may be summarized with the C_P index.

$$\hat{C}_P = \frac{\text{Tolerance}}{6\hat{\sigma}_{TOTAL,ST}} = \frac{22 - 8}{6(3.29)} = .71$$

With an estimate of the overall process average, performance capability is evaluated with the C_{PK} index.

$$\hat{\mu} = \overline{\overline{X}} = 14.89$$

$$\hat{C}_{PK} = \text{Minimum}\left(\frac{\hat{\mu} - LSL}{3\hat{\sigma}_{TOTAL,ST}}, \frac{USL - \hat{\mu}}{3\hat{\sigma}_{TOTAL,ST}} \right)$$

$$= \text{Minimum}\left(\frac{14.89 - 8}{3(3.29)}, \frac{22 - 14.89}{3(3.29)} \right)$$

$$= \text{Minimum}(.70, .72) = .70$$

As performance and potential capability are almost identical, the process is properly centered for average plating thickness. However, because both C_P and C_{PK} are well below their minimum capability requirements, work must begin on slashing process variation. As $\sigma_{W,ST}$ is substantially larger than $\sigma_{P,ST}$, efforts should first concentrate on reducing the variation in plating thickness on each brake hose fitting. Perhaps a more thorough mixing of fittings during plating is necessary, or an increased agitation of the plating bath is needed.

Suppose the specifications were 15 plus 7, minus 5. Capability measures such C^*_P and C^*_{PM} should be selected to appraise this plating process.

$$\hat{C}^*_P = \text{Minimum}\left(\frac{T-\text{LSL}}{3\hat{\sigma}_{TOTAL,ST}}, \frac{\text{USL}-T}{3\hat{\sigma}_{TOTAL,ST}}\right)$$

$$= \text{Minimum}\left(\frac{15-10}{3(3.29)}, \frac{22-15}{3(3.29)}\right)$$

$$= \text{Minimum}\,(\,.51\,,.71\,) = .51$$

τ^*_{ST} is estimated as 3.292.

$$\hat{\tau}^*_{ST} = \sqrt{\hat{\sigma}^2_{TOTAL,ST} + (\hat{\mu}-T)^2} = \sqrt{3.29^2 + (14.89-15)^2} = 3.292$$

From this, C^*_{PM} is computed as .51.

$$\hat{C}^*_{PM} = \text{Minimum}\left(\frac{T-\text{LSL}}{3\hat{\tau}^*_{ST}}, \frac{\text{USL}-T}{3\hat{\tau}^*_{ST}}\right)$$

$$= \text{Minimum}\left(\frac{15-10}{3(3.292)}, \frac{22-15}{3(3.292)}\right)$$

$$= \text{Minimum}\,(\,.51\,,.71\,) = .51$$

Because the process average is so close to the target, the estimate of τ^*_{ST} is almost identical to $\sigma_{TOTAL,ST}$, resulting in the estimate of C^*_{PM} being very similar to that for C^*_P. Again, a conclusion is reached that the process average is properly centered, but process variation must be significantly decreased.

Capability Studies for Piece Averages

In isolated cases, piece averages also have a stated specification. For example, a metal plate used in the construction of nuclear reactors has its surface finish checked in four locations. The measurement at each location must meet a surface finish specification of 35 $\pm$ 6. In addition, the *average* of these four measurements on each piece ($\overline{X}_P$) must fall within a range of 35 $\pm$ 3. In order for a plate to be considered acceptable, both requirements must be met. A component with all four measurements equal to 39 satisfies the first requirement, but not the second, making it a nonconforming part.

Process capability for the specification on individual measurements is evaluated as previously described. For assessing how well this process fulfills the additional stipulation on piece averages, estimates of $\sigma_{P,ST}$ and μ are used in conjunction with the second specification.

$$\hat{\mu} = \overline{X} \qquad \hat{\sigma}_{P,ST} = \frac{\overline{R}_P}{d_{2_n}}$$

In this example, $\sigma_{P,ST}$ is estimated as .43, μ as 36.05, with the piece average specification being 35 ± 3. Measures of both potential $(C_{P,PA})$ and performance capability $(C_{PK,PA})$ are computed as given below, where the subscript "PA" represents "piece average."

$$\hat{C}_{P,PA} = \frac{\text{Tolerance}}{6\hat{\sigma}_{P,ST}} = \frac{38 - 32}{6(.43)} = 2.33$$

$$\hat{C}_{PK,PA} = \text{Minimum} \left(\frac{\hat{\mu} - \text{LSL}}{3\hat{\sigma}_{P,ST}}, \frac{\text{USL} - \hat{\mu}}{3\hat{\sigma}_{P,ST}} \right)$$

$$= \text{Minimum} \left(\frac{36.05 - 32}{3(.43)}, \frac{38 - 36.05}{3(.43)} \right)$$

$$= \text{Minimum} (3.14, 1.51) = 1.51$$

This process has both good potential and performance capability for average surface finish. However, as performance capability is considerably less than potential, centering the process average closer to the middle of the tolerance will help boost the $C_{PK,PA}$ index.

13.6 Features with Geometric Tolerances

Geometric dimensioning and tolerancing (GD&T) is an engineering standard (ANSI Y14.5M-1982) providing a unified terminology and methodology for describing both the geometry of product features and their associated tolerances. For example, Figure 13.18 below illustrates how GD&T generates a tolerance zone for hole location, which is a circle circumscribing the square tolerance zone allowed by the conventional coordinate tolerancing method for this type of feature.

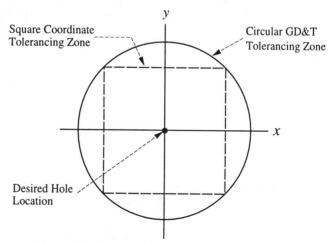

Fig. 13.18 A circular tolerance zone for hole location.

In some cases, feature tolerances are interrelated, making the tolerance zones much more complex in shape. These complicated shapes create challenging problems for estimating a process's ability to satisfactorily produce this characteristic.

Establishing Stability

Individual features are examined for stability with conventional charts, as tolerances play no role in control charting. Although if several characteristics are correlated, one of the multivariate charts described in Section 13.8 (p. 816) may be needed. At times, features with GD&T specifications are checked on simply a pass/fail basis. With attribute-type data being generated in these situations, use either a *np* or *p* chart to help establish stability. Note that these charts are not as effective in monitoring a process as variable-data charts are.

Measuring Process Capability

If the feature's performance is being tracked on an attribute chart, process capability can be measured with either equivalent P^*_{PK} or $ppm_{TOTAL\ LT}$, as explained in Section 9.3 (p. 536).

Most efforts for assessing capability for variable-data characteristics with GD&T have concentrated on circular and spherical tolerance zones. For example, consider the radial distance of a hole center from its target. The *x* and *y* coordinates of 15 actual hole centers were measured and recorded in Table 13.8. The location of these hole centers in comparison to their requirement is displayed in Figure 13.19.

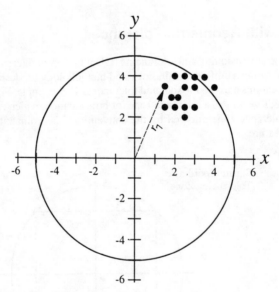

Fig. 13.19 Actual hole location versus target.

The radial distance, r_i, between the target hole center and the *i*th hole is given by this next formula, where x_i and y_i represent the coordinates of the *i*th hole's center, while *a* and *b* are the *x* and *y* coordinates of the target.

$$r_i = \sqrt{(x_i - a)^2 + (y_i - b)^2}$$

For ease of calculation, the target hole center is often assigned coordinates of $0,0$, simplifying the above formula to:

$$r_i = \sqrt{x_i^2 + y_i^2}$$

Distances for all 15 holes were computed with this formula, then recorded in column four of Table 13.8. For example, the radial distance from the target to the seventh hole center, r_7, is calculated as 3.81.

$$r_7 = \sqrt{x_7^2 + y_7^2} = \sqrt{1.5^2 + 3.5^2} = 3.81$$

Before a capability study for hole location can begin, these r values must be checked for stability. Plotting the 15 radial distances on an *IX & MR* chart implies the process was stable during their production as no out-of-control conditions are noted. Because an *IX & MR* chart is chosen, the assumption of normality for the process output distribution must be checked. Due to the relatively small sample size of 15, the Anderson-Darling goodness-of-fit method was chosen. As the A-D statistic is only .302, whereas the critical value for .99 confidence is 1.035 (review page 413), there is no evidence that the process is not normally distributed. Had the r values been non-normally distributed, control limits for the *IX* chart would have needed adjustment, as was explained in Section 7.4 (p. 401).

Table 13.8 Calculation of radial distance for hole location.

Hole #	x	y	r	$x - \bar{x}$	$y - \bar{y}$	r_C
1	3.0	3.5	4.61	.5	.3	.58
2	2.0	4.0	4.47	-.5	.8	.94
3	2.0	3.0	3.61	-.5	-.2	.54
4	3.5	4.0	5.32	1.0	.8	1.28
5	2.5	2.0	3.20	.0	-1.2	1.20
6	2.5	3.5	4.30	.0	.3	.30
7	1.5	3.5	3.81	-1.0	.3	1.04
8	2.5	2.5	3.54	.0	-.7	.70
9	4.0	3.5	5.32	1.5	.3	1.53
10	3.0	4.0	5.00	.5	.8	.94
11	2.5	4.0	4.72	.0	.8	.80
12	2.0	2.5	3.20	-.5	-.7	.86
13	1.5	2.5	2.92	-1.0	-.7	1.22
14	3.0	2.5	3.91	.5	-.7	.86
15	2.0	3.0	3.61	-.5	-.2	.54
Avg.	2.5	3.2	4.10			.89
S_{TOT}	.71	.68	.78			.33

When a spherical tolerance is specified, compute radial distances with this formula, assuming the target center has the coordinates of $0, 0, 0$.

$$r_i = \sqrt{x_i^2 + y_i^2 + z_i^2}$$

Stability and normality are checked in the same manner as described above.

Measuring Performance Capability

To measure the performance capability of this process to correctly position a hole, process parameters for the distribution of the r values must first be estimated.

$$\hat{\mu} = \overline{X} = \overline{r} \qquad \hat{\sigma}_{ST} = \frac{\overline{MR}}{1.128} \qquad \hat{\sigma}_{LT} = \frac{S_{TOT}}{c_4} = \frac{1}{c_4} \sqrt{\frac{\sum_{i=1}^{k}(r_i - \overline{r})^2}{k-1}}$$

Because they have a normal distribution in this example, Kane (pp. 507-511) recommends the following formulas for evaluating capability:

$$\hat{C}_{PK} = \hat{C}_{PU} = \frac{USL - \hat{\mu}}{3\hat{\sigma}_{ST}} \qquad\qquad \hat{P}_{PK} = \hat{P}_{PU} = \frac{USL - \hat{\mu}}{3\hat{\sigma}_{LT}}$$

If the r values are non-normally distributed, which is fairly common in these situations, switch to the equivalent P_{PK} index, where the required percentiles are estimated directly from normal probability paper.

$$\text{Equivalent } \hat{P}_{PK} = \frac{USL - \hat{x}_{50}}{\hat{x}_{99.865} - \hat{x}_{50}}$$

For the previous example, the average of the 15 r values is 4.10, while their σ_{LT} is estimated as .795.

$$\hat{\mu} = \overline{r} = 4.10 \qquad\qquad \hat{\sigma}_{LT} = \frac{S_{TOT}}{c_4} = \frac{.781}{.9823} = .795$$

Given an USL of 5 for radial distance, P_{PK} is found to be only .377, indicating very poor performance capability for locating hole centers.

$$\hat{P}_{PK} = \frac{USL - \hat{\mu}}{3\hat{\sigma}_{LT}} = \frac{5.0 - 4.10}{3(.795)} = .377$$

With an estimated Z value of only 1.13, over 12.9 percent of the hole locations are predicted to be beyond a radial distance of 5 from the target center. This agrees closely with the actual 13.3 percent (2/15) of nonconforming hole centers.

$$\hat{Z}_{USL,LT} = 3\hat{P}_{PK} = 3(.377) = 1.13$$

Had the r values been non-normally distributed, estimate the percentage of nonconforming holes from normal probability paper. For another method of estimating the percentage of nonconforming hole centers, read Davis *et al.*

Measuring Potential Capability

Potential capability is the best a process can do when centered at the target average. This second aspect of process capability is measured by assuming the cluster of hole locations is actually centered at the target, then recalculating radial distances. This is accomplished by using the cluster's actual center as the surrogate target, then computing radial distances to each hole location. These revised radial distances are then compared to the USL, as is illustrated in Figure 13.20.

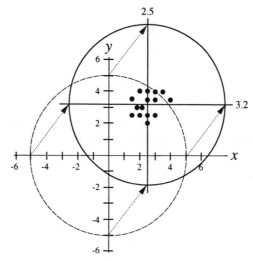

Fig. 13.20   Potential capability is based on radial distances from the cluster's center.

The *x* coordinate for the cluster's center is found by averaging the *x* coordinates for all 15 holes, while the *y* coordinate is the average of the individual *y* coordinates.

$$\overline{x} = \frac{\sum\limits_{i=1}^{k} x_i}{k} \qquad\qquad \overline{y} = \frac{\sum\limits_{i=1}^{k} y_i}{k}$$

To distinguish it from the radial distance to the target, the radial distance from each hole to the cluster's center is labeled r_C and is calculated as follows:

$$r_{C,i} = \sqrt{(x_i - \overline{x})^2 + (y_i - \overline{y})^2}$$

In the above example, $\overline{x}$ is 2.5, while $\overline{y}$ is 3.2 (see Table 13.8). For hole number 7, the r_C distance from the cluster's center is 1.04. r_C values for the remaining 14 hole centers are given in column seven of Table 13.8.

$$r_{C,7} = \sqrt{(x_7 - \overline{x})^2 + (y_7 - \overline{y})^2} = \sqrt{(1.5 - 2.5)^2 + (3.5 - 3.2)^2} = 1.04$$

Plot these radial distances on an *IX & MR* chart to check for stability, then on normal probability paper to check for normality. When these conditions are satisfied, estimate the average and standard deviation of the r_C values as follows:

$$\hat{\mu}_C = \overline{X}_C = \overline{r}_C \qquad \hat{\sigma}_{ST,C} = \frac{\overline{MR}_C}{1.128} \qquad \hat{\sigma}_{LT,C} = \frac{S_{TOT,C}}{c_4} = \frac{1}{c_4} \sqrt{\frac{\sum_{i=1}^{k} (r_{C,i} - \overline{r}_C)^2}{k-1}}$$

Evaluate potential capability with the average and standard deviation of the radial distances.

$$\hat{C}_P = \hat{C}_{PU} = \frac{USL - \hat{\mu}_C}{3\hat{\sigma}_{ST,C}} \qquad\qquad \hat{P}_P = \hat{P}_{PU} = \frac{USL - \hat{\mu}_C}{3\hat{\sigma}_{LT,C}}$$

If the r_C values are non-normally distributed, find the appropriate percentiles from normal probability paper to estimate equivalent P_P.

$$\text{Equivalent } \hat{P}_P = \frac{USL - \hat{x}_{50,C}}{\hat{x}_{99.865,C} - \hat{x}_{50,C}}$$

From the summary of hole location measurements in Table 13.8 (which have close to a normal distribution), the average and standard deviation of the r_C values are estimated as:

$$\hat{\mu}_C = \overline{X}_C = \overline{r}_C = .89 \qquad\qquad \hat{\sigma}_{LT,C} = \frac{S_{TOT,C}}{c_4} = \frac{.33}{.9823} = .336$$

These two process parameters represent what the radial distances would be if the process were centered at the target. Thus, the potential capability of this process to locate a hole may be quantified with the following P_P index.

$$\hat{P}_P = \frac{USL - \hat{\mu}_C}{3\hat{\sigma}_{LT,C}} = \frac{5 - .89}{3(.336)} = 4.077$$

This process has extremely high potential capability ($P_P = 4.077$), but very low performance capability ($P_{PK} = .377$) due to its inability to properly center holes at the target.

When dealing with a spherical tolerance, switch to this formula, where $\overline{z}$ is the z coordinate of the cluster center.

$$r_{C,i} = \sqrt{(x_i - \overline{x})^2 + (y_i - \overline{y})^2 + (z_i - \overline{z})^2}$$

Little developmental work for capability studies has been undertaken for features with more involved GD&T specifications. Interested readers can learn what has been done to date by reviewing Dowling *et al*, Gruner, Hulting, Jackson, Karl *et al*, Littig *et al* (*a*), Phillips and Cho, and Sarkar and Pal (1998).

13.7 Characteristics without Specifications

In rare occasions, no specification limits are given for a variable-data characteristic being manufactured. This situation can happen in process/product development, or in prototype production. Not knowing the "voice of the customer" creates a unique challenge for assessing the ability of a process to satisfy customers.

Establishing Stability

This first step of a capability study poses no problems as specification limits play no role in charting a process. Simply select a suitable control chart, and begin charting. When the process achieves a reasonably good state of control, estimate process parameters from the centerlines of the chart. Keenan describes a methodology for evaluating producibility during the development cycle, which relies on run charts for detecting assignable-cause variation. Levinson and Ben-Jacob provide an example of how SPC techniques were employed during the developmental stage to reduce problems of a photolithography process for manufacturing integrated circuits.

Measuring Process Capability

Even though an absolute level of capability cannot be derived without specifications, improvements in capability can be tracked, which is typically the prime concern when developing new products or processes.

Potential Capability

Bissell (p. 241) suggests a method related to the coefficient of variation, while Gallo provides a similar concept for a problem involving the mixing of dry powders. These indicators of capability improvement assume a target for the process average is known, which is fairly common in product development.

$$C_{RD} = \frac{6\sigma_{ST}}{T} \qquad P_{RD} = \frac{6\sigma_{LT}}{T}$$

If the characteristic's measurements are non-normally distributed, find the appropriate percentiles from normal probability paper to estimate equivalent P_{RD}.

$$\text{Equivalent } \hat{P}_{RD} = \frac{\hat{x}_{99.865} - \hat{x}_{.135}}{T}$$

These dimensionless ratios are interpreted like the C_R and P_R ratios, in that smaller values are better, with the best capability occurring when they equal zero. However, a C_{RD} ratio equal to 1.00 does not mean anything in particular, just that the process is better than when this ratio equaled 1.27 and is not as good as when the process has C_{RD} equal to .89.

Herman proposes monitoring improvements in potential capability by plotting the ratio of the current estimate of the process standard deviation to its initial estimate on a run chart.

$$\text{Plot point} = \frac{\hat{\sigma}_{LT, CURRENT}}{\hat{\sigma}_{LT, INITIAL}}$$

Some practitioners prefer to track the ratio of the variances instead.

$$\text{Plot point} = \frac{\hat{\sigma}^2_{LT, CURRENT}}{\hat{\sigma}^2_{LT, INITIAL}}$$

Beginning at 1.0, either ratio will track changes in process capability over time, as is demonstrated in Figure 13.21. Like C_R, decreases in this measure over time signify improved process capability, whereas increases indicate capability deterioration. Being dimensionless, these ratios may be easily compared to those for other processes or characteristics with similar targets.

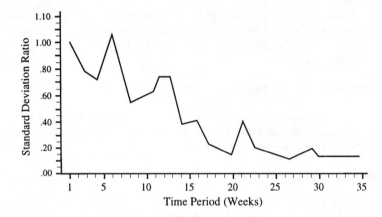

Fig. 13.21 Tracking the ratio of $\hat{\sigma}_{CURRENT}$ to $\hat{\sigma}_{INITIAL}$.

Of course, the simplest method is to just study the estimated standard deviation. Any improvements in potential capability are unveiled as this estimate approaches zero.

Example 13.13 Bolt Torque

In the development of a new line of ceiling fans, a target of 15 newton-meters is set for a critical fastener. After control of the bolt-tightening process is established, σ_{ST} is estimated as 2 nm. The C_{RD} ratio is computed as .80.

$$\hat{C}_{RD} = \frac{6\hat{\sigma}_{ST}}{T} = \frac{6(2\,\text{nm})}{15\,\text{nm}} = .80$$

After several process modifications, σ_{ST} is reduced to 1.5 nm. This improvement in potential capability is reflected by a decrease in C_{RD} to .60.

$$\hat{C}_{RD} = \frac{6\hat{\sigma}_{ST}}{T} = \frac{6(1.5\,\text{nm})}{15\,\text{nm}} = .60$$

Performance Capability

To track improvements in performance capability, a measure must include an estimate of the process average. This can be done as follows for characteristics without specifications:

$$C_{DT} = \frac{6\sqrt{\sigma_{ST}^2 + (\mu - T)^2}}{T} \qquad\qquad P_{DT} = \frac{6\sqrt{\sigma_{LT}^2 + (\mu - T)^2}}{T}$$

If the characteristic's measurements are non-normally distributed, determine the .135, 50, and 99.865 percentiles from normal probability paper and compute equivalent P_{DT}.

$$\text{Equivalent } P_{DT} = \frac{6\sqrt{[(x_{99.865} - x_{.135})/6]^2 + (x_{50} - T)^2}}{T}$$

As the process spread is reduced, indicating a more precise process, both the C_{DT} and P_{DT} ratios decrease, reflecting improvement in performance capability. When the process median is centered closer to the target average, these dimensionless ratios will also decrease, again implying performance capability has improved.

A slightly simpler approach is to compute just the following value:

$$C_{EL} = \sigma_{ST}^2 + (\mu - T)^2 = \tau_{ST}^{*\,2}$$

Directly proportional to the economic loss (*EL*), this alternative measure is very similar to the C_{PG} ratio. Values approaching zero imply overall loss is decreasing and performance capability is improving. However, this index is not unitless and, therefore, should not be used to compare different processes.

Example 13.14 Bolt Torque

In the development of a new line of ceiling fans, a target of 15 newton-meters is set for a critical fastener. After establishing control, μ is estimated as 13 nm and σ_{ST} as 2 nm. The C_{DT} ratio is calculated as 1.13.

$$\hat{C}_{DT} = \frac{6\sqrt{\hat{\sigma}_{ST}^2 + (\hat{\mu} - T)^2}}{T} = \frac{6\sqrt{2^2 + (13 - 15)^2}}{15} = 1.13$$

If the process average is shifted to 14 nm, which is now closer to the target of 15 nm, C_{DT} decreases to .89, indicating improved performance capability. This is true no matter where the specifications will eventually be set.

$$\hat{C}_{DT} = \frac{6\sqrt{2^2 + (14 - 15)^2}}{15} = .89$$

After several process modifications, σ_{ST} is reduced to 1.5 nm. This process improvement is reflected by a decrease in C_{DT} to .72.

$$\hat{C}_{DT} = \frac{6\sqrt{1.5^2 + (14 - 15)^2}}{15} = .72$$

When the process average is finally centered at the target, this ratio drops to just .60, signaling another improvement in performance capability. In fact, it now equals potential capability (recall Example 13.13 on the previous page).

$$\hat{C}_{DT} = \frac{6\sqrt{1.5^2 + (15 - 15)^2}}{15} = .60$$

Unlike most capability measures, no universal rating scale for C_{DT} exists. The above result applies to only this particular process and should never be compared to that for any other.

13.8 Product Capability for Correlated Characteristics

Chapter 12 demonstrated how to combine the capability of several product characteristics into a single measure with Product C_{PK}(p. 730), assuming all characteristics are independent. Unfortunately, there are many products where this assumption is invalid. For example, the weight of a part is dependent on its overall length. These two features are positively correlated: the longer the part, the greater its weight; the shorter the part, the less its weight. In a similar fashion, the strength of a fiber is positively correlated with its thickness.

The distance of a punched hole to the edge of a part is dependent on where the hole is originally located by the turret punch. These two features are negatively correlated: when the hole is nearer the left side, it is farther from the right; when farther from the left side, the hole is closer to the right.

If product features are highly correlated, special charting techniques are required to demonstrate stability, as well as to assess overall product capability.

Establishing Stability

On-line computers can now collect vast amounts of data on many process-related characteristics, both input and output. More often than not, many of these characteristics are correlated to some degree. The conventional approach for control charting a multitude of related features is maintaining a separate chart for every characteristic. Each chart is referred to as a "univariate" chart because just one variable is being monitored. For example, fiber strength is positively correlated with the thickness of a textile fiber. Univariate control charts for these two characteristics are displayed in Figure 13.22 (only the $\overline{X}$ portions are shown), with both demonstrating good control.

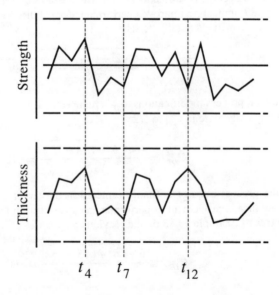

Fig. 13.22 Univariate charts for fiber strength and thickness.

Note how at time t_4 average fiber thickness is high, and fiber tensile strength is also high. At t_7, average thickness and strength are both low. Because these characteristics are highly correlated (see Figure 13.23), the patterns on their charts look very similar and, thus, Kane (p. 506) suggests charting only one. However, a significant and important process change can occur that may not show up on either univariate chart. Consider time t_{12}. Fiber thickness is very high, but fiber strength is quite *low*. The relationship between these two characteristics has dramatically changed, yet neither univariate chart signals an out-of-control condition.

The thickness (x) and strength (y) measurements for each time period are plotted on the scatter diagram shown in Figure 13.23. All points fall on, or are clustered very near, the line, implying that thickness and strength are positively correlated. However, time period 12 is an exception, being located a considerable distance from the other points. This "outlier" point reveals a significant change in the *relationship* between thickness and strength for time period 12. An investigation should be undertaken to discover why thicker fibers now have *lower*, not higher, strength. Note that this vital insight is missed if just the two univariate charts are analyzed, as both indicate the process is in good control.

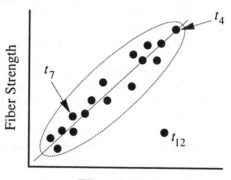

Fig. 13.23 Scatter diagram of fiber thickness versus strength.

A control chart for monitoring the combination of several characteristics is called a "multivariate" chart, with one of the most popular being Hotelling's T^2 chart. Unfortunately, the calculations associated with this chart involve computing covariances between all characteristics, as well as matrix inversion and multiplication, all of which are fairly difficult and time consuming unless appropriate software programs are available.

The concept behind this chart is to reduce information about the j characteristics into a single value, then monitor this single number on one chart. This approach is somewhat similar to that applied to GD&T tolerances, where the x and y coordinates for a hole's location are combined into just one radial distance, which is then monitored on a single *IX & MR* chart. For complete details about this kind of multivariate chart, read Alt (1982, 1985), Hotelling, Kourti and MacGregor (1996), Montgomery (pp. 322-330), or Ryan. A multivariate chart for individual measurements is provided by Tracy *et al*. Mason *et al* (1996) have developed a multivariate charting method when the process average undergoes sustained shifts in level. Multivariate charts have also been developed for continuous and/or batch processes by Kourti and MacGregor (1995), Kresta *et al*, MacGregor and Kourti, and Nomikos and MacGregor.

For other charting approaches, see Fuchs and Benjamini, Liu, Subramanyam and Houshmand, Wheeler (1995, pp. 329-339), and Wright. Wierda (1994) provides a thorough review of multivariate SPC charts. In addition to monitoring the process average, the variance must also be checked for control. Alt and Bedewi, as well as Montgomery (pp. 330-332), describe charts designed especially for this purpose.

Besides the difficulty in calculation, Hotelling's T^2 chart is hard to interpret. If a point is out of control, which characteristic, or *combination* of characteristics, is responsible? The cause may even be attributable to a change in the relationship between some of the characteristics, as was the case in the fiber strength example. For help with interpreting multivariate charts, see Doganaksoy *et al*, Mason *et al* (1995), Murphy, or Timm. If some of the characteristics are autocorrelated, see Woodhall and Ncube. In addition to all the above concerns, as the number of characteristics increases, so does the false alarm rate of the multivariate chart, causing even more interpretation problems.

Measuring Process Capability

Once process control is achieved, the normality assumption must be checked before making any estimates of process capability. Everitt demonstrated that Hotelling's T^2 chart is fairly sensitive to departures from normality, especially skewness. This problem intensifies as the number of characteristics increases. Koziol, Mardia, as well as Royston, provide methods to verify the normality assumption for multivariate distributions.

If the normality assumption is satisfied, multivariate process capability may be assessed by the methods proposed by Boyles, Chan *et al*, Chen, Fuchs and Kenett, Hellmich and Wolff, Littig *et al* (*b*), Niverthi and Dey, Padgett *et al* (for machine capability studies), Pearn *et al*, Taam *et al*, Wang and Chung, Wang *et al*, or Wierda (1993). In addition to Hubele *et al* (1991), Kocherlakota and Kocherlakota offer an approach when only two characteristics ("bivariate") are being studied. Kotz and Johnson offer a detailed explanation, analysis, and comparison of all these proposed methods. A technique not involving matrix algebra is advocated by Hubele *et al* (1992). Unfortunately, all of these methods are rather complicated and computer software is needed to perform the necessary calculations.

13.9 Summary

This chapter explained how both potential and performance capability are measured in the presence of certain unique operating conditions:

- systematic trends,
- set-up variation,
- autocorrelation,
- short production runs,
- within-piece variation,
- GD&T tolerances,
- characteristics with no specifications, and
- correlated features.

When necessary, modified control charting techniques for establishing stability were described for those circumstances where traditional charts did not apply. Procedures for assessing various types of process capability were then given. Some of these methods considered a "worst-case" capability scenario (tool wear, set up, autocorrelated data). This approach always generates a very conservative estimate of process capability, but one that customers can be assured of receiving.

The development of simplified assessment methods is needed for estimating the capability of characteristics with more complicated GD&T tolerances, as well as for the combination of multiple features that are highly correlated.

13.10 Exercises ━━━━━━━━━━━━━━

13.1 Lead Concentration

Brake hose fittings are plated with a layer of lead to prevent corrosion. As fittings pass through a plating bath, the concentration level of lead decreases over time. When the concentration declines to a certain level, lead balls are added to replenish the supply of lead in the plating solution. Due to this procedure, an *IX & MR* chart monitoring the percentage of lead concentration reveals a series of decreasing trends. The SPC implementation team decides to replace the control limits on the *IX* chart with the maximum and minimum SPA limits. With a specification for lead concentration of 11 to 18 percent, an average moving range of .429, and a capability goal of 1.67, compute these maximum and minimum SPA limits, then explain how they are used to help run this plating operation. Estimate the potential capability of this lead-plating process with $C_{P. MOD}$. Given the average of the initial setups is 15.83, while the average of final subgroup averages is 13.12, estimate $C_{PK. MOD}$.

13.2 EDM Setup

An *IX & MR* chart for the setup averages ($n = 7$) of an EDM machine has the centerline of the *IX* chart equal to 124.3, with the average moving range equal to 1.21. With this information, calculate the control limits for this chart. Assuming the setup operation is in control, and the measurements are normally distributed, estimate the standard deviation of the setup operation. Given specification limits of 125.0 ± 4.0, predict $C_{P. SETUP}$, k_{SETUP}, and $C_{PK. SETUP}$.
A Target $\overline{X}$, R chart is chosen to monitor the parts during each run, with R equal to .81 when in control. Assuming variation within each run is similar, estimate the standard deviation for the runs, $C_{P. RUN}$, as well as $C_{PK. CUSTOMER}$. What would be your recommendations for improving this process?

13.3 Adhesive Paper

A thin, continuous strip of paper has an adhesive applied to one side. The height of this adhesive is a critical characteristic and must be between 2.00 mm ± .10 mm. Because the average height slowly cycles up and down, an EWMA chart is selected to monitor this autocorrelated process. The forecast versus actual residuals are plotted on an *IX & MR* chart (special-cause chart) and display good control, with the average moving range equal to .01895 mm. Estimate the standard deviation of these residuals and then the potential capability of this process, assuming the wandering average is maintained at the middle of the tolerance for adhesive height.
From the common-cause chart of forecasted heights, the highest expected average height is 2.031 mm, while the lowest is 1.982 mm. With this additional information, predict the worst-case quality customers can expect to receive from this process (use $C_{PK. AUTO}$). Estimate the $ppm_{TOTAL. ST}$ for this worst-case scenario.

13.4 Control Panels

Two locating holes are drilled in control panels for various types of refrigerators. Correct distance between these hole centers is important when mounting a panel in the refrigerator. The SPC implementation team chose a Short Run $\overline{X}$, R chart ($n = 2$) to monitor this critical characteristic since there are several models of control panels with different distance requirements, all run in small quantities over the same hole-drilling operation. Once control is established, the results listed in Table 13.9 are generated for the four models of control panels. From this summary, estimate both potential and performance capability for these models.

Table 13.9 Results of short-run capability study on refrigerator control panels.

Panel	$\hat{\sigma}_{LT}$	$\overline{X}$	LSL	USL	k	$C_{P,\,GOAL}$	$C_{PK,\,GOAL}$
A	.52	35.61	32.5	37.5	12	1.50	1.50
B	.36	24.98	23.0	27.0	7	1.67	1.67
C	.35	51.02	47.5	52.5	9	1.50	1.50
D	.61	39.83	38.5	41.5	6	1.33	1.33

13.5 Control Panels - Radar Chart

Plot the ratio of actual to desired capability for each control panel model on one radar graph (review Section 12.4, p. 743). Make suggestions on how to improve the capability for each of the four models.

13.6 Control Panels - Confidence Bounds

Calculate lower .95 confidence bounds for both the C_P and C_{PK} indexes of each control panel model listed in Table 13.9. Comment on your findings.

13.7 Integrated Circuits

Integrated circuits (chips) are manufactured on silicon wafers. These wafers are circular disks containing from 20 to as many as 600 chips. Before individual chips are cut from a wafer, a series of tests are applied to determine if each chip is conforming or nonconforming. Because several varieties of chips are produced on this one process, a Short Run p chart is chosen to monitor the chip-making operation. Once control is established, the average percentage nonconforming is found for each of the three types of chips produced, with these results recorded in Table 13.10.

Table 13.10 Results of short-run capability study.

Chip Type	$\overline{p}$	Σn_i	Equivalent $P^*_{PK,GOAL}$
C-127	.00092	7,609	1.10
C-129	.00153	7,190	1.25
F-772	.00008	12,500	1.20

Estimate the equivalent P^*_{PK} index for each chip type, then plot their ratios of estimated capability to the goal on a radar graph. Finally, compute a lower .90 confidence bound for the performance capability of each chip type.

13.8 Solder Paste Height

A process utilizes surface mount technology to locate electronic components on a large printed circuit board. One critical feature is the height of the solder paste applied prior to mounting the components. To monitor this application process, solder paste height is measured at six randomly chosen locations ($j = 6$) on each of two consecutively-produced boards ($n = 2$) every half hour. Measurements are recorded on a 3-D $\overline{X}$, R chart, with control eventually established ($k = 28$). Provided the following averages, calculate control limits for all portions of this chart.

$$\overline{R}_W = .0563 \qquad \overline{R}_P = .0402 \qquad \overline{\overline{X}} = .7091$$

Estimate the short-term standard deviation for the combination of within- and between-board variation for paste height. Given the LSL is .550 and the USL is .850, estimate both potential (C_P) and performance (C_{PK}) capability. What should be done first to help improve process capability? Explain what would have occurred if the average within-board range had been incorrectly used to calculate control limits for the $\overline{X}$ chart.

13.9 Average Solder Paste Height

What is the potential and performance capability for maintaining the average solder paste height at .70 $\pm$ .05 for each board in Exercise 13.8?

13.10 Solder Paste Height with a Target Average

Repeat Exercise 13.8 with a solder paste height of .73, plus .12, minus .18. What action can be taken to improve performance capability?

13.11 Hole Location

The location of a hole drilled in the leading edge of the wing for a military aircraft has a GD&T tolerance, where the maximum allowed radial distance from the target center is 7.5. The radial distances plot as a curved line on normal probability paper, implying the process output has a non-normal distribution. From the NOPP, the median is found to be 2.74, with the 99.865 percentile determined as 5.63. Estimate equivalent P_{PK} for hole location.
Radial distances from the centroid of the actual hole locations are calculated and found to be in control, but non-normally distributed. From this NOPP analysis, the median of the r_C values is estimated as .59, while the 99.865 percentile is projected to be 3.18. Estimate the potential capability for locating holes with the equivalent P'_P index.

13.12 Valve Leak Rate

A new design is proposed for reducing the leak rate of mini-valves for gas cooking ranges to a target of 250. Leak rate measurements on the new design are plotted on an $\overline{X}$, R chart ($n = 3$) and seen to be in control. From this chart, the overall average is calculated as 254.1, while the average range is 6.08. The individuals plot as a straight line on NOPP. Estimate C_{RD} and C_{DT} for this design.
A competing valve design is tested and found to have an average of 251.3, with a short-term standard deviation of 3.97. Which is a better design from a capability viewpoint?

13.13 Kite String Strength

A producer of children's toys is developing a new type of light-weight kite string which is planned to have a breaking strength of 40 kg. The initial string design is tested several times for break strength, with the results plotted on an $\overline{X}$, R chart. Once in control, the individual break strength measurements are plotted on NOPP and discovered to have a skewed-right distribution. With $x_{.135}$ estimated as 28.4, x_{50} as 37.6 and $x_{99.865}$ as 49.8, compute equivalent P_{RD} and equivalent P_{DT}.

For the second design, $x_{.135}$ is estimated as 26.4, x_{50} as 35.2, and $x_{99.865}$ as 46.3. Estimate equivalent P_{RD} and equivalent P_{DT}, then determine if there is an improvement in break strength. In the next design iteration, $x_{.135}$ turns out to be 30.5, x_{50} is 39.2, and $x_{99.865}$ is 50.3. Again, estimate equivalent P_{RD} and equivalent P_{DT}, then comment.

13.11 References

Abraham, Bovas, and Whitney, James B., "Applications of EWMA Charts to Data from Continuous Processes," *44th ASQC Annual Quality Congress Transactions*, May 1990, pp. 813-818

Aerne, L.A., Champ, C.W., and Rigdon, S.E., "Evaluation of Control Charts Under Linear Trends," *Commun. in Statistics - Theory and Methods*, 20(10), 1991, pp. 3341-3349

Alt, F.B., "Multivariate Quality Control: State of the Art," *36th ASQC Annual Quality Congress Transactions*, May 1982, pp. 886-893

Alt, F.B., "Multivariate Quality Control," *Encyclopedia of Statistical Sciences*, Vol. 6, S. Kotz, and N. Johnson, editors, 1985, John Wiley & Sons, New York, NY, pp. 110-122

Alt, F.B., and Bedewi, G.E., "SPC of Dispersion for Multivariate Data," *40th ASQC Annual Quality Congress Transactions*, May 1986, pp. 248-254

Alwan, Layth C., "Autocorrelation: Fixed Versus Variable Control Limits," *Quality Engineering*, Vol. 4, No. 2, 1991-1992, pp. 167-188

Alwan, Layth C., and Roberts, Harry V., "Time-Series Modeling for Statistical Process Control," *Journal of Business & Economic Statistics*, Vol. 6, No. 1, January 1988, pp. 87-95

Anderson, T. W., *The Statistical Analysis of Time Series*, 1971, John Wiley, NY, NY

Arcelus, F. J., and Banerjee, P. K., "Selection of the Most Economical Production Plan in a Tool Wear Process," *Technometrics*, Vol. 27, No. 4, November 1985, pp. 433-437

Arcelus, F. J., and Banerjee, P. K., "Optimal Economical Production Plan in a Tool Wear Process with Rewards for Acceptable, Undersized and Oversized Parts," *Engineering Costs and Production Economics*, Vol. 11, 1987, pp. 13-19

ASQC, *Quality Assurance for the Chemical and Process Industries, A Manual of Good Practices*, 1987, ASQ Quality Press, Milwaukee, WI

ASQC Chemical and Process Industries Division, *ISO 9000 Guidelines for the Chemical and Process Industries*, 2nd ed., 1996a, ASQ Quality Press, Milwaukee, WI

ASQC Chemical and Process Industries Division, *Specifications for the Chemical and Process Industries: A Manual for Development and Use*, 1996b, ASQ Quality Press, Milw., WI

Bingham, Christopher, and Nelson, Lloyd S., "An Approximation for the Distribution of the von Neumann Ratio," *Technometrics*, Vol. 23, No. 3, August 1981, pp. 285-288

Bissell, Derek, *Statistical Methods for SPC and TQM*, 1994, Chapman & Hall, London, England

Bothe, Davis R., "A Powerful New Control Chart for Job Shops," *43rd ASQC Annual Quality Congress Transactions*, May 1989, pp. 265-270

Bothe, Davis R., "A Chart for Monitoring Within-Piece Variation," *44th ASQC Annual Quality Congress Transactions*, May 1990, pp. 899-904

Bothe, Davis R., *SPC for Short Production Runs Reference Handbook*, 1991, International Quality Institute, Inc., Sacramento, CA (916-933-2318)

Boyles, Russell A., "Exploratory Capability Analysis," *Journal of Quality Technology*, Vol. 28, No. 1, January 1996, pp. 91-98

Box, G.E.P., "Bounded Adjustment Charts," *Quality Engineering*, 4(2), 1991-1992, pp. 331-338

Box, G.E.P., and Jenkins, G.M., *Time Series Analysis, Forecasting and Control*, 2nd ed., 1976, Holden-Day, San Francisco, CA

Box, George, and Kramer, Tim, "Statistical Process Monitoring and Feedback Adjustment - A Discussion," *Technometrics*, Vol. 34, No. 3, August 1992, pp. 251-267

Burr, I.W., *Statistical Quality Control Methods*, 1976, Marcel Dekker, NY, NY, pp. 161-168

Burr, J.T., "SPC in the Short Run," *43rd ASQC Annual Quality Congress Transactions*, May 1989, pp. 776-780

Chan, L.K., Cheng, S.W., and Spiring, F.A., "A Multivariate Measure of Process Capability," *International Journal of Modelling and Simulation*, Vol. 11, 1991, pp. 1-6

Chen, H. "A Multivariate Process Capability Index Over a Rectangular Solid Tolerance Zone," *Statistica Sinica*, Vol. 4, 1994, pp. 749-758

Constable, Gordon K., and Hobbs, Jon R., "Small Samples and Non-normal Capability," *46th ASQC Annual Quality Congress Transactions*, May 1992, pp. 37-43

Cook, H.M., Jr., "Some Statistical Control Techniques for Job Shops," *43rd ASQC Annual Quality Congress Transactions*, May 1989, pp. 638-642

Cryer, Jonathan, D., and Ryan, Thomas, P., "The Estimation of Sigma for an X Chart: $\overline{MR}/d2$ or $s/c4$?" *Journal of Quality Technology*, Vol. 22, No. 3, July 1990, pp. 187-192

Davis, R.B., and Woodhall, W. H., "Performance of the Control Chart Trend Rule Under Linear Shift," *Journal of Quality Technology*, Vol. 20, No. 4, October 1988, pp. 260-262

Davis, Robert D., Kaminsky, Frank C., and Saboo, Sandeep, "Process Capability Analysis for Processes with Either a Circular or a Spherical Tolerance Zone," *Quality Engineering*, Vol. 5, No. 1, 1992-1993, pp. 41-54

Deleryd, M. and Vannman, K., "Process Capability Studies for Short Production Runs," *International Journal of Reliability, Quality, and Safety Engineering*, Vol. 5, 1998, pp. 353-401

Dodson, Bryan, "Control Charting Dependent Data: A Case Study," *Quality Engineering*, Vol. 7, No. 4, 1995, pp. 757-768

Doganaksoy, N., Faltin, F.W., and Tucker, W.T., "Identification of Out of Control Characteristics in a Multivariate Manufacturing Environment," *Communications in Statistical - Theory and Methods*, Vol. 20, No. 9, 1991, pp. 2775-2790

Dowling, M., Griffin P., Tsui, K., and Zhou, C., "Statistical Issues in Geometric Verification Using Coordinate Measuring Machines," *1993 Proceedings of the Section on Quality and Productivity*, American Statistical Association, Alexandria, VA, pp. 287-296

Eibl, S., Kess, U., and Pukelsheim, F., "Achieving a Target Value for a Manufacturing Process: A Case Study," *Journal of Quality Technology*, 24(1), Jan. 1992, pp. 22-26

Ermer, D.S., "A Control Chart for Dependent Data," *34th ASQC Annual Quality Congress Transactions*, Atlanta, GA, May 1980

Everitt, B.S., "A Monte Carlo Investigation of the Robustness of Hotelling's One and Two-Sample T^2 Tests," *Journal of the American Statistical Assoc.*, Vol. 74, 1979, pp. 48-51

Farnum, Nicholas R., "Control Charts for Short Runs: Non-constant Process and Measurement Error," *Journal of Quality Technology*, Vol. 24, No. 3, July 1992, pp. 138-144

Fuchs, Camil and Benjamini, Yoav, "Multivariate Profile Chart for Statistical Control," *Technometrics*, Vol. 36, No. 2, May 1994, pp. 182-195

Fuchs, Camil and Kenett, Ron S., *Multivariate Quality Control: Theory and Applications*, 1998, Marcel Dekker, New York, NY

Gallo, Paul P., "Characterizing the Variability of a Class of Mixing Processes," *Journal of Quality Technology*, Vol. 18, No. 1, January 1986, pp. 16-21

Gardiner, Stanley, C., and Mitra, Amitava, "Estimation of Process Standard Deviation Under Autocorrelated Observations with Variable Subgroup Sizes," *Quality Engineering*, Vol. 8, No. 2, 1995-1996, pp. 215-224

Gibra, I.N., "Optimal Control of Processes Subject to Linear Trends," *Journal of Industrial Engineering*, Vol. 18, 1967, pp. 35-41

Gibra, I.N., "Economically Optimal Determination of the Parameters of an $\overline{X}$-Control Chart," *Management Science*, Vol. 17, 1971

Gilbert, K.C., and Kirby, K., and Hild, C.R., "Charting Autocorrelated Data: Guidelines for Practitioners," *Quality Engineering*, Vol. 9, No. 3, 1997, pp. 367-382

Gitlow, Howard S., and Oppenheim, Rosa, "Residual Analysis with Shewhart Charts," *Quality Engineering*, Vol. 3, No. 3, 1991, pp. 309-331

Grant, Eugene L., and Leavenworth, Richard S., *Statistical Quality Control*, 5th ed., 1980, McGraw-Hill Book Co., New York, NY, pp. 298-302

Gruner, Glenn, "Calculation of Cpk Under Conditions of Variable Tolerances," *Quality Engineering*, Vol. 3, No. 3, 1991, pp. 281-291

Harding, A.J., Jr., Lee, K.R., and Mullins, J.L., "The Effect of Instabilities on Estimates of Sigma," *46th ASQC Annual Quality Congress Transactions*, May 1992, pp. 1037-1043

Harris, T. J., and Ross, W. H., "Statistical Process Control Procedures for Correlated Observations," *Canadian Journal of Chemical Engineering*, Vol. 69, 1991, pp. 48-57

Hellmich, M. and Wolff, H., "A New Approach for Describing and Controlling Process Capability for a Multivariate Process," *Proceedings of the Section on Quality and Productivity*, American Statistical Association, 1996, pp. 44-48

Herman, John T., "Capability Index - Enough for Process Industries?," *43rd ASQC Annual Quality Congress Transactions*, Toronto, Canada, May 1989, pp. 670-675

Hill, William J., and Bishop, Lane, "Quality Improvement Approaches for Chemical Processes," *Quality Engineering*, Vol. 3, No. 2, 1990-1991, pp. 137-152

Hillier, F.S., "$\overline{X}$- and R-Chart Control Limits Based on a Small Number of Subgroups," *Journal of Quality Technology*, Vol. 1, 1969, pp. 17-26

Holmes, D.S. and Mergen, A.E., "Measuring Process Performance for Multiple Variables," *Quality Engineering*, Vol. 11, No. 1, 1998-1999, pp. 55-59

Hotelling, H., "Multivariate Quality Control," *Techniques of Statistical Analysis*, Eisenhart, Hastay, and Wallis, editors, 1947, McGraw-Hill, New York, NY, pp. 111-184

Hubele, N.F., Shahriari, H., and Cheng, C-S., "A Bivariate Process Capability Vector," *Statistical Process Control in Manufacturing*, Keats, J.B., and Montgomery, D.C., editors, 1991, Marcel Dekker, New York, NY, pp. 299-310

Hubele, N.F., Montgomery, D.C., and Chih, W-H., "An Application of Statistical Process Control in Jet-Turbine Engine Component Manufacturing," *Quality Eng.*, 4(2), 1992, pp. 197-210

Hulting, F.L., "Process Capability Analysis with Geometric Tolerances," *Proceedings of the Section on Quality and Productivity*, American Statistical Assoc., 1993, pp. 207-216

Hunter, J. Stuart, "The Exponentially Weighted Moving Average," *Journal of Quality Technology*, Vol. 18, No. 4, October 1986, pp. 203-210

Hunter, J. Stuart, "A One-Point Plot Equivalent to the Shewhart Chart with Western Electric Rules," *Quality Engineering*, Vol. 2, No. 1, 1989-1990, pp. 13-19

Jackson, Paul F., "Simple Process Capability?," *Quality*, 40(2), Feb. 2001, pp. 34-38

Jang, J.S., Ahn, D.G., Lee, M.K., and Elsayed, E.A., "Optimum Initial Process Mean and Production Cycle for Processes With a Linear Trend," *Quality Engineering*, Vol. 13 No. 2, 2000-2001, pp. 229-235

Janis, Stuart J., "Is Your Process Too Good for Its Control Limits?," *44th ASQC Annual Quality Congress Transactions*, May 1990, pp. 1006-1011

Juran, J.M., Gryna, Frank M., Jr., and Bingham, R.S., Jr., *Quality Control Handbook*, 3rd ed., 1974, McGraw-Hill, New York, NY

Kalyanasundaram, M. and Balamurali, S., "A Superstructure Capability Index for Short Run Production Processes and Its Bootstrap Lower Confidence Limits," *Economic Quality Control*, Vol. 12, No. 2, 1997. pp. 85-93

Kane, Victor E., *Defect Prevention: Use of Simple Statistical Tools*, 1989, Marcel Dekker, Inc., New York, NY

Karl, Dennis P., Morisette, Jeffery, and Taam, Winson, "Some Applications of a Multivariate Capability Index in Geometric Dimensioning, and Tolerancing," *Quality Engineering*, Vol. 6, No. 4, 1994, pp. 649-665

Keller, Paul A., "Demystifying SPC, Part I: Choosing X Over X-bar Charts," *PI Quality*, Vol. 3, May/June 1993a, p. 26

Keller, Paul A., "Demystifying SPC, Part II: Sampling Period, and Autocorrelation," *PI Quality*, Vol. 3, July/August 1993b, pp. 34-35

Keller, Paul A., "Demystifying SPC, Part III: EWMA Makes It Easy," *PI Quality*, Vol. 3, September/October 1993c, pp. 50-52

Keenan, Thomas M., "A System for Measuring Short-Term Producibility," *49th ASQC Annual Quality Congress Transactions*, Cincinnati, OH, May 1995, pp. 50-56

Kocherlakota, S., and Kocherlakota, K., "Process Capability Index: Bivariate Normal Distribution," *Commun. in Statistics - Theory and Methods*, 20(8), 1991, pp. 2529-2547

Koons, George F. and Luner, Jeffery J., "SPC in Low-Volume Manufacturing: A Case Study," *Journal of Quality Technology*, Vol. 23, No. 4, October 1991, pp. 287-295

Kotz, S., and Johnson, N.L., *Process Capability Indices*, 1993, Chapman & Hall, London, U.K., pp. 179-201

Kourti, T. and MacGregor, J.F., "Process Analysis, Monitoring and Diagnosis Using Multivariate Projection Methods," *Journal of Chemometrics and Intelligent Lab. Systems*, 28, 1995, pp. 3-21

Kourti, T. and MacGregor, J.F., "Multivariate SPC Methods for Process and Product Monitoring," *Journal of Quality Technology*, Vol. 28, No. 4, October 1996, pp. 409-428

Koziol, J.A., "A Class of Invariant Procedures for Assessing Multivariate Normality," *Biometrika*, Vol. 69, 1982, pp. 423-427

Kresta, J., MacGregor, J.F., and Marlin, T.E., "Multivariate Statistical Monitoring of Process Operating Performance," *Canadian Journal of Chemical Engr.*, Vol. 69, 1991, pp. 35-47

Krishnamoorthi, K.S., "On Assignable Causes that Cannot be Eliminated - An Example from a Foundry," *Quality Engineering*, Vol. 3, No. 1, 1990-1991, pp. 41-47

Levinson, Harry J., and Ben-Jacob, Jake, "Managing Quality Improvement on a Development Pilot Line," *Quality Management Journal*, Vol. 3, No. 2, 1996, pp. 16-35

Liao, W.S., Wu, S.M., and Ermer, D.S., "A Time Series Approach to Quality Assurance," *Inspection and Quality Control in Manufacturing Systems*, Vol. 5, 1982, ASME

Littig, S.J., Lam, C.T., and Pollock, S.M. (*a*), "Process Capability Measurements for a Bivariate Characteristic Over an Elliptical Tolerance Zone," *Technical Report 92-42, University of Michigan*, Ann Arbor, MI

Littig, S.J., Lam, C.T., and Pollock, S.M. (*b*), "Process Capability Measurements for Multivariate Processes: Definitions and Example for a Gear Carrier," *Technical Report No. 42, Industrial and Operations Engineering Department, University of Michigan*, Ann Arbor, MI

Little, Thomas A., and Harrelson, C. Scott," Short Run Capability Assessment Using the t Distribution," *47th ASQC Annual Quality Congress Transactions*, May 1993, pp. 181-191

Liu, Regina Y., "Control Charts for Multivariate Processes," *Journal of the American Statistical Association*, Vol. 90, No. 432, December 1995, pp. 1380-1387

Long, Jeri M., and De Coste, Melinda J., "Capability Studies Involving Tool Wear," *42nd ASQC Annual Quality Congress Transactions*, Dallas, TX, May 1988, pp. 590-596

Lucas, James M., and Saccucci, Michael S., "Exponentially Weighted Moving Average Control Schemes: Properties and Enhancements," *Technometrics*, Vol. 32, No. 1, February 1990, pp. 1-12

MacGregor, J.F., and Kourti, T., "Statistical Process Control of Multivariate Processes," *Control Engineering*, Practice 3, 1995, pp. 403-414

Mamzic, C. L., and Tucker, T. W., "SPC in the Process Industries," (part I) *Hydrocarbon Processing*, Vol. 67, No. 11, November 1988a, pp. 132D-132P

Mamzic, C. L., and Tucker, T. W., "SPC in the Process Industries," (part II) *Hydrocarbon Processing*, Vol. 67, No. 12, December 1988b, pp. 34C

Mandel, B.J., "The Regression Control Chart," *Journal of Quality Tech.*, 1(1), 1969, pp. 1-9

Manuele, J., "Control Chart for Determining Tool Wear," *Ind. Quality Control*, 1, pp. 226-233

Mardia, K.V., "Applications of Some Measures of Multivariate Skewness and Kurtosis in Testing Normality and Robustness Studies," *Sankhya, Series B*, Vol. 36, 1974, pp. 115-128

Martin, John E., "SPC Bridges Lab/Production Gap," *PI Quality*, Vol. 1, 1991, pp. 40-41

Mason, R.L., Tracy, N.D., and Young, J.C., "Decomposition of T^2 for Multivariate Control Chart Interpretation," *Journal of Quality Technology*, Vol. 27, No. 2, April 1995, pp. 99-108

Mason, R.L., Tracy, N.D., and Young, J.C., "Monitoring a Multivariate Step Process," *Journal of Quality Technology*, Vol. 28, No. 1, January 1996, pp. 39-50

McNeese, William H., and Klein, Robert A., *Statistical Methods for the Process Industries*, 1991, ASQ Quality Press, Milwaukee, WI

Montgomery, Douglas C., *Introduction to Statistical Quality Control*, 2nd ed., 1991, John Wiley & Sons, New York, NY

Montgomery, D.C. and Mastrangelo, C.M., "Some Statistical Process Control Methods for Autocorrelated Data," *Journal of Quality Technology*, 23(3), July 1991, pp. 179-193

Montgomery, D.C., Keats, B.J., Runger, G.C., and Messina, W.S., "Integrating Statistical Process Control and Engineering Process Control," *Journal of Qual. Tech.*, 26(2), Apr. 1994, pp. 79-87

Morris, R.A. and Watson, E.F., "Determining Process Capability in a Chemical Batch Process," *Quality Engineering*, Vol. 10, No. 2, 1997-1998, pp. 389-396

Murphy, B.J., "Selecting Out of Control Variables with the T^2 Multivariate Quality Control Procedure," *The Statistician*, Vol. 36, 1987, pp. 571-583

Negiz, A. and Cincar, A., "Statistical Monitoring of Strongly Autocorrelated Processes," *Proceedings of the Section on Quality and Productivity*, American Statistical Assoc., 1996, pp. 55-64

Niverthi, M. and Dey, D.K., "Multivariate Process Capability: A Bayesian Approach," *Communications in Statistics - Simulation*, Vol. 29, 2000, pp. 667-687

Nomikos, P., and MacGregor, J.F., "Multivariate SPC Charts for Batch Processes," *Technometrics*, Vol. 37, No. 1, February 1995, pp. 41-59

Ng, H.C., and Case, K.E., "Development and Evaluation of Control Charts Using Exponentially Weighted Moving Averages," *Journal of Quality Tech.*, 21(4), Oct. 1989, pp. 242-250

Padgett, M.M., Vaughn, L.E., and Lester, J.A., "Statistical Process Control and Machine Capability in Three Dimensions," *Quality Engineering*, Vol. 7, No. 4, 1995, pp. 779-796

Palm, A.C., "SPC Versus Automatic Process Control," *44th ASQC Annual Quality Congress Transactions*, 1990, pp. 694-699

Pearn, W.L. and Chang, C.S., "An Implementation of the Precision Index for Contaminated Processes," *Quality Engineering*, Vol. 11, No. 1, 1998-1999, pp. 101-110

Pearn, W.L., Kotz, S., and Johnson, N.L., "Distributional and Inferential Properties of Process Capability Indices," *Journal of Quality Technology*, Vol. 24, No. 4, October 1992, pp. 226-227

Phillips, M.D. and Cho, Byung-Rae, "Quality Improvement for Processes With Circular and Spherical Specification Regions," *Quality Engineering*, 11(2), 1998-1999, pp. 235-243

Proschan, F., and Savage, I.R., "Starting a Control Chart," *Industrial Quality Control*, Vol. 17, No. 3, September 1960, pp. 12-13

Pugh, G. Allen, "On the Application of Individual and Moving-Range Charts to Insulated Wire Production," *Quality Engineering*, Vol. 7, No. 1, 1994-1995, pp. 105-111

Pyzdek, Thomas, *Pyzdek's Guide to SPC, Volume 2: Applications and Special Topics*, 1992, Quality Publishing, Tucson, AZ

Quesenberry, C.P., "An SPC Approach to Compensating a Tool-Wear Process," *Journal of Quality Technology*, 20(4), Oct. 1988, pp. 220-229

Quesenberry, C.P., "SPC *Q* Charts for Start-Up Processes and Short or Long Runs," *Journal of Quality Technology*, 23(3), July 1991, pp. 213-224

Royston, J.P., "Some Techniques for Assessing Multivariate Normality Based on the Shapiro-Wilks W," *Applied Statistics*, Vol. 32, No. 2, 1983, pp. 121-133

Runger, George C., "Robustness of Variance Estimates for Batch and Continuous Processes," *Quality Engineering*, Vol. 7, No. 1, 1994-1995, pp. 31-43

Ryan, Thomas P., *Statistical Methods for Quality Improvement*, 1989, John Wiley & Sons, New York, NY, pp. 215-227

Sarkar, Ashok and Pal, Surajit, "Estimation of Process Capability Index for Concentricity," *Quality Engineering*, Vol. 9, No. 4, 1997, pp. 665-671

Sarkar, Ashok and Pal, Surajit, "Process Control and Evaluation in the Presence of Systematic Assignable Cause," *Quality Engineering*, Vol. 10, No. 2, 1997-1998, pp. 383-388

Schneider, H. and Pruett, J.M., "Control Charting Issues in the Process Industries," *Quality Engineering*, Vol. 6, No. 3, 1994, pp. 347-373

Seder, Leonard A. and Cowan, David, *The Span Plan Method: Process Capability Analysis, ASQC General Publications No. 3*, 1956, American Society for Quality Control, Milw., WI

Shaw, J. A., "Statistical Process Control for the Process Industries," *ISA Transactions*, Vol. 30, No. 1, 1991, pp. 11-34

Shore, Haim, "Process Capability Analysis When Data Are Autocorrelated," *Quality Engineering*, Vol. 9, No. 4, 1997, pp. 615-626

Spiring, F.A., "An Application of Cpm to the Tool Wear Problem," *43rd ASQC Annual Quality Congress Transactions.*, 1989, pp. 123-128

Spiring, F.A., "Assessing Process Capability in the Presence of Systematic Assignable Cause," *Journal of Quality Technology*, 23(2), Apr. 1991, pp. 125-134

Spurrier, J.D. and Thombs, L.A., "Control Charts for Detecting Cyclical Behavior," *Technometrics*, Vol. 32, 1990, pp. 163-171

Subbaiah, P. and Taam, W., "Inference on the Capability Index: Cpm," *Commun. in Statistics - Theory and Methods*, 22(2), 1993, pp. 537-560

Subramanyam, N. and Houshmand, A.A., "Simultaneous Representation of Multivariate and Corresponding Univariate $\overline{X}$ Charts Using Line-Graph," *Quality Engineering*, Vol. 7, No. 4, 1995, pp. 681-692

Taam, W., Subbaiah, P., and Liddy, J.W., "A Note on Multivariate Capability Indices," *Journal of Applied Stat.*, 20(3), 1993, pp. 339-351

Tatum, L.G., "Control Charts for the Detection of a Periodic Component," *Technometrics*, 38(2), May 1996, pp. 152-160

Taylor, Wayne A., *Optimization & Variation Reduction in Quality*, 1991, McGraw-Hill, Inc., New York, NY, pp. 174-176

Thompson, L.A., Jr., "SPC & the Job Shop - Strange Bedfellows?," *43rd ASQC Annual Quality Congress Transactions*, Toronto, Canada, May 1989, pp. 896-901

Timm, Neil H., "Multivariate Quality Control Using Finite Intersection Tests," *Journal of Quality Technology*, Vol. 28, No. 2, April 1996, pp. 233-243

Tracy, Nola D., Young, John C., and Mason, Robert L., "Multivariate Control Charts for Individual Observations," *Journal of Quality Technology*, 24, 2, April 1992, pp. 88-95

Traver, R.W., and Davis, J.M., "How to Determine Process Capabilities in a Developmental Shop," *Industrial Quality Control*, Vol. 18, No. 9, 1962, pp. 26-29

Vander Wiel, S.A., "Monitoring Processes That Wander Using Integrated Moving Average Models," *Technometrics*, Vol. 38, No. 2, May 1996, pp. 139-151

von Neumann, John, "Distribution of the Ratio of the Mean Square Successive Difference to the Variance," *Annals of Mathematical Statistics*, Vol. 13, 1941, pp. 86-88

Wang, F.K. and Chen, J.C., "Capability Index Using Principal Components Analysis," *Quality Engineering*, Vol. 11, No. 1, 1998-1999, pp. 21-27

Wang, F.K., Hubele, N.F., Lawrence, F.P., Miskulin, J.D., and Shahriari, H., "Comparison of Three Multivariate Process Capability Indices," *Journal of Quality Technology*, Vol. 32, No. 3, July 2000, pp. 263-275

Wardell, D. G., Moskowitz, H., and Plante, R. D., "*Control Charts in the Presence of Data Correlation*," Management Science, Vol. 38, No. 8, 1992, pp. 1084-1105

Wheeler, Donald J., *Advanced Topics in Statistical Process Control*, 1995, SPC Press, Inc., Knoxville, TN

Wheeler, Donald J., *Short Run SPC*, 1991, SPC Press, Inc., Knoxville, TN

Wheeler, Donald J., and Chambers, David S., *Understanding Statistical Process Control*, 2nd edition, 1992, SPC Press, Inc., Knoxville, TN

Wierda, Siebrand J., "A Multivariate Process Capability Index," *47th Annual ASQC Quality Congress Transactions*, 1993, pp. 342-348

Wierda, Siebrand J., "Multivariate Statistical Process Control - Recent Results and Directions for Future Research," *Statistica Neerlandica*, Vol. 48, 1994, pp. 147-168

Wise, Stephen A. and Fair, Douglas C., *Innovative Control Charting*, 1998, ASQ Quality Press, Milwaukee, WI

Woodhall, W., and Ncube, M., "Multivariate CUSUM Quality-Control Procedures," *Technometrics*, Vol. 27, 1985, pp. 285-292

Wright, Don, "Multivariable Charting," *Quality*, Vol. 32, No. 3, March 1993, p. 49

Yang, C.H., and Hillier, F.S., "Mean and Variance Control Chart Limits Based on a Small Number of Subgroups," *Journal of Quality Technology*, Vol. 2, 1970, pp. 9-16

Zhang, N.F., "Estimating Process Capability Indexes for Autocorrelated Data," *Journal of Applied Statistics*, Vol. 25, 1998, pp. 559-574

Zhang, Y. and Fang, X.D., "Target Allocation for Maximizing Wear Allowance of Running Fits Based on Process Capability," *Quality Engineering*, 12(2), 1999-2000, pp. 169-176

Chapter 14

Understanding the Six-Sigma Philosophy

Much discussion in recent years has been devoted to the concept of "six-sigma" quality. The company most often associated with this philosophy is Motorola, Inc., whose definition of this principle is stated by Harry in his booklet, *The Nature of Six Sigma Quality*.

> *A product is said to have six sigma quality when it exhibits no more than 3.4 ppm at the part and process step levels.*

There is often confusion about the relationship between 6σ and this definition of producing no more than 3.4 nonconformities per million opportunities. From Appendix Table III, the area underneath a normal curve beyond 6σ from the average is found to be 1.248×10^{-9}, or .001248 *ppm*, which is about 1 part per billion. Considering both tails of the process distribution, this would be a total of .002 *ppm*, as is shown in Figure 14.1. This process has the potential capability of fitting two 6σ spreads within the tolerance, or equivalently, having 12σ equal the tolerance.

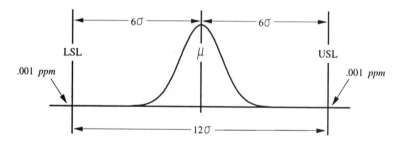

Fig. 14.1   A process with potential capability of 12 σ equaling the tolerance.

However, the 3.4 *ppm* value corresponds to the area under the curve at a distance of only 4.5σ from the process average. Why this apparent discrepancy? It's due to the difference between a static and a dynamic process.

14.1 A Static versus a Dynamic Process

If the process portrayed in Figure 14.1 is *static*, meaning the process average remains centered at the middle of the tolerance, then approximately .002 *ppm* will be produced. But under the six-sigma concept, the process is considered to be *dynamic*, implying that over time, the process average will move both higher and lower because of many small changes in material, operators, environmental factors, tools, etc. Recall from the graph in Figure 2.13 (p. 16) that most small shifts in the process average will go undetected by the control chart. For an *n* of 4, there is only a 50 percent chance a 1.5σ shift in μ is detected by the next subgroup after this change. By the time this next subgroup is collected, μ may have returned to its original position. Thus, this process change will never be noticed on the chart, which means no corrective action is implemented. However, this movement has caused the actual long-term process variation to increase somewhat because between-subgroup variation is greater than within-subgroup variation. Note that estimates of short-term process variation are not impacted because they are determined from only within-subgroup variation.

Based on studies analyzing the effect of these changes on process variation (Bender, 1962, 1968; Evans, 1970, 1974, 1975*a*, 1975*b*; Gilson), the six-sigma principle acknowledges the likelihood of undetected shifts in the process average of up to ±1.5σ. Because shifts in the average greater than 1.5σ are expected to be caught, and σ is assumed not to change, the worst case for the production of nonconforming parts happens when the process average has shifted either the full 1.5σ above the middle of the tolerance, or the full 1.5σ below it. For this worst case, there would be only 4.5σ (6σ minus 1.5σ) remaining between the process average and the nearest specification limit, as is illustrated in Figure 14.2.

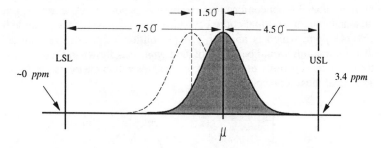

Fig. 14.2   Process average has shifted 1.5 σ higher.

This reduced *Z* value of 4.5 for the dynamic model corresponds to 3.4 *ppm*. When this size of shift occurs, the *Z* value for the other specification limit becomes 7.5, which means essentially 0 *ppm* are outside this limit. Because the process average can shift in only one direction at a time, the maximum amount of nonconforming parts produced is 3.4 *ppm*. Notice that for most of the time, the average should be closer to the middle of the tolerance, resulting in far fewer than 3.4 *ppm* actually being manufactured.

To achieve a goal of 3.4 *ppm*, the process average must be no closer than 4.5σ to a specification limit. Assuming the average could drift by as much as 1.5σ, potential capability must be at least 6.0σ (4.5σ plus 1.5σ) to compensate for shifts in the process average of up to 1.5σ, yet still be able to produce the desired quality level. The required 4.5σ plus this added buffer of 1.5σ create the 6σ requirement, and thereby generate the label "six-sigma."

To counter the effect of shifts in μ, a buffer of 1.5 standard deviations can be added to other capability goals as well. If no more than 32 *ppm* are desired outside either specification, the goal would be to have ±4.0σ fit within the tolerance, assuming no change in the process average. This target equates to a C_p of 1.33 (4.0/3). Under the static model, this potential capability goal translates into 32 *ppm* outside each specification when the average is centered at *M*. But with the inevitable 1.5σ drifts in μ occurring with the dynamic process model, the average could move as close as 2.5σ (4.0σ minus 1.5σ) to a specification limit before triggering any type of corrective action. This change in centering would cause as many as 6,210 *ppm* to be produced, quite a bit more than the desired maximum of 32 *ppm*.

To make allowances for these expected (but undetected) shifts in μ of up to 1.5σ, the potential capability goal must be increased to at least 5.5σ (4.0σ plus 1.5σ), meaning C_p must be at least 1.83 (5.5/3). Achieving this goal allows the process to endure a 1.5σ shift in average and still produce no more than 32 *ppm*.

This crucial difference between a static and dynamic process is summarized in Table 14.1 for several additional values of C_p. The dynamic process assumes unnoticed shifts in the process average of up to 1.5σ may possibly occur when a subgroup size of four is used.

Table 14.1   Effect on *ppm* of a static versus a dynamic process.

C_P	Static Process			Dynamic Process, $n = 4$		
	Z_{MIN}	Z_{MAX}	*ppm*	Z_{MIN}	Z_{MAX}	*ppm*
1.00	3.0	3.0	2,700	1.5	4.5	66,813
1.33	4.0	4.0	63	2.5	5.5	6,210
1.50	4.5	4.5	6.8	3.0	6.0	1,350
1.67	5.0	5.0	.6	3.5	6.5	233
1.83	5.5	5.5	.0	4.0	7.0	32
2.00	6.0	6.0	.0	4.5	7.5	3.4

As Figure 2.14 (p. 17) illustrates, smaller subgroup sizes have less chance of detecting a 1.5σ shift in μ. For example, when a subgroup size of 2 is chosen, the probability of catching a 1.5σ shift with the next subgroup sampled is only 20 percent. In fact, the process average could move by as much as 2.1σ before there is a 50 percent chance of discovery.

Table 14.2   *ppm* levels for a dynamic process with *n* equal to 2.

C_P	Dynamic Process, $n = 2$		
	Z_{MIN}	Z_{MAX}	*ppm*
1.00	0.9	5.1	184,100
1.33	1.9	6.1	28,720
1.50	2.4	6.6	8,198
1.67	2.9	7.1	1,866
1.83	3.4	7.6	337
2.00	3.9	8.1	48
2.20	4.5	8.7	3.4

With this magnitude of change, the *ppm* levels associated with a given C_p value under the dynamic model will be somewhat higher, as is displayed in Table 14.2. Note that Z_{MIN} is now 2.1 less than the static value, while Z_{MAX} is 2.1 higher.

In fact, to have a maximum of 3.4 *ppm*, potential process capability must be 4.5σ plus 2.1σ, for a total of 6.6σ. This requirement necessitates a C_p of 2.20 (6.6/3), as is shown in the last line of Table 14.2. This substantial increase in potential capability is necessary to offset larger undetected movements in the process average due to the smaller subgroup size.

Products with Multiple Characteristics

Extremely low *ppm* levels are imperative for producing high quality products possessing many characteristics (or components). Table 14.3 compares the probability of manufacturing a product with all characteristics inside their respective specifications when each is produced with $\pm 4\sigma$ ($C_P = 1.33$) versus $\pm 6\sigma$ ($C_P = 2.00$) capability. The processes producing the features are assumed to be dynamic, with up to a 1.5σ shift in average possible.

Suppose a product has only one feature, which is produced on a process having $\pm 4\sigma$ potential capability, then from Table 14.1, a maximum of .6210 percent of these parts will be nonconforming under the dynamic model. Conversely, at least 99.3790 percent will be conforming, as is listed on the first line of Table 14.3. If this single characteristic is instead produced on a process with $\pm 6\sigma$ potential capability, at most .00034 percent of the finished product will be out of specification, with at least 99.99966 percent within specification.

Table 14.3 Probability of a completely conforming product.

No. of Char.	With 1.5σ shift	
	$C_P = 1.33$ ($\pm 4\sigma$)	$C_P = 2.00$ ($\pm 6\sigma$)
1	99.3790	99.99966
2	98.7618	99.99932
5	96.9333	99.9983
10	93.9607	99.9966
25	85.5787	99.9915
50	73.2371	99.9830
100	53.6367	99.9660
150	39.2820	99.9490
250	21.0696	99.9150
500	4.4393	99.8301

If a product has two characteristics, the probability both are within specification (assuming independence) is .993790 times .993790, or 98.7618 percent when each is produced on a $\pm 4\sigma$ process. If they are produced on a $\pm 6\sigma$ process, this probability increases to 99.99932 percent (.9999966 times .9999966). The remainder of this table is computed in a similar manner.

When each characteristic is produced with $\pm 4\sigma$ capability (and assuming a maximum drift of 1.5σ), a product with 10 characteristics will average about 939 conforming parts out of every 1,000 made, with the 61 nonconforming ones having at least one characteristic out of specification. If all characteristics are manufactured with $\pm 6\sigma$ capability, it would be very unlikely to see even one nonconforming part out of these 1,000.

For a product having 50 characteristics, 268 out of 1,000 parts will have at least one nonconforming characteristic when each is produced with $\pm 4\sigma$ capability. If these 50 characteristics were manufactured with $\pm 6\sigma$ capability, it would still be improbable to see one nonconforming part. In fact, with $\pm 6\sigma$ capability, a product must have 150 characteristics before expecting to find even one nonconforming part out of 1,000. Contrast this to $\pm 4\sigma$ capability level, where 60.7 percent of these parts would be rejected, and the rationale for adopting the six-sigma philosophy becomes quite evident.

14.2 Short- and Long-term Six-Sigma Capability

The six-sigma approach also differentiates between short- and long-term process variation. Just as in this book, the short-term standard deviation is estimated from within-subgroup variation, usually from $\overline{R}$. The long-term standard deviation incorporates both the short-term variation and any additional variation in the process introduced by the small, undetected shifts in the process average that occur over time. Although there is no exact relationship between these two types of variation that applies to every kind of process, the six-sigma philosophy ties them together with this general equation (Harry and Lawson, p. 6-8).

$$\sigma_{LT} = c\,\sigma_{ST}$$

As c is affected by shifts in the process average, it is related to the k factor, which quantifies how far the process average is from the middle of the tolerance.

$$c = \frac{1}{1-k} \qquad k = \frac{|\mu - M|}{(\text{USL} - \text{LSL})/2}$$

If a process has a C_P of 2.00 and is centered at the middle of the tolerance, then there is a distance of $6\sigma_{ST}$ from the average to the USL. When the process average shifts up by $1.5\sigma_{ST}$, it has moved off target by 25 percent of one-half the tolerance ($1.5/6.0 = .25$). For this k factor of .25, c is calculated as 1.33.

$$c = \frac{1}{1 - .25} = \frac{1}{.75} = 1.33$$

The long-term standard deviation for this process would then be estimated from σ_{ST} as:

$$\hat{\sigma}_{LT} = c\,\sigma_{ST} = 1.33\,\sigma_{ST}$$

1.33 is quite commonly adopted as the relationship between short- and long-term process variation (Koons). This factor implies long-term variation is approximately 33 percent greater than short-term variation. Other authors are more conservative and assume a c factor between 1.40 and 1.60, which translates to a k factor ranging from .286 to .375 (Harry and Lawson, pp. 6-12, 7-6). For a c factor of 1.50, k is .333.

$$1.50 = \frac{1}{1-k}$$

$$1-k = \frac{1}{1.50}$$

$$k = 1 - \frac{1}{1.50} = .333$$

This assumption expects up to a 33.3 percent shift in the process average. With six-sigma capability, there is $6\sigma_{ST}$ from M to the specification limit, a distance that equals one-half the tolerance. A k factor of .333 represents a maximum shift in the process average of $2.0\sigma_{ST}$, a number derived by multiplying one-half the tolerance, or $6\sigma_{ST}$, by .333.

14.3 Capability Goals Under the Six-Sigma Philosophy

In order to meet the short-term capability goals under the six-sigma doctrine, there must be at least $6\sigma_{ST}$ between the process average and either print specification when the process average is at the middle of the tolerance. This goal means at least $12\sigma_{ST}$ must fit within the tolerance, as is presented in Figure 14.3.

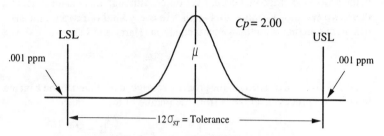

Fig. 14.3 Six-sigma goal for C_P.

If $12\sigma_{ST,\,GOAL}$ equals the tolerance, the six-sigma goal for C_p becomes 2.00.

$$C_{P,GOAL} = \frac{\text{Tolerance}}{6\sigma_{ST,GOAL}} = \frac{12\sigma_{ST,GOAL}}{6\sigma_{ST,GOAL}} = 2.00$$

Because the process average is expected to vary by as much as $\pm 1.5\sigma_{ST}$, the minimum distance allowed between the process average and either specification limit that still achieves the performance capability goal of $3.4\,ppm$ is $4.5\sigma_{ST,\,GOAL}$. Thus, both the following conditions must be met:

$$\mu - LSL \geq 4.5\sigma_{ST,GOAL} \text{ and } USL - \mu \geq 4.5\sigma_{ST,GOAL}$$

These conditions define a "six-sigma" goal for C_{PK}, since this measure of capability considers both σ_{ST} and μ, as well as this minimum allowed distance of $4.5\sigma_{ST,\,GOAL}$ between the average and either specification limit (review Figure 14.2, p. 832).

$$C_{PK,GOAL} = \frac{1}{3\sigma_{ST,GOAL}} \text{Minimum} (\mu - LSL, USL - \mu) = \frac{1}{3\sigma_{ST,GOAL}} (4.5\sigma_{ST,GOAL}) = 1.50$$

Thus, the definition of six-sigma quality for variable-data characteristics may be expressed quantitatively by utilizing these three capability measures:

$$C_{P,\,GOAL} = 2.00 \qquad C_{PK,\,GOAL} = 1.50 \qquad ppm_{TOTAL,\,ST,\,GOAL} = 3.4$$

All three must be met in order to achieve the goal of six-sigma quality.

14.4 Six-Sigma Quality for Attribute Characteristics

The previous definition given for six-sigma quality pertains to product characteristics measured with variable-type data. However, this definition can be easily extended to cover characteristics generating attribute-type data as well:

Six-sigma quality is a product having a $ppm_{TOTAL,\ LT}$ of no more than 3.4.

With the capability measures developed in this book for attribute data, this ppm goal translates into an equivalent $P^{\%}_{PK}$ goal of 1.50. Recall from Section 9.2 (p. 553) that equivalent $P^{\%}_{PK}$ assumes all the nonconforming parts are in the same tail of the output distribution.

$$\text{Equivalent } P^{\%}_{PK,GOAL} = \frac{Z(\ 3.4\ ppm\ )}{3} = \frac{4.50}{3} = 1.50$$

These capability goals are consistent with those developed for variable-data situations.

14.5 Summary

The six-sigma philosophy is becoming more and more popular in the quality field, especially with companies in the electronics industry (de Treville *et al*). Organizations striving to attain the quality levels required with the six-sigma system usually adopt the following three recommended strategies for accomplishing this goal (Tomas offers a six-step approach). Improving an existing process to the six-sigma level of quality would be very difficult, if not impossible. That's why Fan insists this type of thinking must already be incorporated into the original design of new products *and* the processes that will manufacture them if there is to be any chance of achieving six-sigma quality.

1. ***Design Phase***

    a. Design in $\pm6\sigma$ tolerances for all critical product and process parameters. For additional information on this topic, read *Six Sigma Mechanical Design Tolerancing* by Harry (see also Shina and Saigal, and Turmel and Gartz).

    b. Develop designs robust to unexpected changes in both manufacturing and customer environments (see Harry and Lawson).

    c. Minimize part count and number of processing steps.

    d. Standardize parts and processes.

Knowing the process capability of current manufacturing operations will greatly aid designers in accomplishing this first step. And of course, good designs will positively influence the capability of future processes.

Once a new product is released for production, the designed-in quality levels must be maintained, and even improved upon, by working to reduce (or eliminate) both assignable and common causes of process variation. McFadden lists several additional key components of a six-sigma quality program specifically targeted at manufacturing.

2. ***Internal Manufacturing***

    a. Standardize manufacturing practices.

    b. Audit the manufacturing system. Pena provides a detailed audit checklist for this purpose.

    c. Use SPC to control, identify and eliminate causes of variation in the manufacturing process. Mader *et al* have written a book entitled *Process Control Methods* to help with this step.

    d. Measure process capability and compare to goals. Koons' book on capability indices is useful here.

e. Consider the effects of random sampling variation on all six-sigma estimates and apply the proper confidence bounds (review Chapter 11, then read Arthur, Breyfogle, Harry and Schroeder, Pyzdek, and Tavormina and Buckleysee).

f. Reduce variation by continuous process improvement (see Kelly and Seymour, as well as Bothe). Delott and Gupta reveal how the application of statistical techniques helped achieve six-sigma quality levels for copperplating ceramic substrates. Harry (1994) provides several examples of applying design of experiments to improve quality in the electronics industry.

Even if these first two strategies are adopted, a company will never achieve six-sigma quality unless it has the full cooperation, and participation, of all its suppliers.

3. *External Manufacturing*

a. Qualify suppliers.

b. Minimize the number of suppliers.

c. Develop long-term partnerships with remaining suppliers.

d. Require documented process control plans.

e. Insist on continuous process improvement.

Craig shows how Dupont Connector Systems utilized this set of strategies to introduce new products into the data processing and telecommunications industries. Noguera discusses how the six-sigma doctrine applies to chip connection technology in electronics manufacturing, while Fontenot *et al* explain how these six-sigma principles pertain to improving customer service. Daskalantonakis *et al* describe how software measurement technology can identify areas of improvement and help track progress toward attaining six-sigma quality in software development.

As all these authors conclude, the rewards for achieving the six-sigma quality goals are shorter cycle times, shorter lead times, reduced costs, higher yields, improved product reliability, increased profitability, and most importantly of all, highly satisfied customers.

"Quality remains long after the price is forgotten."

H. Gordon Selfridge

14.6 References

Arthur, Lowell J., *Six Sigma Simplified: Quantum Improvement Made Easy*, 2000, available through ASQ Quality Press, Milwaukee, WI

Bender, Art, "Bendarizing Tolerances - A Simple Practical Probability Method of Handling Tolerances for Limit-Stack-Ups," *Graphic Science*, December 1962, pp. 17-21

Bender, Art, "Statistical Tolerancing as It Relates to Quality Control and the Designer, "SAE Paper No. 680490, *Society of Automotive Engineers*, Southfield, MI, May 1968

Bothe, Davis R., *Reducing Process Variation*, 1993, International Quality Institute., Inc., Sacramento, CA, 916-933-2318

Breyfogle, III, F.W., *Implementing Six Sigma: Smarter Solutions Using Statistical Methods*, 1999, available through ASQ Quality Press, Milwaukee, WI

Craig, Robert J., "Six-Sigma Quality, the Key to Customer Satisfaction," *47th ASQC Annual Quality Congress Transactions*, Boston, MA, May 1993, pp. 206-212

Daskalantonakis, Michael K., Yacobellis, Robert H., and Basili, Victor R., "A Method for Assessing Software Measurement Technology," *Quality Engineering*, Vol. 3, No. 1, 1990-1991, pp. 27-40

Delott, Charles, and Gupta, Praveen, "Characterization of Copperplating Process for Ceramic Substrates," *Quality Engineering*, Vol. 2, No. 3, 1990, pp. 269-284

de Treville, Suzanne, Edelson, Norman M., and Watson, Rick, "Getting Six Sigma Back to Basics," *Quality Digest*, Vol. 15, No. 5, May 1995, pp. 42-47

Evans, David H., "Statistical Tolerancing Formulation," *Journal of Quality Technology*, Vol. 2, No. 4, October 1970, pp. 188-195

Evans, D.H., "Statistical Tolerancing: The State of the Art, Part I: Background," *Journal of Quality Technology*, Vol. 6, No. 4, October 1974, pp. 188-195

Evans, D.H., "Statistical Tolerancing: The State of the Art, Part II: Methods for Estimating Moments," *Journal of Quality Technology*, Vol. 7, No. 1, January 1975a, pp. 1-12

Evans, D.H., "Statistical Tolerancing: The State of the Art, Part III: Shifts and Drifts," *Journal of Quality Technology*, Vol. 7, No. 2, April 1975b, pp. 72-26

Fan, John Y., "Achieving 6σ in Design," *44th ASQC Annual Quality Congress Transactions*, San Francisco, CA, May 1990, pp. 851-856

Fontenot, Gwen, Behara, Ravi, and Gresham, Alicia, "Six Sigma in Customer Satisfaction," *Quality Progress*, Vol. 27, No. 12, December 1994, pp. 73-76

Gilson, J., *A New Approach to Engineering Tolerances*, 1951, Machinery Publishing Co., London, England

Harry, Mikel, *The Nature of Six Sigma Quality*, Motorola Univ. Press, Schaumberg, IL, p. 3

Harry, Mikel, *Six Sigma Mechanical Design Tolerancing*, 1988, Motorola University Press, Schaumberg, IL

Harry, Mikel J., *The Vision of Six Sigma: Case Studies and Applications*, 2nd edition, 1994, Sigma Publishing Co., Phoenix, AZ

Harry, Mikel J. and Lawson, J. Ronald, *Six Sigma Producibility Analysis and Process Characterization*, 1992, Addison-Wesley Publishing Co., Reading, MA, p. 6-8

Harry, Mikel and Schroeder, Richard, *Six Sigma: The Breakthrough Management Strategy Revolutionizing the World's Top Corporations*, 2000, available through ASQ Quality Press, Milwaukee, WI

Kelly, Harrison W. and Seymour, Lee Ann, *Data Display*, 1993, Addison-Wesley Publishing Co., Reading, MA

Koons, *Indices of Capability: Classical and Six Sigma Tools*, 1992, Addison-Wesley Publishing Co., Reading, MA, pp. 22-23

Mader, Douglas P., Seymour, Lee Ann, Brauer, Douglas C., and Gallemore, Marian Lee, *Process Control Methods*, 1993, Addison-Wesley Publishing Co., Reading, MA

McFadden, Fred R., "Six-Sigma Quality Programs," *Quality Progress*, Vol. 26, No. 6, June 1993, pp. 37-42

Noguera, John, "Implementing Six Sigma for Interconnect Technology," *46th ASQC Annual Quality Congress Transactions*, May 1992, pp. 538-544

Pena, Ed, "Motorola's Secret to Total Quality Control," *Quality Progress*, Vol. 23, No. 10, October 1990, pp. 43-45

Pyzdek, Thomas, *The Six Sigma Handbook*, 2000, available through ASQ Quality Press, Milwaukee, WI

Shina, S.G. and Saigal, A., "Using C_{PK} as a Design Tool for New System Development," *Quality Engineering*, Vol. 12, No. 4, 2000, pp. 551-560

Tadikamalla, Pandu R., "The Confusion Over Six-Sigma Quality," *Quality Progress*, Vol. 21, No. 11, November 1994, pp. 83-85

Tavormina, Joseph J., and Buckley, Shawn, "SPC and Six-Sigma," *Quality*, Vol. 31, No. 11, November 1992, p. 47

Tomas, Sam, "Motorola's Six Steps to Six Sigma," *34th International Conference Proceedings*, APICS, Seattle, WA, October 1991, pp. 166-169

Turmel, Jeff and Gartz, Larry, ""Designing In" Quality Improvement: A Systematic Approach to Designing for "Six Sigma"," *51st ASQ Annual Quality Congress Transactions*, May 1997, pp. 391-398

Appendix

Statistical Tables

The following statistical tables provide control limit constants, areas underneath the normal curve, as well as probabilities needed to calculate confidence bounds for the various capability measures covered in this book.

Table I. Control Limit Constants for n from 1 to 10, by 1

Table II. c_4 Values For kn from 2 to 500, by 1

Table III. Areas Under the Normal Curve for Z Values from 0.00 to 7.79, by .01

Table IV. Cumulative Areas Under the Normal Curve from -2.99σ to 2.99σ, by .01

Table V. Median Ranks for kn from 10 to 50, by 1

Table VI. Values of χ^2 Distribution for v from 1 to 500, by 1

Table VII. Values of Student's t Distribution for v from 1 to 500, by 1

Table I. Control Limit Constants for $n = 1$ to 10

n	A_2	A_3	D_3	D_4	B_3	B_4	d_2	c_4
1	2.660	3.760	-	-	-	-	-	-
2	1.880	2.659	0	3.267	0	3.267	1.128	.7979
3	1.023	1.954	0	2.575	0	2.568	1.693	.8862
4	0.729	1.628	0	2.282	0	2.266	2.059	.9213
5	0.577	1.427	0	2.115	0	2.089	2.326	.9400
6	0.483	1.287	0	2.004	0.030	1.970	2.534	.9515
7	0.419	1.182	0.076	1.924	0.118	1.882	2.704	.9594
8	0.373	1.099	0.136	1.864	0.185	1.815	2.847	.9650
9	0.337	1.032	0.184	1.816	0.239	1.761	2.970	.9693
10	0.308	0.975	0.223	1.777	0.284	1.716	3.078	.9727

Note that the A_2 and A_3 factors for $n = 1$ assume the moving ranges (or moving standard deviations) are based on two consecutive *IX* measurements.

Table II. c_4 Values for $kn = 2$ to 175

kn	c_4	kn	c_4	kn	c_4	kn	c_4	kn	c_4
1	-	36	.9929	71	.9964	106	.9976	141	.9982
2	.7979	37	.9931	72	.9965	107	.9976	142	.9982
3	.8862	38	.9933	73	.9965	108	.9977	143	.9982
4	.9213	39	.9934	74	.9966	109	.9977	144	.9983
5	.9400	40	.9936	75	.9966	110	.9977	145	.9983
6	.9515	41	.9938	76	.9967	111	.9977	146	.9983
7	.9594	42	.9939	77	.9967	112	.9978	147	.9983
8	.9650	43	.9941	78	.9968	113	.9978	148	.9983
9	.9693	44	.9942	79	.9968	114	.9978	149	.9983
10	.9727	45	.9943	80	.9968	115	.9978	150	.9983
11	.9754	46	.9945	81	.9969	116	.9978	151	.9983
12	.9776	47	.9946	82	.9969	117	.9978	152	.9983
13	.9794	48	.9947	83	.9970	118	.9979	153	.9984
14	.9810	49	.9948	84	.9970	119	.9979	154	.9984
15	.9823	50	.9949	85	.9970	120	.9979	155	.9984
16	.9835	51	.9950	86	.9971	121	.9979	156	.9984
17	.9845	52	.9951	87	.9971	122	.9979	157	.9984
18	.9854	53	.9952	88	.9971	123	.9980	158	.9984
19	.9862	54	.9953	89	.9972	124	.9980	159	.9984
20	.9869	55	.9954	90	.9972	125	.9980	160	.9984
21	.9876	56	.9955	91	.9972	126	.9980	161	.9984
22	.9882	57	.9955	92	.9973	127	.9980	162	.9984
23	.9887	58	.9956	93	.9973	128	.9980	163	.9985
24	.9892	59	.9957	94	.9973	129	.9980	164	.9985
25	.9896	60	.9958	95	.9973	130	.9981	165	.9985
26	.9901	61	.9958	96	.9974	131	.9981	166	.9985
27	.9904	62	.9959	97	.9974	132	.9981	167	.9985
28	.9908	63	.9960	98	.9974	133	.9981	168	.9985
29	.9911	64	.9960	99	.9975	134	.9981	169	.9985
30	.9914	65	.9961	100	.9975	135	.9981	170	.9985
31	.9917	66	.9962	101	.9975	136	.9981	171	.9985
32	.9920	67	.9962	102	.9975	137	.9982	172	.9985
33	.9922	68	.9963	103	.9976	138	.9982	173	.9985
34	.9925	69	.9963	104	.9976	139	.9982	174	.9986
35	.9927	70	.9964	105	.9976	140	.9982	175	.9986

Table II. c_4 Values for $kn = 176$ to 500

kn	c_4	kn	c_4	kn	c_4	kn	c_4	kn	c_4
176	.9986	211	.9988	246	.9990	281	.9991	316	.9992
177	.9986	212	.9988	247	.9990	282	.9991	317	.9992
178	.9986	213	.9988	248	.9990	283	.9991	318	.9992
179	.9986	214	.9988	249	.9990	284	.9991	319	.9992
180	.9986	215	.9988	250	.9990	285	.9991	320	.9992
181	.9986	216	.9988	251	.9990	286	.9991	321	.9992
182	.9986	217	.9988	252	.9990	287	.9991	322	.9992
183	.9986	218	.9988	253	.9990	288	.9991	323	.9992
184	.9986	219	.9989	254	.9990	289	.9991	324	.9992
185	.9986	220	.9989	255	.9990	290	.9991	325	.9992
186	.9986	221	.9989	256	.9990	291	.9991	326	.9992
187	.9987	222	.9989	257	.9990	292	.9991	327	.9992
188	.9987	223	.9989	258	.9990	293	.9991	328	.9992
189	.9987	224	.9989	259	.9990	294	.9991	329	.9992
190	.9987	225	.9989	260	.9990	295	.9992	330	.9992
191	.9987	226	.9989	261	.9990	296	.9992	331	.9992
192	.9987	227	.9989	262	.9990	297	.9992	332	.9992
193	.9987	228	.9989	263	.9990	298	.9992	334	.9992
194	.9987	229	.9989	264	.9990	299	.9992		
195	.9987	230	.9989	265	.9991	300	.9992	335	
196	.9987	231	.9989	266	.9991	301	.9992	to	.9993
197	.9987	232	.9989	267	.9991	302	.9992	385	
198	.9987	233	.9989	268	.9991	303	.9992		
199	.9987	234	.9989	269	.9991	304	.9992	386	
200	.9987	235	.9989	270	.9991	305	.9992	to	.9994
201	.9988	236	.9989	271	.9991	306	.9992	455	
202	.9988	237	.9989	272	.9991	307	.9992		
203	.9988	238	.9989	273	.9991	308	.9992	456	
204	.9988	239	.9990	274	.9991	309	.9992	to	.9995
205	.9988	240	.9990	275	.9991	310	.9992	500	
206	.9988	241	.9990	276	.9991	311	.9992		
207	.9988	242	.9990	277	.9991	312	.9992		
208	.9988	243	.9990	278	.9991	313	.9992		
209	.9988	244	.9990	279	.9991	314	.9992		
210	.9988	245	.9990	280	.9991	315	.9992		

Table III. Areas Under the Normal Curve for Z Values from 0.00 to 2.99

Values in the body of this table are the areas under the normal curve to the right of the Z value (shaded area in the picture at right).

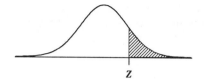

Z

Z	0.00	0.01	0.02	0.03	0.04	0.05	0.06	0.07	0.08	0.09
0.0	5.000-01	4.960-01	4.920-01	4.880-01	4.840-01	4.801-01	4.761-01	4.721-01	4.681-01	4.641-01
0.1	4.602-01	4.562-01	4.522-01	4.483-01	4.443-01	4.404-01	4.364-01	4.325-01	4.286-01	4.247-01
0.2	4.207-01	4.168-01	4.129-01	4.090-01	4.052-01	4.013-01	3.974-01	3.936-01	3.897-01	3.859-01
0.3	3.821-01	3.783-01	3.745-01	3.707-01	3.669-01	3.632-01	3.594-01	3.557-01	3.520-01	3.483-01
0.4	3.446-01	3.409-01	3.372-01	3.336-01	3.300-01	3.264-01	3.228-01	3.192-01	3.156-01	3.121-01
0.5	3.085-01	3.050-01	3.015-01	2.981-01	2.946-01	2.912-01	2.877-01	2.843-01	2.810-01	2.776-01
0.6	2.743-01	2.709-01	2.676-01	2.643-01	2.611-01	2.578-01	2.546-01	2.514-01	2.483-01	2.451-01
0.7	2.420-01	2.389-01	2.358-01	2.327-01	2.297-01	2.266-01	2.236-01	2.207-01	2.177-01	2.148-01
0.8	2.119-01	2.090-01	2.061-01	2.033-01	2.005-01	1.977-01	1.949-01	1.922-01	1.894-01	1.867-01
0.9	1.841-01	1.814-01	1.788-01	1.762-01	1.736-01	1.711-01	1.685-01	1.660-01	1.635-01	1.611-01
1.0	1.587-01	1.562-01	1.539-01	1.515-01	1.492-01	1.469-01	1.446-01	1.423-01	1.401-01	1.379-01
1.1	1.357-01	1.335-01	1.314-01	1.292-01	1.271-01	1.251-01	1.230-01	1.210-01	1.190-01	1.170-01
1.2	1.151-01	1.131-01	1.112-01	1.093-01	1.075-01	1.056-01	1.038-01	1.020-01	1.003-01	9.853-02
1.3	9.680-02	9.510-02	9.342-02	9.176-02	9.012-02	8.851-02	8.691-02	8.534-02	8.379-02	8.226-02
1.4	8.076-02	7.927-02	7.780-02	7.636-02	7.493-02	7.353-02	7.214-02	7.078-02	6.944-02	6.811-02
1.5	6.681-02	6.552-02	6.426-02	6.301-02	6.178-02	6.057-02	5.938-02	5.821-02	5.705-02	5.592-02
1.6	5.480-02	5.370-02	5.262-02	5.155-02	5.050-02	4.947-02	4.846-02	4.746-02	4.648-02	4.551-02
1.7	4.457-02	4.363-02	4.272-02	4.182-02	4.093-02	4.006-02	3.920-02	3.836-02	3.754-02	3.673-02
1.8	3.593-02	3.515-02	3.438-02	3.363-02	3.288-02	3.216-02	3.144-02	3.074-02	3.005-02	2.938-02
1.9	2.872-02	2.807-02	2.743-02	2.680-02	2.619-02	2.559-02	2.500-02	2.442-02	2.385-02	2.330-02
2.0	2.275-02	2.222-02	2.169-02	2.118-02	2.068-02	2.018-02	1.970-02	1.923-02	1.876-02	1.831-02
2.1	1.786-02	1.743-02	1.700-02	1.659-02	1.618-02	1.578-02	1.539-02	1.500-02	1.463-02	1.426-02
2.2	1.390-02	1.355-02	1.321-02	1.287-02	1.255-02	1.222-02	1.191-02	1.160-02	1.130-02	1.101-02
2.3	1.072-02	1.044-02	1.017-02	9.903-03	9.642-03	9.387-03	9.137-03	8.894-03	8.656-03	8.424-03
2.4	8.198-03	7.976-03	7.760-03	7.549-03	7.344-03	7.143-03	6.947-03	6.756-03	6.569-03	6.387-03
2.5	6.210-03	6.036-03	5.868-03	5.703-03	5.543-03	5.386-03	5.234-03	5.085-03	4.940-03	4.799-03
2.6	4.661-03	4.527-03	4.396-03	4.269-03	4.145-03	4.024-03	3.907-03	3.792-03	3.681-03	3.572-03
2.7	3.467-03	3.364-03	3.264-03	3.167-03	3.072-03	2.980-03	2.890-03	2.803-03	2.718-03	2.635-03
2.8	2.555-03	2.477-03	2.401-03	2.327-03	2.256-03	2.186-03	2.118-03	2.052-03	1.988-03	1.926-03
2.9	1.866-03	1.807-03	1.750-03	1.695-03	1.641-03	1.589-03	1.538-03	1.489-03	1.441-03	1.395-03

Table III. Areas Under the Normal Curve for Z Values from 3.00 to 5.99

Values in the body of this table are the areas under the normal curve to the right of the Z value (shaded area in the picture at right).

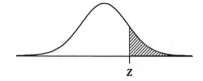

Z	0.00	0.01	0.02	0.03	0.04	0.05	0.06	0.07	0.08	0.09
3.0	1.350-03	1.306-03	1.264-03	1.223-03	1.183-03	1.144-03	1.107-03	1.070-03	1.035-03	1.001-03
3.1	9.676-04	9.354-04	9.042-04	8.740-04	8.447-04	8.163-04	7.888-04	7.622-04	7.364-04	7.114-04
3.2	6.871-04	6.637-04	6.410-04	6.190-04	5.977-04	5.770-04	5.571-04	5.378-04	5.191-04	5.010-04
3.3	4.835-04	4.665-04	4.501-04	4.343-04	4.189-04	4.041-04	3.898-04	3.759-04	3.625-04	3.495-04
3.4	3.370-04	3.249-04	3.132-04	3.019-04	2.909-04	2.804-04	2.702-04	2.603-04	2.508-04	2.416-04
3.5	2.327-04	2.242-04	2.159-04	2.079-04	2.002-04	1.927-04	1.855-04	1.786-04	1.719-04	1.655-04
3.6	1.592-04	1.532-04	1.474-04	1.418-04	1.364-04	1.312-04	1.262-04	1.214-04	1.167-04	1.123-04
3.7	1.079-04	1.038-04	9.974-05	9.587-05	9.214-05	8.855-05	8.509-05	8.175-05	7.854-05	7.545-05
3.8	7.248-05	6.961-05	6.685-05	6.420-05	6.165-05	5.919-05	5.682-05	5.455-05	5.236-05	5.025-05
3.9	4.822-05	4.627-05	4.440-05	4.260-05	4.086-05	3.920-05	3.760-05	3.606-05	3.458-05	3.316-05
4.0	3.169-05	3.037-05	2.911-05	2.790-05	2.674-05	2.562-05	2.455-05	2.352-05	2.253-05	2.158-05
4.1	2.067-05	1.979-05	1.895-05	1.815-05	1.738-05	1.663-05	1.592-05	1.524-05	1.458-05	1.396-05
4.2	1.335-05	1.278-05	1.222-05	1.169-05	1.118-05	1.070-05	1.023-05	9.780-06	9.351-06	8.940-06
4.3	8.546-06	8.169-06	7.807-06	7.461-06	7.130-06	6.812-06	6.508-06	6.217-06	5.939-06	5.672-06
4.4	5.417-06	5.173-06	4.939-06	4.716-06	4.502-06	4.297-06	4.102-06	3.914-06	3.736-06	3.564-06
4.5	3.401-06	3.244-06	3.095-06	2.952-06	2.815-06	2.685-06	2.560-06	2.441-06	2.327-06	2.218-06
4.6	2.115-06	2.015-06	1.921-06	1.830-06	1.744-06	1.661-06	1.583-06	1.508-06	1.436-06	1.368-06
4.7	1.302-06	1.240-06	1.181-06	1.124-06	1.070-06	1.018-06	9.692-07	9.223-07	8.776-07	8.350-07
4.8	7.944-07	7.556-07	7.187-07	6.836-07	6.501-07	6.181-07	5.877-07	5.588-07	5.312-07	5.049-07
4.9	4.799-07	4.560-07	4.334-07	4.118-07	3.912-07	3.716-07	3.530-07	3.353-07	3.184-07	3.024-07
5.0	2.871-07	2.726-07	2.588-07	2.456-07	2.331-07	2.213-07	2.100-07	1.992-07	1.890-07	1.793-07
5.1	1.701-07	1.614-07	1.530-07	1.451-07	1.376-07	1.305-07	1.237-07	1.173-07	1.112-07	1.053-07
5.2	9.983-08	9.460-08	8.964-08	8.492-08	8.045-08	7.620-08	7.217-08	6.835-08	6.472-08	6.128-08
5.3	5.802-08	5.493-08	5.199-08	4.921-08	4.657-08	4.407-08	4.170-08	3.946-08	3.733-08	3.531-08
5.4	3.340-08	3.158-08	2.987-08	2.824-08	2.670-08	2.524-08	2.386-08	2.256-08	2.132-08	2.015-08
5.5	1.904-08	1.799-08	1.699-08	1.605-08	1.516-08	1.432-08	1.352-08	1.277-08	1.206-08	1.138-08
5.6	1.075-08	1.014-08	9.574-09	9.035-09	8.526-09	8.045-09	7.590-09	7.160-09	6.754-09	6.370-09
5.7	6.008-09	5.665-09	5.342-09	5.036-09	4.748-09	4.476-09	4.218-09	3.976-09	3.746-09	3.530-09
5.8	3.326-09	3.133-09	2.952-09	2.780-09	2.618-09	2.466-09	2.322-09	2.186-09	2.058-09	1.937-09
5.9	1.824-09	1.716-09	1.615-09	1.520-09	1.430-09	1.345-09	1.266-09	1.190-09	1.120-09	1.053-09

Table III. Areas Under the Normal Curve for Z Values from 6.00 to 7.79

Values in the body of this table are the areas under the normal curve to the right of the Z value (shaded area in the picture at right).

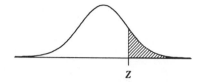

Z	0.00	0.01	0.02	0.03	0.04	0.05	0.06	0.07	0.08	0.09
6.0	9.901-10	9.310-10	8.753-10	8.228-10	7.734-10	7.269-10	6.831-10	6.420-10	6.032-10	5.667-10
6.1	5.324-10	5.001-10	4.697-10	4.411-10	4.142-10	3.889-10	3.652-10	3.428-10	3.218-10	3.020-10
6.2	2.835-10	2.660-10	2.496-10	2.342-10	2.197-10	2.061-10	1.933-10	1.813-10	1.700-10	1.594-10
6.3	1.495-10	1.401-10	1.314-10	1.231-10	1.154-10	1.081-10	1.013-10	9.494-11	8.895-11	8.333-11
6.4	7.805-11	7.310-11	6.846-11	6.410-11	6.002-11	5.619-11	5.260-11	4.924-11	4.608-11	4.313-11
6.5	4.036-11	3.776-11	3.533-11	3.305-11	3.091-11	2.891-11	2.704-11	2.529-11	2.364-11	2.211-11
6.6	2.067-11	1.932-11	1.805-11	1.687-11	1.577-11	1.473-11	1.377-11	1.286-11	1.201-11	1.122-11
6.7	1.048-11	9.785-12	9.137-12	8.531-12	7.964-12	7.434-12	6.939-12	6.476-12	6.043-12	5.639-12
6.8	5.262-12	4.909-12	4.579-12	4.271-12	3.983-12	3.715-12	3.464-12	3.230-12	3.011-12	2.807-12
6.9	2.616-12	2.438-12	2.272-12	2.117-12	1.973-12	1.838-12	1.712-12	1.595-12	1.485-12	1.383-12
7.0	1.288-12	1.199-12	1.117-12	1.040-12	9.676-13	9.005-13	8.381-13	7.799-13	7.256-13	6.751-13
7.1	6.281-13	5.842-13	5.433-13	5.054-13	4.700-13	4.370-13	4.062-13	3.777-13	3.511-13	3.263-13
7.2	3.032-13	2.818-13	2.618-13	2.432-13	2.259-13	2.099-13	1.950-13	1.811-13	1.682-13	1.561-13
7.3	1.450-13	1.346-13	1.249-13	1.159-13	1.076-13	9.992-14	9.270-14	8.593-14	7.971-14	7.394-14
7.4	6.861-14	6.362-14	5.906-14	5.473-14	5.074-14	4.707-14	4.363-14	4.041-14	3.741-14	3.475-14
7.5	3.220-14	2.975-14	2.764-14	2.554-14	2.365-14	2.198-14	2.032-14	1.887-14	1.743-14	1.610-14
7.6	1.499-14	1.388-14	1.277-14	1.188-14	1.099-14	1.010-14	9.326-15	8.660-15	7.994-15	7.438-15
7.7	6.883-15	6.328-15	5.884-15	5.440-15	4.996-15	4.663-15	4.330-15	3.997-15	3.664-15	3.331-15

Table IV. Cumulative Areas Under the Normal Curve from -2.99σ to .00

Values in the body of this table are the areas under the normal curve to the left of the Z value (shaded area in the picture at right).

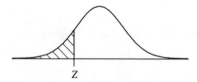

Z

Z	0.09	0.08	0.07	0.06	0.05	0.04	0.03	0.02	0.01	0.00
-2.9	.0014	.0014	.0015	.0015	.0016	.0016	.0017	.0017	.0018	.0019
-2.8	.0019	.0020	.0021	.0021	.0022	.0023	.0023	.0024	.0025	.0026
-2.7	.0026	.0027	.0028	.0029	.0030	.0031	.0032	.0033	.0034	.0035
-2.6	.0036	.0037	.0038	.0039	.0040	.0041	.0043	.0044	.0045	.0047
-2.5	.0048	.0049	.0051	.0052	.0054	.0055	.0057	.0059	.0060	.0062
-2.4	.0064	.0066	.0068	.0069	.0071	.0073	.0075	.0078	.0080	.0082
-2.3	.0084	.0087	.0089	.0091	.0094	.0096	.0099	.0102	.0104	.0107
-2.2	.0110	.0113	.0116	.0119	.0122	.0125	.0129	.0132	.0136	.0139
-2.1	.0143	.0146	.0150	.0154	.0158	.0162	.0166	.0170	.0174	.0179
-2.0	.0183	.0188	.0192	.0197	.0202	.0207	.0212	.0217	.0222	.0228
-1.9	.0233	.0239	.0244	.0250	.0256	.0262	.0268	.0274	.0281	.0287
-1.8	.0294	.0301	.0307	.0314	.0322	.0329	.0336	.0344	.0351	.0359
-1.7	.0367	.0375	.0384	.0392	.0401	.0409	.0418	.0427	.0436	.0446
-1.6	.0455	.0465	.0475	.0485	.0495	.0505	.0516	.0526	.0537	.0548
-1.5	.0559	.0571	.0582	.0594	.0606	.0618	.0630	.0643	.0655	.0668
-1.4	.0681	.0694	.0708	.0721	.0735	.0749	.0764	.0778	.0793	.0808
-1.3	.0823	.0838	.0853	.0869	.0885	.0901	.0918	.0934	.0951	.0968
-1.2	.0985	.1003	.1020	.1038	.1056	.1075	.1093	.1112	.1131	.1151
-1.1	.1170	.1190	.1210	.1230	.1251	.1271	.1292	.1314	.1335	.1357
-1.0	.1379	.1401	.1423	.1446	.1469	.1492	.1515	.1539	.1562	.1587
-0.9	.1611	.1635	.1660	.1685	.1711	.1736	.1762	.1788	.1814	.1841
-0.8	.1867	.1894	.1922	.1949	.1977	.2005	.2033	.2061	.2090	.2119
-0.7	.2148	.2177	.2207	.2236	.2266	.2297	.2327	.2358	.2389	.2420
-0.6	.2451	.2483	.2514	.2546	.2578	.2611	.2643	.2676	.2709	.2743
-0.5	.2776	.2810	.2843	.2877	.2912	.2946	.2981	.3015	.3050	.3085
-0.4	.3121	.3156	.3192	.3228	.3264	.3300	.3336	.3372	.3409	.3446
-0.3	.3483	.3520	.3557	.3594	.3632	.3669	.3707	.3745	.3783	.3821
-0.2	.3859	.3897	.3936	.3974	.4013	.4052	.4090	.4129	.4168	.4207
-0.1	.4247	.4286	.4325	.4364	.4404	.4443	.4483	.4562	.4562	.4602
0.0	.4641	.4681	.4721	.4761	.4801	.4840	.4880	.4920	.4960	.5000

Table IV. Cumulative Areas Under the Normal Curve from .00 to 2.99σ

Values in the body of this table are the areas
under the normal curve to the left of the Z value
(shaded area in the picture at right).

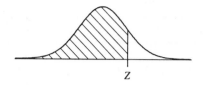

Z	0.00	0.01	0.02	0.03	0.04	0.05	0.06	0.07	0.08	0.09
0.0	.5000	.5040	.5080	.5120	.5160	.5199	.5239	.5279	.5319	.5359
0.1	.5398	.5438	.5478	.5517	.5557	.5596	.5636	.5675	.5714	.5753
0.2	.5793	.5832	.5871	.5910	.5948	.5987	.6026	.6064	.6103	.6141
0.3	.6179	.6217	.6255	.6293	.6331	.6368	.6406	.6443	.6480	.6517
0.4	.6554	.6591	.6628	.6664	.6700	.6736	.6772	.6808	.6844	.6879
0.5	.6915	.6950	.6985	.7019	.7054	.7088	.7123	.7157	.7190	.7224
0.6	.7257	.7291	.7324	.7357	.7389	.7422	.7454	.7486	.7517	.7549
0.7	.7580	.7611	.7642	.7673	.7703	.7734	.7764	.7794	.7823	.7852
0.8	.7881	.7910	.7939	.7967	.7995	.8023	.8051	.8078	.8106	.8133
0.9	.8159	.8186	.8212	.8238	.8264	.8289	.8315	.8340	.8365	.8389
1.0	.8413	.8438	.8461	.8485	.8508	.8531	.8554	.8577	.8599	.8621
1.1	.8643	.8665	.8686	.8708	.8729	.8749	.8770	.8790	.8810	.8830
1.2	.8849	.8869	.8888	.8907	.8925	.8944	.8962	.8980	.8997	.9015
1.3	.9032	.9049	.9066	.9082	.9099	.9115	.9131	.9147	.9162	.9177
1.4	.9192	.9207	.9222	.9236	.9251	.9265	.9279	.9292	.9306	.9319
1.5	.9332	.9345	.9357	.9370	.9382	.9394	.9406	.9418	.9429	.9441
1.6	.9452	.9463	.9474	.9484	.9495	.9505	.9515	.9525	.9535	.9545
1.7	.9554	.9564	.9573	.9582	.9591	.9599	.9608	.9616	.9625	.9633
1.8	.9641	.9649	.9656	.9664	.9671	.9678	.9686	.9693	.9699	.9706
1.9	.9713	.9719	.9726	.9732	.9738	.9744	.9750	.9756	.9761	.9767
2.0	.9772	.9778	.9783	.9788	.9793	.9798	.9803	.9808	.9812	.9817
2.1	.9821	.9826	.9830	.9834	.9838	.9842	.9846	.9850	.9854	.9857
2.2	.9861	.9864	.9868	.9871	.9875	.9878	.9881	.9884	.9887	.9890
2.3	.9893	.9896	.9898	.9901	.9904	.9906	.9909	.9911	.9913	.9916
2.4	.9918	.9920	.9922	.9925	.9927	.9929	.9931	.9932	.9934	.9936
2.5	.9938	.9940	.9941	.9943	.9945	.9946	.9948	.9949	.9951	.9952
2.6	.9953	.9955	.9956	.9957	.9959	.9960	.9961	.9962	.9963	.9964
2.7	.9965	.9966	.9967	.9968	.9969	.9970	.9971	.9972	.9973	.9974
2.8	.9974	.9975	.9976	.9977	.9977	.9978	.9979	.9979	.9980	.9981
2.9	.9981	.9982	.9982	.9983	.9984	.9984	.9985	.9985	.9986	.9986

Table V. Median Ranks for *kn* from 10 to 20

i	10	11	12	13	14	15	16	17	18	19	20
1	6.7	6.1	5.6	5.2	4.9	4.5	4.3	4.0	3.8	3.6	3.4
2	16.3	14.9	13.7	12.7	11.8	11.0	10.4	9.8	9.2	8.8	8.3
3	26.0	23.7	21.8	20.1	18.8	17.5	16.5	15.5	14.7	13.9	13.2
4	35.6	32.5	29.8	27.6	25.7	24.0	22.6	21.3	20.1	19.1	18.1
5	45.2	41.2	37.9	35.1	32.6	30.5	28.7	27.0	25.5	24.2	23.0
6	54.8	50.0	46.0	42.5	39.6	37.0	34.8	32.8	31.0	29.4	27.9
7	64.4	58.8	54.0	50.0	46.5	43.5	40.9	38.5	36.4	34.5	32.8
8	74.0	67.5	62.1	57.5	53.5	50.0	47.0	44.2	41.8	39.7	37.8
9	83.7	76.3	70.2	64.9	60.4	56.5	53.0	50.0	47.3	44.9	42.6
10	93.3	85.1	78.2	72.4	67.4	63.0	59.1	55.8	52.7	50.0	47.6
11		93.9	86.3	79.9	74.3	69.5	65.2	61.5	58.2	55.2	52.4
12			94.4	87.3	81.2	76.0	71.3	67.2	63.6	60.3	57.4
13				94.8	88.2	82.5	77.4	73.0	69.0	65.5	62.2
14					95.1	89.0	83.5	78.7	74.5	70.6	67.2
15						95.4	89.6	84.5	79.9	75.8	72.1
16							95.7	90.2	85.3	80.9	77.0
17								96.0	90.8	86.1	81.9
18									96.2	91.2	86.8
19										96.4	91.7
20											96.6

Table V. Median Ranks for *kn* from 21 to 30

i	21	22	23	24	25	26	27	28	29	30
1	3.3	3.1	3.0	2.9	2.8	2.7	2.6	2.5	2.4	2.3
2	7.9	7.6	7.3	7.0	6.7	6.4	6.2	6.0	5.8	5.6
3	12.6	12.0	11.5	11.1	10.6	10.2	9.8	9.5	9.2	8.9
4	17.3	16.5	15.8	15.2	14.6	14.0	13.5	13.0	12.6	12.2
5	22.0	21.0	20.1	19.3	18.5	17.8	17.2	16.6	16.0	15.6
6	26.6	25.4	24.4	23.4	22.4	21.6	20.8	20.1	19.4	18.8
7	31.3	29.9	28.6	27.5	26.4	25.4	24.2	23.6	22.8	22.0
8	36.0	34.4	32.9	31.6	30.3	29.0	28.1	27.1	26.2	25.3
9	40.6	38.8	37.2	35.7	34.2	33.0	31.8	30.6	29.6	28.6
10	45.3	43.3	41.4	39.8	38.2	36.7	35.4	34.2	33.0	31.9
11	50.0	47.8	45.7	43.8	42.1	40.5	39.0	37.7	36.4	35.2
12	54.7	52.2	50.0	48.0	46.1	44.3	42.7	41.2	39.8	38.5
13	59.4	56.7	54.3	52.0	50.0	48.1	46.4	44.7	43.2	41.8
14	64.0	61.2	58.6	56.2	53.9	51.9	50.0	48.2	46.6	45.1
15	68.7	65.6	62.8	60.2	57.9	55.7	53.6	51.8	50.0	48.4
16	73.4	70.1	67.1	64.3	61.8	59.5	57.3	55.8	53.4	51.6
17	78.0	74.6	71.4	68.4	65.8	63.3	61.0	58.8	56.8	54.9
18	82.7	79.0	75.6	72.5	69.7	67.0	64.6	62.3	60.2	58.2
19	87.4	83.5	79.9	76.6	73.6	70.8	68.2	65.8	63.6	61.5
20	92.1	88.0	84.2	80.7	77.6	74.6	71.9	69.4	67.0	64.8
21	96.7	92.4	88.5	84.8	81.5	78.4	75.6	72.8	70.4	68.1
22		96.9	92.7	88.9	85.4	82.2	79.2	76.4	73.8	71.4
23			97.0	93.0	89.4	86.0	82.8	79.9	77.2	74.7
24				97.1	93.1	89.8	86.5	83.4	80.6	78.0
25					97.2	93.6	90.2	87.0	84.0	81.2
26						97.4	93.8	90.5	87.4	84.5
27							97.4	94.0	90.8	87.8
28								97.5	94.2	91.1
29									97.6	94.4
30										97.7

Table V. Median Ranks for *kn* from 31 to 40

i	31	32	33	34	35	36	37	38	39	40
1	2.2	2.2	2.1	2.0	2.0	1.9	1.9	1.8	1.8	1.7
2	5.4	5.2	5.1	4.9	4.8	4.7	4.6	4.4	4.3	4.2
3	8.6	8.3	8.0	7.8	7.6	7.4	7.2	7.0	6.8	6.7
4	11.8	11.4	11.1	10.8	10.4	10.2	9.9	9.6	9.4	9.2
5	15.0	14.5	14.1	13.7	13.3	12.9	12.6	12.2	11.9	11.6
6	18.2	17.6	17.1	16.6	16.1	15.7	15.2	14.8	14.5	14.1
7	21.3	20.7	20.1	19.5	18.9	18.4	17.9	17.4	17.0	16.6
8	24.5	23.8	23.0	22.4	21.8	21.2	20.6	20.0	19.5	19.1
9	27.7	26.8	26.0	25.3	24.6	23.9	23.3	22.7	22.1	21.5
10	30.9	29.9	29.0	28.2	27.4	26.6	25.9	25.3	24.6	24.0
11	34.1	33.0	32.0	31.1	30.2	29.4	28.6	27.9	27.2	26.5
12	37.3	36.1	35.0	34.0	33.0	32.1	31.3	30.5	29.7	29.0
13	40.4	39.2	38.0	36.9	35.9	34.9	34.0	33.1	32.2	31.4
14	43.6	42.3	41.0	39.8	38.7	37.6	36.6	35.7	34.8	33.9
15	46.8	45.4	44.0	42.7	41.5	40.4	39.3	38.3	37.3	36.4
16	50.0	48.5	47.0	45.6	44.4	43.1	42.0	40.9	39.8	38.9
17	53.2	51.5	50.0	48.5	47.2	45.9	44.6	43.5	42.4	41.3
18	56.4	54.6	53.0	51.4	50.0	48.6	47.3	46.1	44.9	43.8
19	59.6	57.7	56.0	54.4	52.8	51.4	50.0	48.7	47.5	46.3
20	62.7	60.8	59.0	57.3	55.6	54.1	52.7	51.3	50.0	48.8
21	65.9	63.9	62.0	60.2	58.5	56.9	55.3	53.9	52.5	51.2
22	69.1	67.0	65.0	63.1	61.3	59.6	58.0	56.5	55.1	53.7
23	72.3	70.1	68.0	66.0	64.1	62.4	60.7	59.1	57.6	56.2
24	75.5	73.1	71.0	68.9	66.9	65.1	63.4	61.7	60.2	58.7
25	78.7	76.2	74.0	71.8	69.8	67.9	66.0	64.3	62.7	61.1
26	81.8	79.3	76.9	74.7	72.6	70.6	68.7	66.9	65.2	63.6
27	85.0	82.4	79.9	77.6	75.4	73.4	71.4	69.5	67.8	66.1
28	88.2	85.5	82.9	80.5	78.2	76.1	74.1	72.1	70.3	68.6
29	91.4	88.6	85.9	83.4	81.1	78.8	76.7	74.7	72.8	71.0
30	94.6	91.7	88.9	86.3	83.9	81.6	79.4	77.3	75.4	73.5
31	97.8	94.8	91.9	89.2	86.7	84.3	82.1	80.0	77.9	76.0
32		97.8	94.9	92.2	89.6	87.1	84.8	82.6	80.5	78.5
33			97.9	95.1	92.4	89.8	87.4	85.2	83.0	80.9
34				98.0	95.2	92.6	90.1	87.8	85.5	83.4
35					98.0	95.3	92.8	90.4	88.1	85.9
36						98.1	95.4	93.0	90.6	88.4
37							98.1	95.6	93.1	90.8
38								98.2	95.7	93.3
39									98.2	95.8
40										98.3

Table V. Median Ranks for *kn* from 41 to 50

i	41	42	43	44	45	46	47	48	49	50
1	1.7	1.6	1.6	1.6	1.5	1.5	1.5	1.4	1.4	1.4
2	4.1	4.0	3.9	3.8	3.7	3.7	3.6	3.5	3.4	3.4
3	6.5	6.4	6.2	6.1	5.9	5.8	5.7	5.6	5.5	5.4
4	8.9	8.7	8.5	8.3	8.1	8.0	7.8	7.6	7.5	7.3
5	11.4	11.1	10.8	10.6	10.4	10.1	9.9	9.7	9.5	9.3
6	13.8	13.4	13.1	12.8	12.6	12.3	12.0	11.8	11.5	11.3
7	16.2	15.8	15.4	15.1	14.8	14.4	14.1	13.8	13.6	13.3
8	18.6	18.2	17.7	17.3	17.0	16.6	16.2	15.9	15.6	15.3
9	21.0	20.5	20.0	19.6	19.2	18.8	18.4	18.0	17.6	17.3
10	23.4	22.9	22.4	21.8	21.4	20.9	20.5	20.0	19.6	19.2
11	25.8	25.2	24.6	24.1	23.6	23.1	22.6	22.1	21.7	21.2
12	28.3	27.6	27.0	26.4	25.8	25.2	24.7	24.2	23.7	23.2
13	30.7	30.0	29.3	28.6	28.0	27.4	26.8	26.2	25.7	25.2
14	33.1	32.3	31.6	30.9	30.2	29.5	28.9	28.3	27.7	27.2
15	35.5	34.7	33.9	33.1	32.4	31.7	31.0	30.4	29.8	29.2
16	37.9	37.0	36.2	35.4	34.6	33.8	33.1	32.4	31.8	31.2
17	40.3	39.4	38.5	37.6	36.8	36.0	35.2	34.5	33.8	33.1
18	42.8	41.7	40.8	39.9	39.0	38.1	37.3	36.6	35.8	35.1
19	45.2	44.1	43.1	42.1	41.2	40.3	39.4	38.6	37.8	37.1
20	47.6	46.5	45.4	44.4	43.4	42.5	41.6	40.7	39.9	39.1
21	50.0	48.8	47.7	46.6	45.6	44.6	43.7	42.8	41.9	41.1
22	52.4	51.2	50.0	48.9	47.8	46.8	45.8	44.8	43.9	43.1
23	54.8	53.5	52.3	51.1	50.0	48.9	47.9	46.9	46.0	45.0
24	57.2	55.9	54.6	53.4	52.2	51.1	50.0	49.0	48.0	47.0
25	59.7	58.2	56.9	55.6	54.4	53.2	52.1	51.0	50.0	49.0
26	62.1	60.6	59.2	57.9	56.6	55.4	54.2	53.1	52.0	51.0
27	64.5	63.0	61.5	60.1	58.8	57.5	56.3	55.2	54.0	53.0
28	66.9	65.3	63.8	62.4	61.0	59.7	58.4	57.2	56.1	55.0
29	69.3	67.7	66.1	64.6	63.2	61.8	60.5	59.3	58.1	56.9
30	71.7	70.0	68.4	66.9	65.4	64.0	62.7	61.4	60.1	58.9
31	74.2	72.4	70.7	69.1	67.6	66.2	64.8	63.4	62.1	60.9
32	76.6	74.8	73.0	71.4	69.8	68.3	66.9	65.5	64.2	62.9
33	79.0	77.1	75.3	73.6	72.0	70.5	69.0	67.6	66.2	64.9
34	81.4	79.5	77.6	75.9	74.2	72.6	71.1	69.6	68.2	66.9
35	83.8	81.8	80.0	78.2	76.4	74.8	73.2	71.7	70.2	68.8
36	86.2	84.2	82.3	80.4	78.6	76.9	75.3	73.8	72.3	70.8
37	88.6	86.6	84.6	82.7	80.8	79.1	77.4	75.8	74.3	72.8
38	91.1	88.9	86.9	84.9	83.0	81.2	79.5	77.9	76.3	74.8
39	93.5	91.3	89.2	87.2	85.2	83.4	81.6	80.0	78.3	76.8
40	95.9	93.6	91.5	89.4	87.4	85.6	83.8	82.0	80.4	78.8
41	98.3	96.0	93.8	91.7	89.6	87.7	85.9	84.1	82.4	80.8
42		98.3	96.1	93.9	91.8	89.9	88.0	86.2	84.4	82.7
43			98.4	96.2	94.1	92.0	90.1	88.2	86.4	84.7
44				98.4	96.3	94.2	92.2	90.3	88.5	86.7
45					98.5	96.3	94.3	92.4	90.5	88.7
46						98.5	96.4	94.4	92.5	90.7
47							98.5	96.5	94.5	92.7
48								98.6	96.6	94.6
49									98.6	96.6
50										98.6

Table VI. Values of χ^2 Distribution for $\nu = 1$ to 25

Values in the body of this table are the $\chi^2_{\nu,\alpha}$ values for selected α levels and degrees of freedom (ν) used in establishing confidence bounds for capability measures and determining critical values for goodness-of-fit tests.

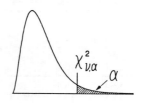

	α		
ν	.99	.95	.90
1	0.00016	0.00393	0.01579
2	0.02010	0.10259	0.21072
3	0.11483	0.35185	0.58438
4	0.29711	0.71072	1.06362
5	0.55429	1.14547	1.61030
6	0.87207	1.63538	2.20413
7	1.23903	2.16734	2.83311
8	1.64646	2.73263	3.48953
9	2.08785	3.32511	4.16816
10	2.55819	3.94029	4.86517
11	3.05349	4.57481	5.57778
12	3.57050	5.22602	6.30378
13	4.10688	5.89185	7.04149
14	4.66038	6.57062	7.78953
15	5.22928	7.26091	8.54675
16	5.81218	7.96163	9.31223
17	6.40768	8.67174	10.08517
18	7.01482	9.39043	10.86493
19	7.63263	10.11700	11.65090
20	8.26031	10.85080	12.44261
21	8.89705	11.59130	13.23960
22	9.54237	12.33801	14.04148
23	10.19561	13.09050	14.84796
24	10.85627	13.84839	15.65867
25	11.52394	14.61141	16.47341

	α		
ν	.10	.05	.01
1	2.70554	3.84147	6.63486
2	4.60517	5.99147	9.21034
3	6.25139	7.81472	11.34489
4	7.77944	9.48773	13.27671
5	9.23635	11.07050	15.08634
6	10.64464	12.59159	16.81190
7	12.01704	14.06716	18.47535
8	13.36157	15.50732	20.09024
9	14.68367	16.91899	21.66602
10	15.98718	18.30704	23.20925
11	17.27502	19.67515	24.72498
12	18.54935	21.02607	26.21697
13	19.81194	22.36204	27.68825
14	21.06414	23.68479	29.14124
15	22.30713	24.99579	30.57792
16	23.54183	26.29623	31.99993
17	24.76904	27.58711	33.40866
18	25.98942	28.86930	34.80531
19	27.20357	30.14353	36.19087
20	28.41198	31.41043	37.56624
21	29.61509	32.67057	38.93217
22	30.81328	33.92444	40.28936
23	32.00690	35.17246	41.63840
24	33.19625	36.41503	42.97982
25	34.38159	37.65248	44.31411

Table VI. Values of χ^2 Distribution for $v = 26$ to 50

Values in the body of this table are the $\chi^2_{v,\alpha}$ values for selected α levels and degrees of freedom (v) used in establishing confidence bounds for capability measures and determining critical values for goodness-of-fit tests.

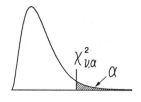

	α				α		
v	.99	.95	.90	v	.10	.05	.01
26	12.19805	15.37915	17.29188	26	35.56317	38.88514	45.64168
27	12.87839	16.15139	18.11390	27	36.74122	40.11327	46.96294
28	13.56457	16.92785	18.93923	28	37.91592	41.33714	48.27824
29	14.25633	17.70837	19.76773	29	39.08747	42.55697	49.58789
30	14.95331	18.49266	20.59921	30	40.25602	43.77297	50.89218
31	15.65536	19.28053	21.43356	31	41.42174	44.98535	52.19140
32	16.36208	20.07190	22.27058	32	42.58475	46.19426	53.48577
33	17.07348	20.86649	23.11019	33	43.74518	47.39988	54.77554
34	17.78907	21.66428	23.95224	34	44.90316	48.60237	56.06091
35	18.50894	22.46499	24.79664	35	46.05879	49.80185	57.34208
36	19.23258	23.26857	25.64329	36	47.21218	50.99846	58.61922
37	19.96017	24.07492	26.49209	37	48.36341	52.19232	59.89250
38	20.69131	24.88390	27.34292	38	49.51259	53.38354	61.16209
39	21.42605	25.69537	28.19577	39	50.65977	54.57223	62.42812
40	22.16411	26.50928	29.05051	40	51.80506	55.75848	63.69074
41	22.90554	27.32554	29.90708	41	52.94851	56.94239	64.95007
42	23.64993	28.14403	30.76542	42	54.09021	58.12404	66.20624
43	24.39758	28.96466	31.62545	43	55.23019	59.30351	67.45935
44	25.14784	29.78747	32.48709	44	56.36854	60.48089	68.70951
45	25.90105	30.61222	33.35038	45	57.50531	61.65623	69.95683
46	26.65707	31.43896	34.21516	46	58.64054	62.82962	71.20140
47	27.41581	32.26758	35.08143	47	59.77429	64.00111	72.44331
48	28.17686	33.09804	35.94911	48	60.90661	65.17077	73.68264
49	28.94046	33.93030	36.81820	49	62.03754	66.33865	74.91948
50	29.70653	34.76421	37.68862	50	63.16712	67.50481	76.15389

Table VI.   Values of χ^2 Distribution for $\nu = 51$ to 75

Values in the body of this table are the $\chi^2_{\nu,\alpha}$ values for selected α levels and degrees of freedom (ν) used in establishing confidence bounds for capability measures and determining critical values for goodness-of-fit tests.

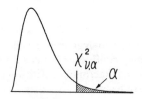

ν	α .99	.95	.90
51	30.47500	35.59984	38.56035
52	31.24567	36.43708	39.43338
53	32.01835	37.27586	40.30758
54	32.79332	38.11618	41.18300
55	33.57043	38.95795	42.05961
56	34.34947	39.80121	42.93732
57	35.13042	40.64588	43.81611
58	35.91335	41.49193	44.69600
59	36.69821	42.33927	45.57695
60	37.48483	43.18794	46.45889
61	38.27314	44.03785	47.34179
62	39.06326	44.88901	48.22569
63	39.85488	45.74136	49.11052
64	40.64828	46.59486	49.99629
65	41.44336	47.44956	50.88292
66	42.23992	48.30533	51.77046
67	43.03822	49.16222	52.65884
68	43.83776	50.02020	53.54804
69	44.63897	50.87923	54.43809
70	45.44150	51.73926	55.32892
71	46.24559	52.60031	56.22055
72	47.05090	53.46226	57.11293
73	47.85757	54.32526	58.00604
74	48.66559	55.18915	58.89996
75	49.47499	56.05406	59.79454

ν	α .10	.05	.01
51	64.29540	68.66930	77.38596
52	65.42241	69.83217	78.61576
53	66.54820	70.99346	79.84334
54	67.67279	72.15322	81.06877
55	68.79622	73.31149	82.29212
56	69.91851	74.46833	83.51343
57	71.03972	75.62375	84.73277
58	72.15985	76.77780	85.95018
59	73.27894	77.93053	87.16571
60	74.39701	79.08195	88.37942
61	75.51409	80.23210	89.59135
62	76.63021	81.38102	90.80154
63	77.74539	82.52873	92.01003
64	78.85964	83.67526	93.21686
65	79.97300	84.82065	94.42208
66	81.08549	85.96491	95.62572
67	82.19711	87.10808	96.82782
68	83.30791	88.25016	98.02840
69	84.41787	89.39121	99.22752
70	85.52705	90.53123	100.42519
71	86.63544	91.67024	101.62144
72	87.74305	92.80827	102.81632
73	88.84992	93.94534	104.00984
74	89.95605	95.08147	105.20203
75	91.06146	96.21667	106.39293

Table VI. Values of χ^2 Distribution for $\nu = 76$ to 100

Values in the body of this table are the $\chi^2_{\nu,\alpha}$ values for selected α levels and degrees of freedom (ν) used in establishing confidence bounds for capability measures and determining critical values for goodness-of-fit tests.

ν	.99	.95	.90
76	50.28545	56.91974	60.68985
77	51.09731	57.78642	61.58584
78	51.91036	58.65393	62.48251
79	52.72438	59.52226	63.37984
80	53.53993	60.39142	64.27780
81	54.35628	61.26147	65.17643
82	55.17427	62.13223	66.07572
83	55.99303	63.00382	66.97560
84	56.81285	63.87623	67.87607
85	57.63369	64.74936	68.77715
86	58.45572	65.62328	69.67880
87	59.27896	66.49784	70.58102
88	60.10291	67.37322	71.48382
89	60.92790	68.24928	72.38718
90	61.75399	69.12598	73.29109
91	62.58095	70.00343	74.19552
92	63.40865	70.88153	75.10045
93	64.23780	71.76034	76.00595
94	65.06764	72.63972	76.91192
95	65.89808	73.51982	77.81842
96	66.72968	74.40053	78.72541
97	67.56205	75.28182	79.63284
98	68.39558	76.16377	80.54083
99	69.22956	77.04629	81.44921
100	70.06472	77.92946	82.35813

ν	.10	.05	.01
76	92.16617	97.35097	107.58255
77	93.27018	98.48438	108.77092
78	94.37352	99.61693	109.95807
79	95.47619	100.74862	111.14402
80	96.57820	101.87947	112.32880
81	97.67958	103.00951	113.51241
82	98.78033	104.13874	114.69490
83	99.88046	105.26718	115.87627
84	100.97999	106.39485	117.05655
85	102.07892	107.52174	118.23575
86	103.17727	108.64789	119.41390
87	104.27504	109.77331	120.59102
88	105.37225	110.89801	121.76711
89	106.46890	112.02199	122.94221
90	107.56501	113.14527	124.11632
91	108.66059	114.26788	125.28947
92	109.75563	115.38980	126.46166
93	110.85016	116.51105	127.63291
94	111.94417	117.63165	128.80325
95	113.03769	118.75161	129.97268
96	114.13071	119.87094	131.14122
97	115.22325	120.98964	132.30888
98	116.31530	122.10774	133.47568
99	117.40689	123.22523	134.64162
100	118.49801	124.34212	135.80672

Table VI. Values of χ^2 Distribution for $\nu = 101$ to 125

Values in the body of this table are the $\chi^2_{\nu,\alpha}$ values for selected α levels and degrees of freedom (ν) used in establishing confidence bounds for capability measures and determining critical values for goodness-of-fit tests.

	α				α		
ν	**.99**	**.95**	**.90**	ν	**.10**	**.05**	**.01**
101	70.90055	78.81314	83.26746	101	119.58867	125.45842	136.97101
102	71.73735	79.69747	84.17726	102	120.67888	126.57415	138.13447
103	72.57450	80.58234	85.08749	103	121.76865	127.68931	139.29714
104	73.41293	81.46774	85.99814	104	122.85798	128.80391	140.45901
105	74.25197	82.35372	86.90922	105	123.94688	129.91796	141.62011
106	75.09143	83.24022	87.82074	106	125.03536	131.03146	142.78044
107	75.93209	84.12731	88.73269	107	126.12342	132.14443	143.94002
108	76.77318	85.01487	89.64510	108	127.21107	133.25686	145.09885
109	77.61528	85.90300	90.55784	109	128.29831	134.36878	146.25694
110	78.45809	86.79162	91.47103	110	129.38514	135.48019	147.41431
111	79.30159	87.68064	92.38458	111	130.47158	136.59107	148.57096
112	80.14569	88.57036	93.29853	112	131.55763	137.70147	149.72691
113	80.99068	89.46046	94.21283	113	132.64329	138.81137	150.88216
114	81.83601	90.35110	95.12757	114	133.72858	139.92078	152.03672
115	82.68235	91.24217	96.04270	115	134.81349	141.02971	153.19061
116	83.52923	92.13366	96.95815	116	135.89802	142.13816	154.34382
117	84.37639	93.02579	97.87398	117	136.98220	143.24615	155.49638
118	85.22482	93.91819	98.79018	118	138.06601	144.35367	156.64828
119	86.07344	94.81122	99.70673	119	139.14946	145.46075	157.79954
120	86.92292	95.70460	100.62362	120	140.23257	146.56736	158.95017
121	87.77320	96.59847	101.54087	121	141.31533	147.67353	160.10016
122	88.62392	97.49267	102.45845	122	142.39774	148.77926	161.24955
123	89.47518	98.38747	103.37638	123	143.47982	149.88456	162.39831
124	90.32722	99.28261	104.29464	124	144.56156	150.98944	163.54647
125	91.17960	100.17811	105.21324	125	145.64297	152.09388	164.69403

Table VI. Values of χ^2 Distribution for $v = 126$ to 150

Values in the body of this table are the $\chi^2_{v,\alpha}$ values for selected α levels and degrees of freedom (v) used in establishing confidence bounds for capability measures and determining critical values for goodness-of-fit tests.

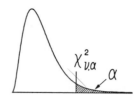

	α					α		
v	.99	.95	.90		v	.10	.05	.01
126	92.03280	101.07411	106.13215		126	146.72406	153.19791	165.84100
127	92.88649	101.97053	107.05139		127	147.80482	154.30152	166.98739
128	93.74081	102.86735	107.97094		128	148.88526	155.40472	168.13320
129	94.59548	103.76460	108.89081		129	149.96538	156.50752	169.27845
130	95.45071	104.66222	109.81101		130	151.04520	157.60992	170.42313
131	96.30666	105.56014	110.73151		131	152.12471	158.71193	171.56726
132	97.16314	106.45857	111.65228		132	153.20392	159.81355	172.71083
133	98.02043	107.35740	112.57341		133	154.28282	160.91478	173.85386
134	98.87770	108.25655	113.49475		134	155.36142	162.01563	174.99635
135	99.73615	109.15612	114.41649		135	156.43973	163.11610	176.13831
136	100.59466	110.05597	115.33843		136	157.51775	164.21621	177.27974
137	101.45357	110.95628	116.26075		137	158.59549	165.31594	178.42066
138	102.31310	111.85689	117.18330		138	159.67294	166.41530	179.56106
139	103.17349	112.75789	118.10613		139	160.75011	167.51431	180.70095
140	104.03400	113.65929	119.02918		140	161.82699	168.61296	181.84034
141	104.89522	114.56101	119.95263		141	162.90361	169.71125	182.97923
142	105.75693	115.46299	120.87630		142	163.97995	170.80920	184.11763
143	106.61900	116.36552	121.80022		143	165.05603	171.90680	185.25554
144	107.48188	117.26820	122.72442		144	166.13183	173.00407	186.39297
145	108.34497	118.17132	123.64887		145	167.20737	174.10098	187.52992
146	109.20866	119.07478	124.57359		146	168.28265	175.19757	188.66640
147	110.07231	119.97852	125.49858		147	169.35768	176.29382	189.80241
148	110.93698	120.88263	126.42381		148	170.43244	177.38975	190.93796
149	111.80178	121.78699	127.34928		149	171.50695	178.48536	192.07305
150	112.66720	122.69175	128.27500		150	172.58122	179.58064	193.20769

Table VI.   Values of χ^2 Distribution for $\nu = 151$ to 175

Values in the body of this table are the $\chi^2_{\nu,\alpha}$ values for selected α levels and degrees of freedom (ν) used in establishing confidence bounds for capability measures and determining critical values for goodness-of-fit tests.

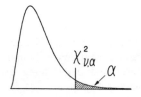

ν	.99	.95	.90	ν	.10	.05	.01
151	113.53320	123.59676	129.20097	151	173.65523	180.67560	194.34188
152	114.39942	124.50212	130.12726	152	174.72899	181.77025	195.47562
153	115.26638	125.40782	131.05371	153	175.80252	182.86459	196.60893
154	116.13355	126.31381	131.98046	154	176.87581	183.95861	197.74180
155	117.00126	127.22003	132.90740	155	177.94885	185.05233	198.87424
156	117.86896	128.12671	133.83460	156	179.02166	186.14575	200.00625
157	118.73756	129.03351	134.76199	157	180.09424	187.23887	201.13783
158	119.60629	129.94076	135.68969	158	181.16659	188.33170	202.26900
159	120.47557	130.84827	136.61750	159	182.23870	189.42422	203.39975
160	121.34542	131.75604	137.54568	160	183.31058	190.51646	204.53010
161	122.21546	132.66412	138.47397	161	184.38225	191.60841	205.66003
162	123.08616	133.57242	139.40258	162	185.45369	192.70007	206.78957
163	123.95706	134.48108	140.33137	163	186.52491	193.79145	207.91870
164	124.82845	135.38997	141.26037	164	187.59591	194.88254	209.04743
165	125.69986	136.29916	142.18953	165	188.66669	195.97336	210.17577
166	126.57208	137.20863	143.11901	166	189.73726	197.06391	211.30373
167	127.44443	138.11837	144.04865	167	190.80761	198.15418	212.43130
168	128.31729	139.02834	144.97846	168	191.87776	199.24418	213.55848
169	129.19060	139.93856	145.90853	169	192.94769	200.33391	214.68530
170	130.06417	140.84909	146.83885	170	194.01743	201.42338	215.81172
171	130.93831	141.76000	147.76930	171	195.08694	202.51258	216.93778
172	131.81265	142.67099	148.69999	172	196.15626	203.60152	218.06348
173	132.68709	143.58242	149.63084	173	197.22538	204.69020	219.18881
174	133.56224	144.49400	150.56198	174	198.29429	205.77863	220.31377
175	134.43780	145.40581	151.49329	175	199.36301	206.86680	221.43838

Table VI. Values of χ^2 Distribution for $v = 176$ to 200

Values in the body of this table are the $\chi^2_{v,\alpha}$ values for selected α levels and degrees of freedom (v) used in establishing confidence bounds for capability measures and determining critical values for goodness-of-fit tests.

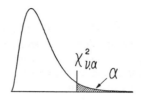

| | α | | |
v	.99	.95	.90
176	135.31349	146.31779	152.42476
177	136.18967	147.23025	153.35637
178	137.06615	148.14285	154.28828
179	137.94298	149.05566	155.22037
180	138.81989	149.96875	156.15256
181	139.69782	150.88204	157.08499
182	140.57568	151.79559	158.01765
183	141.45369	152.70942	158.95050
184	142.33235	153.62345	159.88349
185	143.21114	154.53767	160.81660
186	144.09038	155.45223	161.74998
187	144.96980	156.36690	162.68356
188	145.84966	157.28190	163.61728
189	146.72961	158.19700	164.55112
190	147.60986	159.11250	165.48519
191	148.49106	160.02799	166.41948
192	149.37210	160.94396	167.35388
193	150.25311	161.85993	168.28845
194	151.13530	162.77632	169.22319
195	152.01738	163.69284	170.15815
196	152.89955	164.60955	171.09320
197	153.78222	165.52634	172.02850
198	154.66500	166.44360	172.96388
199	155.54813	167.36093	173.89950
200	156.43158	168.27850	174.83521

| | α | | |
v	.10	.05	.01
176	200.43153	207.95472	222.56263
177	201.49985	209.04239	223.68652
178	202.56799	210.12981	224.81007
179	203.63593	211.21698	225.93327
180	204.70368	212.30392	227.05613
181	205.77124	213.39060	228.17864
182	206.83861	214.47705	229.30082
183	207.90581	215.56327	230.42266
184	208.97281	216.64924	231.54417
185	210.03963	217.73499	232.66534
186	211.10627	218.82050	233.78619
187	212.17274	219.90578	234.90672
188	213.23902	220.99082	236.02692
189	214.30513	222.07565	237.14681
190	215.37106	223.16025	238.26638
191	216.43682	224.24463	239.38562
192	217.50241	225.32878	240.50456
193	218.56782	226.41272	241.62319
194	219.63307	227.49644	242.74151
195	220.69815	228.57994	243.85953
196	221.76306	229.66323	244.97724
197	222.82780	230.74630	246.09467
198	223.89238	231.82917	247.21178
199	224.95679	232.91182	248.32860
200	226.02102	233.99423	249.44507

Table VI. Values of χ^2 Distribution for $v = 201$ to 225

Values in the body of this table are the $\chi^2_{v,\alpha}$ values for selected α levels and degrees of freedom (v) used in establishing confidence bounds for capability measures and determining critical values for goodness-of-fit tests.

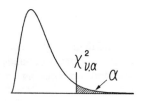

| | α | | |
v	.99	.95	.90
201	157.31572	169.19619	175.77110
202	158.19933	170.11425	176.70723
203	159.08471	171.03229	177.64339
204	159.96877	171.95068	178.57977
205	160.85381	172.86943	179.51635
206	161.73904	173.78814	180.45314
207	162.62447	174.70718	181.38993
208	163.51007	175.62616	182.32690
209	164.39584	176.54566	183.26407
210	165.28255	177.46508	184.20141
211	166.16862	178.38500	185.13893
212	167.05643	179.30483	186.07641
213	167.94356	180.22496	187.01426
214	168.82999	181.14537	187.95206
215	169.71815	182.06586	188.89021
216	170.60559	182.98663	189.82830
217	171.49477	183.90747	190.76674
218	172.38319	184.82856	191.70510
219	173.27085	185.74971	192.64379
220	174.16023	186.67110	193.58240
221	175.04882	187.59253	194.52133
222	175.93913	188.51462	195.46038
223	176.82863	189.43652	196.39953
224	177.71899	190.35864	197.33888
225	178.60851	191.28098	198.27833

| | α | | |
v	.10	.05	.01
201	227.08509	235.07646	250.56131
202	228.14905	236.15851	251.67725
203	229.21281	237.24034	252.79291
204	230.27642	238.32196	253.90829
205	231.33989	239.40339	255.02337
206	232.40319	240.48463	256.13820
207	233.46636	241.56567	257.25272
208	234.52935	242.64650	258.36699
209	235.59219	243.72714	259.48097
210	236.65490	244.80759	260.59468
211	237.71745	245.88786	261.70811
212	238.77984	246.96792	262.82128
213	239.84210	248.04781	263.93418
214	240.90419	249.12749	265.04683
215	241.96616	250.20701	266.15919
216	243.02798	251.28632	267.27131
217	244.08966	252.36546	268.38317
218	245.15119	253.44442	269.49476
219	246.21256	254.52319	270.60610
220	247.27382	255.60178	271.71719
221	248.33492	256.68019	272.82802
222	249.39589	257.75844	273.93861
223	250.45673	258.83649	275.04894
224	251.51742	259.91437	276.15903
225	252.57799	260.99208	277.26887

Table VI. Values of χ^2 Distribution for $\nu = 226$ to 250

Values in the body of this table are the $\chi^2_{\nu,\alpha}$ values for selected α levels and degrees of freedom (ν) used in establishing confidence bounds for capability measures and determining critical values for goodness-of-fit tests.

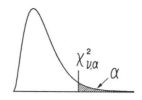

		α				α	
ν	.99	.95	.90	ν	.10	.05	.01
226	179.49889	192.20333	199.21798	226	253.63842	262.06962	278.37847
227	180.39011	193.12587	200.15760	227	254.69871	263.14699	279.48782
228	181.28046	194.04884	201.09752	228	255.75887	264.22418	280.59692
229	182.17164	194.97156	202.03740	229	256.81889	265.30121	281.70581
230	183.06366	195.89491	202.97756	230	257.87879	266.37807	282.81445
231	183.95476	196.81822	203.91779	231	258.93855	267.45476	283.92285
232	184.84668	197.74171	204.85808	232	259.99818	268.53128	285.03100
233	185.73943	198.66516	205.79864	233	261.05770	269.60764	286.13894
234	186.63121	199.58921	206.73925	234	262.11707	270.68384	287.24663
235	187.52380	200.51297	207.67990	235	263.17631	271.75987	288.35411
236	188.41631	201.43734	208.62081	236	264.23545	272.83574	289.46135
237	189.30961	202.36141	209.56186	237	265.29446	273.91144	290.56836
238	190.20280	203.28607	210.50295	238	266.35333	274.98700	291.67516
239	191.09679	204.21041	211.44417	239	267.41207	276.06239	292.78173
240	191.98975	205.13535	212.38564	240	268.47070	277.13762	293.88806
241	192.88348	206.06041	213.32712	241	269.52921	278.21270	294.99418
242	193.77798	206.98536	214.26872	242	270.58761	279.28760	296.10009
243	194.67234	207.91066	215.21056	243	271.64585	280.36237	297.20576
244	195.56654	208.83583	216.15228	244	272.70402	281.43698	298.31123
245	196.46151	209.76158	217.09422	245	273.76203	282.51144	299.41647
246	197.35629	210.68742	218.03639	246	274.81994	283.58574	300.52151
247	198.25183	211.61313	218.97854	247	275.87774	284.65991	301.62633
248	199.14719	212.53893	219.92091	248	276.93540	285.73390	302.73094
249	200.04234	213.46528	220.86338	249	277.99294	286.80776	303.83534
250	200.93918	214.39171	221.80593	250	279.05041	287.88148	304.93952

Table VI.   Values of χ^2 Distribution for $v = 251$ to 275

Values in the body of this table are the $\chi^2_{v,\alpha}$ values for selected α levels and degrees of freedom (v) used in establishing confidence bounds for capability measures and determining critical values for goodness-of-fit tests.

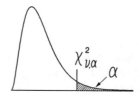

v	.99	.95	.90
251	201.83485	215.31798	222.74857
252	202.73126	216.24455	223.69141
253	203.62744	217.17143	224.63420
254	204.52434	218.09812	225.57732
255	205.42099	219.02535	226.52038
256	206.31836	219.95239	227.46362
257	207.21449	220.87972	228.40705
258	208.11328	221.80733	229.35042
259	209.01082	222.73472	230.29409
260	209.90807	223.66264	231.23768
261	210.80601	224.59034	232.18157
262	211.70464	225.51855	233.12537
263	212.60297	226.44653	234.06946
264	213.50098	227.37502	235.01347
265	214.40067	228.30326	235.95775
266	215.29903	229.23175	236.90219
267	216.19805	230.16049	237.84653
268	217.09775	231.08948	238.79101
269	217.99812	232.01845	239.73564
270	218.89709	232.94741	240.68041
271	219.79673	233.87660	241.62532
272	220.69702	234.80576	242.57023
273	221.59797	235.73540	243.51527
274	222.49853	236.66501	244.46032
275	223.39869	237.59484	245.40575

v	.10	.05	.01
251	280.10772	288.95503	306.04349
252	281.16491	290.02844	307.14726
253	282.22203	291.10171	308.25083
254	283.27899	292.17483	309.35419
255	284.33588	293.24780	310.45735
256	285.39265	294.32064	311.56031
257	286.44928	295.39333	312.66307
258	287.50583	296.46587	313.76562
259	288.56225	297.53829	314.86798
260	289.61855	298.61056	315.97012
261	290.67478	299.68269	317.07209
262	291.73088	300.75468	318.17387
263	292.78688	301.82654	319.27543
264	293.84275	302.89826	320.37681
265	294.89855	303.96982	321.47800
266	295.95423	305.04129	322.57900
267	297.00979	306.11260	323.67980
268	298.06527	307.18376	324.78041
269	299.12064	308.25482	325.88084
270	300.17589	309.32572	326.98108
271	301.23105	310.39651	328.08114
272	302.28611	311.46716	329.18101
273	303.34106	312.53767	330.28070
274	304.39593	313.60807	331.38022
275	305.45067	314.67835	332.47954

Table VI. Values of χ^2 Distribution for $\nu = 276$ to 300

Values in the body of this table are the $\chi^2_{\nu,\alpha}$ values for selected α levels and degrees of freedom (ν) used in establishing confidence bounds for capability measures and determining critical values for goodness-of-fit tests.

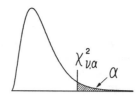

	α		
ν	.99	.95	.90
276	224.30054	238.52463	246.35104
277	225.20092	239.45463	247.29645
278	226.10194	240.38457	248.24197
279	227.00360	241.31498	249.18761
280	227.90482	242.24533	250.13336
281	228.80666	243.17588	251.07922
282	229.70805	244.10636	252.02518
283	230.61115	245.03730	252.97125
284	231.51270	245.96815	253.91741
285	232.41486	246.89893	254.86354
286	233.31763	247.83015	255.80975
287	234.22101	248.76129	256.75620
288	235.12390	249.69287	257.70273
289	236.02740	250.62435	258.64934
290	236.93039	251.55573	259.59604
291	237.83398	252.48755	260.54282
292	238.73816	253.41925	261.48967
293	239.64070	254.35139	262.43660
294	240.54494	255.28341	263.38360
295	241.44978	256.21559	264.33080
296	242.35406	257.14763	265.27808
297	243.25780	258.08038	266.22527
298	244.16211	259.01272	267.17281
299	245.06813	259.94547	268.12026
300	245.97244	260.87809	269.06776

	α		
ν	.10	.05	.01
276	306.50534	315.74847	333.57868
277	307.55989	316.81848	334.67765
278	308.61434	317.88835	335.77643
279	309.66870	318.95811	336.87503
280	310.72296	320.02774	337.97348
281	311.77712	321.09726	339.07173
282	312.83120	322.16663	340.16982
283	313.88516	323.23590	341.26772
284	314.93905	324.30504	342.36546
285	315.99282	325.37406	343.46301
286	317.04651	326.44295	344.56040
287	318.10011	327.51173	345.65763
288	319.15361	328.58037	346.75467
289	320.20700	329.64890	347.85156
290	321.26033	330.71732	348.94828
291	322.31355	331.78561	350.04483
292	323.36667	332.85379	351.14121
293	324.41972	333.92186	352.23743
294	325.47267	334.98982	353.33349
295	326.52552	336.05764	354.42937
296	327.57830	337.12535	355.52510
297	328.63099	338.19295	356.62066
298	329.68357	339.26044	357.71607
299	330.73607	340.32781	358.81132
300	331.78849	341.39508	359.90640

Table VI. Values of χ^2 Distribution for $v = 301$ to 325

Values in the body of this table are the $\chi^2_{v,\alpha}$ values for selected α levels and degrees of freedom (v) used in establishing confidence bounds for capability measures and determining critical values for goodness-of-fit tests.

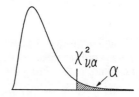

v	α .99	.95	.90
301	246.87732	261.81112	270.01547
302	247.78276	262.74400	270.96322
303	248.68877	263.67700	271.91116
304	249.59302	264.61014	272.85915
305	250.50014	265.54339	273.80703
306	251.40550	266.47706	274.75524
307	252.31141	267.41056	275.70349
308	253.21786	268.34387	276.65178
309	254.12369	269.27759	277.60010
310	255.03006	270.21172	278.54844
311	255.93697	271.14565	279.49696
312	256.84442	272.07939	280.44550
313	257.75122	273.01383	281.39422
314	258.65736	273.94806	282.34295
315	259.56522	274.88239	283.29200
316	260.47241	275.81681	284.24076
317	261.38013	276.75163	285.18984
318	262.28716	277.68592	286.13892
319	263.19593	278.62090	287.08801
320	264.10400	279.55566	288.03741
321	265.01137	280.49081	288.98682
322	265.91925	281.42572	289.93622
323	266.82764	282.36101	290.88562
324	267.73654	283.29638	291.83517
325	268.64471	284.23181	292.78487

v	α .10	.05	.01
301	332.84082	342.46224	361.00133
302	333.89304	343.52928	362.09610
303	334.94521	344.59620	363.19071
304	335.99728	345.66302	364.28517
305	337.04927	346.72973	365.37947
306	338.10116	347.79631	366.47361
307	339.15298	348.86280	367.56760
308	340.20471	349.92918	368.66145
309	341.25633	350.99545	369.75514
310	342.30791	352.06163	370.84868
311	343.35937	353.12768	371.94206
312	344.41076	354.19364	373.03530
313	345.46208	355.25949	374.12839
314	346.51332	356.32521	375.22131
315	347.56447	357.39086	376.31410
316	348.61552	358.45639	377.40674
317	349.66649	359.52180	378.49923
318	350.71741	360.58713	379.59159
319	351.76824	361.65235	380.68379
320	352.81899	362.71748	381.77586
321	353.86965	363.78250	382.86776
322	354.92023	364.84740	383.95953
323	355.97075	365.91222	385.05117
324	357.02117	366.97694	386.14266
325	358.07151	368.04155	387.23402

Table VI. Values of χ^2 Distribution for $\nu = 326$ to 350

Values in the body of this table are the $\chi^2_{\nu,\alpha}$ values for selected α levels and degrees of freedom (ν) used in establishing confidence bounds for capability measures and determining critical values for goodness-of-fit tests.

ν	.99	.95	.90	ν	.10	.05	.01
326	269.55339	285.16730	293.73456	326	359.12180	369.10606	388.32521
327	270.46257	286.10286	294.68439	327	360.17200	370.17049	389.41628
328	271.37225	287.03848	295.63436	328	361.22213	371.23480	390.50720
329	272.28119	287.97415	296.58432	329	362.27217	372.29903	391.59799
330	273.19061	288.91019	297.53426	330	363.32214	373.36314	392.68865
331	274.09927	289.84597	298.48433	331	364.37203	374.42717	393.77916
332	275.00969	290.78211	299.43454	332	365.42184	375.49109	394.86954
333	275.91933	291.71830	300.38488	333	366.47157	376.55493	395.95977
334	276.82946	292.65453	301.33503	334	367.52123	377.61867	397.04989
335	277.73880	293.59080	302.28548	335	368.57084	378.68231	398.13985
336	278.64990	294.52744	303.23605	336	369.62034	379.74585	399.22969
337	279.56021	295.46378	304.18658	337	370.66979	380.80930	400.31939
338	280.47099	296.40049	305.13724	338	371.71917	381.87266	401.40895
339	281.38097	297.33754	306.08786	339	372.76847	382.93591	402.49840
340	282.29141	298.27431	307.03876	340	373.81769	383.99909	403.58771
341	283.20233	299.21110	307.98962	341	374.86683	385.06216	404.67688
342	284.11372	300.14824	308.94044	342	375.91593	386.12515	405.76593
343	285.02428	301.08540	309.89153	343	376.96492	387.18805	406.85483
344	285.93530	302.02225	310.84258	344	378.01386	388.25084	407.94362
345	286.84679	302.95977	311.79357	345	379.06273	389.31355	409.03228
346	287.75874	303.89698	312.74483	346	380.11151	390.37617	410.12080
347	288.66984	304.83453	313.69604	347	381.16026	391.43869	411.20920
348	289.58139	305.77209	314.64752	348	382.20889	392.50114	412.29746
349	290.49340	306.70966	315.59894	349	383.25748	393.56348	413.38560
350	291.40587	307.64756	316.55030	350	384.30599	394.62574	414.47363

Table VI. Values of χ^2 Distribution for $\nu = 351$ to 375

Values in the body of this table are the $\chi^2_{\nu,\alpha}$ values for selected α levels and degrees of freedom (ν) used in establishing confidence bounds for capability measures and determining critical values for goodness-of-fit tests.

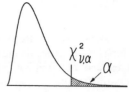

| | α | | | | α | | |
ν	.99	.95	.90	ν	.10	.05	.01
351	292.31746	308.58547	317.50193	351	385.35445	395.68790	415.56152
352	293.22949	309.52338	318.45349	352	386.40281	396.74999	416.64928
353	294.14333	310.46128	319.40532	353	387.45112	397.81197	417.73691
354	295.05492	311.39952	320.35708	354	388.49935	398.87387	418.82444
355	295.96695	312.33775	321.30876	355	389.54754	399.93570	419.91184
356	296.88078	313.27597	322.26071	356	390.59564	400.99742	420.99911
357	297.79234	314.21417	323.21258	357	391.64367	402.05907	422.08625
358	298.70570	315.15271	324.16471	358	392.69165	403.12062	423.17328
359	299.61951	316.09123	325.11676	359	393.73953	404.18209	424.26018
360	300.53238	317.02972	326.06907	360	394.78736	405.24348	425.34696
361	301.44569	317.96854	327.02129	361	395.83514	406.30479	426.43362
362	302.35943	318.90734	327.97343	362	396.88284	407.36598	427.52018
363	303.27361	319.84612	328.92582	363	397.93048	408.42713	428.60661
364	304.18683	320.78521	329.87830	364	398.97804	409.48817	429.69291
365	305.10048	321.72427	330.83085	365	400.02556	410.54912	430.77909
366	306.01456	322.66330	331.78349	366	401.07300	411.61000	431.86516
367	306.92906	323.60264	332.73602	367	402.12036	412.67079	432.95111
368	307.84259	324.54195	333.68881	368	403.16768	413.73151	434.03694
369	308.75654	325.48121	334.64150	369	404.21493	414.79214	435.12265
370	309.67091	326.42079	335.59426	370	405.26209	415.85268	436.20826
371	310.58569	327.36033	336.54709	371	406.30922	416.91314	437.29375
372	311.50090	328.29982	337.49999	372	407.35626	417.97354	438.37910
373	312.41510	329.23962	338.45313	373	408.40326	419.03384	439.46437
374	313.32972	330.17937	339.40617	374	409.45019	420.09406	440.54951
375	314.24475	331.11906	340.35945	375	410.49704	421.15420	441.63452

Table VI. Values of χ^2 Distribution for $\nu = 376$ to 400

Values in the body of this table are the $\chi^2_{\nu,\alpha}$ values for selected α levels and degrees of freedom (ν) used in establishing confidence bounds for capability measures and determining critical values for goodness-of-fit tests.

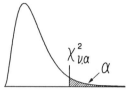

| | α | | | | | α | | |
ν	.99	.95	.90		ν	.10	.05	.01
376	315.16019	332.05906	341.31261		376	411.54384	422.21426	442.71944
377	316.07460	332.99936	342.26566		377	412.59059	423.27423	443.80424
378	316.99086	333.93925	343.21913		378	413.63726	424.33415	444.88893
379	317.90609	334.87943	344.17248		379	414.68389	425.39395	445.97351
380	318.82172	335.81955	345.12589		380	415.73043	426.45372	447.05797
381	319.73775	336.75997	346.07953		381	416.77692	427.51336	448.14233
382	320.65273	337.70032	347.03305		382	417.82337	428.57297	449.22657
383	321.56957	338.64097	347.98662		383	418.86974	429.63245	450.31071
384	322.48535	339.58154	348.94043		384	419.91603	430.69190	451.39471
385	323.40153	340.52205	349.89411		385	420.96228	431.75124	452.47863
386	324.31810	341.46284	350.84802		386	422.00849	432.81051	453.56244
387	325.23359	342.40356	351.80179		387	423.05462	433.86972	454.64614
388	326.14948	343.34457	352.75580		388	424.10067	434.92885	455.72972
389	327.06723	344.28549	353.70967		389	425.14669	435.98789	456.81321
390	327.98241	345.22671	354.66376		390	426.19266	437.04686	457.89657
391	328.89946	346.16784	355.61790		391	427.23855	438.10575	458.97984
392	329.81689	347.10888	356.57190		392	428.28436	439.16456	460.06300
393	330.73322	348.05021	357.52612		393	429.33013	440.22333	461.14605
394	331.64993	348.99144	358.48057		394	430.37587	441.28198	462.22900
395	332.56702	349.93296	359.43487		395	431.42152	442.34059	463.31184
396	333.48450	350.87438	360.38940		396	432.46713	443.39910	464.39457
397	334.40235	351.81608	361.34377		397	433.51267	444.45755	465.47722
398	335.32059	352.75768	362.29836		398	434.55816	445.51592	466.55975
399	336.23768	353.69918	363.25280		399	435.60357	446.57423	467.64216
400	337.15515	354.64096	364.20746		400	436.64895	447.63244	468.72448

Table VI. Values of χ^2 Distribution for $\nu = 401$ to 425

Values in the body of this table are the $\chi^2_{\nu,\alpha}$ values for selected α levels and degrees of freedom (ν) used in establishing confidence bounds for capability measures and determining critical values for goodness-of-fit tests.

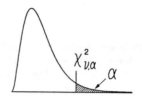

	α		
ν	**.99**	**.95**	**.90**
401	338.07299	355.58263	365.16215
402	338.99121	356.52458	366.11686
403	339.90826	357.46680	367.07161
404	340.82722	358.40891	368.02638
405	341.74501	359.35091	368.98137
406	342.66317	360.29317	369.93638
407	343.58169	361.23532	370.89141
408	344.50058	362.17774	371.84647
409	345.41827	363.12043	372.80154
410	346.33789	364.06260	373.75683
411	347.25631	365.00504	374.71194
412	348.17508	365.94775	375.66726
413	349.09421	366.89072	376.62279
414	350.01370	367.83356	377.57814
415	350.93197	368.77627	378.53350
416	351.85217	369.71924	379.48907
417	352.77114	370.66207	380.44466
418	353.69047	371.60516	381.40024
419	354.61014	372.54812	382.35604
420	355.53017	373.49133	383.31164
421	356.44894	374.43479	384.26745
422	357.36966	375.37811	385.22326
423	358.28912	376.32129	386.17908
424	359.20892	377.26472	387.13509
425	360.13069	378.20840	388.09111

	α		
ν	**.10**	**.05**	**.01**
401	437.69428	448.69061	469.80671
402	438.73953	449.74867	470.88882
403	439.78475	450.80668	471.97083
404	440.82988	451.86463	473.05276
405	441.87498	452.92250	474.13457
406	442.91999	453.98027	475.21626
407	443.96497	455.03800	476.29787
408	445.00990	456.09566	477.37940
409	446.05476	457.15323	478.46081
410	447.09957	458.21076	479.54210
411	448.14435	459.26817	480.62331
412	449.18904	460.32555	481.70444
413	450.23370	461.38286	482.78545
414	451.27831	462.44009	483.86636
415	452.32285	463.49726	484.94719
416	453.36734	464.55435	486.02790
417	454.41176	465.61137	487.10853
418	455.45613	466.66832	488.18906
419	456.50047	467.72521	489.26947
420	457.54476	468.78204	490.34980
421	458.58898	469.83878	491.43005
422	459.63315	470.89547	492.51020
423	460.67725	471.95208	493.59025
424	461.72134	473.00863	494.67020
425	462.76535	474.06510	495.75005

Table VI. Values of χ^2 Distribution for $\nu = 426$ to 450

Values in the body of this table are the $\chi^2_{\nu,\alpha}$ values for selected α levels and degrees of freedom (ν) used in establishing confidence bounds for capability measures and determining critical values for goodness-of-fit tests.

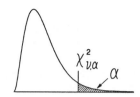

| | α | | | | | α | | |
ν	.99	.95	.90		ν	.10	.05	.01
426	361.04956	379.15193	389.04713		426	463.80929	475.12153	496.82982
427	361.97039	380.09571	390.00314		427	464.85322	476.17789	497.90950
428	362.89157	381.03933	390.95936		428	465.89708	477.23418	498.98907
429	363.81145	381.98321	391.91557		429	466.94089	478.29039	500.06856
430	364.73167	382.92692	392.87177		430	467.98462	479.34656	501.14795
431	365.65223	383.87088	393.82817		431	469.02836	480.40263	502.22726
432	366.57312	384.81468	394.78436		432	470.07201	481.45866	503.30646
433	367.49435	385.75873	395.74075		433	471.11559	482.51463	504.38559
434	368.41591	386.70302	396.69733		434	472.15916	483.57052	505.46460
435	369.33615	387.64714	397.65369		435	473.20265	484.62633	506.54354
436	370.25671	388.59150	398.61025		436	474.24610	485.68210	507.62239
437	371.17928	389.53570	399.56701		437	475.28951	486.73779	508.70113
438	372.10050	390.48013	400.52353		438	476.33287	487.79343	509.77981
439	373.02206	391.42439	401.48026		439	477.37618	488.84900	510.85838
440	373.94226	392.36889	402.43696		440	478.41943	489.90453	511.93686
441	374.86447	393.31363	403.39386		441	479.46264	490.95998	513.01527
442	375.78700	394.25819	404.35074		442	480.50577	492.01535	514.09355
443	376.70818	395.20256	405.30738		443	481.54890	493.07067	515.17179
444	377.62967	396.14759	406.26443		444	482.59195	494.12592	516.24991
445	378.55148	397.09244	407.22145		445	483.63496	495.18113	517.32795
446	379.47362	398.03710	408.17845		446	484.67793	496.23625	518.40591
447	380.39607	398.98200	409.13543		447	485.72083	497.29134	519.48377
448	381.31885	399.92712	410.09259		448	486.76371	498.34634	520.56155
449	382.24023	400.87206	411.04994		449	487.80652	499.40130	521.63923
450	383.16364	401.81723	412.00705		450	488.84929	500.45619	522.71685

Table VI. Values of χ^2 Distribution for $v = 451$ to 475

Values in the body of this table are the $\chi^2_{v.\alpha}$ values for selected α levels and degrees of freedom (v) used in establishing confidence bounds for capability measures and determining critical values for goodness-of-fit tests.

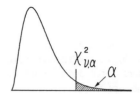

| | α | | |
v	.99	.95	.90
451	384.08566	402.76263	412.96434
452	385.00798	403.70783	413.92181
453	385.93062	404.65283	414.87904
454	386.85357	405.59849	415.83646
455	387.77683	406.54395	416.79405
456	388.69867	407.48920	417.75140
457	389.62255	408.43468	418.70892
458	390.54501	409.38039	419.66663
459	391.46951	410.32589	420.62430
460	392.39258	411.27161	421.58193
461	393.31595	412.21756	422.53974
462	394.23962	413.16329	423.49750
463	395.16184	414.10924	424.45523
464	396.08612	415.05542	425.41313
465	397.00893	416.00138	426.37121
466	397.93382	416.94711	427.32903
467	398.85723	417.89351	428.28701
468	399.78094	418.83968	429.24518
469	400.70495	419.78563	430.20329
470	401.62926	420.73224	431.16158
471	402.55206	421.67862	432.11983
472	403.47696	422.62477	433.07779
473	404.40215	423.57159	434.03615
474	405.32584	424.51772	434.99446
475	406.24981	425.46451	435.95294

| | α | | |
v	.10	.05	.01
451	489.89202	501.51103	523.79437
452	490.93471	502.56580	524.87180
453	491.97732	503.62048	525.94914
454	493.01992	504.67512	527.02641
455	494.06245	505.72972	528.10359
456	495.10493	506.78424	529.18068
457	496.14738	507.83872	530.25770
458	497.18975	508.89312	531.33463
459	498.23211	509.94746	532.41146
460	499.27440	511.00176	533.48821
461	500.31668	512.05598	534.56488
462	501.35888	513.11016	535.64147
463	502.40105	514.16428	536.71798
464	503.44317	515.21831	537.79441
465	504.48525	516.27231	538.87075
466	505.52726	517.32625	539.94702
467	506.56926	518.38012	541.02318
468	507.61118	519.43394	542.09929
469	508.65306	520.48769	543.17529
470	509.69494	521.54140	544.25123
471	510.73674	522.59507	545.32709
472	511.77850	523.64867	546.40286
473	512.82022	524.70218	547.47855
474	513.86189	525.75565	548.55415
475	514.90349	526.80908	549.62969

Table VI. Values of χ^2 Distribution for $\nu = 476$ to 500

Values in the body of this table are the $\chi^2_{\nu,\alpha}$ values for selected α levels and degrees of freedom (ν) used in establishing confidence bounds for capability measures and determining critical values for goodness-of-fit tests.

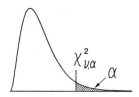

| | α | | | | | α | | |
ν	.99	.95	.90		ν	.10	.05	.01
476	407.17407	426.41107	436.91136		476	515.94509	527.86247	550.70515
477	408.10044	427.35784	437.86973		477	516.98661	528.91578	551.78052
478	409.02347	428.30437	438.82827		478	518.02812	529.96901	552.85582
479	409.94860	429.25156	439.78697		479	519.06956	531.02221	553.93103
480	410.87402	430.19806	440.74539		480	520.11099	532.07534	555.00616
481	411.79974	431.14522	441.70398		481	521.15235	533.12844	556.08123
482	412.72390	432.09214	442.66273		482	522.19366	534.18147	557.15622
483	413.64835	433.03926	443.62143		483	523.23494	535.23445	558.23111
484	414.57492	433.98614	444.58028		484	524.27621	536.28736	559.30594
485	415.49994	434.93322	445.53908		485	525.31740	537.34024	560.38069
486	416.42523	435.88051	446.49781		486	526.35855	538.39304	561.45537
487	417.35081	436.82754	447.45670		487	527.39966	539.44582	562.52996
488	418.27667	437.77478	448.41576		488	528.44073	540.49852	563.60448
489	419.20095	438.72223	449.37475		489	529.48172	541.55116	564.67893
490	420.12737	439.66988	450.33367		490	530.52271	542.60376	565.75329
491	421.05220	440.61726	451.29276		491	531.56366	543.65630	566.82759
492	421.97919	441.56486	452.25177		492	532.60456	544.70879	567.90179
493	422.90457	442.51218	453.21094		493	533.64543	545.76121	568.97594
494	423.83023	443.45971	454.17004		494	534.68626	546.81360	570.05000
495	424.75616	444.40744	455.12954		495	535.72700	547.86592	571.12398
496	425.68237	445.35538	456.08873		496	536.76775	548.91821	572.19789
497	426.60886	446.30304	457.04808		497	537.80845	549.97042	573.27174
498	427.53561	447.25090	458.00735		498	538.84911	551.02260	574.34551
499	428.46074	448.19896	458.96678		499	539.88970	552.07473	575.41918
500	429.38805	449.14675	459.92613		500	540.93028	553.12678	576.49281

Table VII. Values of Student's t Distribution for ν Equal 1 to 50

Values in the body of this table are the $t_{\nu,\alpha}$ values for selected α levels and degrees of freedom (ν) used in establishing confidence bounds for various capability measures.

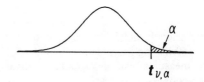

$t_{\nu,\alpha}$

ν	α 0.10	0.05	0.01	ν	α 0.10	0.05	0.01
1	3.078	6.314	31.821	26	1.315	1.706	2.479
2	1.886	2.920	6.965	27	1.314	1.703	2.473
3	1.638	2.353	4.541	28	1.313	1.701	2.467
4	1.533	2.132	3.747	29	1.311	1.699	2.462
5	1.476	2.015	3.365	30	1.310	1.697	2.457
6	1.440	1.943	3.143	31	1.309	1.696	2.453
7	1.415	1.895	2.998	32	1.309	1.694	2.449
8	1.397	1.860	2.896	33	1.308	1.692	2.445
9	1.383	1.833	2.821	34	1.307	1.691	2.441
10	1.372	1.812	2.764	35	1.306	1.690	2.438
11	1.363	1.796	2.718	36	1.306	1.688	2.434
12	1.356	1.782	2.681	37	1.305	1.687	2.431
13	1.350	1.771	2.650	38	1.304	1.686	2.429
14	1.345	1.761	2.624	39	1.304	1.685	2.426
15	1.341	1.753	2.602	40	1.303	1.684	2.423
16	1.337	1.746	2.583	41	1.303	1.683	2.421
17	1.333	1.740	2.567	42	1.302	1.682	2.418
18	1.330	1.734	2.552	43	1.302	1.681	2.416
19	1.328	1.729	2.539	44	1.301	1.680	2.414
20	1.325	1.725	2.528	45	1.301	1.679	2.412
21	1.323	1.721	2.518	46	1.300	1.679	2.410
22	1.321	1.717	2.508	47	1.300	1.678	2.408
23	1.319	1.714	2.500	48	1.299	1.677	2.407
24	1.318	1.711	2.492	49	1.299	1.677	2.405
25	1.316	1.708	2.485	50	1.299	1.676	2.403

Table VII. Values of Student's *t* Distribution for ν Equal 51 to 100

Values in the body of this table are the $t_{ν,α}$ values for selected α levels and degrees of freedom (ν) used in establishing confidence bounds for various capability measures.

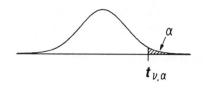

$$t_{ν,α}$$

		α					α		
ν	0.10	0.05	0.01		ν	0.10	0.05	0.01
51	1.298	1.675	2.402		76	1.293	1.665	2.376
52	1.298	1.675	2.400		77	1.293	1.665	2.376
53	1.298	1.674	2.399		78	1.292	1.665	2.375
54	1.297	1.674	2.397		79	1.292	1.664	2.374
55	1.297	1.673	2.396		80	1.292	1.664	2.374
56	1.297	1.673	2.395		81	1.292	1.664	2.373
57	1.297	1.672	2.394		82	1.292	1.664	2.373
58	1.296	1.672	2.392		83	1.292	1.663	2.372
59	1.296	1.671	2.391		84	1.292	1.663	2.372
60	1.296	1.671	2.390		85	1.292	1.663	2.371
61	1.296	1.670	2.389		86	1.291	1.663	2.370
62	1.295	1.670	2.388		87	1.291	1.663	2.370
63	1.295	1.669	2.387		88	1.291	1.662	2.369
64	1.295	1.669	2.386		89	1.291	1.662	2.369
65	1.295	1.669	2.385		90	1.291	1.662	2.368
66	1.295	1.668	2.384		91	1.291	1.662	2.368
67	1.294	1.668	2.383		92	1.291	1.662	2.368
68	1.294	1.668	2.382		93	1.291	1.661	2.367
69	1.294	1.667	2.382		94	1.291	1.661	2.367
70	1.294	1.667	2.381		95	1.291	1.661	2.366
71	1.294	1.667	2.380		96	1.290	1.661	2.366
72	1.293	1.666	2.379		97	1.290	1.661	2.365
73	1.293	1.666	2.379		98	1.290	1.661	2.365
74	1.293	1.666	2.378		99	1.290	1.660	2.365
75	1.293	1.665	2.377		100	1.290	1.660	2.364

Table VII. Values of Student's *t* Distribution for ν Equal 101 to 150

Values in the body of this table are the $t_{\nu,\alpha}$ values for selected α levels and degrees of freedom (ν) used in establishing confidence bounds for various capability measures.

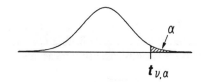

$t_{\nu,\alpha}$

	α				α		
ν	0.10	0.05	0.01	ν	0.10	0.05	0.01
101	1.290	1.660	2.364	126	1.288	1.657	2.356
102	1.290	1.660	2.363	127	1.288	1.657	2.356
103	1.290	1.660	2.363	128	1.288	1.657	2.356
104	1.290	1.660	2.363	129	1.288	1.657	2.356
105	1.290	1.659	2.362	130	1.288	1.657	2.355
106	1.290	1.659	2.362	131	1.288	1.657	2.355
107	1.290	1.659	2.362	132	1.288	1.656	2.355
108	1.289	1.659	2.361	133	1.288	1.656	2.355
109	1.289	1.659	2.361	134	1.288	1.656	2.354
110	1.289	1.659	2.361	135	1.288	1.656	2.354
111	1.289	1.659	2.360	136	1.288	1.656	2.354
112	1.289	1.659	2.360	137	1.288	1.656	2.354
113	1.289	1.658	2.360	138	1.288	1.656	2.354
114	1.289	1.658	2.360	139	1.288	1.656	2.353
115	1.289	1.658	2.359	140	1.288	1.656	2.353
116	1.289	1.658	2.359	141	1.288	1.656	2.353
117	1.289	1.658	2.359	142	1.288	1.656	2.353
118	1.289	1.658	2.358	143	1.287	1.656	2.353
119	1.289	1.658	2.358	144	1.287	1.656	2.353
120	1.289	1.658	2.358	145	1.287	1.655	2.352
121	1.289	1.658	2.358	146	1.287	1.655	2.352
122	1.289	1.657	2.357	147	1.287	1.655	2.352
123	1.288	1.657	2.357	148	1.287	1.655	2.352
124	1.288	1.657	2.357	149	1.287	1.655	2.352
125	1.288	1.657	2.357	150	1.287	1.655	2.351

Table VII. Values of Student's *t* Distribution for ν Equal 151 to 200

Values in the body of this table are the $t_{\nu,\alpha}$ values for selected α levels and degrees of freedom (ν) used in establishing confidence bounds for various capability measures.

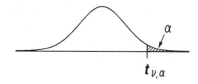

ν	α 0.10	0.05	0.01	ν	α 0.10	0.05	0.01
151	1.287	1.655	2.351	176	1.286	1.654	2.348
152	1.287	1.655	2.351	177	1.286	1.654	2.348
153	1.287	1.655	2.351	178	1.286	1.653	2.347
154	1.287	1.655	2.351	179	1.286	1.653	2.347
155	1.287	1.655	2.351	180	1.286	1.653	2.347
156	1.287	1.655	2.350	181	1.286	1.653	2.347
157	1.287	1.655	2.350	182	1.286	1.653	2.347
158	1.287	1.655	2.350	183	1.286	1.653	2.347
159	1.287	1.654	2.350	184	1.286	1.653	2.347
160	1.287	1.654	2.350	185	1.286	1.653	2.347
161	1.287	1.654	2.350	186	1.286	1.653	2.347
162	1.287	1.654	2.350	187	1.286	1.653	2.346
163	1.287	1.654	2.349	188	1.286	1.653	2.346
164	1.287	1.654	2.349	189	1.286	1.653	2.346
165	1.287	1.654	2.349	190	1.286	1.653	2.346
166	1.287	1.654	2.349	191	1.286	1.653	2.346
167	1.287	1.654	2.349	192	1.286	1.653	2.346
168	1.287	1.654	2.349	193	1.286	1.653	2.346
169	1.287	1.654	2.349	194	1.286	1.653	2.346
170	1.287	1.654	2.348	195	1.286	1.653	2.346
171	1.287	1.654	2.348	196	1.286	1.653	2.346
172	1.286	1.654	2.348	197	1.286	1.653	2.345
173	1.286	1.654	2.348	198	1.286	1.653	2.345
174	1.286	1.654	2.348	199	1.286	1.653	2.345
175	1.286	1.654	2.348	200	1.286	1.653	2.345

Table VII. Values of Student's *t* Distribution for v Equal 201 to 250

Values in the body of this table are the $t_{v,\alpha}$ values for selected α levels and degrees of freedom (v) used in establishing confidence bounds for various capability measures.

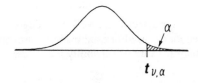

$t_{v,\alpha}$

	α					α		
v	0.10	0.05	0.01		v	0.10	0.05	0.01
201	1.286	1.652	2.345		226	1.285	1.652	2.343
202	1.286	1.652	2.345		227	1.285	1.652	2.343
203	1.286	1.652	2.345		228	1.285	1.652	2.343
204	1.286	1.652	2.345		229	1.285	1.652	2.343
205	1.286	1.652	2.345		230	1.285	1.652	2.343
206	1.286	1.652	2.345		231	1.285	1.651	2.343
207	1.286	1.652	2.344		232	1.285	1.651	2.343
208	1.286	1.652	2.344		233	1.285	1.651	2.342
209	1.286	1.652	2.344		234	1.285	1.651	2.342
210	1.286	1.652	2.344		235	1.285	1.651	2.342
211	1.286	1.652	2.344		236	1.285	1.651	2.342
212	1.286	1.652	2.344		237	1.285	1.651	2.342
213	1.286	1.652	2.344		238	1.285	1.651	2.342
214	1.286	1.652	2.344		239	1.285	1.651	2.342
215	1.286	1.652	2.344		240	1.285	1.651	2.342
216	1.285	1.652	2.344		241	1.285	1.651	2.342
217	1.285	1.652	2.344		242	1.285	1.651	2.342
218	1.285	1.652	2.344		243	1.285	1.651	2.342
219	1.285	1.652	2.343		244	1.285	1.651	2.342
220	1.285	1.652	2.343		245	1.285	1.651	2.342
221	1.285	1.652	2.343		246	1.285	1.651	2.342
222	1.285	1.652	2.343		247	1.285	1.651	2.342
223	1.285	1.652	2.343		248	1.285	1.651	2.341
224	1.285	1.652	2.343		249	1.285	1.651	2.341
225	1.285	1.652	2.343		250	1.285	1.651	2.341

Table VII. Values of Student's *t* Distribution for ν Equal 251 to 300

Values in the body of this table are the $t_{v,\alpha}$ values for selected α levels and degrees of freedom (ν) used in establishing confidence bounds for various capability measures.

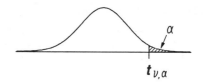

$$t_{v,\alpha}$$

	α				α		
ν	0.10	0.05	0.01	ν	0.10	0.05	0.01
251	1.285	1.651	2.341	276	1.285	1.650	2.340
252	1.285	1.651	2.341	277	1.285	1.650	2.340
253	1.285	1.651	2.341	278	1.285	1.650	2.340
254	1.285	1.651	2.341	279	1.285	1.650	2.340
255	1.285	1.651	2.341	280	1.285	1.650	2.340
256	1.285	1.651	2.341	281	1.285	1.650	2.340
257	1.285	1.651	2.341	282	1.285	1.650	2.340
258	1.285	1.651	2.341	283	1.285	1.650	2.340
259	1.285	1.651	2.341	284	1.285	1.650	2.340
260	1.285	1.651	2.341	285	1.285	1.650	2.340
261	1.285	1.651	2.341	286	1.285	1.650	2.339
262	1.285	1.651	2.341	287	1.285	1.650	2.339
263	1.285	1.651	2.341	288	1.284	1.650	2.339
264	1.285	1.651	2.341	289	1.284	1.650	2.339
265	1.285	1.651	2.341	290	1.284	1.650	2.339
266	1.285	1.651	2.340	291	1.284	1.650	2.339
267	1.285	1.651	2.340	292	1.284	1.650	2.339
268	1.285	1.651	2.340	293	1.284	1.650	2.339
269	1.285	1.651	2.340	294	1.284	1.650	2.339
270	1.285	1.651	2.340	295	1.284	1.650	2.339
271	1.285	1.650	2.340	296	1.284	1.650	2.339
272	1.285	1.650	2.340	297	1.284	1.650	2.339
273	1.285	1.650	2.340	298	1.284	1.650	2.339
274	1.285	1.650	2.340	299	1.284	1.650	2.339
275	1.285	1.650	2.340	300	1.284	1.650	2.339

Table VII. Values of Student's *t* Distribution for ν Equal 301 to 500

Values in the body of this table are the $t_{ν,α}$ values for selected α levels and degrees of freedom (ν) used in establishing confidence bounds for various capability measures.

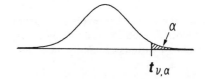

$t_{ν,α}$

<table>
<tr><td></td><td colspan="3">α</td></tr>
<tr><td>ν</td><td>0.10</td><td>0.05</td><td>0.01</td></tr>
<tr><td>301
to
308</td><td>1.284</td><td>1.650</td><td>2.339</td></tr>
<tr><td>309
to
328</td><td>1.284</td><td>1.650</td><td>2.338</td></tr>
<tr><td>329
to
335</td><td>1.284</td><td>1.649</td><td>2.238</td></tr>
<tr><td>336
to
368</td><td>1.284</td><td>1.649</td><td>2.337</td></tr>
<tr><td>369
to
408</td><td>1.284</td><td>1.649</td><td>2.336</td></tr>
</table>

<table>
<tr><td></td><td colspan="3">α</td></tr>
<tr><td>ν</td><td>0.10</td><td>0.05</td><td>0.01</td></tr>
<tr><td>409
to
418</td><td>1.284</td><td>1.649</td><td>2.335</td></tr>
<tr><td>419
to
435</td><td>1.284</td><td>1.648</td><td>2.335</td></tr>
<tr><td>436
to
458</td><td>1.283</td><td>1.648</td><td>2.335</td></tr>
<tr><td>459
to
500</td><td>1.283</td><td>1.648</td><td>2.334</td></tr>
<tr><td>∞</td><td>1.282</td><td>1.645</td><td>2.326</td></tr>
</table>

Index

A

A/P graph, 245-247, 343, 742-744, 750

a'_4, 372

A_2, 12, 18, 760, 773-774, 794, 802, 804

A^2, 412

A^{2*}, 412-413

a_3, 369-370, 418-419

A_3, 18, 98

a_4, 372, 418-419

accumulated frequency, 359, 386
(*See also* estimated accumulated frequency)

accuracy, 12, 113, 227, 230, 234, 237, 244-246, 249, 298, 321
(*See also* k and C_A)

action limits, 788

actual frequency, (f_A), 405-410

AIAG, 7, 42, 45, 69, 220, 582

aim, *see* accuracy

α, *see* alpha risk

alpha risk, 403, 406-409, 413, 569-571, 598-609, 628, 634, 642, 648, 692, 701-705

ANSI, 508, 589, 807

ARIMA, 786-788

ASME, 59, 589

ASQC, 39

ASTM, 44

assignable-source variation, *see* variation

*, 115, 139, 159, 198

attribute data, 521-555, 569, 686, 730-732, 750, 789, 836-837

attribute-data charts, *see* control charts

autocorrelation;
capability for, 781-792
coefficient of, 785

average;
loss, *see* $\overline{L}_{ST}$ and $\overline{L}_{LT}$
loss per part, *see* $\overline{L}$
of moving ranges, *see* $\overline{MR}$
of nonconformities per subgroup, *see* $\overline{u}$ and u'
number of nonconforming units, *see* $n\overline{p}$ and np'
number of nonconformities, *see* $\overline{c}$ and c'
percentage nonconforming, *see* $\overline{p}$ and p'
of population, *see* μ
of ranges, *see* $\overline{R}$
sample standard deviation, *see* $\overline{S}$
of subgroup, *see* $\overline{X}$
of subgroup averages, *see* $\overline{\overline{X}}$

Average C_{PK}, 722-728, 750

Average P_{PK}, 722-728, 750

B

b, 672-681, 683, 710-711

B_3, 18

B_4, 18

bell-shaped curve, 352

benchmarking, 66

best-fit line, 363, 381, 391, 392, 762

between-subgroup variation, *see* variation

bias correction factor, 138-139

biased estimator, 41, 138, 254

bilateral specification, 55, 74, 94, 174, 220-221, 266, 269, 366, 392, 378, 499, 534, 729

bimodal distribution, 352, 355

binomial chart, 521, 536, 694

binomial distribution, 521-522, 536, 547, 686

b_{kn}, 138-139

Boeing, 59, 596

bounded distribution, *see* distributions

brainstorming, 60

buffer, *see* safety margin

C

c (loss), 280-282, 285-287, 319, 323-324

c (number of nonconformities), 527-530

c (relationship between σ_{LT} and σ_{ST}), 835

$\overline{c}$, 527-528, 540, 542, 686-688

c', 527, 688

c_4, 18, 36, 37, 39, 41, 91, 97, 98, 99, 113, 130, 170, 250, 277, 381, 389, 392, 403, 433, 463, 566, 587, 614-617, 619, 640, 763, 810, 812

$C'_{P.MOD}$, 767

$C'_{PK.MOD}$, 767-768

C'_{PL}, 144-149, 212-213, 237-239, 571, 643

$C'_{PL.RUN}$, 776

$C'_{PL.SETUP}$, 775

$C'_{PM.SUB}$, 768

C'_{PU}, 144-147, 155-156, 571, 643-644

ABOUT THE AUTHOR

Davis R. Bothe is the Director of Quality Improvement at the International Quality Institute, a consulting firm for quality and productivity improvement with a worldwide client base. He previously worked as a systems analyst for NASA, as a senior quality engineer for General Motors, and as an adjunct professor of Statistics at Eastern Michigan University.

A Fellow of the American Society for Quality, Mr. Bothe is an ASQ certified quality and reliability engineer and is listed in the 1st edition of the *International Who's Who in Quality*.

With over 25 years of industrial experience, he has published numerous articles and is the author of various books and booklets, including *Industrial Problem Solving*, *SPC for Short Production Runs*, and *Reducing Process Variation*. Call 262-375-8868 for ordering information or visit www.I-Q-I.com.